Sir Rudolf Peierls

Selected Private and Scientific Correspondence Vol. 1

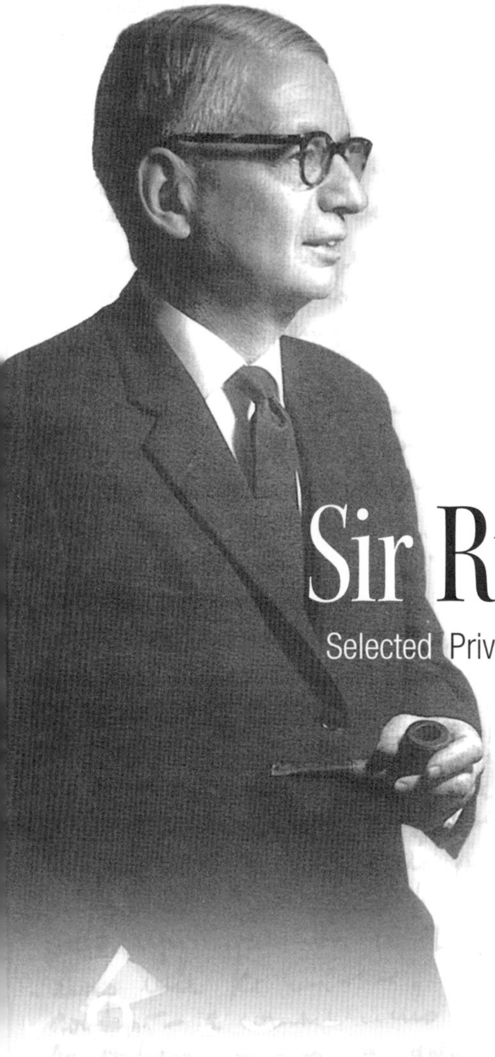

Sir Rudolf Peierls

Selected Private and Scientific Correspondence Vol. 1

Sabine Lee
University of Birmingham, UK

World Scientific
NEW JERSEY • LONDON • SINGAPORE • BEIJING • SHANGHAI • HONG KONG • TAIPEI • CHENNAI

Published by

World Scientific Publishing Co. Pte. Ltd.

5 Toh Tuck Link, Singapore 596224

USA office: 27 Warren Street, Suite 401-402, Hackensack, NJ 07601

UK office: 57 Shelton Street, Covent Garden, London WC2H 9HE

British Library Cataloguing-in-Publication Data
A catalogue record for this book is available from the British Library.

SIR RUDOLF PEIERLS
Selected Private and Scientific Correspondence
Volume 1

ISBN-13 978-981-256-503-7
ISBN-10 981-256-503-5

Printed in Singapore by World Scientific Printers (S) Pte Ltd

Contents

Foreword

These letters are a record of the thoughts of a remarkable man and his remarkable friends and family, living through a historical period of exceptional violence and danger. They illuminate the larger scene of the Hitler years as well as the intimate scene of the Peierls family. They show us how quiet courage and sanity could bring a family safely through the storm and build a solid future upon the ruins of the past.

The two chief characters in the story are the German physicist Rudi Peierls and his Russian wife Genia. The two most dramatic incidents are the meeting of Rudi and Genia in Odessa in 1930 and the first calculation of the critical mass of a nuclear bomb by Rudi and Otto Frisch in 1940. Luckily for us, Rudi and Genia were separated for a year after their first meeting, so that we have an authentic record of their epistolary courtship in the magnificent series of letters exchanged between Leningrad and Zurich in 1930–1931. The Peierls–Frisch calculation of the critical mass was the decisive event that started large-scale projects to develop nuclear weapons both in Britain and America. The correspondence of 1940–1943 gives us an inside view of the fumbling efforts to get these projects organized in both countries, before they were finally and efficiently combined in 1943 at Los Alamos.

I had the good fortune to be a boarder in the Peierls home in Birmingham for the academic year 1949–1950 while I was working as a research student in Rudi's department. Thanks to Rudi's skills as a teacher and administrator, his department of theoretical physics was the best in Britain, having left Oxford and Cambridge far behind. Unmarried research students were welcome to stay in the Peierls home, where we were well fed and entertained by Genia. The home was a warm and wonderful chaos of teenagers and babies. The four Peierls children were Gaby aged 16, Ronnie 14, Kitty 2 and Joanna six months, no two of them alike, each having individual needs and schedules. Genia managed them all with her unique combination of loving heart and loud voice. Food was then still rationed in England, and gas-fires were regulated by shilling-in-the-slot meters which required an infinite supply of shillings. House-keeping for a big family was complicated. But Genia made light of all difficulties. She always had time left over from her family to take care of personal problems for me and other student boarders. When any of us complained about anything, she would say, "What you have to complain, compared with Jew under Hitler or prisoner in Russia?" When I expressed surprise that she had another baby so soon after Kitty, she said, "There's never convenient time for having baby". When one of the students made an unsuitable marriage or bought an unsuitable motorcycle, she said, "Worst thing for health is to be an idiot". Since the definite article does not exist in Russian, she considered that it was also superfluous in English. Having myself been raised in a proper well-ordered British home without much noise or excitement, I fell in love with the uninhibited Russian-Jewish way of life that Genia brought to Birmingham.

From time to time, Rudi and Genia would throw a big party at the house for their numerous friends and students. Massive quantities of food and drink would appear and the boarders would be put to work peeling potatoes and apples. A frequent visitor on such occasions was Klaus Fuchs, who had been a close friend and collaborator of Rudi in the bomb project. Klaus was fond of the Peierls children, and Genia said he had been their favorite baby-sitter when they were together at Los Alamos. He sat quietly at the parties and spoke only when spoken to. I talked with him mostly about Harwell, the British Atomic Energy Research Establishment, where he was head of the Theoretical Division. He was enthusiastic about Harwell and the prospects for peaceful development of nuclear energy. In February 1950 we were flabbergasted to learn that Klaus had been arrested and had confessed to being a Soviet spy. At first Genia did not believe it, but Rudi went to visit Klaus in jail and confirmed that it was true. For Genia, with her long experience of living in fear of the Soviet police, the key to survival was to have friends that one could trust, and the unforgivable sin was the betrayal of that trust. For once, she was speechless with anger. After Klaus was convicted, she too visited him in jail and gave him a piece of her mind.

While Genia was welding the families of students and staff into a tight community of friends, Rudi was building his Department of Theoretical Physics into a first-class international center of research. He attracted bright young people from all over the world. During my time in Birmingham I shared ideas with Gerald Brown from America, Frank Nabarro from South Africa, Jens Lindhard from Denmark, Stuart Butler from Australia, Wladek Swiatecki from Poland and Tony Skyrme from England, besides many others whose names I have forgotten. The atmosphere of youthful exuberance was similar to the atmosphere at the Institute for Advanced Study at Princeton where I had spent the previous year. But in many ways Rudi was a better leader than Oppenheimer for a group of young people. Rudi was a platoon-leader while Oppenheimer was a general. Rudi was friendlier and more personally involved with the troops. He was actively engaged in research as well as teaching. During my two years in Rudi's group, the most important discovery was made by Rudi himself, who found a new and elegant method of deriving quantization rules for any classical field theory. Rudi's method gave for the first time a clear and intuitive meaning to the uncertainty principle in quantum field theory.

In summer 1951 I said good-bye to Rudi and Genia and went to start a new life at Cornell. Richard Feynman had moved from Cornell to Caltech and I was invited to take Feynman's place, an offer I could not refuse. But I left Birmingham with a heavy heart. Having once been a member of Rudi's team and Genia's incomparable household, I knew that I had reached childhood's end. For the rest of my life I would be making my own way in the world, without Rudi's wisdom and Genia's warmth to fall back on. Many years later, when my own children were growing up, we spent a summer in Seattle together with the Peierls family, and my children got to know Rudi and Genia. By that time, Rudi had become Sir Rudolf and Genia was Lady Peierls. But they had not changed. Genia still gave us good advice in her uniquely Russian version of English, and my children called her "The Loud Lady." In these letters I can still hear her voice.

Freeman Dyson
Institute for Advanced Study, Princeton

Acknowledgements

On a sunny June afternoon in 1992, when I was in the middle of writing up my doctoral thesis in contemporary history, a college friend suggested that I accompany him to a public lecture in physics. I had always been interested in the sciences, and I would have called myself mathematically literate, as I had completed an undergraduate degree in mathematics. But I very much doubted that my knowledge of physics would be sufficient to follow a lecture — public or not. My friend, a theoretical physicist himself, assured me that I need not worry, and I reluctantly went along finding myself in the presence of many eminent scientists and even more keenly interested students who had come to listen to the 1992 Dirac Memorial Lecture at Cambridge University. The lecture on 'Broken Symmetries' was delivered by Rudolf Peierls. I was mesmerized by a scientist who, well into his eighties, was not only still actively pursuing research, but who was making a point of communicating his findings to an audience of experts and non-experts alike. I was fascinated by the clarity of the exposition, and I was intrigued to find out more about the man behind the lecture, who was evidently more than just 'another physics professor'. Little did I know at that time that this lecture had triggered an interest that would occupy me for years to come and that would ultimately lead to this edition of two volumes of correspondence of Sir Rudolf Peierls. (Little, also, did I know that the friend who encouraged me to attend the lecture would continue to observe and comment on my work on Rudolf Peierls initially as an 'expect adviser' and later as my husband.)

Not long after the completion of my Ph.D. I joined the University of Birmingham to become a lecturer in modern history. Birmingham, of course, was the place where Rudolf Peierls had spent the longest part of his distinguished career as a researcher and teacher of theoretical physics. It was not long before I made contact with Sir Rudolf, who, after a characteristically thorough investigation of my credentials as a 'numerically literate' historian gave me permission to research his life and work and granted access to all his papers, both the archival and privately held collections. My original plan to write a biography soon gave way to the current project of editing a selection of his vast collection of private and scientific correspondence in order to make this most valuable source available to a much wider audience.

Most of the letters reproduced in volume 1 of the Selected Correspondence of Rudolf Peierls are held in the Bodleian Library, Oxford. Sir Rudolf had passed his papers on to the Contemporary Scientific Archives Centre in two instalments in the mid 1970s and mid 1990s. They were catalogued and duly deposited at the Department of Western Manuscripts of the Bodleian Library, Oxford. I am grateful to Colin Harris and his colleagues in the Western Manuscript Department for all their support and advice, which greatly facilitated the completion of the project.

Many of the early letters exchanged between Rudolf Peierls and his wife and her family remain in family possession. Some of the material is reproduced here with kind permission of the Peierls family. Other material was collected from the Library of Congress (Oppenheimer Papers), the Pauli Archive at CERN, the Sommerfeld Archive at the Deutsches Museum, Munich, the Division of Rare and Manuscript Collections, Cornell University Library (Bethe Papers), the Churchill Archives Centre, Cambridge (Chadwick Papers, Dirac Papers), the National Archives, Public Record Office, London, and the Niels Bohr Library of the AIP Center for History of Physics. In all these archives and many other libraries I have received valuable assistance in assembling this collection and researching the historical background of the material and the work described therein.

Material authored by Rudolf and Genia Peierls, Heinrich Peierls, Hermann and Annie Krebs, Siegfried and Jenny Peierls and Nina and Alfred Peierls is reproduced by permission of the Peierls family; letters written by James Chadwick are reproduced by kind permission of the Master and Fellows of Churchill College; letters authored by Wolfgang Pauli are reproduced by permission of the Pauli Archive, CERN; letters written by Niels Bohr are reproduced by permission of the

Niels Bohr Archive, Copenhagen; the letter written by Carl-Friedrich von Weizsäcker is reproduced by permission of the Carl Friedrich von Weizsäcker-Gesellschaft e.V. Some of the letters reproduced in this volume have previously been published. This applies to some of the letters exchanged with Wolfgang Pauli,[a] Niels Bohr[b] and V.A. Fock.[c]

Among other copyright holders who have granted permission to reproduce letters in this collection are: Mrs. Rose Bethe, Prof. Jochen H. Heisenberg, Dr. I. Landau, Dr. Olga Cherneva, Frau Monika Baier, Mrs. A.F. Crampin, Prof. Gustav Born, Christian Suter, F. Anthony Placzek, the Fowler family, Mrs. Jane Gatrell, Dr. Susan K. Martin, and Mrs. K.D. Baxandall.

Research for this project, in particular archival work in Russia and the United States, has been supported by the British Academy and the Royal Society. Furthermore, the Royal Society contributed to the funding of translations of Russian material into English. The work was facilitated by a term's research leave from the University of Birmingham and an AHRB grant which permitted the extension of this leave by a further term.

Much of the early correspondence between Rudolf and Genia Peierls was in Russian. This material was translated by Vladimir Kviatkovsky, St. Petersburg. To my great relief, he agreed to working from the manuscripts. Those with first hand experience of Genia Peierls' handwriting will appreciate the magnitude of his task.

My thanks go to the Peierls family, and in particular to Jo Hookway (née Peierls) who has supported this project with enthusiasm, insider knowledge and much common sense. She greatly facilitated access to material which had not been archived and is still in family possession, she read and commented on parts of the book, and gave useful insights into many aspects of the Peierls family life which would have been difficult, if not impossible, to gain in any other way.

When Jo Hookway spoke at the opening of the Rudolf Peierls Centre for Theoretical Physics in Oxford in 2003, she jokingly asked whether

[a] Wissenschaftlicher Briefwechsel.
[b] Collected works.
[c] J. Frenkel, Memoirs, letters, documents, "*Science*", 1986: 401–405.

those in charge knew what they were doing, when they chose to name the building after Peierls. Having lived with the name PEIERLS herself for 21 years, she admitted that one of the joys of being married was having an easy-to-pronounce surname. My own children are probably just about the only 10- and 11-year old children in this country, who can spell and pronounce the name Peierls to perfection. Having watched their mother work on this edition almost from the moment they were born, (and, in the case of my daughter, having spent a lot of time and effort reading out letters and helping me transcribe them) they deserve a special thanks. It is to them and to all other budding scientists that I dedicate this book.

Summer 2007

Sabine Lee
University of Birmingham

A Tennis Player — Not a Golfer

> ... Several times when I was in Europe French, German and other scientists commented how unjust it was that he had never received the Nobel Prize for physics. Maybe it was his breadth which was not helpful — nuclear and solid state physicists were surprised that he had been active in other fields — each had thought he had done plenty in one field. Others thought that his involvement with nuclear fission had been controversial. Whatever the verdict — he should have received it as one of the great thinkers in 20th-century physics.

Thus Gillian Gehring's[1] recollection and assessment of Rudolf Peierls, expressed in a letter written to Peierls' youngest daughter Jo shortly after her father's death in September 1995.[2] Gillian Gehring referred to herself as one of the 'intellectual grandchildren' of Rudolf Peierls; all her teachers, Sam Edwards, Brian Flowers and Walter Marshall[3] had been Peierls' pupils. She is one of many theoretical physicists in this country,

[1]Gillian Gehring (1941–), studied physics at Manchester and Oxford, where she obtained her D. Phil in 1965 working with a former Peierls student, Walter Marshall. After research and lecturing at Oxford, she became Professor of Condensed Matter Theoretical Physics at Sheffield University in 1989.

[2]Letter Gillian Gehring to Jo Hookway, 21.9.1995; family possession.

[3]Samuel Frederick Edwards (1928–), studied at Cambridge and Harvard before moving to Princeton for research. After posts at Birmingham and Manchester, Edwards became Professor of Theoretical Physics, Plummer Professor and later Cavendish Professor of Physics at Cambridge. Brian Hilton Flowers (1924–), studied physics at Cambridge. After work at the Anglo-Canadian Atomic Energy Project at Montreal, he joined the UK Atomic Energy Research Establishment (AERE) at Harwell and did research at Birmingham University after which he returned to the AERE; later he became chairman of the Science Research Council (1967–72), rector of Imperial College London (1973–1985), vice-chancellor of the University of

and indeed abroad, who can trace their intellectual ancestry directly or indirectly, to Sir Rudi, or 'Prof', as numerous of his students and junior colleagues would call him with respect and affection. Many thought of him as an impressive researcher; as many or more remember him for his great contribution to theoretical physics through his teaching.

To adopt his wife Genia's well-rehearsed characterisation of Rudolf, he was an intellectual tennis player, not a golfer.[4] He needed partners in his research; in the early days fellow students, then colleagues, or later his own more advanced students. He thrived on bouncing ideas off and receiving the return from others, or conversely, returning balls that had been played into his court and thereby moving 'the game' on to higher levels. It is not surprising, therefore, that some of Peierls' important achievements have occurred as a result of direct and intense contact with others. Among the more widely-known are Bethe and Peierls' attempt to develop a theory of the deuteron photo-effect in 1933. In a conversation with Chadwick in Cambridge, the latter had challenged the two friends to develop a theory of this phenomenon. On the train, returning from Cambridge to Manchester, where Bethe and Peierls were based at the time, doubtless after intense conversations and with the help of numerous backs of envelopes, they succeeded in developing a consistent theoretical approach.[5] More significant, if not scientifically then politically, was another one-to-one effort, namely the collaboration, in early 1940, with Otto Frisch[6] that led to the calculation of the critical mass of uranium 235, a finding that suggested the theoretical feasibility of an atomic weapon based on a nuclear self-sustaining

London (1985–1990) and chancellor of Manchester University (1994–). Walter Marshall, received his Ph.D. in physics from Birmingham University after only two years of research training; he went on to work at the AERE with Brian Flowers, and later held important government advisory positions under Labour and Conservative governments.

[4]Genia Peierls to Mr Raison (journalist working for the *New Scientist*), 16.7.1961, vol. II, chapter 8 of *Peierls Correspondence.*

[5]Rudolf Peierls, *Bird of Passage. Recollections of a Physicist*, (Princeton: Princeton University Press, 1985), p. 103; see also letter Hans Bethe to Rudolf Peierls, 8.9.1995, vol. II, chapter 11 of *Peierls Correspondence.*

[6]Robert Otto Frisch (1904–1979), received his doctorate from the University of Vienna in 1926; he continued research in Berlin, Hamburg and Copenhagen before moving to Birmingham in 1939. Here, Frisch and Peierls calculated that the fissioning

chain reaction.[7] And these two examples are merely the tip of the iceberg.

Being an efficient collaborator, however, does not equate with being an enthusiastic or a good teacher. For Rudolf Peierls, his own experiences as a student in the late 1920s, at a time of great excitement, stimulation and achievement in physics, were a key to his own passion for the subject as well as his approach to communicating it. He had learnt his trade — theoretical physics — in an environment with a belief in intellectual exchange as an essential ingredient of scientific progress. The most influential among his teachers, Sommerfeld in Munich, Heisenberg in Leipzig, Pauli in Zurich and Bohr in Copenhagen, in their distinct ways, created settings which would provide for a spirit of collaboration and communication. Sommerfeld had created what justifiably became known as the 'Sommerfeldschule', a school of theoretical physics that set standards and acted as a nursery for the spread of this growing and thriving discipline.[8] Sommerfeld was not only a superb, imaginative and creative scientist but also an outstanding teacher and facilitator of academic discourse at his institute at Munich and beyond.

Heisenberg, a Sommerfeld pupil, imported some of the seminar experiences he had gained at Munich to Leipzig, when he took up his appointment as professor of theoretical physics there in 1927. Apart from regular seminars, he organised, together with Peter Debye, what became known as the 'Leipziger Festspiele', meetings dealing with a particular research theme of current interest.[9]

Wolfgang Pauli, another Sommerfeld pupil, did not make quite the same impact as a teacher generally or as an advisor to undergraduate

---

of uranium had the potential to create a chain reaction which could be used to develop an atomic weapon. Frisch later joined the Manhattan Project. After the war he worked for the AERE at Harwell before taking up the Jackson Chair of Physics at Cambridge in 1947.

[7]'The Frisch-Peierls Memorandum of 1940', in R.H. Dalitz and R.E. Peierls (eds.), *Selected Scientific Papers of Sir Rudolf Peierls*, London: World Scientific, 1997, pp. 277–282.

[8]Michael Eckert, *Die Atomphysiker: eine Geschichte der theoretischen Physik am Beispiel der Sommerfeldschule*, Braunschweig/Wiesbaden: Vieweg, 1993; esp. ch. 2.

[9]The meetings were popular as a focal point of theoretical research in Germany, and the proceedings were published by Peter Debye as *Leipziger Vorträge*.

students in particular. But he had great influence on talented young scientists through his collaboration with his research students, assistants and colleagues, in personal discussions and in particular in correspondence. Peierls later summed up Pauli's role when he wrote: 'To discuss some unfinished work, some new or speculative ideas with Pauli was a great experience, because of the clarity of his understanding and his standard of intellectual honesty which would never let a slipshod or superficial argument get by.'[10]

Another influential mentor of Peierls', Niels Bohr, of course, was at the centre of much of the theoretical physics research in the early 20th century, and scores of young, promising physicists undertook their regular pilgrimages to Denmark to spend some time absorbing the spirit of 'science in the making' when attending the Bohr festivals at his institute in Copenhagen. As Peierls himself acknowledged later, 'Bohr's influence on modern physics should not be viewed merely through his own contributions, great as they are: an important part of it consisted of the inspiration, encouragement and guidance he gave to others.'[11]

Rudolf Peierls was among the generation of physicists who spent their 'Lehr- und Wanderjahre', years as traveling apprentices, at the research centres across Europe, at Munich, Göttingen, Berlin, Leipzig, Zurich, Copenhagen, Leningrad, Kharkov, Leiden, Cambridge, Paris or Rome. He was among those who would go on to move the boundaries of science even further. Others included Landau, Gamow, Weisskopf, Bethe, Placzek, Kramers, Rosenfeld, Born, Bloch and many more. While the majority of those members of the community of theoretical physicists played an important role in scientific progress through their own research, few felt quite as passionate about and had such an immense influence on teaching the discipline as Rudolf Peierls had. During the post-war years at Birmingham, he built, virtually from nothing, what arguably became the best department of theoretical physics in England, if not in Europe.

---

[10] R. Peierls, 'Wolfgang Ernst Pauli, 1900–1958', *Biographical Memoirs of Fellows of the Royal Society*, **5** (1960), pp. 175–87, here, p. 183.

[11] R. Peierls, 'An Appreciation of Niels Bohr', *Proc. Phys. Soc.* **81** 793–99 (1963), here p. 798.

When looking back at his own experiences as a student, Peierls fondly remembered the warm and friendly atmosphere of the Bohr institute in Copenhagen. He commented that Bohr's keen interest in people turned the personal relations in his institute into a family-like atmosphere and in fact an extension of the Bohr family, into which members of the institute 'were allowed to intrude at any time'.[12] The same could be said about the Peierls household in Birmingham and later in Oxford. What impressed his numerous students and junior colleagues about the 'Peierls experience' was the way in which departmental affairs were allowed to be extended into the 'Peierls family'. There was no boundary between home and other portions of Rudolf Peierls' life,[13] and to many students and postdocs the Peierls family became 'their family' for the time of their stay at Prof's institute and sometimes beyond. This was partly due to the fact that Rudolf Peierls took a genuine interest in his students, but it was also due to the fact that this deep-rooted concern was shared by his wife Genia, who felt responsible for her nuclear family as much as the 'adopted' one.[14]

Rudolf and Genia Peierls' shared a remarkable relationship. They first met at a physics conference in Odessa in 1930. At the time Rudolf was Wolfgang Pauli's assistant at the ETH in Zurich and Genia had just completed her physics degree in Leningrad. After the conference and a short holiday thereafter, their ways parted again until Rudolf returned to Leningrad in March 1931 to give some lectures. In the course of this trip to Russia, he married Genia. Genia had to remain in Russia waiting for her visa which allowed her to join Rudolf in Zurich in the autumn of 1931. At first sight the two were an unlikely couple. Among the reminiscences compiled by friends for Genia's 70th birthday was one contribution which contemplated defining 'a genia' as a unit of 'loudness, big-heartedness, self-confidence, loving concern, bossiness, generosity, stubbornness, unbreakable English, compassion, bad verse,

[12]Peierls, *Bohr. An Appreciation*, p. 799.

[13]Comment made by Ronnie Peierls, Rudolf Peierls' eldest son, in his father's funeral address. Bodleian, *The Correspondence and Papers of Sir Rudolf Ernest Peierls.* Supplementary Catalogue. A 207. Hereinafter cited as *Peierls Papers.* Supp.

[14]Gerry Brown, talk given at the naming of the Rudolf Peierls Centre for Theoretical Physics, Oxford, 2.10.2004.

fantasy, hypnotic gaze and irresistible kindness', and it concluded that whatever the final verdict, Genia stood for something 'larger than life: the milli-genia should be perfectly adequate for normal purposes'.[15] In contrast, Rudolf was unassuming and modest, quiet, and, at times, even shy. However, he was no less determined than his wife, and the two complemented each other in many important aspects. The marriage lasted for 55 years until Genia's death in 1986. The intense affection of their early long-distance relationship remained strong throughout their married life. To Rudolf, Genia was a friend and companion, but she was also a counsel. As one of their close friends would later recall, Rudolf would often listen to a student's or colleague's problem, promise to think about it, go home and discuss the problem with Genia only to return the next day with an answer, undoubtedly often provided by Genia's input.[16] Genia was not 'merely' the professor's wife, she herself was the initiator of many of the not strictly speaking scientific aspects of departmental life, ranging from entertainment to housing, from student counseling to health and safety advice.

Rudolf Peierls' role as a physicist, as a teacher, as an administrator and as a friend to many scientists and non-scientists cannot be grasped without an understanding of his early academic experiences, and these, similarly, cannot be understood without an inkling of the close bond and intensity of the relationship between him and his wife Genia. Fortunately for the historian of science, Rudolf Peierls' correspondence, ranging from the early 1920s to his death in 1995, allows us an unusually precise insight into these aspects as well as into the more directly science-related parts of his career. This has been the core consideration in the decision to include a substantial part of private letters in this edited selection of the Peierls correspondence.

The correspondence

Rudolf Peierls was a prolific letter writer; and he was also a meticulous collector of papers of all description including his own correspondence.

---

[15] *Genia Peierls, 25.7.1978*, private possession, contribution of Denys and Helen Wilkinson.

[16] Editor's interview with Gerry Brown, 2.10.2004.

Upon his retirement from the Wykeham Chair of Physics at Oxford, he made available his papers to be catalogued[17] and later deposited at the Bodleian Library, Oxford.[18] The collection, consisting almost entirely of correspondence written and received, covers all aspects of his career to 1974, including research, committee activities, work on advisory boards, appointments, overseas visits, and some biographical material. In 1994/5 the vast majority of the remaining correspondence, consisting of a large amount of family correspondence, and much additional material relating to Rudolf Peierls' retirement, was made available, and duly deposited at the Bodleian. Only a small proportion of largely family related documents have been retained by the Peierls family. The Peierls papers consist of several thousand letters, and only a small fraction of these can be reproduced here. These are supplemented by some additional material (mostly letters written by Rudolf Peierls and held by the recipients and therefore deposited elsewhere).[19] However, the core of the correspondence, consisting of letters received by Peierls and carbon copies of letters sent by him, is held at the Bodleian. The material is remarkable for a number of reasons.

- The earliest letter dates from 1922, when Rudolf Peierls was still at school, and regular exchanges of letters commence in 1926, when Peierls began his studies at Munich; some of these are continued until Peierls' death in 1995 (e.g. with his school friend Heinz Rudolph or his closest friend Hans Bethe, whom he met in Munich in 1926).

- For the early years, the core of the material consists of letters received by Peierls. However, while at university, Peierls was

[17]Contemporary Scientific Archive Centre, CSAC 52/6/77 and National Cataloguing Unit for the Archives of Contemporary Scientists, NCUACS 57/6/95.

[18]Bodleian, *The Correspondence and Papers of Sir Rudolf Ernest Peierls.* Ms.Eng.misc.b197ff; and *The Correspondence and Papers of Sir Rudolf Ernest Peierls.* Supplementary Catalogue.

[19]These include Pauli papers at CERN/Geneva, Bethe Papers at Cornell, Oppenheimer Papers at the Library of Congress/Washington, Sommerfeld Papers at the Deutsches Museum/Munich, Chadwick Papers at the Churchill Archives Centre/Cambridge, Dirac Papers at Florida State University.

given his ‘friend Corona’[20] a typewriter which he used for letter writing as well as for academic work, and therefore, increasingly frequently, carbon copies of letters written by Peierls are collected, too.

- A significant part of the correspondence covers the early phase of the acquaintance of Genia and Rudolf Peierls between their first meeting in the summer of 1930 and their marriage in March 1931, and from May 1931 until Genia was finally allowed to join Rudolf in Switzerland in the autumn of 1931.

The material selected for this edition relates to the life and work of Rudolf Peierls, and by implication presents a balance of personal letters and scientific correspondence. Volume one covers the years 1926 to 1945 (with one additional letter from 1922), volume two covers the years 1945 to 1995.

Several criteria played a part in selecting the material for inclusion. Material was included if it was judged to be

- significant for an understanding of Rudolf Peierls’ personal and academic development, for his family background and the fate of his family throughout the Third Reich (mostly correspondence with Genia, and with Else and Heinrich Peierls)

- of historical interest in view of the specific situation of the Peierls family as an assimilated German Jewish family in interwar Berlin (largely family correspondence with Heinrich Peierls (father) and Else Peierls (stepmother) and siblings)

- historically significant for an understanding of scientific developments, especially in solid state and nuclear physics and quantum mechanics (e.g. early correspondence with Bethe, Placzek, Landau, Heisenberg, Pauli and Dirac; later correspondence relating also to particle physics e.g. correspondence with Brown, Dalitz, Dyson, Edwards)

---

[20]Letter [58].

- interesting with view to the historical development of the situation of Jewish and non-Jewish refugees and particularly refugee scientists (mostly correspondence with Bethe, Placzek and Bohr)
- of significance for an understanding of the circle of young Russian theorists around Landau, Gamow, and Ivanenko (mostly correspondence with Genia, Landau, but also Bohr and Placzek)
- of significance for an understanding of Britain's role in the development of nuclear weapons (mostly correspondence with Chadwick, Frisch, Akers, Simon, Fuchs)
- of significance in relation to Rudolf Peierls' role in the development of teaching and research of theoretical physics in post-war Britain in general, and at Birmingham in particular
- of significance for an understanding of the Fuchs case and recurring espionage allegations
- of significance where Rudolf Peierls contributed to an understanding of science historical issues as one of the key figures of early quantum mechanical developments

The majority of material consists of letters either written or received by Rudolf Peierls. Exceptions are two pamphlets written by scientists at a conference in Odessa in 1930 which capture the spirit among theoretical physicists at a crucial time of its international development, a letter written by Genia to Klaus Fuchs in early 1950, a memorandum about lessons to be learnt from the Fuchs case, a memorandum about dealings with German scientists during the post-war era, a letter written by Genia in preparation for a biographical sketch of her husband in the *New Scientist* in 1961, and some items relating to Rudolf Peierls death in 1995.

Editorial comments

1. The letters are listed chronologically and, unless stated otherwise, reproduced in toto.

2. The transcription is limited to the text content, not to layout or graphic details.

3. Letters are reproduced in the original language, with the exception of Russian letters which appear in English translation. Where the original mixes language (which happens frequently in correspondence between Genia and Rudolf Peierls,) the passages not reproduced in the original are italicised.

4. Typos and misplaced letter sequences are corrected.[21]

5. Reproduced letters are held in various locations. These are not noted as comments to the letters but can be ascertained from a list in the appendix.

6. Where letters are misdated in the original, the correct date is reproduced with an explanation in the footnote.

7. Where manuscripts do not contain date or location, these are added in square brackets, if they can be deducted from the context and content.

8. Handwritten additions, where included, are identified as such.

9. Editorial additions are made in square brackets.

10. Deleted text is not transcribed.

11. The reproduced material makes no distinction between single and multiple underlining.

12. Gaps arising from punching or filing are filled and only noted in square brackets, if there are doubts. Abbreviations are only completed in square brackets where necessary.

[21]This applies in particular for correspondence in the late 1980s and 1990s. As a result of failing eyesight in later life, many of Rudolf Peierls' word-processed letters written in those years contain a large number of typographical errors which have been corrected.

13. Duplication of $\overline{n}$ and $\overline{m}$ is done without being noted.
14. Illegible words are noted as [??], with [?] for one word, [??] for two words and [???] for three or more words.
15. Footnotes in the document are marked with asterisk and added at the end of the relevant page. Editorial notes appear at the bottom of the page.
16. Repeated words at the page break are transcribed as one word.
17. Handwritten additions are marked as such.

1. Lehr- und Wanderjahre:
The Traveling Apprentices of the Second Quantum Generation 1922–1930

A 'mountain air compressor with integrated temperature control' was the remarkable invention about which Rudolf Peierls communicated with his friend Heinz Rudolph in the earliest surviving letter of his long and extensive correspondence.[22] Heinz commented that the idea, while 'nice', served better as a source of amusement than a serious invention, not least because it would not be very good value for money; nevertheless it indicates that Rudolf Peierls, from a young age was interested in science and engineering. He had always been a bright child, who found school work easy and was keen to probe further into areas that interested him most: the sciences. His friends remember many an occasion when he would leave their play in order to 'think' only to return once he had solved whatever problem puzzled him at the time.[23] Rudolf's original plan had been an engineering career. His interest in all things mechanical went beyond the usual child's fascination with automobiles, telephones and transistors. But his family, doubting his practical abilities, persuaded Rudolf to settle for physics instead. It appears that father Heinrich's lack of manual dexterity was generally assumed to have been passed on to his younger son. Heinrich, even in adulthood could not tie a knot and had to have his boots fitted with special hooks on which to anchor the laces without having to tie them. Nor could he shave himself without causing himself bodily harm and he had a barber doing the job for him daily.[24] The rather unpersuasive

[22]Letter [1].

[23]Letter Waltraud Krause to Rudolf Peierls, 27.8.1989, vol. II, chapter 11 of *Peierls Correspondence.*

[24]Rudolf Peierls, *Bird of Passage. Recollections of a Physicist*, Princeton: Princeton University Press, 1985, p. 5.

arguments against engineering, namely his supposed clumsiness and his poor eyesight failed to convince the young Rudolf. Yet, he had little choice but to obey his father's wishes.

No lesser than Fritz Haber,[25] an old family friend, advised Rudolf to consider studying experimental physics, and while Rudolf would have preferred to study elsewhere, he bowed to parental pressure and enrolled for a course in experimental physics at Berlin. He was soon to discover that first-year students were prevented from taking any practical courses due to overcrowding, and thus he became a theoretician by default.[26] This accidental choice was to be a decisive career move, as Rudolf firmly established himself in the theoretical field in subsequent years.

Not unlike many young men of his generation, Rudolf fitted into the mould of his protective middle-class assimilated background.[27] The third child of Heinrich and Elisabeth Peierls, Rudolf Peierls had been brought up in a materially comfortable position. His father (1867–1945), who had joined the Allgemeine Elektrizitätsgesellschaft (AEG) in 1888 had worked his way up to become director of the factory in Berlin-Oberschöneweide (in succession to Erich Rathenau). Eventually he became a member of the managing board (1908) and later a member of the Supervisory Board (1929). His first wife, Elisabeth, died in 1921, and Heinrich remarried. His non-Jewish second wife, Else (née Hermann) was the daughter of a famous actor and the sister-in-law of the playwright Ludwig Fulda, which added a stronger cultural dimension to the Peierls household.

Throughout his student life at Munich, Leipzig and Zurich, then as a research assistant to Pauli at Zurich, and later while occupying research

[25]Fritz Haber (1886–1934), studied chemistry at Heidelberg and Berlin. After research at Jena, Zurich and Karlsruhe, he became professor of Physical Chemistry at Karlsruhe in 1906 and Director of the Institute for Physical Chemistry and Electrochemistry in Berlin in 1911, from where he was forced to retire in 1933.

[26]Peierls, *Bird of Passage*, p. 18.

[27]Sebastian Haffner (born as Raimund Pretzel (1907–1999)) a German Jewish journalist growing up in similar circumstances, gives an insightful first-hand account of the scenario which bears remarkable similarities with Rudolf Peierls' experiences. Sebastian Haffner, *Geschichte eines Deutschen*, Stuttgart: Deutsche Verlags-Anstalt, 2000.

and lecturing positions at Rome, Cambridge, Manchester and Birmingham, Rudolf Peierls exchanged letters with his father and stepmother regularly. A small number of these is reproduced here. They provide an interesting insight into the way in which their family relationship developed. Rudolf, who had initially been at the receiving end of advice increasingly emancipated himself from parental influence and eventually reversed the dynamics by providing assistance in later years, when his father and stepmother finally decided to emigrate from Nazi-Germany, years after their children had taken similar steps to settle in the United States and the United Kingdom.[28]

While a student at Berlin, Rudolf Peierls encountered some of the leading figures in scientific research: Max Planck, Walther Nernst, Walther Bothe to name but a few.[29] The first semesters of academic study had also given him the first taster of a new exciting discipline that was to dominate his work in years to come: quantum physics. The short account of his first university year at Berlin in his autobiography describes Rudolf as a conscientious student who, despite overburdening his time-table with a large number of lectures outside his main discipline found time for an active social life. Many of his friends, interestingly mostly from an assimilated Jewish background, shared his interest in mathematics and the sciences, and were probably also active in the mathematical and physics society, MAPHA,[30] a student club which Rudolf Peierls later recreated at Munich.[31]

The first decisive step into a more independent life was Peierls' move to Munich in the autumn of 1926, when he became a student at Arnold Sommerfeld's institute. From the turn of the century most German universities offered some lectures in theoretical physics, and almost everywhere students could choose to study topic in this field at doctoral level. Yet, although there were already some centres of theory such as Göttingen or Berlin, it was Arnold Sommerfeld in Munich who as professor from 1906 established the first school of theoretical

[28]Letters [264], [276], [291], [295], [303], [304], [319], [321].

[29]Studienbuch, *Peierls Papers* Supp. A.31.

[30]Mathematisch-Physikalische Fachschaft. For details see *Peierls Papers*, Supp., A.8.

[31]Letter [5].

physics in Germany, a scientific community from which the 'spirit of Sommerfeld's teaching' was to spread across Germany, Europe and the world.[32]

Unlike Planck at Berlin whose research genius did not find expression in his teaching, Sommerfeld deservedly had the reputation of being a superb communicator and great teacher. His lectures were a model of clarity. It is no coincidence, that the list of his students and assistants includes virtually everybody who made his name in (quantum) physics in years to come: Pauli, Heisenberg, Bethe, Peierls, von Laue, Kossel, Ewald, Lenz, Herzfeld, Wentzel, Heitler, Houston, Eckart, Rubinowicz, Pauling, Laporte, Brillouin, Condon, Fröhlich, London, Landé and many others.

Being introduced to quantum mechanics in such an inspiring manner and being confronted by Sommerfeld with the topical question of the electron theory of metals proved to be important for Peierls' short-, medium- and long-term physics career. As significant was his acquaintance with Hans Bethe, a fellow student one year his senior. They shared an interest in and a passion for physics which resulted in a lifelong friendship that went far beyond the research-related acquaintance. The correspondence between Bethe and Peierls, which commenced with an amusing postcard written from Bethe to Peierls in the summer of 1927 in his customary witty and insightful manner,[33] would conclude almost seventy years later with an equally insightful resumé of seventy years of friendship and collaboration in science, written at the eve of Rudolf Peierls' death and never received by him.[34] It is evident in the letters, that the two young men shared many interests which, incidentally, they had in common with many young scientists of their generation. They were keen hikers and skiers, they were interested in literature, theatre and music, they enjoyed traveling, but above all they

[32]Michael Eckert, *Die Atomphysiker. Eine Geschichte der theoretischen Physik am Beispiel der Sommerfeldschule*, Braunschweig/Wiesbaden: Vieweg, 1993, pp. 37ff.

[33]Letter [9].

[34]Hans Bethe to Rudolf Peierls, 8.9.1995, vol. II, chapter 11 of *Peierls Correspondence*.

were keen to share the latest developments in their discipline, especially those driven by their own teachers and friends.[35]

Peierls' move to Munich was a liberating experience for the nineteen-year old. Although he had had an active social life in Berlin and had enjoyed the rich cultural and intellectual life in the Golden Twenties at its centre, the German capital, it was not until his settling in Munich that he gained a degree of independence. But even then, his parents took more than just a superficial interest in his academic and social life. Heinrich Peierls expected to be informed about his son's whereabouts in some detail, and he continued to demand a say in the organisation of the latter's spare time, his travels and his holidays.[36] The tone of the letters between father and son and stepmother and stepson changed over the years, and as time went on, pride in the intellectual achievements increasingly mingled with a willingness to allow Rudolf to take decisions himself.[37]

Peierls' years at Munich, the autumn of 1926 until the spring of 1928, were a time of rapid personal and scientific development. Basic ideas of quantum mechanics had already been worked out by de Broglie, Schrödinger, Heisenberg, Pauli, and Dirac, but the new formalism had not been tested widely on the problems which had defeated the old quantum theory of Bohr and Sommerfeld. In 1923, Arthur Compton had discovered the quantum nature of x-rays, thus confirming photons as particles;[38] 1924, de Broglie had proposed that matter had wave properties;[39] in early 1925, Pauli had formulated the exclusion principle for electrons in atoms;[40] and in the summer of the same year Heisenberg completed his famous paper widely accepted to be the breakthrough in quantum mechanics;[41] early in 1926, Schrödinger developed a wave-

[35]Letters [9–10], [12], [14], [17], [25], [31], [37], [40], [53].

[36]Letters [21], [23].

[37]Letters [33–34], [39].

[38]Arthur H. Compton, 'A Quantum Theory of the Scattering of X-rays by Light Elements', *Phys. Rev.* **21**, 483–502 (1923).

[39]L. de Broglie, *Recherches sur la théorie des quanta*, Thesis, Paris, 1924.

[40]W. Pauli, 'Über den Zusammenhang des Abschlusses der Elektronengruppen im Atom mit der Komplexstruktur der Spektren', *Z. Phys.* **31**, 765–83 (1925).

[41]W. Heisenberg, 'Über die quantentheoretische Umdeutung kinematischer und mechanischer Beziehungen', *Z. Phys.* **33**, 879–93 (1925).

mechanical description of quantum physics which included the formulation of the Schrödinger equation;[42] in the same year Fermi and Dirac found that quantum mechanics required a different way to count particles, the Fermi-Dirac statistics, which opened the way to solid state physics; and in the spring of 1927 Heisenberg formulated his Uncertainty Principle.[43]

This was the context of Peierls' first seminar paper.[44] Dirac and Jordan, independently but concurrently had proposed a theory for the description of measurements in quantum mechanics, which was to become known as transformation theory. Peierls was to report on the papers of Dirac and Jordan[45] to Sommerfeld's seminar. This was a difficult first assignment, but one which provided a useful learning experience, as Peierls recalled later.[46]

In anticipation of Sommerfeld's plan to travel the world in the summer of 1928, a cultural and scientific journey that would take him to the United States via the Far East,[47] Peierls followed his teacher's advice and continued his studies in Leipzig from mid-1928. Another of Sommerfeld's most promising pupils, Werner Heisenberg had been appointed to the chair there in 1927 (at the time Germany's youngest full professor) and had started a school of theoretical physics of some promise. In stark contrast to the elderly 'Geheimrat Sommerfeld' who was about to turn sixty, Heisenberg represented the new generation of theoreticians. 'Aufbruch in das neue Land'[48] is the title of the chapter on the discovery of quantum mechanics in Heisenberg's autobiography.

---

[42]His ideas were developed in a series of papers published in *Annalen der Physik* in 1926. *Ann. d. Phys.* **79**, 361–376; 489–527; 734–56 (1926); *Ann. d. Phys.* **80**, 437–90 (1926); *Ann. d. Phys.* **81**, 109–139 (1926).

[43]W. Heisenberg, 'Über die Grundprinzipien der Quantenmechanik', *Z. Phys.* **43**, 172–98 (1927).

[44]Letter [14]. See also *Peierls Papers*, Ms.Eng.misc.b197. (A.2).

[45]P.A.M. Dirac, 'The Quantum Theory of Emission and Absorption of Radiation', *Proc. Roy. Soc.* **A112**, 243–265 (1927); P. Jordan, 'Zur Quantenmechanik der Gasentartung', *Z. Phys.* **44**, 473–480 (1927).

[46]Rudolf Peierls, interview by J.L. Heilbron, 17./18.6.1963, Niels Bohr Library, American Institute of Physics, College Park, MD, USA.

[47]Eckert, *Atomphysiker*, pp. 119–20.

[48]Departure into a new land.

Heisenberg, for many, personified this expeditionary spirit of a new generation of intellectual adventurers.

When Peierls joined Heisenberg in Leipzig the latter was working on the theory of ferromagnetism. Although Peierls was still occupied with lectures and some experimental work,[49] he was given his first research assignments. After an unsuccessful attempt to work out a theory explaining the observation that spectral lines emitted by streams of fast-moving positive ions, so-called canal rays, were wider than expected,[50] Peierls turned to a rather more complex problem related to the doctoral work of Heisenberg's first Ph.D. student Felix Bloch. It also meant a return to the more familiar territory of the electron theory of metals which he had studied under Sommerfeld in Munich. The application of quantum mechanics to metals was one of the prime examples of the fast route from the discovery of first principles to the solution of problems which had puzzled scientists for a significant period of time and for which classical physics could not provide adequate answers. According to Pauli's exclusion principle and the Fermi-Dirac statistics electrons had to locate themselves in such a way that no energy state was taken up by more than one electron. This, the Sommerfeld School occasionally referred to as 'Wohnungsamt' (housing office) of electrons.[51] Encouraged by early results in the quantum mechanical approach to the electron theory of metals, Sommerfeld turned it into one of the central research areas of his institute.[52] While Sommerfeld's theories, based on quantum mechanics, explained a number of previously paradoxical experimental data,[53] they had not explained the conductivity in metals. Any metal

---

[49]See note 46.

[50]Peierls, *Bird of Passage*, pp. 33–34.

[51]Pauli to Wentzel, 5.12.1926, A. Hermann, K. v. Meyenn, V.F. Weisskopf (eds.), *Wolfgang Pauli. Wissenschaftlicher Briefwechsel mit Bohr, Einstein, Heisenberg u.a.*, vol. 1: 1919–1929, New York/Heidelberg/Berlin: Springer, 1979, p. 361. Cited hereinafter as Pauli, *Wissenschaftlicher Briefwechsel*, with volume number.

[52]This is evident, among others, in the fact that he lectured on the topic to his advanced students in the summer semester of 1927. Eckert, *Atomphysiker*, pp. 127ff.

[53]See in particular A. Sommerfeld, 'Zur Elektronentheorie der Metalle auf Grund der Fermi'schen Statistik. 1. Allgemeines, Strömungs- und Austrittsvorgänge', *Z. Phys.* **47**, 1–32 (1928).

of reasonable purity would show a level of conductivity which suggested that electrons could travel many times the distance between the atoms before colliding. Moreover, experiments demonstrated that this average 'mean free path' (the distance traveled without collision) became larger at lower temperatures.

Felix Bloch tackled some of these questions in his doctoral thesis.[54] He investigated the wave functions of electrons in a lattice by looking at the behaviour of electrons not in the individual atoms or molecules but in the overlapping electric fields of the atoms making up a crystal. Working on the assumption of a high potential in which the electrons were tightly bound to the lattice, he discovered that the energy states were not, as previously thought, discrete, but that continuous bands of allowed energies existed. Bloch's explanations did not take into account the fact that electrons in metals repel each other, as they were electrically charged. In his theory, electrons were treated no differently whether they were near an atom with a full complement of electrons (and thus neutral) or an atom which had an electron missing and was therefore positively charged.

Werner Heisenberg encouraged Peierls to consider this problem by assuming the exact opposite, namely that all atoms in the lattice had their full complement of electrons and that, therefore, the electrons could only change places.[55] Again, as in Peierls's first assignment in Leipzig, his research did not bear fruit in the form of a positive, publishable result. Taking the lead from work by Heitler and London who had investigated a similar problem in molecules,[56] Peierls concluded that the model suggested no conductivity at all in this case, as all the electrons would have to 'jump' simultaneously, a phenomenon the probability of which was negligible.

Although these initial studies in Leipzig appeared a frustrating phase in the scientific progress of the young physicist, they proved to be of great value beyond the issue of immediate 'output'. Firstly, they fa-

[54] F. Bloch, 'Über die Quantenmechanik der Elektronen in Kristallgittern', *Z. Phys.* **52**, 555–59 (1928).

[55] Peierls, *Bird of Passage*, p. 35.

[56] W. Heitler and F. London, 'Interaction of neutral atoms and homopolar binding according to the quantum mechanics', *Z. Phys.* **44**, 455–72 (1927).

miliarised Peierls with research at the cutting edge of physics. They deepened his understanding of the revolutionary science of the day — quantum mechanics. His work with Heisenberg also made him witness a completely different style of mathematical physics compared with that practiced by his first 'master', Sommerfeld. Sommerfeld's approach to theoretical physics was best summarised in his own words. He once addressed a group of students by saying: 'If you want to be a physicist, you must do three things — first, study mathematics, second, study more mathematics, and third, do the same'.[57] In contrast, Werner Heisenberg, relied more heavily on his brilliant intuition.[58]

Peierls spent much of the summer of 1928 in England. He saw his stay as an opportunity to improve his English, but also as a chance to make new academic contacts across the Channel. He renewed his acquaintance with Dirac,[59] whom he had briefly met at Leipzig[60] and he was introduced to R.H. Fowler,[61] who invited Peierls to give a talk at a meeting of the Kapitza Club, an informal discussion forum for theoretical physics in Cambridge.[62] His working vacation was something of a coming of age in his studies as well as in his personal development. His father and stepmother were beginning to accept his judgements as those of a maturing young man[63] and they increasingly allowed him to make his own choices (while still offering welcome and sometimes less welcome advice).[64]

[57]Daniel J. Kevles, *The Physicists. The History of a Scientific Community in Modern America*, Cambridge/Mass. and London: Harvard UP, 1995, 4th printing, p. 200.

[58]For the contrast between Sommerfeld, Heisenberg and Pauli see David Cassidy's description of Heisenberg's time in Munich, in David C. Cassidy, *Uncertainty. The Life and Science of Werner Heisenberg*, New York: W.H. Freeman and Co., 1992, pp. 91–109.

[59]Letter [22].

[60]R. Peierls, 'Dirac' in J.G. Taylor (ed.), *Tributes to Paul Dirac*, IOP Publishing, 1986, pp. 35–7, here p. 35.

[61]Ralph Fowler (1889–1944), studied mathematics at Trinity College Cambridge; after the First World War, under the influence of Rutherford, (whose daughter Eileen he was to marry in 1921) he turned his attention to physics. In 1932 he took up the post of the newly created Plummer Chair of Theoretical Physics.

[62]Peierls, *Bird of Passage*, pp. 35–6.

[63]Letters [21], [23].

[64]Letter [27].

On his return from England and prompted by Heisenberg, Peierls began the research project which would lead to his first published paper, an examination of the theory of galvano-magnetic effects,[65] a study of the anomalous or positive Hall effect. As early as 1879, E.H. Hall had tried to determine whether the force experienced by a current carrying wire in a magnetic field was exerted on the whole wire or whether it was only exerted on what would later be called the moving electrons in the wire. Hall suspected the latter.[66]

The phenomenon of the Hall effect is largely analogous to the deflection of cathode rays on a magnetic field. In some metals, however, it produces a positive sign as if the current was carried by positive carriers. An explanation of this paradox was impossible as long as electrons were visualised as freely-moving in the metal. Bloch's theory of conductivity explained that conductivity was caused by jumps of the electrons from atom to atom where their energy could be less than the maximum potential barrier between atoms. There was no classical analogue for the process and, as pointed out in the Peierls' paper, with the new understanding, it no longer represented force-free motion. The positive Hall-effect was then explained in terms of a concept of holes, when Peierls showed that electrons could give an anomalous sign of the Hall coefficient in a regular lattice.[67]

Peierls' time at Leipzig was stimulating because of Heisenberg's impressive presence, but also because of a remarkable array of exceptional scientists around him. Gregor Wentzel, who held the chair in mathematical physics at the time of Peierls' arrival at Leipzig, was replaced by Friedrich Hund on moving to Zurich later in the year. Peter Debye,

---

[65]R.E. Peierls, 'Zur Theorie der galvanomagnetischen Effekte', *Z. Phys.* **53**, 255–266 (1929).

[66]E.H. Hall, 'On a new action of the magnet on electric currents', *Am. J. Math.* **2**, 287–92 (1879).

[67]In a second paper, which was based on a report to the regional association of the German Physical Society in Leipzig, Peierls further elaborated on the results and demonstrated that the current was proportional to dE/dk (E = energy; k = wave vector) and that the rate of change of k in an electric or magnetic field was the same as for a free electron. These results were published in his second paper published in the same year. (R.E. Peierls, 'Zur Theorie des Hall-Effekts', *Phys. Z.* **30**, 273–4 (1929).

another Sommerfeld pupil, headed the experimental side of the physics institute. Guido Beck was Heisenberg's assistant[68] and through him Peierls met Georg Placzek, who became a close friend and with whom he was to collaborate closely until Placzek's untimely death in 1955.

In the spring of 1929, when Heisenberg accepted an invitation to lecture in the US, Japan, China and India, Peierls moved on to Zurich to work with yet another Sommerfeld pupil, Wolfgang Pauli.

Before embarking on new projects at the ETH, Peierls had to complete a task of a different nature. A Leipzig publisher had invited him to translate de Broglie's book on wave mechanics[69] from French into German, and Peierls agreed to take on the work. Despite this extra burden, he also quickly settled down to the research which would, in the summer of 1929 earn him a Leipzig Ph.D., a study of thermal conductivity in crystals with its recognition of the importance of the so-called 'Umklapp-process' at low temperatures.[70]

In the autumn of 1929, Peierls took up Pauli's offer to become his assistant in succession of Felix Bloch. Heisenberg had envisaged a regular exchange of advanced students and young researchers between Zurich and Leipzig, not unlike Bloch and Peierls.[71] This idea, in fact, was later put into practice by Rudolf Peierls and Hans Bethe who, in the 1950s and 1960s oversaw an active exchange of graduate students and research fellows between Birmingham and Oxford on the one hand and Cornell on the other.[72]

As before in Munich and Leipzig, Peierls time in Zurich was characterised as much by his contact with his academic teacher as with his new (and old) academic friends. Among the many friendships he forged, one stood out not merely because of its immediate impact but because of its long-term implications. In the autumn of 1929, the young Russian

[68]Cassidy, *Uncertainty*, p. 269.

[69]De Broglie, L. *Einführung in die Wellenmechanik*, translated by R. Peierls, Leipzig: Akademische Verlagsgesellschaft, 1929.

[70]R.E. Peierls, 'Zur kinetischen Energie der elektrischen und thermischen Leitfähigkeit von Metallen', *Ann. d. Phys.* **4**, 121–48 (1930); 'Zwei Bemerkungen zur Theorie der Leitfähigkeit', *Ann. d. Phys.* **5**, 244–6 (1931).

[71]Heisenberg to Pauli, 1.8.1929, Pauli, *Wissenschaftlicher Briefwechsel*, I, p. 517.

[72]See letters exchanged betwen Bethe and Peierls in vol. II of the *Peierls Correspondence*, chapters 6–9.

theoretician, Lev Landau, visited Pauli's institute. Landau had come on a Soviet government scholarship. Despite the brevity of his initial visit, Landau and Peierls initiated a deep and lasting friendship as well as an intense working relationship that was to result in a number of important (and controversial) publications in the following years.[73] Peierls was immediately impressed with the depth of his knowledge and his impressive intuition. Half a year younger than Peierls, Landau already had what his German friend judged to be a 'very mature understanding of physics'. Born in Baku in 1908, the young Landau displayed an early interest in and aptitude for physics and mathematics. At the age of fourteen he became a student of those two subjects at Baku University, before moving to Leningrad two years later where he studied under A.F. Joffe, the 'father of Soviet physics'.[74] During Landau's first stay at Zurich, in 1929, the most vigorously debated subject was quantum electrodynamics. Prior to his European tour, Landau had completed some work on the diamagnetism of metals by using quantum mechanics. At Zurich, however, he moved on to a collaborative study with Rudolf Peierls. The two investigated the limitations posed on the measurability of physical quantities in the relativistic quantum region. Landau and Peierls looked at light quanta (photons) in space, and wrote a wave equation for photons not unlike Schrdinger's equation for electrons. From this they derived sequences of equations for different numbers of photons, but as the two would recognise later, the results were not only complicated but physically nonsensical.

Landau was forced, by visa regulations, to limit his first stay in Zurich, but Peierls and he met several times during Landau's 'European tour' in 1929/30,[75] and they spent part of the summer at a conference in Odessa, a conference which was to prove fruitful for Rudolf Peierls as it initiated important contacts with the Russian physics community, but even more so on a personal level, because it was on this occasion that he first met his future wife, Genia Kannegiser.

---

[73] See introduction to chapter 2 below, p. 99.

[74] See E.M. Lifshitz, 'Lev Davidovich Landau (1908–1968)', in I.M. Khalatnikov (ed.), *Landau. The Physicist and the Man. Recollections of L.D. Landau*, Oxford: Pergamon Press, 1989, pp. 7–29.

[75] Letters [43] and [45].

[1] Heinz Rudolph[76] to Rudolf Peierls

Essen, 1.8.[19]22

Lieber Rudi!

Du fragst in Deinem letzten Brief wann ich in Berlin, d.h. Oberschweineöde, eintreffen werde; sicher erst nächste Woche; genaueres weiss ich noch nicht. Ich werde Dich noch davon benachrichtigen. Dein "Gebirgsluftkompressor" (D.R.P.) mit "Temperatur-Regler" (P.R.P.) ist ein sehr schöner Gedanke, der den Witzblättern schon jede Menge Stoff geliefert hat; ob aber die Sache nicht zu teuer wäre? Man würde dann nur in besonderen Fällen davon eine Anwendung machen.[77] Du musst bedenken, dass Du Gebirgs- oder sonst irgendeine Luft verfrachten oder besser gesagt verpflanzen willst, mit dem Unterschiede, dass sie nicht "blühen oder sich gar vemehren" soll. Man würde eine ziemliche Menge dafür gebrauchen. Dein Fahrradmantel leidet daran, dass er nicht genügend federn kann; der Blechmantel würde den Schlauch an den Seiten durchquetschen lassen, würde sich bald verbeulen durch die Stösse und würde dadurch dass das Rad gewaltsam gedreht würde, den Schlauch bald durchscheuern. Deine Fabel ist sehr gut geschrieben, nur hat sie das eine dass sie noch zu "weich" ist; sie müsste etwas besser durchgebildet sein, meiner Meinung nach. Auf die Anfrage, warum ich immer nur die Antwort zu dem schreibe, was Du vorher angefragt hast, habe ich eigentlich keine richtige Antwort; ich könnte Dir eine Unmenge schreiben, was sehr interessant ist, aber was nun eigentlich? Ausserdem ist das Schreiben für mich etwas unangenehm.

Bis auf weiteres

Heinz

---

[76] Heinz Rudolph was Rudolf Peierls' closest friend at school. Peierls recalls frequent discussions on contemporary topics during their grammar school years with the two friends holding "similar views on many topics". (Peierls, *Bird of Passage*, p. 12.) Rudolph and Peierls kept in touch during their students years but lost contact after Peierls had moved to Manchester in 1934. From 1978 onwards they communicated sporadically and stayed in contact until the Peierls' death.

[77] Heinz Rudolph was the more practical among the two friends. He later became a successful engineer. See letters exchanged between R. Peierls and H. Rudolf in *Peierls Correspondence*, vol. II, chapters 10 and 11.

Es wird in dem Brief nicht alles fehlerfrei sein. Du entschuldigst dieses, wenn Du erfährst dass ich immer während des Schreibens aus meinen Gedanken gerissen werde. Von der Schrift erwähne ich nichts als das, was dieser Satz sagt!

[2] Else Peierls[78] to Rudolf Peierls

[Berlin], 8.11.[19]26

Lieber Rudel!
Du hast uns durch Deine reichlichen ausführlichen Nachrichten + hübschen Briefe so erfreut, daß es mir ordentlich leid tut, wenn Du länger, als ich es gewollt, auf eine Antwort warten mußtest.[79] Na inzwischen hast Du ja auch die beiden Pakete bekommen. Vati insbesondere findet es sehr nett von Nernst, daß er Dir gleich antwortete;[80] außerdem freuen wir uns, wie gut Du Dir Deine Zeit einteilst + wie gewissenhaft Du Dich in die verschiedensten Zeitungen vertiefst, um gerecht urteilen zu können. Hoffentlich geben die Collegs das, was Du Dir vorgestellt hast — an Inhalt + Schwung in Auffassung + der Art vorzutragen, so daß es eine Lust ist zuzuhören. Ich wünsche Dir natürlich auch netten Anschluß, + glaube sicher, daß Ihr Beide, Ihr "Rudis" viel Gemeinsames habt + unternehmen werdet.[81] Für die Besorgung des Caviar besten

[78] Rudolf Peierls' mother died of Hodgkin's disease in 1921, when Rudolf was only 14. His father Heinrich married again not long after. His second wife, Else Hermann, was the daughter of a famous actor, and she herself was a talented musician. Her sister Helene was married to the poet and playwright, Ludwig Fulda.

[79] After studying at the University of Berlin for one semester, Rudolf went to Munich to study under Arnold Sommerfeld. His father had been reluctant to allow him to move away from home earlier, as he was sceptical about Rudolf's ability to look after himself. (Peierls, *Bird of Passage*, pp. 16, 23) The letters during the following years frequently demonstrate concerns of his father and stepmother.

[80] Peierls had attended the lectures of Walther Nernst, who taught the introductory physics course at Berlin university in 1925/6. Rudolf followed his father's advice of remaining in contact with some of his former teachers when moving to different universities, a practice he would continue with Sommerfeld, Heisenberg, Pauli and others later. See e.g. letter [33].

[81] The other Rudi is Rudi Friedländer, the son of a family friend.

Dank. Hoffentlich triffst Du bei Bernsteins viel nette Jugend, wenn sie Dich mal auffordern.

Bei uns ist alles gesund, nur haben wir reichlich viel Einladungen vor + hatten Sonnabend 12 Gäste, die erst um 1 1/4 Uhr verschwanden, so lustig war's. Dein früherer Chef "Kubi" war auch da + Dr. Preuß versuchte Dich zu vertreten + Cigarren mit mir herumzureichen.

Gestern früh waren wir i[n] d[er] Automobilausstellung, die grandios ist. Besonders elegant der Stand der M.W.O.,[82] mit den herrlichen Stangen + Rohren von Metallen. An Wagen: Maybach!! Das wäre was für Dich gewesen. Alfred[83] bombardiert uns fast täglich mit Postkarten aus allen möglichen amerik[anischen] Städten, die er sehr schön findet. Er macht eine Studienreise als Führer einiger Directoren. Wie ist d[er] Kursus f[ür] Leibesübungen. Deine Mirabelle[84] scheint ja sehr nett zu sein, hoffentlich ist ihr Töchterchen wieder gesund. Was hat Dir denn vom Deutschen Museum am meisten imponiert? Sonnabend sind wir in Dahlem zu Muttis Geburtstag.

Daß die Skier gespannt werden mußten, wußte ich leider nicht, hoffentlich kommen sie wieder in Ordnung. Dazu mußt Du aber Skyschuhe haben, die sind ohne Nägel. Die Bergschuhe kannst Du ja zum Wandern gut brauchen. Friedländers wissen einen guten Schuster für Bergschuhe etc., falls Du sie nicht fertig bekommen solltest. Den Krebsen geht es gut.[85] Ilse[86] dito. Annie will Dir selbst schreiben, sie waren gestern Abend bei uns! Mußt Du das Bad extra bezahlen + habt Ihr überhaupt eins? Wie ist d[as] Essen i[n] d[er] Kantine? Vati + ich haben Donnerstags eine s[ehr] famose Engländerin hier + machen in Conversation,

---

[82]As part of the company's diversification programme, the AEG had acquired, in 1901, the Allgemeine Automobil-Geselleschaft and transformed this into the Neue Automobil-Gesellschaft (NAG). The production of cars, largely designed by a member of the board of the AEG, Georg Klingenberg, was located in the Kabelwerke Oberspree, the director of which was Heinrich Peierls. See Manfred Pohl, *Emil Rathenau und die AEG*, Frankfurt/M.: Verlag v. Hase & Koehler, 1988, p. 164.

[83]Alfred, Rudolf's brother, who was eight years his senior, was an engineer and became an expert in electric condensers. In the mid–1920s he spent a number of years working in the United States. Peierls, *Bird of Passage*, p. 21.

[84]A reference to Peierls' landlady whose name was Mirabeau.

[85]Peierls' older sister Annie was married to Hermann Krebs. Within the family they were often referred to as 'die Krebse'.

[86]Ilse Krebs, daughter of Annie and Hermann Krebs.

ich lese viel engl[isch] Shaw! fein. Der Charly[87] ist herzig + wird immer hübscher; hoffentlich Deine Studentinnen auch!!! Vati läßt schön grüßen, ich sende den Brief ohne seine Anschrift fort — damit er vorher mitkommt. Also alles Gute + viel Spaß.
Herzliche Grüsse

D[eine] Else

[3] Heinrich Peierls to Rudolf Peierls

Berlin, 23.11.1926

Lieber Rudi.
Für Deine Mitteilungen vom 20. d[iesen M[onats], die wir mit grossem Interesse gelesen haben, vielen Dank. Ich freue mich darüber, dass Du mit einem geringeren Etat wirtschaften kannst und werde die definitive Summe bei Deiner Anwesenheit in Berlin feststellen.[88] Inzwischen habe ich RM 300.— an Dich abgeschickt, damit Du die Reise bezahlen und noch einiges Geld in der Tasche behalten kannst.

Dass Du zu Weihnachten zu uns kommen willst, freut uns sehr. Wenn wir Dir die Entscheidung in dieser Frage überlassen haben, so geschah das nur, um Dich nicht, falls das Wetter anders wird, in der Ausübung eines gesunden Sportes zu stören. Wir sind aber natürlich mit der Teilung Deines Aufenthaltes, wie Du sie vorschlägst, sehr einverstanden.

Du hattest uns seinerzeit mitgeteilt, dass Du Frau Justizrat Bernstein besuchen willst. Wir haben aber nichts über diesen Besuch gehört, und ich bitte Dich, dies nachzuholen.
Herzliche Grüsse

Dein Vater

---

[87]Else's pet.

[88]Heinrich Peierls gave his son an allowance during his student years. The correspondence regularly refers to financial matters, especially with regard to extra expenses such as holidays. See in particular also letters between Rudolf Peierls and his parents March 1927 and April 1929. (*Peierls Papers* Supp., A.98–A.100) Peierls was meticulous in his book keeping. Many of his notes from his early student days well into his professorial times at Birmingham (in family possession) demonstrate this.

Lieber Rudel!
Dein letzter Brief hat uns s[ehr] erfreut, ebenso daß Du Weihnachten kommst. Ich habe eine kleine Verschönerung Deines Zimmers vorgenommen, gegen die Du sicher nichts einzuwenden haben wirst. Eben habe ich Gretel Fischer condoliert + gehört, wie gern Du in München bist + wie gut Du aussehen sollst. Der kleine Rudi Friedländer hat nur Angst, daß Du Dich bei ihm langweilst, weil Dich nur schwierige mathemat[ische] Probleme interessieren, beweise ihm mal, daß das nicht so ist. Wie ist die Turnstunde + hast Du schon Sky Kursus im Trockenen? Viel Spaß

Gruß Else

[4] Heinrich Peierls to Rudolf Peierls

Berlin, 1.12.1926

Lieber Rudi.
Herr Dr. Hamburger, der Neffe von Emil Ludwig,[89] der augenblicklich in Genf Privatdozent und im Völkerbundsbüro beschäftigt ist, schreibt mir den abschriftlich beifolgenden Brief. Ich hatte ihn vor einiger Zeit persönlich gebeten, sich für Deine englischen Studien zu interessieren.[90] Du brauchst auf den Brief nicht zu antworten; ich habe dies bereits direkt getan und freue mich doppelt, dass Du zu Weihnachten hier bist, damit ich Euch bekanntmachen kann. Er kann Dir für Deine Absichten sehr von Nutzen sein. Wir wollen in Berlin dann weiter über die Angelegenheit sprechen.

Die Reise nach England ist nur dann ausführbar, wenn Du inzwischen englische Studien treibst. Da Du bisher davon nichts erwähnt

[89]Emil Ludwig (originally named Emil Cohn 1881–1948), writer and journalist from Breslau, Heinrich Peierls' home town. At the time he lived in Switzerland. His abhorence of the National Socialist regime later led him to work for the U.S. government and, from 1940, to write anti-fascist pamphlets.

[90]Peierls had intended to improve his English by spending his university vacations in England in the summer of 1927. Instead he went in the summer of 1928; see letters [19–20], [22].

hast und auch in Deinem Etat hierüber nichts vermerkt, hast, so nehme ich an, dass dies bisher nicht erfolgt ist.
Herzliche Grüsse

Dein Vater

[5] Else Peierls to Rudolf Peierls

Vati grüßt vielmals. Berlin 23.2.[19]27

Verehrter Herr Vorsitzender.[91]
Zunächst beglückwünschen wir Dich zur Gründung Deiner Fachschaft. Der Erfolg ist wirklich fein, ich wünsche Dir viel Freude + Anregung + menschlichen Gewinn + Anschluß! Schade, daß Du nicht mal mit mir den Fasching am Rhein kennen lerntest — wenn er Dir dort eben nicht liegt, hast Du ja recht, nicht mehr hinzugehen. Ich bin nur froh, daß Du sonst einige neue Bekannte hast + hoffentl[ich] nette Zerstreuung. Es ist s[ehr] lieb von Dir, Dich um mein Wohl und Wehe zu kümmern; ich hatte eben gleich nach + an dem schönen Festtag solche Schmerzen, daß ich zum Professor mußte, der feststellte, daß meine Gallenblase sehr geschwollen + entzündet ist + mir sofort Compressen, Carlsbader Cur, strenge Diät + — keinen Tropfen Alkohol — oh weh! verordnete. Ehe ich nicht richtig in Carlsbad war, was aber nicht vor Mai sein kann, werde ich nicht gesund sein können. Das sind alles nur Hilfs- und Ersatzmittel. Na, Unkraut vergeht ja nicht — + das bin ich trotz Deiner schönen Ansprache.[92] Vati spürt Gott s[ei] D[ank] nichts v[on] d[er] Grippe, arbeitet nur s[ehr] viel. Carli[93] bestand gestern glänzend ist seelig + stürzt sich in "Mulus-Orgien", wie er's nennt. Morgen hat Alfred Geburtstag. Fuldas sind in Teneriffa! Es freut mich, daß Du es mit Erich so nett hattest. Er war eine Woche als Logiergast hier. Doris

[91]During his year at Berlin, Peierls had been involved with the MAPHA, a student-run mathematics and physics society. He initiated the foundation of a similar group in Munich and became the first chairman.
[92]Reference to Peierls' speech at his father's 60th birthday party in February 1927.
[93]Carli Fulda, son of Ludwig and Helene Fulda.

Sietzkowski hat sich in St. Moritz mit d[em] Mitinhaber einer Bankfirma Stier verlobt. Wir gratulieren Sonntag. (Adr[esse] Dahlem! Podbilskiallee 35) falls Du 1 Glückwunschkarte schreibst, ev[entuell] auch an [?]. Vati hat ein Herrenessen ich werde nun engl[isch] lesen.
Viel Spaß. Gute Nacht

Else

[6] Heinrich Peierls to Rudolf Peierls

Berlin, 16.7.1927

Lieber Rudi.
Ich freue mich, aus Deinem Schreiben vom 13. d[iesen] M[onats] zu ersehen, dass Du, wenn es Deine Zeit erlaubt, Dich der körperlichen Erholung widmest. Warst Du zu Fuss auf der Zugspitze oder benutztest Du die Bahn?

Es interessiert mich sehr, dass Herr Hagelberg, den Du durch Alfred kennengelernt hast, der Bruder von Frau Dr. Levi ist. Wie alt ist er? Wird es ein Verkehr für Dich werden? Die Bilder vom Chiemsee haben uns sehr interessiert; bei dem einen habe ich zuerst allerdings geglaubt, dass es sich um einen bereits leblosen Menschen handelt.

Ich freue mich, dass Du von Deinem Guthaben etwas für das Hilfswerk Sächsische Unwetter-Katastrophe tun willst. Ich werde aber zunächst nur RM 25.- einzahlen, da ja leider Gottes die Unglücke ziemlich häufig sind. So kannst Du später noch etwas Gutes tun.[94]
Viele herzliche Grüsse

Dein Vater

[94]Heinrich Peierls himself supported many good causes himself (see e.g. [23], Waltraud Krause to Rudolf Peierls, 27.8.1989, vol. II, chapter 11 of *Peierls Correspondence.*). While being generous, he was also cautious in his donations and loans, and as he suggested to Peierls in this letter, tried to keep support well-balanced. See e.g. letter [47].

Lieber Rudel!
Deine Sportvergnügungen leiden natürlich auch unter dem wechselnden und ungünstigen Wetter. Jedenfalls bin ich froh, daß Du nicht nakkend im Englischen Garten herumtollst, sondern dafür in die heimischen Hochgebirgshöhen kommst. Wie war d[ie] Zugspitz-Fahrt? Am 15. hatte Ludwig Fulda Geburtstag, + Sonnabend, d[en] 23. wird Ilse 1 Jahr alt. Hermann fährt dazu für 2 Tage n[ach] Norderney. Heut schrieb Frau Krebs,[95] wie glänzend es Ilse + Annie gehe. Letztere geht ohne Schmerzen, badet tägl[ich]. Sie [?] Ilse + Anne + schlafen alle tüchtig.[96] So hoffen wir endl[ich], diese große Sorge los zu werden. In d[en] Zeitungen stehen viel Unglücke, man freut sich doppelt über jede gute Nachricht.

Eben kam beifolgende Einladung. Kürzlich regneten wir mit Motorboot kolossal ein + vorgestern ging es Hermann, der Feinsteins besuchte, ebenso.

Weiter viel Vergnügen + Anregung i[n] d[er] Fachschaft + in d[er] Natur.
Viel Liebes von

Anschl[97] + Else

[7] Heinrich Peierls to Rudolf Peierls

Berlin, 3.8.1927

Lieber Rudi.
Dein Schreiben vom 2.8. ist in unserem Besitz. Einerseits freue ich mich, dass Du auf Wanderschaft gehen willst, andererseits bedauere ich lebhaft, dass Du keinen Anschluss gefunden hast. Ich kann mir nicht recht denken, dass es ein grosser Genuss für Dich sein wird, allein zu reisen,

---

[95]Mother of Hermann Krebs.

[96]Annie Krebs had had a skiing accident in the spring and spent some time on Norderney to recover.

[97]Else's mother Anna Hermann.

aber ich weiss auch nicht, in welcher späteren Zeit Dir der Anschluss sicher gewesen wäre. Wenn Du glaubst, Dich auch auf eigene Faust gut unterhalten zu können, so habe ich gegen die Reise nichts einzuwenden. Anstatt aber von Bozen weiter südwärts zu gehen, möchte ich vorschlagen, dass Du Fuldas in Karersee aufsuchst. Sie haben, wie sie uns schrieben, Platz, da sie Absage von Bekannten bekommen haben, die zu ihnen kommen wollten. Ich glaube, sie würden sich freuen, wenn Du auf einige Tage hingingest. Solltest Du den Eindruck haben, dass Du störst, was ich aber nicht glaube, so kannst Du ja jederzeit weiterwandern. Ich lasse Dir jedenfalls per Postanweisung RM 150.– zugehen, bitte aber, da wir, wie Du weisst, verreisen, Herrn Juliusberg, Kabelwerk Oberspree, Oberschöneweide, gleichzeitig mitzuteilen, wohin Du Deinen Monatswechsel haben willst. Herr Bernhardt ist gleichfalls auf Urlaub.

Was uns betrifft, so fahren wir am Donnerstag, den 11.8. abends mit dem Schlafwagenzug nach München und kommen früh, 8.10 Uhr in München an, steigen aber gleich um, um nachmittags in Gastein zu sein. Ich nehme aber nach Deinen Mitteilungen an, dass Du um diese Zeit nicht mehr in München sein wirst.

Unsere Briefadresse ist bis etwa 2. September Bad Gastein, Hotel Kaiserhof. Wir bleiben dann noch drei bis vier Tage unterwegs und fahren dann wieder nach Hause. Ich bitte Dich dringend, uns immer genauere Adressen für Depeschen zu geben, so dass Du erreichbar bist. Besten Gruss

Dein Vater

Lieber Rudel!
Es tut mir sehr leid, dass Du keine Reisegesellschaft gefunden hast. Vati schreibt heute an Fuldas, die Dich sicher für einige Zeit gerne aufnehmen. So kannst Du dort mit ihnen bekannten Führern noch schöne Hochtouren machen, die Dir neu sind. Ein beneidenswerter Zustand. Was gäb ich drum, wenn ich wie früher klettern dürfte. So werden wir s[ehr] brav + beschaulich in Gastein leben + hoffentlich gute Luft haben. Gib um Gottes Willen oft und vernünftige Nachricht. Da Fuldas nur bis 1. Sept[ember] dort bleiben, beginnst Du viell[eicht]

mit Bozen + wanderst durchs Eggenthal hinauf, was früher oft Freunde taten. Alles Gute + Schöne
herzl[ich]

Else

[8] Hans Bethe to Rudolf Peierls

Bad Gastein, 27.8.1927[98]
(postcard)

Schöne Gegend kann man sehen
(Bitte Karte nur zu drehen!)
Doch darin ein Häuserblock
So von fünf bis dreizehn Stock,
Drinnen Menschen über achtzig
(Was nicht hindert, dass man macht sich
Zur Promnade äusserst fein)
Dieses nennt der Mensch Gastein.
Wünsche Ihnen viel Vergnügen
In Carezzas Sonn zu liegen.[99]

[98]This postcard is the earliest piece of the correspondence between Hans Bethe and Rudolf Peierls. It is followed by a more or less regular exchange of letters that would continue until just days before Peierls' death in September 1995.

[99]Hans Bethe spent the holidays with his mother in Bad Gastein. The little verse indicates both his wit and his dislike for the place Gastein, which he regards as an architectural disaster, made worse by the 'typical tourists' as described in his poem.

[9] Hans Bethe to Rudolf Peierls

Seite 1055[100]

Frankfurt/M, 19.9.1927

Lieber Peierls!
Ich fühle mich entsprechend schuldbewußt, und dabei rege ich mich normalerweise sogar noch auf über mangelhafte Schreibtätigkeit bei anderen Leuten, und behaupte, ich schriebe gern Briefe. Wofür ich als Beweis nur die Zahlen oben in der Ecke anführen kann (gerechnet weder von der Erschaffung der Welt noch von Christi Geburt an, sondern von meinem Auszug nach München am 26 IV 26). Aber angesichts des vorliegenden Tatbestandes nehme ich alles zurück, was ich in den letzten 40 Jahren gegessen und getrunken habe und behaupte das krasse Gegenteil. (Was Sie anstelle einer Entschuldigung nehmen sollten, denn die sind doch nie wahr). Um historisch anzufangen, belastete der Schöpfersche Balken mein Gemüt noch längere Zeit in Baden-Baden, bzw. lag mit im Magen (wo er wohl in der Magensäure schwamm), wo er aber wenig Platz fortzunehmen schien, da sich für das vorzügliche Essen der Selighofs in Baden noch genügend Platz fand.[101] Als ich dort meine Warendepots genügend aufgefüllt hatte, liess ich mich vier Tage in Frankfurt sehen, und fuhr dann mit meinem Vater über München, wo ich Frau Höchstetter mit meinem Besuch beglückte, nach Gastein, dem baulich scheusslichsten Ort, den ich kenne, und der um so scheusslicher wirkt, weil seine zehnstöckigen Häuser in einer fabelhaften Umgebung liegen. Dementsprechend habe ich dann auch eine Autotour mit Herrn und Frau Prof. Sachs[102] gemacht, zum Glocknerhaus bei strahlendster Sonne und entsprechender Stimmung, und zum Ausgleich mit meinem Vater eine Hochtour bei Nebel und Erkältungsentwicklung. (war aber

---

[100]Hans Bethe was in the habit of numbering his letters and pages thereof. This first letter is p. 1055 of letter 217. Peierls later used a numbering system in his correspondence with his future wife, Genia, when every letter contained a number with date and the same information about previous letter.

[101]Bethe's parents had divorced; his mother lived in Baden-Baden, his father in Frankfurt.

[102]Parents of their friend Werner Sachs.

natürlich doch sehr schön, und vor allem sehr übend, weil auf dem Grat ein Windchen wehte, das zwar ohne Schnee und Kälte, aber sonst von ähnlicher Güte war wie das an der Zugspitze, und infolgesdessen die Standfestigkeit übte). Als drittes Ereignis leistete ich mir noch eine Tour allein auf einen Aussichtsberg, der aber durch einen Steingrat ohne gebahnten Weg + Neuschnee interessant gemacht wurde. Sonst bestand der Gasteiner Aufenthalt meist aus Jausen mit Familienbeigabe, die weniger lohnend als auf Dauer langweilig waren, und aus anderen Jausen mit Herrn und Frau Sachs, die die "bessere Sorte repräsentierten". Zum Schluss machte ich es Ihnen nach, in dem ich als der "bessere junge Mann" über die Hauptkette der Alpen, in diesem Fall durch den Tauerntunnel, für zwei Tage an die Kärntener Seen fuhr. Zuhause hatte ich dann die angenehme Aufgabe, 460 Renditen für die Wirtschaftskurve auszurechnen, und dann meine diversen trauernden Witwen zu besuchen, letzteres in Ermangelung augenblicklich hier vorhandener männlicher Wesen, die mit mir befreundet sind. Allmählich will ich sogar jetzt anfangen, etwas zu arbeiten. (es wird Zeit!) und ich habe vorläufig zwei Arbeiten studiert, die über das Thema meiner Doktorarbeit gehen, die sich aber beide als relativ ungefährlich erwiesen.[103] Womit sich aber nicht erreichen liess, dass meine Arbeit bisher erschienen ist. Was damit passiert, sage ich Ihnen nicht, aber es ist ja wohl anzunehmen, dass die Geschichte bald mal rauskommen wird. Ich werde jedefalls etwa am 10. X. nach München fahren, weil experimentell erwiesenermassen das Arbeiten dort immer noch wesentlich besser geht als hier. Dann werde ich Berlin die Ehre geben (wann sagt die Redaktion Ihres Blattes mir vielleicht). Worauf vermutlich der Semesterbeginnkater wegen Zuweniggearbeitethaben folgen wird. Das wäre also das Programm. Ich glaube zwar noch nicht, dass dieser Brief eine normale Länge hat, aber es ist nachmittags 17h und da kann man noch nicht richtig schreiben, das geht erst ab 21. In diesem Sinne wünsche

[103] Bethe was one year Peierls' senior; after two years of studying physics at Frankfurt, he had joined the 'Sommerfeld school' in Munich in the spring of 1926. By the summer of 1927, Sommerfeld had suggested to Bethe to develop a theory of diffraction of electrons by crystals. See interview N. David Mermin with Hans Bethe, 25.2.2003, http://ifup.cit.cornell.edu/bethe/.

ich Ihnen dass Sie sich ebenso gut erholt und ebensowenig gearbeitet haben, aber nicht ebenso lang mit der Antwort warten.
Herzlichst Ihr

Hans Bethe

[10] Hans Bethe to Rudolf Peierls

Frankfurt/M, 1.10.1927
Inliegend meine Arbeit!
212/1063

Lieber Peierls!
Vielen Dank für das Fachschaftsprogramm, das mich zur Umstossung meines Programms veranlasste, indem und dadurch dass ich meinen Plan, in München im Oktober zu arbeiten, aufgab, und statt dessen etwa am 16. nach Berlin kommen will, und gleich nach der Tagung nach München fahre. Ihre nette Einladung, bei Ihnen zu wohnen, kann ich vorläufig noch nicht annehmen, da Werner Sachsens Verwandte noch "ältere Rechte" auf mich haben. Seien Sie mir also bitte nicht böse, wenn ich Ihnen zunächst einen Korb gebe; unser Zusammensein braucht aber dadurch in keener Weise gestört zu werden, da ich sowieso ja wohl den ganzen Tag in der Stadt sein werde.

Sie werde nicht darum herumkommen, mir eingehend die "scheene Jejend zu erklären". Hier ist momentan Physiologenkongress, bei dem ich väterlicherseits beteiligt bin.[104] Immerhin hält sich diese Beteiligung, von einem Vortragsvormittag abgesehen, an dem mein Vater selbst sprach und als Attraktion einen Film vorführte, in den relativ engen Grenzen der Teilnahme an allgemeinen Abspeisungen und nach-

[104]Bethe's father was professor of physiology at Frankfurt University. Hans was close to his father (See Silvan Schweber 'Writing the Biography of a Living Scientist: Hans Bethe', Special Collections, Oregon State University. Libraries: www.oregonstate.ed/ dept/Special-Collections/subpage/ahp/1995symposium/ schweber/html.) and occasionally joined him for conferences.

folgenden Tanzvergnügen. Im übrigen war ich im Wesentlichen mit Werner Sachs und Eltern zusammen, obwohl ich zunächst erklärte, mit einem schnurrbärtigen Menschen, der aussieht wie ein unsympathischer Filmheld,[105] nicht befreundet sein zu können.

Gearbeitet habe ich natürlich wieder nichts. Nichtsdestotrotz auf Wiedersehen in dem trauten Städtchen Berlin.
Herzlichst Ihr

Hans Bethe

[11] Heinrich Peierls to Rudolf Peierls

Berlin, 15.12.1927

Lieber Rudi.
Ich freue mich, aus Deinen Zeilen vom 12. d[ieses] M[onats], zu ersehen, dass Du für die nächste Zeit in angenehmer Weise beschäftigt bist. Das avisierte Paket ist angekommen, aber wir werden es erst am nächten Sonntag öffnen, an welchem Tage wir Weihnachten vorfeiern werden, da wir Montag abend abreisen wollen. Wir fahren zunächst nach Lugano, Parkhotel, und wollen es vom Wetter abhängig machen, ob wir dort bleiben oder weitergehen. Jedenfalls bitte ich, dorthin kurz zu schreiben, und mir Deine Adresse anzugeben, damit ich immer weiss, wo Dich Briefe erreichen.[106]

Was den von Dir erwähnten Kälte-Techniker betrifft, so ist es nicht ausgeschlossen, dass wir gerade einen solchen brauchen können.[107] Du kannst ihm anheimstellen, sich unter Bezugnahme auf mich schriftlich unter Darlegung seiner bisherigen Tätigkeit an Herrn Direktor Rosenberg, AEG Fabriken Oberleitung, Friedrich-Karl-Ufer 204 zu wenden.

---

[105]Reference to Werner Sachs' new hairstyle and moustache.

[106]Peierls intended to spend the Christmas holidays with student friends in the Gerlostal, in the Alps.

[107]From time to time, Rudolf Peierls enquired about the possibility of friends or acquaintances being employed at the AEG. As indicated in this letter, his father insisted on using the correct channels of application, but was generally supportive, if he judged the matter to be potentially successful.

Ich bemerke ausdrücklich, dass er vielleicht gerade seinem Spezialwissen entsprechend beschäftigt werden könnte.
Nochmals viele Grüsse und Feiertagswünsche

Dein Vater

17.12.27

Lieber Rudel!
Leider komme ich etwas verspätet zum Anschreiben. Der Brief soll gleich fort. Also nochmals recht frohe Festtage + gutes Schiwetter + Glück auf in jeder Beziehung.
Von Herzen

Else

[12] Hans Bethe to Rudolf Peierls

Baden-Baden, 24.12.1927
(postcard)

Lieber Peierls, lieber Thorner![108]
Sie werden jetzt irgendwo im wilden Gerlostal als wackere Männer auf die rauhen Berge steigen, aber da ich "Ihnen jetzt nicht sage", wo Sie sind, schreibe ich Ihnen nach München. Sie haben sich so rührend in München um mich gekümmert, dass ich Ihnen trotz des ungewissen Schicksals dieser Karte gleich dafür danken muß.[109] Ich bin also glücklich hier angekommen, von Oos nach Baden im Rollstuhl per Gepäckwagen und dann teils per Auto, teils per Schultertransport. Jetzt bin ich in der Pflege meiner Mutter, die trotz des Schrecks, den sie

[108]Hans Thorner, medical student and friend of Peierls' living in Heidelberg.
[109]After a weekend skiing trip, Bethe suffered from frostbite causing an infection in his foot. This was to trouble him well into the New Year, so that he did not return to Munich until mid-February 1928; see letter [14].

zuerst bekam, froh ist, mich hier zu haben, und eines Arztes, der alle 2 Tage kommt. Der linke Fuss heilt schon und dem rechten geht es etwas besser. Also jetzt nochmals herzlichen Dank für Alles; fröhliche Tage und Ski-Heil! Herzlichst Ihr

Hans Bethe

[13] Heinrich Peierls to Rudolf Peierls

Berlin, 25.1.1928

Lieber Rudi.
Ueber Deinen Bericht vom 18.1. habe ich mich sehr gefreut und übermittle Dir meine besten Wünsche für Deine weitere Vortragstätigkeit.[110] Soweit Du Geld für Chemikalien brauchst, bitte ich um spezifizierte Aufgabe und werde dann das Weitere veranlassen.
Von uns ist nichts neues zu berichten.
Herzliche Grüsse

Dein Vater

Lieber Rudi!
Wie fein, daß Dein ersten großer Vortrag so wohl gelang, so daß sogar Dein Prof. Sommerfeld sich derartig lobend äußerte. Am Ende bist Du doch dazu bestimmt, ein kleiner, resp[ektive] großer Professor zu werden. Kannst Du Dich mit Prof. Sommerfeld mal über Deine Zukunft beraten lassen? Sicher nimmt er doch viel Interesse an Dir. Erich war kurz hier, wohnte bei uns + meinte Ihr wolltet eine Osterwanderung gemeinsam machen. Ich schlage Euch d[as] entzückende Salzburg vor, auf dessen bergigen Höhen ringsum Ihr Euch wundervoll erholen könnt.

[110]Rudolf Peierls had given his first seminar paper in Sommerfeld's seminar on transformation theory based on two papers by Dirac and Jordan. P.A.M. Dirac, 'The Quantum Theory of Emission and Absorption of Radiation', *Proc. Roy. Soc.* **A114**, 243–65 (1927); P. Jordan, 'Zur Quantenmechanik der Gasentartung', *Z. Phys.* **44**, 473–80 (1927).

Alfred reist endgültig am 2. Februar v. New York ab. Wir haben am selben Tag hier eine gr[oße] Gesellschaft von 44 Pers! Meine Freundin aus Amerika singt, + H[ermann] Weigert begleitet, hoffentlich kommt nix dazwischen. Wann kommst Du nach Hause, wann beginnt Göttingen? Lene Ullmann hat Dich dort schon warm empfohlen. Von wann — wann gehst Du nach England, darüber kannst Du Dich ja mit Karli verständigen. Ich weiss nicht, ob Dir bekannt ist, dass Werner Milch seinen Vater plötzl[ich] an Herzschlag verlor. Die Trauerfeierlichkeiten sollen großartig gewesen sein. Werner kommt demnächst her, + i[n] d[er] II. Hälfte Februar wird Edith Honigmann aus Wien bei uns wohnen. Es freut mich so, dass Du so netten Verkehr + so viel Spaß hast.
Also so weiter!
Herzl[ich]

Else

[14] Hans Bethe to Rudolf Peierls

Baden-Baden, 28.1.[19]28

Lieber Peierls!
Ich liege immer noch kontinuierlich auf der Matratze, während Sie darüber vortragen. Zur Abwechslung, und zur Vervollständigung des Lazaretteindrucks hat sich meine Mutter jetzt auch ins Bett gelegt, mit einer Angina, aber nicht sehr. Dafür mache ich schüchterne Versuche, auch mit dem rechten Bein aufzutreten, verfüge aber unter dem kleinen Zehlein (das ich jedoch nicht falte), noch über eine offene Stelle. Da ich bisher die vom Doktor genannten Heilungszeitem immer versechsfachen musste, könnte ich eigentlich nicht vor dem 1. IV. erwarten, nach München zu kommen. Aber ich löse Kreuzworträtsel mit Tinte, d.h. ich bin Optimist und schrieb demgemäss an Unsöld,[111] ich käme

[111] Albrecht Unsöld (1905–95), studied physics under Sommerfeld, completed his Ph.D. at the age of 21 in Munich, where he became Sommerfeld's assistant. In 1932, aged 27, he became professor of theoretical physics at Kiel. He later made his most important contributions to the theory of stellar atmospheres.

about 13 II, es wäre mir aber sympathischer, erst am 20. auf dem Podium erscheinen zu müssen. On verra.

Da ich keine Tiegel quantitativ kaputtschlagen kann (Ihre diesbezüglichen Ausführungen haben mich übrigens sehr an mein eigenes quantitatives Arbeiten erinnert) begnüge ich mich mit dem Erwerb mehr oder weniger tiefgründiger mathematischer Kenntnisse aus Bieberbachs[112] Differentialgleichungen. Wobei ich wieder meine mangelnde Begabung für die reine Mathematik sehe, in dem ich meine, der Mond geht unter und der Beweis wäre zu Ende, wenn der strenge und kluge Mathematiker noch drei grosse Haken findet. Ich könnte, glaube ich nie solche Beweise selbst bis ins Letzte herauskriegen, d.h. so, dass Bochner[113] es glauben würde. Ganz so schlimm wie Ihre principia mathematica sind aber Differentialgleichungen nicht; die Probe, die Sie mir davon gaben, scheint mir auf eine sinnvolle Mischung des "grossen Lalula" von Morgenstern und von "Fisches Nachtgesang vv-v" zu deuten nach Belieben garniert mit Pfeilen $\downarrow$ abwärts (Konfektionsabteilung, Trikotagen, Erfrischungsraum).

Meine Hauptbeschäftigung ist jetzt, solange ich nicht gerade darüber nachdenke, dass ich es noch beliebig lange im Bett aushalten würde, und meine eigene Geduld lobe, zu lesen, und ich finde diese Beschäftigung a) sehr angenehm, b) sehr zeitraubend, c) sehr notwendig von wegen Spezialwissenschaft und Allgemeinbildung, d) sehr unangenehm, weil alle guten Gedanken schon gedacht worden sind. Als letztes habe ich in vier Tagen die "Kristin Lavranstochter"[114] besiegt, und bin sehr begeistert davon. Das ist ein Buch, in dem so viele Freuden und Qualen der Menschen geschildert sind, so wie wir sie selbst erleben und um uns sehen, wie ich kaum ein zweites kenne. Das sind wirklich tiefe Menschen, die da vor einen hingestellt werden, und ich verstehe nicht recht, wie

---

[112] L. Bieberbach, *Theorie der gewöhnlichen Differentialgleichungen.* Berlin: Springer, 1923.

[113] Salomon Bochner (1899–1982). At the time lecturer of mathematics in Munich working mostly in harmonic analysis. In 1933 he emigrated to Princeton where he stayed until his retirement.

[114] Kristin Lavransdatter, a trilogy set in 14th century Norway by Sigrid Undset, for which she received the Nobel Prize for Literature in 1928.

z.B. die "Forsyte Saga",[115] die ein stellenweise gutgeschriebener Roman von uninteressanten Menschen ist (ich kenne übrigens nur den ersten Band), damit an Qualität verglichen wird. (Wobei ich eben die ekelhafte Angewohnheit mancher Menschen nachahmte, von jedem anderen die Kenntnis aller "Neuerscheinungen" vorauszusetzen.)

Dazwischen bin ich mal ganz klassisch geworden, und habe gefunden, dass der alte Shakespeare auch was gekonnt hat. Trotzdem muss ich (zu meiner Schande?) gestehen, dass mir die modernen Sachen doch sehr viel näher liegen, und die Probleme von Shakespeare oft ganz fernstehen. Schiller kann ich ja überhaupt nicht mehr lessen.

Ihrem und Unsölds gemeinsamen Schluss, dass die Welt nicht existieren kann, weil sonst die Quantenmechanik gelten müsste, muss ich zustimmen. Aber das dürfen wir nicht weitersagen, denn sonst entzieht man uns die Gelder, und was soll dann aus all den Physikern werden? Aber vielleicht finden Sie auf diese Feststellung hin eine Anstellung bei Lenard,[116] und werden mit dem Uräther verwandt.

Sehen Sie Thorner eigentlich mal? Oder vertieft er sich so sehr in Examensvorbereitungen? Und kann man Ski laufen? D.h. bei Arosa kann mans, wie mir Erwin Strauss[117] Karte mitteilte. Und ich kanns diesen Winter nicht mehr. Wann ich komme, sage ich Ihnen also noch nicht, und bin für Post stets gern Nehmer.
Herzlichst Ihr

Hans Bethe

---

[115] John Galsworthy's trilogy, Forsyte Saga, chronicling the lives of three generations of a monied, middle-class English family at the turn of the century, would be awarded the Nobel Prize in 1932.

[116] Philip Lenard (1862–1947), Nobel laureate and professor of physics at Heidelberg University until his retirement in 1931; well-known and outspoken supporter of national socialism.

[117] Fellow student and friend of Bethe's and Peierls' from Munich.

[15] Heinrich Peierls to Rudolf Peierls

Berlin, 9.2.1928

Lieber Rudi.
Ich habe mich sehr gefreut, wieder einmal Deine Stimme zu hören und danke Dir herzlich für Deine Glückwünsche. Ebenso danke ich Dir für Dein Geschenk, das uns grosse Freude macht, und das Du bei Deiner späteren Anwesenheit wirst sehen und hören können.

Deine Berichte haben uns natürlich lebhaft interessiert. Bevor man sich für Göttingen entscheidet, wird die Frage akut werden, ob man, falls Prof. Sommerfeld für Dich besonderes Interesse haben sollte, dies als genügenden Grund ansehen kann, um für München zu optieren oder ob sachlich Göttingen auf jeden Fall vorzuziehen ist. Ich nehme aber an, dass wir uns darüber noch werden aussprechen können, zumal ich glaube, dass Du jedenfalls nach Berlin kommen wirst, ehe Du alle Brücken in München abbrichst.[118]
Alfred erwarten wir morgen abend oder übermorgen früh.
Herzliche Grüsse

Dein Vater

10.2.1928

Lieber Rudi!
Es war sehr schneidig von Dir anzurufen. Wir feierten Vati, der gerade an diesem Geburtstag viel zu tun hatte, indem wir schnell köstliche Forellen + Obst aßen + dann mit den Krebsen i[n] d[ie] Oper fuhren: Barbier v[on] Sevilla, die entzückend + s[ehr] drollig ist. Heut früh

[118]Arnold Sommerfeld left Munich in the spring of 1928 to spend a year in America. Peierls considered moving to Göttingen, but in the end decided to join Heisenberg's school of theoretical physics in Leipzig for the summer semester 1928. N.b. in interviews he later mentioned that he never considered Göttingen as a possbile place to study, but the correspondence shows unambiguously that he did. See e.g. Rudolf Peierls, interview by John Heilbron, 17.6.1963, Niels Bohr Library, AIP.

kam ein Telegramm von Alfred aus Hamburg, das meldete, daß Alfred dort glückl[ich] landete. Die Krebse sind z[um] Empfang dort, + heute abend 11 Uhr treffen sie alle hier ein. Wir müssen vorher eine Abendgesellschaft bei Jacobsohns über uns ergehen lassen![119] Was nun mit Deinem Studium wird, ist ja sehr interessant, letzten Endes wirst Du selbst entscheiden können, welchen Weg Du wählst. Dr. Hamburger, (Augenarzt), der sich nach Dir erkundigte, meinte, wenn ein Prof. wie Sommerfeld sich für Dich interessiere, sei das sehr wertvoll + Du würdest wohl am besten in München bleiben. Fuldas sind auf ihrer Orientreise. Das Wetter ist heut mal wieder schön + d[ie] Sonne scheint, ich wanderte mit Charly im Tiergarten. Viel Spaß für den Frühling.
Addios!
Viele Grüße

Else

[16] Heinrich Peierls to Rudolf Peierls

Glion, 2.6.[19]28

Lieber Rudi.
Innige Geburtstagswünsche. Diesmal wirst Du mündig + darfst zu mir "altes Kamel" oder dergleichen sagen. Hoffentlich zeigt sich Dir das neue Lebensjahr gewogen + Du bleibst vergnügt und wirst in Deiner Arbeit erfolgreich.
Im Übrigen:

Willst Du mir was zu Deinem Geburtstag schenken, mußt Du stets an Nägel und Ohren denken.

In diesem Sinne grüßt Dich herzlich

Dein Vater

---

[119] Relatives of the Peierls. The son, Franz, remained a good friend of Rudolf Peierls throughout their lives. (See letters [62], [75].)

[17] Hans Bethe to Rudolf Peierls

302/1571
Zug Baden-Frankfurt, 5.6.[19]28

Lieber Peierls!
Sie sehen, ich reise mal wieder spazieren, und meinen, ich hätte es gut. Aber ich bin darüber noch nicht ganz meiner Meinung. Denn es ist eigentlich sehr traurig, warum ich jetzt nach Frankfurt fahre: Die Mutter einer Freundin von mir ist gestorben, und ich will sehen, ob ich nicht ein bisschen helfen kann. Das ist da wohl nicht so ganz einfach. Aber manchmal tut einem wohl Aussprache und Gesellschaft in solchen Momenten sehr wohl. Und manchmal braucht man auch jemand, der einen ein wenig aufmuntert, etwas zu tun, um den Grübeleien zu entgehen, die ja die grösste Gefahr für einen Menschen nach einem Schmerz sind. Sie scheint übrigens ziemlich ruhig zu sein.

Vorher war ich in Baden-Baden bei meiner Mutter, um mich zu erholen. Dafür ist Baden sehr geeignet, denn der Wald ist ein angenehmer Aufenthalt, und man meint immer, wenn man dort ist, man wäre bei sich selbst zur Sommerfrische. Ausserdem bietet die Küche bei meiner Mutter eine rein angenehme Abwechslung gegenüber dem unter den verschiedenen Namen auftretenden Münchener Kalb- und Rindsbraten. Immerhin – ich bin nicht ohne Entschuldigung in Baden gewesen sondern habe mich vorher – offenbar um einen Grund zu haben nach Baden zu fahren, zehn Tage lang ins Bett gelegt, mit Fieber beliebiger Qualität (ich habs bis 40,3 gebracht!) und einer selbst duch Kapazitäten des Schwabinger Krankenhauses nicht näher erklärten Krankheit. Also sagen wir Grippe. Aparterweise habe ich infolgedessen im Schwabinger Krankenhaus gewählt, aber die Niederlage der Demokraten nicht verhindern können.[120] Dafür habe ich bei der bayrischen Landtagswahl "gesiegt", indem ich in weiser Voraussicht der

[120]The early letters between Bethe and Peierls contain primarily comments on physics, cultural issues or personal remarks. This is one of the few examples of expression of political opinion. These only resurfaced in the mid-1930s when both had settled outside Germany, Bethe in America and Peierls in England.

Tatsache, dass die demokratische Partei Bayerns keine hervorragende Bedeutung im Landtag Bayerns haben wird, sozialdemokratisch gewählt habe. Wenn meine Tante in Stettin das wüsste, würde sie mich enterben, da sie aber sowieso nichts hat, macht das nichts. Im übrigen bin ich mit dem deutschen Volk relativ zufrieden, ausser mit der Zunahme der Kommunisten. Nach Münchener Plakatsäulen zu urteilen, müsste übrigens der Reichstag mindestens 200 Nationalsozialisten enthalten: Hitler liess nicht weniger als 10 Plakate los, die ein guter Karrikaturist hätte illustrieren sollen. Welch schönes Bild könnte man z.B. zu den Worten machen: "Der internationale Börsenjude triumphiert".

Im Institut ist der Personalbestand beinahe unverändert, nur die Amerikaner sind bis auf unseren Freund four times u (WW) Sleator entschwunden.[121] Der ist aber nicht magerer geworden, obwohl er gerade eine Arbeit publiziert, über den Halleffekt[122] und noch einen anderen Effekt mit wohlklingendem Namen. Ich selbst schwelge darin, dass mein Name in Arbeiten zitiert wird, und habe ausserdem nach den Satz "jeder sein eigener Amanullah" meine Doktorarbeit abgeschrieben (in Baden-Baden). Über die Länge bin ich erschrocken. 70 Seiten, und Sommerfeld wird gewiss dasselbe tun. Gelernt habe ich noch immer Null, aber habe jetzt den festen Vorsatz und wenig Zeit, um ihn auch auszuführen.[123]

Letzten Samstag war ich mit Thorner zusammen, der, umgeben von Frl. Ster, die dick aber nett und witzig war (das sind Mädchen selten), von einer Schwarzwaldtour mit 3h-Aufstehen und derlei Stuss nach Baden-Baden kam zum Neurologentag. Er schien von

---

[121]W.W. Sleator, American student of Arnold Sommerfeld's until 1928.

[122]In Munich already, Rudolf Peierls had been interested in physical interpretations of quantum mechanics. In February 1928 he had given a paper entitled: 'Neuere Arbeiten zur physikalischen Deutung der Quantenmechanik (prinzipielle Unschärfe der Beobachtung)'. At Leipzig, where he joined Heisenberg's institute in the summer of 1928, the quantum theory of magnetism was being developed and Peierls published his important paper on the Halleffekt. R. Peierls, 'Zur Theorie der galvanomagnetischen Effekte', *Z. Phys.* **53**, 255–66 (1929). W.W. Sleator, *Phys. Z.* **29**, 628 (1928).

[123]Hans Bethe was preparing for his doctoral examination which was due to take place on 24th July 1928.

Krehl[124] recht befriedigt und hatte sich offenbar mit dem Thema seiner Arbeit schon etwas befreundet.

Es freut mich, dass Ihnen das Arbeiten in Leipzig so gut gefällt. Heisenberg muss rührend sein, wenn er um 11 abends noch Unterhaltungen über Arbeiten führt. Wie ist denn Wentzel?[125] Und Sie sind sehr experimentell interessiert mit Arbeiten über Hg.[126] Ich glaube, Sie sind seit langem der erste Physiker, der gleichzeitig experimentiert und theoretisert. Mit meinen Geistesprodukten sind Sie hoffentlich glücklich geworden! In diesem Sinne herzlichst Ihr

Hans Bethe

[18] Heinrich Peierls to Rudolf Peierls

Berlin, 20.7.1928

Lieber Rudi.
Vor Abgang meines Schreibens erhalte ich Deine Mitteilung von gestern, die mich natürlich sehr interessierten.[127] Ich glaube, dass Du unbedenklich das Anerbieten Deines Professors annehmen kannst. Das einzige,

---

[124]Hans Thorner was interested in general medicine as well as phsychosomatic problems. At Heidelberg he was working with Ludwig von Krehl (1861–1937), director of the Heidelberg Medical School and founder of the so-called 'Heidelberg School' which stood for a more anthropological approach to medicine.

[125]Gregor Wentzel (1898–1978), studied at the Universities of Freiburg and Greifswald before moving to Sommerfeld's institute in Munich where he took his doctorate in 1921 and his habilitation a year later; 1926 professor of mathematical physics at the University of Leipzig, 1928 appointed to the University of Zurich as successor to Erwin Schrödinger. After the war, Wentzel became professor at the University of Chicago where he taught between 1948 and 1970.

[126]While at Leipzig, Peierls took part in the advanced practical laboratory, a rather unusual step for a theorist. One of his tasks in the lab was the measurement of the fine structure of the mercury lines by using a simple spectrometer and a parallel-plate interferometer. See Peierls, *Bird of Passage*, p. 38.

[127]Refers to an offer by Heisenberg that Peierls could take on more teaching and administrative duties. The subsequent suggestion to use the stipend to support a talented poor student is characteristic of Heinrich Peierls' charitable attitude and activities. He supported several good causes in Berlin and elsewhere.

was dagegen spräche, wäre, dass Deine Studien etwa durch Verwaltungstätigkeit gehindert werden könnten, was aber nach Deinen Mitteilungen ja sicher nicht der Fall ist. Was das Honorar betrifft, so könntest Du ihm ja mitteilen, dass Du den gleichen Betrag für ein Stipendium an einen vermögenslosen Studenten verwenden würdest.
Nochmals herzl[iche] Grüsse

Dein Vater

[19] Heinrich Peierls to Rudolf Peierls

Berlin, 13.8.1928

Lieber Rudi.
Deine Karte von vorgestern haben wir erhalten und Deinen Bericht mit grossem Interesse gelesen. Anbei eine für Dich bestimmte Karte.

Wir haben hier eine sehr aufregende Zeit hinter uns. Anschel hat am Sonnabend einen schweren Schlaganfall erlitten, der eine vollkommene rechtseitige Lähmung zur Folge hatte. Es lässt sich zur Stunde noch nicht sagen, welche Wendung die Krankheit nehmen wird. Eine augenblickliche Lebensgefahr scheint nicht vorhanden zu sein, aber es lässt sich nicht mit Bestimmtheit sagen, ob und wieweit die Lähmung zurückgeht. Ihr Zustand ist augenblicklich sehr traurig. Bitte, verständige Karli, da ich nicht weiss, wieweit er von Fulda's orientiert ist. Mit letzteren habe ich vereinbart, dass sie vorläufig in Karersee bleiben, da augenblicklich ein vorzeitiges Zurückkommen keinen Zweck hätte.
Viele Grüsse

Dein Vater

Lieber Rudi! Dank f[ür] D[einen] Brief. Ich kann eben nichts mehr schreiben. Es ist zu trostlos, Mutti so leiden zu sehen, + am schlimmsten, wenn sie sprechen will + nicht kann. Bitte teile Carli alles mit. Er soll mir s[eine] Adresse geben.

Kuss Else

[20] Heinrich Peierls to Rudolf Peierls

Berlin, 25.8.1928

Lieber Rudi.
Besten Dank für Deine uns sehr interessierenden Zeilen vom 22. d[iesen] M[onats]. Hoffentlich ist auch der Rest Deines Aufenthaltes vom guten Wetter begünstigt. Ich nehme davon Kenntnis, dass Du etwa am 1. September nach London gehen wirst, und habe mir Deine neue Adresse vorgemerkt. Wie ich schon schrieb, würde ich nichts dabei finden, wenn Du noch länger an der See bliebst.

Ich bin sehr damit einverstanden, dass Du ungefähr am 15. Oktober in Berlin eintriffst. Was die Rückreise betrifft, so wäre vielleicht zu erwägen, ob Du nicht der Abwechslung halber und um es kennenzulernen, mit einem der grossen Ozeandampfer von Southampton nach Hamburg bezw. Bremen fährst. Die Kosten sind nicht sehr gross, und wenn man sich die Verpflegung anrechnet, ist es nicht teurer, als wenn man die Route über Harwich wählt. Ich überlasse es aber besser Deiner eigenen Entscheidung. Eventuell erhälst Du ja auch in London durch die Reisebüros alle nötigen Informationen.

Anschels Zustand hat sich nicht sehr verändert. Wir müssen im Augenblick damit zufrieden sein, dass wenigstens keine Verschlimmerung eingetreten ist.
Herzliche Grüsse

Dein Vater

Viele herzl[iche] Grüße Else.
Siehst Du Dir Ventnor[128] an?

Wegen London möchte ich noch folgendes sagen: Es kann sein, daß zwar die Wahl der Lancasterschen Wohnung an sich empfehlenswert ist, daß aber 1) die Entfernung von der City besonders groß ist 2) ein boarding house Dir mehr Bekanntschaften internationaler Art vermittelt. In

[128]Village on the Isle of Wight.

diesem Falle würde ich an Deiner Stelle den Aufenthalt teilen, mir ein günstiges boarding house suchen + den grössten Teil des Aufenthaltes dort verbringen.
Nochmals grüßend

D[ein] V[ater]

Lieber Rudi.
Wir sind gerade immer im geeigneten Moment bei den Eltern, wenn ein Brief an Dich abgeht und ich benütze gern die Gelegenheit Dir eine schöne Zeit zu wünschen. Du wirst ja als vollendeter Engländer zurückkommen. Also mach's gut.
Viele Grüsse

Deine Hermannie[129]

[21] Rudolf Peierls to Paul Dirac

Hampton, 14.9.1928

Dear Mr. Dirac,
You certainly do not know my name but you will remember that we have been in the Leipzig Opera together.[130]

Excuse my troubling you with a little entreaty: I have now been in England for six weeks, not mainly for working, but for learning the language and meeting with English people and English life. But I begin to get tired of being lazy, so that I should like, if possible, to stay the first two or three weeks of October beginning with Oct. 3, in Cambridge doing a bit of work in the libraries. Would you mind telling me whether

[129]Hermann and Annie Krebs used this abbreviation of their first names when signing joint letters.

[130]In the summer of 1928 Dirac visited Leipzig, and Peierls had to accompany him to the opera, where he was terrified to find that Dirac would not abide by the standard rule of leaving his coat and hat at the cloakroom. See J.G. Taylor (ed.), *Tributes to Paul Dirac*, London: IOP Publishing, 1986, p. 35.

this is possible without making difficulties to anybody? I should be very much obliged to you.
Your sincerely

Rudolf Peierls (Leipzig)

[22] Heinrich Peierls to Rudolf Peierls

Berlin, 27.9.1928

Lieber Rudi.
Deine ausführlichen Nachrichten vom 24. d[iesen] M[onats] haben wir erhalten. Deine Begründung, warum Du noch einmal nach Cambridge möchtest, leuchtet mir ohne weiteres ein. Ich hoffe, dass Du Deine Absichten dort verwirklichen kannst. Inzwischen haben uns Dr. Freund's, die sich sehr gefreut haben, mit Dir zusammen gewesen zu sein, Deine Grüsse telefonisch übermittelt. Dass Du offenbar von Deinem englischen Aufenthalt sehr profitierst, freut mich ungemein. Von Deiner Abhebung bei der AEG habe ich Kenntnis genommen.

Hier ist nichts Neues. Meine Indisposition hat sich wieder gegeben; auch Fulda's geht es gut.
Herzliche Grüsse

Dein Vater

Dear Rudi! I enjoyed your letter very much. Your life must be very interesting + the people you meet too. Mrs Freund told me by tel[ephone] that you are like an Englishman, perfectly in language. What impression did you get of Shakespeare's house, or did you never see it? I hope you will make some notices for yourself about everything you see in England. What about photos? Excuse me, I think I made many mistakes, but my head is not yet in order + I can't write as I should like to do.
Use your time + enjoy yourself!
Ever yours

Else

[23] Rudolf Peierls to Paul Dirac

Leipzig, 15.10.1928

Dear Mr. Dirac,
Unfortunately I had to leave Cambridge somewhat earlier than I expected. You will not think me impolite if I say you good bye this way only. I also wish to express you my thanks for your kindness which made it possible for me to have a very interesting time in Cambridge. I hope it will not have been the last time I spend there.
With many thanks
Yours very sincerely,

Rudolf Peierls

[24] Hans Bethe to Rudolf Peierls

F[rankfurt], 2.12.[19]28

Lieber Peierls,
Ich bin also jetzt schon so alt und dumm geworden, dass ich für den Staatsdienst geeignet erscheine. Immerhin ist die Art Staatsdienst, die ich hier bei Madelung[131] zu leisten habe, recht harmlos, und meine Befürchtungen, mit dem Doktor wäre die bequemste Zeit des Lebens vorbei, war unrichtig. Im Grunde genommen verlangt Madelung nichts weiter von mir, als dass ich jeden Morgen etwa zwei Stunden mit ihm schmuse, teils über vernünftige Dinge wie Quantenmechanik, teils indem ich einige philosophisch-physikalische Aperçus anhöre, die mir meistens sehr in ihrer Geistesfülle imponieren, aber doch nicht ernstzunehmen sind. M[adelung] ist an sich ein ziemlicher Skeptiker im Gegensatz zu Sommerfeld, und es macht ihm mehr Vergnügen als die ganze Physik, wenn er aus einpaar sowieso schon vagen Hypothesen ein Gebäude von

[131] Erwin Madelung (1881–1972). Succeeded Max Born as professor of theoretical physics at Frankfurt University in 1921.

kühnen Aphorismen aufbaut, das er dann sozusagen von unten her ansehen, bestaunen und sich darüber moquieren kann. Dazwischen macht er verblüffend einfache Erfindungen für physikalische Apparate, Elektrometer, Viskosimeter, Bolometer usw., die er mit Vorliebe so erklärt, dass man erst ganz am Schluss merkt, worum es sich im Grunde handelt, und dann natürlich umso erstaunter ist über die Einfachheit des Prinzips.[132] Den grössten Teil des Tages bin ich allein im Institut, was mir nach dem Münchener Betrieb etwas komisch vorkommt, und kann, wenn ich will, einen Kaffee oder einen dancing tea veranstalten. Das Alleinsein hat sich vorläufig übrigens nicht bewährt, ich arbeite dabei weniger als wenn ich mit mehreren zusammen in einem Zimmer sitze, gehe im Zimmer spazieren und denke mitunter garnicht an Physik bzw. Fourierintegrale, mit denen ich mich momentan beschäftige. Es kommt aber recht wenig bisher heraus. Ich möchte für praktische und ev[entuell] auch theoretische Zwecke eine vereinfachte Methode zur Fourieranalyse eines willkürlich gegebenen Kurvenzugs herausbekommen, und ausserdem das Fourierspektrum gewisser willkürlicher Punktanordnungen berechnen, Aber, wie gesagt, es wird noch nichts, und ich bewundere die Leute, die jeden Monat ihr Arbeits-Ei legen.

Daneben muss ich das physikalische Anfängerpraktikum leiten, und staune, trotz minimalster Erwartungen – über die Unkenntnis der einfachsten Dinge. Als neulich ein Mädchen nicht einmal die Hebelgesetze kannte, und zwei lineare Gleichungen mit zwei Unbekannten nicht auflösen konnte, nützte bloss noch Anschreiben. Als ich mitten dabei war, kamen aus dem Nebenzimmer so etwa 10 Leute hereingestürzt, als ob der Teufel hinter ihnen her wäre: Zwei Mädchen hatten Wasser im zugekorkten Gefäss gekocht, und dem Wasser wurde

---

[132]The remarks demonstrate that Bethe was not very impressed with Madelung. In an interview he later summarised his time at Frankfurt: "Madelung didn't suggest anything. He suggested that I should paint in India ink some wiggles on paper which one could then use as a grating and Fourier analyse light going through it. It couldn't have worked. It didn't work. And the other thing Madelung was interested in was the different interpretations of Dirac's equation. In fact, he wrote a couple of papers about it and he talked to me about it. It didn't seem to me then, and it does not seem to me now that it added anything to Dirac's equation. So I was entirely on my own." Hans Bethe, interview by Lillian Hoddeson, 29.4.1981, Niels Bohr Library, AIP.

Mazedonien zu klein. Ausserdem stellte man bei der Bestimmung des spez[ifischen] Gewichts nach dem Archimedischen Prinzip regelmässig das Wassergefäss, in das der Körper eintaucht, mit auf die Waage. Der Chef von der Experimentalphysik[133] ist übrigens nicht klug.

Frankfurt hat immer noch den Vorzug, dass ich hier nette Bekannte habe. Für das Arbeiten ist das allerdings ein Nachteil, selbst wenn nach den drei Jahren, die ich weg war, bloss noch drei Leute übriggeblieben sind – die aber umso intensiver. Dafür ist aber auch zur Abwechslung ganz schön, mal bis $\frac{1}{2}$ 5h morgens zu tanzen, wie ich das gestern tat, besonders wenn die Gesellschaft trotz viel Frankfurt-W so nett und amüsierlich ist wie diese es war. Man ist zwar am nächsten Tag ziemlich unfähig, aber schön wars doch. Auch habe ich gefunden, dass Theater mit Freundin schöner ist als allein, und bin derartig snobistisch geworden, dass ich in jede Premiere gehe – Dreigroschenoper, Krankheit der Jugend, Lederköpfe von Georg Kaiser etc.[134] Manche Sachen sind sehr reizend gewesen, davon z.B., "Es liegt in der Luft", was bei Ihnen in Berlin schon im Sommer in der Luft lag. Die Dreigroschenoper hat mir in ihrer Originalität auch sehr imponiert; sie ist etwas so ganz anderes als das Gewohnte und wurde sogar hier in Frankfurt einigermassen annehmbar gespielt. Die "Krankheit der Jugend" (Bruckner) ist schon ein schwieriger Fall, besonders weil das Stück, das wirklich von recht krankhaft veranlagten jungen Menschen handelt, von der Frankfurter Kritik meiner Ansicht nach ganz falsch aufgefasst wird als symptomatisch für die "Jugend von heute". Nicht etwa im anklagenden Sinn der Leute, die den verlorenen Krieg, die böse Inflation und die heutige Jugend als die Grundübel unserer augenblicklichen Lage sehen, und die ich neulich auf einer Gesellschaft in grosser Schar beobachten konnte, wie sie erst die Probleme der heutigen Jugendverderbnis erörterten und sich ach so einig waren, um sich nachher über den Streitpunkt "Auto oder Fussgänger?" gründlichst zu verkrachen. Sondern die Kritiker sind hier auf die Idee gekommen, dass gerade die Besten, Begabtesten

---

[133]In 1925 Karl-Wilhelm Meissner succeeded Walter Gerlach as professor of experimental physics in Frankfurt, before becoming professor of astronomy in 1928.

[134]Refers to the Kurt Weill's *Three Pennies Opera* which had its premiere in 1928, and two recent plays, Ferdinand Bruckner, *Krankheit der Jugend* (1926) and Georg Kaiser, *Die Lederköpfe* (1928).

und Nachdenklichsten jungen Leute in der Auseinandersetzung mit den Problemen des Lebens besonders mit der Sexualität zu Grundegehen und beim Selbstmord enden. Da kann ich nicht mit. Ebensowenig wie bei Kaisers Lederköpfen, die mir physischen Ekel erregt haben und das will allerhand heissen. Ausserdem war die Aufführung schlecht wie hier meistens.

Die Geschichte vom Hilbert[135] und dem trostlosen jungen Mann, die Sie mir schrieben, hat auch mich einen Tag lang erheitert. Er muss überhaupt ein Original sein: Eines Abends hatte er Gesellschaft, und sollte sich noch eine andere Krawatte anziehen. Nach einer halben Stunde, als alle Gäste da waren, war er noch nicht zurück. Seine Frau suchte ihn und fand ihn – im Bett. Der Reflex des Krawatteausziehens hatte in ihm die Vorstellung "ins Bett gehen" erzeugt. Und das tue ich jetzt auch.
Herzlichst Ihr

Hans Bethe

[25] Hans Bethe to Rudolf Peierls

Frankfurt, 17.2.[19]29

Lieber Peierls!
In Göttingen war ich nicht, wie Sie wohl festgestellt haben.[136] Ich wäre wahrscheinlich auch nicht hin, wenn ich nicht erkältet gewesen wäre und in erbittertem Kampf mit der Grippe läge, die ich zu vermeiden versuche. Denn wenn dieser Kampf nicht wäre, hätte ich wahrscheinlich wacker die rauhen Berge des Taunus mit Skiern bestiegen, was sich trotz verbreiteter gegenteiliger Ansichten mangels besserer Tätigkeitsfelder

[135] David Hilbert (1862–1943). At the time professor of mathematics at the University of Göttingen where he stayed until his death in 1943. He had a reputation of being the stereotypical absent-minded professor, and many anecdotes to that effect were being recounted by his students and younger colleagues.

[136] Peierls had been to a conference in Göttingen, which Bethe also had intended to attend.

bewährt hat. Der Schwarzwald ist allerdings schöner; jedenfalls schien es mir vom 1.–5. Januar dieses Jahres auf dem Feldbergerhof also, obwohl Petrus mir fünf Sonnentage schuldig blieb. Die liefert er jetzt bei -22 deg, also quasi in entwerteter Valuta, nach (ob wohl eine Aufwertungsklage da Erfolg hätte?).

Trotzdem war Göttingen sicher sehr interessant, d[as] h[eißt] die Sitzungen, Göttingen selbst wohl weniger. Viele Sachen schlugen sogar in mein Spezialgebiet, wie Elektronenbewegung, Austrittsarbeit, und so. Aber Göttingen ist weit, und die Anziehungskraft von Kongressen ist direkt proportional den Reizen der Stadt und umgekehrt dem Quadrat der Entfernung.

Meine Produktivität ist inzwischen auf einem Minimum angelangt.[137] Die Fourieranalyse unvollkommen periodischer Funktionen habe ich abgeschlossen, ohne etwas herausbekommen zu haben, was des Publizierens wert wäre. Jetzt versinke ich in Fermischer Statistik, und komme nicht über die unangenehme Feststellung hinaus, dass andere Leute gute Arbeiten schreiben, z[um] B[eispiel] Bloch, und ich nichts herauskriege.[138] Vor allem möchte ich mein mittleres Potential in Ordnung bringen, das Herr Aalpern mir nicht glauben will. Dass es bei hoher Ordnungszahl nicht stimmt, liegt einfach daran, dass die Paulingschen Abschirmungszahlen falsch sind. Also beschäftige ich mich erst einmal mit Abschirmungszahlen, für die ich die Fermische Statistik

[137]Bethe had taken up a post at the institute of his former teacher Madelung in Frankfurt. He found the research atmosphere stale and unstimulating, and used the time to learn about current developments in physics. (See J. Bernstein, *Prophet der Energie*, Stuttgart: S. Hirzel Verlag, 1988, pp. 18–19.) Nevertheless, his own assessment of his work in the letter to Peierls is difficult to sustain. In this supposedly unproductive phase he published a paper on the passage of cathode rays through electric fields formed by grids ('Über den Durchgang von Kathodenstrahlen durch gitterförmige elektrische Felder', *Z. Phys.* **54**, 703–10 (1929)), a paper on the comparison of the distribution of electrons in the helium ground state ('Vergleich der Elektronenverteilung im Heliumgrundzustand nach verschiedenen Methoden', *Z. Phys.* **55**, 431–36 (1929)) and a substantial article on the splitting of terms in crystals ('Termaufspaltung in Kristallen', *Ann. Phys.* **3**, 133–208 (1929)) which was widely used by physical chemists.

[138]Felix Bloch (1905–1983). At the time working with Heisenberg in Leipzig. He had just completed his doctoral thesis which promulgated a quantum theory of solids that provided the basis for understanding electric conduction.

verwenden will. Dann schwebt es mir vor, auch die Leitfähigkeit etwas quantitativer zu berechnen, als das Bloch getan hat. Aber all das ist nur Programm, und mit einem rohen Ansatz brachte ich bisher z[um] B[eispiel] für das mittlere Potential im Metall immer das Doppelte des beobachteten Wertes heraus. Nebenbei bemerkt, habe ich selbst schon daran gedacht, den Berechnungsindex für Elektronen durch Absorption zu erklären. Das gibt aber immer Indizes < 1, und die Abweichung von eins beträgt ungefähr ein Prozent des beobachteten Betrages. Ich glaube kaum, dass Herr Aalpern damit glücklicher wird.

Mit dem wissenschaftlichen Niedergang ist auch ein persönlicher verknüpft: Ich amüsiere mich neuerdings auf Maskenbällen. Allerdings stand mir ein Aufgebot verschiedener guttanzender und netter Mädchen aus dem Institut meines Vaters zur Verfügung, zu denen ich immer flüchten konnte, wenn es im Gewühl nicht mehr ging. Aber ich hatte den immerhin erstaunlichen Mut, fremde Mädchen aufzufordern, und sogar die bisher immer fehlende Fähigkeit, sie für einige Zeit an meine liebenswürdige Persönlichkeit zu fesseln, sodass ich am Schluss der Unternehmung prophezeien konnte, dass ich mit 30 Jahren die Wissenschaft des Maskenballs beherrschen würde. Im übrigen betätige ich mich nicht viel gesellschaftlich, sondern meist nur im "trauten Kreise" von beiderlei Geschlechts, was Frau Neuberger,[139] die Ihnen ja dem Namen nach bekannt ist, dazu veranlasst, ihren Sohn und mich als unnormal zu bezeichnen, weil wir in der Faschingszeit zu Hause schmusen. Diese Tätigkeit ist mir aber – trotzdem ich heruntergekommen bin – doch immer noch lieber als Faschingsbälle.

Im Sommer gehe ich, wie ich Ihnen vielleicht schon schrieb, doch nach Stuttgart zu Ewald[140] – als Assistant. Werden Sie in diesem Semester fertig? Und was gedenken Sie dann anzufangen? Eigentlich haben Sie ja im Vergleich zu mir noch viel zu kurz studiert, um schon den Doktor machen zu dürfen. In diesem Sinne mit herzlichen Grüssen. Ihr

Hans Bethe

[139]Mother of a colleague and friend of Hans Bethe's.

[140]Paul Peter Ewald (1888–1985). Professor of theoretical physics at Stuttgart University. Hans Bethe's father Albrecht was a good friend and former assistant to Ewald's uncle in Strasburg. In 1939 Hans Bethe married Ewald's daughter Rose.

[26] Heinrich Peierls to Rudolf Peierls

Rome, 18.4.1929

Lieber Rudi.
Ich müßte lügen, wenn ich behaupten wollte, daß Dein Brief vom 13. große Begeisterung bei uns verursacht hätte. Zunächst habe ich den Eindruck, daß Du Deine Arbeiten in letzter Zeit unnötig vernachlässigt hättest. Um nur einen Deiner Gründe herauszugreifen: Die ausgefallenste Porzellantafel hätte ich Dir immer in kürzester Zeit in der AEG oder in Rosenthal beschaffen können. Wir sind fachlich weiter als die staatliche Porzellan-Manufaktur. Privat-Affären dürften Dich auch in Deinem Arbeiten nicht aufhalten; auch ich bin jung gewesen, hatte aber keinen sorgenden Vater mehr + war einzig um existieren zu können, gezwungen, mich in meinem Arbeiten nicht irritieren zu lassen. Auch ich habe mich mehrfach sterblich in Deinem Alter verliebt, habe aber niemals diese Gewalt über mich gewinnen lassen. Du spielst mit dem Feuer. Wenn ich Dir auch gern glaube, dass Du Dich nicht "verlobt" hast, so unterschätzt Du die Schwierigkeit des Loskommens, zumal wenn die Liebe, wie in Deinem Falle, + wie sehr oft bei warm empfindenden weichen Menschen das Mitleid zur Basis hat.[141] Selbst nur durch eine sorgfältige Correspondenz bindest Du Dich menschlich fester und fester + schließlich schlidderst Du in die Sache, auch ohne es positiv zu wissen, wie Deutschland in den Weltkrieg. Nebenbei mußt Du Dir darüber klar werden, daß Du Dir einen Beruf ausgesucht hast, der Dir selbst bei günstigstem Verlauf — mit großem Glück + großem Fleiß — nicht gestatten wird, Dir einen Haushalt ohne Zuschuß zu gründen. Rechne hierbei nicht zu stark auf mich, denn wenn ich auch zu leben habe, so weißt Du genau, daß eine veränderte Position mir auch großen Einbußen an meinen Einkommen zugesteht. Laß Dein Herz nicht mit Deinem Verstand durchgehen. Die Folgen

[141] Heinrich Peierls refers to Peierls' friendship with Marga Vogt, a young gymnastics teacher from Halle whom he had met in Leipzig. In the spring and summer of 1929 the two corresponded regularly, and the letters sent by Marga Vogt to Peierls indicate that she was unstable and probably suffered from some kind of depression. *Peierls Papers*, Supp., A.32.

können für Dich ein verpfuschtes Leben bedeuten. Nun zu dem materiellen Teil: An sich spielen die 100 M[ark] einmalig keine große Rolle. Sie sind erledigt und es bedarf keiner Rückzahlung. Ich ersuche Dich aber, mir jetzt einen genauen Plan über Deinen Etat in Zürich zu machen + Dich, wenn er von mir gutgeheißen worden ist, genau daran zu halten. Übrigens, sollte Dir Herr Bickel nicht einmal aushelfen, sondern Du solltest überhaupt Deinen Wechsel durch ihn beziehen, was ich mit ihm besprochen hatte und was mir bequemer ist als direkte Überweisung. Sollte Dir dieses Verfahren unvernünftig erscheinen, so bitte ich um sofortige Nachricht. Die Idee mit der Übersetzung ist — wenn sie auch auf aus anständigem Grunde entstanden ist — durchaus krumm.[142] Wenn Dir jemand M[ark] 1500 zahlt, so muß es eine solche Riesenarbeit sein, daß Du dabei Deine Hauptarbeit vernachlässigst. Und das geht nicht. Meine Bitte geht dahin:

1. Gieb die Correspondenz mit dem Mädchen auf.

2. Widme Dich mit Energie Deiner Arbeit.

3. Erhole Dich trotzdem nach Herzenslust in Deiner freien Zeit, ohne Übersetzungen zu machen.

4. Veranschlage Deinen Etat reichlich, denn es ist unser Wunsch, daß Du anständig lebst, aber komme mit Deinem festgesetzten Etat auch wirklich aus!

Alles Weitere wollen wir Ende Juni persönlich besprechen. Wir kommen dann nach Zürich oder ersuchen Dich, nach Constanz zu kommen (wahrscheinlich zwischen 20 + 25 Juni). Enttäusche das Vertrauen nicht, das ich in Dich setze + handele verständig. Bitte schreibe aber jetzt keine Entschuldigungsschreiben, denn wir haben ja später Gelegenheit zur Aussprache. Für mich, oder richtiger gesagt für Dich handelt

[142] Peierls was working on a translation of de Broglie's new book on wave mechanics. It was published first in English and German, and only after the award of the Nobel prize to de Broglie in 1930 it came out in French. De Broglie, L. *Einführung in die Wellenmechanik*, translated by R. Peierls, Leipzig: Akademische Verlagsgesellschaft, 1929.

es sich nicht um die Vergangenheit, sondern um die Zukunft. Das ganze Leben liegt vor Dir, versaue es Dir nicht![143]

Alfred hat, wie Du gehört haben wirst, auf unsere Anwesenheit an seinem Hochzeitstage keinen Wert gelegt, + heiratet übermorgen.[144] Wir bleiben bis 23. hier, Adresse Bolzano, Hotel Laurin.
Herzliche Grüße

Dein Vater

Lieber Rudi!
Seit langer Zeit endlich mal ein Brief von Dir, der leider auf mich denselben Eindruck machte, wir auf Vati. Nimm Dir bitte seine Ratschläge zu Herzen + sei vielmals gegrüßt.

D[eine] Else

[27] Heinrich Peierls to Rudolf Peierls

Berlin, 14.5.1929

Lieber Rudi.
Mit bestem Dank bestätige ich den Eingang Deines Schreibens vom 8. d[iesen] M[onats] — Uns geht es weiter gut. Wir beabsichtigen, am 20. d[iesen] M[onats] nach München zu fahren, wo ich während 2 Tage verschiedene Verbandssitzungen mitzumachen habe. Am 23. d[iesen] M[onats] wollen wir uns bei dieser Gelegenheit das Walchenseewerk

[143]Heinrich Peierls' remarks must have made some impression on his son. Although he did not stop his correspondence with Marga immediately, there is no indication that he was not focussing on his work. Not only did he settle into a new environment in Zurich, he also completed the de Broglie translation, received his dissertation topic and completed his doctoral dissertation in less than four months.

[144]Alfred married a Russian woman, Nina, who had been a secretary in the Russian trade delegation in Berlin. She already had a 4-year old daughter, Vera. Heinrich and Else Peierls evidently had reservations against the marriage, and later correspondence prior to Peierls' marriage to the Russian Genia Kannegisser, indicate, that they objected among others to her nationality. See letter [112].

ansehen. Schade dass Du nicht mehr dort bist. Wir sind also vom 20. Mai bis einschliesslich 23. Mai in München, Regina-Palast-Hotel und vom 24. Mai an wieder zu Hause.
Viele Grüsse

Dein Vati + Else

Es liegt Vati daran, bald Adressen für Ankäufe für Villen in der Schweiz zu bekommen.[145]

Ich schrieb vor einiger Zeit an Herrn Bickel,[146] ohne Antwort zu erhalten. Ist er verreist?

[28] Heinrich Peierls to Rudolf Peierls

Berlin, 13.6.1929

Lieber Rudi.
Wir fahren nunmehr bestimmt Sonntag früh nach Konstanz, wo wir Sonntag abend ankommen — Adresse Insel-Hotel. Wir bleiben dort bis Donnerstag, den 20. Juni früh. Ich habe Herrn Bickel, der mir für die Fahrt nach Zürich ein Automobil anbot, geschrieben, dass ich es nur annehmen würde, wenn er mir 1.) seine Kosten berechnet und 2.) wenn er es entbehren kann, und habe ihn gebeten, das Auto ev[entuell] so zu schicken, dass es um 10 Uhr früh am Insel-Hotel ist. Wir würden also in diesem Falle am 20. d[iesen] M[onats] gegen Mittag in Zürich sein, wo ich ein Zimmer im Savoy-Hotel (Baur en ville) bestellt habe und würden Dich dort zum Mittagessen erwarten.*

Zu Deiner Orientierung teile ich Dir mit, dass ich bis Sonntag, den 23. Juni nachmittags bleibe, zu welcher Zeit ich nach Berlin zurückfahren muss.

---

[145] Heinrich Peierls was nearing retirement and he was contemplating to settle in Switzerland. In this context he asked his son to look for appropriate properties.

[146] Herr Bickel helped with financial transactions such as Heinrich Peierls' monthly payments to support his son.

Else bleibt aber bis zum 28. Juni mittags in Zürich, wo ich sie dann abholen würde und mit ihr gemeinsam nach Vulpera fahre.

Ich hoffe, dass Du diesen Brief nach Rückkehr von Deiner Freiburger Reise vorfindest, wo Du hoffentlich Onkel und Tante besucht haben wirst. (Ich höre inzwischen, daß sie in Sonthofen sind.)

* Soeben Dein Schreiben erhalten. Selbstverständlich werden wir uns sehr freuen, Euch beide im Auto zu fahren.

Viele Grüsse

Dein Vater

Auf frohes Wiedersehen! E[lse]

[29] Else Peierls to Rudolf Peierls

St. Moritz, 23.7.[19]29[147]

Lieber Rudel!

Hurray kleiner Doctor! Von Herzen beglückwünsche ich Dich hocherfreut! War's sehr stramm, die Prüfung noch dazu bei der Hitze? Bleibst Du noch in L[eipzig] + hast Du noch zu tun? Es wäre sonst sehr nett, wenn Du Dich mit uns treffen könntest. Wir bleiben bis zum 6. Aug[ust] hier, dann Autotour-Lenzerheide wo wir 1–2 Tage bleiben, dann nach Luzern für 4–5 Tage, wenn die Conferenz nicht gerade dort tagt. Also gieb bald Nachricht über D[eine] Pläne. Übrigens sind i[n] d[er] Roonstr. alle Mädels wieder zurück + können für Dich sorgen.

Also nochmals herzl[ichen] Glückwunsch —

Kuss

D[eine] Else

[147]The letter is misdated as 29.7.1929, but it must have been written before the letter of 26 July, which refers to the telegram mentioned in this letter. The reference to Ilse's birthday indicates that the letter was written on 23 July 1929.

Lieber Rudi, Sohn, Freund + College!
Als wir heute Abend von einer größeren Tour zurückkamen, erhielten wir zu unserer größten Freude Dein Telegramm. Lasse Dich zunächst brieflich umarmen, sei herzlichst beglückwünscht, daß Du die erste Etappe Deiner von Dir gewählten Laufbahn glücklich geschafft hast. Ist jetzt Dein Züricher Engagement perfekt? Wie lange hast Du noch in Leipzig zu tun? Was hast Du denn vor? Wie ist es mit einem Treffpunkt in der Schweiz? Oder willst Du uns in Berlin erwarten, wo wir Mitte August einzutreffen gedenken? Ich nehme an, daß briefliche Neuigkeiten an uns unterwegs sind + daß Du Alfred und Annie informiert hast. Hast Du Onkel Ludwig nach Carezza zum 15. Juli gratuliert? Vielleicht teilst Du es ihm und Tante Lala auch mit. Ilse hatte heute ihren Geburtstag. Ich umarme Dich innig.

Dein Vater

[30] Heinrich Peierls to Rudolf Peierls

St. Moritz-Dorf, 26.7.1929

Lieber Rudi!
Gestern erhielten wir Deinen mit großer Spannung erwarteten Brief. Nun können wir Dir nochmals zum Prädikat: summa cum laude herzlich gratulieren. Wir sandten heute, um unserer großen Freude Ausdruck zu geben, 2 kleine Packetchen an Dich ab, eines mit Pralinés, das andere mit candierten Früchten. (Idee und Auswahl von Else). Zoll ist bereits bezahlt.

Wir werden etwa am 16.8. früh in Berlin sein. Nach Deinen Angaben würden wir Dich dort treffen oder jedenfalls in Kürze erwarten können, worauf wir uns sehr freuen. Dann wird aber nachgefeiert. Abrechnung ist unnötig, sie wird Dir immer, wenn Du ein Examen machst erlassen. Herzliche Grüße

Dein Vater

Lieber kleiner, neugebackener "summa cum laude" Doctor. Nun ist die Freude erst ins Riesengroße gewachsen. Nochmal gratulor. Du darfst Dir daheim eine Feier nach D[einem] Geschmack arrangieren. Nun hat d[ie] Hitze etw[as] nachgelassen + so wirst Du — Süßigkeit im Munde + im Herzen, sicher flott weiterarbeiten. Vergiß d[ie] Erholungspause nicht! Also auch viel Spaß! Tausend liebe Grüße

Else

[31] Hans Bethe to Rudolf Peierls

Zug Stuttgart-Heidelberg
27.7.[19]29

Lieber Peierls,
also zuerst mal meinen herzlichsten Glückwunsch, dass Sie offenbar Examen und die damit zusammenhängenden Störungen der Behaglichkeit, wie Lernen und so, hinter sich gebracht haben, und jetzt nicht nur berechtigt sind, A-Assistant zu werden, sondern auch alles wieder zu vergessen, was Sie gelernt haben, und das ist mindestens 500 frs pro Monat wert.

Ich scheine gestern meine Tätigkeit als A-Assistant (vielleicht ausser ein paar Tagen im September) abgeschlossen zu haben. Der Abschluss war sehr feierlich und wurde durch Sortieren von einigen hundert Separatabdrucken von Ewald aus den Jahren 1900–1929 festlich begangen, eine Beschäftigung, die lange genug dauerte, um uns ein Bewußtsein äusserster Pflichterfüllung zu verschaffen.[148] Ab 1. Oktober hab ich mich dann an Sommerfeld verkauft[149] – also doch! – und bekam dafür einige Bonbons in den Mund gesteckt, nämlich 1) ein Reisestipendium von einem Jahr, wovon ich nur $\frac{1}{4}$ Jahr nach New York zu gehen brauche –

---

[148]Bethe had been working as Ewald's assistant in Stuttgart during the summer semester of 1929.

[149]Bethe enjoyed his semester at Ewald's institute, which he later recalled to be 'among the happiest times of his youth' (Bernstein, *Prophet der Energie*, p. 19), but Sommerfeld was keen to see him return to Munich and offered him a very attractive package, both financially and in terms of career prospects.

Sommerfeld hat sich also doch meiner Abneigung gegen längeren Aufenthalt in experimenteller Umgebung nicht verschliessen können und im übrigen nach Cambridge und wahrscheinlich Rom. 2) Habilitation zum nächsten Sommer; nachdem ich erklärt hatte, an Stuttgart reizte mich u.a. die Aussicht, mich baldmöglichst habilitieren zu können, d.h. am Ende des SS1930, blieb S[ommerfeld] nichts anderes übrig als das noch ein halbes Jahr früher zu versprechen, 3) die nächste freiwerdende Assistentenstelle in München; das letzte Versprechen ist zwar nichts wert, weil wahrsheinlich so bald nichts frei wird und weil man ausserdem derartige Versprechen vergessen kann. Immerhin war ich ganz zufrieden mit dem, was ich für 130 DM pro Monat erhandelt habe, so sehr, dass ich meine Arbeit über die Austauschenergie bei Ionenkristallen nicht weiter angeschaut habe[150] und infolgedessen auch Ihr Zitat nicht so bald brauche. (Denn jetzt sind Ferien und überhaupt).

Ich habe übrigns Ihrem Mut bewundert, mit so einfachen Annahmen die Ionisierungsspannung zu berechnen! Ich hätte es nie riskiert, die Störungsenergie einfach mit Wasserstofffunktionen zu berechnen, sondern hätte (warum denn einfach, wenns auch kompliziert geht) sicher erst mal versucht, ein self-consistent field auszurechnen usw. Es ist aber eigentlich sehr hübsch, dass man so leicht durchkommt! Ausserdem habe ich mich bei Ihrer Arbeit überzeugt, dass man etwas "leicht mit Gruppentheorie aber noch leichter ohne sie"[151] sehen kann, eben die Koeffizienten der einzelnen Integrale $\mathfrak{J}$ und $\mathfrak{K}$. Ich habe mich übrigens inzwischen auch bemüssigt gesehen, eine "Ionisierungsspannung" auszurechnen, aber eine "negative", nämlich die Elektronen-Affinität des Wasserstoffs nach der Methode von Hylleraas,[152]

[150]At Stuttgart Bethe met Douglas Hartree who had recently developed the so-called Hartree method which facilitated the approximate determination of properties of atoms. Bethe had already developed an interest in the quantitative description of atomic structures, and now he applied Egil Hylleraas' calculations of electron energy to the negative hydrogen ion.

[151]At the time, theoreticians began to talk about "Gruppenpest", as many were tempted to use group theory more extensively than necessary. Interview with Léon Rosenfeld. Interview by Thomas Kuhn and John L. Heilbron, 1.7.1963 and 19.7.1963, Niels Bohr Library, AIP.

[152]E.A. Hylleraas, 'Neue Berechnung der Energie des Heliums im Grundzustande, sowie des tiefsten Terms von Ortho-Helium', *Z. Phys.* **54**, 347–66 (1929).

im Gegensatz von den Resultaten von Brück und Pauling (negative El[ektronen]-Affin-ität] von 80 bezw. 2 kcal.) finde ich +17 mit einem Fehler von ±1 kcal pro Mol H^- (d.h. Grundterm des H^- gleich 1,053 Rh). Haben Sie übrigens gelesen, was der gute Brück neulich für einen, gelinde gesagt Unsinn, über Hyperfeinstruktur fabriziert hat?[153] Es kommt das z.B. das Axiom vor, dass 2 Vektoren vom Betrag $\frac{3}{2}$ und $\frac{1}{2}$ zur Resultierenden 0 zusammensetzen können.

Momentan fahre ich mal auf zwei Tage nach Heidelberg, um den Werner Sachs als neuen Doktor begrüssen zu können. Dann ev[entuell] 1 Tag nach Ludwigshafen zu Mark,[154] der Elektronen beugen und mich zu disem Zweck interviewen will.[155] Ein paar Tage zu meiner Mutter nach Baden-Baden, und von da aus mit ihr in eine sehr verlockend aussehende Gegend, am unteren Ende des Aletsch-Gletschers (Südhang der Jungfraugruppe) wahrscheinlich Hotel Riederalp. (Aber wenn Sie das Bedürfnis nach schriftstellerischer Tätigkeit haben, schreiben Sie nach Baden, Gunzenbachstr. 27). Die Prospekte von da sehen wunderbar aus!

Und was machen Sie bei 308° K (d.h. momentan sind hier höchstens 295) in der mancherlei Reize bietenden Stadt Leipzig? Eben intoniert der Männergesangverein Bruchsal ein sehr lautes Abschiedslied. Da kann ich Ihnen vor Schreck nur noch herzliche Grüsse schicken!
Ihr

H. Bethe

---

[153]Hermann Alexander Brück studied at Kiel, Bonn and at Munich where Arnold Sommerfeld supervised his doctoral studies on the wave mechanics of crystals. He later turned to astronomy, moved to Cambridge and Dublin before becoming Astronomer Royal of Scotland in 1957.

[154]Hermann Francis Mark (1895–1992), "father of polymer science"; at the time director of the research laboratory of highmolecular compounds at IG Farben in Ludwigshafen.

[155]Mark's work with Bethe was used in H.F. Mark and R. Wierl, 'Die Ermittlung von Molekularstrukturen durch Beugung von Elektronen an einem Dampfstrahl' *Z. Elektrochemie* **36**, 675–6 (1930); and in H.F. Mark and K.H. Meyer, *Der Aufbau der hochmolekularen organischen Naturstoffe*, Leipzig: Akademische Buchgesellschaft, 1930. Bethe, in turn, used Wierl's and Mark's results on the determination of the atomic form factor with electrons in his article on the theory of the passage of fast corpuscular rays through matter ('Zur Theorie des Durchgangs schneller Korpuskularstrahlen durch Materie', *Ann. Phys.* **5**, 325–400 (1930)).

[32] Heinrich Peierls to Rudolf Peierls

Berlin, 7.10.1929

Lieber Rudi.
Deine Karte sowie Deinen ausführlichen Brief vom 4.d[iesen] M[onats] haben wir erhalten und freuen uns, dass Du ein passendes Zimmer gefunden hast.[156]

Von Deiner Abhebung nehme ich Kenntnis und bin selbstverständlich einverstanden. Ebenso bin ich natürlich damit einverstanden, dass Du gegen Ende des Monates noch einmal Geld abhebst. Du kannst mir ja später eine Generalabrechnung geben.

Der Tod von Stresemann ist ein grosser Verlust für Deutschland.[157] Du hast ja inzwischen erfahren, dass Curtius,[158] der sehr geschätzt wird, provisorisch an seine Stelle tritt. Es ist aber noch nicht sicher, ob diese Berufung in ein Definitivum umgewandelt wird. Ev[entuell] sind noch Hoesch[159] und Schubert[160] Kandidaten für diesen wichtigen Posten.
Herzliche Grüsse

Dein Vater

Lieber Rudel.
Wie schön, daß Du so ein luftiges Zimmer auf der Höh gefunden hast + mit Pauli oft zusammenkommst. Wir hörten die g[anze] Trauerfeier gestern per Radio + sahen uns d[en] Reichstags Platz an, der prachtvoll stilvoll geschmückt war. Es ist ein Jammer um Stresemann, aber ich

---

[156]Peierls had moved back to Zurich to begin his work as Pauli's assistant at the ETH.

[157]Gustav Stresemann (1878–1929), German foreign minister who had worked for reintegration of Gemany into the international community after the First World War by negotiations with the victorious powers. His reconciliation with France led to the award of the nobel peace prize to Briand and Stresemann in 1926.

[158]Julius Curtius (1877–1948), had served as economics minister between 1926 and 1929, before succeeding Stresemann as foreign minister from 1929–1931.

[159]Leopold von Hoesch, German Ambassador to Paris.

[160]Carl Theodor von Schubert, Staatssekretär (minister) in the German foreign ministry.

glaube man kann seinem Werk nicht mehr schaden. Die Richtlinien sind gegeben + so muß es sich vollenden. Ein Glück, daß er einen schnellen ahnungslosen Tod fand. Jahrelang hat er genug gelitten.

Wir haben eben einen reizenden Logierbesuch: Frl. Torranke aus Heidelberg. Sie übersiedelt Donnerstag in ihr eigenes Zimmer in d[er] Berlinerstr. War dann 14 T[age] bei uns. Unser s[ehr] unangenehmer Wiener Besuch E[dith Theiner] ist gestern (nach 6 Wochen Hiersein) abends abgedampft!!! Hast Du Alfreds geschrieben? Lugano, Hotel Meister, Schweiz. Frl. Toni regiert i[n] d[er] Wohnung + hütet das Kind.

Viel Liebes

Else

[33] Heinrich Peierls to Rudolf Peierls

Berlin, 15.10.1929

Lieber Rudi.

Besten Dank für Deine Nachrichten vom 11. d[iesen] M[onats], die mich sehr interessierten. Ich freue mich, dass die Gehaltsfrage zu Deiner Zufriedenheit geregelt ist und erkläre wiederholt mein Einverständnis damit, daß Du bis 100.- Frs. monatlich bei Bickel abhebst.

Was Deine Doktorarbeit betrifft, so bitte ich um Zusendung eines Exemplars, sobald Du die Druckerlaubnis erhalten hast.

Was Heisenberg betrifft, so glaube ich auch, dass er Dir Deine Entschließung nicht übel nehmen kann. Immerhin würde ich die Fühlung mit ihm zu behalten suchen. Vielleicht kannst Du Dich auch, wenn Du mal von Zürich nach Berlin fährst, bei ihm ansagen und Dich auf der Durchreise dort aufhalten. In persönlicher Aussprache lassen sich manche Dinge besser klären.

Alfred's haben, wie aus ihren neueren Berichten hervorgeht, besseres Wetter als zuerst gehabt.

Dein Vater

Sie haben übrigens sofort ihr Hotel gewechselt + wohnen nicht "Meister", sondern Hotel Bellevue kommen aber schon Dienstag daheim an. Eben kommt die Karte. Es ist auch hier sommerlich. Eben 3 Besucher zum Wohnen hier, aus Palermo.
Eiligst viele Grüße

Else

[34] Heinrich Peierls to Rudolf Peierls

Berlin, 14.11.1929

Lieber Rudi.
Deinem Schreiben vom 11. d[ieses] M[onats] entnahmen wir die Aussicht von Deinem Zimmer, die beneidenswert ist. Wenn aber Deine jetzige Wohnung Schwierigkeiten wie Du sie schilderst, hat, so wird man Dir raten müssen, bald Aussicht auf eine andere Wohnung zu suchen.

Wir haben nun definitiv beschlossen, unsere Reise nach Zürich aufzugeben und zwar deswegen, weil wir am 22. d[iesen] M[onats] für 8–10 Tage nach Paris fahren wollen, wo ich Eugene Weigert[161] gerne sprechen möchte. Da Else Paris noch nicht kennt und ich für diese Zeit abkommen kann, haben wir uns entschlossen, so lange zu bleiben. Wir würden dann vorläufig nicht nochmals wegfahren und unsere nächste Reise erst Mitte März nächsten Jahres anlässlich der geplanten Mittelmeerreise antreten.

Da wir uns nun auf diese Weise bis Ostern nicht sehen würden, — übrigens fallen die Osterfeiertage sehr spät; Ostersonntag ist am 20. April — so wäre es schon ganz schön, wenn Du Weihnachten für eine paar Tage herkommen würdest. Einen Zwang soll es nicht bedeuten. Selbstverständlich würden wir vollkommen damit einverstanden sein, dass Du nur einen kleinen Teil Deiner Ferien hier verbringst und Dich im übrigen, wenn das Wetter hierzu einlädt, dem gesunden Wintersport widmest.

[161] Relative of Heinrich and Elisabeth Peierls.

Deinem Urteil über Atlantic kann ich nicht beipflichten.[162] Ich habe den Eindruck, dass der Tonfilm noch nicht weit genug ist, um sich mit einer so schweren Aufgabe, wie der Schilderung von Menschen, die voller Verzweiflung untergehen, zu befassen. Hier haben die Worte geradezu ernüchternd gewirkt. Hiermit will ich natürlich nichts prinzipielles gegen den Tonfilm sagen, von dem ich sogar überzeugt bin, dass er eine grosse Zukunft hat; aber er muss noch sehr stark vervollkommnet werden.

Wir hoffen, noch vor dem 22. d[iesen] M.[onats] eine Zeile von Dir zu haben.

Viele herzliche Grüße

Dein Vater

Lieber Rudel!

Es war sehr lieb von Dir, an Muttis Geburtstag zu denken. Vielen Dank. Ich stelle ein paar weiße Rosen an ihr Bild, da draußen leider alles erfriert. Inzwischen hat Fredi Cohn-Oesterreicher, Osnabrückerstr. 21, Berlin-Wilmersdorf einen furchtbaren Schicksalsschlag erlitten. Seine Frau Anneliese, Annies Freundin, hat sich unter die fahrende Untergrundbahn geworfen + war sofort tot. Sie war nervenleidend, das arme Ding, + ging schon länger mit Selbstmordideen um. Gestern war die Beerdigung. Alle sind s[ehr] erschüttert. Vielleicht schreibst Du ihm ein paar Worte. Alle Bekannten freuten sich über die netten Briefe; alles Gute, bleib gesund!

Kuss Else

[35] Heinrich Peierls to Rudolf Peierls

Berlin, 2.1.1930

Lieber Rudi.

Wir freuen uns sehr, aus Deiner Karte zu ersehen, dass es Dir gut geht und dass Ihr gute Schneeverhältnisse habt. Hoffentlich ist es bei der

---

[162] *Atlantic* (EA Dupont), produced at Estree (London) in German and English in 1929.

grossen Höhe, in der Ihr Euch befindet, so geblieben; wir haben hier das scheusslichste Tau- und Regenwetter.

Wenn sonst kein Anlass vorliegt, werden wir nicht mehr nach Arosa schreiben, sondern unsere nächsten Briefe nach Zürich richten.
Beste Neujahrsgrüsse

Vater

Prosit Neujahr, lieber Rudel, bleibe gesund + fidel + genieße d[en] herrlichen Wintersport. Bitte schreibe öfter 1 Kartengruß. Die Karte aus Arosa kam gerade Sylvester abends 7 Uhr ehe wir i[n] d[ie] Scala fuhren, während Vati sich schon wieder Sorgen machte. Wir waren sehr vergnügt mit Fischer's i[n] d[er] Scala + jeder bekam dann hier bei mir um 11 Uhr 1 Hummer, Sekt, Punsch + Karpfen, so erreichten wir spielend das neue Jahr.
Hoffentlich geht's im Lande + bei uns seelisch + geistig aufwärts!
Viele Grüße auch an Thorner

D[eine] Else

In Breslau verlief übrigens der Abend reizend.[163] Es wurden 6 famose Reden gehalten. Darunter war die von Vati natürlich die Witzigste. Am 6. Januar ist Lalas Geburtstag; 1 Karte würde sie sehr erfreuen!

---

[163] 60th birthday of Heinrich Peierls' brother, Franz.

[36] Heinrich Peierls to Rudolf Peierls

Berlin, 28.1.1930

Lieber Rudi.
Dein Schreiben vom 21. d[iesen] M[onats] habe ich erhalten, und Deine Doktor-Dissertation und auch die Diplom-Kopie ist in unseren Besitz gelangt.[164] Verstehen kann ich allerdings nur die letztere, denn die Schrift ist mir ein Buch mit sieben Siegeln. Indessen genügt es mir, dass Deine Professoren sie vestehen. Ich freue mich, dass Du zusammen mit Professor Pauli eine Arbeit planst und würde mich freuen, sie seinerzeit zu erhalten, auch wenn ich sie nicht verstehe.

Von hier ist nichts neues zu melden. Oskar Weigert,[165] der übrigens vor einiger Zeit Ministerialdirektor geworden ist, wird möglicherweise von Genf für 1 Tag nach Zürich kommen. Ich habe ihm jedenfalls Deine Adressen sehr ausdrücklich mitgeteilt.
Herzliche Grüsse

Vater

Lieber Rudolf!
Wolltest Du Dich nicht als Dr so nennen? Ich habe mich ebensosehr wie Vati über die Beglaubigungs-Scheine gefreut. Übrigens hat Dein Freund Friedländer in München das Schriftliche gut überstanden + für D[einen] Freund Juliusburger sucht Vati heftig etwas, s'ist aber schwer, da er nur genügend hat.[166] Die Braut hat uns s[ehr] gefallen, ist s[ehr] resulut.

---

[164] Peierls' doctoral thesis had been published at the end of 1929 as R. Peierls, 'Zur kinetischen Theorie der Wärmeleitung in Kristallen', *Ann. Phys.* **3**, 1055–1101 (1929).

[165] Heinrich Peierls' first wife, Elisabeth (née Weigert) was his cousin. Oskar Weigert was a relative of both.

[166] This refers to Hans Juliusberger, a friend of Peierls, from Oberschöneweide. Against the wishes of his parents he got engaged to his long-standing girl-friend. Peierls' parents were open-minded and tried to bridge the gap between parents and children in the Juliusberger family, not least by helping Hans Juliusberger to find a job. See letters [39], [72].

Nun kommt's drauf an, ob die beiden — wenn er eine Stellung hat — wirklich aus eigener Kraft, so bescheiden mit 1 Zimmer + kl[einem] Zuschuß auskommen. Dies bezweifeln seine Eltern sehr (ihre Eltern haben nichts!) da sie dazu bis jetzt allen Grund hatten. (siehe gr[oßer] Wechsel, der nie reicht.) Na, nach 2 Stunden kann man noch kein genügendes Urteil fällen. Jedenfalls wollen sie bald heiraten.

Mit dem Eislaufen ist's dieses Jahr nix. Mild und nebeliges Wetter. Ich habe viel zu arbeiten, da das Buch meines Vaters für die neue Auflage umgearbeitet — modernisiert wird, was der Leiter der Reinhardt Schauspielschule mit mir gemeinsam macht. Es macht viel Freude, mehr als "feucht Gebabbel" + Mischpoke.
Vergiß Vaters Geburtstag nicht.

Kuss Else

[37] Hans Bethe to Rudolf Peierls

Korrektur folgt baldigst! (Inliegend zum Trost zwei Marken, mit denen ich nichts anfangen kann)

Zug Baden-Baden – München 30.1.[19]30

Lieber Peierls,
Sie werden schon auf mich geschimpft haben, weil ich Ihre Korrektur so lange nicht zurückgeschickt habe.[167] Dafür habe ich Sie populär gemacht, d.h. in München und Stuttgart über Ihre Arbeit im Seminar vorgetragen, die mir durch besondere Freundlichkeit des Verfassers bereits in der Korrektur zugänglich gemacht wurde. Mein hoher Chef ist nach anfänglichem Kopfschütteln von der Wichtigkeit

[167]Peierls had asked Bethe to comment on a paper on the kinetic theory of thermal conduction in crystals based on his Ph.D. See letter [36], note 165.

der Umklapp-Prozesse überzeugt[168] und trauert mit dem ganzen Institut dem Wiedemann-Franzschen Gesetz nach.[169] Dass das nicht herauskommt, ist der einzige Fehler Ihrer Theorie, die sonst sehr schön ist, dass man sie sogar beinahe glauben könnte. Aber wo ein Fehler stecken soll, sehe ich nicht. Im übrigen hab ich "for Spass" überall, wo mir ein Druckfehler zu sein schien, eine entsprechende Korrektur vorgenommen – ohne Garantie, aber vielleicht können Sie es doch für die zweite Korrektur brauchen!

In Arosa war es noch sehr hübsch,[170] obwohl ich am Tag Ihrer Abreise bei der Pasern-Tour, die Ihnen vielleicht ein Begriff ist, (Langwies-Watterscheide (2600m) Küblis (400m)) Minkos bekam. Die Abfahrt war nämlich dermassen von laufenden Leuten bevölkert, dass alle zwei Minuten mindestens drei Leute an einem vorbeifuhren, und zwar alle so schnell, dass man schon alleine vom Luftdruck hinfiel, und dann, wenn man sich mühsam erhoben hatte, gleich nochmal vor Schreck über diesen Vorgang sich hinsetzte. Wenn man davon aber abstrahiert, war diese berühmte Tour sehr schön – eine zweite Tour aufs Matlieshorn auf ganz anderem Weg aber noch schöner, während das Parpaner Rothorn uns infolge des Nebels und Schneefalls versagt blieb.

Eben komme ich von einer Tour ohne Skier nach Stuttgart zu meinem früheren Chef, bei dem ich drei Tage lang ungefähr so ununterbrochen geredet habe, wie Sie neulich in Arosa. Hauptsächlich sollte

[168]In his paper, Peierls looks at thermal resistance in crystals as result of the anharmonic part of the force between atoms. He refined Debye's earlier results (P. Debye, *Vorträge über die kinetische Theorie*, Leipzig/Berlin: Teubner, 1914), and investigated the coupling between otherwise independent waves of different lengths and directions caused by these anharmonic forces. While the wave vectors in a continuous medium are conserved exactly, in a crystal they can change by multiples of a Bragg vector. Peierls called this process 'Umklapp' process, a term which remained in use.

[169]The Wiedemann-Franz laws had postulated the correlation between thermal and electric conductivity. As the Umklapp-processes as described by Peierls become very rare at low temperatures, the theory predicted an exponential rise of thermal conductivity at low temperatures which was not consistent with the Wiedemann-Franz laws.

[170]Bethe had joined Thorner and Peierls on 5th January in Arosa. Peierls had to leave earlier than the other two because of work commitments in Zurich.

ich Ewald eine Arbeit von Rosenfeld über optische Aktivität erklären,[171] wobei ich auch wissen sollte, was R[osenfeld] eigentlich für ein Mann sei und ob er irgendeine besondere Funktion in Zürich hätte.[172] Da ich es nicht wusste, soll ich es eruieren. Anschliessend Abstecher nach Frankfurt und Baden-Baden, sowie einen moralischen Kater wegen 5 Wochen Ferien.
Herzliche Grüsse Ihr

Hans Bethe

Grüssen Sie Ihren neuvermählten Chef![173]

[38] Lev Landau to Rudolf Peierls

Leipzig, 6.2.1930

Hochverehrter Herr Doktor!
Entschuldigen Sie bitte für meine verzögerte Antwort welche leider keineswegs etwas mit irgend welcher Untersuchung der Gleichungen zu tun hat. Sie haben natürlich vollständig recht, dass die von uns gewonnene Lösung den Gleichungen keineswegs genügt. Dass habe ich auch schon am ersten Tag auf eine Bemerkung Heisenbergs hin feststellen müssen, aber wegen einer ungeheuren Schreibfaulheit Ihnen nicht gleich mitgeteilt. Ich bin auch ganz Ihrer Meinung, dass es überhaupt nicht lohnt irgend welche Lösungen der Gleichungen zu suchen welche so wie so wegen der + - Übergänge keine einzige konkrete Frage richtig beantworten könnten.

---

[171] L. Rosenfeld, 'Über die Gravitationswirkungen des Lichts', *Z. Phys.* **65**, 589–99 (1930).

[172] Léon Rosenfeld (1904–1974), had completed his Ph.D. in Liège in 1926. After studying in Paris (1926–7) and Göttingen (1927–9), he came to Zurich in the summer of 1929 where he was working mostly on Heisenberg-Pauli's quantumelectrodynamics. (See Pauli, *Wissenschaftlicher Briefwechsel*, II, pp. 74–5.) Later, he taught in Liège, Utrecht, Manchester and Copenhagen, where he was also a close collaborator and personal assistant of Niels Bohr.

[173] On 23 December 1929, Pauli had married the Leipzig performer Margarete Käthe Deppner.

Also schicken Sie die Geschichte mit dem Falle entsprechenden tiefsinnigen Bemerkungen über die Naturgeheimnisse ruhig dem Geh.Reg. usw. ab, und begeben Sie sich an die Bekämpfung des $-m$. Ich habe mich schon etwas damit beschäftigt aber leider noch nur allgemeine Momente herausgefunden.

Hoffentlich geht in Zürich auch ohne meiner hohen·Persönlichkeit alles gut. Ihre Bitte alle hiesige langweilige Leute zu grüssen konnte ich leider nicht erfüllen, weil denen ungeheuer viele sind. Beck[174] stellte sich noch etwas dümmer heraus als ich es von Berlin den Eindruck hatte und Halpern[175] komprommettiert mit seinem agressif langweiligen Gesicht Ihren Geschmack. Hund[176] erschien mir dagegen ganz nett.

Grüssen Sie meinerseits Ihren verehrten Professor wie auch seine Gemahlin, und good-bye bis aufs weiteres.

Yours truly,

L. Landau

[39] Heinrich Peierls to Rudolf Peierls

Berlin, 10./11.2.1930

Lieber Rudi.

Für Deine freundlichen Wünsche vom 4. d[iesen] M[onats] danke ich Dir herzlich, ebenso für das bei uns pünktlich eingetroffene Buch, von dem ich glaube, dass es mich ganz besonders interessieren wird.

[174]Guido Beck (1903–1988), Austrian-born physicist who despite Peierls' critical remarks here became a good and respected friend. Peierls first met him in Leipzig where he was Heisenberg's assistant. Beck later worked in Odessa and Paris, before emigrating to Latin America where he worked at the National Astronomic Observatory in Cordoba and at the Centro Brasileiro de Pesquisas Fisicas in Rio de Janeiro.

[175]Otto Halpern (1899–1982), at the time at Leipzig, later emigrated to the US and taught at Los Angeles and later at New York University.

[176]Prof. Friedrich Hund (1896–1997) had studied mathematics, physics and geography at Marburg and Göttingen, taught at Göttingen before becoming professor of physics at Rostock (1927) and Leipzig (1929).

Ich freue mich, dauernd gute Nachrichten von Deinem Ergehen zu haben und kann Dir von uns das gleiche berichten. Unsere Reise werden wir erst am 12. März antreten, doch muss ich vorher noch einmal geschäftlich verreisen. Da unser Haus Roonstr. verkauft worden ist, ziehen wir am 30. Juni um. Näheres teilt Dir Else mit.
Herzliche Grüße

Dein Vater

Lieber Rudi!
Nach kurzem heftigen Suchen fanden wir eine wunderschöne 7-Zimmer-Wohnung am Reichskanzlerplatz 10. Es gibt viel vorzubereiten vor der Reise.
Juliusburger's Hans hat durch Vati eine Praktikantenstelle bei einer Osram-Untergesellschaft für monatl[ich] 250.- M[ark] bekommen, tritt Montag an!
Viel Liebes

Else

[40] Lev Landau to Rudolf Peierls

Berlin, 6.3.1930

Lieber Peierls,
$\frac{\sqrt{\frac{hc}{K}}}{m}$ Entschuldigungen. Zuerst das Geld. Aber die russische Regierung war in diesem Falle so unakkurat, dass ich überhaupt die ganze Zeit auf Pumpen lebte und wahrhaftig nichts tun konnte. Ich hab es Ihnen natürlich am selben Tag wo ich die Überweisung aus Moskau bekommen telegraphisch geschickt. Hoffentlich sind Sie deswegen in keine unangenehme Situation geraten und sind mir auch nicht böse.

Besonders aber die Briefe. Schreibfaulheit ist wirklich absolut katastrophal geworden. Hoffentlich geht es weiter besser. Besonders blöde ist es weil ich Sie für Paris durchaus gewinnen will (sonst sehen wir uns ja beliebig lange nicht). Schreiben Sie mir daher möglichst schnell über Ihre Ferienpläne. Ich bin im Moment in Berlin, wo ich noch ungefähr eine Woche bummeln möchte um dann, wenn es mit dem französischen Visum alles in Ordnung ist (bis jetzt habe ich es noch nicht!) nach Paris zu gehen. Wenn Sie sich beim Empfangen dieses Briefes in Berlin befinden so rufen Sie mich (Zehlendorf 32-13) an.

Über Theorie und auch andere damit nichts zu tun habende andere Sachen könnte ich Ihnen verschiedenes schreiben aber im Moment ohne entsprechender Stimmung kann ich es nicht und noch auf das Abschicken des Briefes zu warten wäre noch blöder. Hoffentlich sehen wir uns doch bald und besprechen alles mündlich. Also seien Sie mir wirklich nicht böse.

Ihr

L. Landau

[41] Heinrich Peierls to Rudolf Peierls

Berlin, 12.3.1930

Lieber Rudi.

Im Begriff abzureisen, erhalte ich soeben Deine Zuschrift vom 10.d[iesen] M[onats] und teile Dir mit, dass ich nicht das Geringste dagegen einzuwenden habe, wenn Du sowohl nach Holland wie nach Kopenhagen fährst.[177] Voraussetzung ist aber, dass Du Dich nicht überanstrengst und dass Du tatsächlich wertvolle Anregungen hierdurch erwartest. Du kannst selbstverständlich das nötige Geld und auch den Vorschuss auf Dein Gehalt bei Bickel erheben.

Unser Programm bleibt unverändert bestehen; wir werden Dir von Venedig aus nochmals schreiben.

Herzliche Grüsse

Dein Vater

Else grüßt bestens.

---

[177]Peierls had been invited to join Pauli for a visit of Ehrenfest in Leiden and Hendrik Kramers in Utrecht, in mid-March. Letter Pauli to Casimir, 23.3.1930, refered to in Pauli, *Wissenschaftlicher Briefwechsel*, II, p. 11 footnote 1. Peierls and Pauli, on this occasion signed the famous wall behind the sliding blackboard.

[42] Lev Landau to Rudolf Peierls

Berlin, 20.3.[19]30

Lieber Peierls!
Also nach Holland geh ich sobald das Visum da ist, was wohl (ich habe vor zwei Tagen einen telegraphischen Gesuch eingereicht) in einigen Tagen sein mag. Schreiben Sie mir bitte, wo Sie in Holland die nächsten Tagen treffbar sein werden, wenn Sie überhaupt noch da länger bleiben. Herzliche Grüsse auch an Bloch.
Ihr

L. Landau

P.S. Wo und wann ist Pauli?

[43] Heinrich Peierls to Rudolf Peierls

Berlin, 8.5.1930

Lieber Rudi.
Besten Dank für Deine freundliche Zuschrift vom 3. d[iesen] M[onats]. Wir sind gut hier angekommen und haben uns wieder eingerichtet. Die Arbeiten am Reichskanzlerplatz gehen langsam vorwärts, da verschiedenes geändert werden muss. Wir hoffen, in der zweiten Hälfte Juni umziehen zu können.[178]

Dein Vetter Edgar[179] hat sich mit Fräulein Betsy Vogel in New York verlobt. Der Vater ist Direktor einer großen Bank und der Großvater ist der grosse amerikanische Mäzen Henry Goldmann,[180] der u[nter] a[nderem] die bekannte Sängerin Helena Gerhard finanziert. Die Famile wird also nicht für ihn sammeln müssen.

[178]Heinrich and Else Peierls were moving to a flat at the Reichskanzlerplatz. See letter [39].
[179]Edgar Peierls, son of Siegfried and Jenny Peierls.
[180]Henry Goldmann was a co-founder of the investment bank Goldmann & Sachs.

Inzwischen ist Edgar, der ja Bankier werden soll, kurz nach der Verlobung nach Hamburg abgefahren, um dort bei Warburg's und später auch in London und Paris sich über das europäische Bankwesen zu informieren.

Bereits vor der Ankunft des Dampfers in Hamburg hatte Edgar Unterleibsschmerzen, und als er in Hamburg ankam, musste er Hals über Kopf in das Krankenhaus, um am Blinddarm operiert zu werden. Die Operation ist gut gelungen und er scheint ausser Gefahr zu sein. Er muss aber, da es höchste Zeit war und offenbar das Bauchfell schon angegriffen war, einige Zeit im Krankenhaus liegen und vorläufig eine Kanüle tragen. Onkel Adolf[181] war gestern bei ihm und ich werde heute hinfahren; über Sonntag wird ihm Alfred Gesellschaft leisten. Vielleicht schreibst Du ihm eine Zeile, mit der Du ihm gleichzeitig zu seiner Verlobung gratulieren kannst, nach Hamburg, Krankenhaus Bethanien.

Den Ausschnitt aus der Dänischen Zeitung schicke ich Dir anbei zurück.[182]
Sonst nichts Neues. Herzliche Grüsse

Dein Vater

Lieber Rudel!
Als Strohwitwe sitze ich hier. Vater hat das Wichtigste berichtet, gestern feierten wir Hermann's Geburtstag im Familienkreise bei ihnen, Gerda Mutter und Ur-Omi,[183] wir + Nina's. Es war s[ehr] gemütlich. Was sagst Du zur Edgars Neuigkeit?

Im Juli kommen alle Amerikaner mit der neuen Familie her!!! Ich stehe in Umzugsberatungen und Arbeiten. Fuldas sind in Karlsbad in der hübschen Königsvilla. Dein Zeitungsbild ist s[ehr] ulkig. Hast Du viel zu tun + gern?
Viel Liebes

Else

[181]Heinrich Peierls' brother-in-law.
[182]Article with photo about the Copenhagen colloquium.
[183]Hemann Krebs' mother and grandmother.

[44] Lev Landau to Rudolf Peierls

Haarlem, 17.5.1930

Lieber Peierls,
Die 10^{20} Entschuldigungen bin ich Ihnen natürlich auch diesmal schuldig, jetzt haben Sie aber auch einen Teil der Schuld. Mit aller Mühe kann ich Ihren Lichtquantenbeweis einfach nicht verstehen.[184]

Die Korrektur ist natürlich schon beliebig lange bei (ungefähr am Tage von Gamovs Brief) beim Geh.Reg.Rat.Prof.Dr.

Mit dem Feldoperator bin ich in meisten Behauptungen einverstanden. Da ich in der letzten Zeit auch ungefähr zu den selben Vorstellungen gelange, bin ich geneigt, sie als zwingend aufzufassen. Ihr Glaube an die Möglichkeit einer vernünftigen Orthogonalisierung der Eigenfunktionen "wirklichen Feldes" scheint mir aus eigener Erfahrung beliebig aussichtslos und unabhängig von den Schwierigkeiten im Mehrkörperproblem, die ja doch sekundärer sind (denken Sie an die klassische Elektrodynamik). Ich würde deshalb die Sache so kurz formulieren. Das "messbare"Heisenberg-Paulische Feld entspricht nicht der physikalisch interessanten Raumeigenschaft welche prinzipiell unmessbar ist sondern ist teilweise an die notwendigen Eigenschaften des Messkörpers zurückzuführen. Deswegen scheint es mir "praktisch" gänzlich uninteressant was man besonders an dem Falle $\mathfrak{E}$ ersieht, was ja absolut verrückten physikalischen Bedingungen entspricht. Vernünftig wird es nur bei $|\mathfrak{E}| \gg \frac{\sqrt{hc}}{v^{2/3}}$ (v Messungsvolumen) was aber als $h \to 0$ interpretiert werden kann also einfach der klassischen Elektrodynamik entspricht. In der Quantenelektrodynamik ist diese Eigenschaften der Theorie von den Schweinereien unabhängig der Feldbegriff einfach nicht zu gebrauchen. Da auch die Koordinaten der Lichtquanten unvernünftig zu sein scheinen so würde ich geneigt zu glauben dass die Impulse der Lichtquanten einzigen einfachen sinnvollen Grössen sein dürften. Von Ihrer Behauptung über die unendliche Selbstenergie habe ich leider kein

[184] Landau and Peierls were investigating the behaviour of light quanta in space, writing a wave equation for photons similar to Schrödinger's equation for electrons.

Wort verstanden wozu mir auch Bloch nicht verhelfen konnte. Die unendliche Selbsternergie hat ja nichts mit irgendwelchen Eigenschaften des Feldbegriffes zu tun, sondern ist eine Eigenschaft der Gleichungen. Das man in diesem Fall keine sinnvollen Aussagen machen kann ist ohne weiteres zuzugeben.

Die schlechten Aussichten meiner zweiten Wette sehe ich leider auch aus theoretischen Gründen ein, da ihr theoretischer Hintergrund absolut unhaltbar sein dürfte. Wie ist übrigens Paulis Stimmung.

Den Diamagnetismus werde ich also doch zusammenschreiben. Gegen das periodische Feld seien Sie nicht so vorsichtig und erzählen Sie natürlich darüber wem und was Sie Lust haben. Ich halte die Bemerkung eigentlich für ziemlich unwichtig. Mir geht es eigentlich bis auf einige unendlich grosse Faktoren sehr gut. Den Rockefeller habe ich doch gekriegt und bleibe also im Ausland ungefähr bis Januar. Die Woche in København war wegen der in person erschienenen Dänin und Schwedin ziemlich amüsant. Bohr ist reizend. Ich würde ihm aber nicht mehr als Z23 Parameter zutrauen was meine Behauptung über Unendlichkeit genialer Theoretiker glatt widerlegen dürfte. Seit zwei Wochen bin ich in Holland. Das RijksMuseum gehört zu den 7 bisherigen Bilderbetrunkenheiten meines Lebens. (Eremitage und Russisches Museum in Leningrad, Tretjakowkaja Gallerie in Moskau, Kaiser Friedrich Museum und Kronprinz Palais in Berlin, Gemäldegalerie in Dresden) Besonders Frans Mals habe ich viel schöner gefunden als ich es mir nach anderen Bildern vorstellte. Das Mauritshuis[185] find ich dagegen ziemlich mittelmässig. Mit der Holländischen Landschaft bin ich eigentlich zufrieden.

Mit Bloch vertragen wir uns ganz gut was aus der ersten Begegnung in Zürich keineswegs vorauszusagen wäre. Die scheinenden Selbstzufriedenheitsmomente haben keinen reelle Bedeutung was Sie nach Bloch auch eingestanden haben, und sind wahrscheinlich einfach durch einige Reste der Schweizerischen Erziehung (siehe auch Schiller und C$^{\underline{o}}$) formuliert worden, die Bloch hoffentlich bald loswerden wird. Ehrenfest hat sich etwas an mich gewohnt und fürchtet sich nicht mehr so stark wie

[185] Museum in The Hague housing the Royal Cabinet of Painting, consisting of masterpieces of the Dutch Golden Age.

es am Anfang deutlich zu spüren war. Ich hatte eigentlich zuerst etwas Lust ihn zu necken, dazu ist er aber zu wehrlos. Von den Holländern gefiel mir am besten Wiersma. Fokker[186] ist ein Witz.

Am 19.V. fliege ich nach England. Schreiben Sie nach Dr. L. Landau, c.o. Dr. G. Gamov, Cavendish Laboratory, Cambridge.
Ihr

L. Landau

P.S. Wenn es Ihnen nicht sehr schwer wäre, schreiben Sie mir die Liste der Leute, Sie Separata von unserer Geschichte geschickt haben, dass ich wissen kann, wer es hat.

[45] Heinrich Peierls to Rudolf Peierls

Berlin, 28.5.1930

Lieber Rudi.
Dein Brief vom 25. d[iesen] M[onats], für den ich bestens danke, haben wir mit großem Interesse gelesen. Wenn Du im nächsten Semester Vorlesungen halten kannst, wenn auch zunächst nicht unter eigenem Namen, so wäre dies sehr zu begrüssen.[187]

Was Deinen Geburtstag betrifft, so glaube ich, dass es keinen Zweck hat, von hier aus ein Paket nach dem Ausland zu schicken, falls Du nicht besondere Wünsche hast, die in Deutschland erfüllbarer sind. Bitte äussere Dich umgehend dazu.

Von Hans Juliusburger habe ich zufällig gehört, dass er bereits am 16. Mai ohne alle Formalitäten geheiratet hat. Wenn Du ihm nachträglich gratulieren willst, so ist seine Adresse: Dr. Bleichröder, Berlin Pankow, Breitestr. 33.

---

[186] Adriaan D. Fokker (1887–1972), Dutch physicist who derived the Fokker-Planck equation in 1913.

[187] Peierls gave a lecture course on wave mechanics in Pauli's name. See letter Peierls to Genia Peierls, 8.10.1930.

Bei uns sieht es jetzt schon etwas wüst aus. Wir werden wahrscheinlich am 12. Juni ziehen. Neue Adresse: Berlin-Charlottenburg 9, Reichskanzlerplatz 10.
Herzliche Grüsse

Dein Vater

Lieber Rudi!
Es freut uns ja sehr, daß Du bald Vorträge halten darfst + so schnell vorwärts kommst. Der Geburtstagsmonat ist angebrochen! Vera hat am 2. Juni, Nina am 15. Juni Geburtstag. Die amerikanische Sippe naht scharenweise! Wir gehen jetzt ins Adlon, Frau Martin, Jenni's Schwester[188] zu begrüßen. Edgar und Braut werden wohl in diesen Tagen eintreffen.

Wir hatten gestern einen großen erhabenen Eindruck durch das Genie Toscanini im Concert d[er] Philharmonie mit d[em] Orchester v[on] New York. Sonst schufte ich nur + habe dazwischen Besuch. Viel Spaß, gutes Wetter! Was macht Dein Herz? Vous-êtes libre?
Ou non?
Viel Liebes

Else

[188] Jenni Peierls, wife of Siegfried.

[46] Werner Heisenberg to Rudolf Peierls

[Leipzig], 3.6.1930

Lieber Herr Peierls!
Aus Ihrem und Paulis Brief[189] sehe ich, daß Sie beide meinen Versuch einer Umschreibung der bisherigen Gleichungen noch nicht richtig verstanden haben.[190] Ich möchte, wie Sie auf der letzten Seite Ihres Briefes schreiben, dem Operator Gesamtimpuls genau dieselbe Bedeutung lassen, wie bisher. Man bekommt also in der Tat genau die selben Gleichungen wie bisher, nur mit der Nebenbedingung $-i\alpha_k\pi_k + \alpha_i\alpha_k f_{ik} = 0$. Genauer möchte ich das so schreiben: Für eine kräftefreies Elektron soll $-i\alpha_k\pi_k + \alpha_i\alpha_k f_{ik} = 0$ gelten und ich hoffe, daß aus dieser Bedingung folgt, dass $\int_V div\pi$ die Eigenwerte $0, \mu$ hat, wobei das $\int div\pi$ über irgendein endliches Volumen zu erstrecken ist. Im Falle von Elektronen + Lichtquanten kann nicht mehr $-i\alpha_k\pi_k + \alpha_i\alpha_k f_{ik}$ verschwinden. Wenn man dies nicht fordert, so hat sicher $div\pi$ zunächst kontinuierliche Eigenwerte. Man kann aber hoffen, dass es nur für solche Werte von $div\pi$ Lösungen mit endlicher Selbstenergie gibt, für die es im kräftefreien Fall solche, d.h. Lösungen mit der Nebenbedingung $-i\alpha_k\pi_k + \alpha_i\alpha_k f_{ik} = 0$ gibt. Man müßte also auch im Fall von Elektronen + Lichtquanten fordern, daß $\int_V div\pi$ die Eigenwerte 0 und μ hat.[191] In diesem Fall kann man eine Hilfsvariable $\underline{q}$ einführen nach der Gleichung $div_p\pi = \mu\delta(\underline{q}-\underline{q}_p)$

[189] The letter has not survived, but its contents and the definitions and notations used below can be deduced from Heisenberg, W., 'Die Selbstenergie des Elektrons', *Z. Phys.* **65**, 4–13 (1930). This paper contains all the definitions of terms used in this letter, e.g. α_k are spin matrices, Π_k are the components of the electric field.

[190] One of the consequences of the uncertainty relation was that it was not possible to determine the exact location as $l = h/MC$ (M =Protonenmasse); therefore Heisenberg wanted to introduce a fundamental unit l, a so-called 'universal length' and a lattice structure by dividing space into cells of the volume of $(l = h/Mc)^3$. While the one and two dimensional cases could be reconciled with known theory, the three-dimensional extension caused difficulties, and energy impuls conservation laws remained valid only in approximation. Pauli continued to investigate the β-decay, but correspondence on the matter with Werner Heisenberg was largely via Peierls. Peierls, however, at the time, was more interested in field quantisation, work he had engaged in in collaboration with Lev Landau.

[191] μ was defined as $\mu = 4\pi e/\sqrt{2hc}$.

und die Katze fällt wieder auf die alten Füße. Das Wesentliche an meinem Versuch erscheint mir nur, daß für die Selbstenergie eine einfache und durchsichtige Formel resultiert $\frac{1}{2}\int dV(\delta\rho_2\mathfrak{E}+\rho_3\mathfrak{G})^2$, und daß man daraus genauer sieht, welchen Bedingungen das Feld zu unterwerfen ist, damit die Selbstenergie verschwindet.[192] Die ganze Frage ist also nach wie vor die, ob es in der bisherigen Theorie Lösungen ohne Selbstenergie gibt; ferner, für welche Werte von $\int div\pi$ dies der Fall ist. Die Korrespondenz scheint mir dagegen völlig trivial, wie ich immer an Pauli schrieb, da ja an den Gleichungen nichts geändert ist.

In der Frage der Nullpunktsenergie bin ich dagegen sehr viel skeptischer als Sie; ich muß gestehen, daß ich den Operator $\sqrt{\Delta}$ als etwas so fürchterliches ansehe, daß ich zu seiner Anwendung nur mit Gewalt gezwungen werden kann. Wie man aus dem Operator Δ die Wurzel zieht, hat gelegentlich der Spintheorie Dirac gezeigt. Ob es immer durch einen solchen Witz geht, weiß ich nicht.[193] Aber jedenfalls bin ich so nicht zufrieden. Ich sehe auch nicht, wie in Ihrem $\mathfrak{E}_1$ und $\mathfrak{H}_1$ nach Umstellung der Faktoren im Energieausdruck die Selbstenergie noch als Quadrat oder etwas ähnliches geschrieben werden kann. Geht das????

Eine weitere fundamentale Schwierigkeit scheint mir darin zu liegen, daß die Größe $\int_V div\mathfrak{E}dV$ in der bisherigen Theorie überhaupt keine Eigenwerte haben kann, da $\int_V div\mathfrak{E}dV = \int \mathfrak{E}_n df$; es handelt sich also hier nicht mehr um räumliche Mittelwerte von $\mathfrak{E}$, sondern um solche auf einer Fläche und dafür divergiert (vgl. Paulis Brief vor einigen Wochen)[194] das bisherige Verfahren.

---

[192]See Peierls R. and L. Landau, 'Quantenelektrodynamik im Konfigurationsraum', *Z. Phys.* **62**, 188–200 (1930). On the basis of Heisenberg and Pauli's derivation of the electromagnetic field (W. Heisenberg and W. Pauli, 'Zur Quantendynamik der Wellenfelder', *Z. Phys.* **56**, 1–61 (1929) and W. Heisenberg and W. Pauli, 'Zur Quantendynamik der Wellenfelder. II', *Z. Phys.* **59**, 168–90 (1930)), Peierls and Landau described the electromagnetic field and its interaction with matter by a Schrödinger equation in the configuration space of the light quanta, obtaining identical results to those of Heisenberg and Pauli.

[193]In the above mentioned publication on light quanta in configuration space, (note 192), Peierls and Landau used the operator $\sqrt{\Lambda}$ to which Heisenberg objected.

[194]Pauli's letter could not be located.

Ich glaube also z[ur] Z[eit] (Dienstag abend), daß die ganze Feldquantelung noch in Unordnung ist und daß sich diese Unordnung unter anderem in der Nullpunktsenergie und in der Undefiniertheit von Größen à la $\int \mathfrak{E}_n df_{\sqrt{\Delta}}$ äußert. An Pauli richte ich die dringende Bitte, den Operator $\sqrt{\Delta}$ ohne Fourierzerlegung zu definieren, wie mit Bleistift in Ihrem Brief angedeutet.
Sonst die herzlichsten Grüße von Institut zu Institut.
Ihr

W. Heisenberg

P.S. Vom 5.–15.6. bitte ich Sie oder Pauli, Briefe nach Helgoland postlagernd zu schicken.[195]

[47] Heinrich Peierls to Rudolf Peierls

Berlin, 10.6.1930

Lieber Rudi.
Deinen Brief vom 6. d[iesen] M[onats] haben wir erhalten und freuen uns zu hören, dass Du Deinen Geburtstag angenehm verbracht hast.

Wenn Du vom 22. bis 27. Juni in Leipzig bist,[196] so werden wir uns sehr freuen, Dich am 28. Juni in Berlin zu sehen, damit Du Dein neues Zimmer inspizieren kannst. Unser Umzug findet programmässig übermorgen statt, und Du kannst Dir denken, dass unsere alte Wohnung nicht mehr sehr einladend aussieht.

---

[195]Heisenberg suffered from hayfever and therefore spent the early summer away from Leipzig on Heligoland.

[196]Peierls took part in the so-called 'Leipziger Festspiele' a colloquium organised jointly by Heisenberg and Debye. The third such colloquium, in June 1930, dealt with electron interference, and Peierls, who stepped in for Pauli, gave a paper about his work on magnetism. Pauli, *Wissenschaftlicher Briefwechsel* II, p. 16, note a.

Was die Angelegenheit Hirsch betrifft,[197] so bin ich der Meinung, dass Du ihm diesmal helfen solltest. Ich würde ihm an Deiner Stelle sagen, dass dies nur einmal und ausnahmsweise geschehen könnte. Ich bitte Dich ausdrücklich, ihm zu sagen, dass Du mir davon nichts schreiben konntest und dass Du Dir das Geld geborgt hättest, um ihm behilflich zu sein und es also zu einem bestimmten Termin zurückgeben müsstest; anderenfalls würde nach meiner Meinung mit einer Rückgabe des Geldes nicht zu rechnen sein. Ich stelle Dir anheim, ihm entweder das Geld von dort direkt zu schicken, nachdem Du die Summe bei Bickel erhoben hast oder mir aufzugeben, wohin ich das Geld in Deinem Namen schicken soll. Vielleicht hat auch die Angelegenheit Zeit, bis Du zu Hause bist.

Else ist, wie Du Dir denken kannst, in den nächsten Tagen sehr beschäftigt und sendet Dir herzliche Grüsse; das gleiche tut

Dein Vater

Sieh zu, am 28. möglichst pünktlich hier zu sein, da ich am 29. Mittag oder Abend nach München fahren muß.

[48] Werner Heisenberg to Rudolf Peierls

Helgoland, 13.6.1930

Lieber Herr Peierls!
Vielen Dank für Ihren Brief; ich glaube, daß wir jetzt so ungefähr einig sind über den Sinn der Nebenbedingung. Für den materiefreien Fall bedeuten meine Gleichungen ein System, das aus dem Strahlungshohlraum und einem ungeladenen Partikel besteht. Es muß also Lösungen der Gleichungen geben, die sich als Produkt von zwei Faktoren schreiben lassen, von denen der erste einer (physikalisch nicht vorkommenden) Partikel mit der Ladung 0 entspricht, der zweite dem

[197] Peierls had enquired about lending some money to a friend, Kurt Hirsch. See also letters Heinrich Peierls to Rudolf Peierls, 15.6.1930, 4.2.1931 and 11.2.1931. *Peierls Papers*, Supp., A101–102.

ladungslosen Hohlraum. Die Existenz dieser Lösungen ist wieder trivial, wenn man in der von Ihnen angegebenen Weise Faktoren umstellt und statt (I) $\delta_1\rho_2\mathfrak{E} + \rho_3\mathfrak{H}^2$ etwa (II) $(\delta_1\rho_2\mathfrak{E}^{(1)})(\delta_1\rho_2\mathfrak{E}^{(2)})$ einführt und $(\delta_1\rho_2\mathfrak{E}^{(2)}) + (\rho_3\mathfrak{H}^{(2)}) = 0$ fordert. Es ist aber zu bemerken, was Sie ja auch schon schreiben, daß I und II nur dann ersetzt werden kann (d.h., nach Integration über den Raum, bis auf Umstellungen identisch ist), wenn $div\mathfrak{E} = 0$. Sobald longitudinale Schwingungen auftreten, so ist I von II wesentlich verschieden. Dieser Umstand scheint mir zu zeigen, daß Ihre Beseitigung der Nullpunktsenergie durch Umstellung von Faktoren noch nicht die endgültige Lösung darstellt. Man möchte doch den Fall $div\mathfrak{E} = 0$ nicht wesentlich anders behandeln als den Fall des Elektrons. Ihre Umformung der Nebenbedingung in eine Summe von drei Ausdrücken scheint mir nicht sehr vorteilhaft, weil aus dem Verschwinden der Summe ja noch gar nicht das der einzelnen Glieder folgt; Ihre Gleichungen verlangen also wohl zu viel.

Die Einführung des Elektronenortes scheint mir sehr einfach in folgender Weise möglich. Es sei eine Lösung von

$$H = \alpha_n I_n + \frac{1}{2}\int(\delta_1\rho_2\mathfrak{E} + \rho_3\mathfrak{H})^2 dV$$

gegeben; d.h. etwa ein Funktional $\Phi_\rho(\mathfrak{E})(= \Phi_\rho(\pi))$, das außerdem die Nebenbedingung $\int div\mathfrak{E} = \mu$ erfüllt. Dabei ist es zunächst unwesentlich, ob $(\delta_1\rho_2\mathfrak{E}+\rho_3\mathfrak{H}) = 0$ ist oder nicht; d.h. ob man ein kräftefreies Elektron hat oder nicht. Dann ist das Funktional

$$\Psi(\underline{q}_1\mathfrak{E}) = \delta\left(\underline{q}_x - \frac{1}{\mu}\int div\mathfrak{E}\cdot xdV\right)\delta\left(\underline{q}_y - \frac{1}{\mu}\int div\mathfrak{E}\cdot ydV\right)\delta(\cdots{}^2)\frac{\Phi(\mathfrak{E})}{\delta}$$

eine Lösung der Ausgangsgleichung

$$H = \alpha_k[p_k - \mu\chi_k(q)] + \int(\mathfrak{E}^2 + \mathfrak{H}^2)dV\,,$$

wie man durch direktes Einsetzen verifiziert. Notwendig ist hierbei nur, daß $\int div\mathfrak{E}dV = \mu$. (III) gibt also allen notwendigen Aufschluß über die Bahn des Elektrons.

Im ganzen glaub' ich aber doch nicht daran, daß mit den bisherigen Vertauschungsrelationen eine vernünftige Lösung der Gleichungen möglich ist. Ich möchte in ähnlicher Weise, wie wir in Kopenhagen besprochen haben, doch Ungenauigkeiten der Ortsko-ordinaten einführen; aber diesmal ohne neue universelle Länge, sondern etwa nach dem Schema $\{\chi_i, x_k\} = \delta_{ik}$ oder $\{f_{ik}|x_r x_s\} = \delta_{ik}\delta_{ks}$.

Mir ist es bisher nicht gelungen, solchen Vertauschungs-Relationen einen vernünftigen mathematischen Sinn zuzuordnen. Dagegen wären die zugehörigen Ungenauigkeitsrelationen äußerst vernünftig; auch wären die Vertauschungs-Relationen relativistisch invariant und es gäbe beliebig kurze Wellen. Fällt Ihnen oder Pauli nicht vielleicht etwas über den mathematischen Sinn solcher Vertauschungs-Relationen ein?

Ich freue mich sehr auf Ihr Kommen zu den Leipziger Festspielen[198] und hoffe dringend, daß Sie auch Ihren Chef mitbringen.[199] Ihnen und Pauli also die herzlichsten Grüße von
Ihrem

Werner Heisenberg

[49] Heinrich Peierls to Rudolf Peierls

Berlin, 4.7.1930

Lieber Rudi.
Meine Karte aus München hast Du wohl erhalten, ebenso wie wir die Deinige und Deinen letzten Brief.
Ich bin sehr dankbar, daß Du die Annonce aufgegeben hast[200] und bitte, mir diejenigen Meldungen, die nicht von vornherein ungeeignet erscheinen, baldigst zugehen zu lassen.

Alfreds sind nach Adelboden, Colm-Hotel, abgereist; von Annie kommen weiter gute Nachrichten. Die Hitze hier ist enorm.

---

[198]See letter [47], note 196.
[199]Pauli did not attend, but Niels Bohr, who was to receive the Plack-medal in Berlin on 20 June 1930, briefly visited Leipzig on that occasion.
[200]See letter [27], note 145.

Ich bitte Dich, am 10.d[iesen] M[onats] eine Glückwunschdepesche zu der Hochzeit von Edgar zu schicken: Adresse: Peierls, Hotel Plaza Athénée, Paris.
Herzliche Grüsse

Dein Vater + Else

[50] Hans Bethe to Rudolf Peierls

München, 11.7.1930

Lieber Peierls!
Schönsten Dank dafür dass Sie Herrn Frank[201] eine falsche Theorie geschrieben haben![202] Ich bin nämlich dadurch auf die Erklärung der anormalen Grössenordnung der magnetischen Widerstandsänderung (Koeffizient B von Frank) gekommen.

Die vollständig exakte Durchrechnung Ihres Falles – Elektron im elektrischen Feld F_x und magnet[ischen] Feld H_z ohne Beeinflussung durch das Gitterfeld, aber mit endlicher freier Weglänge l – finden Sie anliegend; es ergibt sich ein Widerstand, der vollkommen unabhängig ist von Magnetfeld.[203] Dabei ist angenommen, dass alle

[201] Nathaniel Hermann Frank (1903–84), postdoctoral fellow from MIT, he later returned to Princeton. Frank was working on electron theory of metals.

[202] Peierls had been working on the increase of electric resistivity of metals caused by a magnetic field. Experiments showed that after an initial gradual rise, resistivity became proportional to the strength of the magnetic field, but all the theories determined the additional resistivity as proportional to the square of the field.

[203] Peierls believed to have found a way out by arguing that in strong magnetic fields the electrons which follow spiral orbits in the magnetic field, would complete several turns and therefore the quadratic law would cease to exist. Shortly before the Leipziger Festspiele, where he was to present his ideas, Peierls realised that, mathematicaly, the different parts cancelled each other and would not serve to explain the situation. See R. Peierls, 'Das Verhalten metallischer Leiter in starken Magnetfeldern', *Leipziger Vorträge: Elektronen Interferenzen*, Leipzig: Hirzel, 1930, pp. 75–85. In writing up the paper, Peierls could insert a note with the correct explanation which he had found, independently of Bethe's calculations in this letter.

Elektronen die gleiche Geschwindigkeit v_0 besitzen, und dass nach jedem Stoss alle Bewegungsrichtungen des Elektrons gleich wahrscheinlich und unabhängig von der Bewegung vor dem Stoss sind (Mittelung über ϕ). Der Fehler bei Ihnen war offenbar, dass Sie den Stoss als gleichwahrscheinlich an jedem Punkt der Bahn des Elektrons betrachteten, das bedeutet aber Vernachlässigung von Gliedern der Ordnung $\frac{\beta^2}{h^2} = \left(\frac{\text{Krümmungsradius im Magnetfeld}}{\text{freie Weglänge}}\right)^2$; infolgedessen müssten Sie in Ihrer Formel $\rho = \frac{\frac{1}{\alpha}H^2}{1+const.\cdot h^2}$ konsequenterweise auch die 1 im Nenner gegen das Glied mit H^2 streichen, und bekommen demnach sicher keine Aussage über den Widerstand in der Nähe der Wendetangente (d.h. $\beta \approx h$).

Dass der Widerstand bei fester Elektronengeschwindigkeit streng unabhängig vom Magnetfeld wird, entspricht der Tatsache, dass die magnetische Widerstandsänderung in der "ersten Näherung" von Sommerfeld verschwindet. Eine Widerstandsänderung ist erst dann zu erwarten, wenn mehrere Elektronensorten verschiedener Geschwindigkeiten[204] sich an der Leitung beteiligen, man bekommt dann in meinen Bezeichnungen für kleine Felder ($h \ll \beta$)

$$\mathfrak{J}_x = enf\int \frac{\beta(v)}{\beta^2(v)+h^2}w(v)dv = enf\cdot\left(\int\frac{w(v)dv}{\beta(v)} - h^2\int\frac{w(v)dv}{\beta^3(v)}\right)$$

$$\mathfrak{J}_y = enf\left(h\int\frac{w(v)}{\beta^2(v)}dv - h^3\int\frac{w(v)}{\beta^4(v)}dv\right)$$

wobei $w(v)$ die Anzahl der Leitungselektronen von der Geschwindigkeit v ist. Rechnet man dasselbe etwas strenger nach Sommerfeld durch, so kommt man zum gleichen Resultat; und es wird

$$dv\cdot w(v) = v\frac{df_0}{dv}d\Omega$$

[204]Handwritten addition: besser gesagt mit verschiedenen $\beta = \frac{v}{l}$.

in den Bezeichnungen von Sommerfeld. Wenn nun

$$\overline{\beta^n} = \int \beta^n w(v) dv$$

ist, so folgt in unserer Näherung (d.h. solange man nach H entwickeln darf!):

$$\sigma = \frac{\mathfrak{F}_{x^2} + \mathfrak{F}_{y^2}}{F\mathfrak{F}_x} = \frac{enf}{F} \frac{\left(\overline{\beta^{-1}} - h^2\overline{\beta^{-3}}\right)^2 + \left(h\overline{\beta^{-2}}\right)^2}{\overline{\beta^{-1}} - h^2\overline{\beta^{-3}}}$$

$$= \frac{e^2}{m} n \frac{\overline{\beta^{-1}}^2 + h^2\left(\overline{\beta^{-2}}^2 - 2\overline{\beta^{-1}\beta^{-3}}\right)}{\overline{\beta^{-1}}\left(1 - h^2\frac{\overline{\beta^{-3}}}{\overline{\beta^{-1}}}\right)}$$

$$= \frac{e^2}{m} n\overline{\beta^{-1}} \cdot \left(1 + h^2 \cdot \frac{\overline{\beta^{-2}}^2 - \overline{\beta^{-1}\beta^{-3}}}{\overline{\beta^{-1}}^2}\right)$$

Der mit $H^2 \sim h^2$ proportionale Teil enthält also eine typische Schwankungsgrösse: $\overline{\beta^{-2}}^2 - \overline{\beta^{-1}} \cdot \overline{\beta^{-3}}$. Bisher hat man nun die Schwankungen von $\beta^{-1} = \frac{l}{v}$ immer nur soweit in Betracht gezogen, als sie durch die Fermische Verteilung bedingt sind, d.h. durch die kleine Variation der Energie der Leitungselektronen. In Wirklichkeit wird aber, weil die Elektronen sich im Gitterfeld bewegen, die freie Weglänge bezw. β nicht eine reine Funktion der Energie der Elektronen sein, sondern von der Bewegungsrichtung abhängen (von C in der Bezeichnung von Bloch).[205] Das macht nichts aus für die Berechnung des Widerstandes selbst $(\overline{\beta^{-1}})$, aber alle Schwankungsgrössen werden natürlich enorm vergrössert — und das ist gerade das, was wir zur Erklärung der magnetischen Widerstandsänderung brauchen: β schwankt bei fester Energie viel stärker als vermöge der kleinen Energiedifferenzen unter den Leitungselektronen. Eine quantitative Ausrechnung scheint mir ziemlich hoffnungslos, aber man kann leicht sehen, dass man bei Annahme

[205]For Bloch's notation see: F. Bloch, *Über die Quantenmechanik der Elektronen in Kristallgittern*, Dissertation Leipzig, July 1928. A published version can be found in *Z. Phys.* **52**, 555–600 (1928).

von vernünftigen Schwankungen der β-Werte für die Leitungselektronen auch zu vernünftigen Grössenordnungen für die Widerstandsänderung kommt. Vielleicht hilft die neue Arbeit von Morse (Phys. Rev. Juni)[206] zu einer etwas quantitativeren Behandlung solcher Fragen. Der Punkt, wo das quadratische Gesetz zu gelten aufhört, wird durch die Nicht-Freiheit der Elektronen nicht berührt, weil er nur davon abhängt, dass die freie Weglänge selbst vergleichbar wird mit der Bahnkrümmung im Magnetfeld, und nicht den Schwankungen der mittleren freien Weglänge bei den verschiedenen Leitungselektronen. Infolgedessen stimmt Franks C und nicht sein B.

Es tut mir leid, dass auf diese Weise nicht nur Frank, sondern auch ich Ihnen ins Handwerk pfusche. Aber man kann nichts dagegen machen, dass einem bisweilen eine Idee kommt!

Übrigens war ich in Leipzig schon ziemlich gespannt auf Landaus Theorie des Diamagnetismus im Elektronen-Kasten.[207] Könnten Sie ihn wohl veranlassen, uns hier etwas darüber zu schreiben? Oder können Sie das selbst tun? Mein Chef interessiert sich glaube ich auch sehr dafür.

In Leipzig verschwand ich ohnehin so sang- und klanglos in Hunds Arbeitszimmer, sodass Sie mir dort nicht bloss den Diamagnetismus sondern auch irgendeinen herrlichen Witz schuldig blieben, den Hund nicht hören sollte. Was machen Sie sonst? Hier ist augenblicklich eine erträgliche Temperatur ausgebrochen, sodass sich leben lässt.
Herzliche Grüsse
Ihr

Hans Bethe

---

[206]Philip M. Morse, 'The Quantum Mechanics of Electrons in Crystals', *Phys. Rev.* **35**, 1310-24 (1930).

[207]Lev Landau spent 18 months travelling working among others in Germany, Switzerland, and with Niels Bohr in Copenhagen. When he arrived in Zurich in 1930 he had just completed his theory on diamagnetism which was of great interest to Pauli who himself had developed a theory of paramagnetism due to electron spins.

[51] Hans Bethe to Rudolf Peierls

location unspecified, 24.7.1930

Lieber Peierls,
Entschuldigen Sie, dass ich Ihren Sommerfrischen-Seelenfrieden in Karersee störe — aber da Herr Guth[208] mir im Verlaufe einer unendlichen, nicht konvergenten, Reihe von Mitteilungen u[nter] a[nderem] auch sagte, dass Sie sowieso Material zum ixen[209] mitgenommen haben, ists vielleicht doch nicht so schlimm.

Also herzlichen Dank für Ihren Brief, aus dem ich entnehme, dass Sie im gleichen Moment die gleiche Erklärung für die Widerstandsänderung im Magnetfeld gefunden haben wie ich.[210] Anscheinend hatte ich mich in meinem mehr langen als durchdachten Brief so unklar wie irgend möglich ausgedrückt, sodass ich nochmal den Versuch mache, Ihnen zu sagen, was ich damals sagen wollte und was Ihnen sowieo bekannt ist, also:

Bei gebundenen Elektronen ist die Energie nicht mehr eine Funktion der Geschwindigkeit allein, sondern hängt ausserdem von der Richtung der Bewegung des Elektrons ab. Infolgedessen lässt sich auch die Geschwindigkeitsverteilung der Elektronen im elektrischen Feld nicht mehr in der Form $f_o(\varepsilon) + \xi \cdot \chi(\varepsilon)$ (ε = Energie, ξ = Komponente der Geschwindigkeit in der Feldrichtung) schreiben, sondern χ hängt auch wieder von allen drei Geschwindigkeitskomponenten einzeln ab. Dasselbe tut die Sommerfeldsche freie Wellenlänge, die ja direkt proportional zu χ ist (nach Definition). Sie wird also bei gebundenen Elektronen "richtungsabhängig", d.h. hängt ausser von ε von der Bewegungsrichtung ab. Sie sehen, dass meine Erklärung mit Ihrer identisch ist.

[208] Eugène Guth (1905–), Viennese physicist who was working on the application of the Thomas-Fermi model to positive ions with Peierls. They later published a paper on this problem; R.E. Peierls and E. Guth, 'Application of the Thomas-Fermi model to positive ions', *Phys. Rev.* **37**, 217 (1931). Guth later moved to the University of Notre Dame and worked mainly on polymer physics.

[209] Terms used by the young physicists for tedious calculations.

[210] See letter [50], note 203. See also Peierls, *Bird of Passage*, p. 59.

Da Sie sich schon immer mit magnetischer Widerstandsänderung befasst haben, während ich die Sache eigentlich bloss per Zufall fand, ist es unbedingt richtig, wenn Sie die Sache publizieren. Andererseits möchte allerdings Sommerfeld gern, dass in der Z[eitschrift] f[ür] Physik im Anschluss an die Franksche Widerstandsänderung bei grossem H auch gleich die Richtigsstellung der absoluten Grösse der W[iderstands] Ä[nderung] bei kleinen Feldern erscheint, entweder von Ihnen oder von mir oder eventuell gemeinsam. Was denken Sie dazu? Es sollte nur eine kurze Notiz sein, die die Sache qualitativ erklärt, ohne tiefe Theorien.[211] Franks Arbeit ist fertig.[212]

Ich reise am 1. VIII nach Frankfurt, Blumenstr. 12, und am 7. mit meiner Mutter in die Schweiz, vermutlich nach Fetan (Adresse Baden-Baden, Gunzenbachstr. 27). Gute Erholung und herzliche Grüsse
Ihr

Hans Bethe

[52] Heinrich Peierls to Rudolf Peierls

Berlin, 25.7.1930

Lieber Rudi.
Deinen Brief aus dem Unterengadin haben wir erhalten und uns darüber gefreut, dass Ihr gutes Wetter hattet. Ich muss allerdings befürchten,

[211]Peierls published a paper, though not in *Z. Phys.*, as Sommerfeld had wished. R. Peierls, 'Zur Theorie der magnetischen Widerstandsänderung', *Ann. Phys.* **10**, 97–110 (1931).

[212]Frank was working on a joint paper with Arnold Sommerfeld. See Correspondence Sommerfeld-Bethe-Frank 2.11.1930, 27.11.1930 and 15.12.1930, Michael Eckert and Kark Märker (eds.), *Arnold Sommerfeld. Wissenschaftlicher Briefwechsel, vol. 2, 1919–1951*, Berlin/Diepholz/Munich: GNT, 2004, pp. 313–16, (hereinafter cited as Eckert, *Sommerfeld Briefwechsel*) and www.lrz-muenchen.de/~Sommerfeld/PersDat/02980.html. Arnold Sommerfeld and Nathaniel H. Frank, 'The Statistical theory of thermoelectric, galvano- and thermo-magnetic phenomena in metals', *Rev. Mod. Phys.* **3**, 1–42 (1930) and N.H. Frank, 'Über die metallische Widerstandsänderung in starken Magnetfeldern', *Z. Phys.* **64**, 650–56 (1930).

dass der Rest Eurer Reise bei weniger günstigem Wetter verlief. Hier sind augenblicklich ständige Regengüsse an der Tagesordnung.

Wir haben nun inzwischen unsere Reise fortgesetzt: Wir fahren am nächsten Mittwoch, den 30. d[iesen] M[onats] abends hier ab, werden uns voraussichtlich, falls Weigerts[213] in Freiburg zu Hause sind, dort einige Stunden aufhalten und am Donnerstag, den 31. Juli gegen Abend in Luzern eintreffen. Sollten Weigerts nicht zu Hause sein, was ich inzwischen ermittle, so sind wir schon Donnerstag mittag in Luzern. Unsere Adresse ist dort Hotel Schweizerhof.

Wir haben deswegen zunächst Luzern gewählt, weil wir Zürich genügend kennen und wir in Luzern, falls das Wetter sich bessert, geeignete Ausflugsmöglichkeiten haben.

Ich schlage vor, dass Du zunächst auf einen Tag nach Luzern kommst. Wenn Du wochentags dazu nicht Zeit hast, so wäre der Sonntag geeignet. Es wäre uns angenehm, hierüber eine Zeile von Dir im Schweizerhof vorzufinden, damit wir danach disponieren können. Wir gedenken etwa 5 Tage in Luzern zu bleiben und dann per Auto weiterzufahren. Voraussichtlich gehen wir nach Klosters und treffen uns am 15. August mit den Amerikanern in St. Moritz. Wieweit Du uns begleiten kannst, können wir ja dann in Luzern besprechen.

Jedenfalls hoffen wir, Dich zunächst in Luzern zu sehen. Bis dahin viele Grüsse, auch von Else,

Dein Vater

[53] Hans Bethe to Rudolf Peierls

Zuoz, 12.8.[19]30
(post card)

Lieber Peierls,
Vielen Dank für Ihren Brief + Manuskript. Ich bin sehr damit einverstanden, und werde in der Arbeit von Frank, deren Korrektur ich

[213] A retired judge, Felix Weigert, who was an uncle of Elisabeth and Heinrich Peierls (who had been first cousins).

bekomme, einen entsprechenden Hinweis zufügen.[214] Ich nehme an, Sie haben noch ein zweites Korrektur-Exemplar und schicke meines an Sommerfeld. Ich hoffe, das ist Ihnen recht.
Herzlichst Ihr

Hans Bethe

Viel Spass in Odessa.[215] Hier kann man auf mehr oder weniger rauhe Berge steigen und sich dabei aufs Abendessen freuen. Adresse bis zum 21. VIII hier, dann postlagernd Meran.

---

[214]See letter [51]. Bethe communicated with Sommerfeld about the calculations. See letter Hans Bethe to Arnold Sommerfeld, 2.11.1930, Eckert, *Sommerfeld Briefwechsel*, II, pp. 313–16.

[215]The 7th All-Union Congress, which took place in Odessa in August 1930, attracted a great number of scientists from the Soviet Union and other countries.

2. Russia: A Scientific Adventure and Romance

The 'main attraction of the summer of 1930' for Rudolf Peierls and a number of younger and more established Western and Soviet physicists was the 7th All Union Conference at Odessa. For many Westerners it was the first exposure to the Soviet Union, and for many Soviet scientists it was a rare opportunity to encounter non-Soviet scientists.

Soviet physics had undergone remarkable changes in the preceding decade. After the end of the Civil War, scientists had pressed for a reestablishment of scientific relations with the West. By the mid-1920s, foreign contacts had become an integral part of Soviet research and an important source for scientific equipment.[216] By the end of the 1920s, at the eve of collectivisation and rapid industrialisation, Soviet physics had rebuilt enough of such scientific links to feel part of the international community of scholars. Among the signs of the successful return of Soviet science onto an international platform was the attendance of Western physicists at the Odessa Conference in August 1930. Among those were more senior figures such as Sommerfeld, Pauli, Simon and Dirac, but also the relatively junior Rudolf Peierls.

For Peierls delight about the invitation to attend the conference was spurred by a variety of factors. Any young scientist would have been glad about any invitation to a conference, especially one where the host would cover all expenses. Moreover, it was a Soviet national conference with only a small number of foreign scientists being invited. It was evident that Peierls was beginning to make a name for himself.[217] Last, but not least, Peierls had been intrigued by this country which was

[216]Paul R. Josephson, *Physics and Politics in Revolutionary Russia*, Berkeley: University of California Press, 1984, ch. 4.

[217]He suspected that he owed his invitation to Yakov I. Frenkel, who was interested in some of his work.

evidently ruled by an oppressive regime but which had, on the other hand, produced some outstanding scientists. He had met and had been impressed by Lev Landau, and he had encountered the work of Frenkel, Fock, Tamm and many others. The Odessa Conference was promising to bring him into closer contact with many of these physicists, most notably with Frenkel and Tamm, and some of the remarkable younger colleagues around his friend Landau.

At the conference, and during a subsequent boat trip across the Black Sea, Rudolf Peierls met many of Lev Landau's Leningrad colleagues, a close-knit community of exceptionally gifted young men (and woman) not unlike the community which Rudolf had known in Sommerfeld's institute. They were known as the 'Jazz Gang',[218] a group formed around George Gamow (Jonny), Dmitry Ivanenko (Dymus) and Landau (Dau), also called the 'three musketeers', with Genia Kannegiser and Matvei Bronstein (Abbot) also playing an active role.[219] No doubt facilitated by his friendship with Lev Landau, Rudolf Peierls was accepted into the inner circle of this group of young scientists from Leningrad University, and as is evident in the pamphlet 'Am Morgen nach der Schlacht'[220] he took an active role in the social as well as scientific aspects of the meeting in Odessa and beyond. The pamphlets, edited by Matvei Bronstein, give an inkling of some of the heated discussions arising from what was perceived by Bronstein and others to be the 'new crisis in the theory of quanta'[221] (the old crisis being the that which caused Bohr's theory to be replaced by quantum mechanics in the mid-1920s). In the contribution 'Frühstück von Pauli', Bronstein sums up some of the perceived problems. The negative energy states in Dirac's equation, the infinite nature of the electron energy and the uncertainties about the nucleus itself gave rise to a variety of views about the

[218]Gennady Gorelik and Victor Ya. Frenkel, *Matvei Petrovich Bronstein and Soviet Theoretical Physics in the Thirties*, Basel: Birkhäuser Verlag, 1994, pp. 20–22.

[219]Apart from these five, A. Anselm, V. Kravtsov and I. Sololskaya were members of the Jazz Band.

[220]See items [54–55].

[221]M. Bronstein, 'Novi krizis teorii kvant', *Nauch slovo*, 38–55 (1930). This was a popular science journal edited among others by Ioffe.

measurability of the 'inner electronic order'. Heisenberg,[222] Ivanenko and Ambartsumyan[223] had independently developed ideas based on a quantisation of space, which would result in a theory containing a unit of the 'smallest possible length'. This inevitably led to a replacement of the Eukledian continuous space with a discrete entity of points that would form a cubic lattice,[224] and this lattice geometry was at the heart of many discussions aboard the *Grusija.*[225]

For Rudolf Peierls, quite apart from the buzz of the scientific discussions, the Odessa conference held in store other exciting developments. By far the most important new acquaintance of the summer of 1930 was the only female member of the Jazz Band: Eugenia (Genia) Nikolaevna Kannegiser. As Peierls recalls in his autobiography, she 'seemed to know everybody, and was known to everybody',[226] and throughout the meeting and during the travels thereafter, Rudi and Genia got to know each other better. With their German and Russian backgrounds, neither of them could converse in the other's mother tongue and the only common language spoken sufficiently well by both to communicate reasonably comfortably was English. After journeying across the Black Sea to Yalta and Batum, a smaller number of travelers continued by train to Tiflis, and by car to Vladikavkaz, from where Genia and Rudi traveled on to Kislovodsk, a village in the high mountains. This trip became infamous because Rudolf Peierls was forced to make the overnight journey in the wooden luggage rack due to a lack of available seats.[227] In Kislovodsk they parted, with Genia continuing her travels through Georgia and Rudolf returning to the Russian centres of science in Moscow and Leningrad.[228]

---

[222]See letter [46], note 190.

[223]Viktor Ambartsumyan (1908–1996), studied astronomy at Leningrad and as such was attached to both the university and the observatory at Pulkovo. He wrote the first Russian textbook on theoretical astrophysics. Later he cofounded and directed the Byurakan Astrophysical Observatory and served as president of the Armenian Academy of Sciences (1947–1993).

[224]See Gorelik and Frenkel, *Matvei Bronstein*, p. 40.

[225]The name of the boat that took the physicists across the Black Sea.

[226]Peierls, *Bird of Passage*, p. 63.

[227]Ibid., p. 64.

[228]See letters [56–57].

When Rudolf Peierls, after his summer adventures in Russia, returned to Zurich his physics endeavours found themselves in serious competition with a long-distance romance which resulted in an extensive correspondence between Leningrad and Zurich. The letters exchanged between Rudolf and Genia are an exceptional source. They were written by two young people who had grown up in different countries, different social environments, spoke different languages, and knew little about each other or each other's worlds, when they parted just three weeks after their first meeting. But they were determined to get to know each other through their letters. As a result, the early correspondence (September 1930 to February 1931) reveals much about Rudolf and Genia themselves as they attempted to explain themselves to each other and engage in discussions on mutual perceptions, interests, work and social life. At the same time, the letters show the dynamics of their increasingly intense friendship, especially in their long exchange about the nature of friendship, relationships, love and 'adventures'. In addition, the letters contain interesting details about the different members of the Jazz Band, about the Russian way of life as experienced by Genia and her family as well as information about the Zurich physics community and Rudolf Peierls' close friends.

That the relationship was serious right from the beginning is evident not merely in the volume and intensity of the communications, but also in the fact that Rudolf Peierls, almost immediately, started to learn Russian,[229] and from October onwards Russian words, expressions and sentences are included in the correspondence, and from early 1931 onwards parts of letters or whole letters are exchanged in Russian.[230]

Another result of the meetings in the Soviet Union was the collaboration of Peierls and Landau (who continued on his 'European tour' after the Odessa meeting) with Vladimir Fock. Landau and Peierls revisited the unresolved questions of infinite self-energy of electrons, and they felt that Heisenberg's uncertainty relations for nonrelativis-

[229]See letter [62].

[230]The Russian parts of the letters are included in English translation and reproduced in italics.

tic quantum mechanics needed extension in the relativistic field.[231] In particular, they discussed the issue of measurability of momenta and accuracy of measurements of the intensity of electric and magnetic fields, issues which also interested Fock.[232] Landau and Peierls, partly because of the logistic complications of collaborating between Zurich and Leningrad, and partly because they had further developed Fock's initial ideas, wanted to publish their results independently of Fock, which caused slight misunderstandings between the three.[233] But by far the more significant upset was caused by the content of the publication itself. Much of the Landau-Peierls had been done at various stays in Copenhagen, most notably their visit in February/March 1931.[234] Niels Bohr did not agree with their key conclusions,[235] in particular the conclusion that it was impossible to measure accurately electromagnetic fields. As he put it in a letter to Pauli later: 'Although I could not find an error immediately in Peierls' and Landau's arguments concerning the measurability of electromagnetic field quantities, I always felt uneasy about their criticism of the foundations of the formalism.'[236] Apparently, Bohr was so upset about the publication of the paper that he did not want to be acknowledged in it.[237]

But in late 1930 and early 1931, Rudolf Peierls was not only busy with the Landau-Peierls paper. As Pauli's assistant he was deeply involved in discussions with Pauli surrounding the latter's thought processes, and at the time they evolved around β-decay. In the late 1920s it was observed that when a β-particle (electron) was emitted from a

---

[231]Letters [93], note 380 and [103].

[232]Letters [76], [86].

[233]Letter [97].

[234]Peierls had stayed in Copenhagen from 24 to 28 February 1931; Lev Landau from 25 February to 19 March 1931.

[235]See letter [103].

[236]Letter Niels Bohr to Wolfgang Pauli, 25.1.1933, Pauli, *Wissenschaftlicher Briefwechsel*, II, p. 153. Original in Danish, translation by the editor.

[237]About Bohr's reaction to the Landau-Peierls paper see Abraham Pais, *Niels Bohr's Times. In Physics, Philosophy and Polity*, Oxford: Clarendon Press, 1991, pp. 359–61. See also Viktor Weisskopf, interview by John Heilbron, 10.7.1963, Niels Bohr Library, AIP; Christian Møller, interview by Thomas S. Kuhn, 29.6.1963, Niels Bohr Library, AIP.

nucleus, there was generally some energy and momentum missing, a grave violation of the laws of conservation. Rather than allowing these laws to be discarded, Pauli proposed in late 1930 that the missing energy and momentum was carried away from the nucleus by some particle (later named the neutrino by Enrico Fermi) that was uncharged and had little or no mass and had gone unnoticed because it interacted with matter so rarely that it was nearly impossible to detect. Rudolf Peierls was the one to communicate these ideas in the famous open letter to the 'radioactive ladies and gentlemen' in December 1930, and he told Genia in some detail the essence of Pauli's thought, although he rather doubted their accuracy.[238]

While some of the correspondence between Rudi and Genia was concerned with their scientific work, most of the letters were taken up by exchanges about day-to-day events, social and cultural experiences and the development of their mutual relationship. In late September 1930 already, Rudolf contemplated the possibilities of Genia joining him in Zurich.[239] He came to the conclusion that, on financial grounds, this would not be feasible, as his position was not secure, and Genia would find it difficult to get a job. Considerations of this kind were a recurring theme of the letters, and although no solutions presented themselves, it is clear from the correspondence, that both Genia and Rudi regarded this as a goal to work towards.

Despite their distinct backgrounds, Genia and Rudolf shared a common European cultural heritage of which they were both very aware. Both were interested in music and enjoyed going to concerts and ballets. Rudi's stepmother had been a musician, a profession she had had to abandon due to ill health. But her influence as well as her brother-in-law's connections within literary circles[240] and had impacted on Rudi,

---

[238]See letters [81–83].

[239]Letter [59].

[240]Ludwig Fulda, the husband of Else's sister Helene, was a playwright who took an active part in cultural activities in interwar Germany. Among others, he was, between 1925 and 1932) the first president of the German PEN-Club (Poets, Essayists and Novelists), an international organisation which had established itself as a voice speaking out for persecuted and suppressed writers. As such he had plenty of contacts, both in Germany and internationally in the literary world.

just has Genia's stepfather's profession as a translator, mainly of literary works, had instilled into her an interest in books, both prose and poetry. How intense the interest in literature was among Genia and her friends, is evident in an anecdote related by science historian Gennady Gorelik about the origin of Matvei Bronstein's nickname 'Abbot'. It was taken from Anatole France's book *La Rôtisserie de la Reine Pédauque*, which had been translated into Russian by Genia's stepfather. While reciting passages from the book on a train trip to Pulkovo, Genia and her friends were struck by the similarities between one of the main characters, Abbot Jêrome Coiquard, and Matvei Bronstein who shared many of the Abbot's remarkable characteristics such as 'profound mind, wide knowledge, balanced scepticism, kindheartedness and tolerance.'[241] Rudi's and Genia's fascination with literature is evident in much of the correspondence, as they repeatedly commented on their current reading.[242] Similarly, both evidently enjoyed the cultural life in Zurich and Leningrad respectively with frequent and knowledgeable reports of visits to concerts, ballets and operas and the cinema.[243]

The correspondence took on an even more involved tone at the turn of the year. Rudolf had joined a group of friends for a skiing holiday in Arosa, and on meeting a girl from Berlin, Eva Gäbeler, he had reason to consider his willingness or otherwise to engage in an 'adventure', the term used by Rudolf and Genia for close friendships.[244] Expecting a 'lecture on adventures' from Genia[245] Rudolf is relieved to read in her reply about her lack of jealousy.[246] The exchange is very enlightening, and it is an indication of how Genia's and Rudolf's relationship would develop. There is no doubt that the foundation of their partnership was a combination of deeply-rooted love and affection, combined with an acceptance of the personality and way of life of the partner. The letters throughout the half century of their marriage demonstrate, that

[241]Gorelik and Frenkel, *Matvei Petrovich Bronstein*, p. 25.
[242]E.g. letters [64], [69], [71–72], [77], [79], [81].
[243]E.g. letters [67–71].
[244]Letters [90], [92].
[245]Letter [93].
[246]Letter [94].

the cornerstones of their relationship was honesty and a willingness to share every last detail of their experiences, pleasant or painful. Just as Rudi had told Genia about his previous friendship with Marga Vogt, a young woman in Leipzig[247] and the relationship with Lilly Fenigstein, a Swiss violinist,[248] Genia related to Rudi her previous liaison with Dumus Ivanenko.[249]

Much of the correspondence of the first few months of 1931 was taken up with preparations for Rudolf's trip to Russia in March. His parents had assumed that he went to Leningrad primarily to teach. But when Rudi changed his story about the necessity of his renewed Russian trip, they became suspicious, and guessed correctly that the driving force was of a private nature, namely 'chercher la femme'.[250] This concern could not be dispelled by a meeting between Rudi and his parents shortly before his departure for Leningrad, and indeed, on 15 March 1931, Rudolf and Genia, much to the dismay of his parents, got married in Leningrad. Heinrich Peierls' reaction, as expressed in his letter to Rudi on receiving the news a few days later, was most revealing. Concerns about his son's insecure financial position were mingled with worries about his wife's nationality, which he feared to be a real hindrance to securing an academic position in Germany. Moreover, married life at this stage in Rudolf's career appeared to affect his flexibility of having to take on even badly-paid academic positions abroad, a sacrifice which was often essential on the way to a permanent academic post. Heinrich Peierls' feelings were summed up in his comment: 'If you cannot find a suitable partner in all of Germany, then at least test yourselves before making a permanent commitment and get married when you have the necessary financial means to support a wife.'[251]

Rudolf's and Genia's change in marital status had little impact on the logistics of their relationship initially, as visa regulations forced them to live apart until the late summer of the year, when Genia was allowed

[247]Letter [79]. See also letter [26], note 141.
[248]Letter [59].
[249]Letter [68].
[250]Letter [107].
[251]Letter [111].

to leave Russia to join her husband in Switzerland.[252] Consequently, the two continued their frequent and detailed exchange of letters throughout the spring and summer.

As Peierls remarked later in his autobiography, he had little recollections of this semester in Zurich, partly because his mind was preoccupied with his distant wife, but also because Pauli attended the summer school at Ann Arbor and therefore left Zurich well before the end of term. As a consequence, Peierls spent some time in Leipzig at Heisenberg's institute, and this provided the opportunity to revive his acquaintance with some of the German physicists.[253]

Despite Pauli's well-recorded aversion to solid state physics, Peierls' next research topic again was within this field. For his habilitation, the qualification necessary to teach independently at tertiary level in Germany and Switzerland, he investigated the question of electrical resistance at small temperatures.[254] With Pauli being in the United States for the best part of the summer, Peierls had to rely on written correspondence to get his tutor's advice and commentary. Three elaborate letters from Pauli written within three days at the beginning of July 1931[255] demonstrate his dislike of the both 'Größenordnungsphysik' and the 'Dreckeffekte',[256] but they are also witness to the astute and constructive criticism which he was able to offer. And despite extensive travels throughout the summer of 1931, Peierls completed his studies in connection with his habilitation in October of that year.[257]

Pauli's long absence from Zurich in the summer of 1931 allowed his assistant Peierls not only a visit to Leipzig but also a more extended stay in Berlin. En route from Leipzig to Leningrad, Peierls had to wait for his Russian visa and therefore had the opportunity to spend two weeks with his father and stepmother. It turned out to be the last longer stay in Berlin. After 1933, travels to Germany for Jews who resided abroad

[252]Peierls, *Bird of Passage*, pp. 70–74.
[253]Letters [134–135], [137].
[254]Letter [125], note 465.
[255]Letters [125–127].
[256]Letter [125].
[257]Pauli, *Wissenschaftlicher Briefwechsel*, II., p. 87, note d.

became increasingly risky, and apart from very brief visits in the mid-1930s, Rduolf Peierls chose not to take this risk. Later, following his wartime experiences, he did not visit Germany for many years, and, in fact, did not visit his home in East Berlin until after German unification in 1990.[258] Being back in Leipzig and in Berlin, allowed Rudolf Peierls to gain some more first-hand experience of the volatile political and financial situation in Germany in the early 1930s. The Great Depression, which had left its mark globally, hit Germany harder than most other countries, because of the huge financial obligations in connection with reparations after the First World War. Although Peierls claimed that he 'preferred thinking about physics',[259] he could not escape the realities of friends and relatives losing jobs and facing financial hardship. And the changing climate, both political and economic, became one of the regular themes of letters exchanged between him and his father in subsequent years.[260]

However, the summer of 1931 was occupied predominantly with his trip to Russia, with his attempts to finalise visa arrangements for Genia's move to Switzerland, and on arrival in Switzerland with the completion of the Habilitation.

[258]See letters in chapter 11 of vol. II of *Peierls Correspondence*.
[259]Letter [132].
[260]See *Peierls Papers*, Supp., A. 102–109.

[54] Pamphlet by Some Physicists Who Had Attended the Odessa Conference

Boat Grusija, 26.8.1930

AM MORGEN NACH DER SCHLACHT[261]

Die Schwierigkeiten der Quantenelekrtodynamik
von W. Pauli jr und J. Frenkel[262] (erste gemeinsame Arbeit)
Angesichts des merkwürdigen Sachverhaltes, dass manche Physiker immer noch nicht wissen, dass aus der Quantenelektrodynamik nichts herausgekommen ist und auch nichts herauskommen wird, hat die Kongressleitung beschlossen eine besondere quantenelektrodynamische Abteilung einzurichten. Ein kompetenter Physiker des Auslandes wurde besonders gerufen, um die russischen Physiker von der Richtigkeit zu überzeugen. In der Tat hat sich auch gezeigt: wer andern im kontinuierlichen Raum zwischen zwei diskreten Gitterpunkten eine Grube gräbt, fällt selbst hinein. Nachdem sich weiter gezeigt hat, dass die Dichte der Lichtquanten an einem Punkt nicht berechnet werden kann, wurde die Diskussion über die genannte These geschlossen, da diese von allen Teilnehmern einstimmig angenommen wurde.

Eine lange Diskussion entstand jedoch über die Frage, ob die Quantenelektrodynamik überhaupt existiert. Um diese Frage zu erörtern müssten erst die folgenden Fragen geklärt werden:
a) Folgt aus dem Stattfinden von Vorlesungen über ein Wissensgebiet die Existenz des Wissensgebietes?

[261] The participants of the Odessa conference were taken by boat across the Black Sea, and during the cruise, some of them (Bronstein as editor with Ambarzumian, Dorfman, Fock, Frenkel, Mandelstam Jr., Pauli, and Sommerfeld and others as contributors) wrote what they called an international newspaper in two parts. The pamphlet was written in Russian and German. Some of the articles are included here to give a flavour of the spirit of 'scholarship' among the thriving community of theoretical physicsts at the time. Later, in a similar spirit, many of the Bohr pupils and collaborators, wrote a number of pamphlets called Journal of Jocular Physics, to which RP also contributed. See http://www.nbi.dk/nbi-history.html.

[262] Yakov Ilich Frenkel (1894–1952), head of the department of theoretical physics at the Ioffe Leningrad Physical Technical Institute from 1926 until 1952.

b) Folgt aus dem Vorhandensein von Arbeiten über ein Gebiet die Existenz des Gebietes?

Die Konferenz entschied diese beiden Fragen dahin, dass aus den geannten Prämissen nur der Wunsch der Vortragenden, dass das Gebiet existieren möge, bzw. der Wunsch der Verfasser, das Gebiet begründen zu können, gefolgert werden kann.

Sodann wurde auch die Hauptfrage der Existenz der Quanteneletrodynamik verneint und man schob die Schuld den unglückseligen Elektronen zu.

Kleine Nachrichten

Verbotene Bücher

Bei einem zweifelhaften Individuum der 3-ten Klasse (Physiker?) wurde im Verlaufe des gestrigen Nachmittags ein Ex[emplar] des verbotenen Buches von Born und Jordan vorgefunden. Der Obercensor W. Pauli hat die sofortige Konfiskation angeordnet, da eine Durchbrechung des sogenannten Pauli(jr)-Verbotes das schwache Gebäude der heutigen theoretischen Physik um-zu-stos-sen droht.

Bronstein[263] in Angeregtem Zustande

Ein seltener Fall eines verbotenen Übergangs aus einem symmetrischen in einen antisymmetrischen Zustand ist heute auf unserem Schiffe beobachtet worden. Es ist zu vermuten dass man es mit einer Resonanzerscheinung zwischen Paulis Eigenschwingungen und Bronsteins periodischen Seelenbewegungen zu tun hat, trotz der nicht unerheblichen Massenverschiedenheit. Wie Dirac in seinem unlängst erschienenen Traktat betont hat, bietet die theoretische Erklärung dieser rätselhaften Erscheinung grosse Schwierigkeiten dar, da man von Pauli keineswegs annehmen kann, (nach Experimentan von Peierls), dass er ein Loch im Passagierraum darstellt.

[263]Matvei Bronstein was a talented young physicists among a group of Russian theoretitians, including Lev Landau, George Gamow, Genia Kannagiser, Dimitri Ivanenko. He was not only a gifted researcher but also a great populariser of science, which is evident in this pamphlet. Bronstein later perished in a Soviet camp. See Gorelik and Frenkel, *Matvei Bronstein.*

Frühstück von Pauli
Gedicht von M. Bronstein

Herr Pauli ist Theorienfresser
Und es ihm keineswegs gefällt,
Dass der Elektronendurchmesser
Die kleinste Länge in der Welt.
Der Raum, denkt er, ist kein Punktgitter
Von Unsell und Ambarzumian.
Die ganze Theorie ist ein Flitter
Und kaum verbessert werden kann.

[55] Pamphlet by Some Physicists Who Had Attended the Odessa Conference

Boat Grusija, 28.8.1930

AM MORGEN NACH DER SCHLACHT

Zur Quantentheorie der Seekrankheit
Eine internationale Zeitung für die Leute, die vom Kongress übriggeblieben sind.
von J. Dorfmann und R. Janus, z.Z. auf "Grusija"
(Vorläufige Mitteilung)
Wie bekannt, ist bisher noch keine befriedigende Theorie der Seekrankheit vorhanden. Die klassischen Theorien können nicht die Tatsache erklären, dass bei der Seekrankheit eine langwellige Grenze beobachtet worden ist. Nun kann man aber leicht mit Hilfe der Wellenmechanik eine qualitativ und quantitativ befriedigende Theorie aufbauen. Wir betrachten die Wirkung periodischer Schwingungen von der Frequenz v auf eine Anzahl von Menschen, die in einem Dampferhohlraum eingeschlossen sind. Wir betrachten ferner die Passagiere wie ein entartetes Gas, in welchem die einzelnen Individuen in Paaren sich bewegen. Unter der Einwirkung periodischer Schiffsschwingungen wird ein Teil der Individuen aus den Doppelzellen entfernt und nähert

sich dem Rande des Schiffes, wobei verschiedene unangenehme Effekte auftreten, die in der Praxis eine bedeutende Rolle spielen können. Es wird gezeigt, dass das Auftreten des unangenehmen Seeeffektes durch die Bedingung

$$W - Wo = hv$$

bestimmt wird. Dieses Gesetz ist dem Gesetze des photoelektrischen Effektes ganz analog. Und zwar der Seeeffekt besitzt eine langwellige Grenze, die durch die Bedingung

$$hv_0 = W_0$$

bestimmt wird. Durch eine genaue Durchrechnung der einzelnen φ-Werte konnte man leicht die entsprechenden v_0 im Einklang mit den Experimenten berechnen. Die Anzahl der Seekranken und die Intensität des Effektes ergibt sich aber zu klein. Wir haben aber den Austauscheffekt nicht in Betrracht genommen. Eine vorläufige Rechnung zeigt, dass dadurch ganz plausible Werte erhalten werden können.

Zuletzt möchten wir unseren herzlichsten Dank an Hern Prof. Frenkel aussprechen.

[56] Genia Kannegiser to Rudolf Peierls[264]

Gagry, 13.9.1930

Rudi,
Dearest, it is the seconde edition, the first I wrote this morning on the seaside and on cliff — a lot of foam and all was covered with a big wave, it was very jolly, but it was so long, long letter.[265] ... Darling,

---

[264] At the Odessa Conference Peierls met the young Russian physicist Yevgenia Kannegiser. The two corresponded for the next six months until Peierls visited Russia again. On the occasion of this second visit, he married Genia. See Peierls, *Bird of Passage*, pp. 61–73.

[265] Genia's letters (in English and German) contain many errors and odd phrases. Some of the spelling mistakes have been corrected, but no attempts have been made to alter the oddities of phraseology which remained typical of Genia's written and spoken English. The writing on a number of letters has faded, and passages are illegible.

I am afraid that this will be also very long, and I beg your pardon for your hard labour. It will be so difficult for you to read and to understand something, but Rudi I cannot do anything else. I have no post paper, no dictionary and even no ink and pen and now it is the evening. I do not know what o'clock it is, because I live without watch, but I think that something like eleven, and the moon is yet under mountain and I have so bad lamp. So, darling, once more I beg your pardon. I received yesterday your letter, and I thank you for it very seriously, very very seriously, because I believe now that I was not for you only "adventure" — one of many other "girls" that you know that you will know. Certainly, I thought that before also, because if I have not, I should never give you my hands and my lips, they are too "clean" for "adventure", but I was not sure and that way very sad, some minutes. Now all is all right. I am afraid that in such long discussions you did not understand my meaning of "serious". I will try to explain it once more. "Serious" is not so awfully how you think it my darling.

Serious — it is when the person is "dear" for you, when you love not for your own pleasure or happiness, not for yourself, but for other, when you love not because you have not something other to do, but because you cannot to love not, when your love is not an adventure, not "pour passé le temps" but an happiness, a great and deep thing, a "serious" thing and not only a play. But serious is only first step to marriage. Always — is a fiction, but sometime, now — is "always" — when you want to say "moment — stop". Certainly, one cannot know what will be tomorrow, but you can want such and such future, and your want to love me always is the greatest thing I can ask you. Darling you write about my meaning of love — marriage and for you it is not so? Certainly, marriage is a logical development of love. If I love a man it is necessary for me to be together with him, to know all about his life and work to show him all that interests me, to see him when I want and so on and so on. It would be for me quite insupportable to eat when he is hungry to have something that he want and has not. etc., etc. and all this is marriage — when I cannot live without a man I marry him. I know you only two weeks, but I love you and it is so awfully that you are not here.

I have such a lot to tell you, I want to know about you a thousand things. Six months is very long, but perhaps with the letters it would be better will you, shall we write to each other very often (twice a week or something in this kind) and about all our life? Work, sport, books theatre, cinema, dances, new people, etc. You know now my people, my town, my family, and my home. I know about you less, certainly, but a little I know, so it would be not so difficult to write and we shall live a little together. So, darling, answer this letter, and write all that happens with you in Berlin, especially how goes the thing with Simon's institute.[266] I shall relate you about *Gagry* in next letter, because that's too long. You know the letter from *Leningrad* with dictionary would be better I believe. It is awfully late now, and I must go up to morrow very early and go arrange my ticket to *Leningrad*. Good night, my dearest boy, I wanted to write you a lot of "philosophical" things more, but it is so tedious – the philosophy. Yes? And now is so beautiful: the sea under the moon and big big stars fall down into the sea. Good night!

14.9.[1930]

You know that I cannot stop my letter, if it remains a bit of clean paper.[267] I think that this side is quite well for my pencil and "beautiful" handwriting. So I go on and you must suffer an half an hour more. You are not angry? write to me your preferable format and which I will try not to write so time some letters well? Now I am after my dinner which I get not without difficulties, such a crowd in restaurant! But it was splendid and I am in very "lyrical" state.

Darling, here is splendid this morning I swim very fare, so fare that I cannot see the people on the beach. *Great*?! Certainly I was not alone, because I am [?] a little with me were two very good swimmer (two

---

[266]Franz Simon (1893–1956) studied physics in Munich (1912–14) and Berlin (from 1919) and completed his Ph.D. under Nernst in 1921. He continued research in Berlin and became associate professor there in 1927. Simon had suggested to Peierls to join him in Berlin in 1931, but he was appointed himself to the chair of Physical Chemistry in Breslau in 1931. In 1933 he accepted an invitation by F.A.Lindemann to work at the Clarendon Laboratory in Oxford where he stayed until his death.

[267]The letter is written on the back of four empty telegram forms. Part of the reverse of the forms do not contain writing; Genia refers to this in the letter.

ingeneer, one of *Moscow* and other of *Kharkov*). But they were not necessary at all, and I swim, swim, swim in blue blue sea under the sun, and I think that I can go till Constantinopol.

Fabelhaft!! But I was tired in the return! My heart is not well at all. You know in the mountain near my home lives bears, charcalsm and wolves, here is a *cave* (I forgot the word, you know this thing in which we were sitting in *Kislovodsk* when it rains so hard). In this *cave* in the 6th century lives a saint-*holy*. Now I was there, very interesting. In five minutes I will go to see underground river(!) and to morrow in the mountains to see the waterfall! It will be very interesting, even klm from my home.

Rudi, you can be jealous (are you jealous? I am not at all.) I am quiet charmed with one turkish merchant named *[?]*. He sells the fruit on the *Basar* of *Gagry*. He is as heavy as two Pauli, and so splendidly lazy! I think that he is charmed also, he gives me best peaches and laugh when I come in the shop. So if I will be in some harem, don't be afraid. Oh, the last leaf. Rudi, good bye! Till the March! Write the answer in *Leningrad*. I will be there 26. Have you received my first letter in *Leningrad*? Darling, I love you very very strong and am very happy. I am an optimist. It would be yet better that it was. Good bye.

Genia

[57] Rudolf Peierls to Genia Kannegiser

Berlin, 24.9.[19]30

I was so glad to get your letter, dear, and of course it was not too long, and if you write still longer letters the only change will be that I can only read them three times instead of 6 times. Also it is "fishing for compliments" when you say it is difficult to read your letters. I really understand every word in spite of the bad paper, pencil, language, philosophy and so on. I am afraid it will not be so easy with my letters, but when I get back to Zurich I will have my little friend again, my little Corona, and then letter-writing is much quicker and you can read them easier. (If you are being jealous, I must tell you that "Corona" is a portable typewriter.) As for jealousy, I am not jealous at all, and I should like to [think] I will never be, but so far one must be careful: "One must never say 'never' ". Theoretically, jealousy is very silly, because you never l[o]se anybody by the influence of somebody else, but if somebody else has such an important influence, it only shows that you have lost that person before.

Now I am already in the middle of philosophy which, as you say, is very tedious. Still, I am afraid I cannot help bringing a lot of philosophy into this letter, but first I will speak about more agre[e]able things.

My last days in *Leningrad* were very interesting again. Did I tell you how much I liked *Eremitage*? I had one more evening in the Theatre with Mrs Frenkel,[268] *Olga Gregorevna* and Simons,[269] saw the Bhuddist temple with *Bronstein* and some things more. Then with some difficulties I got my steamer tickets, and I had a very nice time on the sea. I was glad that the sea was rather rough, because then most people could not eat their dinner and we got double portions. On the steamer, I met a very nice man, a professor of social hygiene from *Leningrad* who will come to Berlin in one of these days.

In the train from Stettin to Berlin, I suddenly saw some relatives who could tell me that my sister had born another child, a son this time,[270] and I was very anxious to get back to Berlin and see her. The child is really charming and I have never seen such an intelligent-looking child of 4 days.

Now my other niece is not quite well, so she is not allowed to stay in the same house with the little one, and had to go to my parents and they put me to my brother's flat. I like that very much because I had never had much time for my brother and my sister-in-law and now I can see them very often. Also I have quite a flirt with my brother's step-daughter (aged 6 years) so there is another reason for you to be jealous.

---

[268] Sara Isaakovna, wife of Yakov Il'ich Frenkel.

[269] Franz Simon and his wife.

[270] Günther Krebs. See letter Heinrich Peierls to Rudolf Peierls, 10.10.1930, *Peierls Papers*, Supp., A101.

I like to talk with my sister-in-law, Nina, because she can understand so many things and I think she is the only person of my family whom you would not call "european". She is also the only person of my family who has a slight idea that somewhere in *Leningrad* there is some girl . . . .

Besides her, also Barbara Zarniko knows a bit — that is that girl of Simon's institute I told you of and one of my best friends. She saw the picture I took of you and liked you so much — the picture really gives one little image of all the things I love in you — that I could not help telling her that she was not alone with her opinion. Generally I do not like to speak to other people about such things, but this time I could not help it. Are you angry about it? You are not. *Yes?* I had a splendid time here, of course, and had to see a lot of people, all of them were interesting to hear about my travel and I had to talk the same story three times a day.

I met all my friends again, but I cannot tell you about all of them, because then this letter would become a big book. But besides this, I did not do very much here and I am afraid it will be very hard to start working again, when I get to Zürich. On my way back I shall stop in Heidelberg to see my best friend who works in a hospital there[271] and in Freiburg to see an old uncle of mine,[272] a concession which I make to my parents.

25.9.[1930]

Darling, if marriage is such as you describe, of course love means marriage. But then marriage does not mean family as you once told me. "Family" is a responsibility which I do not yet feel able to undergo. Marriage in the sense of your last letter, and I was glad to read it — means a relation between two persons, but family means so much more (and less in certain respects), I do not want family, perhaps I shall want it later, perhaps never.

---

[271]Hans Thorner. See also letter [60].

[272]Felix Weigert and his wife, see letter Heinrich Peierls to Peierls, 25.7.1930, *Peierls Papers*, Supp. A101.

There is another word which astonishes me in your letter: "to be too clean for adventures". Do you really think adventures immoral? I can understand if you do not want an adventure in a special case. I could perhaps understand if you say you don't want adventures at all. But why generally say they are immoral? Of course this has nothing to do with our relation, because I do not want an adventure with you and even if I would it would be immoral to have what Landau calls an unsymmetrical situation. I mean that two persons mean a different thing and simply let things go without trying to get a solution. I realize that I speak dreadfully philosophical, i.e. ununderstandable, I only want to say: if you should want love and I want an adventure, it would be quite immoral. But adventures in general can be quite clean, why not? I told you that in many respects, I do not yet know my way and this is one of the points I will have to find out. Do I want adventures, and if so is it possible to have them without l[o]sing more important things? *Genia*, I am afraid you do not understand a single word of this letter. But if I would try to explain more, it would become more and more philosophical and you would understand less and less. So I better stop it.

You say I know much of you, but I do not. Which way do you think to go? Do you do your work now because you feel it impossible not to do it, have you serious relation to it, or is it just because you want some occupation? If you could do what you like, what would you do?

I had to ask you many more things, but if one only writes it, everything sounds so abstract and so silly.

Darling, I shall be so glad if you write to me very often, but you must not, through these letters l[o]se the persons you live with in *Leningrad.*

It is sure now, that I come next March, if nothing extraordinary happens in the meantime. So, *good bye.* In my hands I have yours as in that car on the *Grusinskaja* (do you remember?) and my lips ... well I better tell nobody what they are doing!

Your Rudi

Give my regards to *Nina*[273] and tell her how unhappy I was I could not say good-bye to her, but poor *Bronstein* telephoned her one hour without getting her.

Your next letter to Zürich, Physikalisches Inst. d.E.T.H., Gloriastr. 35.

[58] Genia Kannegiser to Rudolf Peierls

Leningrad, 26.9.1930

My very very dear!
I don't know why, but this three days I so awfully want to see you to speak with you, that it is rather hard to write a letter the answer on which I shall receive only in two weeks, and it is the minimum. Ou, *Rudi*

I received your "angry" letter five minutes before I start of moto car from *Gagry* to *Sochi* and I have read it in this "fabelhaft" way. Do you remember *Bidari* in *Crimea* and the way from *Sochi* to *Riga*. *Gagry* is 10^5 better: the high mountains covered with curled forests, roses, palm trees and high high pins right under your feet — the sea — blue and green and white foam of tide. *Great!*

I was sitting near the moto driver (as is the best place) and certainly began to speak with him a little and to ask him to go quicker; if you know what a scandal began between the people in the moto car! Ladies cried, of horror and told me that I must immediately cease to speak because the way is dangerous and so I was obliged to be silent all the way! Poor girl! Yes? I was awfully tired because I packed all the night my luggage; think only I brought with me 30 klgr(!) of fruits and sweet things! And all this except the wine grapes was packed in my bag and my trunk (do you remember them, they are not so big at all!) But they were a marvell of packing — the peaches and pears were in mine dresses and after five days of way (I remained 2 days in *Moscow* were in such a

[273]Nina Kannegiser, Genia's sister, a biologist whom Peierls had met on his visit to the Soviet Union in the summer.

beautiful state how immediately from the garden. But the state of my dress wanted to be better).

I made from my cushion a volume of $5cm^3$! *Great.* But if you know how it was heavy to carry all this things! In *Sochi* and in *Moscow* I had the porters, but in *Gagry* it was very difficult indeed to manage with these 4 things two of which were 15 klgr each. But *Nina* and my mother were in right enthusiasm when they saw all this treasures.

I was two days in *Moscow* and walked all the time in the streets and was in *Tretjakov* and in the Museum of Western Art, Masereel[274] is very *good.* Do you remember the man with the little dog near him? And his autoportret — fabelhaft.

I awfully like Renoir, but why all his women resemble themselves so much, but they are really charming with blue eyes and such beautiful lips. Yes?

I was very sorry because on the place of Picasso and new french and german painters, (do you know Kissling? We have there only one picture of him, a girl in rose dress, but I like it awfully) they were an exhibition of Rabinranath Tagore.[275] I don't like the dilletants and he is one in the painture, but this little museum is charming and I prefer it to all other *Moscow* museums. This time I was in a museum of old painters, something of the kind of *Eremitage* but very little one, and there I saw a "Madonna" of early italian primitifs, it is one of the best picture I saw in all my life, and I want to sent you a post carte with it but they were all sold. You know long, long, hands and light hair. When you look on her very beautiful thing I am always a little sad, and you?

Now I am at home and to morrow will go in the institute. This night we go at the theater with a lot of people. I think that the company is of 8 persons! I am so sorry, that I was not in *Leningrad* when you were here, but I am so glad that you were charmed with it. I think *Lening[rad]* is one of the beautifullest town of the world. Such a lot of water and gardens, and such unity in style. I will show you a lot of things in March.

---

[274]Frans Masereel (1889–1972), Belgian-born woodcut artist who widely exhibited in communist countries.

[275]Rabindranath Tagore (1861–1941), Calcutta-born writer and poet, who also painted.

Darling, I cannot find my dictionary and the mistakes are awful; I beg your pardon for them and for the tedious letter also I want to say you such a lot and it is so difficult! And it is so sad when perhaps it is not interesting for you at all. Between your last letter and this are two weeks and it is all the time from Odessa to *Kislovodsk.* I hope that you send me the photograph from Berlin; my mixt letter with Bubi[276] was very cool my dear sir, but it was mixt and you understand the conditions. Yes? I sent you my big photo. It is not very nice, but the others are far and I want to post the letter immediately by air post. Sent me your big one. Well?

Genia

P.S. Rudi, I forgot to tell you: Dymus Ivan.[277] told me that Sack[278] was awfully shocked that you remained in *Kislovodsk.* Is it true?

Write me often, because if you will not I should think that I am not interesting for you at all and it is very disagreeable to think.

P.P.S. *Rudi*, you know that one month from six is passed. Hooray!!! $\frac{1}{6}$ is on! Oh, one sensation! Sara Isaakovna told me that Houtermans[279] will remain in *Leningrad.* It will be very jolly in March. Yes?

---

[276]Sergei Mandelstam, son of the well-known Soviet physicist Leonid Isaakovich Mandelstam. Sergei himself was a physicist and later specialised in optics. He was known among his friends as Bubi. A short letter from Genia to Rudolf (23.9.1931), which is not reproduced here, had been a joint letter with Sergei Mandelstam.

[277]Dmitri D. Ivanenko (1904–1994), Russian physicist and close friend of Landau's. He was a member of the three/four musceteers. For details see Gennady Gorelik and Frenkel, *Matvei Bronstein*, chapter 2.

[278]Heinrich Sack, physicist, married to the chemist Charlotte Sack.

[279]Friedrich Georg Houtermans (1903–1966), son of the chemist Elsa Wanek and the banker Otto Houtermans, completed his Ph.D. at Göttingen under James Franck in 1927 and worked as Gustav Herz's assistant in Berlin 1928–1933, when his left-wing leanings forced him to leave Germany. He did not stay in Leningrad long-term in 1930, as Genia speculated in this letter, but in 1935 he took up a position at Kharkov. He was among the victims of Stalinist purges and was arrested in December 1937, extradited to Germany in 1940 and arrested by the Gestapo. After his release, he continued to work in Berlin, Göttingen (1945–1952) and Bern (1952–1966). In 1950, Houtermans, in collaboration with K.F. Schleppa, wrote a book about experiences in Soviet camps. (F. Beck and W. Godin, *Russian Purges and the Extraction of Confession*, Trans. Eric Mosbacher and David Porter. New York: Viking, 1951.)

[59] Rudolf Peierls to Genia Kannegiser

Berlin, 29.9.[19]30

My dear *Genia*,
It is my last day in Berlin, and I must quickly write you a bit of a letter. In the last days I thought very much about the possibilities of your coming to Zürich and living with me, but the result is rather bad. First of all I decided after some calculations, that is absolutely impossible for us both to live on my salary which amounts to 400 Swiss francs (150rb) a month. This is very much for one person, but not far enough for two. Of course it could be possible that you have a profession also, and anyhow, you would not like sitting around the whole day and doing nothing, but earning money with scientific work is much more difficult in Europe than in Russia now, and the probability of getting this position in the same town I am living in is very near to zero. And even if it were greater, it would be a great risk as you had to come here without knowing how to live. The worst thing is that I have no hope to improve my position. If for instance, this offer with Simon's Institut converges,[280] and it seems to converge, as far as I can see, although it is still too early to decide, as it cannot be before next winter, even if this converges, I shall not get more money than now. Nobody knows how long it will take till I get a better position, but don't forget that I am terribly young.

I tell you all this, dear, not to make you sad, but that you do not lay too many hopes in this marrying possibility.

I am ashamed to confess that I was not a single time in the theatre during these two weeks in Berlin, but we have no good earnest plays now, and I am tired of comedies, because they have so plenty of them here or in Zürich and always the same kind of thing. So I spent all evenings with friends or with my family. I am a bit dubious about the marriage of my brother and Nina now. They are very happy now, but she has too much time, she has a servant, and as she is very clever, housekeeping only takes a very small part of her time. My brother is

[280]See letter [56].

about the whole day and he is very tired when he gets home. Of course, she ought to find an occupation, if not a position, but she cannot make up her mind. I am very doubtful what will become of it.

I am curious what will happen in Zürich. You remember that evening on *Grusija* when I told you I would only stay a short time and then go back to Zürich where there is some girl. You know I think differently now about this, and the girl became rather unimportant to me, so unimportant that I did not even tell you the details. Still, she does exist, and I have to do something about it when coming back. She is a very nice girl, much younger than I am, violinist, a very good violinist, as I know, and awfully well educated, so that she cannot do anything without asking her parents.[281] My relation to her was very young and on the way of developing, and so we had not spoken very much about our relation. But now, when I came back from Leningrad, I wrote her a letter, and told her that in the meantime something had happened which could not disturb or affect my relation to her, but showed me what feelings and how intense feelings were possible besides this relation. Now perhaps everything will be all right, and this will stay as it was; a very good friendship — surely not more for many reasons — but as she is very "European" it is also possible that she will not want to see me again. In this case, I would be sorry — but only sorry. We will see what happens.

On September 6, Gamov got the following telegram: "Sensationelle Flucht Jenny Peierls nach Gagry falsch oder trivial Dymus Kiebel". and Gamow sent it to me with the remark: "What does that mean". I explained as far as I understood it myself and laughed very much about it. Perhaps you better do not tell *Ivanenko* that I told you this.
Please read the last lines of Gamow's "Fine structure of α rays" Nature, September 13, 1930.[282]
Very sincerely yours

Rudi

Answer soon!

---

[281] Lilly Fenigstein (1911–1999), Swiss violinist who later married the English-born Herbert Herz. See www.gbnf.com/genealog2/herz/html/d0006/I574.HTM.

[282] G. Gamow, 'Fine Structure of Alpha-Rays', *Nature* Sep 13, 397 (1930). Gamow had signed the paper at Piz de Daint. See letter [64], note 298.

[60] Rudolf Peierls to Genia Kannegiser

Zürich, 2.10.[19]30

My dear and distant *Genia*,
It is really awful with the long time between a letter and an answer, because I cannot remember why you call my letter from Leningrad an "angry" one. But now I really ought to be angry with you. You apologize for everything, for writing long letters, for writing tedious letters, for writing letters with mistakes, for writing letters with Bubi *Mandelstam*[283] and so on, and I only wonder why don't you apologize for being born! Do you really not know how glad I am about these long and "tedious" letter which, of course, instead of being tedious are the more amusing the longer they are. Don't you know that I like these letters with all mistakes, that I like them more the more mistakes they contain, because I can hear you speak the more distinctly. Or did you expect I would not understand why the style of a "mixed" letter must be different? In the contrary I was very glad about this letter because there were some very little and very fine traces of true meaning in it, which, of course, nobody else could understand. Sometimes one can be nearer to each other than ever, if one is not alone and cannot speak frankly. But of course you know all this and you only wanted me to tell you once more.

As for the time it takes for you to get an answer, the minimum is much less than two weeks: you will even get this letter earlier although yours had to wait here for me and although I could not answer it yesterday, as I was busy the whole day, and although I have not yet the time-table of the airplanes.

The distance from Berlin to Zürich seemed very short to me now. The two hours in Heidelberg were very nice,[284] but very short, because this friend, Thorner — I told you about him, do you remember, as my friend in München, he is now a doctor working in internal medicine and neurology — I had not seen since last Christmas. Last winter it

[283]See letter [58].
[284]See letter [57].

happened that I thought with whom could I go skiing to the mountains. So I sent a post-card to this Thorner. When I returned from the letter-box there was a post-card from him: "What do you think about making a ski-tour together?" Landau came with us, then, and I think they liked each other, though they are very different types and quarreled at every occasion. Since then I had not seen this Thorner, and we wrote letters very seldom, so we had a lot of things to tell. We talked during every second during these two hours, till my train left. Then I went to Freiburg and payed a visit to an old uncle of my father. He and his wife are really charming people, although it is very sad, as he is now entirely blind, and begins to become deaf also, although his mind is still quite fresh and young (He is aged 85.).

Then I came here, on a splendid evening with stars, mountains and so on, one can see the mountains from my street! The whole day I had to do a lot of things, speak with Pauli and discuss with him the letters which arrived in the meantime, and so on. Our relation is very strange; usually he says: "Here is a letter of Mr. So and so who has calculated something, but I cannot yet see through it. I am sure there must be a mistake." Then usually I can show him the mathematical proof that he is right, and that there is actually a mistake. So Pauli always says that although I am not very interesting physically, I am very comfortable and a very good help for his laziness. "The ideal assistant".

But I was terribly tired last night, because I am no more used to do something reasonable from morning to evening!

Bloch is here now. He had a bad accident in the mountains. He was climbing with a friend and suddenly lost a step and fell down. He was on a rope, of course, so his friend could keep him from falling down entirely, but he had a broken leg, and had to wait there two days in a bad position till they could come and carry him down with many difficulties.

I telephoned with Lilly Fenigstein (the girl I wrote to you about in my last letter)[285] and it seems as if she is not offended at all, which would mean that I was less important to her than I feared. I am glad about it; we will probably be good friends.

---

[285]Letter [59].

I think it nearly impossible that Sack was shocked,[286] perhaps he was a little bit angry that I let him go to *Moscow* alone. Dymus seems to take everything 100 times more important than it is. Sacks photos came out well.

Also, Mandelstams photos are very nice, in spite of the "negress". About your photo I was very glad, but send me a better one occasionally, I am sorry I have no large one of me, I shall try and get one and send it to you.

But it is terribly late now and I must go to the institute. Before, I will look for somebody who can give me Russian lessons.

How do you enjoy working again, and how do you enjoy being in *Leningrad* again with so many people and with all the pleasures of a big city?

Regards to everybody and my love to you, darling!

Your *Rudi*

I am anxious to get your letter!

I do not send this letter *recorded delivery* because then it goes quicker. I hope it will not be lost.

[61] Genia Kannegiser to Rudolf Peierls

Leningrad, 3.10.1931

I am awfully pride that you understand my letters, darling, but you are really very clever boy because they are many people who cannot understand even my russian letter and write so angry answers that I am quite ashamed when I read them, and I certainly understand your every word. So all is quite right.

I want to say you so many things, that I don't know when I shall finish this letter. Now it is evening, I am a little tired because I worked all day very hard, but at home is very nice and my cat is sitting near me and looks on our toys with very sweety eyes and sings, sings, sings.

[286] Letter [58].

You asked me if I like my work, or it is only "some occupation". I love the physics very much and I work now with "clean physics". I like to think.

4.10.1930

Rudi, on this word I fall asleep and when I opened my eyes it was late and Nina slept so well that I went right to bed and dreamed all night about you. (What a material for Freud and his prophet Bronstein!) Dear, I go on, so I like to think and sometimes I have some ideas but I am not of high opinion about me. It is true that I understand all very easily, perhaps too easily, but I am too "woman", too "surfacing" not "deep" (you understand?) and so lazy, so awfully lazy! But certainly I prefer this work to any other and it will be for me very tedious without it. I write to you in my laboratoria all the people is going away I am alone (it is 9 o'clock in the evening) and awfully angry – my Kohlhörster (you know such electroscope for penetrating rays) became quite mad, to day I measured the first time after vacances, and after his traveling on *[?]* with my patron and in the same condition he show now 21.0 instead of 11.5 before summer. I cannot understand what is the matter with him and dance around 11 hours without any result. I am afraid that my patron dirtied him and some radioactive products I cannot find other reason and if it is so it is awful!

I don't know what will be my way, now I am on the quite path of "laboratoria girl" and I enjoy it very much, but Rudi I am an adventurist in the high sense of the word and too young also, so it is a great probability that I shall have many adventures and perhaps many professions. You know, dear, it is so tedious and even so sad to be sure in your future. Future must be unknown and bright. Yes? When you know your future you are big slave in present and it is immoral. Rudi, it is too philosophical but you understand not only the words but their deep sense and it is so easy to speak with you. You know I write not only the funny verses but and lyrical (?) also, I am sad that you cannot understand them, I don't read them to anyone (there were some exceptions of the rule) but sometimes I like them, your opinion interest me very much, but it is nothing to do. I think now a little about a novel, but I am too lazy to write it and it is too difficult.

If I can do what I want, perhaps I will travel and write books about towns, peoples and stars. Sometimes I want to be a writer. And there are such moments when I want to be an ingeneer, not an ingeneer of fabrique near the big city, but somewhere in the forest or in the mountains with wilde people and wilde nature around. In such conditions you can very well think and see and write. About "adventures" – you distinguish between "love" and "adventure", so it is easy to explain my point of view on this thing. Is an "adventure" immoral? What is moral? I understand only self-moral, if remembering some thing you do b-r-r-r (if you are uncomfortable in the middle) it is immoral, so can exist moral adventure, such after which you are quite well, but it is "unhonest" it is "profanation" of love, of words, of kisses, it is fox-trott under 7 symphonie of Beethoven (you understand). And the love for the money – it is "dirty" also in this sense — "profanation" of love and of woman. Ou how late, I must see for the last time my Kohlhörster. Till to morrow, dear.

6.10.1930

When I returned at home the 4th, I found your letter from Berlin[287] on my dining table and yesterday your Luftpost from Zürich[288] awakened me in the morning. It was so nice, just as you entered in the room. Darling I was glad of both your letters, but really I had not want compliments with "long, tedious, mixed etc. letter" but they are awful in comparison with that which I have in my head, I am so sad I cannot write you russian yet, I am a little afraid of it because I remembered the little tale of Maupassant: a frenchman fell in love with an english girl and married her she was charming and spoke an awful french and they were very very happy until she learned french quite well. Then he saw that she was stupid and vulgar and fell into deep spleen and was unhappy all his life. It is not our "cas". No?

My dear, dear boy — I had not many hopes in such nice and simple end of our story — as life with you in Zürich, but I am an optimist and, it is a long time to wait that it can happened a lot of things in the style

[287]Letter [59].
[288]Letter [60].

of moral novels you found 100.000 from on the street, or the death of American uncle bring you a house in New Orleans and so on. Yes?

No, "seriously", darling, when a man want something very strong, he always get it. Rudi, I want to see you! It will be splendid in March.

Now is very beautiful here — "the gold autumn", and the town with blue rivers and gild and red gardens is charming. I walk sometimes very long time through the streets and you know it is such a pleasure to look around when the sky is high and blue and dry leaves under your feet. How I love *Leningrad* if you know how I love it! Why you are not here now, we should go in *Detskoye Selo*[289] and in *Pavlovsk.*[290] The park of *Pavlovsk* is one of the best in the whole world and in autumn there is really fabelhaft.

But in early spring (March and April) is also very beautiful another style certainly but you shall see it.

I related you about *Moscow* in last letter but I forgot to tell you about my *Moscow* friend. Do you remember this evening in *Kislovodsk*, our first tête-à-tête when I tell you my story (it is true that you were not very interested in it). And I relate you that last summer when I was in *Crimea* I meet there one "not ingeneering ingeneer". He is forty two, traveled very much, likes verses and pictures and understood very easily many things. I walked with him and we spoke many hours and afterwards we wrote letters to one another sometimes. I was very pride that such aged man (42) is so interested in me. Now in *Moscow* we walked and spoke again and you know I thought it before and now I am sure that he loves me awfully, why I think so I do not know because we never spoke a single word about love, but there is something in eyes and laugh that make sure that there is a big and sad love — the "lost love".

I don't love him at all, he is only good friend, very intelligent and more aged, but I am very serious in such things and I don't like "to play" with love, but it is not my fault. I never want that he loves me.

---

[289]The town was founded in the 18th century as Tsarskoye Selo (Royal Village); it was renamed Detskoye Village (Children's village) in 1918. In 1937 the name was changed again into Pushkin to commemorate the centenary of Pushkin's death.

[290]Town 25 km south of Leningrad, famous for its grand palace and parks.

So I am sorry with it, but for other side. Love is such splendid thing, not perhaps it is better a not divided love that not at all. I don't know.

Rudi, dear, I forgot to thank you for photo. They are very nice, the best I think are Pauli in the train and a view of *green* with pin and palm tree and my portret.

Don't you find that all our story is awfully romantic? Such your plane of love as *from the Caucasus to the Black Sea*! When I remember all I am quite astonished how it can happen when we were alone only two last days.

Do you remember Sack with two trunks in our wagon? And the first evening in the *Grusija* when came quite afraid Bronstein thinking that we fell into the water! It was only two weeks but we saw such a lot of things and it passed so many between us that it seems to me much longer time.

Good night my darling, I love you so strong

Yours, Genia

I write this letter 3 days, but yesterday I cannot write a word it was my free day and all the day came all sorte of friends. I was not very charming with them at the end. Nina sent you her regards and beg you to sent her a face of Pauli of you have it.

[62] Rudolf Peierls to Genia Kannegiser

[Zürich], 8.10.30

Genia,

Dearest, *when will I receive your letter? I have waited so long!*

And I cannot even be angry for I kept you waiting much longer for my letter from Berlin.

My life here begins to become more "stationary" without any exciting events. I am in the institute the whole day (which does not mean, of course, that I am working all the time) only sometimes. I am with Bloch, who has not yet quite recovered from his accident. It must have been a dreadful thing, lying 8 hours alone on a small place on a rock, where he could not move without falling down, and then he had to stay another two nights in this place (3000 meters high) and only within the next two days they could bring him down with the help of about 12 guides. But he was very brave all the time, although it must not be a pleasure to be carried down the rock with a broken leg.

Yesterday I was careless enough to show him the pictures of *Caucasus* and he was quite upset and if there is an opportunity, he will surely climb one of the summits there.

This week I started my Russian lessons, and I very much enjoy them. My teacher is a student of Laws, a very clever girl. If you like, you may sometimes write me *in Russian*, but in the first time, I will have to read it together with my teacher — you understand!

This is all my life here now; not very much. I make many efforts to work, but this is very *hard*, if you are so used to laziness and if there is nobody to control you. Most of my time is spent in preparing a lecture on wave mechanics, which I shall give in Pauli's place this winter. This is *very interesting.*

What about Houtermans? In a post card to Pauli it seemed they would go back to Berlin? Did you hear the details of their marrying story?[291]

In Germany there is a very good talking film now, "Die Affäre Dreyfuss". If it comes to *Leningrad* you <u>must</u> go there! I had never thought they could make such a good earnest but not sentimental film in Germany.

I spoke with Pauli, and I can stay away from Zurich from first of March to first of May, so I shall have at least seven weeks in *Leningrad.* But until then there are still nearly 5 months.

Genia, do you know how often I think of you, talk with you, dream of you! Aren't we silly, you and I? There are so many boys and girls in the world, and just this one in *Leningrad* and this one in Zurich ... But who can help it? *I love you very much.*

---

[291] Fritz Houtermans and his girl friend, the physicist Charlotte Riefenstahl, decided to get married in Tiflis and asked Pauli and Peierls to be their witnesses. See I. B. Khriplovich, 'The eventful life of Fritz Houtermans', *Physics Today* (July 1992), 29.

But I promised to Pauli I would read some papers before he comes and he will be here "early morning" today (i.e. $13\frac{1}{2}$ o'cl.) *Good bye, my dear!*

Your *Rudi*

The included pictures are the best ones I could find.

[63] Rudolf Peierls to Genia Kannegiser

[Zürich], 9.10.[19]30

My dear child,
If only you knew how happy I was when I got your letter. I knew I would get it to-day, because I had yesterday posted a letter for you. I think I could see you when you wrote it so much of you is in this letter. I am sorry I cannot write letters like that but my letters are so terribly cool and do not express what I want to say. And probably I say everything twice, because I speak so often with you, that I never know have I written you this in a letter or only told you while you were not here.

My questions what would you do if you could do what you want, was not purely curious, I very often think about the very similar question, what would you do if you came here and live with me, in the both possible cases, if I would have enough money for both of us or if I would not. But as yet I do not see a solution, not a clear one for the first case and none at all for the second.

Of course it is terrible to know about your future, but sometimes you have to decide for so long a time. So for instance I promised Pauli I would tell him half a year before I go away from Zurich, and as Simon needs the same time to find a position or more, I had to write now to Simon that I should like to go to Berlin next winter if possible.[292]

Dear, I think I am much more lazy than you are, at least I am terribly lazy. Fancy what we would do together!

---

[292] See letter [56].

We seem to have the same idea about adventures and "morals" but I do not understand what you mean by "profanation"? Is not this highly conventional? Of course there is no need to discuss love-for-money. But there the point is not that it is dirty to buy something you want, but that you really will never get what you want. But if two people strongly want to be together a short time and have each other entirely, body and soul, and they are honest to each other, I mean tell each other the truth, where is the profanation? But stop discussing theories.

Do you really believe that the Maupassant story applies to us?[293] I think not, although there is no doubt, that the many feelings of strangeness between us are very important. But this is a strangeness of the whole person, of character, of opinion and so on, and the language is not important. In the contrary, I am so sorry you cannot speak Russian with me, because I enjoy so much hear you speak Russian. You have such a personal style of speaking that I could notice it although I only understand 10^{-3} of the words. Seriously, this is one of my greatest anxieties for the case you could ever come to Germany. To whom could you speak then? And I do not know whether it is possible at all to learn a foreign language, that one can speak it like you speak Russian. In March I hope I will know a bit more of the language, so that I will probably understand 0.3 of the words and perhaps already then your Maupassant story could happen, but will it? I think in one year I shall know enough Russian to understand even verses. But I have no great sense for verses. I like them very much, but it is not that I could not live without them, you understand?

Of course I remember that evening in *Kislovodsk*[294] and I was very silly then, I must confess you now. It is nonsense of course when you believe I was not interested in the story. But I was terribly tired. You remember that we were in the railway the night before and that I went down to *Vladikavkaz* and back in the afternoon? But I did not want to be tired. I wanted so much to hear your story, and then I thought it would be so silly to sleep just when it was the first time we were together

[293] See letter [61].

[294] On their journey through Georgia, Rudi and Genia decided to visit Kislovodsk, a resort in the high mountains; Peierls, *Bird of Passage*, p. 64.

first time that we could speak as we liked, and such a short time — and to spend this time with sleeping! But of course I could not help being tired, and I was silly enough not to tell you. And so some parts of your story are now for me like seen through fog, and you must tell me some more when we see us again (writing such things is so difficult).

With your engineer you are probably right, that it is better to love in vain than not to love at all. But still it is always sad; most sad, if he has some little hopes, which one takes so easily from a smile or a word, which did not mean anything. I seldom was so furious as on this evening in *Grusija* and the morning in the train you speak of. This probably was the reason why I was not very polite to Sack then, and therefore again he probably was angry with me, and this was what *Ivanenko* understood as "being shocked".

By the way have you much time to read books and do you like to read? I should sometimes like to send you some books which you do not know and which would probably interest you. (At least I believe you told me you did not know them; for instance "Les climats" by Maurois (French) and "A Farewell to Arms" by Hemingway and "Liebe" by Wildgans (German) and many more).

The included pictures are for *Nina Nicolaeva.* Tell her that Pauli has many faces but I send some of the nicest.

Now good night, *Genia* sleep well — oh I suddenly remember that it is now 2 o'clock in *Leningrad* so that you probably are long long asleep. Never mind — sleep well my dear!

Your Rudi

[64] Genia Kannegiser to Rudolf Peierls

Leningrad, 10.10.1930

To day is my free day and in the morning I was in *Eremitage* with Nina and two my cousins *very nice, isn't it*? I was not there five months and it was so agreeable to look again on all "ensemble". You were charmed with it, but certainly one visit is nothing. What schools and painters do you prefer? I like very much van Dyck and *Rubens* — the portret (Do you remember these of *Eremitage* but to all I prefer the early germans and italians, von Eyck and his school and fra Bueto angelico. From italian I love Sandro Boticelli awfully love his faces which are on the line between beauty and ugliness and Peruqino Madonnas — mothers and little girls together with childish eyes and wise lips and such touching trees — like big branches, and more genial painters of all peoples and all times — Leonardo da Vinci, but we have here only one his picture and I know him from books and reproduction only. But before Rafael I remain quite cool, more than cool I dislike him too quiet (*calm*, too "easy", too sweet) and I do not like Rembrandt, why I don't know, but he is nothing for me. We have now many new pictures — new french painters from *Moscow* one very nice Monet and Dégas and Renoir. I like him very much I wrote you that in one of my letters. We will go together there in March. Yes?

After dinner we were sitting with Nina on our divan and we write verses about "*Grusija*" and *our trip*, some of them quite funny, Oh, *Rudi* if you will see here "Herr Dr. Sack" or write to him tell him that I am very angry, because he promised me to send photo to me and did not and more than that, I wrote him a letter with Bubi together and he did not answer. Perhaps he is angry that I did not see him before he went from *Kislovodsk* but I begged his pardon in the letter and really I was very sorry with that. Do you remember? On the railway station. So I am angry.

The day before yesterday I have a great walk quite alone in the Ostrova.[295] (You were there with Bronstein, it is near the Buddist temple). It was splendid gold parks and the lakes as looking glaces in the sun set and the sea white and rose two stone lions with wise eyes, and I was alone quite quite alone. Only the moon came after and laughed on the sky. *Nice.* I like sometimes to walk alone, this time I thought about you darling and wanted so much that you were near me.

And yesterday began the winter in the way to my laboratory. I fell into the snow storm and it was such wind that I cannot stay on my feet when I arrived at last in my room I was a bit of ice. To day is stormy and it rains and it is very high water in Neva — a typical autumn paysage, and it is cold ... I speak so much about this "weather", there because

[295]Island in Leningrad, known as Kamenny Ostrov.

my warm overcoat will be ready only in three weeks(!) and it is rather disagreeable to wait it such long time, but we must buy many things and the dressmaker is occupied. Oh, oh, oh!

But it will be splendid.

Yesterday I was half mad, because — every few minutes I read a splendid criminal novel from Edgar Wallace "The face in the Night". Do you like Sherlock Holmes style? I awfully. In the evening I soupped with one of my friends (this Kiebel of this incomprehensible wire)[296] the first time after my traveling in restaurant I remembered our last supper in *Kislovodsk* do you remember the gray sky and musik in the garden?

We finished it with a glass of chocolate, ou I adore it! Kiebel is very good boy, really very good, but a little "strange" too often too sad and don't know what to do next. When he is with me — all right in five minutes he begin to be unhappy again. It was a sad love story with him, but I think that it is simple "melancholic character". You know about this wire; all the town gossip about us. I make an ununderstanding face and big eyes, but *Dumus* made all he can. For me it is nothing at all, but for you?[297]

I got the abstracts from Gamov, from Nature. The last was with photo of Piz de Daint with you and Rosenfeld on the snow. Very well indeed, is it yours? You know that it was the Gamov's dream to inscribe under his work some mountain. Now I think he is quite happy.[298] I must write to him, but if you know how many letters I must write! Now it is 2^h30^m and to morrow I must go up at 7^h30^m. So good night my very dear. I wanted to write you one thing, but you would say it is for compliments. Good night! How many things passed and it is only less than 1.5 months. I am very glad of your letter. Very.

Yours, Genia

[296] Letter [59]. I.A. Kibel, known as Kib, was among a group of astronomer friends of Genia's, a group which also included Ambartsumyan, Gamow and Bronstein.

[297] See letter [58].

[298] Peierls, Gamow and Rosenfeld went hiking in the mountains where Gamow conceived the idea for the paper. He wrote the paper in a mountain hut, took it and signed it on the top. See George Gamow, interview by Charles Weiner, 25.4.1968, Niels Bohr Library, AIP.

[65] V. Fock to Rudolf Peierls

Charkow, 11.10.1930

Lieber Herr Peierls!

Ihren freundlichen Brief mit dem Manuscript unserer gemeinsamen Arbeit habe ich am Tage meiner Abreise nach Charkow (2.X.) erhalten, und deshalb hat sich meine Antwort verzögert.[299]

Ausserdem wollte ich mir noch die ganze Sache gründlich überlegen, und bin jetzt in Zweifel geraten.

Man kann nämlich Wellenfunktionen aufbauen, die der Divergenzbedingung genügen, aber trotzdem nur in einem beliebig kleinen Bereich um den Nullpunkt von Null verschieden sind.

Es sei z.B.

$$F_x = -\frac{y}{\rho}\delta(\rho - a)\delta(z)$$

$$F_y = -\frac{x}{\rho}\delta(\rho - a)\delta(z)$$

$$F_z = 0 \qquad\qquad (\rho = \sqrt{x^2 + y^2})$$

Man wähle $\underline{a}$ beliebig klein und approximiere dann $\delta(\rho - a)$ und $\delta(z)$ durch stetige Funktionen. Man bekommt auf diese Weise ein punktförmiges Wellenpaket. Einen durch diese Wellenfunktion beschriebenen Zustand kann man als Anfangszustand betrachten, und ihn den Maxwellschen Gleichungen gemäß zeitlich fortsetzen. Der Anfangszustand entspricht der Messung des Ortes eines Lichtquants.

Der Ort wäre danach doch meßbar!

Ich bin sehr neugierig zu wissen, was Sie darüber denken.

Bis sich diese Schwierigkeiten nicht geklärt haben, möchte ich abwarten und unsere Arbeit nicht endgültig abfassen.

---

[299]This refers to work of Landau and Peierls in which they tried to extend the uncertainty principle to relativistic quantum theory in the hope that unmeasurability of wavemechanical quantities would help to modify current theories. There was some misunderstanding about the publication of the research results between Landau, Peierls and Fock. See letters [76], [86], [93], note 380.

Ich bin jetzt mit P. Jordan und einigen russischen Kollegen in Charkow, bleibe aber hier nur noch einige Tage. Am 19. Oktober bin ich wieder in Leningrad. Schreiben Sie mir also dorthin!

Es ist mir eine Freude, an unsere Zusammenkunft in Rußland zu denken.

Mit vielen freundlichen Grüssen
Ihr herzlichst ergebener

V. Fock

[66] Genia Kannegiser to Rudolf Peierls[300]

Leningrad, 14.10.1930

Dearest,
I write you very high on the tower, where I work every day because it is of wood and glace and my Kohlhörster is too sensitive for the strong walls, my writing table is below, but I must to measure every ten minutes and I wrote you with a pencil on a book in intervals. This tower stay right on the bank of the Neva and is the highest point of the district. So I can see very far around, and it is beautiful — three bridges, blue river, with fabelhaft buildings around, and on the horizon sea and forests. It is quite a pleasure to sit here.

Rudi, dearest some days before I wrote you a long letter, but when I came to post office I find that in my pocket I had enough money only for the simple letter and I send it so, I think that you will get it together with this.

The day before yesterday and yesterday I got your both letters.[301] I was awfully glad. Awfully! You know, darling, I read in them many things I wanted to write you, it is very curious indeed. You know that every evening in my bed I wrote you long long letters, so long and complete that I also forgot always what I had write to you in my head and what on the paper.

[300]The letter was written in pencil and some passages are illegible as a result.
[301]Letters [62–63].

Rudi, every day, every hour, it becomes [?] to wait. I thought that it will be better, but ... Really it is too hard to speak on such distance every evening! The Russian people says that love is how fire and distance how wind, it kills a little love and blows up a big one. It is true, I think. *Yes?* You wrote that your letters are cooly and I just thought why you have never tell me such "warm" things as you write! Perhaps, you simply haven't a time. Yes? Oh, *Rudi* how warm, hot. I think now, and now it's difficult to write! Yes, about "profanation" simply I forgot all less "pathetical" words and so I put profanation into " " to make it more "at home" but you understand.

Your Russian progress are "fabelhaft". Really, in two weeks I shall write you russian letters for practice, but I am afraid that they will be too lyrical for a teacher!

Yes, I received yesterday a letter from Sack with four photos, two of them you in *Tiflis*. Sack wrote one half of his very short letter in Russian. and really very well, so well that I think that [?] wrote 3/4 of this half. Thanks for your photo. I enjoy them very much. It is your Zürich house? To morrow is my free day (it is so nice in the *USSR* every fifth day you are free, poor boy with all your laziness you have the moral right to do nothing only each seventh day!) and I will photo myself and sent you in next letter quite new portrets.

I have been at the opening of the season of our *concert hall.* Our "Generalmusikdirektor" is nothing in comparison with Bruno Walter,[302] Kleiber,[303] or Klemperer,[304] but the *orchestra* is very nice and a splendid first violin. It was *Skriabin*[305] evening. Do you like it? I don't know him very well, but I love the things I know very much. The "poem of extase" is splendid, really splendid with its ill violins in the middle and joyful trumpets at the end. Skriabin is free and "windy". Yes? I

[302]Bruno Walter (1876–1962), at the time conductor of the Leipzig Gewandhausorchester.

[303]Erich Kleiber (1890–1956), at the time Generalmusikdirektor of the Berlin State Opera. In 1934 he resigned in protest against the Nazi regime and emigrated to South America.

[304]Otto Klemperer (1885–1973), at the time director of the Staatsoper am Platz der Republik, a branch of the Berliner State Opera.

[305]Aleksandr Nikolaevich Skriabin (1872–1915), Russian composer and pianist.

don't like Wagner very much, only few things the "fly of *Walküre*", for instance, but only once in my life I fall asleep in the theatre and it was on Tangeuser.[306] It is true that I was little girl then, but it is true also that remembering it now I close my eyes. When he begin something he can not stop, fifty times the same thing! *You know that my motive* of my love for you is Carman of Schumann's "Carneval". I don't know why, but perhaps because when I in the first time thought of you very seriously, someone played "Carneval" in the ladies room in *Grusija.*

I was so glad to read "*you*"[307] in your letter, it is stupid this english with his "you" in all conditions.

Good bye, till to morrow! I must go downstairs and work without intervals till the evening and then go home — dinner, bath myself and sleep. Not a very interesting programme! Yes good bye, darling.

[67] Rudolf Peierls to Genia Kannegiser

Zürich, 14.10.[19]30

My dearest,
Now the winter begins, and that means that the airplane Berlin-*Leningrad* no longer goes. So the distance between you and me increased by some days. *That's terrible, isn't it?* On the other hand, it means that 0.25 of the time till 1. March has passed.

I just came from a *concert of Stravinski, very interesting*, but not easy to understand. Still I liked it very much, especially one piece where he played the piano himself. A very light "capriccio".

There is one man more now in our institute: Waller from Upsala[308] (Scattering of X-rays, theoretically). He is a sympathetic man, but terribly silent. I am usually not talking much either, but he is 1000 times worse. Still one can walk with him and so we did last Sunday. It

[306] Tannhäuser.

[307] In Russian, there is a distinction between the formal and less formal address of you. Rudi, in his letter to Genia, used the informal term.

[308] Ivar Waller (1898–1991), theoretical physicist and crystallographer; later professor of mathematical physics at Upsala University.

is very nice, walking in the hills around Zurich, especially in this time of year. The fog lies on all valleys and covers them, so that one can only see the tops of all hills. In the background of course the higher mountains, which already are covered with snow.

I wrote you the other day of Heisenberg's letters in which he usually writes very wild theories to Pauli. Some time ago he had written a paper on the self-energy of the electron and connected it with the other difficulty, the zero-point energy of electromagnetic field.[309] We showed that if one solves the one difficulty, then the other will be solved, too. Pauli wrote to him, then, that this paper was like the old proverb: "If my aunt had four wheels, she would be an omnibus". Now some days ago, Heisenberg wrote: "I think the aunt is an omnibus", and gave a solution of both difficulties. But we did not like this solution, it has no physical idea in it, and we tried to find a mistake. And really to-day I found it. Fancy the face of the post-official, when he had to receive the telegram: "Tante ist doch kein Omnibus, Fehler gefunden, Hurra". Pauli was so glad about it that he promised me a bottle of champagne, which we will drink when he is over back from the Solvay Congress.[310]

I am so tired, *Darling, good night.* Write *soon*!

Your *Rudi*

How are you and how is your Kohlhörster?[311]

[68] Genia Kannegiser to Rudolf Peierls

Leningrad, 15.10.1930

It is 11 1/2 o'clock now and I go up to only 15 minutes before. Quite a "Pauli's" inclinations!

I am awfully jolly today, darling. It is so nice in the street — sun and blue sky, and warm wind. I still go at post office and after to *islands*

[309] W. Heisenberg, 'Die Selbstenergie des Elektrons', *Z.Phys.* **64**, 4–13 (1930).

[310] Wolfgang Pauli was attending the 6th Solvay Congress in Brussels between 20 and 25 October 1930 where he spoke about the 'magnetic electron'. W. Pauli, 'Les Théories quantiques du magnetisme: l'électron magnétique', in *Le magnétisme*, Paris: Gauthiers-Villars, 1932, pp. 175–238.

[311] See letter [61].

what I shall do in the evening, I don't know yet. ("It is better not to know the future" from the letter of ... (I don't know the date) October).[312] Yesterday was really nice, also. I returned from my laboratoria on foot, and by the way had climbed on the top of St. Isaac's church' it is awfully high and difficult climbing indeed, but it is "fabelhaft". Will you, we shall go there together.

Rudi, if you knew how afraid I was this evening in *Kislovodsk* when I let you go alone in *Mineralnaja*! But I was so tired, so dreadfully tired that I thought that if I went with you you would be obliged to carry not only three trunks but me also.

My story, really, is not very interesting, but you must to know it — for to know me. It is difficult to write such things but I will try. I don't remember you or not, but I told you this evening that you are my second love, four or five men loved me, but I loved only once before this summer and my first love was *Ivanenko*. I was a little girl then, this story began four years ago and ended 2 1/2 years, and now seems to me that I read it in some book. But it was very "serious" and I awfully growed up during that time. It began with pity and something like mother's feelings. (He was very unhappy then) afterwards came the classical "first love" and it ended from both side in the same time. So I was *Dymus* first bride near one year. Bride not wife, you understand? Ouf! Rudi, perhaps it is very uncomprehensible but I shall relate to you the details when we shall see one another again.

Now we are in very cool relation, more than cool, because *Dymus* has an awful character and was so angry when this story ended, although it had a very "symmetrical" end, that we did not speak for half a year.

Do you remember the way from *Beslan* till *Mineralnaja*, it was very strange my first and second love in one car!

Rudi, I will be very glad to receive the books that you like. But perhaps it is too expensive. You forgot that I relate you my high opinion of "Les climats". I read them one year ago, and was charmed. The hero is disgusting-like man, but this book has a very important thing — his own "tune", it is like in musical things. Do you know Eugène

[312] Letters [61], [63].

Fromentain[313] and his "Dominique" a very nice book, a very classical one but with such beautiful "imprèssionist" peysages!

I will sent you the Russian book when you will understand them, certainly you have not read *Bodin, Sredin* and many many other new writers. Some one are very interesting, and for you will be easier to understand them after your traveling, and before another one.

Rudi, I forgot to tell you that I started speaking German. Oh! Es ist fabelhaft, ich spreche mit meinem Stiefvater und meinem Vetter.[314] I speak awfully with a lot of english words but I can write that, Ich liebe dich sehr stark, sehr sehr stark!

Nina wrote these lines below before I begin this side.

So good bye, I want to talk with you more, but it will be without end and you will never receive this letter.

I write you so long, long, long letters that you can write me not only answers. Yes? Ou, *Nina* telephoned that she is waiting for me so at last good bye.

Rudi, my dear. I miss you terribly. Good bye. Kiss you. Love you strongly.

Hello. M Peierls! How do you do? Thank you very much for yours photos. Pauli sitting in the boat is really charming. I shall be very glad to see you here next March, then more. *I hope* till this time you shall speak Russian fluently. *Once again thank you very much.* Do you understand?

N. Kannegiser

[313]Eugène Fromentain (1820–1976), French artist, better known for his paintings, also published literature. "Dominique" was first published in 1862 in *Revue des deux mondes.*

[314]Genia's father, a well-known gynecologist had died when Genia was a little girl. Her mother remarried Isaak Mandelstam, a translator and talented writer. *Peierls Papers*, Supp. A118.

[69] Rudolf Peierls to Genia Kannegiser

Zurich, 16.10.[19]30

My Dearest!
How is that possible? I thought you were so much used to be teased, and take such theories humorously, and now you seem to be offended by that harmless joke about "fishing for compliments"! Isn't that silly? *No, seriously,* you will write me what you meant in your letter without being afraid of jokes I would possibly make about it? *Yes?* You will get much more training from the side of Dymus, and so on, and it must not be agre[e]able for you. I, of course, do not mind it, I am far from *Leningrad* and then, much talk usually is less agre[e]able for a woman than for a man. But it must be terrible for you and I am so sorry to be the cause of such inconveniences for you. And I am afraid that your situation will be even worse in March and afterwards. Afterwards — a funny word; really I cannot think of the time after I will have left *Leningrad* again, it is quite impossible to imagine it now, that there is some time "afterwards". I agree with your taste as far as Raphael is concerned, I think him very tedious and pedantic. But I am strongly fond of Rembrandt. I think his pictures are the kind I like most. Then I like also to see the impressionists. Did you see Slevogt?[315] or Lesser Ury?[316]

Then they are also beautiful things amongst the Dutch (Rembrandt is quite apart from all the other), although I generally dislike Rubens. (In *Eremitage* I saw some pictures of his who are quite different from all I have seen and very beautiful). There is for instance Vermeer van Delft and many others (I have no memory for names unfortunately). I like many of the Italian painters (Giorgione, Tizian, most of Leonard), I saw too little I have no impression. Generally I do not understand the very early painters with one exception: Konrad Witz,[317] of whom

---

[315] Max Slevogt (1868–1932), German Impressionist painter.

[316] Lesser Urey, (1861–1931), German painter, painted among others the well-known 'Portrait of Dr. Rathenau'.

I saw splendid things in Basel. Also some Holbein. This is the — very unsystematic — report of my taste. Of course we will be in *Eremitage* in March and will see many things.

Speaking of taste I must confess that I sometimes love criminal stories. But Wallace is terrible. He is the worst thing a writer of criminal stories can be: *boring*. The most attractive thing of this kind of literature is that one sees a complicated and fine logical development by which the hero of the story comes to find out the truth. But in Wallace's stories the detective comes and simply says: "This man is the murderer" and nobody ever knows why.

To-day I had my Russian lesson again. *It is good to speak Russian.* My teacher is Russian, but she left Russia very early, she was only a child when her parents took her out from Russia. But she loves this country so much. To-day I stayed there for two hours after the lesson and had to tell her about Russia and she became quite upset when I showed her pictures and told her many things I had seen there, things that made me also think of Russia with a kind of homesickness. She knows a lot of Russian people here who live here now for many years, are married here and have their profession and so on, but they all say that they can never feel at home here, that their home is still *in Russia.*

But my room is not heated. *It is very cold.* (How cold will it be for you in *Leningrad* and without winter coat!) So I will go to bed and go on writing to-morrow. Good Night *Genia. In Leningrad* it is now 2 o'clock so you already will be fast asleep.

Sleep well!

I am so glad your letter only went 5 days. So after all it is not so far to *Leningrad*!

I know that Gamow laid so much stress on undersigning his paper in the mountains, and he wanted by all means a mountain with snow on it. Every time when we planned a trip he asked what was the name of the place and was there snow. I have the suspicion that he only came with us for that reason! But it is a pity that we could not go on "Piz Quatter Vals" the name is so much more thrilling, but it was too

---

[317]Konrad Witz (ca 1400–1445), German-Swiss painter.

difficult. Gamow had difficulties with his foot (he had had some accident in England before) and Rosenfeld's shoes went entirely to pieces and we got into fog and so on. But Piz Da Daint was also very beautiful.[318]

To-morrow probably my friend Thorner comes from Heidelberg for a weekend. In those two hours in Heidelberg we spoke also about "philosophy" you know: causality.[319] Heisenberg's uncertainty principle, and so on. He is a doctor of medicine, but one of the few people who could possibly understand these things without being a physicist. But of course it will be hard work and one must always be very careful that these biologists do not take anything they want and claim it as "a consequence of the new physical theories". I will be very glad, for personal reasons, too, if he comes for there are not very many people in Zurich with whom one can speak reasonably. I am so sorry, I wanted to have a photograph of me for you as you wanted it and I made Waller take it. But it came out so dreadful, that I did not send it to you.

O dearie.

So as I said, Thorner was here and it were very nice days.[320] Last night we went to the theater to see "How to become rich and happy", a very harmless but nice comedy, well played and well made up. Before and afterwards we had long discussions about the meaning and the use of the notion "time" in biology. He had some plans, some problems and ideas he was very fascinated by and he wanted to discuss them a long time, there remained no problem, no idea and he had to confess that he had had no clear thought. So in the end the effect of our discussion is only destructive. But besides this we had a splendid time. This morning we went with an early train to Einsiedeln, a place near the Vierwaldstätter See, climbed a 1900 meter mountain there, the most beautiful, although unknown point of this part of Switzerland, the great "Mythen". This is very sharp and steep rock, quite near the lake, and as it is the highest of the surrounding mountains, you have a beautiful view on the whole chain of the mountains behind. Beneath you you have many lakes in narrow valleys. To-day like always in autumn, the valleys

[318]See letter [64].
[319]See letter [60].
[320]Letters [60], [62].

were filled with fog but on the higher mountains the sun was shining and one looked down on the fog like a sea with many islands, the other mountains above a dark blue sky. I had never seen this spectacle before so beautifully. I tried to make some pictures and when they are ready I hope I can give you a little impression of the beauty.

The end of the beautiful day was a bit exciting, because our train back to Zurich was late, and he had to get here the train to Heidelberg, as he must be in his hospital again to-morrow morning. Now his train to Heidelberg waited for the other one but the man to whom he had given his bag (which he could not carry to great Mythen and back) was not there. So he got his train all right, but his bag stayed here and I had to send it on to him and there were all kinds of complications.

Now it is quite a general theorem that every letter has an end, and even this one must have it, although it sounds very improbable. I think often of you, dear, I dream often of you (far too often!) and I am longing for March.

Good bye, my Genia

Your Rudi

One must always be careful with general theorems! I just got your letter of 14./15. and I was so happy.[321] Now this letter would certainly not have an end if I had the time. But sometimes one must work a little bit! I am longing for you, Du Liebes!

[70] Genia Kannegiser to Rudolf Peierls[322]

Leningrad, 22.10.1930

Rudi,

It is a too long monologue! Think only, that it is the third letter and I have not the answer to the first one. I know dear, that it is not your fault, but there is now two weeks between an letter and an answer. Horrible! Why aeroplanes cannot fly in the autumn? It is so stupid

[321] Letter [66].

[322] The letter is written in pencil and some if it is barely legible.

that the air post is only in the summer. Yes? But nothing to do. I think that I shall have your letter in four or five days, so the next one will be a little more "diagonally".

Rudi I don't think that I killed somebody — this red spots are only Fe_2O_3, I work with it now and I am tired to wash my hands every 3 minutes, it is awfully difficult with it so fine fine pulver, and I am now quite an Indian girl. Really!

We all at home are mad this week with Busch.[323] Do you like him? I awfully. I don't know what is better, verses or illustrations. I read "Max und Moritz", "Plisch und Plum", when I was a little girl, four years old, but I read it in russian and now I try to understand the German text and we have a very thick book now with poems quite unknown for me. So a great pleasure. Yesterday I was at the "Klavierabend" — Chopin and Lizst. It is very strange but I can not receive pleasure from a thing I know too well, for instance this "Marche Funebre" of Chopin, a splendid thing, but I heard it so many times that I was now a stone and around me a lot of ladies cried with very hot tears. I never cry in "public places" only two times I was very near of it. Once on the 7th symphonie of Beethoven with Klemperer and the second time it was the ouverture to the Marriage of Figaro of Mozart and I am astonished till now with this second occasion. Perhaps very beautiful things, too beautiful make me sad.

Rudi, tell me the things without which you "cannot live". For me verses, Musik, sunny and windy days and I awfully love the "white nights" perhaps you will see a bit of them in the April. At the concert I saw a lot of people between others Mrs Frenkel who was in her blue dress of *Grusija*, Ivanenko with his wife. Dymus is here only few days and will return very soon in *Kharkov*. In *Kharkov* two weeks were Jordan, Fock and *Ambartsumyan* I like him (*Ambarts.*) very much, such a nice boy without any *tricks*, and such a lot of people has them, Landau, Ivanenko, Bronstein, this all are a little "mad" (too complicated). Only Gamow is always in good state of mind and it is very easy with him. Yes? *Ambartsumyan* relate me very many funny things, and I was glad

---

[323]Wilhelm Busch (1832–1908), German author and illustrator, most famous for his satirical drawings and his illustrated stories, mostly for children.

to see him. Do you like him? He has a splendid head not only for the [?] but for all. You understand? He is a very clever boy and very "original". Rudi, a boat goes under my window and a long tail of white foam is behind it how beautiful it was three nights on *Grusija.* Do you remember?

It is so sad to write when you want to see and speak. Oh, *Rudi*, it is soon two months, one third only, a half of a year it is such long time. Think and you and I can be quite different people. We can "grow up". This letter is too dirty, but if I copy it it will be too clean and unnatural. So "I beg your pardon"(!!!) and sent it so.

What do you now? How is Bloch? How are you with this girl? How is Pauli? It seems to me that I will never receive a letter from you! I am very interested now in cosmical rays, and of absorption of hard γ and so on. I think that you can have from absorption coef. an desirable wave length very easily. Yes? Here is only ".." Yesterday I was looking in "Berichte" and meet your name there hundred times. I was glad to see your name, if I cannot see you. Good bye, darling! *Good bye.*

Genia

Dear Rudi,
I was so happy when I received your letter and the photos. Your write Russian fabelhaft. *Serious* fabelhaft. I think that with my english you are not so satisfied. Yes? (It is not "fishing.") Darling, if I could be near you only half an hour! But half an hour is a lot of time. Do you remember the train between *Vladikavkaz* and *Beslan*, so many things passed in it, and it was only 40 minutes. In half an hour I could tell you at least five letters, because you know that I can speak 10^3 times easier than write. Many people says that I speak too easy. But I never thought that it was difficult for you to speak, at contrary I remember how you screamed writing letter. Rudi, do you know that in one of these days I was very near of "death" or of serious wounding. I fall into the tram-accident. My tram jostle with another one and I was standing because it was near my stop and I wanted to go away. It was an awful shock I fall down on my back and knock my head really very hard. Then began an dreadful panic, because all the glass was broken, (half of

them was on me) and all the people with screams, crys etc. run away. They runned on me and one lady even lost her galottes on my breast, so it was not too agreeable to lay under all these legs on glaces and on dirty floor: Brrr! At last I rose, gathered all the things that this stupid people lost in the tram and in the street and recovered them to their owners. But if you know in what lamentable state I was: dirty(!!) all covered with glaces (I have cuts even on the stomach)) and on the 3/4 covered with bruises. Only to day I am better and can to sit and to lay without suffering. And I was returning from splendid walk in *Pavlovsk* (I wrote you about his parks) I had going there for the services affair, but on the way had many pleasures. The sun set, the high sky and the lakes covered with dry leaves. We returned (I was with one man of my laboratoria) in the night under stars and moon in quite desert and dark park, my companion was afraid a little of this too romantic surrounding, but you know, certainly that I am afraid only of cows. Even in this tram I was calm.

Darling, this story with your friend is really too "european". But the position of these people for my sight is not so bad. Why, they are young, healthy, loving, they had some position. You say that they can lost it, but in this world all is very uncertain. And, you know, for me the money is not a great thing. I am so habituated till the early childhood to the difficulties with money! Worn and homekeeping, but 0.9 of the women do so, and I think that they were very happy when all these stories turn their end and they were together. But poor girl, it is awfully hard to marry a man against his family, because now she certainly think that she must be for him all — and family, and position, and money (you understand?) and if she is not sure, that he is happier now with her and without all those things, than before, she certainly will be very unhappy. If he is a clever boy, he must be very careful and tender, because it is always easier to give something that to take and he gives her to many. But you know, although it seems to me very "unnatural" I can understand the parents (and you?), and perhaps they are the unhappiest in this story. Think only they loved their boy, they loves him also certainly they think that they are right, that he would be happier if he made all that they wanted. That now his life is spoiled, that he did not love them etc. etc. It is an awful thing to think such things (!!!). But the boy is right, because in 23 man must live with his own head, he knows in general what it is better for him and what is more honest for him. (Ouf, how many philosophy can one write, you are quite tired now my poor boy.)

Dear Rudi, dearest Darling, what I must do? I write this few lines in russian because they are not very interesting for you and you can pass them if it is difficult for you to read awful handwriting in strange language. Rudi, what a splendid little girl of five months I saw yesterday! She is a little daughter of one of my acquaintance, she has big round eyes, quite round and dark grey, she is rose, rose, and her nose is turned up and so small! She laughs all the time and begin to "speak" and she has curled hair. How I love the little children. I think that even more that boys. How are your nieces and nephew? Dear I think that the position of uncle and aunt is the best in the world. We have all the pleasures and none of the disagreements and responsibility. Yes? I have many little nephews (the children of my cousins) but they all live far away. And always the uncle for the child is the most splendid thing. Because he is always in good humour and brings nice and sweet things in his pocket, I remember how I awaited one of my uncles!

Darling, I want to walk with you, to day in sun and blue river with boats in so near!

Thorner is a very nice boy, but very very boy. He is younger than you. Yes? How are your lectures? Last winter I read the lectures for the worker, who wanted to go into the high school. I learned with them mathematics and physics. If you knew how I was afraid the first time, the youngest of them was 27 minimum! And amongst them were quite old men (35!) But they were very nice and learned with great pleasure. Afterwards I had another group, with more younger people. They were very lazy and I was awfully severe — awfully! You don't know how severe I am!?

Darling, and you with Stradoba is very nice. I don't know why but old people and little children always like me, with the middle age it is not as easy, but with old one if I don't tease them it is always all right. I did not know that you have bad character. At contrary, I think that you have a very nice one in comparison with Landau for instance or even with Sack. Do you remember how easy he became unhappy, every insect desolated him and with you it was very easy!

Dear, I was very astonished in *Odessa* with you. You were so "uneuropean" and I like in you so much that you can make "nonsense", they are some people who cannot. Sack cannot certainly, and Mr Simon, and Ramsauer(!) and *Bursian* (do you know him?) and even Dorfman,[324] but I think that Pauli can, perhaps Bothe, and any russian physicist — Gamow, Frenkel, Landau, Bronstein, Krutkow, Fock. You understand what I mean under "nonsense".

3.11.

Darling, I think that I will never stop this letter. But I must end it because it will lost any interest in one week. So good bye my dear, dear, dear! *I kiss you.*

Genia

P.S. I beg your pardon, because this letter is so dirty (It is not fishing, really I am very ashamed with it), but I cannot help it because I write it in free places, and in my work room also with dirty hand. *Good bye.*

Genia

[71] Rudolf Peierls to Genia Kannegiser

Zurich, 24.10.[19]30

My dearest,
I am so angry with the time, it runs so slowly! I could think that the winter term must soon be finished and I on my way to *Leningrad* and to you, but it has not yet begun. *Terrible*! At least it seems to begin now. My lecture on wave mechanics will start next week and yesterday I already had a meeting of the people who will hear this lecture to arrange the time and so on.[325] It is a very strange feeling, if you are

[324]Yakov Grigorevich Dorfman (1898–1974), physicist and later head of the Leningrad magnetic laboratorium.
[325]See letter [62].

in some way responsible for what you say. I very often gave lectures, so that is nothing new to me, but it is new that it is a regular lecture and that people sit there with open diaries and write down nearly every word I say. It was somewhat funny, because I am not yet allowed to announce lectures (being not yet a "Dozent") and so it was announced as a lecture of Pauli. Yesterday some people asked me: "Please will you give the lectures or Pauli?" and when I said I would, they made a face like "What is the use of coming here, then?"

I hope that at least some people will remain. Now, to-day Stodola telephoned me. This is an old professor of engineering, the greatest "Bonze" here. (If you do not know what a Bonze is, ask for instance *Bronstein*.) He is so pedantic, that when he heard a concert, he strictly avoids to hear one tone of music for some days, because he must digest this concert. He is conscientious enough to begin now to learn relativity and quantum mechanics although he is already about 70 years of age. So he had heard of my lecture and wanted me to explain him a lot of things and I already to-day spent two hours with him. He is a very strange man, think for instance that he has an organ in his dining room, no harmonium, but a real organ. Do you know that I have a very bad character? So I tried to make a good impression on this man and I entirely succeeded. I started with refusing when he offered me money for the "lessons" I will give him, which as he said was very "generous", but in truth a good calculation since the high opinion of a man like him is more useful than money. Then I tried to understand his complicated way of thinking and to explain to him the things he had difficulties with, and he was charmed.

Besides this I have a lot of other things now, and if I would do them all, I ought to work the whole day (this of course, is arguable. You do not know much psychology, what do you think, is the pleasure of being lazy if you have the moral right to do so?) [???] I thought I could see you sleeping in "Tannhäuser"! We once had, as a joke, a Jiddish-German dictionary There was: Zoff = das Ende; ka Zoff = kein Ende; gar ka Zoff = die Götterdämmerung". (Gar kein in German means: not at all)

Yesterday I sent you some books. Please do not take them as very important or typical. They are just two books I like, and you will probably like too, but not the books I like most or something similar. You understand?

These are also some photos. The last ones were not taken in Zurich but one on a trip from Arosa and the other one in the train.

Genia. My dear. Could you be here only one minute!

25.10.

I got many letters of friends in the last time, and by and by I must tell you about all these people. Only there is an awful lot of them and each with a complicated story. Some of these stories will be very strange for you and show you some parts of European life which you would probably not believe to exist. There is for instance Hans Juliusberger, a very good friend of mine in Berlin.[326] He is the son of quite a wealthy director of a factory. He studied engineering and his father wanted him to take the state career which wants some more studying than the usual one. Now he met a girl, fell in love with her and practically always was together with her during all his studies. She is really one of the nicest girls I have ever met, charming, intelligent, beautiful and understands many things. Now, instead of being glad about it, his parents tried to separate these two. Of course, these two did not dream of marrying (formally or actually, you understand) before he had finished his studies, and if this was right or wrong — anyhow it was a terrible situation. The parents wanted him to give her up, because she "disturbed him so that he could not work". He again thought that the continuous interruption of his parents disturbed him. He is not very strong and in the end he failed in his examination. Then she went to Vienna to stop all this talk, he entirely broke with his parents, and worked hard and the second time he passed his examination. Then he succeeded in getting a little position in the office of a newspaper and they married. Now they both work very hard, and live very modestly. (Fortunately they can stay in the house of her parents who are more reasonable than his.) But even now, his parents refuse to speak even one word to them. I admire the ability of these two people, mostly of the young woman, who comes also from

[326]See letter [36].

quite a wealthy family and now at the same time can work the whole day and after all manage the little household with quite a personal style, and always with good humour, although this is not easy, because you know that in Germany now, everybody now must be afraid of losing his job. If this would happen to either of them, it would be terrible. His parents of course could be a security for him, but they seem to be perfectly crazy.

Personally, their situation is one of the best possible. Symmetrical. And if perhaps he is a little too weak, she is very strong. I like these people very much, and you would probably like them too. But if you will ever see them???????

Darling send me your eyes that I may look into them half an hour; send me your hands that I may hold them in mine for half an hour, send me your lips that I may kiss them half an hour; I shall return everything half an hour later, well packed, *recorded delivery and post-paid.* No you better send them not, for I would not return them, I would keep them and you would be uncomfortable in *Leningrad* without eyes, without hands and without lips.

Television is not yet installed between Zurich and *Leningrad* and even if it was, what about the hands and the lips then? So nothing can be done, except writing things which one cannot write. *Good night.*
Yours
Good night, Genia!

Rudi

[72] Genia Kannegiser to Rudolf Peierls

[Leningrad], 28.10.1930

My dear and silly boy, do you really think that I was offended with "fishing"? Certainly I am not and I even forgot what I wanted to tell you in that letter. Two weeks is such a long time and I want to tell you such a lot of things! Do you know this old anecdote about the crocodil which is from head to tail longer than from tail to head? Now I think that it is true because the letter from *Leningrad* to Zürich goes five or even four days and from Zürich to *Leningrad* seven or eight. What is the matter with it? I cannot understand, but it is so. Perhaps you can explain this phenomenon with the principle of relativity?!

Aunt — omnibus is fabelhaft.[327] I related this story to Bronstein, and he was quite charmed with it and especially with the Hurra in the end. He said that this hurra is all Pauli — the "Theorienfresser".[328]

Darling, Zürich is a bit tedious after Berlin, yes? But it is splendid to go out Sundays in the mountains and such a lot of water around. I cannot imagine a beautiful town without water. Do you remember *Tiflis*, a very nice town, especially in the evening, but only one river — *Kura* and it is interesting, it is all that you want, but not beautiful, and *Moscow* also, I do not like it at all, because after *Neva any other river* is nothing at all. And we have such beautiful towns — *Kiev* for instance, on the *Dnepr* and the towns on *Vodin*, but I have not seen the last ones and only in one night from *Leningrad* (7 hours) it is the most charming little town I even see — *Novigrod.* Think only — a town of 12th century on the blue *blue* river with green green meadows around and the churches in the green gardens and big towers and the corners of old walls and around white cloisters with green and gold heads and on the meadows herds (?) of gees quite like snow. I was there this summer with *Ambartsumyan* it was very nice and very jolly but we did not sleep to nights and I fell from the *bench on the tram* when I go in the morning to the laboratoria, so strong I was asleep. It was my birthday. Perhaps we will go in *Novigrod* in the April, it is awfully interesting, as you can see from my description (!!) Rudi, I think you are a year older than I, so your birthday is somewhere in the summer. Yes? *Stravinski* is sometimes very interesting[329] and soon some more than that. Do you heard "*Sacre du Printemps*? It is the morning of the world, first men, first labour, first gods, first love and strong young earth. It is wildly and large and awfully nice. *Ansermé (he is Generalmusikdirektor in Paris)*

[327]See letter [67].

[328]Matvei Bronstein had written a poem about 'Pauli, der Theorienfresser' in the Odessa Pamphlet [54]. The copy of the pamphlet located in the Peierls collection carries a handwritten dedication "Dem furchtbaren Theorienfresser überreicht vom Herausgeber. Leningrad, den 6. Oktober". *Peierls Papers.* Supp., A 140.

[329]See letter [67].

manages Stravinski fabelhaft well. And *Petrushka* is a ballet and there are such places in it, that you cannot not to laugh. Once near me there was sitting an old, old man and he laugh also with all people around. It is so jolly!

If you like "criminale litterature" perhaps you have read "Der Meister des jüngsten Tages" of Chapet[330] — very interesting. But Wallace you are right is tedious, first at once too long, but he has an humouristical laugh of his own affairs and it is nice.

I am very interested would you like Flaubert, or you would not. He is awfully objective and cool, but Mme Bovary is one of the greatest litteral expressions. It is so stupid that all the best books you read before you are twenty years old. And for all the rest of your life you have only a little bit of contemporary writers and books you occasionally have not read before. I am in best position because I know the German literature only very bad, so I have many pleasures in perspective. I like very much Schnitzler, sometimes Wasserman, but he read too much Dostojevsky. Zweig is too sentimental and too *Freudian*, also, Kellermann — tedious and others, I also forget all the names, I don't know so well to have an idea of whole writer.

And if you will learn Russian you shall have such a lot of quite new, quite original books to read, and classical and moderne writers I want to send you so many things. My laboratoria is awfully far from my home, 40 minutes *on the tram*, so I every morning have a long time to read, but in tram I can only read the "stupid" novells, so I do. The last was "a girl from about" of Woodhouse it is funny, very funny.

Rudi, I also cannot, cannot imagine the time after March. It is the "end of the world", but it is such a long time till this end! I think that I used a bottle of ink and now only two months are passed and how many letters wrote with pencil! Four months — 120 days — 24 free days! Awful, awful, awful. Think only — 24 free days!

Darling, I quite forget an very important thing. Are you quite well in the sense of tuberculose? Because March and April are the worst months for weak breasts, and if you have some inclination of that kind,

[330]The novel was actually written by Leo Perutz in 1923.

you can fell ill, I am a little, but am habituated and after mountains our wet spring can be very dangerous. So dear, you must think about it. It is midnight now, and my eyes closes too often. After dinner I go on foot very far (five kilomenters in one end) so I walked nearly two hours and I am very tired. (But splendid evening!) You know, with new photo-100 catastrophes, so I sent you the old ones, on one of them I am awful, but my cat is very nice indeed. It is named "Chimmie". Have you seen him? *Rudi*, I want to know so many things about you? Why it is cold in your room? Where live you? With whom live you? etc. etc. [331]

[73] Heinrich Peierls to Rudolf Peierls

[Berlin], 1.11.1930

Lieber Rudi,
Mit dem Bild, das Du uns mit Deinem Schreiben vom 28. d[iesen] M[onats] geschickt hast,[332] hast Du uns eine grosse Freude bereitet. Wir finden es ausgezeichnet, und man sieht Dir ordentlich an, dass Du Dich recht wohl fühlst.

Deine Mitteilungen interessieren uns gleichfalls sehr. Fahre nur fort, uns recht viel von Deinem Ergehen zu erzählen; es interesseirt uns alles, was Dich betrifft. Evas Adresse ist: Breslau 16, Fischerau 14. Es scheint, dass ich es übersehen habe, Dir das frohe Ereignis mitzuteilen.

Dass Du Dir bei der Schweizer Bank-Gesellschaft ein Sparkassen-Konto eingerichtet hast, ist sehr richtig. Es ist mir sogar sehr lieb, dass Du es nicht auch bei der Kreditanstalt eingerichtet hast, denn nach meiner Erfahrung finden sehr viele Verwechslungen statt, wenn bei einer Bank mehrere Konten des gleichen Namens sind. Mir ist es auch sehr angenehm, dass hierdurch die Entnahmen bei Bickel nicht mehr nötig sind. Am besten wäre es, wenn ich Deinem C[onto] von einem bestimmten Zeitpunkt an regelmässig 100 Frs. monatlich überweisen lasse. Da ich nicht mehr genau weiss, von wann an dieser Betrag fällig ist, so bitte, ich um gelegentliche Aeusserung hierüber.

[331] The end of the letter is missing.
[332] *Peierls Papers*, Supp. A.101.

Unser Besuch bei Juliusberger hat einen gewissen Erfolg gehabt. Vor einigen Tagen sind die Kinder das erste Mal wieder bei ihren Eltern gewesen. Ich habe aber noch keinen Bericht hierüber, wie die Rücksprache ausgefallen ist.[333]

Toni und Werner Milch[334] sind auf wenige Tage hier, weil ein Onkel von ihnen seinen 70. Geburtstag feiert, und wohnen bei uns.
Viele Grüsse

Dein Vater

Lieber Rudi-Rudi! Bild + Brief haben uns s[ehr] gefreut! Eben wirst Du eingerahmt. Dein Prof. Stodola[335] ist ja s[ehr] interessant. Da er Bach-Schwärmer ist, kennt er vielleicht den Erfinder des Bechstein-Moor Flügels, der ein Bachfanatiker ist. Wir lernten ihn, Herrn Moor, kennen. Er hat den Bechstein Flügel verbessert, indem er eine 2. Tastatur über der I. angebracht hat, sodaß das Spiel etw[as] Orgelartiges + Orchestrales bekommt. Seine Frau — eine große Pianistin (Schottin) gastiert mit diesem Flügel mit Bruno Walter zusammen + Euer I. Züricher Musl Dirigent (Namen vergessen) schrieb ihm begeistert über diese Erfindung. Vielleicht fragst Du mal gelegentlich. Weiter viel Spaß! Gruß

Kuss Else

[74] Rudolf Peierls to Genia Kannegiser

[Zurich], 4.11.1930

Meine liebe *Genia*!
Isn't it a terrible method of conversation to which we are condemned? Suppose only we would talk together and only after four questions always one would hear the answer to the first one! It is impossible to

[333] See letter [36].
[334] Relatives from Vienna.
[335] Professor of engineering at the ETH. See letter [71].

remember exactly what one has written two weeks ago. Now, the situation is not so bad as it could be, because you always answer to my letters before you receive them and I who read in your letters every time questions to which I have answered a week ago. We must try and develop this system further and then it will be splendid.

These two months went quicker as I thought. Think only, already 0.3333 ... of the time is passed! What is 0.333 ...? It is a factor, which the order of magnitude 1 and I as a theorist neglect such factors. Or, you can, for instance, take the cubic root $\sqrt[3]{\frac{1}{3}}$ is already 0.69! Of course I know the story with the crocodile (only some weeks ago I read it again in the very nice article of Frenkel an "Am Morgen nach der Schlacht").[336] But you quoted it the wrong way: it is shorter from head to tail, and the simple explanation is that from head to tail it goes downhill. But with Zürich and Leningrad it is the other way round; for Zürich is higher than *Leningrad.* But the reason is perhaps that you always send your letters *recorded delivery* and I do not. I thought the registration would want time, and so recorded letters go slower. But now I will try it. Please write me at what time you get this letter.

Now I am quite filled with splendid harmonics and with computing integrals and developing the most impossible things into power series, only to treat a very simple case of my "change of resistance".[337] This kind of work is more tedious than the other one, thinking about physical explanations or methods: But it is so much more concrete and you know what you have done.

Then I have the lecture, which was not quite as bad as I expected.[338] But what the hearers expected and what they really thought of it I not yet know. But I will know it day after tomorrow, when I count the number of people who come there for the second time.

---

[336] Refers to pamphlet written on Grusija; see items [54–55]. Only the Russian version of Frenkel's contribution could be located and is found in *Peierls Papers*, Supp. A 140.

[337] Peierls was working on a theory of the magnetic change of resistance. This was published in 1931. R. Peierls, 'Zur Theorie der magnetischen Widerstandsänderung', *Ann. Phys.* **10**, 97–110 (1931).

[338] Peierls started lecturing a course on wave mechanics in the autumn of 1930. See [62].

I get on *well.* There is, for instance, no such characterisation of a situation as: "Musik wird störend oft empfunden, weil stets sie mit Geräusch verbunden" or many other things.

You want to know about my life here; there is not much to tell, for my life here is simple and tedious. I live in a beautiful little house with a Swiss family. They are quite nice, but I never see them. My room is about as large as yours and fortunately not the usual "chambre meublie". All the house is made up with a good and original taste. The best of my room is the view from the window — I include the picture — and also the view from the street if I walk to the institute. Then if the weather is clear, one sees the whole chain of the mountains. Now, they are already quite white (I like this view not only for its beauty, but also with respect of the ski-possibilities). Sometimes are clouds above us, so that it is here quite dark, but the mountains are bright and sunny ... impossible to describe that impression. Or for instance, if it is night, clear and cold and silent, and under your feet you have all the lights of the town. (It reminds me a little of *Tiflis.*)[339]

I am at home usually only in the evening, and even then I am often out, but I generally have supper at home (or at least something which I call supper, for I make it myself) and I am bourgeois enough to spend half an hour or even an hour on this occupation, reading my newspaper and so on. From about 9 till 7 o'clock I am in the institute with half an hour or a bit more interruption for lunch. I do not do very much besides. It can happen that I do not speak a word for some days, except to the maid who brings my breakfast and the housekeeper in the institute and perhaps some words with Pauli, that "there is nothing new to-day, physics are as tedious as before, papers in ZS f.Phys. are 'falsch oder trivial' but there is a new film in the cinema". From time to time I go to this quoted cinema, or to the theater, but both are not very interesting now. Then I have the Russian lessons, or I see one of the people in the institute after supper, on Wednesday after our colloquium I have to drink with Pauli and Wentzel. So you see there is nothing exciting. This week we shall have great festivities, as there is the 75. *jubilee* of the

[339] After the Odessa Conference, as part of the group journey, Peierls and the other physicists visited Tiflis. Peierls, *Bird of Passage*, p. 64.

Technical High School (to which I belong) and for two days everybody will think of nothing else. Friday night they make a *torch procession* and Saturday will be great ball in the Hochschule with I think about 2000 people. It might be quite amusing. In the *Stravinski* concert I heard "sacre du printemps" this is probably the same as you mean. It was awfully interesting, but I think it is impossible to understand it first time. I should very much like to hear it again. They played it not too well. The director of the orchestra here is rather tedious — and the whole orchestra was upset by *Stravinski.* Lilly Fenigstein plays also in the orchestra and she told me she also thought *Stravinski* is no music. (You know they are all Swiss people).

There is one more of my collection of friends: Franz Jacobsohn a cousin of mine in Berlin.[340] It is again a story with parents, and you know that the problem parents $\leftrightarrow$ children is still very important and sometimes complicated in Germany. I think in Russia these things are much simpler. The story is this: He was already as a boy interested in all kinds of sciences, botany, zoology, and so on. In his room he had white mouses and birds and frogs and guinea-pigs and so on, so that nobody dared come into his room. And as is usual with such people he was at the same time interested to help other people. Everybody who knows him only superficially would think him a very great egoist. But really he is the contrary. One day he met a very strong woman. She is about 10 years older than he (or even more) and when he met her she had one illegitimate child and expected her second. The father was a student, a very mean subject who did not care for the fact that she had lost her family for him, and left her quite alone without help and only sometimes he came and tried to get money from her. So she was in a terrible situation. Now this Jacobsohn, who was then a school boy, began to take the responsibility for everything, got money for her for the time of the birth (I do not know from where) and organized everything. Then his parents knew about the affair and said that it was bad for him to be together with "such person" and it would spoil his morals and so on. He was a bit sentimental then and they persuaded him to promise

[340]See letter [15].

that he would not see her again. He really did not till once he met her on the street. And then he noticed how helpless she was without him and that she would probably be lost (she already had made an appointment with some man who had spoken to her on the street and this usually means to give up herself entirely) and he also noticed that he liked to see her again, and so he saw quite clearly how dishonest the "morals" of his parents were and, as it is necessary for him to live with his parents he managed to do everything without letting them know. This goes now since about 6 years. She writes now for newspapers and he can help her very much as he has a good taste, and he sometimes writes himself, but he cannot publish it under his name as he is a doctor of medicine now and therefore it is against the "unwritten laws" to write in newspapers. But she has a nice little flat and he is most of the time with her. It is very strange, because she is about 35 now and she looks like 40, and he is in my age. He is a "beautiful organisator". For instance, once in the holidays, he made a journey of about two months through: Switzerland, the North Italian lakes, Venedig, Florence, Rome, Neaple, Sardinia, Corsica, Riviera, South France, Hispania with all great towns like Gibraltar, Tanger, back to South France, and to Paris, and home. All this together with his "wife" and with the money he had saved in a few months. When he came back, though, he had lost about 10kg of his weight. As all medical people, he is very interested in the public institutions, public health, and so on and very critical, so I like to talk to him, and I learned many things from him.

He was also a member of the debating evenings which we had in the first years of my studies in Berlin. We were about 15 people of different professions and discussed about once a week on everything: education, politics, psychoanalysis, and so on. Amongst these were also Hans Juliusburger — I wrote you about him some weeks ago — and Barbara Zarniko, the physicist from Berlin.[341] It was quite interesting, not so much for the results of the discussions, but psychologically. I never knew before so clearly how strongly a man is influenced by his profession, and how clearly one can, if you hear people discuss on any

[341] See Peierls, *Bird of Passage*, p. 12; see also letter [57].

subject, tell at once, who is a student of medicine, who is interested in laws and so on.

I was so glad to get your pictures, especially the large one, where one looks just into your eyes. Since some time I try to remember their colour, but, strange enough, I cannot. So once more I must say you good night, pack a lot of kisses and many more unrespectable things into this letter and let it go on its long journey.
Good night, my darling!
Write soon and write more about you!

Rudi

[75] Genia Kannegiser to Rudolf Peierls

Leningrad, 6.11.1930

My Darling,
I have not replied for a long time because I was in the lab all day long and I had to sort out that nobody would disturb me: guests, concerts, cinema. But my last letter was so awfully long, and my life also is so uninteresting that nothing exceptionally new and interesting did not happened during this five days. Only the concert was very nice. Brahms, I like him very very much. Yesterday I wrote you a letter but all in Russian and so ununderstanding, because I wrote very quickly, that I was afraid that you will not understand a bit so to day I wrote you a 'mixt' one. And you know till now I am habituated to speak with you english I always think english of you and even in dreams our conversations are english. Yesterday was my free day, the worst free days of all that time, think only I was obliged to get up awfully early (for the free days at 10 o. clock and to go in the *public library* to look about a journal. *I was looking for it for two hours, but of course I could not find it. After the public library* I had a rendez vous with *Nina and Mama* at my dressmaker, they were certainly late half an hour (the women always are late, I am the only exception) and the worst of all my dressmaker was not at home, and she lives awfully far and the weather was as bad as possible. And I was hungry (!!!) and wet and angry ... After dinner I lost my chess party with my stepfather, I washed my head and was obliged to stay at home

all evening, because *mother said I'd catch a cold if I went out, and at night the wind opened our window and while Nina and I were arguing whose turn it was to shut it I got so chilled that I think I have a cold anyway.*

Something awful instead of a free day. *Yes*? And I am angry till now because I want a new dress, long dress, all my old ones are very short. *Sorry. I forgot to tell you — I've got a new overcoat. Hooray!!!!* my old one I had 8 years(!!) and I could not to look at it once more. The new one is long I look at least 30 years old in it, and I am awfully glad. Think only 8 years the same coat! It is true that Nina has not yet a new one but she will in some days. So all will be all right. Dear, my life resemble yours very much. Only I live in the family so the dinner is the time when all are together and we speak very much during it, and you have only new-papers, poor boy, and in laboratoria, I speak also all the time I think that I should dead, if I did not speak all the day. Rudi, I don't think that you are a good cook. And in your letter I found lunch and supper (?) but no dinner. Why? *I always marvel at how thin you are in comparison to, say, Sack, not to mention Pauli. It's because you don't eat enough. You must eat lunch, and a lot of it, alright, darling?*

Yes, a pleasant news. I got the translation from english journals for one institut. I am very glad because *it is very well paid, and I'll have some spare cash now. When you come I'll have enough to go to Novgorod and even to Moscow.* and I want to buy many things before — ski for instance and *skates.* It is not sure yet, but very probable. It is true that I will be very occupied because in the evenings I will translate ... but every things has two sides.

I want to know so much your cousin,[342] awfully nice boy, and more then boy a "man". I don't understand is he a husband of this woman, or simply a friend, if the second is true is it ... I know a boy of 22, who married a woman with three children, the eldest is only 4 years younger than he. So it is not strange, I know two stories of that kind. If it is love, it is easy to make many things, but if only friendship, so it is more difficult, 10^3 more.

[342] Franz Jacobsohn. See letter [74].

This is very "european", you know we have not "illegitimate" children and it is very nice. European laws are awfully severe and stupid severe, because all fall on the woman and the man is "'all-right", and without these laws all falls on the woman, all harsh and sad of love and woman is weaker side. All our laws were on the mother and child side and it is true. In all these stories, man is guilty and if you can say that woman know what she do the child is quite the *guilty* that he is born and he will grow up not only without father, but and without parents also, because very often I think the family is not kind for girl-mother, end even laws are contre him. And the laws contre *abortion*? I think that every woman will prefer a child for these operations, if she can have it, and it is too hard to kill young child, and it is too sad, and such laws as yours make only that the women are obliged to be operated in such conditions that not only child is lost for the world but mother also is very often dead of dirty and such things. Stupid! Oh, how many philosophy! *I won't do it again.*

Darling, I've read a wonderful book about Everest, about Mallory and Irwin, their expedition and their sherpas, about 8,500 meters above sea level. Fantastic. Very fantastic.

I want to climb on the snow mountains, will you come in July and we shall go. O.k.? But I am afraid that it will be too difficult for me, because I have a bad heart, but perhaps I can, I will try. Darling, I wrote you in laboratoria, and it is time to go home, perhaps today I will begin translation so I must have all the evening for that. So good bye.

Rudi, when I answer your answer to this letter, it will be half way to March 1. Good bye. It's so far to stretch my hands and lips, but I am still kissing you.

Genia

My regards to Barbara Zarniko. I think that she knows me, for you and your picture. I was very glad to see the things that you see every day. Thanks for photo. *You take wonderful pictures.*

[76] V. Fock to Rudolf Peierls

[Leningrad], 8.11.1930

Lieber Herr Peierls,
Ihren freundlichen Brief vom 22. Oktober habe ich erhalten, und bin damit einverstanden, daß es gut wäre, wenn Landau seine Überlegungen ganz kurz in unsere Arbeit einfügen könnte.

Ich habe über die physikalische Bedeutung der nicht-Vollständigkeit des Funktionensystems nachgedacht und habe meine Überlegungen in die Arbeit eingeschaltet. Diese zerfallen in zwei Abschnitte: 1) Die Folgerung, daß es Zustände geben würde, für welche die Wahrscheinlichkeit jedes denkbaren Meßresultats verschwindet. 2) Über die Art der Singularitäten eines vollständigen Funktionensystems. Den Abschnitt 1) halte ich für besonders wesentlich, und möchte ihn unbedingt in die Arbeit einschalten.

Außerdem habe ich noch einige wesentliche Änderungen und Kürzungen im Text vorgenommen, und hoffe, daß Sie und Landau damit einverstanden sind. Das Manuskript der Arbeit sende ich mit diesem Briefe.

Falls Landau wirklich nach Zürich kommt, wird das hoffentlich zur Beschleunigung des Erscheinens der Arbeit beitragen. Sie könnten dann mit ihm die Arbeit endgültig abfassen und mir 1 Exemplar davon senden. Wahrscheinlich werde ich mit Ihrem Text einverstanden sein. Jedenfalls werde ich Ihnen sofort antworten, und falls keine Änderungen mehr nötig sind, das Manuskript direckt an Scheel[343] senden, so daß unser schwieriges Unternehmen nunmehr zu einem glücklichen Ende kommen wird.

Very good!

Ihre Aufnahmen, die Sie an Schubnikow[344] nach Charkow geschickt haben, habe ich ihm übergeben (er ist immer noch hier in Leningrad). Tamm ist durch seine Kugelsymmetrie ganz entzückt.

---

[343]Karl Scheel, (1866–1936), editor of *Zeitschrift für Physik*, president of the Deutsche Physikalische Gesellschaft, co-editor of the Handbuch der Physik.

[344]A.V.Shubnikov (1887–1970), Russian crystallographer and first director of the Russian Institute of Crystallography in 1944.

Wenn Landau in Zürich ist, grüßen Sie ihn, und bitten Sie ihn, daß er mir schreibt. Auch möchte ich Pauli und allen Kollegen einen Gruß übergeben.
Ihr herzlich ergebener

V. Fock

[77] Rudolf Peierls to Genia Kannegiser

[Zurich], 16.11.[19]30

My dear,
It seems that our system of correspondence works very well. Yesterday I said to our housekeeper: "At 12 o'clock to-day, please bring me a letter from Leningrad", and he really did. And when I opened it, the beginning (your "theorem" about free time) was nearly a literal translation of what I said in my last letter!

I very often think of the first days in *Odessa*: No you were not at all "exotic" to me, on the contrary, there was something like the feeling I had known you long before. Everything was so "selbstverständlich" and so I did not philosophise much about you in the first days. You simply were there and I liked you to be there and did not bother to analyze anything. Had I done it, I probably had thought you a bit superficial and it was only later that I saw you better and then I was astonished how it could be possible to be so serious and know so much and understand so much and at the same time to be so easy and bright. I never dreamt of comparing you to any other girl I have seen but I do not think that I know any other like you (although I know many other "uneuropean" girls!) It is curious that you are at the same time typically Jewish (and therefore familiar to me) and typically Russian (and therefore strange). But all this is only the frame, which of course is unseparable from you. When you speak or laugh you are sometimes like an exploding firework, only the important difference is that there are so many things left when the firework is burnt out. May-be I had a wrong impression of you in the first days but then (and also from the letters) I know you better so that it would make no difference when you are more dressed and less sun-burnt, less tired and less in a hurry. There are some pictures of Sack where one only sees one arm of you or one leg or only a curl of your hair, and this is entirely sufficient. I think if I would meet you on the street with a Mohammedanian costume, where one only sees your eyes, I should be sure to recognize you. (If of course I would look around me at all; it may happen that I go on the street without seeing anybody and afterwards I shall tell you a funny story about it). In my memory, however, it belongs to you to be in a hurry a little for it suits you so well. I shall never forget the whirlwind you were like in those ten minutes on *Mineralnaja* station before your train left.

Darling, could you think I would have an adventure without telling you? It is not impossible still, that I shall have one, but as yet there is no danger. In Zurich there are only two girls I know more intimately. One is Lilly Fenigstein, I told you about my relation to her, and then *Fania Moskovskaja*[345] with whom I shall certainly be very good friend, but no more (or, if you want, no less!) She really is very clever and it is easy to talk with her. A week ago I had a discussion with her about adventures (theoretically) which may perhaps interest you. There is a superstition that if your cigarette burns only on one side, it means your husband will betray you. She told me that and added that her husband would never do that — as long as he loved her. "Don't you think it is possible to betray a person one loves?" I asked. "Of course it is possible but he simply would not do it. I, for instance, think I would, or at least I could do it." I said this made no difference. "No it makes a lot of difference whether you really do it or not, but you are right that the only interesting question is, could you do such thing." She is right in this point. As for myself, I do not know yet, whether I could, or not. But for the present, there is no danger in this respect and anyhow, be sure that I would tell you. Else, it would not be "clean".

I am very glad you like Hemingway, I think you found the same things in this book as I loved. First time I read this in a newspaper and I was quite thrilled by it, although this is a barbaric way of reading, if you are obliged to read two pages every day!

[345]Rudolf Peierls' Russian teacher.

As for Wildgans, it is long ago that I read him and it may be poss[ible] that many things and observations I found in this book and which then were quite new to me then, would seem trivial to me also now. Still I remember the language very clearly and I like this even now. As for the philosophy, I do not read plays for philosophy. But I noticed that we have a different way of reading books: You always tell whether the persons are sympathetic to you and I can only feel sympathy or antipathy for the writer. Perhaps I take books more "academically" than you.

It seems also that we use the word "character" in a different sense. If somebody is always in good humour, and if he is careful not to displease other people, it does not mean that he has a good character. The first only means that he has a good temper and the second that the opinion of other people is important to him. But it does not matter, because you also have a very good character in my sense of the word and I am worse in your sense than you believe me. Do not forget that I also was not in my "normal state" that time, and I can also be very boring to other people, very tedious or so cool that people who start making jokes, become earnest and the atmosphere "icy"; I can have very bad tempers. But why tell you, you will soon enough notice yourself.

I have to go now; I promised Waller to have lunch with him. Therefore good bye, my dear!

— Last night I heard a concert of Bach-quartett. 3 pieces of Beethoven: *healthy!* They played very well and I love Beethoven. Friday I was, as I told you, invited by my landlady. *That was very boring!* These people are quite interesting, as they have seen very much, been in all parts of the world, India, Japan, China, America, also in Java, Egypt, and I don't know where. But they understand to talk about this so tediously, to emphasize with so long phrases the most unimportant things and to take themselves so important, with one word: Swiss people!

The new students' house is really splendid. I eat out now every day and it is very nice, one can meet reasonable people.

My lectures are going on; this week I put them questions for exercise and when they give the solution I can see whether they understand anything or not. I am still reading *Turgeniev* and still the same story

I began with. It goes very slowly because I want to learn words and write down every new word and learn it. So very soon will be a holiday, when I learn the 1000th word.

This week I will try to get M[adame] Bovary from the library, and I shall also look for a new book by Hemingway. Yes, Landau on skies is a funny picture and I remember how once he was in some place and could neither forward nor backward, because he declared he would fall down (it was a rather steep place). Then Thorner became very angry and called him a coward and then with little, little steps he managed to get through. Thorner was so rough that I was afraid Landau would be cross, but evidently he was not. I am sure that we will go skiing together in *Leningrad*. Tell me, do you believe it possible to hire a pair of skies there? It would be terribly complicated to bring mine from here.

I promised to tell you a funny story, here it is:

Some days ago I went on the street with a chemist, who speaks very well Russian, and we talked Russian together. By this I was so absorbed that I did not see Lilly Fenigstein coming behind us with a friend and speaking very loud. She run against me several times and finally gave me such a push to my bag that I nearly lost it. Then they walked immediately before us for a certain time; all that was so obvious that my companion noticed and laughed very much, but I did not feel that anything happened and only knew it some days later when she told me! So at least in this respect I am fit to be a professor.

I think the position of a school master in Germany is much worse than you imagine. Not only tedious, but you have a hard struggle against your colleagues and against the system if you want to do anything reasonable at all. The girl, for instance, I told you about was blamed by an old teacher, because "she occupied herself with the children too much after lessons" (one ought to appreciate this as a sign of interest) and this "spoilt her authority". Then she has not only 8 hours but a lot of other work and a long way to her school, she is supposed to hear lectures and so on. Of course there is some time left, but she wants very much to do mathematical work, and for this she is too tired. Of course you are right one ought to be happy in spite of all this and I know people who have much less money and much less free time and still are always happy and always find time to do what they want. But one can understand her a bit, can't you?

This week I scarcely got further in my calculations, but at least now I see "how to do them"; they first were so complicated that I had no hope to get through without errors, but I could simplify them a little. Even now it is terribly complicated, with numerical coefficients large like telephone numbers. In German one has a beautiful expression for this occupation: "ixen", and one might try to translate it by "to ex".

Again I must leave to have tea with *Fania Moskovskaja* (this letter is a good illustration of your "general theorem"'). So good-bye till to-night or to-morrow)!

Rudi

[78] Genia Kannegiser to Rudolf Peierls

Leningrad, 17.11.1930

Darling,
if you knew only how anxious I was these days, think two weeks without any word from you and I know that you are too "european" for left a letter without answer. I thought that you fell from some mountain and broke your leg like Bloch, or even worse that this was your head broken, or that you are ill and are lying alone in Zürich, or that something happened with your *family* in Berlin etc. etc. The "best" things were that you fell in love with some girl on that ball and quite forgot all other world and that you have left the letter in your pocket, or forgot to put the number of the house in the address, the last solution I think was the simplest and right as I can see from your letter of 16 that nothing extraordinary has happened with you. *Rudi, my darling, darling, I felt so sorry for you, when I thought of you lying there all alone. with nobody to care for you and look after you.* Now all is all right. So and I have not write to you a lot of time, but really I could not because I thought that you can be in Berlin or in love.(!)

Now it is winter, darling, Neva is covered with ice and the streets also are and all the people fall down; every moment somebody is lying under your feet. Sometimes it is yourself. I think that in some days we shall go on the ski in the first time this year. *Ice rinks opened yesterday, I will start skating. I've never tried before, and it is most likely to be quite a funny scene — my vain attempts to stay vertical. Among the audience there will be photo cameras and first aid materials.* These week I was twice in the theatre, now we have the *Moscow theatre.* In *Moscow* they are the best theatre of the *USSR* and perhaps of the world also. Have you heard of *Goset* (the state Jewish theatre) it was *very good, terribly good.* Think, the little Jewish *village* little village) in Prussia before the Revolution an old half mad Jew think himself the king of Jewish people and goes with his friend to search of Jerusalem. He thinks that he is very fare, but they walked all the time around their native place and have all sorts of adventures. It is all. But they played that *very well* (!!!) and the decorations, and the musik, it was the most touching spectacle that I see. It was very difficult to understand the words, because they speak Jew and I understand only 0.1 of them, but it was not necessary. You know the first time on my life I feel how deep Jew I am. I was so sorry that you have not see that, darling. I was also in the *Kamerini Theatre.* It was very jolly — an operette *[?]* I laughed there awfully, but it is a little rough in some places. But really I was quite charmed with one of the actors, he danced charlestown — fabelhaft, more than fabelhaft. In the second act there fly the birds of paper, (you know how children in schools make them always) in the interact Bronstein and *Ambartsumyan* went to avant-scene and brought a lot of them and we throw them during all the third act into the public it was very very funny, we had many acquainted in the theatre. Yesterday I was in *Pavlovsk* with Nina and one of my cousin, who is here in the first time, and I show him all of our "beauties", it was very nice weather, and very beautiful parks, and castles. We will go there in March. About "adventures" — I think that man never knows what he could do, or could not. You think that you could do anything and in reality you don't make it, and in the contrary you make the things that you think quite impossible to do. You are too "theoretical", darling, really. *Yes*? For me, I think that I could not betray a person. I love, simply because one person is quite sufficient for me, but it is a pure theory, it is possible that I would do it in some conditions. You must never tell "never". *Yes*? I don't know why you think that you are lazy. Dear, you work *very hard.*

It is awfully late, darling, and I am quite asleep. Good night, once more. How many good nights I will write you till I can say it to you?

You can write wonderfully well in Russian, only with the tenses of the verbs you're in a mess, terribly "european" future and imperfectum you translate always to have or haben, it is not necessary at all.

My darling absent-minded professor. You know once I set off for the university not wearing a skirt. And Landau when asked what colour were his socks, could not give an answer. But not to notice being attracted by a young lady — is truly remarkable, Rudi, I want to see you.

I want to post this letter tomorrow morning and because it is so short. *Good night.*

Genia

[79] Rudolf Peierls to Genia Peierls

Zürich, ! 30.11.[19]30 !
half time

My dear,
Do you know that today is half time? Three more months, still a long time, although February only has 28 days. The last months were so long, and if I remember today everything that has happened, then it seems to me as though it has been more than a year ago. What will it be like in March, what will I be like, what will you be like? One cannot imagine it. I am a great optimist, I think everything will be fine. I am glad that we have the letters. This way there was always a Genia close to me. But was it the same Genia who was in Odessa and whom I shall see again? The same Genia whom I am longing to see (and not only see, as you know, also kiss and caress, talk to and listen to, feel her hand in my hair, and more, more!) I am believing in it, and I am optimistic again. I think, my dear, that since I have met you I have changed into a different person. It is dreary to wait, but I know that you exist, even in the moments when I do not think about you, and that gives me great and deep joy. I am longing for your next letter, the last one was nine days ago. But this is a just punishment: my last letter was also very

late and you will probably also wait for me. We always act in the same way: It seems that this is a secret connection between us (cosmic rays maybe? You must know).

Dau wrote that he will be here next week. He is very lazy and he only answered after I had written him three letters, and only a postcard on which there was nothing but a big "?". I had to know whether he would be renting, as I was supposed to book a room. Now everything is o.k. Landau, Thorner and I will be in Arosa, in the same house where we stayed last year.

Yesterday I read Madame Bovary. I could not stop and read the whole book without interruption. Very good.

He knows a lot about love, he knows a lot about men. What he knows is not always pleasant, but true.

Only Homais is too schematic; you can meet him in may books and when he appears for the first time in a book you know everything he is going to do.

Next week I will not have any Russian lessons. Fania Moskovskaja will be in London to meet a friend. I envy her a lot!

I wrote a letter to Barbara Zarniko and passed on your regards. She has already replied and also passed regards on to you. Tomorrow her friend will be here, a very clever English girl, whom I have already met in Berlin. Today is Saturday and therefore I do not have time to finish this letter; in the afternoon "Dancing" in the students' house and in the evening I am invited by a Yugoslav chemist.

Good bye my dear!

I returned very early from the dancing, as it was *terribly boring* and now I had guest in my room, and even a female, although my landlady explicitly and strictly forbids ladies' visits. When I came home to-day I heard somebody cry heartbreakingly and found the little niece of my landlady, who was quite alone at home, the servant having promised to be back by 5 o'clock and being late. And the little girl was awfully afraid in the dark house. So I took her in my room and had a charming tête-à-tête. (Don't be afraid of an "adventure", she is only 5 years old.)

1.12.

Dearest my poor little girl,
Did you really believe I would not write to you anymore if I fell in love with somebody else or if I had to make a journey? And dangers are not so great here, you know, even if I would go to the mountains, I would never go alone, and also one cannot get lost in the mountains. Of course I wrote you a long letter, just on the morning after the ball and posted it *recorded delivery* with correct address in German and Russian. It is very silly that it got lost. If once more you will get no answer, and will be afraid that something happened, you will simply send me a telegram and within one or two days you can have the confirmation that I am still alive. (In this case, please telegraph to Hadlaubstrasse 51, my home address, as a telegram in the institute always causes great sensation).

I tried very hard to remember the content of the letter and fortunately I have your letter to which it was the answer. *But it is so difficult to write everything again!* I spoke for instance of my friend Juliusberger, you remember and quite agreed with you that the main difficulty lies in him sacrificing himself for her and that he will probably not understand this but he is very young, and also this will be her duty, she will entirely lead him in these things and I think she will do it well. My parents, by the way, spoke to his with the effect that there will be another meeting between parents and children, the result of which I do not know yet.[346]

I also spoke of your "melancholy" friend. How did you get on with him in the meantime? It is always so sad, if one sees somebody suffering and cannot help him. The only thing he can do is waiting. There will be people who love him, but the only question is will he have the patience to wait for it?

Thorner, I also told you, is no boy, he is 3 or 4 years older than I, and not only in years but in mind too. He has made a lot of things already, took his doctor degree in psychology and in medicine, and he knows much about men. When he speaks to people only some minutes, he quickly knows where are there difficulties and what are their weak points, a quality very important for a doctor.

[346]See letter [71].

And I also told you, how glad was I to know I had a better character than Landau and more "sense for nonsense" than Ramsauer.[347]

Then there was a report of this ball, which was in spite of its gigantic dimensions, very nice and on which I was, as you probably guessed from my next letter with *Fania Moskovskaja*, my Russian teacher.

The rest of my wise remarks is probably lost forever. Maybe I am theoretical in many things, I admit, but I think there are some things which I know not only theoretically and one of this is, what would I be able to do. Formerly I had wrong views in this point; for instance I would never have believed before that I could act as I did with this girl from Leipzig[348] — you remember? although now I am not sorry for it at all. But if you know yourself a little bit, and mainly know why do you do things and why not, (The real reasons never are the simple ones.) you can know what you are able to do. Which of course does not mean that you will do it. Last night I talked a lot of philosophy with that chemist; he is a very nice and intelligent man, only too weak and he has a terrible wife: *stupid* and "grande dame", a most frightening combination.

To-morrow, Landau will come and I am very glad, I have so many things to speak with him that probably we will have to talk 48 hours without interruption. You in the university without gown must be fabelhaft. Exactly the same things once made a friend of my sister, only it was not in the university but in a concert. And she was brave enough to put on her overcoat and stay at the concert! The same girl once came home by underground and instead of giving her ticket to the man at the *counter* she gave him her key of the house. The man had sense for houmour as every such man in Berlin and said: "The key is of no use for me, if you do not give me your address!"

How are your measurements going?

Your Rudi

---

[347]See letter [70].

[348]This refers to Marga Vogt, the lady with whom Peierls exchanged letters in the spring and summer of 1929. See letter [26], note 141. Correspondence in *Peierls Papers*, Supp. A32.

[80] Werner Heisenberg to Rudolf Peierls

Leipzig, 5.12.1930

Lieber Peierls!
Vielen Dank für Ihren Brief![349] Ich verstehe immer noch nicht, weshalb man Ihre Nebenbedingungen noch explizit braucht. Wenn man alle vier Variable $\mathfrak{E}_{(1)}\mathfrak{E}_{(2)}\mathfrak{H}_{(1)}\mathfrak{H}_{(2)}$ in the Hamiltonfunktion und den Impuls einführt, so folgen aus den Vertauschungsrelationen die Wellengleichungen für die $\mathfrak{E}_1$ u[nd] s[o] w[eiter]. Diese Wellengleichungen enthalten allerdings automatisch den Operator $\frac{1}{\sqrt{\Delta}}$, das scheint mir aber unwesentlich. Die Gleichungen sagen nun ziemlich dasselbe aus, wie Ihre Nebenbedingungen. D[as] h[eißt] man bekommt etwa:

$$\dot{\mathfrak{E}}_1 = \frac{1}{2}(\dot{\mathfrak{E}} + \frac{rot}{\sqrt{\Delta}}\dot{\mathfrak{G}});$$

ebenso aus dem Impuls

$$\frac{\partial}{\partial x_i}\mathfrak{E}_1 = \frac{\partial}{\partial x_i}[\frac{1}{2}(\mathfrak{E} + \frac{rot}{\sqrt{\Delta}}\mathfrak{H})].$$

Aus den Wellengleichungen folgen also Ihre Nebenbedingungen wenigstens bis auf eine (räumliche und zeitliche) Konstante. Ich sehe also nicht ein, wozu man noch explizit eine Nebenbedingung braucht.

Mit dem Faktor 2 in den Vertauschungsrelationen können Sie natürlich sehr leicht recht haben. Dagegen seh' ich wieder nicht ein, weshalb im Falle $div\mathfrak{E} \neq 0$ irgendetwas geändert wird. Die Vertauschungs-Relationen zwischen den $\mathfrak{E}^{(1,2)}$ und $\mathfrak{H}^{(1,2)}$ bleiben doch ungeändert, die zwischen $\mathfrak{A}$ und den $\mathfrak{E}$ ebenfalls. Alles andere folgt wieder aus den Vertauschungs-Relationen. Der longitudinale Teil von $\mathfrak{E}$ wird sich dabei in gleichen Hälften auf $\mathfrak{E}^{(1)}$ und $\mathfrak{E}^{(2)}$ verteilen.

---

[349] The preceding letter from Peierls to Heisenberg does not survive. But undoubtedly it referred to Peierls' and Landau's attempts to extend Heisenberg's uncertainty principle to relativistic quantum theory.

Ich habe also deutlich den Eindruck, daß Sie sich das Leben unnötig schwer machen.
Viele herzliche Grüße

Ihr W. Heisenberg

[81] Genia Kannegiser to Rudolf Peierls

Leningrad, 6.12.1930

Darling, I have in my mouth a very nice — "fabelhaft" bonbon. So I think my letter will be very sweet. I am not angry at all. Seriously, if you want to go to sleep, or to 'ex' *or something else like that you can answer carelessly. I won't get hurt.* Rudi, for me it is very difficult to write, when I want to speak, and now I want to speak (it would not be very distinctly of course, bonbon! but ...)

Dear, you have seen L'Aiglon, do you like Rostand?[350] I do not very much; L'Aiglon was the night-mare of my youth. When I was thirteen we (with Nina) had French lessons with an old, angry, etc. "madame" as we read l'Aiglon on two voices, it was "fabelhaft" how you can imagine! The monologues seemed to me without end, and I read only every fourth line, it was so tedious, and so "unending" that even Madame had not bemerken it. Nina had another "method" she read every fourth page. And even with all this we read it 2 months. I prefer the comedies: "Les Romanesques", "Cirano", but not "Chantequelais" it is right that french language is made for hens, the que, qui, q'est, quand etc. but it is too much if you have 140 pages of clucking, *Yes?*

Certainly, I read "L'Ame enchantée", but only one half of it, because I fall asleep in the middle of the second volume.[351] Romain Roland is the germaniest of all french writers, as Schnitzler is the most french of all german ones. Romain is German in the bad sense of the word, he

---

[350]Refers to Edmond Rostand (1869–1918) and his tragedy L'Aignlon.

[351]Romais Rolland (1866–1944), French writer and nobel laureate, wrote *L'Âme enchantée*, a work in seven volumes between 1922 and 1933.

is caricature even French is strange to you, and to me at contrary very familiar sometimes; I love Anatole France[352] and Regnier.[353]

7.12.

On that place came Ivanenko and his wife and one other boy, so I was obliged to stop till to day. Jules Romain[354] and Mourroi.[355] I love "l'ironique" and "la clarité française", but "pathetique" brrrrr. (Hugo, for instance, or "l'Aiglon").

Every man, if he is even a little interesting, have many "faces". I have three groups of them. First — to everybody — you know it very well certainly, secondly — only for very few persons, I think that I show you all, that I could in such a short time, of this second face, and the third nobody knows, even I do not know it till the end, if I shall write about it. I think you will understand nothing, it is very difficult even to speak about such things, and to write ....

In two words — I think I am awfully "girl" (perhaps even "boy"!) to everybody, and in the middle I am awfully "woman". You understand? But "faces" are really "faces", I do not "play" simply I am so, and it is all. *You must be fed up with all this humanistic philosophy, poor boy. I will try not to philosophise anymore.*

Darling, all this week it rains, rains, rains, all the snow become dirty and water, you cannot even think about ski, and it seems to me that it is still October, and not December and that it is till the 1 March five months and not 3, even less — to day is the 7/XII. I write you in laboratory and I think that it is late now, all the people goes home now, and I have not the watch, but I am very hungry and my appetite is the best time choose to me.

The days run by, as I do not [?] what but awfully quick, I am awfully occupied, so since "general theorem" I do nothing, all the day to day I run around the house and I think that I jumped so many stairs that

[352]Anatole France (1844–1924), French novelist and nobel laureate.

[353]Henri de Régnier (1864–1936), French poet best known for his symbolist poetry.

[354]Jules Romains (1885–1972), French writer and philospher who became known as an exponent of unanimism.

[355]Reference unclear.

the total height is Youngfrau[356] minimum, somewhere I lost my *purse* with all my money (not very much 1 *rouble*), but I must go in the tram and sent this letter to you, all the people are at home now, so it is the question "where I can get this 50m, oh I know I will take them from the "portier", (Hourrrra!) and the key from my laboratory chamber door. So you can see what a wind is in my head.

Darling, these days I was in the theatre and saw a comedie, one of the heros was quite Houtermans, all, all — hairs, eyes, costume, but all. It was very amusing, because he played a role very *unsuitable* to Houtermans, I laughed awfully. To-morrow I will go to the ballet and one of the nearest days in cinema to see a new film, on [?] very interesting. Something in the kind of [?], but it is Russia the far West and North. I like very much films with animals. And you?

I am hungry(!!!) but it is not so bad to be hungry, when you think about dinner. You know, never in my life I have forgot to eat, especially I like the sweet things, do you remember [?] in *Odessa* and [?] I eat them only when I was quite little girl — six years old, because after the war we have not them in *Leningrad*. I remember them but not very well. Bring. *yes?*

Soon it will be New Year. [???] Or the records in the skiing? Or something else? I will think about this subject, dear. I am awfully curious about Pauli's idea, if it is the seconde "Verbot" it is *"cheers"!*

Rudi, are you curious? I am — "nee plus ulta" You know that till now I think about this box in ... (do you remember?) and that I become quite ill when I think about "Masque de Fer" (you know this story) and the end of the last Dickens novel[357] which he did not finish, it is awful that exists thing that nobody knows and never will know. I do not know why but I think that to day or to morrow I will get your letter.

In my last free day my cousin made of me and Nina 7(!) pictures 5 (!) of which were spoiled for different reasons. Next free day we will try another series and I will sent you me in new coat perhaps in new dress. I will have a new dress long, long till my end.

[356]Refers to a mountain in the Alps called Jungfrau.

[357]Refers to Charles Dickens (1812–1870), *the Mystery of Edwin Drood*, his final novel which was left unfinished at the time of his death in 1870.

Good bye my dear.

Once more, if you are not in the state of writing letters don't write them. I will not be angry, at all. Write when you want without thinking about answers. *Ruuuudi* I want to see you. Good bye! Very very "yours".

Genia

[82] Rudolf Peierls to Genia Kannegiser

On the train: Zürich — Tübingen, 6.12.[19]30

My dear Genia,

Today and tomorrow there is a physics meeting, where Lise Meitner,[358] *Geiger*[359] *and many others will be. I am supposed to go there and talk to them. Pauli wants to know something, and cannot go himself, because there is a big ball of the Italian students and he will have to be there.*[360]

Well, I am going with pleasure, I love traveling. Ewald will also be there, a very nice man, and Ruhemann, a friend of Barbara Zarniko.[361]

The only unpleasant thing was that the train left in the middle of the night! I had to get up at $6\frac{1}{2}$ *in the morning. And this was particularly unpleasant today. Last night I was in the "Kanne" with Pauli and Dau*[362] *and we drank very much.*

The "Kanne" is a very funny place, perhaps the most "uneuropean" place in Zurich, a little business bar with very good drinks and with a

---

[358]Lise Meitner (1878–1968), Austrian-Swedish physicist, close collaborator of Otto Hahn, with whom she studied radioactivity. In 1918, they discovered the element protactinium, and in the late 1930s their work would provide the evidence for nuclear fission.

[359]Hans Geiger, 1882–1945, inventor of the Geiger counter, an instrument to detect and count alpha particle, later refined by him and Walter Müller.

[360]The Tübingen meeting is the now famous gathering at which Wolfgang Pauli, in his open letter to his colleagues, addressed to 'Dear Radioactive Ladies and Gentlemen!', announced his neutrino proposition. The letter is printed in Pauli, *Wissenschaftlicher Briefwechsel*, II, pp. 39–40; an English version is published in Enz, *No Time to be Brief*, p. 215.

[361]Martin Ruhemann, experimental physicist from Berlin, later to marry Barbara Zarniko.

[362]Nickname of Lev Landau.

landlord who every night is drunk and then gives long speeches in mixt German, Schweizerdytsch, and English, very good philosophy mixed with indecent jokes. Therefore he does not allow ladies to come to his bar except the barmaid who is used to his speeches. He generally does not allow everybody to come to his "place" and if anybody displeases him, he is thrown out by manual force. Pauli is a very great friend of his and it is very funny to watch him in this milieu.

Now there is some life in the institute, ardent discussions about everything and if you would come near my room you would hear a dreadful noise, because at any time of the day there will be some people of different opinion trying to convince each other. All this *of course* is due to Landau. Of course I did not do any work this week (About your idea that I work like two horses, it might be said that I do, but like two horses who pull on the same car in different directions.) Excuse my very bad handwriting, I now changed my train and am in a dreadfully shaky car.

It is now 12 o'clock. This is the time when the letters from Leningrad arrive. Perhaps your letter is now in the institute, and I will only get to read it the day after tomorrow. Impatient! Next week, there will be a Japanese theatre in Zurich, and also a Rachmaninow concert. Both will be good.

Between 24.12. and 5.1. I will be in Arosa. Therefore you will have to write to Pension Alpenblick, Inner-Arosa (Kt. Graubünden), Switzerland after the 19.12. I am very glad, because it has been very warm, but now it has snowed a lot.

In the evening

The meeting is very interesting and many clever people are here.

Geiger is splendid and the most clever organisator. I have never seen a meeting where every lecture started and finished at the time it was supposed to etc. And then they also showed us the institute: people were divided into groups, each group got a special guide, and in every room was the man who worked there and showed his apparatus, which was illustrated by schematical pictures on the walls; every important part of the apparatus could be projected on a screen by a little lamp, and so on. All the experiments were with the Geiger-Müller Zählrohr, which they use here for almost everything. Regener[363] also gave a lecture and told how he measured the cosmic rays in the depth of the Bodensee with such a counter. It is really very nice to hear single α-particles or single particles of the cosmic radiation give a noise in a loudspeaker and we just discussed that there would [be] no principal difficulty against making an α-particle fire a gun or any such thing.

7.12.

Now I have spoken with Lise Meitner and Geiger and my errand is fulfilled.[364] But it seems that Pauli's idea was wrong and not in agreement with experiments. My train to Zurich only goes in the evening, so I sit in a little cafe in this funny little town, whose population by a great part consists of students, and what students! I hope I can find Gerlach in the afternoon,[365] and talk with him about ferromagnetism. Then there will be much time also to go on with my *Turgeniev* with the story I have not yet finished.

To-day's session was interesting again. Regener is a very nice fellow but Kohlhörster awfully stupid.

Such a German little town is very strange. The houses, the streets, the surroundings are very nice and "lovely". But the atmosphere! It is a dreadful thought that one perhaps would have to live in such a town. Everybody knows everybody else, the people you meet on the street are either bourgeois of worst kind or students in "couleur", i.e. coloured caps and with a behaviour that can only be explained by the assumption that only a man who has studied at the university and especially one who is a member of a "corporation" drinks much beer and has cuts in his face can be regarded as a human being. Now also

---

[363] Erich Regener, professor of experimental physics at Stuttgart.

[364] Pauli's letter to the meeting had been addressed to Lise Meitner and Hans Geiger, and Pauli had instructed Peierls to obtain feedback from these two scientists on his ideas. See note 360.

[365] Walter Gerlach (1889–1979), studied physics in Tübingen where he received his doctorate in 1912. After working in Frankfurt and Tübingen he accepted a chair Munich in 1929, where, with some interruptions, he remained until his retirement in 1957. During the war Gerlach became "Beauftragter des Reichsmarschalls für Kernphysik" and from 1944 he was responsible for the German uranium project.

the influence of the stupid national socialists is very great. It is difficult to choose between the average Swiss and these people, these being — with Landau's terminology — aggressively tedious, and the Swiss only harmlessly tedious. Fortunately one meets a lot of reasonable people in Switzerland and this would be very difficult here. But — terrible thought! — the greatest part of German professors lives like that!

Terrible. I talked to Gerlach much too long and missed my train. Now I have to wait here for one night. Well, this is not too bad, I am very tired and so I can sleep very much. I would like to post the letter here, because you will get it earlier. I kiss you,

Your Rudi

[83] Genia Kannegiser to Rudolf Peierls

[Leningrad], 11.12.1930

Darling, darling, darling.
It is evening now, the winter is here again the wind is blowing through our balcony. High winter sky is looking in our windows, with bright stars. I have just washed my hair and now I'm drying it near the air heater, some hairs fall on my eyes and prevent me from writing. I do so much want to see you that tears are welling up and make writing difficult, too. It is terribly hard to wait when you can't count days, as they are too many, but you have to count weeks or, rather, five-day weeks, because the days-off are like milestones on the road from September to March. There remain 16 days-off, just as many letter and lots, lots of dreams and evenings when I lie down and speak to you. One of these days I had a dream as if you were sitting on our wardrobe (!) and I was feeding you with buttered bread and honey. It was terribly awkward for me to reach up with buttered bread; honey flowed down on my fingers, they get sticky and sweet, sweet. It sometimes seems to me that I have known you not only for 3 months, but for many, many years. So many things that I have told you about, so many places that I have shown you.

Straight away you seemed to me an old acquaintance. Sack, for example, was like an inhabitant of Mars for me. He was so much different from anybody I had known before; I had to explain to him everything and still he didn't much understand.

Rudi, do you remember we were in Vladikavkaz and you asked me, "Do you want me to stay in Kislovodsk, too?" and I answered, "I do want and I don't want: it will be too sad to part. Now that I haven't got used to you it wouldn't be so difficult, but later..." It proved to be even more difficult than I imagined — 1000 times more difficult. But I'm glad, more than glad now that you went to Kislovodsk.

I can't help crying, but I'm very happy. Everything will be all right when you come, you only do that as soon as you can. If only they could invent the "Time Machine" (remember Well's book!): I'd press the button and we would have March.

My hair is almost dry, my eyes are almost shut, and I almost see you in my dream. Good night, and what are you doing now? In your sleep you wave your hands and laugh — I was afraid you'd fall down from the third berth in that train between Beslan and Mineralnye Vody.

Good night. I love you so much and won't see you for so long.

12.12.1931

Darling — pencil because I am sitting on this awfully tedious lecture, if I will write you this letter the time will not be lost. Yes? To day is cold ... I look into window, the sky is rose and red — the sunset, and *Neva* begins to freeze the first lights on the other side and the ships on the river — very beautiful. I want to go home on foot and I must hear the things in which I am not interested at all. Stupid! I am very glad that you like "Mme Bovary". I read it twice and both times without interruption, the best places are the death of Mme Bovary and the ball in the castle. "Education sentimental"[366] is very nice, splendid also, there they are. Two women and quite different and awfully "living" both. Flaubert's people are always so living that you think that you have seen them somewhere. But "Salambo"[367] the most "classique"

[366]L' Education Sentimentale (1869). Flaubert's last novel to be published in his lifetime and one of the most influential 19th century French novels.

thing of Flaubert, I could not read till the end so tedious it is. It is interesting how the point of view changes. Mme Bovary was "verboten" by censure, and poor Flaubert had a lot of disagreements, and now ... I wrote you about the ballet, where I had to go. It was "The Swan's Lake" perhaps you have heard the musik — *Tchaikowski.* I have seen it two times before, and every time I received more pleasure. Awfully beautiful — and musik and dances. I even don't know what is better, perhaps the best is "ensemble" of music and dance. It is the most "classique" ballet. Think 50 women all in white — white swans and 10 — all in black — black ones. It is beautiful that really you can see it every day with same pleasure. I was also in the cinema in that film I wanted to see so much, but it was not as nice as I thought. But some places were very interesting. A tigre in the Ice in *Aldan.*(!!!) (such big river in Siberia) and so on. Now I read a biography of Heine, he was charming. Yes? Certainly I do not want even a good friend like he; poor man, such illness, such long death only a very strong man can bear, and think that the best things like "Bimini" for instance he wrote half cadavre.[368] I think that "Bimini" is one of the most ingenious lyrical things in the world. *Yes*?

Darling, I am angry as a tigre. Today I tried all the morning to put a *"draht"* in galvanometre. It was the first time in my life, and diameter of *"draht"* is 0.003mm(!!) You know they are in *"Silberhülle".* You must [?] the silver in [?] etc. It is difficult as to an camel to go through a needle, but I did all, all till the end and in the very end when all was finished I wanted to make a little netter and teared this damned "draht" into five parts. It is true that no man in the world made this thing from the first time, but I was so near. *Yes* darling, "ce meilleur est l'enemie du lieu". Ou, this idiot will finish in 3 minutes All is all right all the people think that I write this lecture. Hoorray! It is all.

Now is 10 o'clock and I am still in my laboratoria; I am tired — because from the 13 hours which I am here I worked 11 1/2 (1 hour this lecture) and 1/2 hour lunch) and it was such long time ago this lunch!

[367] Salammbô (1862). Fantasy novel by Gustave Flaubert set in 3rd century BC Carthage.

[368] Poem by Heinrich Heine, written in 1953/4, while he was already seriously ill.

But I want to post the letter on my way home, because if I would not do it, I will never finish and such a lot of things I want to tell you to day, awfully "philosophical" things so better I will not begin. You are tired, I think, to read this time, because one half is in russian and the seconde so dirty and uncomprehensible that I myself can not understand any word. To morrow I will go to the doctor and I will not have the tickets to the opera comique. I will go into the hospital 16.-17. I think that I wrote you about this operation. *That's enough.* But when the pain is "reasonable" I bear it very well. But I am very disagreeable when I had "stupid" pains — the headache or something of that kind. The worst is that I will be obliged not to speak five or four days (!!!) I cannot imagine what I will do. And not to eat, only drink milk, I like it very much (you know), but really I am not now in the age of five months.

So not very bright perspectives. *Yes*? Your land lady is not a nice woman, to leave a child of five years alone, quite alone in the whole house, an aunt must to have not a bit of imagination, or to be a bad subject. ... really the ladies. Visits are forbidden for you? It is too "european". *Yes*?

It is awfully cold to day and not a bit of snow and the sky problem is dark and sad. You know that the dream of my life was the "winter-sport" in style of "bobsleys" and skiing.

You know now after your last letters, I begin to think about you and to speak with you Russian on which language think you about me? Now you have three. *Rudi, Good night! Genia*

Oh yes, darling, please tell to *Dau* that if it not difficult to him, he brings me a little thermos (one glass) it is for one of my parents, he has *"stomach ulcers"* and some days can only drink the hot milk, and his thermos is broken.
Good night!

Yours, Genia

[84] Werner Heisenberg to Rudolf Peierls

Leipzig, 12.12.1930

Lieber Peierls!
Ich komme mir bald so vor, wie ein altes Waschweib, was immer das letzte Wort behalten will. Auch ist unsere Diskussion reichlich uninteressant; aber trotz alledem sehe ich nicht ein, wie in aller Welt Ihr geliebter Operator $\frac{1}{\sqrt{\Delta}}$ in den Vertauschungs-Relationen der (sämtlicher) $\mathfrak{E}_1, \mathfrak{E}_2, \mathfrak{H}_1, \mathfrak{H}_2$ auftreten soll. Ich habe doch damals sämtliche V[ertauschungs-]R[elationen] an Pauli geschrieben und der Operator $\frac{1}{\sqrt{\Delta}}$ ist mir nirgends begegnet. Also seien Sie doch bitte so nett und schreiben mir, auch wegen der Rechenfehler, sämtliche V[ertauschungs-]R[elationen] der vier Größen $\mathfrak{E}_1, \mathfrak{E}_2, \mathfrak{H}_1, \mathfrak{H}_2$, auf einer Postkarte. Dann will ich auch nie mehr über $\frac{1}{\sqrt{\Delta}}$ schimpfen.
Viele Grüße

Ihr W. Heisenberg

[85] Rudolf Peierls to Genia Kannegiser

[Zürich], 14.12.1930

My dear!
My poor girl! I could not read what they did to you in hospital, I could only read that you will be there and I can imagine how unpleasant that will be. Or rather: was, because I hope that by now everything is o.k. again.

Was is very painful, my poor girl? The "cosmic rays" were the best again: we both write about two "Genias" and you even know three. The first two I even know very well, and I had already wondered, when I met the second. But the third I have never met. But I know that, in fact, you were not a girl but a "woman" and I love that. But why do you think that your philosophy is boring? When you think about something and you want to talk about it — that will never be boring. And it appears to me that you do not philosophise very much. Dau philosophises all day long and that is perhaps too much. Now I spend almost all day arguing with him about theories of physics or theories of films or many more.

Last week I went out a lot. Monday I was in the Japanese theatre: very interesting. Exceptional actors, with a strange style, although we could not understand the meaning. Too many people die or make "Harakiri". The language is also strange. Tuesday a bad cinema film, Wednesday Rachmaninov-concert: he played brilliantly! Schubert and Chopin very good; Schumann's and Rachmaninov's compositions a bit boring. But what he played best was Liszt.

Yesterday I went to the theatre with Fanya Moskowskaya, and we saw an opera of a young Czech composer: "Schwanda the Bagpiper" by Weinberg.[369] *It is a fairytale opera and the music is fantastic. Very simple and the story goes that he has stolen everything from Czech folk songs. One scene was excellent: Schwanda says: "If I did this, the devil should come and get me!" At that moment he drowns and the devil literally takes him. You would have liked that too. It was just a shame that the playing and singing was awful. Afterwards we met Dau and we drank until 3 o'clock. Not very civilised, is it?*

This morning, Dau and I looked at the paintings of an Italian-Swiss painter: Giacometti.[370] *I had never heard his name before, but his pictures are very nice. Beautiful bright colours.*

Tonight there is a performance of the actors "Rote Hilfe Berlin" *(German morgue).*[371]

Last week in Tübingen I suddenly noticed that I did not have any money left; there was not even enough to pay for my breakfast and I had to send money, when I arrived back in Zurich. That is the reason why I could not send the letter to you from Tübingen. Everything for the professor!

I only know very little about French literature. I know neither Anatole France, nor Regnier, nor Romains[372] *and I know that I should*

---

[369] Czech opera, first performed in 1927.
[370] Alberto Giacometti (1901–1966), Swiss surrealist painter and sculptor who became a close friend of Jean-Paul Sartre.
[371] Communist welfare organisation under the auspices of the Communist Party.
[372] See letter [81].

read them. By Rostand I love Cyrano de Bergerac: Do you know it? And I also like Romain Rolland very much. L'ame enchantée. Jean Christophe and above all Pierre et Luce.

My poor girl, I think it is more difficult for you to read my Russian than my English.

I will rather go on in English. If you are so curious about Pauli's idea,[373] I am the more sorry to tell you that it probably is wrong. I can tell you now about it, if you will not talk about it. (It is no use to let so many people know about unpublished ideas, which probably are false). He thought that one could explain the difficulty with the nitrogen nucleus and with continuous β-rays if one assumes that nuclei not only consist of electrons and protons but also of neutrons of a very small mass. By this one could understand why N-nucleus has wrong symmetry, as the symmetry only depends on the total number of particles in the system and in radioactive elements always a neutron and electron would leave the nucleus at the same time and as the neutrons, which one cannot easily see, may have any energy, the continuous β-spectra can easily be understood. The difficulty is only that if one does not make very crazy assumptions about these neutrons, their ionizing power would be so great that one ought to have seen them already, which is not the case.

By the way, I forgot to tell you that I saw an experiment in Tübingen, which will probably interest you. They are going to measure directly the scattering of cosmic rays by the Bothe-Kohlhörster coincidence method. They put three Geiger-Müller counters over each other. The third is not in a straight line with the first ones, and between the second and the third there is a piece of lead:

If they get a coincidence in 1 and 2, they know that a ray with certain direction passed. And if they have a current in all three counters at once, they know that this ray was deflected in the lead by a certain angle. Of

[373]The neutrino hypothesis which Pauli had communicated to his colleagues in an open letter in early December. See letter [82], note 360.

course it takes a long time till they get enough triple coincidences, but it seems possible, only as far as now they have not yet any results. *When Dau sees you in a long dress, he will be very annoyed, he is very much against long dresses.*

Good bye, my dear, dear Genia,

Your Rudi

[86] V. Fock to Rudolf Peierls and Lev Landau

Leningrad, 20.12.1930

Liebe Landau und Peierls!
Ich war sehr erstaunt, daß Sie unsere gemeinsame Arbeit aus dem Grunde aufgeben wollen, daß Sie deren Idee weiter entwickelt haben. Das ist doch ein sehr seltsamer Grund!

Jedenfalls wäre es besser gewesen, wenn Sie mir das früher mitgeteilt hätten.
Ihr ergebener

V. Fock

[87] Heinrich Peierls to Rudolf Peierls

Berlin, 20.12.1930

Lieber Rudi,
Wir danken Dir für Deine Karte aus Tübingen und nehmen an, dass Du inzwischen wieder völlig in Zürich eingelebt hast. Wenn wir auch ein grosses Weihnachtspaket in diesem Jahr nicht abschicken wollen, so haben wir Dir doch erneut einige Proben deutscher Schokolade als Muster ohne Wert zugehen lassen und wünschen Dir guten Appetit.

Bitte teile uns doch baldigst mit, wohin Du reist und gib ab und zu ein Lebenszeichen von Dir.
Herzliche Weihnachtswünsche

Dein Vater

Ich bin gestern in die hiesige Handelskammer gewählt worden.

Die Chocolade ißt Du ja gern, so sandten wir Dir gestern 3 Hamann-Päckchen als Proviant f[ür] D[eine] Skitouren. Ein Buch, was ich bestellte, folgt leider erst später. Feiere recht vergnügt hoffentl[ich] in schöner Natur bei gutem Wetter.
Viele herzl[iche] Grüsse
v[on] D[einer] Else

[88] Rudolf Peierls to Genia Kannegiser

Zürich, 21.12.1930

My dear,
Now your operation is over and you are healthy again and you can speak again. My dear, how is someting like this possible? Not to talk for 5 days? And it is impossible to make up for this, because there is no faster speed than c (3×10^{10}), *and this is your normal speed. I am very sorry for you, write soon, when everything is well again.*[374]

You see, time moves faster and faster. When you receive this letter, it is almost New Year's Day and already 4 of the 6 months over. But I am longing, honestly. There is a Well's time machine and this is a task. But that machine does not work at night, and at night one could think that it will never be March.

Genia, I desperately want to see you. I cannot imagine, what it will be like in March. Too good! I am often thinking about you, English, German, Russian and often without language. But I am still speaking Russian too slowly, and therefore it is still difficult to talk to you in Russian. Writing is not too difficult, but the pronunciation .... Its was useful to speak Russian with Dau, but we are now talking frequently and much about physics and we both speak German so quickly that it would take a week to say in Russian what we say in German in one day.

We are both very much occupied with the "uncertainty relations". Every day we are improving the experiments, which we have done the previous day. This is very entertaining, but there are few results.

[374]Genia was troubled by recurrent tonsillitis. See correspondence throughout 1931.

It is late, dearest, and I can write quicker in English. This week I did not do very much extraordinary; once we were in the cinema, but saw nothing very interesting. Wednesday we all were invited by Wentzel and his wife and there again Dau made great impression. The main point why he makes such an impression on "bourgeois" people who are not in a safe psychological situation is not mainly that he has different and revolutionary views. This is not seldom and people are used to defend themselves with all arms of their tradition. But he discusses not at all with reasons, he simply puts against them a system which is closed and as "obvious" as theirs. He does not try to be objective but is subjective so they are only with different prejudices and this amazes them awfully much. It is really an interesting experiment to notice the effect he makes on different people, for instance, Pauli, Waller, Wentzel, *Fania Moskovskaja* and many others. If someone is not shocked he is surely in a very safe position.

To-night we had long "philosophic" discussions. This was the result of the fact that we had nothing to do, the cinema programme being terribly bad and so on, and we without appetite for working. So we talked all the evening. I am not so sure that you are right with your opinion of his character, even in your sense of the word.

Last night I danced with Lilly Fenigstein, all the evening without interruption, in a queer place, "Kursaal". This is a dancing place with terrible people. But we had quite a jolly time.

Amongst others, I saw a Turkish girl, who was in this Kursaal, every time I had been there. Obviously it was her profession to be there every night, like many other girls (you imagine the "profession"). One year ago she looked quite nice but this time so dreadfully broken and spoiled that I was very astonished. I knew that this is the quickest way to ruin her, but I did not imagine it was so quick! *Well, it is very late now, good night my dear.*

22.12.[1930]

Today is the shortest day, very cold and one can feel the snow in the air, how beautiful will it be in the mountains today! Today Dau and I finished our discussion of the "uncertainty relations" and now one only has to write up everything.[375] *This will be terribly difficult, because we always want to use different expressions. I will take my typewriter, and perhaps we'll write the article in Arosa. It will be very nice there, I only fear that there will be many people of an acquaintance. Everybody whom you ask, "Where will they all be?" answers, "In Arosa."*

Dau sends you his regards, he is much too lazy for writing or, rather, writing is too difficult an occupation for him. You must not expect any letters from him before he returns. *All the best for the New Year, my dear, many challenges, many tasks, much fun and, above all, health (so that you can talk as much as you like!) Once again good night! When I am asleep I'll fidget with my hands, probably, but nobody will tell me and I will not hit anybody. Good bye,*
Your Rudi

[89] Genia Kannegiser to Rudolf Peierls

Leningrad, 23.12.1930

Darling, once again my hair is wet and my translation unfinished, but still I am writing to you because I won't have time tomorrow and I want you to receive this letter by the New Year. I found this letter with Arosa's address in Byron's work, I read him now with great pleasure. "Don Juan" a little long but good verses. Three days ago I was at last at this doctor, the story is more disagreeable I thought before – it is 8–10 days in hospital and as, with me always happens, all sorts of complications. I am afraid what it will be two weeks. I want to be at home in New Year Night and I have all sorts of affairs in my service, translations etc. to end. So I will be in the hospital in the very beginning of January (2–3) and remain there till 12–14 and not speak all this time! Awful! It is the operation in the throat. At home, every dinner we have "discussions" on this theme. Mama is certainly (mothers always are) afraid, and don't want that I do it now, but I prefer always to do the disagreeable things as soon as possible. It is

[375]The paper was published as R. Peierls, L. Landau, 'Erweiterung des Unbestimmtheitsprinzips für die relativistische Quantenmechanik', *Z. Phys.* **69**, 56–69 (1931).

true that it is a little dangerous, because I will lost two glass of blood and my temperature can begin (you know I have a little toubercoulose) and sometimes temperature jumps up till 38.2) but it is not necessary, at all and I hope that all will end in this 10 days.

Darling, if you don't know France and Romain, read as soon as you can: "La rotisserie de la reine Pédogue" and "Le Chat Maigre" of France and "les Copains" of Jules Romain. You will receive the great great pleasure, really. Very funny and awfully good written things. I was sure that you would like Mme Bovary, with these ones I am not, but I think you will. What a lot of theatres, concerts and *cinema* in one week?

Japan theatre was *"kitsch". Yes?* ... I awfully love the Japan peinture, and you? You know these portrais signed with 2–3 "trait" and peysages and animals. And the Japan tales are charming, they were the preferables tales of my childhood and now I like them awfully. Do you know ???

To day is ten days that I am without Musik, and not interesting concert till the New Year and afterwards I will be in hospital, so one month without concert. Stupid! In March and April we shall have a lot of yours "Generalmusikdirektor"

Darling, really I am woman in the middle. Sometimes it seems to me that I am awfully old, old, would and that all the people are children and that they are unhappy like children and that all will pass if you will take them in the knees (you understand?)

"Woman" perhaps with a big letter, is a very strange thing, always a little "'mother". *Yes?* It is awfully ununderstandingly and I am myself half asleep of that.

Darling, I do not know also what you prefer, my english or my russian. What is easier for you? How we will talk in March, I think that I will be "mixt" but we are habituated now. *Yes?* In some days it will remain only $\frac{1}{3}$, the worst is passed. But it will be very complicated "technically" in March, such a lot of "friends" and only evenings, and such long distance between all the places. Tête-à-tête in such condition is really very difficult. But "tête-à-tête" or without it I want to see you, Rudi.

Ou, all the paper is in my writing table, and Nina is fast asleep. So good night, darling. Happy New Year! And jolly skiing. Take the pictures and sent me them. I want so much to be in Arosa and not to lie in hospital. *Nina wishes you and Dau a Happy New Year.*

Genia

[90] Rudolf Peierls to Genia Kannegiser

Arosa, 25.12.[1930]

Cherished and dear Genia!
Today it is really great and it is a shame that you cannot be together with us here. We arrived yesterday and already the journey was good. In Zurich the weather was bad, but when the train ascended, we were at first in and then above the clouds. There was sun, snow, high mountains and a bright blue sky. You know, Arosa is a big holiday resort with big hotels, but we are staying in a little village 200 metres above the town. A small and very nice house. The people here are not very interesting apart from one interesting worker whom we have already met last year.[376]

Today we went skiing for the first time. That was strange, because I had not been skiing all year. Very strange when the skies all of a sudden move apart. Dau skies very badly and when we go together I always have to wait for him for a long time. Then you get very cold. But what shall I do with him? It is almost impossible to teach him.

There is also a very nice girl here, Marli Heiman *whom we also met last year.*

27.12.

I could not finish Thursday because they had a great dinner and so on for reason of Christmas. And yesterday, my dear darling, I could not write to you because we were on the skier all the day and in the evening danced all the time in our pension: The people are not too interesting,

---

[376]Peierls and a group of friends (Thorner and Bethe) had been to Arosa the previous Year.

but they dance quite well. I am already brown like an Indian. You cannot imagine how beautiful this is in the sun and the snow. How beautiful also the night when we very far see all the houses of Arosa, with all their little lights and above all the mountains which by night look so small that you think you could touch them all with your fingers. To-day I met this Miss Heimann and we will have a great trip on a summit of nearly 3000 meters together with some people of our pension. Now there will be supper and I must stop writing again. So good bye my dear. *Good night.*

29.12.

The tour yesterday was really splendid although we got into fog and had to return before we reached the top. Also going down was very funny in the fog as one never can see whether the way is going up or down and where one is going. In the next time we will try again to get on this "Parpaner Rothorn"! Dear, this letter is awfully silly, I know, but we all are here in such a silly state of mind, all the day messing around with the skies dancing and similar intelligent occupation. By the way, dear, I posted to you from Zurich a book of Huxley which you perhaps may like. It is a bit too much of a "salad" but it contains really good scenes and many interesting thoughts.

1.1.31

My dearest I did not write you all these days although I had practically nothing to do (or rather "because" I had not). Darling I must tell you now a complicated a very complicated and very strange story and I do not know whether I can tell it so that you understand, but yet it is necessary to tell you. I said that the people in our pension are not too interesting. There are two exceptions, a mathematician from Göttingen, a very nice fellow, but very young (in mind) and not very active — he is generally called "Bübchen" (little boy). The second exception is Eva Gäbeler, a girl from Berlin, gymnastic teacher, quite good looking with an easy character (as you would say), not very "deep" but with a quick mind. Now the situation is this. She seems to be very much in love with the above mentioned mathematician. He is very cold and it seems that he is not very much interested in her. But this is not complicated enough, originally she loves a man in Berlin and she promised him to be "treu" to him. But she cannot be without men and she cannot resist. So she is always here with this "Bübchen", a not too pleasant situation as you will imagine. And furthermore there is the beginning of an "adventure" with me. You understand that she is not in love with me, I even less in love with her, also nothing "serious" has happened except some kisses. Can you understand this, my dear? Will you be angry? I am very anxious of adventures, because I fear that they could trouble the relations between you and me — now at least I know that adventures of this kind have nothing to do with my feelings for you, that after all you are as dear to me as before (or even more?) Can you understand this? I am afraid this will make you angry, you will take it more seriously than it is. I ask you for very much you can understand this, my darling? I cannot help having adventures if I am here and you so far and this one even will possibly develop into an "adventure" in the normal sense of the word and if it does not the reason will be that I do not find a way of success, but not any thought of "purity". I am not sorry for it — but what if this brings a distance or a coolness between you and me? *Genia* for no adventures in the world I would like to lose one single point of what is between you and me. Write me dearest what you think about all this and whether you are angry about your foolish boy or not.

The situation here is the more complicated as this Eva herself does not know exactly what she wants. Sometimes she is very sorrowful and dreams of the promise she has given to that man in Berlin. At other moments she is able to say things like: "Thorner", (Thorner is here since two days) "you may come a little nearer, because first it will be warmer then, and second you will like to stay longer and not say you will go to sleep." Such things are, of course, not serious at all, they are, by the way, not at all "European". They give a certain atmosphere which you surely know; I do not know if you like it.

It is impossible to relate you everything, for instance the dramatic events when we were near of being found together by this "Bübchen" or by Landau and so on. Perhaps all these things are due to the fact that this girl is the only reasonable girl in our house (will you laugh here or will you be shocked?) when we go for a dance, for instance, she has to dance with all the present men (she really dances well). Now you must not think that all our occupation is dancing and flirting, most of the time we are in the snow and there is no such problem. Do you know how beautiful it is amongst all the mountains when you feel the speed of your power over the skies. I conceded even in teaching to Landau a bit of the art how to run down a little hill without falling down at every step. One must simply treat him as a little boy and say him "now do this" and "now that" and be angry if he makes mistakes. Only he is so slow, so slow! My dearest I was so glad to get your letter,[377] for which I had to wait some time, as I did not get it in Zurich and it had to be forwarded here. Of course it makes no difference if you lose my Arosa address, I will only get your letters one day later. But if you are too busy, darling, do not spoil your sleep with writing, even if *Nina* does not protest, just take a scrap of paper and write two words to show that you are still alive. I am so ashamed you are so busy and we are so lazy all the time here. Yesterday at 24^h O.E.Z.= 23^h M.E.Z. I knew that you were drinking for me and I did the same, and I thought of all good good wishes for you and the New Year and mainly (egoistic as usual!) for its third and fourth month. Now there remain only 60 days!

I am very fond of the Dreigroschenoper, I like mainly the songs (the music as well as the verses, I do not know whether it is possible to translate the verses into Russian) The end is a bit silly and cheap.

Dau will be in Leningrad in March, a little later than me.

My dearest write me soon on a few words whether you are angry about my foolishness! Kisses, many kisses!
Good night, good bye my dear.

Your Rudi

[91] Genia Peierls to Rudolf Peierls

[Leningrad], 1.1.1931

The New Year — hurrah!! First of all I beg your pardon ("fishing?") for the paper not "trop chique" really. Now it is 18 o'cl[ock], but I am

[377]Letter [89].

quite asleep and *Mother is pacing to and fro and saying, "Little Genia, my girl, go to bed!" Rudi, today is the first day of the New Year, and the night was the New Year's night; I didn't sleep a wink. At 8 a.m. I went out together with our guests and started directly for the laboratory, from where I have just returned.*

I'd like to go on dancing, singing, and drinking (!) too. What on earth, I'm in marvelously high spirits now! But my appearance is quite melancholic: two 5-mm-wide slits instead of the eyes, the hair (!!!), and what is the most important, Rudi, my voice, my penetrating voice is ruined by singing and drinking. Since the very morning I can only speak in whispers, and when I try to speak aloud, terrible sounds are uttered. What a shame, what a shame! Yes, seriously speaking, I caught a cold somewhere. I'm afraid the operation may be postponed because of this, and if it goes on in this manner, you'll have to go to hospital to see me, when you come at the end of April. Will you? The party was very hilarious. Great!!! Never before did I drink so much and was so high-spirited with wine. It was according to the equation I = I0ex, where x is the number of wine-glasses, ∼100, I0 — my normal state. It was a great pity you weren't by my side, darling. I drank to your health a full glass of hock-cup. That was at 2 a.m. of our local time, which means it was midnight in your time zone. Didn't you feel your ears burning? (Do you know the saying: having hot ears means somebody is thinking of him). As it seems, I went beyond the allowable limit in my glee. At any rate, Mother was scolding me today at dinner for my shocking behaviour. In my opinion, it is a prejudice. I was so gleeful that other people must have become merry too. Is that right? So, to my mind, everything was all right. By the way, I drank Bruderschaft *4 (!) times. One of them, we drank solemnly with Bronstein; following to the accepted rite, I even broke the wine-glass and we kissed. (I kissed each time, but broke the glass only once, as Mother was obviously against this version of the ritual). Bronstein was tolerably drunk too, he is really charming when he is in this state. Exuberant! Darling, you and I didn't drink Bruderschaft; only in our letters we accepted first-name relations, using Russian "thou". Oh, this idiotic* "you"*!*

Darling, are you very tired? Such a lot of Russian and such awful handwriting. Thank you for the books. I have not yet read them (had no time) I shall take them into the hospital and read there. Hospital Brrrr. I was once a fortnight with my appendix (... the first days I was so "ill" that I did not pay any attention to the surrounding, but after ...) Darling you know that if there are three women together (the social position, profession, education do not play any roles), they will all the time speak about their children. (Illnesses mostly also have "childish" origin.) In three days I had such feeling as I have had minimum five children and three husbands. All from the very beginning, till the children of children. Ooooo!! You know that I love the children awfully that I will be very happy to have them, but such "approfondation of question" is quite unbearing. I was half mad after that. Now, I am afraid will be the same thing. ... Dau, I am glad that you like him. I love him very much, quite "seriously", he is a very very "good boy" ...

1.2.

I am ill today darling high temperature etc. and I am in bed now because of that and my handwriting is even worse than normal. So, Dau, I am sure that I will never quarrel with him, because I think he is a little boy, quite, quite little boy and so I do not take seriously all that he says, all that he does. It is too childish you can only laugh, as you laugh when a boy of ten years speaks about "Weltproblem", "love" or something like that. He has the "heart" (you understand) of a man but everything else of a child, even an awfully theoretical child; he has the theories of all cases of life. But he is charming — and I love him awfully — like a "little brother" perhaps. I cannot bear when he is unhappy. Darling, it is so difficult to write in such conditions. I am afraid that you will not understand a word. I sent you my last portait. It is a little like "the details of Rembrandt picture" old without his monkey. I am awfully monkeyish here, but all the people say that the "likeness is perfect". Perhaps!

You are now in Zürich, darling, I hope that you have not broken your feet in Arosa. I am ill now, but awfully jolly and translate a German comic song on Russia. Very difficult, because all is the play of words, but I think I will do it at last.

I was in the opera in the morning we have splendid ones, really. It was the tale of Pushkin musik of Rimski-Korsakov[378] *and all the theater was filled with children, they were quite little, it was their first theater visit ever and they were awfully funny. And the music was charming, and costumes also. You know this old Russian style. Darling, I want you to be here now.*
Good-bye.

Genia

Let March come soon!

[92] Rudolf Peierls to Genia Kannegiser

Arosa, 6.1.[19]31

Dear, my dear Genia!
I did not realize your operation was so complicated as that and that the tuberculosis affected you so much. My poor girl! Now while I am writing to you, you are probably already in the hospital and I imagine how painful all this will be. Poor dear child! Just while I am writing, I sit before the window and outside one sees Arosa quite below, with all the strong and snowy mountains around, the morning sun is brightly shining on all and dark blue sky above. Very cold today, the frost crackling, everything is very beautiful.

The things I told you in my last letter became much simpler in the meantime. There is nothing at all "serious" in them. My relations to this Eva are only a play, for me as well as for her. And truly — a harmless play, impossible to be called an adventure. There is no doubt now that she is seriously in love with Bübchen and that she gave up the man in Berlin. Bübchen is more difficult to understand, he is much cooler, but it seems to be what Dau would call a symmetric situation and so everything seems simple and clear.

[378]Nikolai Rimski-Korsakov (1844–1908), Russian composer and music teacher, who composed among others fifteen operas many of which are based on Russian history and folklore.

As for myself, I am not sorry for anything — not only for anything I did, because I actually did nothing at all, but also for anything I thought. There was no adventure, but there was an opportunity to learn that I would probably take adventures if they would come. But I am not even sure about this. Adventures do not simply come so that you must take them, you must really want them. And in the present case I do not know whether I really wanted it, i.e. did all to get it. Well, anyhow, the situation is now clear.

Some days ago we made a trip to the "Weisshorn" a peak about 2800 m high. We were very tired then and got to the "Weisshornhütte" a house on the edge of this peak, about 2500 m. Quite near of the house, a boy of 15 years fell and damaged his arm. Thorner immediately proved to be a good doctor. But he could not do anything without ether because the adjustment of the elbow was too painful. So the beforementioned Bübchen and I went down to Arosa to *the pharmacy* to get ether and so on. As the patient had bad pains, we hurried and wanted only two hours to go down to a height of 1700 m and back. In the meantime, Thorner had asked the people in the house to help him for the operation and for bringing the patient down. There were a lot of spokesmen, but nobody answered except three very nice English people, a man and two ladies. We all were terribly upset by all this. Later we met some friends of Thorner and they said: "O you have been there when the accident happened? Yes we had also seen it, but we went away as soon as possible, we cannot stand such dreadful view!" So when Bübchen and I came up again — without breath of course, we had to assist for the operation and then we four (Thorner, the English man, Bübchen and I) took all our skies down. It was already dark and at every step we sank into the snow up to the knees or even to the hips. But at the end we got down safely, though late, and next morning, the patient could take his train home. Thorner made this operation very well (afterwards he confessed that it was the first time). He is really a good doctor. But personally the relations between Thorner and me are a bit cooler than before. This is mostly due to Dau. Not in such a way that he "separated" us, I would never give him this influence, but he is a very good "Reagenspapier". He makes people speak and tell their ideas, and although his own ideas are in many ways silly (for my taste) it is easy to use him and find out what other people are like. And Thorner's ideas are really dreadful, especially in the question of relation between men and women. He has the ancient opinion of the two classes of women. One of them "decent" and the others have to be regarded as "Freiwild". It would be not too bad, if he had such views theoretically, but he uses them also practically. So for instance there is in Arosa a friend of that Eva Gäbeler. She is a type of girl I do not like very much. She plays consciously with men. But she does it so openly that of course a man knows what he does if he joins the play. After all these plays are quite harmless. Now one day she complained that men behaved terribly against her. For instance one man had invited her for New Year's night and payed for her against her protest. Afterwards she evidently did not do everything he wanted and then he became angry and made her feel always again that he had payed for her and she was "ingrateful". Instead of being upset Thorner said that the man was not so perfectly wrong, that if a girl behaved like she did, she could assume that [s]he was not a "decent girl" and that he could behave to her as he liked. He even made herself feel that opinion. This of course made a great break in the atmosphere here.

I told you perhaps that Thorner in last time has close connections to France, he had very often been there and he likes French culture very much. Now this is a typical example of French culture and that is also why I do not like it.

As for the books you wrote me about, I wanted to read them long ago, but I never could make up my mind. But as soon as possible I will really do it.

Evening

Last night we invited those three english people we had met on the Weisshorn and arranged a large party for them, it was quite nice and lasted quite a long time. So we were terribly lazy to-day and lay in the sun all the morning. (O my poor conscience, when I think how you work in the same time!) In the afternoon we had a very nice trip, I have never seen such a beautiful light in the mountains. Only it was terribly cold. To-morrow it is finished and we must go back, because to-morrow afternoon I must speak in the colloquium in Zurich.

My darling, I am so strongly looking forward to your letter which will show me how you accepted my last one. My dear, do you believe me that nothing at all is lost between us?

Good night, I kiss you so strongly! You cannot even speak but probably you will be able to kiss and I will try to take off some of your pains, poor girl.

Good Night.

Your Rudi

[93] Rudolf Peierls to Genia Kannegiser

[Zürich], 10.1.[19]31

My dearest girl from Odessa,

Thank you for your card. It was really a nice card and I like it very much. But this is not the girl from Odessa, but my Genia, how I remember her and more than I remember a poor girl! Once again pain and not again in hospital.

If I was with you I would be gentle and would be caring and I would do whatever I could so that you can forget your pain.

But I am not there and instead I am writing to you such unpleasant things! Now look! Now I am very good: You wrote me a letter lying in bed with a fever, and immediately I am answering in the same condition. But luckily I do not have a lot: a slight inflammation of the mucus "stomatitis aftosa", I don't know why! But it is not very bad, I just have to stay at home for a few days.

As for Dau, dearest, I think one can take him a bit more seriously than you do. Of course you are right that he is very childish and what you say about his "heart" is even more true. But all this does not prevent that a lot of his ideas not only in physics but also in life are very interesting and important. I do not say, of course, that I usually agree with them, but it is difficult to say anything against them and that shows that they are new and necessary. Many people act very silly against him, because if he makes them angry (what he claims to be his main occupation or at least his main purpose) they bring the discussion from theory to practice and let him feel how little experience he has in life and especially in love affairs and they say (what is very easy) a lot of things which make him very unhappy. It is their good right, of course, as a defence, but it does not say anything against his arguments.

Your New Year's Night must have been splendid. Yes, I think that nobody can stay serious when you are jolly. I probably heard it in my ears when you drank for me but did not believe it as I expected you to use East European time; and furthermore, 0^h0 MEZ is only 1^h0 OEZ in winter.

11.1.

Last night I had a very high temperature, and that's why I stopped writing. But today I am very cheerful, and everything is well, although I have to write a very unpleasant letter. Do you know the name Guido Beck. He is a theoretician, he used to be Heisenberg's assistant and now he is in Cambridge.

He has never been a clever physicist. But now Pauli received an article from him which is completely nonsensical, so that we are worried that something is wrong with him, and therefore I have to write to Cambridge.[379] *Isn't that strange?*

I have still to relate you a very funny event of our trip back from Arosa. We went down to the station very punctually but our landlord was not yet there with the luggage. We were very angry and looked for him everywhere but did not find him. Then the time came and the train started. And afterwards we found that the man had just come in the last moment and thrown our things into the train. Thorner had also jumped in and Dau and I were standing there lonely without our things. We telephoned down to Thorner (who could not wait for us as he had to catch a train to Heidelberg) left our things in Chur in *hand luggage locker.* He did and when we came to Chur I saw our things lying there but there was no *ticket.* We had only ten minutes and in the afternoon I had to speak in the Kolloquium in Zürich, so I could not miss this train

[379] G. Beck, 'Bemerkung zur Zerstreuung harter γ-Strahlen', *Naturwissenschaften* **18**, 896 (1930). Landau showed that Beck's paper was based on wrong theoretical assumptions and published his reply in L. Landau, 'Remarks on scattering hardness gamma rays', *Naturwissenshaften* **18**, 1112 (1930).

and no *ticket*. As I know later, Thorner had left his *ticket* for us in the buffet, but this I did not know. So I went everywhere to the head of the station and so on and one minute before this train left we got our things and got to Zurich just in time.

The auditory of my lecture is really decreasing; *the lecture is probably too boring.* But I can't help it. *When I was ill, my landlady proved to be very nice. She does everything to make life pleasant for me, and she says nothing when she has too much work because of me. But the strange thing is: personally she is always very nice to me, but in all things in which tradition comes into play, she is terribly unmodern. I have already told you that visits are impossible there, and it is even worse how she treats her servants. They have to work all day and when they have finished they are not allowed to go out, but have to sleep so that they can work better the next day.*

Once a week, they are allowed to go out, but they have no key and the family wait until they have returned etc.

Of course the staff do not like that, and last year there were eight different servants. I don't understand that. She does not have a bad character, but she simply does not reason, and she thinks that she is right. And she herself works more than the staff. Of course, frequently there are no servants at all, and she looks after the whole household herself. She is a typical Swiss lady.

Today the people from Arosa — Bübchen and Eva — are in Zurich. Annoying that I cannot go out. But in the afternoon they will be my guests together with Dau. (Don't be jealous, everything will be decent!)

Now somebody will go to the post office and will take your letter. So, good bye my dear!

We are really going to see each other ~ *in 48 days. I kiss you.*

Your Rudi

12.1.

Nobody went to the post office yesterday and so the letter is still here, darling. Also my guests did not come in the afternoon, and I was quite alone all the time. It is so stupid to be alone when one is ill, isn't it. I was going to say: Could you be here! It would be somewhat strange, however, if you were here just now, you lying in one bed in one corner and I in the other and I would not even be allowed to take your hand without precautions (danger of infection) not to speak at all from kissing! But still: could you be here!

Eva and Bübchen came in the late evening yesterday, they had also missed their train in Arosa! But in the evening my temperature is always high and I am not a very nice fellow. To-day they will come again I hope, and so I'll have some society. But it is so tedious to be ill. It is even better if one is seriously ill, then one at least sees why. But such a little and unnecessary thing, which cannot become worse and does [not] become better, silly!

Look here, darling, do you see now that your boy has a very disagreeable character even in your sense of the word? He is so much grudging and complaining if he is a little bit ill!

By the way did I tell you that I planned to write a paper together with Dau and *Fock*? I started it in *Leningrad* with *Fock* and when Dau came here he had such a lot of ideas loosely connected with the subject that it was necessary to give up the idea and start a new, much more general paper, in which the old theme was only a little special case.[380] Of course this had to be done without *Fock* for simplicity, and so we wrote him that we had to give up the "Dreimännerarbeit". Now he answered very angrily and acted as if we had stolen one of his ideas (we did not, of course, the first idea was his and ours independently, and about the new things he did not dream at all). I wrote again to reconcile him I should really be sorry if he is offended, he is a very nice man.

[380] Peierls and Landau were collaborating over the extension of the uncertainly principle to relativistic quantum theory. A joint paper, which they completed in January 1931 (L. Landau and R. Peierls, 'Erweiterung des Unbestimmtheitsprinzips auf die relativistische Quantenmechanik', *Z. Phys.* **69**, 56–69 (1931)) had been discussed with Fock and Jordan in 1930, and during Jordan's visit to the Physico-Technical Institute at Kharkov, Jordan and Fock had come to slightly different results which they published in P. Jordan and V. Fock, 'Neue Unbestimmtheitseigenschaften des elektromagnetischen Feldes', *Z. Phys.* **69**, 206–209 (1931).

I am so much expecting your letter, dear with a lecture on "adventures"! I am not forbidden to kiss you by letter and so I do with all my force.

Your Rudi

[94] Genia Kannegiser to Rudolf Peierls

Leningrad, 15.1.1931

... "Ich grolle nicht"! Certainly I am not angry. What rights have I to be? I am sorry — or better "unhappy" (...) but angry — not.

You know my point of view on "adventures", but you must to know that for me the chief thing is "to want" and not "to do". I even prefer when the thing is done. If you, for instance, would not kiss (or more) a girl on such "reasons" as "purity", "public opinion", the fear to lose somebody, you will be a simple "philister", the thing I dislike the most in the world. And to want or not to want is the matter of "instincts" on something of that kind and not at all of reason. So if even I had the rights I could not be angry.

Certainly, I would prefer to be loved so "strongly"(?) that you will not "want" to kiss anybody else, but ... nothing to do! Perhaps even it is too "subjective" and different men love differently. I do not know. Thank you very much for your frankness, and I believe you, Rudi, that from your side all is the same ("or even more?")(!) My dear boy, nothing cannot be lost, nothing cannot be obtained. Life, love, they are "vectors". They have not only magnitude but and the direction also. And it is dynamic not static, every moment you lost something and obtain another and every moment the directions are changed. *Yes?* These "changes" are more or less big, certainly. So, my darling, after your last letters, because I got two letters together (13.1. (!!) — this post!)[381] I will meet you in March not quite the same way, as if I did not receive them but it will not be less or more it will be a little "another style". I think that you will even prefer it. (It will be easier for

[381] Letters [92–93].

you, you egoistical boy!). And certainly it will be different from *March* this sea of ink between us and these 6 months, it means something I believe! *Rudi*, we are very good friends. I am sure of that now. *Yes*? Friendship is more difficult than love, I think that between us both it is very well. No, seriously, do you love me? "Adventures" I am glad that you wrote me, I prefer to write to you and not to some "idealisation" I like "la clarité". Perhaps I will prefer if you were a little another, but I cannot, for instance, change your eye into grey or something of that kind. You are so and so and it is all. My, enough of philosophy, it is not too comprehensible, and I beg you pardon, darling of dirty paper and handwriting, but I am ill and lazy.

Rudi. I am unhappy — such nice time now and snow and sun and I was fortnight in the bed and the skies are pure abstraction. Awful! Dreadful! Disgusting! And the operation is pure abstraction also, because I am now in the middle of my tuberculotical temperature, every evening 37.6-38.0 ... and I can not even think of loss of two glaces of blood (of^3!) Of! I was lying all the time, nevertheless I was once in the ballet, right from the bed to the theater. ...*?* And it was the best ballet I never saw before: "*Raymonda*", the music of *Glazunov* and what musik! and the dances ...! *Rudi* I jumped from pleasure so high that one German who was sitting beside me was quite shocked. But quite! You can imagine very easily what "*Wut*" was at home and what pleasant words sayed me Mama, "the foolish girl" was the most gentle. But I told that "Pairs vaut une messe" and that such Ballet worth an inflexion of lungs, but all passed very harmless. We had a party at home this week; one of Nina's friend parted to *Moscow* and all her laboratoria came and it were some people more. For all the other it was very jolly, but for me not so as the New Year night, think only, I was strictly forbidden for me: to drink, to sing and even to dance! (I danced, certainly, but not too much.) Poor girl! *Yes?*

I am very glad that you had a nice time in *Arosa* (perhaps too nice). *Young* Thorner with his operation. Bravo! But all these tourists quite "european" — very typical! I think that Thorner was right with this stupid girl. He must not certainly to show his opinion, very unpolite, why for? But, really, darling, if that man of this girl is pure heart, she is without a grain of brains in her head, and very vulgar.

1. You must know the person with which you have to supper in New Year's Night, that to know can only the "girl" the stupid girl, I think that your is the second class.
2. If you have such unluck to fall in such story — to relate it in the company of unknown persons! It is too vulgar and stupid also.

And perhaps the opinion of Thorner will be a good lesson for her. I dislike the people who cry on every breast. Brrr! Yes, Rudi, I believe that I am unknown to Eva, and will be to any other of your adventures, I beg you that very seriously; Very *Good?* Darling, you have "carte blanche" in your future serious story (perhaps even present because between letter and answer is between 2–3 weeks) and I do not know what you are doing now) but for adventures "strictly forbidden". Yes? Oh, it is time to finish this unendly letter, and too philosophical, I beg your pardon once more (really, it is not "fishing") but I am ill and a little sad. I do not nothing now, but absolutely nothing. Reading of words and speaking with visitors is not an occupation. All are charming with me, bring me bonbons, flowers, books. But I was lying and waiting your letter and did not know, who are wrong, you or post. Now I think you are waiting mine. On my table stay such white flowers they are a little faded and very beautiful as the sails of yawl when the still beginning. Oh darling, I quite forget we go in the big company at the ballet concert this evening and I have 37.8 but really it is impossible to stay all the time at home! Good bye! To day it is 3/4 of time. Only 45 days more. It is really nothing.

... Genia

[95] V. Fock to Rudolf Peierls

[Leningrad], 15.1.1931

Lieber Herr Peierls!
Ich habe keinen Augenblick geglaubt, daß Sie und Landau unsere gemeinsame Idee für sie allein gebrauchen wollen. Das stand in meinem Brief auch zwischen den Zeilen nicht.[382] Ich weiß ja, daß Sie und Lan-

[382]Letter [86].

dau auf die Idee der Unmeßbarkeit des Elektronenortes ganz unabhängig von mir gekommen sind. Daß Sie aber unsere fast fertige Arbeit ohne genügenden Grund aufgegeben haben, war mir unangenehm, und ich habe mich auf Sie geärgert. Aber man kann ja auch auf einen Freund böse sein. Das Wort "Feindschaft" paßt hier überhaupt nicht. Und jetzt bin ich auch nicht mehr böse.

Also — alles ist gut!

In Ihrer Bemerkung über die Arbeit von Jordan und mir[383] (Ihr Brief vom 9. XII)[384] sprechen Sie von der Möglichkeit, durch eine mehrfache Ladung die Genauigkeit der Feldmessung zu steigern. Diese Möglichkeit haben Jordan und ich selbstverständlich im Auge gehalten. Aber der n-fach geladene Kern hat doch eine komplizierte Struktur, und man möchte glauben, daß der Begriff des Feldes durch einen denkbar einfachsten Probekörper definiert werden muß. Außerdem bedeutet eine Vergrößerung der Ladung eine nicht kontrollierbare Störung des Feldes. Jedenfalls fängt unsere Arbeit mit einem "wenn" an.

Es würde mich sehr interessieren, wenn Sie mir eine ausführliche Ableitung der Ungenauigkeitseigenschaften $\triangle f > \frac{\sqrt{hc}}{(c\triangle t)^2}$ mitteilen könnten. Kann man auch zeigen, daß diese Relation relativistisch invariant ist? Auf den ersten Blick scheint das nicht der Fall zu sein.
Mit besten Grüßen

Ihr V. Fock

Grüßen Sie auch Landau von mir und sagen Sie ihm, daß
ich recht böse auf ihn war, daß das aber eben nichts weiteres bedeutet.

[96] Rudolf Peierls to Genia Kannegiser

[Zurich], 18.1.[1931]

My dear,
How are you? Where are you now? In hospital already? Or has the operation already been completed and you are already at home? I hope

[383]P. Jordan and V. Fock, 'Neue Unbestimmtheitseigenschaften des elektromagnetischen Feldes', *Z. Phys.* **69**, 206–209 (1931).
[384]Letter could not be located.

this is the case, and you are completely well again and it is no longer necessary to think about any illnesses! I was well again on Wednesday already. This doctor who told me that I had to stay at home for at least ten days was a silly man. I was very glad that I could give my lecture on Thursday, because already last week there were few people there and it was a bit difficult and boring. If I had not given my lecture this week, next week probably nobody would come.

I am working with Dau all the time, but progress is very slow. Whenever we think we have understood everything, there is a mistake and we have to start all over again. At the moment there are difficulties, and we don't know what to do. Dau is rather melancholic at the moment — I don't know why and he does not seem to either. It is quite frequently like that with him. They are now showing the film "Im Westen nichts Neues" [385] *which is banned in Germany.*[386] *It is supposed to be a very good film. Tonight I will watch it with Lilly Fenigstein. It was very difficult to get tickets, there was a long queue. The German ban was the best advertisement.*

But Zurich is a terrible city. When Dau and I are together, we are talking about everybody in Russian. It is so convenient that nobody understands us. Last night we went on the tram, and we talked loudly about two girls, who sat opposite us. We were of different opinion and discussed them for a long time. Suddenly one of them said: "Strange that there are people who think that they are the only ones to understand Russian." Very embarrassing, isn't it?

Today it was cold and everything was white. A lot of snow has fallen and it is still snowing. I could have gone on skies from the house to the institute. This is very nice and a lot better than "Fön" (warm wind), which we have had all the time.

Time is running, darling, there are only 6 weeks now till March. It is, however, a little bit more till my arrival in *Leningrad*, as the time in the train cannot be neglected and I can probably only leave here on March 1 or last February. Now, when I shall be there, I hope it

[385] All Quiet on the Western Front.

[386] Lewis Milestone's 1930 adaptation of Remarque's novel was banned in Germany because of its pacifist sentiments.

will not be so bad with our seeing each other as you picture it. First of all there are the free days and as far as I know your free days are the same as those of *G.F.T.R.I.*[387] and you will probably be able to make yourself sometimes entirely free on a free day. I even dream of sometimes going away from *Leningrad* for one day so that we can be free from all "friends" and relatives and so on. Then you must not forget that I shall probably have a room in a hotel which is quite near from *Mochowaja*[388] so that you can just come to me if you have one hour free in the evening. Of course time will always be too short (it would even be too short if we had all the day to ourselves) but it is not at all so bad. The only danger is the great hospitality of all the people who will probably think that I am a stranger and therefore it is their duty to accompany me at every time and always to organize entertainments for me. So in every special case, it will be hard work to get rid of them, and under normal condition I should be very much obliged to them for it. But I have some practice in those tasks and also this will not be too difficult.

When shall I get your letter, my dear, you know why I am so strongly waiting for it. Our housekeeper in the institute is slowly becoming angry because I too often ask him for letters. Did I ever tell you about this housekeeper, Herr Hartmann? He is a very funny man, real Swiss, and so "solid". He will never speak to you without using your title and he knows exactly the difference how to behave exactly against a professor, an assistant or a student. He can be very comically discret, so for instance when a lady is asking for Scherrer[389] on the phone (which rather often happens) and he is absent, Hartmann leaves a paper on Scherrers table: "Man hat angerufen und lässt sagen, ..." He is the greatest authority in the house and if Scherrer wants something to be done very quickly, he leaves a paper on the table of some assistant and undersigns it "H" (Hartmann). But it is of no use, as he cannot copy Hartmanns

[387]Genia was working at the Gosudarstvennyi Fiziko-Technicheskii Rentgenologicheskii Institut (GFTRI) and not at the university itself; hence the issue of 'common' free days arose. In Russia, every worker had a free day every five days.

[388]Street where Genia and her family lived.

[389]Paul Scherrer, (1890–1969), professor of experimental physics at Zurich.

handwriting, it is too clean.

Next week I must go to Stodola again. I already had some long speeches with him by telephone. He never speaks less than half an hour. Yesterday, Scherrer passed and saw I was telephoning. Five minutes later he came back and saw I was still telephoning. So he said "Ah, you are speaking with Stodola!" And so I was. My dear dear girl I want to know how you are and how your illness. I want to know whether you are cross with me (you are not, *yes?*) and I want to see you. *Good bye.*

Your Rudi

[97] Genia Kannegiser to Rudolf Peierls

Leningrad, 22.1.1931

"How sorrowful it is
to be dying so young
— of tuberculosis"
(Violetta's aria from 'La Traviata') (!!!)

It is terribly difficult for one to act the parts that are contrary to one's character, so I can't do Violetta. I've tried to have my photos taken showing my high-coloured cheeks and a "gentle" smile on my pale lips, but the result was a suspicious-looking person, a far cry from Violetta.

Rudi, my evening temperature is still 38° C, and they make me swallow all kinds of stuff and Mother makes me eat something every 20 minutes. "There is no question" about skiing because I get tired after a short walk. I was palpated, checked over, X-rayed, in a word, I underwent all sorts of humiliation. If it goes like this, one of these days I'll go to the graveyard to choose a proper place. And I'll write my last will and testament! But enough is enough! Assez! My patience is pushed to the limit. In spite of doctors, medicines, etc., my spirits are up inappropriately, and I feel happy. If I keep on feeling like this, then, as it seems, at my funeral I'll rise up in the coffin and order to play a merry fox-trot instead of the Marche funèbre. All of this is nonsense, of course. In the meantime I gave up working. Two days ago, though, I began to work a little, regardless of the doctor's anger. This job is at *another Institute, much nearer to my home (the weather is now awfully cold: $-25^o C$ (!)). My working time does not last long: 3–4 hours only. The rest of a day is spent on reading, medical treatment (!!!) (my word of honour, it takes a far longer time than scientific research), and even on playing the piano (?!!). In a word, it is an ideal lifestyle (!!). Well, all this cannot be very interesting. My poor dear, my own boy, how are you doing now? I pity you so much, you being ill and alone — it is really awful.*

And I am sure that you forget to take your medicine and do the things you must not to do, and why you write the desagreable letters, when you are ill and have the high temperature. Oh? *Rudi, Rudi* — what I can do with your illness when I am so far! Stupid!

You know that I am very "independent" and can very well arrange all sorts of "life" things, but I can be ill only at home with Mama. When I become ill and Mama is not here, I am quite desolated and do not know what to do. It is very strange, I cannot even understand what is the matter here, but when I am ill I am quite little girl. Last year it happened with me so funny story, when I returned home from *Crimea?*. I will relate you it, it is so long to write. *Grrr.* What an awful night I had on *Grusija* sea-sick, gropes, fever, temperature, and headache — all together and it rained. My only consolation was the little dog, my vis-à-vis who suffered even more and who cried all night. I did not certainly. In morning I was a "very sad picture", but it was better. You know, that in the first time I thought that I am not for you a simple "adventure" when I met your eyes on that wet deck on that grey morning. And now you are alone, and I cannot help you. Did your landlady all you want, and what eat you when you are ill? *You know the ballet.* It was a very nice ballet concert, but we had such a lot of adventures in the way (without " "!) that I thought some minutes that we shall be there when all will be finished. Two "rendez-vous" were *moved.* Two!! The tickets were left at home, the hour of beginning was not right etc. etc. At the end we were only a little late, but I never saw such "bad organisation" it was seven persons and some of us had the meetings in their institute and so on, but I really too much adventures. To morrow I will go somewhere, but I really do not know where. Think — three interesting things in one evening and very far from one another.

The concert is of one famous *ballerina* the splendid ballet in the theater and very interesting concert — old french songs. I think that I will see this *ballet* but I want all these three together. Awful!

Darling, the train adventures is your "specialité de la maison", or is it Thorner's? I think that it is yours, do you remember *Kislovodsk* or even *Mineralnije* where only the last 0.5 minutes all was ready. It is always happens with your something in the railway station, or only this year? *Rudi*, yesterday I saw a very pretty french woman in the tram, she was with her husband very typical french bourgeois and did not shut for a moment her mouth. She was quite delighted with town and I was really proud. Now it is really very beautiful all the town all the trees, all the wires are quite white ... deep, deep snow — and the sun red as blood and pink sky. Awfully pretty. I hope that in March we shall have some very cold days and you will see all this.

Darling, you are a *cheeky one*, "the lectures on "adventures" — (!) I hope that you have received my last letter. It was a lecture?!!!? I remember it very well — "you know my point of view on adventures" and it was all about "adventures". It was a lot of philosophy there and I even asked your pardon for that, but about "adventures" only these five words. Are you ashamed? I hope you are. *Yes?* Now I am sure that you think that I am a Marjorie of Huxley's novel.[390] *Yes?* Thank you for that high opinion. *Rrrrudi* — if you have not be ill when you wrote your last letter I have been furious as to ... If you will act in such manner in future I will not left you in my testament anything. Now I think you are afraid, and will be a [??] You know that the first syllable of your family name is read on Russian [??] and [???] on children's language means [???] So you must not spoil the good impression of your name, having such opinions of your old friends (yes, old, because five months are a very long time ~ 2% of your life my dear sir !!!)

What is the matter with *Fock*? Awfully strange? It is quite the style of *Ivanenko*, who always thinks that he is stolen by all the modern physiks, but *Fock*? Perhaps he did not understand your letter?

My dearest, and what if you are still ill? Oh, *Rudi*,

[390]Refers to Aldous Huxley's 1928 novel *Point Counter Point.*

Darling, it is near now the 1. III. but I cannot imagine yet how it will be. I remember you very well, very. Always I forget the faces very quickly, but they are lying somewhere underneath and when I met a man I always know, who he (she) is and where I have seen him (her). But I tried to remember you and some of your faces are awfully bright in my mind and with spectacles and without them. Are you changed in that time? I am not I think, or a little. We knew one another so little that we even will not notice this change. It is funny! Good night. *Good night, darling* It remains 4 or 5 letters to be written more, your answer on that will be the last I shall answer. What will be after 1.V.? Perhaps they will be the last letters between us. Very interesting such "near" future and we do not know anything. But I am an optimist also. Darling, I want to tell you many things, very "deep" things, but it is very difficult and 20 days!!! I do not know will you be interested in them, when you will receive this letter!

Here is some words for Dau, you can read them certainly, and I am even interested will you understand them or not. Darling, *I kiss you tenderly, all the more if you are unwell.*

Genia

[98] Heinrich Peierls to Rudolf Peierls

Berlin, 22.1.1931

Lieber Rudi.
Unter Umgehung sonstiger Korrespondenz bitte ich Dich, mir sobald wie möglich mitzuteilen, ob Du die Absicht hast, den Professor in Leningrad zu vertreten bezw. ob Du Dich hierzu verpflichtet hast.[391] Nach Empfang Deiner Antwort werde ich Dich auf Verschiedenes aufmerksam machen.
Inzwischen herzliche Grüsse
Dein Vater

[391]Peierls had told his father about his intentions to travel to Russia again, explaining the need to by his intention to take over some lecturing duties for a Russian colleague.

[99] Werner Heisenberg to Rudolf Peierls and Lev Landau

Leipzig, 26.1.1931

Lieber Peierls u[nd] lieber Landau!
Vielen Dank für Ihre interessante Arbeit, deren Tendenz mir außerordentlich sympathisch ist.[392] Es ist mir durchaus klar — und von Ihnen sehr schön herausgearbeitet worden — daß man in der bisherigen Quantenmechanik der relativistischen Effekte viel zu viel "unbeobachtbares" in die Theorie hereinsteckt und sich über die verschiedenen Unglücke eigentlich nicht beklagen kann. Allerdings hab' ich gegen Ihre Arbeit einen Einwand, den ich auch oft gegen Bohr geltend gemacht hab': Ich glaube stets gerne, daß sämtliche bisherigen Begriffe versagen. Was mich aber interessiert, sind die ganz wenigen Relationen, die nicht versagen: so etwas, wie 1924 die Burger-Dorgeloschen Summenregeln.[393] Also, um meine Kritik spießig (im Sinne Landau) einzukleiden: "Δος μοι πον στω[394]"! Aber ich will Ihre Arbeit noch genau studieren, ob wirklich alle die Begriffe versagen, von denen Sie's behaupten; und ich hoffe, daraus zu lernen.

In Ihrer Arbeit steht übrigens noch ein ganz unwesentlicher, aber elemantarer Fehler an der Stelle, wo Sie die Ungenauigkeitsrelation der Quantenelektrodynamik kritisieren, wie ich mit Schadenfreude bemerkte. Es gibt in der Quantenmechanik einen schönen Satz, demzufolge die zeitlichen Mittelwerte sämtlicher Operatoren (i.e. ihre Diagonalelemente!) vertauschbar und daher gleichzeitig meßbar sind.[395] Dies gilt auch für $\mathfrak{E}$ und $\mathfrak{H}$ und daher ist es völlig trivial dass bei Messungen

---

[392]See note 380.

[393]Hendrik Berend Dorgelo (1894–1961) and Herman Carel Burger (1893–1965) had worked out empirical sum rules for intentities in a line spectrum. See comments in Pauli, *Wissenschaftlicher Briefwechsel*, I, pp. 157–8.

[394]This quote is attributed to Archimedes when he attempted to move a ship with the help of block and tackle.

[395]Pauli criticised this point in his handbook article of 1933, when he pointed out that impuls and energy could be measured exactly and therefore the Landau-Peierls uncertainty argument needed further qualification. See W. Pauli, 'Die allgemeinen Prinzipien der Wellenmechanik', in *Handbuch der Physik* ed. by Geiger and Scheel, 2nd edn. vol. 24, part 1, pp. 83–272, here p. 257.

über lange Zeiten Δt, die ja Mittelwerte von $\mathfrak{E}$ und $\mathfrak{H}$ bestimmen, die Ungenauigkeitsrelationen meines Buches nicht herauskommen können und nicht gelten und nicht behauptet wurden. Diese beziehen sich vielmehr auf Momentanwerte $\Delta t = 0$, wie sämtliche Vertauschungs-Relationen und Ungenauigkeitsrelationen der bisherigen Theorie. Man kann aber sagen, daß die Ungenauigkeitsrelationen meines Buches auch dann noch gelten müssen, wenn Δt zwar $\neq 0$ aber noch $\Delta t < \frac{\Delta l}{c^2}$, weil in solchen Zeiten die Mittelwerte von $\mathfrak{E}$ und $\mathfrak{H}$ über das Raumgebiet Δl^3 sich nicht merklich ändern können, also durch die Zeitintegration nichts herausgemittelt wird. (Die hohen Frequenzen sind ja durch die Raumintegration herausgemittelt.) Für $\Delta t < \frac{\Delta l}{c^2}$ gelten aber die Ungenauigkeitsrelationen meines Buches trivialerweise nicht mehr, (was auch nirgens behauptet wurde), ebenso wenig, wie etwa zwischen den Zeitmittelwerten von p und q in der Quantenmechanik irgendeine Ungenauigkeitsrelation besteht; beim Wasserstoff ist z.B. $\overline{x} = 0$ und $\overline{p_x} = 0$. Von statischen Feldern zu sprechen, ist deswegen falsch, weil es die ebensowenig gibt, wie 'statische' Koordinaten beim Oszillator.

Dieser ganze Punkt ist aber zu uninteressant, um ihn ausführlich zu diskutieren. Hoffentlich schreiben Sie also nächstens eine Arbeit über die Begriffe, die nicht versagen. Herzliche Grüße, auch an Pauli!

Ihr W. Heisenberg

[100] Genia Kannegiser to Rudolf Peierls

Leningrad, 27.1.1931

Thank you for the pictures, darling, for the New Year's photos. Why is Dau wearing white trousers? Your question at the back, "I am awful, yes?" is genuine "fishing", isn't it? Oh, no, you are terribly nice, much better than you really are! (Get it, for "fishing".) Dau is majestic, standing on skis in the middle of the snow desert, =(Pat+Don Quixote)/2 -Rossinant, but on skis, instead. (Pat is the Danish cinema comedian; I think you know who he is?) Eva is very nice (how old is she? 18?) and Bübchen, as he appears on the photo, is a clear-looking boy (an egoist!! (?)), and if Eva marries him, he will keep her under *his hoof and rule her with an iron fist (refer to the dictionary for these "technical" expressions). The boy is very lively. Is that right? For the past $1\frac{1}{2}$ years Dau has put on flesh and acquired an "adult" look. Yes, he actually looks much more adult than he did before his departure.*

Thorner also grew up a lot compared to the picture that you sent me in the autumn. Or he simply looks displeased with something on these photos. With what?

I'm so happy to know you got well; I pitied you a lot and in my dreams at night I saw various misfortunes that befell you. Are you quite, quite well? Are you? Don't smoke out-of-doors, if the weather is cold. That is the reason why people catch colds. You won't, agreed? Rudi, only one month to wait (a bit longer, indeed). It is great, really great, yes? Oh, we'll have to elaborate a lot on techniques in our struggle against friends' hospitality. I'm accustomed to this; it is very complicated, but sometimes it is very funny. Oh, there is a ring at the door, someone to see me, I think (now it starts!!).

Yes, that is right. Good-bye, my darling.

They are gone, these two friends. I like one of them very much, but in small portions. You may remember, I wrote about him: the melancholy boy; in fact, not a boy at all — he is 26. This time, luckily, he was in high spirits and rather nice. As to the other, I have known him for a very long time, since childhood, but he is one of the most boring people I'm acquainted with. (To tell the truth, Sack is a bit more boring). When I speak with him I feel my stomach clench (a very strange sensation, you have never experienced this). You know, they are typical "engineers", who speak about anything as though it was a tram ...

Jan. 28

At that moment Mother put me to bed. Now the daily routine has become very strict. I'm early to bed and late to rise; besides, they force me to lie down in the afternoon. Oh, Rudi, I'm bored to death with all this, I have been ill for a whole month. Yesterday, it is true, I had 37.1° *instead of* 37.8°. *I was so hopeful, but this morning again I have* 37.3°. *Ouch!!! Soon I may go to Detskoye Selo, to the Scientists' Holiday Home.* Do you remember we have been in ones in *Kislovodsk* and in *Crimea*, I show you "*Gaspra*" the big palace where we drink water and I wounded my leg. Do you remember roses, and fountains, and cold cold water? In *Detskoye* it is very nice also: a little house in a garden, very quiet and a big *Verandah* on which I shall lie all the day in (n+1) blankets and warm things. *Detskoye* is very near from *Leningrad* 35 km only, but it is much higher and the climate there is better, not so wet. So I hope that there in 2–3 weeks I shall be all right. I am not sure yet, that I will go there because, you know, I am in the state service, and so I must receive my *leave of absence* for such long time and it is connected with some formalities state doctor's commissions and so on, but I think I will obtain it because I have too high temperature for go to work. X-rays showed that I have: "pleuritis, the toubercoulosis of the glands, very advanced and the beginning of toubercoulosis of lungs. But all this is nonsense and I hope that in 3 weeks I will be quite quite all right.

Darling, my good character begin to spoiled. I think but it is always so with me when I am ill. I am really disgusting there. But I think that for 1 March I will be in a good state of mind, again. *Ach, Rudi* I quite foget to tell you that I fell in love with a ballet girl! (Shocking!) No, seriously, I wrote you last time that I shall go to the concert of a famous Moscow ballerina. I was and have been delighted, she is quite charming, dances ...!!! and is so jolly that everybody laugh when she smiled, what a charming girl! (She is very young 24–25.) *Rudi* I am very sad. 2.II. will come Bubi Mandelstam for some days. All the people will have skies, parties, dances, theater and I will lie alone in *Detskoye* Oh, oh, oh! No, not so bad. Not alone there certainly will be some people and I awfully like the new ones. It is always interesting. *Yes?* I hope that my next letter will be from *Detskoye*. Write me on the old adress *Mokhovaya* [street] it will be quicker and shall have the relations with home, so I will have your letters immediately. Write soon and often and long letters, *Agreed?*

I am grumbling like a 80 years old ruin. I beg your pardon for such a tedious letter (not at all fishing). *I kiss you affectionately.* It is not dangerous, because I do not have the toubercoloses bacilli!

Good-bye, my darling, again. Your housekeeper is charming.

Your Violetta (!!!) Ah!

[101] Rudolf Peierls to Genia Peierls

[Zurich], 28.1.[1931]

My dear sick girl!
What only is the matter with you? You have been ill for such a long time. This not funny any more. But seriously, Genia, you have to get better. You even claim that I am doing everything that is not good for me, and what do you do? Go to the theatre, the ballet, even work etc.

My dear, I understand so well, how boring it is to stay at home all the time. But wouldn't you get better more quickly, if you did everything so many doctors told you?

You don't have to feel sorry for me. You know that I was only ill briefly. Even if it had been worse, I am not on my own here. My landlady did everything that was necessary (of course I could also eat at home) and Dau came by every now and again. You know I have already been away from home for more than five years, and therefore I am used to life without parents.

It appears that you are terribly angry about the "lectures on adventures". *Surely you know that it was not meant to be serious? Of course you know that and therefore one does not have to talk about the topic any more. But why do you talk about Marjorie*[396] *so frequently?* It appears as though she made a great impression in you. If the reason for this should be that you discovered similarities between you and her, I must assert that *you are much better than you think and that you are not at all similar to her!*

Silly thoughts! It is true that we are already "old friends" *and I like these 2% of my life very much and do not want to lose them.*

Everything is o.k. with Fock now,[397] *I wrote him a letter from Arosa and I have received an answer*[398] *which shows that he has made peace with us.*

Last week, Dau and I finished the paper. I am very glad that it is completed now. Even Pauli, "der Theorienfresser" *did not discuss a lot*

---

[396] See letter [98].
[397] See letter [93].
[398] Letter [95].

and Heisenberg even wrote that he agrees.[399] *But, I don't know why, it is a bad state, when a task is finished. One day I am glad about it, but then everything is boring, and I have to think about something new. And that is not so easy. Currently I am thinking about something, that I had started before and never finished. I discovered that the method was bad, and everything I had done with it in vain. And this causes me to be in a bad mood. Dau, too. Do you remember that I wrote to you that Dau, in my opinion, is not only a "child", but that you have to take him more seriously than you do? I think this difference is not difference between you and me, but between Dau as he was a year ago and as he is now. There is a big difference, he believes in himself much more. But I have not changed at all and you see the same boy, whom you saw in Kislovodsk, with and without glasses.*

Today my landlady forgot again to close the windows and it is terribly cold in my room. What to do? I am going to bed. Good bye my darling, my dearest Marjorie, *good night!*

29.1.

Good morning dearest. *I am in the institute and should work, but I am too lazy. Isn't that awful? [???] Beautiful piano.* Beethoven, Bach, Chopin *and many more. Now my dear, tell me what you think. You want to tell me something serious, but you do not know whether I am interested in it? Tomorrow you will say you want to write me a letter but you do not know whether I am going to read it. Don't you know, how much I am waiting for your letters and how important everything that you write is to me, especially serious things and "philosophies" (even* "lectures on adventures" *I would read with pleasure).*

By the way, in spite of your critical remarks our cosmic rays seem to work again. At least all the questions of your letter are answered already two weeks ago, aren't they?

At the moment I am reading Tchechov in Russian. That's great. And I am reading a lot faster and more easily than before and I can almost read without a dictionary.

[399] Letter [99].

If your answer to the letter was your last letter, this would mean that I have been without news for 14 days. How mean!

This afternoon is my lecture, and I don't know yet, what I am going to talk about. I will have to think of something witty (as Dau would say). Now, good bye. I hope that I shall read in your next letter that you are completely healthy again and that you have sent away all the doctors. My dear girl. I kiss you strongly. 30 more days!

Your Rudi

[102] Genia Kannegiser to Rudolf Peierls

Leningrad, 2.2.1931

I write you in my laboratoria! *It is great, darling, isn't it great? The thing is, it was a lot of bother being ill, and I didn't succeed to go [to sanatorium] in Detskoye. I was not allowed to go on a long holiday, etc.; so I made up my mind to begin working a little.* So it is now. I am still in the middle of my temperature 37.8–38.0 every evening, but it is in the evening and in the morning 37.2 only, and I will try to remain in the "health"–position (*It is such a bother, my dear, such a lot of bother to stay unwell for a whole day long.*) I will be very tired or if my temperature will spring on the higher degree I will receive ten days and will go to *Detskoye* in the middle of this month. To tell you the very very truth, I will stay in *Leningrad* this time, because we shall have such nice theatre that it is quite impossible not to go. But really it will be better to go in *Detskoye* in some time. I cannot to be ill any more, and in the fortnight I will be very glad to lie and to do nothing. My dearest, dearest, dearest, in that 4 days I wrote you a minimum of 10 letters (9 were in mind and 1 I began on the paper, but it was so uncomprehensible, that I had no courage to sent it to you) and sent one telegram, have you got it, dear and did you recognize the author, I did not know how many cost the words in Swiss and had not money enough for signature. Rudi, I want to tell you such a lot of things, such a lot that it is simply impossible to write. Only one month now, but I want that it will be this evening. Darling, certainly we shall organise our first "meeting". But how? Perhaps I will meet you on railway station, or something another? Write, what you think about it. What happened interesting in these 4 days? Yes, I saw a splendid film, darling, but really splendid. The Ost-Afrika — all the kinds of beasts, the lions, *giraffes, elephants* awfully interesting. Splendid! Rudi, will you we shall go together in the desert. It will be splendid. Yes? Yes, before the cinema we, I with one of my "friends" go on *the Islands* and farther in *a taxi.* It was very nice, indeed, think only, the moon and stars *and there was no one to be seen,* and quite quite *cloudless*, and snow, all covered with snow, the bushes, the trees but all. We were very far from town, *suddenly our automobile made "fr-r-r-r"* forever and stop. We were in snow!!! And Mama did not know that I go there, because she will be very anxious (my temperature!) so it was quite awful she will think that I am killed or something of that kind, and we begun to take the auto from snow, we worked 20 minutes (in snow) but at last all was in order and we could even go in the cinema. But the perspective to remain in snow all the night was very amusing! *Yes?*

Dear, thanks for the letter and for the love: 3 and 4 pages are love in my sense of the word. With me it is the same thing. You know the morning when I got your letter, I wanted to see you so awfully, I wanted you to be here so strong, that I think that I *I would burst into 1000 parts.* Rudi, I dream every night that morning is March. My, good by darling. I am tired to day *(after the relaxation I've had!). I kiss you most affectionately and love you so much.*

G

It has not hurt in reality. I "theoretically" wanted to be. You understand.

[103] Rudolf Peierls to Genia Peierls

[Zurich], 3.2.1931

My dear girl,
I was so happy when I received your telegram on Sunday evening. Many many thanks!

Now, how are you? You are still not well? And you even cannot go to Detskoye Selo? You poor, and you talk about "bad character". What do you think, who would not be angry?

Your telegram got here very quickly, already on Saturday, 31.1. in the evening, but I only received it on Sunday, because we were in the mountains.

Saturday afternoon we were in Davos and Sunday we went on a very nice tour. It is impossible to imagine how nice it was. This tour is one of the nicest in all Switzerland, you have to climb up only two hours and then it goes down ... two hours running down with great speed. Only it is too well known how beautiful it is and there is a great "traffic". You can buy special tickets for the railway from Zürich to Davos and back from Kütters (the other end of the tour) to Zürich, and on Sunday crowds of people are going there. *It was astonishingly beautiful weather, sunshine, blue sky. I was with Waller and Dau. Dau tried all day to think of a nickname for me, but he could not find one. When he heard "pay", he was thrilled. This was exactly what he had been looking for.* Really it is a splendid name!

Last night was the guest performance of an English theatre company; they played "Arms and the man" *by Bernhard Shaw. Brilliant. A super performance. I love this play. I am sure you know it. Yes?*

You know, Dau and I sent our paper to Bohr.[400] *He did not answer for a long time, but now we have heard that he disagrees with everything. We just do not know what he does not believe; it is very difficult for him to write. Therefore I'll probably leave on the 22.II. and will be in Copenhagen for a week.* Do you remember I told you about my plans with Simon and Berlin? They are completely spoiled now, because Simon is going to be professor in Breslau. But on the same day there was a new and more concrete possibility that I could perhaps get a position in the Technische Hochschule in Berlin, working with R. Becker. Becker is not too interesting, and so I am not sure I shall take it, but I am thinking about it. Economically it would not be better than my position here, but Berlin and Zurich!

[400]See letter [95]. See also [155], note 624.

Last week there was a concert of Edwin Fischer;[401] It was really splendid, especially one Chopin and one Bach (although generally I cannot understand Bach).

You are right that Thorner was very melancholic all the time in Arosa, and it can easily be seen on the pictures. But I could not find out why. He was very tired, it is true, but surely this is not the only reason.

Eva is not 18 but 25 or 26. But I am glad that you like her. Bübchen *is an egocentric, honestly, but he is not very strong. If he married Eva (which is never going to happen, I think he has already forgotten her), Eva would be wearing the trousers.*

I think I did not tell you about Bretscher[402]*; he is a physicist in our institute, a very nice boy. He was the friend of Bloch who was with him in the mountains when that accident happened and who did admirable things then. He is only very nervous and always doubts about himself and so it is a bit difficult with him sometimes. Now he has fallen in love with a student of mathematics, a very nice girl (very nice) but very shy and silent. It is really a pleasure to see these two, they fit so well to each other. Now, unfortunately, they are formally "engaged" (with rings and all such things) which made Dau very furious. And really it is awfully European, isn't? Marry or not marry, but engagement ... !*

Next week my sister and her husband will be in St. Moritz and I shall probably see them when they come through Zurich. This will be very nice.

Now I have to work a little. Good bye, my dear, dear girl.

4.2.[1931]

I have an awful lot of time now, and that's why I am doing nothing. Yesterday I had to talk to the assistant for a very long time. Stodola is a very stupid assistant. It is quite difficult to explain anything to Stodola,

[401] Edwin Fischer (1884–1960), Swiss pianist who was most famous for his interpretations of Bach and Beethoven.

[402] Egon Bretscher (1901–1973) studied chemical engineering at the ETH Zurich, and obtained a Ph.D. in Chemistry at Edinburgh, before returning to Zurich in the late 1920s.

but when he sends an assistant who does not understand anything and I have to explain something to him and he has to explain it to Stodola this is a nightmare.

Today Dau, Fania Moskovskaja and I went to the cinema and we watched a terrible movie, a German "detective film". I have rarely seen a fim more moring.

I will probably not travel via Berlin, but by boat Kopenhagen-Helsingfors. The ship arrives in Helsingfors on 2.III in the morning, and I'll be in Leningrad on the 2nd or 3rd. And then I'll see you! Another 3^3 days. That is too much! I am longing for your letter which I will receive tomorrow or the day after.
So get well as quickly as you can or even more quickly! Good night my Genia. See you very very soon!
I kiss you,

Yours Rudi

Between 22.II and 28.II my address will be: Universitetets Institut for teoretisk Fysik, Blegdamsveg 15, København Ø, Denmark.

[104] Heinrich Peierls to Rudolf Peierls

Berlin, 4.2.1931

Lieber Rudi.
Für Deinen Brief vom 3. d[iesen] M[onats] und Deine Karte herzlichen Dank. Wir freuen uns, daraus zu ersehen, dass Du am Sonntag Sport treiben konntest.

Was Deine Reise betrifft, so mache ich zunächst darauf aufmerksam, dass Du Dich offenbar geirrt hast, dass Du nicht den 2. bez[iehungsweise] 3. März, sondern den 2. bez[iehungsweise] 3. April meinst. Unter dieser Voraussetzung kann ich Dir bestätigen, dass Du uns in der genannten Zeit in Berlin antreffen wirst, wenn nicht unvorhergesehene Ereignisse eintreten. Die Mitteilungen, die ich Dir zu machen hatte, kann dann persönlich erfolgen.

Wie ist es übrigens mit der Rückzahlung des Darlehens an Hirsch?[403] Ich frage nicht des Geldes wegen, sondern weil ich glaube, dass man Ordnung halten und verlangen sollte, dass der Betreffende sich äussert, wie er sich die Rückzahlung denkt.
Herzliche Grüße

Dein Vater

Lieber Rudel!
Herrlich Deine Skitouren. Ich war 2 mal auf dem Eis + hoffe, es geht weiter so.

Wir sind im Begriff, uns den Mont Blanc Film anzusehen[404] + so Hochgebirgsluft zu atmen! Darum addio.
Viele Grüße

E[lse]

[105] Genia Kannegiser to Rudolf Peierls

Leningrad, 7.2.1931

My dear boy,
why is your letter so sad? What has happened? Do you feel offended by me for some reason? What is it? That will be funny, if the cosmic rays are in action and in 3 to 4 days I'll receive the answer in your next letter. Or it was only my impression of your being upset? Behind me I can hear the coals cracking in the stove (it is so cold, that central heating is unable to make the room warm and we have to heat the stove). I like very much to watch the embers in the stove; I imagine towns, castles, ships appear and disappear .... Nice thoughts come into my mind when I'm by the fireside, and daydreams too. Is that right? Do you like stoves? I do so much want you to be here — even if we weren't talking.

---

[403] Peierls had lent some money to a friend. See letter [47].
[404] *Stürme über dem Montblanc*, first sound film with Leni Riefenstahl and director Arnold Fanck, 1930.

8.2.[1931]

Unfortunately, to my silent call there came — not you (here the cosmic rays for you!), but Bubi, so I had to stop writing. Yesterday I did so much want to sit together with you by the fireside; at night I dreamed of big fires and you in most fiery images, a mix of fireman and Phoenix! Last night there was a jolly party, some 15 guests came to meet Bubi. There was a lot of dancing, drinking and yelling: my poor parents had terrible headaches — I'm always amazed at their favourable attitude to our behaviour. By the way, the guests nearly broke Bronstein's neck, as they threw him up to the ceiling and then he was passed like a ball around the room; he tolerated it with heroic patience. I didn't expect him to be so courageous; to tell the truth, he was a little drunk. At 2 h 30 m I felt the pricks of conscience (oh, my parents' heads!) and the voice of reason. All of us were to get up early in the morning; Bubi and I with enormous efforts forced this horde down the staircase. I've never seen such confusion of galoshes and overcoats as happened then. As a result, Ambarts went away wearing somebody else's two galoshes, both for the left foot. There was an amusing youth, with whom we sat side-by-side at the table and then danced a lot — he is an outstanding dancer! His parents live in Detskoye Selo, he often visits them, and he promised "to wipe my tears away", when I would feel miserable in the sanatorium (I think I'll have to go there, because the temperature doesn't get lower, and at 4 p.m. I feel rather bad). Yesterday I drank, of course, only one wine-glass, but for me it was insignificant. Poor Bubi, who sat next to me on my other side — his ear-drum must have burst, at least he stated this; he is a disrespectful nephew, isn't he? Today our big company is going to see a ballet, a new performance with music by Shostakovich, a very young and a very talented composer. Tomorrow Bubi is leaving to return to Moscow, and I go to the concert of Rosenstock:[405] *he is on tour for the first time, I hope it will be great. Such is my lighthearted lifestyle. Are you indignant with it?*

I'm awfully glad that you read Chekhov; what story are you reading now? By all means read his "The House with the Attic", "The Lady

[405] Joseph Rosenstock (1895–1985) conductor.

with a Doggy" and "The Black Monk", and also his plays. I hope, in March the Moscow Arts Theater will come on tour (you must have heard about it and about Stanislavsky)[406]*, and we shall go to see "The Cherry Orchard". This is the best dramatic play I've seen in all my life. It is impossible to imagine how good it is. On the 15th I'll go to the matinee "The Brothers Karamazov" by Dostoyevsky (you must have read the book). They dramatized the novel and the people say it is even better than "'The Cherry Orchard". I look forward to seeing it. Nowadays a lot of interesting concerts and performances are on, but I'm able to see only 25%, because at home I'm considered to be unwell, and after my daily work Mother wishes to put me to bed until morning. That is terrible! Now I'm getting better, at least I can do something at home. Darling, you write that in the near future I won't write any letters, because I'll think you won't read them. You know, sometimes my pen stops and I start thinking, how funny it would be, if on getting my letter Rudi pulls a wry face and say, "What on earth, again!" I trust you (really and seriously, I trust you), you are a straight man, but for purely technical reasons you receive a letter written to you such, as you were at least 2 weeks before, and in two weeks a lot may pass. You made my acquaintance, became my friend and fell in love with me during 13 days, and how many days does it take to fall out of love and forget? I think, not much more. Thus, my dear; sometimes I begin to feel depressed in the middle of the letter.*

You are quite right, when you say that people may mean different things when using the same words, especially such words as "mind", "love", etc. I think we agree as to the sense of the word "love". I love you very much, and it is unbearable to be separated by 3 frontiers and lots of kilometers. I look forward to March with such longing that it seems never to come. Though see, Rudi: there remain only 20 days, i.e. 1/9th part — but that is the most difficult period. Just as when going by train, the most boring are the last 2–3 hours. Here is the same: the hardest stretch will be the last 10 days. I'm out of my wits

[406]Constantin Stanislavsky (1868–1938), Russian theatrical director, teacher and actor; co-founder of the Moscow Art Theatre. His outstanding productions include many of the Chekhov plays.

end, imaging what is going on to take place in these 2 months, and especially after these 2 months. We haven't learned enough about the nature of each other, we had no time to get accustomed to each other, and it is so difficult to wait for the letters, and so sad to be so far apart. In addition we arranged a rendez-vous after April, unless we'll quarrel. Well, we have a large space of time till then, we shall think of it, shan't we? This may be called "the irony of fate": the probability of the symmetrical combination is $\leq \frac{1}{4}$ *and we are parted 1000 kilometers apart, no* Luftpost, *no telephone. Ah, darling, this isn't the last letter, there will be another 2 or 3; you'll have time to write the reply to this one. Good-bye, I kiss you affectionately,*

Genia

Yesterday Bubi shot our photos in the most "fantastic" postures. If something good comes out of it (he took a large exposition with electric light), I'll send you the photos.

[106] Rudolf Peierls to Genia Peierls

[Zurich], 9.2.[1931]

Genia,
I am terribly tired and I don't know why. I have talked with Guth *all day (I don't know whether I have told you about Guth, a dreadfully boring person) and with many stupid people.*

But yesterday it was great. I went skiing with Dau, Waller and Motschan (a nice chemist) on the "Hochstuckli" *not far from the* Vierwaldstättersee. Impossible to tell you how beautiful all this, running down with great speed about one hour on splendid snow. I am feeling more and more sure on my skies and even Dau gradually learns to manage them.

We came back early and in the evening we went to the theatre. A very good play "Voruntersuchung". *The author is a famous Berlin solicitor called Alsberg.* It is long ago since I saw an equally interesting and well-written play; and it was very well performed. Last week we were twice in the cinema (You know Landau seduces everybody and when he is present it is impossible not to go to the film) but saw very bad films.

It would be awfully nice if you could meet me at the station in *Leningrad*, but I am afraid that *Dorfman, Bronstein etc. will also be there. They are all very nice people, but ... And Dorfman has to know when I arrive, because he has to book a room for me.*

10.2.[1931]

Darling, I fell asleep in the middle of writing the letter! I'll probably arrive on 3.III. at 13.15h. From the station I'll go to my room and there will be a letter telling me where and when I can see you, yes?

You still have to tell me what I need to bring (apart from bananas) — But when I arrive you will probably be disappointed, because I speak Russian a lot worse than I write.

My dear girl, what a question whether I recognised the author of the telegram?[407] Of course I thought it came from *Fock* or Mrs. *Frenkel*!

Well, Genia, how foolish to drive by car through the snow, and even at nighttime. Your mother was right to be angry!

I hope that you have now got the holiday which you need. It will be good if you can lie in Detskoye Selo and you will have fully recovered after your stay there. My dear, I can understand very well, how awful all this is and that you want to be healthy and that you want to work! Poor you!

My dear girl, there are only three weeks now, it is practically nothing! Ten days in Zurich, and ten days traveling and then I'll be with you for two months! I cannot yet believe it, I always think that there must be something in the last minute and I cannot leave here or cannot get to *Leningrad* or anything of that kind!

11.2.[1931]

In these last days here, there will be a dreadful lot of work. I must get ready with that stupid change of resistance and there are some complications in the paper of Dau and me[408] which have to be solved before

[407] See letter [103].
[408] See letter [88], note 375.

we come to Kopenhagen. Then all the people here, Waller, Guth, etc. cannot get on with their problems and as Pauli is too lazy, practically everything must be done by Dau and me.

This evening we were with Schneider an architect whom we met in Arosa, he is a very nice fellow, very heavy, very "Swiss", but not at all *narrow-minded middle-class. We drank very much, too much!*

Today, Dau gave a paper in our "Kolloquium". It was brilliant. He spoke about the new paper by Milne,[409] which is complete nonsense. And this is what he reported in each and every sentence he used the word "pathologisch". Some of the Bonzen thought he was "shocking" and they protested; everything was brilliant.

My dear, this is so unfair: The last five months passed really quickly, and now that there is only a very short time left, it passed so slowly. I wish it was March already and I was with you. Good bye, my Genia, and good night!

Rudi

[107] Heinrich Peierls to Rudolf Peierls

Berlin, 11.2. 1931 abends

Lieber Rudi.
Eben kommt Dein Brief von gestern. Dein Zimmer wird am 22. F[ebruar] bereit sein; bitte gieb rechtzeitig die genaue Zeit Deiner Ankunft an. Annies Adresse ist: St. Moritz — Dorf, Pension Bosshard. Else geht es besser + sie wird nächste Tage wieder ausgehen dürfen. Die Angelegenheit Hirsch kann bis zu Deiner Rückkehr bleiben, wenn ich sie erwähnte, so geschah dies nur, damit sie nicht in Vergessenheit gerät. Die Voruntersuchung[410] haben wir schon gesehen und sind ebenso entzückt davon, wie Du. Ich schreibe Dir heute in einer gewissen Sorge, Dein Hin- + Herschwanken in der Angabe Deiner Gründe für Deine erneute russische Reise. Else erinnerte sich genau wie ich,

[409] Probably W.E.Milne, 'The Numerical Determination of Characteristic Numbers', *Phys. Rev.* **35**, 863–67 (1930).

[410] 1931 film with Gustav Fröhlich, banned in 1933. See also letter [106].

dass Du früher von der Vertretung eines nach Amerika reisenden Professors gesprochen hast, während Du es jetzt bestreitest — bringt mich zu der bestimmten Überzeugung, dass Du uns die richtigen Gründe verschweigst. Daraus wiederum folgere ich logischerweise, daß diese Gründe nichts, oder jedenfals nicht ausschlaggebend — mit Deinem Beruf, sondern mit Deinem Privatleben zu tun haben. Also: cherchez la femme. Die größte Sorge macht mir, Du könntest Dir jetzt schon eine Kette an Dein Bein binden, wobei Du weder die Tragweite eines solchen Entschlusses richtig würdigst und die erforderliche Menschenkenntnis besitzest. Auch materiell gibst Du Dich Selbsttäuschungen hin, über die Dir am besten Dein Bruder Alfred etwas bekunden kann.[411] Ich kann Dich nicht zwingen, mir die Wahrheit zu sagen, aber ich rechne darauf, daß Du es tust. Jedenfalls warne ich Dich vor unüberlegten Schritten.
Herzliche Grüße

Dein Vater

[108] Genia Kannegiser to Rudolf Peierls

Leningrad, 13.2.1931

Darling,
18 days — one tenth only. Hip Hip, *Hooray. This is the last letter that you get in Zurich, for the last time I'll write on the envelope "Gloriastr. 35"; shall I write some time "d'E.T.H."? Now we have so closely approached March 3, that I begin to think of the time after May 1. I'm thrilled, awfully thrilled by these future 2 months and later. Isn't it really interesting? Yes? When I'm thinking about it, each time it seems to be something different; of these things, to tell you the truth, one must not think. Simply, whatever will be, will be; and one can neither foretell nor foresee anything. Well, my dear, there goes the second ring, the*

[411] Alfred had married a Ukrainian lady who had a daughter from a previous marriage.

curtain will be up soon and the third act of this ... (tragedy, drama, comedy, tragicomedy, farce???!) begins. What would you prefer? As to me — I can't tell, it depends, but not tragedy, I'm too humorous in nature to take part in a tragedy. Rudi, this letter is going to be a mix of "highly lyrical" and "business" things. Let us begin with business. If I were quite convinced about the worldly wisdom of Dau, I'd have to write much less. But as Dau is well-known for his helplessness, I'll have to go into detail; so I begin:

1. *You may know that in March we may have −25 [C] and in April +20 [C], so take all your warm things, woolen socks, pullovers, warm gloves, a winter overcoat, if you have one, etc. If you take a winter coat, you should also take a spring overcoat. Vous comprenez? Two coats, and if you haven't one for winter, then take sweater.*

2. *In March and April the whole lot of snow, that has fallen and still will fall, begins to melt; so there will be pools of water! You can't imagine how much water there is going to be — an ocean! Therefore you'll have to take galoshes. Oh, yes, galoshes, though you have never put them on. To conclude: don't forget to take warm things and galoshes!!! Then, Rudi, if this isn't much trouble to you (too expensive or too hard), please bring me some bananas, a small pine-apple and a slab of Swiss chocolate. (These are the nicest reminiscence of my childhood: bananas and chocolate.) Well, only if it isn't too expensive and no trouble!*

Oh, I quite forgot to write about Rosenstock: from "so-so" to "very good" in some places (the 2nd part of 5th by Beethoven). There was a fine programme including 8th and 5th symphonies, also 3rd concerto for piano and orchestra. I hadn't listened to it before; they usually play 4th and 5th concertos. I was thrilled so much, I thought I'd go crazy. It still sounds in my ears, and I can't help laughing of the pleasure. Also I have seen a "modern ballet". The music was amazing, especially the arrangement of "Tahiti trott".[412] *Some curious scenes represented all kinds of*

[412] Jazz Suite by Dmitrii D. Shostakovich.

sport, as football and jumps, but best of all was swimming. A wonderful tap-dances by 20 Negroes. However, it is rather a pantomime and not a ballet. Rudi, today I'm going to see "The Servant of Two Masters" by Goldoni; tomorrow there is a concert (Mozart and Beethoven), and on next day "The Brothers Karamazov". Shame on me! (?!!!) On the 17th I'm going to Detskoye for 10–12 days. I am still ill, and if I'm not in the laboratory or theatre, I should stay in bed (it is Mother who insists on staying in bed; I hate it). By the 3rd I'll come back to the city.

Dearest, I don't know how it is going to be in future, but up to now it was wonderful, wasn't it?

Thank you very much for your letters and pictures, and for your love — great thanks! I'd like to write more, but people are running about and around, there is noise of the motor and soon it will be time to go home and on my way I'll mail this letter.

By the way, you wrote some time ago, that when you receive my letters, our relations are reflected on your face. What Dau may know? He lacks the power of observation, even worse than Bronstein. But still, what does he know? I have a purely "technical" interest. And what did he gossip to Kit (Gamov, we called him Kit — Walfisch*). Incidentally, if you'll meet Kit, pass my regards to him; tell him I'm going to write to him soon and send various new poems.*[413] *Rudi, what the hell, there are less than three weeks [left] and when you get this letter you will be sitting on your suitcase [ready to go]. How shall we arrange it: I'll meet you at the station if I am in town; if I am still at Detskoye, you will come to see me. It takes half an hour by local train, and here is my address: 4 Shirokaya Street, the "Holiday Home of Scientists", a 3-minute walk from the station. Well, good-bye once more, the next letter will be the last. I'll write to Copenhagen. And you please write to me too, right? From Denmark it takes 5–6 days for letters to arrive. I kiss you affectionately.*

G.

[413] Both Genia and George Gamow enjoyed writing poetry and they sent their oevres to each other.

[109] Rudolf Peierls to Genia Kannegiser

[Zurich], 14.2.[1931]

What's the matter, Genia? You think that I am annoyed about you? I cannot even remember what I wrote in the letter,[414] *it is dreadful that so much time passes between letter and reply, isn't it? No, of course I was not angry and I do not remember that I was in a bad mood, either. Perhaps I was tired or something like that. It is probably not good to write sad letters, and if it was really like this, please forgive me, darling. You see, I do not have such a good character as you thought. (In a conversation with Dau I found out that "bad character" in Russian means something different than in German, and therefore I now understand why you argued about it).*

Gosh, you are leading a shoddy life, but I cannot say that my life is much better. Tonight I am going on a big fancy-dress ball. Dau will also come, for the first time in his life. It will be remarkable. It is a Böser-Buben-Ball *where we all have to dress up as little children.*

15.2.[1931]

The Ball was great. A shame that you did not see Dau dressed up as a little boy. There were lots of people and some magnificent costumes. The general atmosphere of such a mask ball is very strange as it is a very primitive and rough form of erotic. Most of the people are kissing each other, more or less publicly and never mind very much who is it. I do not like this kind of occupation, and it is of course much worse in such towns as Zurich, where "educated people" in general have so good morals. If then there is such an opportunity like a mask dance, where you can do what you like without being "shocking" it will be all the worse. All this is awfully European, *yes? But luckily, not all the people at the ball are like this, and there are always some with whom you can have fun, dance, laugh, cry etc. And this is what we did. I met a language student, a Slav girl, not stupid and not very pretty who danced very well. But the Swiss people are to Swiss, even the girls are so narrow-minded. For*

---

[414] Letter [103].

example, this girl had never been to a student ball, because people might think that she only went there to meet students! And the same is true for Lilly Fenigstein. Whenever there is a good film showing in one of the smaller out-of-town cinemas, she cannot come. She is afraid that her friends in Zurich would talk about her being in such a terrible place. And it is true, they would gossip.

I have to go for lunch now, and in the afternoon we'll be with Fania Moskovskaja. I want to take photos of everybody, especially Dau in his costume. Good bye my love!

In the evening

It was remarkable. We took very many funny photos, and if they turn out well, it will be fantastic. But darling, do you really think that I could forget you so quickly? I did not talk about "adventures" *(this has no longer been the case, and it is was necessary to fear that) You know the difference, don't you? But that you think I could forget in 15 days only … What a thought! But perhaps you are right and it is not good to write about the interesting things. Writing is so difficult, and writing about philosophy is the most difficult. But soon we'll have the opportunity to talk! And we shall talk, talk very much! Another 15 days! The time that passed was so short, but these 15 days will be so terribly long! What will it all be like? I cannot imagine it. I will probably not get an answer to this letter. There will be another 3 or 4 letters and then: no more letters. I do not need to look at post cards any more, I do not have to write to you about kissing, I will be able to kiss you, and I do not have to talk to you in my sleep, I can really be together with you. It is terrible, though, that we cannot be together the entire time, from morning to night, isn't it?*

Good bye once more, *my dear dear girl! I am kissing you tight (unfortunately only in the letter!). Get well quickly.*

Your Rudi

[110] Genia Peierls to Rudolf Peierls

Peterhof,[415] 18.2.1931

This is my last letter, darling, and it will be very short because I left the paper at home and now have to write on some scraps. It is snowing, and right before my eyes high fir trees are swaying, very high and all covered with snow. In the distance some dogs are barking, and I can hear the yard-keeper cutting firewood. I'm lying on the verandah thrust in a fur sack and covered with a blanket. Only my nose and eyes are outside, and at this moment also the right hand with which I'm scribbling this. I arrived here yesterday in the afternoon (for there was no free accommodation in Detskoye). It is so wonderful here! A white house located in the woods; the interior being decorated in the style of the English Modern. *You know these wooden panels in the lower part of the wall with damask in the upper part. There is wooden ceiling in the dining hall and the smoking room. Amazing armchairs and sofas, a big piano, carpets — all this very cozy. The bedrooms accommodate several people each, so I had to share my bedroom with other women; only one of them snores in her sleep, not too loudly, so it is quite tolerable. The food is!! There are a lot of people, about 50. I haven't made my acquaintance with everybody yet; one of the boys is quite nice, he is from Moscow. There is also a pleasant gentleman and two young ladies who are very nice, too, they are enough for me. Everybody plays dominoes — which I don't like, and other games; yesterday I won them all with brilliance. Tonight they are about to arrange a dance-party. Oh, great news: the local doctor discovered I had endocarditis (a heart disease), that's the limit! If they go on and on, they may well find puerperal fever and melancholic insanity!!!*

Today I'm feeling much better and look at the skies with great interest. Everybody here is skiing and makes me envious. Oh, Rudi! Here they have an outstanding cat — he is red, with green eyes, fluffy fur and tail; then an Alsatian dog called Grizzly. My little darling, I'll be in town for one day on 26th and then come back here. March 1 is quite close,

[415]Now Pedrotvorets, suburb of Leningrad.

oh Rudi, 12 days! It is good we aren't in our usual surroundings: you are in Copenhagen and I'm here — it is easier this way. I was waiting for your letter on February 16 before my arrival here — and didn't get it. Well, they will forward it to me. The last good-bye, my darling, I kiss you for the last time on paper, very affectionately.

G

P. S. Take along cheese and sausage for us.

[111] Heinrich Peierls to Rudolf Peierls

Berlin, 19.3.31

Lieber Rudi,
Wenn ich auch das Misstrauen, das ich in meinem Einschreibebriefe nach Zürich zum Ausdruck brachte,[416] nicht völlig besiegen konnte, so muß ich doch gestehen, dass der Inhalt Deines Briefes vom 13.[417] meine schlimmsten Befürchtungen weit übertroffen hat u[nd] in stärkstem Gegensatz zu dem steht, was Du mir in unserem letzten Gespräch sagtest. Ich will nicht nochmals alle meine Argumente gegen eine so zeitige Bindung aufzählen, da Du sie genügend kennst. Aber ich hätte Dir nicht eine solche Verantwortungslosigkeit zugetraut jetzt gar eine Ehe eingehen zu wollen, ehe Du irgend etwas u[nd] ehe Du eine Frau, wenn auch in allerbescheidenster Weise ernähren kannst. Wie kannst Du als vernünftiger Mensch Deine Entschließung davon abhängig machen, dass Du, falls Du nicht jetzt heiratest, später Schwierigkeiten haben könntest, Deine zukünftige Frau aus Russland herauszubekommen? Siehst Du denn nicht ein, dass ihr beide völlig proletarisieren müsstet + daß es mit Deiner kaum begonnenen Carrière bald völlig aus sein dürfte?

[416]See letter [107]. As in his letters to Rudolf during the previous spring regarding Rudolf's liaison with Marga Vogt (see letter [26]), father Heinrich again warns his son about an early commitment. His main reservation, apart from what he perceives a premature step, is Genia's nationality, which he regards as an obstacle to Rudolf's future in German academia.

[417]Letter could not be located.

Kennst Du denn die Auffassung der für die Universitäts-Berufungen in Frage kommenden Instanzen so wenig, dass Du glaubst, man würde Dir eine ausländische Frau nicht ankreiden? Wo bleiben Dein Leben, Deine Pläne mit Amerika + d[ergleichen]? Willst Du Deine Frau zurücklassen, oder die Kosten für ihre Reisen alle mittragen?

Du kannst natürlich eine derartig große Dummheit machen, die Dein Leben ruiniert, ohne dass ich es Dir verbieten kann.[418] Ich aber halte mich verpflichtet, alles, was in meinen Kräften steht, zu tun, um einen derartigen wirtschaftlichen Selbstmord zu verhindern.

Ich muß Dir daher auf das Allerbestimmteste erklären, dass ich Dir nicht nur keinen weiteren Zuschuß geben, sondern den bisherigen Beitrag zurückziehen müßte.[419] So schwer es mir auch wird, muß ich Dir ferner sagen, dass Du unter keiner Bedingung auf einen Pfennig bei etwa (d.h. sicher) eintretenden wirtschaftlichen Complicationen zu rechnen hättest. Willst Du in Dein Unglück rennen + bist Du vernünftigen Ratschlägen nicht zugänglich, so zerschmeiße alles Porzellan, lasse alle Kultur beiseite + leere den Kelch bis zur Neige. Es kann Dir dann keiner helfen. Sollte in ganz Deutschland keine passende Lebensgefährtin zu finden sein, so prüfe Dich wenigstens eine Zeit lang, ehe Du an unabänderliche Beschlüsse herantrittst und heirate, sobald Du wirtschaftlich dazu in der Lage sein wirst. Wie kopflos Du zu sein scheinst, geht daraus hervor, dass Du uns Ende April in Berlin sehen willst, obwohl ich Dir hier mehrfach sagte, dass wir in dieser Zeit verreist wären. Wir fahren am 10. April zunächst nach Lugano, Park-Hotel, späteres Ziel unbestimmt. Jedenfalls sind wir Ende April oder Anfang Mai noch nicht zu Hause. Meine Depesche hat Dich wohl schon auf unseren Standpunkt vorbereitet. Und nun tue, was Dir nach ruhiger +

[418]Peierls' older brother Alfred, in the absence of his father and stepmother, had married in 1929, a step which Heinrich Peierls objected to strongly. See letter [26], note 144.

[419]It is evident from later correspondence, that Heinrich Peierls did not carry out this threat. On the contrary, he repeatedly urged Rudolf and Genia to accept his financial help, within the limits allowed by the increasingly restrictive German legislation. See letters throughout the 1930s in Bodleian, *Peierls Papers*, Supp. A102–109.

nüchterner Überlegung Dein Verstand, der doch nicht plötzlich völlig abhanden gekommen sein kann, rät.
Besten Gruß

Dein Vater

Lieber Rudel!
Dein plötzlicher Entschluß macht uns großen Kummer. Du scheinst in einem solchen Stadium von Verliebtheit zu sein, daß Du alle Deine Principien, Deine Carrière + ganze Zukunft über den Haufen wirfst. Wenn man so begabt ist wie Du, hat man Pflichten gegen sich selbst + wenn man ein solchen Vater hat, der eine reiche Lebenserfahrung besitzt, nicht kleinlich + altmodisch ist sondern gütig, großzügig u[nd] klarsehend,[420] so meine ich, sollte man seinen väterlichen Rat befolgen! Lasse Dir Zeit und komme erst zur Besinnung. Wir alle haben schon etwas erlebt, worüber wir hinwegkommen mußten. Durch ein solches Erlebnis wird man nur reicher, wenn man es überwindet + sich nichts zumutet, was man nicht halten kann.
In wohlmeinender Liebe

Else

[112] Heinrich Peierls to Rudolf Peierls

Berlin, 31.3.1931

Lieber Rudi.
Deine beiden Schreiben vom 21. und 26. d[iesen] M[onats] sind in unseren Besitz gelangt.

[420]Else tended to be full of praise for her husband's judgement; however her description here is accurate. Peierls described his education as civilised and liberal without an excessive emphasis on discipline. Rudolf Peierls, interview by John Heilbron, 17./18.1963, Niels Bohr Library, AIP. His open-mindedness was also evident in his support of Hans Juliusburger, his attitude towarsd equality (See Peierls, *Bird of Passage*, chapter 1), and his generosity towards his own family and others in need.

Unsere Reisedispositionen sind also jetzt bestimmt die folgenden: Wir fahren am 10. April nach Lugano, Park-Hotel, wo wir, falls das Wetter nicht so ungünstig wird, 2-3 Wochen bleiben wollen. Da uns Nachrichten von Dir angesichts der Tatsache, dass Briefe 5-6 Tage gehen, voraussichtlich nicht mehr in Berlin erreichen, stelle ich anheim, an die eben genannte Adresse zu schreiben. Es ist möglich, dass wir auf der Rückreise, die etwa zwischen 8. und 10. Mai stattfinden wird, für einen Tag nach Zürich kommen. Aus diesem Grunde wäre es wünschenswert, wenn Du uns nach Lugano über die Zeit Deiner Rückkehr nach Zürich Genaueres mitteilen würdest.

Für den Fall, dass Du bei Deiner Rückreise über Berlin kommst, und hier übernachtest, so würde diesmal Dein Wohnen am Reichskanzlerplatz nicht in Frage kommen. Ich stelle Dir anheim, Dich in diesem Falle über Dein Logis mit Alfred und Annie rechtzeitig zu verständigen, die entsprechend informiert sind.

Daß Dein Heirats-Entschluß, (zwei Tage, nachdem Du mich über meine Ansicht befragtest) hier wie eine Bombe einschlug, brauche ich Dir nicht erst zu sagen. Deine Auffasung, daß wir eine solche Aktion wichtig und feierlich auffassen, nehme ich zur Kenntnis. Übrigens hat Else keineswegs gemeint, Du solltest auf mich Rücksicht nehmen. Wenn wir beide über diese Geschichte sehr außer uns sind, so denken wir nicht an uns, sondern lediglich an Dich. Mitteilungen an die Verwandten brauchst Du nicht ergehen zu lassen, das wird sich schon zeitig genug herumsprechen. Im Übrigen hat es keinen Zweck mehr, über geschehene Sachen zu discutieren. Jeder ist seines Glückes, aber auch seines Unglückes, Schmied.

Besten Gruß

Dein Vater

[113] Hans Bethe to Rudolf Peierls

23.4.31

Lieber Peierls,

Sie haben es aber eilig! Meinen allerherzlichsten Glückwunsch! D.h. vorerst hat mir meine Mutter Ihren Brief noch gar nicht nachgeschickt, sondern nur den Inhalt mitgeteilt. Nicht nur verlobt, sondern gleich geheiratet ist eine etwas zu grosse Überraschung. Russin? Ich bin jedenfalls sehr auf "sie" gespannt.

Inzwischen habe ich meinen Wohnort, wie Sie sehen, in den Frühling verlegt, und ich bin begeistert von Rom.[421] Die Stadt kann man immer wieder von neuem ansehen, und findet immer noch etwas Schönes. Ich entwickele dabei einen gewissen Sinn für das Altertum und sehr wenig für Renaissance und Barock-Kirchen, die gar nicht dem entsprechen, was ich mir unter einer Kirche vorstelle. Herrlich (und sehr wesentlich) das Essen. Auch auf der Speisekarte kann man Entdeckungsreisen machen, und 53 Abarten von Maccaroni, Tintenfische, Artischoken schmecken entschieden besser als mutton and cabbage in Cambridge. Der Clou von Rom ist aber unbedingt Fermi, der meiner Ansicht nach in der physikalischen Welt noch lange nicht genug geschätzt wird. Bewundernswert sein Überblick über jedes x-beliebige Problem: er sagt unmittelbar, ob eine Arbeit, die er eben zum ersten Mal gesehen hat, sinnvoll ist oder nicht, und sein Urteil über theoretische wie experimentelle (!) Literatur ist eigentlich unfehlbar. Ausserdem ist er der erste meiner Chefs, die sich richtig um mich bzw. meine Arbeit kümmert, Sie haben es bei Heisenberg und Pauli vielleicht besser gehabt. Jedenfalls habe ich hier wieder Lust an der Arbeit bekommen, während mir in Cambridge vorkam, als ob die noch ungelösten Probleme der Physik uninteressant oder unlösbar wären. Es hat mich gefreut, dass Sie die 2·137 1 Notiz richtig fanden.[422] Manche Leute, vor allem Sommerfeld,

[421] Hans Bethe spent six months of a Rockefeller fellowship in Cambridge, and the subsequent six months in Rome. Two years later, Rudolf Peierls chose the same two locations for his year abroad as part of the same fellowship, but he judged that he improved significantly on the arrangement by swapping the order and staying in Rome in the winter months and moving to Cambridge in the spring.

[422] Hans Bethe, together with Guido Beck and W. Riezler, who had been reseach fellows at Cambridge together, had listened to Eddington's talk about about the 'number 137'. Eddington claimed that he could derive the value of the fine structure constant to be $1/\alpha = \hbar c/e^2 = 137$, starting from Dirac's theory which used a 4×4 matrix for an electron. Eddington assumed that there should be a 4×4 matrix pertaining a proton, and on the basis of the assumption of the two then known elementary particles, the electron and proton, that one would need a 16×16 matrix which, if symmetric, would have 136 distinct matrix elements to which 1 was added

fanden, solche Scherze gehörten nicht in ernste Zeitschriften, und es sei eine Herabwürdigung der Naturwissenschaften. Nichtphysiker, denen ich davon erzählte, sind meist derselben Meinung. Immerhin hat es den meisten Physikern gefallen; nur fürchte ich, dass der Erfolg gleich Null sein wird.

Bei der Widerstands-Änderung habe ich in Cambridge ein paar sehr formale Rechnungen gemacht, aus denen ich sah, dass die Sache sehr vom speziellen Modell abhängen muss. Darauf gab ich die Geschichte auf, nachdem ich mich überzeugt hatte, dass aus der Blochschen Theorie bei Entwicklung nach Potenzen von H die Temperaturabhängigkeit

$$\frac{\Delta R}{R} \sim \frac{1}{R^2}$$

folgt, die merkwürdig gut mit den Experimenten stimmt.[423] Sogar Kapitza[424] gab das zu, obwohl er die Ableitung für verkehrt hielt und die ganze Leitfähigkeitstheorie für Unsinn erklärt. Sie sind offnbar doch einiges weiter gekommen als ich. Ich habe, wie gesagt, keinerlei Ambitionen, spezielle Modelle zu rechnen. Momentan habe ich die Termaufspaltung in Kristallen wieder aufgenommen und zunachst die quantitative Berechnung der Terme von Kristall von dem Fall eines Elektrons

---

for the orbital motion of the electron around the proton. The resulting number 137 was close to the then known value of $1/\alpha$. Bethe, Beck and Riezler published 'On the Quantum Theory of the Temperature of Absolute Zero', *Die Naturwissenschaften* **19**, 39 (1931), which made fun of the talk and Eddington's papers on the subject. In a later issue of *Naturwissenschaften*, (6.3.1931) the journal editors published the follwing 'correction': The note ... was not meant to be taken seriously. It was intended to characterize a certain class of papers in theoretical physics of recent years which are purely speculative and based on spurious numerical agreements. In a letter received by the editors from these gentlemen they express regret that the formulation they gave to the idea was suited to produce misunderstanding.'

[423] Bethe returned to the problem and published his results in H. Bethe, 'Change of resistance in magnetic fields', *Nature* **127**, 336 (1931).

[424] Pyotr Leonidovich Kapitza (1894-1984), Russian physicist who worked in Cambridge between 1921 and 1934, when after his annual visit of the Soviet Union he was prevented from returning to England. He later founded the Institute for Physical Problems of the Russian Academy of Sciences, Moscow. His refusal to join the Soviet nuclear weapons programme led to his dismissal in 1946; in 1955 he was resinstated as director of the Moscow institute.

ohne Berücksichtigung des Spins auf mehrere Elektronen mit Spin ausgedehnt. Dabei kamen nebenbei einige komische Beziehungen zwischen den Multipletts von Atomen der gleichen Übergangsgruppe, z.B. der seltenen Erden, heraus, die einige Punkte der Salpeterschen Multiplettarbeit noch klarer machen.[425] Dann fand ich dass das von Becquerel beobachtete Absorptionsspektrum aus verbotenen Linien besteht, weil nur Elektronen einer Schale sich umgruppieren; die Übergänge kommen erst durch das Kristallfeld zustande, und die Intensität, die man mit dieser Annahme abschätzt, stimmt mit dem Experiment. Dann fand ich, dass die Zeeman-Aufspaltung bei kubischen Kristallen für gewisse (4-fach entartete) Terme von der Richtung des Magnetfeldes abhängt, ebenso die Polarisatioen aller Zeemankomponenten, was mit neuen Experimenten in Jena stimmt. Jetzt sehe ich zu, wieweit die Annäherung berechtigt ist, die Ionen des Kristalls einzeln zu behandeln, und finde, dass die Ionen mit unabgeschlossenen Schalen nicht zu nahe aneinander liegen dürfen. Dadurch scheiden Metalle von vornherein aus, was vernünftig ist. Ausserdem mache ich in Paramagnetismus der Kristalle. Aber nochmals herzlichen Glückwunsch und beste Grüsse

Ihr Hans Bethe

[114] Genia Peierls to Rudolf Peierls

No. 1
Detskoye, 2.5.[1931]

There are quite large leaves on bushes and buds on the birch-trees. The park appears to be covered with the green dust. You know, the sort of things that can be seen on the first spring days. Now it is morning, I have already had the breakfast; I'm sitting in the garden and I am about to take a walk. Yesterday I was walking in the parks all day long; in the evening we arranged charades and some dances in the garden. As

---

[425] See earlier paper H. Bethe, 'Zur Theorie des Zeemaneffektes an den Salzen der Seltenen Erden', *Z.Phys.* **71**, 205–26 (1930).

to the dinner, it is outstanding here, with ice-cream for dessert — dairy and orange ones mixed — fabelhaft! *Tonight I'm leaving for the city: I have to go to the Observatory tomorrow, etc. I had brought the book here, but couldn't read a line, to tell the truth. It is so sunny and the air is so fresh that I only want to walk and walk, and even bathe. I'm wearing a thin blouse and feel hot. The well-known law: the less you have to do, the less time you have, now is in effect. Yesterday I was going to write you a letter — and nothing was done. But as you are very busy, it won't be too difficult for you to write, will it?*

In my memory I can see your hand holding a hankerchief. Two days have already passed, or only two days. Today you have just arrived in Berlin, and you relate, relate, relate. I so much wonder, how it happens. I don't ask questions, for you will write everything, won't you?

I didn't succeed in writing a philosophic letter. I acquired the habit to be philosophic at night-time, and now I don't want any philosophy. I want to speak to you, walk in the sun, catch frogs — there are lots of them. When, Rudi, oh when?

On the 30th in the evening there was great fun — of course, I forgot to take soap with me, and discovered that only late at night when going to bed. My bedroom is located in an outhouse, away from the main house. The room is a double room, but I stayed there by myself. I lay down and started reading "The Party Dress", where this unfortunate Nina is bathing for the length of 8 pages. I was reading and crying (!); all of a sudden the door opened and my good acquaintance came in (she works at the Roentgen Institute). I cried, "Olga, where's the soap?!" She said, "Oh, it's in my suitcase, and I cannot open it, as I have broken the key!" Great, that is the irony of fate. Next morning I ably opened the lock with a nail, a penknife and a small stick, so at last everything was in order.

Darling, it is now 6 o'clock, I've been walking the whole day long, my face is feverish (I got tanned). I feel the dust on my teeth, my left stocking is in rags, and I have to change for supper. Till tomorrow! I wonder what they will give us for supper?

3.5.[1931]

Now I'm writing you in the Observatory. Everybody went out to have lunch and I had sandwiches with me, which I have already eaten up.

I'm writing with a fountain-pen, that I stole from Jonny[426] *long ago; a friend of mine just fixed it, but still it is scratching the paper (stolen things don't benefit the thief) — that is why my handwriting is horrible. Now I know how difficult it is for you to read my letters. I beg your pardon, but if I draw each letter properly, it will be too dull. (This "anti-Semite" is especially bad, so I write with a hopeless pen). Yesterday I returned home very late and found the wire on my table. I was awfully happy. The whole day before I was trying to guess whether it would come or not. You are in Zurich now and must be having dinner with Fania Moscovskaya and Mochan, and you are drinking water like a duck. This is the first day I am at work after your departure, and it is so hard for me to write, because I keep on thinking that I'll see you at home. R-r-r-r-r!*

Yesterday after I changed my stocking and went out into the garden, I saw Dau with a mathematician who came to talk to Kibel about some hydrodynamic equations. Dau was fulminating against stagnation and, happily, changed over to summer uniform, that is — no collar and no tie (!). Looking at him now, no idea would come to mind about hairdressers or sunrises. We played catch nicely and had a short walk, so I came back to the city dog-tired, slept like a log, and when I woke up it was so strange to feel the empty space on the sofa, your head wasn't by my side, I was holding an edge of the blanket and not your finger...

Erik took our picture from the balcony at the moment when we were leaving. The photo is very funny: a tiny, tiny cab and our hats. I'll send it to you in my next letter, as I left the photo at home and want to mail this letter on my way back. At home everything is all right, everyone is in good health, and Chimmi catches the flies enthusiastically. All send you their regards.

I felt so lonely this morning, having breakfast without you, I almost didn't touch anything because of my distress. See how I love you! I fear you are bored reading this endless letter with repetitions of the same things, but you know I can't stop, so you'll have to be patient till the end of the page.

[426] Nickname of George Gamow.

You don't understand why I took offence for "I did not say that ..."? *It is really quite clear and simple. You told me that you had learned from your father to repeat and explain each thing five times in order your interlocutor didn't mix up anything. But the things we talked about aren't to be mixed (or are mixed by intimate friends only), and then such things aren't discussed at all — it sounds too formal, they are unbearable to listen to and sting the heart; it is in the style of "commercial correspondence" and too "rude". Well, do you understand it?*

Now I haven't felt offended for some time, but I still mind it!

Und, jetzt, die erste deutsche Wörte. Meine lieber "Kleiner", es ist so schrecklich ohne dich und es ist so traurig, und ich will neben dich seien so schwer! Heute der Himmel ist grau, grau und kaltes norder Wind. Ich glaube, dass es ist Ladoga [*Lake*] Eisergegangen (?!!) (*floating of ice*). Ich wollte sehr, sehr eines "gutes Weibe" für dich zu seien. War ich? Ich weis nicht, aber ich möchte, alles was ich möchte dass du glücklich war. Warst du?

Es ist sehr interessant, hast du eine Worte verstanden? Ich habe alle dieses Zeilen ohne Wörterbuch und ohne kein Grammatik geschrieben und ohne hilfe. Ich Küsse dich, lieber und sehe in deinen augen. [?] Grüssen.

Genia

That isn't so important, so don't you try to grasp everything. This is letter No. 1. I beg pardon for the bad handwriting and stain; I wrote this on my lap, surrounded by lots of people.

Oh, Rudi, how much I wish to see you! I kiss you affectionately!!

[115] Genia Peierls to Rudolf Peierls

Leningrad, 6.5.1931

My own, one week has passed since your departure and I begin to get used to the situation when you are no longer by my side at dinner, and in the morning I no longer reach for your hair to smooth it ... Again stop-watch is ticking on, again the tube of electrometer is before my eyes. Outside the sky is deep blue and scattered with clouds — the springtime sky; the tree before the window is covered with small leaves like at the landscape picture by Perugino[427] *(we saw that in Moscow, do you remember the Madonna with round hazel eyes?).*[428] *Still sometimes I have a feeling that I only fancied all, that there was no March 1, that it was only a day-dream, just as the ones I had had in January and February ... This feeling overwhelms me only for a second and then I come to my senses — that is not because of my wedding ring, the summer overcoat, and chirping of sparrows, but because I have greatly changed in the past 2 months. I grew up and, what is the most important, now I miss you in quite a different way: much deeper, much stronger and in a "calmer" way than before. You see? I was much pondering on you during these 7 days; it seems, now I know you extremely well, I know myself too — once more I understood, that I have known myself since long ago ... I miss you terribly, just as I would feel a missing hand or foot, but still it is so nice! It is nice to feel this ring on my finger. What are you doing now? During the last 3 days there occurred nothing of interest except for your letter from Ostrov*[429] *(What a chivalry! Are you astounded?). Thank you so much for it, I was terribly happy. Now I'm waiting for the next one. Even if it had been posted in Riga, it wouldn't come earlier, than today. I received an agreeable letter from Nina addressed to us both, but since it remained at home and is very short, I don't send it to you; besides, you saw Nina already. The letter of 25th was delivered on May 4. Great! So our only hope is* Luft *— the "airborne" hope! Oh! Our friends visited us lately and gifted an ancient cup and a vase with flowers. Both things are exquisitely beautiful; unfortunately, you cannot appreciate them at such a long distance. Yesterday Nina and I had a walk on the Islands. It was wonderful: picturesque sea, ponds and canals with frogs and the fresh leaves around ... Amazing. We were walking from 4 till 9, and at about 7 o'clock I was trying to lie down on some of those idiotic backless benches, but Nina found it "*shocking*" and didn't let me lie calmly; she pulled me by the collar or pushed me down to the*

[427] Pietro Perugino (1445–1523) Italian Renaissance painter.
[428] Presumably Peregino's *Madonna and Child Enthroned* at the Pushkin Museum of Fine Arts.
[429] Town in Pskow Region.

grass. She sometimes has fits of being comme il faut. *Finally I found wonderful pinewood planks, laid down on them and looked into the sky. We were terribly hungry, and when we dropped in at the place of our acquaintances at Lopukhinskaya St., which is quite close to the Islands, I ate up a whole box of jelly-sweets. What shame?!!!*

Soon I'll be very busy, as two days from now my German lessons begin. They will take place 4 times every 10 days; besides, I have a new social duty that is rather pleasant and beneficial: to train three young men, the technical workers of our Institute. I should prepare them for entering a college or a university. They are agreeable and bright boys, I'll have kind of classes with them twice in 10 days. So, many of my evenings will be occupied. Soon I'm going to finish my measurements here and continue this work at the Radium Institute. I'll have to deal with hard γ*-radiation, they have a lead gun there! On the 10th I'm going to be driven to Petershof by the chief's car, in a small company. Great! If only the weather is fine and the car won't break. I want to drive the car a little, though I doubt whether they'll let me.*

"The Party Dress" *is rather boring, isn't it? I have read a book by a young Russian writer about the lepers (*lepreux*). The book is rather bad, though the theme is interesting. I have never visited a leprosarium and know very little about leprosy.*

The whole flat smells of turpentine, naphthalene, etc.: they store away the winter clothes. I hate these "all hands' jobs". Mother and Nastya are angry, Chimmi goes on sneezing. Rudi, aren't you bothered by letters yet? As the Luft *functions only from the May 5, and on the 3rd I mailed you a letter by ordinary post, I'm afraid you'll receive two letters together — and so thick ones. You must be bothered with reading, you can't decipher my handwriting and you suffer, my poor boy. Have you had a good night? And where did you spend it? Could it be that you swallowed up [while yawning] your little nephew in Berlin!!!! Tonight I had a dream that I was in Berlin and you couldn't find me. It may well be that on coming home I'll find the paper from the Foreign Department, "please come on May 7 to get the answer". That would be wonderful, and if I find your letter, that would be wonderful, too. Will you write one?*

I'm an exemplary "gutes Weib". *I'm as hungry as a thousand devils, it is 8 o'clock now and I'm writing the 4th page. You were lucky,* Herr Doctor, *with your wife. This is genuine "fishing", but* quand même! *I'm so hungry, that I gulp not only some letters, but whole words and very soon I'm going to gulp lines. Before this stage begins I say you good-bye. See you soon, is that right? And I kiss you as firmly as I can, for I'm hungry and weak; I kiss your nose and whatever you like. Good-bye! Your wife, virtuous and hungry, yours from head to foot,*

Genia.

All send their regards.
P. S. Certainly, I haven't got your marks, but that is quite normal, as money is delivered slower, than letters.
I beg pardon for the astoundingly bad handwriting, the stupid philosophy and a boring letter — my thinking is impaired. G

[116] Genia Peierls to Rudolf Peierls

Leningrad, 10.5.[1931]

No. 3
No. 1 mailed on May 3
No. 2 mailed on May 6

Rudi, my darling, how enraged I was on the 6th of May, when at the post-office they told me that the mail isn't sent by Luftpost, *and no one knows when it will. As the postcard from Lugano was delivered only on the 11th day and the letter from Berlin, on the 9th day,*[430] *I calculated with probability 0.95 this letter would arrive only on the 14th or 15th of May; I nearly got crazy. And today all of a sudden I received your letter from the 7th. And such a nice letter! Great! Is there a sofa in the room? Have you rented the room for a short while, or taken me into consideration? Rudi, after your letter I imagined Zurich to be quite close, and its inhabitants too, even Guth!*

[430] The letters written by Peierls in the three months after their marriage could not be located.

Today is the day off and we had two alternative plans: to go either to Petershof by car, with the boss and two fellows from the Observatory; or to Toksovo (called Russian Switzerland) with Dau, Abbot[431] *et al. But both projects fell through, as the temperature is only +2 and it is so windy, that one cannot say whether there is any sunshine. The sky quickly changes from sunlit to cloudy. No car for you. It is very stupid. But if I went, I should have got your letter only in the evening, and think the whole day long about you, and answer only on the 12th. Because tomorrow I have the Observatory, then a lesson with these young men, then a German lesson, I'll be through only at 21 h. Here you are.*

Today I read the book once more. I understand everything except one German word. I study German grammar: it is rather disgusting. Changed your marks, which we received only today; on my way I dropped in to see Dau. He says that [?] nurses a grudge and flatters. Dau received me in his disgusting room, rather dark, looking into the backyard, near the kitchen, etc. But it is his aunt's flat, so he lives with relatives who feed and keep him. He was busy calculating some astronomical data with exponents as 85/11 and 112/13 — disgusting! Dau was very nice, a bit lonely, I made him laugh and we wrote a letter to Kit. He really forgot how it is done — writing letters; that is terrible. In the evening we have visitors such as Ambarts, Dau, Gleb and someone else. We'll try and get to some cinema. Abbot has lost some ardour with respect to Ulanova.[432] *Dau thinks he was in love and tried to prove it theoretically. But as his data are fantastic, he cannot get any results.*

Yes, he saw my wedding ring at last, only after Nina said, "Genia, you're going to drop the ring!" Dau exclaimed, "How? You are wearing the ring? That is a bourgeois habit, how vulgar, etc." but he soon calmed down. On the 14th we are going to the Jewish theater to see a comedy by Sholom-Aleikhem.[433] *It seems full of merriment. On the 13th there is going to be a German soiree again, on the 12th evening*

---

[431] Nickname of Matvei Bronstein.

[432] Galina Ulanova (1910–1998), famous Russian ballerina; from 1944 prima ballerina of the Bolshoi in Moscow.

[433] Sholom Aleikhem (1859–1916), Ukrainian-born Jewish short-story writer, dramatist and humorist. Widely recognised as the best Jewish writer who lived in Russia and wrote in Yiddish.

I must be dictating the book, and so on. Up to the 15th the days are all occupied, just like the ones that have already passed. I'm afraid it will be wearisome for you: so many things to do these days. Too much for one man. The preceding days were marvelous, it was warm, but not too much; amazingly bright colours and merry winds. The season for car driving was opened on the 8th. The boss drove me right from the observatory, through all the embankments, far to the Islands. The speed was high with the wind blowing in my face, and we didn't have any troubles or accidents. On the Neva and Fontanka rivers and on the canals many boats can be seen already (I haven't yet had any boat trips), the trees are still leafless and are beginning to grow green. I want to watch with you the red sunsets and the white nights over the Neva River. Oh, how much I miss you! I may live on and on without any special feelings — and all of a sudden it cuts like a dagger across my heart.

*I didn't tell you when you were leaving, that you were especially nice the last days; you didn't say "terrible" things. You might think I didn't notice that — I really did. I'm awfully happy it is quite all right with the parents. I pitied you and was thinking constantly about it, how unpleasant it would be for you. After getting the postcard I forwarded to you, I saw everything in good order. And you were quite right concerning the money. I think I cannot be in Zurich in July; and how are you going to stay in Pauli's flat? Will his maid remain for that time too? You'll be unable to tidy the flat yourself. What did Lily Fenigstein say? We have a German violinist on tour here. I think I haven't got time to go and listen: I'm too busy; besides, in the good weather I prefer to go for a walk. A couple of days ago I finished reading "*Party Dress*". I've been reading it in trams. Now Tram 24 is on the line again, so it became easier to get to work. The end of the book is better than the beginning, though all this is so utterly stupid, it is enough to mention all those Greek deliberations. Besides, I don't like vulgar symbolism, and the torn dress is a very cheap symbol.*

Yesterday I took out three encyclopedias and read about leprosy. That was terribly exciting and now, it seems, I know everything, that a layman has to know. As I learned lately a remedy has been discovered. In Nature *the paper on "cosmic rays" is published. Oh, Lord, how long will it go on? He supposes the rays originate from the alpha particle and two electrons, transformed into energy. What other combinations will get into their minds? The Bohr principle is an idea of a genius, as a matter of fact. There are always convincing results.*

At home, no news. Only Mother has been feeling pain in the breast for unknown reason for several days. Is it the heart? Lungs? We don't know. You are asking about my temperature — I don't measure it. It is a bother. And then, what is the interest in measuring the temperature?

Well, it is time to go to the post-office, otherwise everybody will come and the letter won't be mailed. Soon, two months will have passed since we got married. The time flies and crawls. Rudi, when shall we see each other? Will it be only on Sept. 1? Good-bye, my dear, my beloved. I kiss you affectionately and love you so much.

G

I answered Nina's letter.
All send their regards.

[117] Genia Peierls to Rudolf Peierls

Detskoye Selo, 10.5.[1931], 21 h

Right now the Moon is rising, all round, shining and a bit confused, because there is enough light, and it wears an embarrassed smile. A white cat with a gray tail is climbing the tree, and there are lots of different sounds in the air: tapping of croquet balls, barking of dogs, croaking of crows on the lime trees. The background of these sounds is the distant music of the street-organ. I'm sitting in a straw wicker-chair in the garden, I do so want to see you, that the street-organ sounds excite me to tears. Rudi, I stay at the Scientists' Rest Home till the morning of May 13. Isn't it marvelous? At home I would have been more bored. When your hand disappeared from view and we turned back, we saw Dau, all confused, at the entrance to the platform. He overslept and was late some $1\frac{1}{2}$ *minutes. He had a desperate look — had shaved himself in a great hurry and the impression was as though he had been pricked with a spear. Horror! Even Abbot, who usually doesn't notice anything, was worried. He didn't take notice of my wedding ring, though I almost hung it on his own nose. At the [Vitebsky] station we met Kravtsov,*[434] *who was also going to worship the relics of Kibel (by the way, he looks much better than before), and we boarded the train with a great noise. Philistines in the carriage made a row around the doggy. It was* fabelhaft *funny! Then we had dinner and went for a walk to Pavlovsk. After seeing them off, I went to the city to get formal papers, necessary for the Holiday Home, and also my tooth-brush, etc. At home I found a postcard from Lugano and forward it to you. I could understand everything without looking into the dictionary. Aha! The postcard is very nice, it seems to me they aren't angry any longer. Is that right? There are many people here; there is one lady of 2.5 class, another of 3 class,*[435] *all the rest are rabble. Then there are a lot of young ladies together with their husbands — I envy them! I know it isn't good, but ... Rudi, tomorrow I'm going to write you a philosophic letter, and now — good night, my darling, darling, darling, my darling little boy. Where are you now? Have you received our telegram? I'm taking off your glasses and kiss your eyes.*

Genia

Doctor, how much I wish to kiss you really!
I cannot help crying, but I shouldn't, as I'm looked at by some 15 eyes (7 people plus 1 in profile). I must keep self-control.
Write! Wire!

[118] Genia Peierls to Rudolf Peierls

[Leningrad], 21.5.1931

My dear,
I have such a lot of things to tell you, I haven't written you for such a long time (5 days!!), that I really don't know what I should begin

[434]T.P. Kravtsov, Soviet physicist and friend of Landau's.
[435]Rudi, Genia and their Leningrad friends had sorted friends and acquaintances into 'classes' according to their personal characteristics and qualities.

with. Well, I'll begin with unpleasant things: it is better to learn about them in the first turn, and then the rest. Yes? Rudi, I visited the Foreign Department on the 17th. The chief wasn't in, there was only the assistant chief, a lady; she was rather polite — you may remember I talked to you about her — she told me to go to "our" nice lady, the one who had taken our papers and whom we bothered a lot. They could only check whether the papers were dispatched to Moscow or not, as almost all papers are usually sent there. Well, I went to the office of the nice lady who recognized me at once, asked about your departure, sympathized with me greatly (and with you, too), but when she looked through the cards where the transfer of papers is marked, it turned out, that our papers are still here. I nearly burst into tears: that means minimum 2 months waiting! It is such a distress, my dear! So you will have to come here again; if you could do that in August, it would be excellent. First, it means a month less to wait; secondly, Teberda: it will be awfully nice there! Well, here are all unpleasant things for you. Don't wait for the wire, my boy, that is most distressing when you wait and don't get it. Better think of your other trip here. I haven't written to you for 5 days because the only day, when I had time for writing, was the 19th and I felt hurt for I didn't receive your letter. It was delivered only yesterday. In general, Luftpost *functions marvelously in this direction: your letters arrive in 2–3 days. Isn't it great? In my last letter I didn't mention the Jewish Theater and the ballet.*[436] *That was really great. At the Jewish Theater I was with Isay Benedictovich and Abbot. It was overwhelmingly funny, really, though even I. B. could only half understand; as to me, I only caught odd words, as usual the unexpected ones, for example* teich *— the pool (I was very proud with this word). The acting was superb, especially Michoels; amazing music and dancing. Besides, I had read that novel by Sholom Aleikhem, which was dramatized, so everything was* fabelhaft. *The adventures of a poor Jew from a little town: he arranges a thousand "combinations", but all his commercial ideas somehow or other come to nothing. Among other businesses, he speculated at the Stock Exchange in Odessa, and Odessky Boulevard was reproduced on the stage, and Jews, and Langeron*[437] *—*

---

[436] See letter [116].

[437] District in Odessa.

terribly funny! I was sorry you couldn't see the play. The ballet was wonderful. Abbot showed he was a renegade, as he was fascinated by Lukom,[438] *who danced the main part. It is true, in one act she was brilliant, her legs could even form the angle of 120 and she was as light as a fly, but Ulanova is still better. In the first act there was a miraculous dance of Troyanovskaya:*[439] *do you remember this tall and handsome ballerina, who danced with Chabukiani.*[440] *The latter jumped just like the toy clown that you bought. Everything was wonderful except the last short act, where the ship is sinking and the scene of shipwreck looks very natural. The waves rolled and the ship was tossed, I thought I'd be seasick. Br-r-r! What a loathsome thing!*

After the ballet I went to visit my boss; we danced there a little — there were a few people. Then some girl, I, and he with his dog went for a drive. It was too late or too early, about 2 a.m., it began to dawn. In the East the sky was green and deep, and above was black, with stars. We took the girl home, she lives at the opposite side of the city, in an industrial district. Then we drove back across the city, it was exquisitely beautiful scenery. The black outlines of the smoke-stacks and houses against this green sky, quiet canals and deserted streets. The boss was at the wheel and I sat at the back; beside me sat the enormous black dog, keeping its paw on my lap. I did want you were beside me, a desperate wish ...

Well, on the 17th I had a drive again, in a taxi with Kib. We had a sitting, even two sittings on the same day; they lasted till 10 in the evening. Our heads were heavy and we drove to Sestroretsk[441] *through the Islands and along the Gulf coast. There was enough light at 10 o'clock, and even at 11.30, when we were getting back, the sky was light, and pine-trees were clear-cut black. The road was marvelous, it ran*

---

[438] Elena Lukom, ballerina with the State Academic Theatre of Opera and Ballet in Leningrad.

[439] Taissia A. Troyanovskaya, Russian dancer with the State Academic Theatre of Opera and Ballet in Leningrad and later dance teacher at the Perm State Ballet School.

[440] Vakhtang Chabukiani (1910–1991), Georgian ballet-dancer and ballet master.

[441] Seaside resort near Leningrad.

either by the sea coast or by the woods. Now it is so nice in the woods, there are small sticky leaves on the birch-trees, the sounds of the cuckoo far away, and the violets. We were driving terribly fast, and at pits we were tossed up, once I was nearly thrown out of the car, and another time I bit my tongue and had to be silent for 20 minutes. But I like high speed and the wind blowing in my face. The sea was dotted with small lights: those were lanterns on the fishing boats; and white sails of yachts were also seen. That isn't everything: the past week was marked by a lot of pleasant things. But as the book and the German lessons were also unavoidable, I went to bed extremely late: at 2–3 o'clock.

On the 18th I visited one of my acquaintances from the Institute. He isn't young at all, being about 43; he wears a small beard (what would Dau say!), bright red lips, black hair and black eyes. His wife is a very agreeable and talented lady (she is a mathematician), she isn't young either and this is her second marriage. She behaves rather funny, playing "enfant terrible". Many unmentionable things she speaks with a special emphasis; that is why her gossiping sparkles with ill wit. But as she is a kind of eccentric person, this seems rather agreeable. He is also witty, good-humoured, and he seems to have a penetrating mind. Well, I visited them and he showed me his collection of paintings — very nice, ah, how I admired them. He possesses rather many paintings by modern Russian artists — of my favourite style, you know what it is. I simply fell in love with one of the portraits by Serebryakova,[442] *and then there were some of her sketches. Also he has two small pictures by Ostroumova*[443] *— the Pavlovsk Park in the spring and in the autumn. The autumn picture is especially nice: transparent air, red leaves and a glassy pool. Great! And there was one unfinished painting by the artist who portrayed Meierkhold*[444] *(you may remember it, exhibited in the museum, and I sent you the picture-postcard). A company of women, all drunk, must be a brothel or something similar, impressive roses in the carpet pattern, green wall paper, the dresses of special cut, and faces*

[442] Z.E. Serebryakova (1884–1967) Russian painter who moved to Paris in 1924.

[443] A.P. Ostroumova-Lebedeva (1971–1955), Russian painter.

[444] V.E. Meierkhold (1874–1940) Russian actor and producer. The portrait mentioned is probably by A.Ya. Golovin (1863–1930), Russian painter.

— so nice, so nice! I nearly laughed with admiration. I returned very late at night by a wayward tram with a friend of mine. The route of the tram was along riverside, the city was so picturesque. The impression was, that a picture painted by genius was unrolled before one's eyes. Well, that isn't all! (Are you bored?)

Yesterday was the day off, and I took Dau to Toksovo[445] *— everybody else was busy. We had a pleasurable trip. The weather was warm, even hot — I had a terrible urge to bathe. There are marvelous lakes, hills and woods here. But as it was only 20th (!) of May, my prudence took the upper hand: I didn't bathe and only walked barefoot. Oh, what a delight! Dau philosophised a lot, and bit-by-bit developed the theory of our "situation". Very funny. I taught him a lot of things: for example, to recognize the voice of cuckoo, to eat the oxalis — do you know these white little flowers with very sour and tasty stem? We spent the whole day lying under various trees, climbing various hills, etc. He even got a little cheerful. In the evening our company of 10 (!!!) people went to the cinema. We watched Monty Banks*[446] *— an old film, rated at least as 2nd class. I split my sides with laughter. As we were sitting in a row (I was on one end, and Dau, Ambarts and Nina were on another one), we had to yell in order to exchange our impressions. We were nearly asked to go out. The documentary showed the Moscow Conference on the planning of scientific research. When Ioffe appeared, we began to applaud, the whole audience followed our example, and a maddening ovation resulted. It is so easy to become famous! Today I'm going to the concert of a German conductor Weissbach*[447] *— Beethoven will be performed. Such a lot I told you about. Are you quite bothered? Darling, thank you for the letter, for the pictures. Good photo paper means a lot! On one of the pictures I'm good-looking, and at another one I'm very poetic: "Genia d'Arc" — looking into the sky, with a meek smile. Ah! The most striking photo in your life, it is where you sit on the lion. The pictures on skis and wearing fancy-dress are less striking. It was me,*

[445] Village north of Leningrad.

[446] Moutny Banks (1897–1950), comedian and film director.

[447] Hans Edgar Weisbach (1885–1961), German conductor and director of the Leipzig Radio Orchestra who flourished in the 1930s, especially with recordings of Bach.

who shot the best photo! Rudi, I'm about to finish dictating the text, one séance is left. The book will come out soon — about 3 [printer's] sheets, maybe $3\frac{1}{2}$. Tomorrow I'll inscribe the formulas, so everything will be ready by the 25th, or the 26th.

Tomorrow the parents are going to Petershof, to the place where I stayed in winter. I'm very glad they managed to arrange this. They are very tired, I. B. even cannot sleep well. I hope they'll have a good holiday. Well, darling, good-bye, I'm so sad without you — I can't help crying, Rudi! You'll get this letter in the evening and read it at night, your eyes will be closing. Good night. I kiss your eyes when they are already closed. Good night!

G

I quite forgot! You know, in a month I'll probably go on an expedition in a motor-boat! The route is: Neva River — Ladoga Lake — Onega Lake. The chief will measure the emanation in the air. One other person will go, too; he'll also take his dog and shotgun for game. Isn't it great? For a week or 10 days, we'll sleep on board the motor-boat.

Next news: the first, I bought myself a pair of shoes for 30 rubles! The second, Abbot studies Spanish! One more kiss, good night, my darling, once more.

[119] Genia Peierls to Rudolf Peierls

[Leningrad], 24.5.1931
No. 7 (I forgot to put a number in two preceding letters)

"I don't know what the time is" — this is a line from my poem; and in fact, the time is unknown, because I have only a stop-watch here and I'm unwilling to go somewhere. I think optimistically that it isn't so late, and I have enough time to write a long letter to you before 5 o'clock, when I leave for home. Today is a remarkable day — taking the bath! Therefore I have to be punctual. Mother is away, it is true, the parents are in Petershof, but Nastya is just as strict. Nina and I have been for $2\frac{1}{2}$ days on our own — we manage the household, etc. In the morning we order the dinner and something for night meal. It is a touching sight:

the two of us are at breakfast, the two of us speak over the phone, the two of us are at dinner (today, though, Abbot is invited). It is a pity you cannot appreciate how virtuous I am in family life! The main items on our menu are: sauerkraut, dried mushrooms and jam (!). Besides, we have consumed the supply of sweets, left by Mother as a kind of "consolation" for us. We have enough for tonight, but tomorrow ... In general, it is delicious (I mean the jam!) and economical. The marvel of science and technology! Here is my program for today: washing my hair and taking the bath, dinner in the company of Abbot, dictating the book — because the typist fell ill on the 22nd and destroyed all my plans. Luckily, she got well, so if I'll be 2–3 days late, there are solid reasons; I'm not to blame. To find a typist presents a great problem: everyone is employed and must be persuaded. According to the contract it is tolerable to hand in the manuscript a month later, so no worry. Really and truly I'm not to blame. Tomorrow I'm going to spend the whole day at Petershof on a visit to the parents. I hope ice-cream will be served at dinner. I have phoned Mother, they seem to be very happy.

25.5. 0.30 a.m.

My dear, it was certainly late and I had no time to write anything. The typist is ill with a sore throat. Unexpectedly we made up our minds to go to the cinema with Dau, Abbot and Kib. We saw Harold Lloyd that can be rated 3rd class, with a good scuffle. Now I'm back, it is windy outside (tomorrow I'll have to go to the suburb!), and I'm in low spirits — so lonely, on the verge of tears, feeling unwell. Rudi, what are you doing just now? Having a drink at some American bar with the Congress mates? Boating? Writing a letter to me? That would be funny! Yes? I'd like to philosophise right now, but while sitting on your lap, or in the dark, with your hand beneath my head. I think I'd be more cheerful now if I had a child or was in a family way. You know, sometimes I fancy I have a child, very clearly. A boy with tiny hands and round eyes. How I'd nurse him and love him. But now — no, I'd feel more lonely with a boy in my arms. Tonight I feel very small myself, I'd like someone to nurse me, to speak sweetly. I want you to kiss me till I get cheerful. Rudi, it is so windy and the sky is black. A stupid letter, isn't it? I write *with a bad pen, it is hard to read the letter. I like to receive letters, it is a pleasure to open an envelope. Do you want some more philosophy, or, to say exactly, sniveling? It seems funny, but I cried very often during those 2 months, when you were here — some 5 times. Don't think, that I always keep crying. Usually I cry rarely, very rarely (today, for example, I nearly cried, but didn't). Our affair was very hard for me, really. And all happened so quickly. You know, I didn't think we would marry. I considered you as "European", as too "considerate" man (you see?), for to marry being only 23-years-old. And when I wrote you, expecting the 1st of May, I couldn't imagine, what would happen. You really can do "crazy things", and you have grown 1010 times in my opinion. I can do "crazy things", too, and like it, and don't fear. You know, I quite realize, how complicated it was for you, what a difficult situation you found yourself in! Your position was more complicated, than mine (thinking of your parents), you also took on the responsibility for me — a fearful thing. But you had an advantage, even double advantage: first, I'm a perfect "domestic animal", and even now I cannot imagine, how I shall leave my home forever! Second, such things aren't to be written and even talked of — I was quite inexperienced, I haven't been kissed before. My nature radically changed. You see? You have no reasons to reprove me for being a hypocrite. I have a simple and natural attitude, don't I? If I'm in love with a man, I'd prefer him to kiss me really, rather than only <u>want</u> to kiss. It is no good <u>to want</u> to do something. Do you remember in Kislovodsk ... Then I understood that the borderline existed, which is hard to cross, unbelievably hard. However we advanced in our relations, the borderline remained. "Theoretically" I agree it is stupid, but "practically" — one must be in great love, and believe that the other person loves you as much, especially the first time. Then the border disappears. These matters are trivial. Well, go on listening to the "history of the issue". I just started coming to my senses, when — bang! — your "declaration" followed. Rudi, even now I'm writing about it with heavy heart, but I'd better continue. I'm going to be very — perfectly — frank. Agreed? When you said this, I believed you and felt aversion, you ceased to exist for me; do you remember how utterly calm I was looking through the lecture? I didn't know exactly what to do, but I was sure to find a way out. I felt like a slap in the face, felt* *heaviness, loathing. Then I glanced up and saw your expression while you were looking at me. I nearly got crazy. I saw that you told a lie, I was quite sure it was a lie. The first person always knows it. Then the most dismayed 10 minutes of my life followed. It was urgent to do something — what was to be done? Rudi, you see, such words are never forgiven — let them not be true — they are not forgiven to anybody. Only to a husband, as I knew later. During the first minute I thought, well, let it be untrue, let it be bad for him, he would be sorry later. But the next moment I saw it was "childish", the case was too serious, the husband is too intimate a man; I must step over my pride, the woman's pride. And I did it, though it was so hard as to make my heart burst.*

Rudi, I often felt so much pain later, even now I feel pain sometimes. I wouldn't like you to think, that I fell on your neck, that I wouldn't go away at once, if I believed you. The first person knew it was a lie, but the two others I repeated to myself that I'm "impudent" in a way, let it be lie or true, for what reason should I impose myself on you? That is why I cried so much then, I didn't know what to do. The little girl who was offended, and the woman who knows, nearly tore me in half. Rudi, sometimes, even now, it seems to me the girl was right, and then I feel extreme shame and discomfort. You see, I don't want you ever to say, that I bound you, don't want you ever to regard me as a "heavy suitcase". Darling, you were very honest, may be even too honest, and that was good. I mean you are completely free, free even now; I won't be angry and blame yourself. Nobody is to blame in such affairs. I say this for the last time, and then I'll erase from my life those few days at the end of March, I'll stop thinking about them. Only I don't want "fragments", you understand, you should tell me whether you take me just as I am, whether you really want me, or tolerate me because "you must". You don't owe me anything. You know I won't be hurt, I won't be angry. Well, it is 3 o'clock already, and I have to leave early tomorrow. Good night.

I beg your pardon for such a tedious letter — it is simply "a psychological study"!!! If you were only aware of my past scares! Now I don't think anything like that. Now I trust you and know you love me and miss me; as to me, I miss you so much that I cannot help crying, and tears are just about to drop on this sheet.

This American lady — what a funny event! You may make acquaintance with any attractive girls or ladies of all the 5 classes. I'm not jealous at all. And you? If you answer at once, the letter may arrive on the 31st or the 1st. Write me, will you? How lonely I am without you, lieber*! I kiss you most affectionately. Good night. Pardon me for this letter and don't feel hurt by my question. I cannot forget. Good-bye. I registered myself for the accommodation in Teberda.*

G

[120] Genia Peierls to Rudolf Peierls

[Leningrad], 28.5.1931
No. 8 (7?)

Darling,
yesterday I was trying to guess whether there would be a letter from you. When on returning home I found the envelope on my table, I became awfully happy. Thank you! Pauli is a swine; to my mind, you shouldn't have talked to him about the two days. He might have thought you would leave each week somewhere to climb mountains or court young ladies. But for one occasion! In general you are free on Sundays, you may leave and then be late for the train or loose your way. Thus you may lose your way in Stuttgart. My poor darling, you will be lonely and sad in Zurich, even without Pauli.[448] *I'd so much like to come and amuse you. Every day I imagine how I come back home and find the envelope from the Foreign Department, but it is only in my imagination. I registered myself for accommodation in Teberda just in case we may go from August 10 to 25. This time can be postponed by 5–10 days; anyway I don't know whether I'll get the accommodation. Well, there is a quite a long space to wait, everything may change in the meantime. Darling, about a month has passed since you left, there remain max. 2.5 months. Oh, Rudi, what a bother — this "co-ordination" and counting of days! The year has passed since I started to count: last June I counted the*

[448] Pauli was leaving Zurich early to teach in the US in the summer.

days remaining till the holiday, because I was unbearably tired (in 1929 I had only a 12-days holiday; unluckily, on my last day in Crimea I ate too much, and was ill on my way back and the next week at home). Then you and I counted the days and hours (what a humiliation!) that remained till our parting; then these 6 months — 180 days. Ah! Then you came and two weeks later we started to calculate the days remained till your departure and when we'll get the papers, etc. Now I go on calculating when I may receive your letter, when you will come. Shall we have to count days after August 15? Good Lord, this arithmetic will make me crazy! Rudi, can you fancy that we shall live without counting? Live "today" — not n days ahead? You know, I'm a person living in the present, and when I have to think a lot about the future, I don't feel I live at all, but sleep or dream. I lose the sense of reality, you see? I have written a very stupid letter No. 7 (of May 24)[449] *— could you understand a thing in it? It was so late at night and I was feeling so sad, that the result was a lot of rubbish. Rudi, on the 25th Nina and I went to visit our parents, and we were terribly glad to find them so healthy-looking after a 3 days of leisure, especially Isay Ben[edictovich], that was great! They are in very high spirits and walk long distances, as though they earn money for each kilometer — like a taxi! Horror! We had such a long walk that Nina and I began to think of a chaiselongue, while they went on and on. The nature over there is really magnificent: a merry late spring. Everything is in bloom: bird-cherry trees, lilac, fruit trees. Flowers spring from the earth and the sun is bright; then there is blue sky and sea, and very high-thrown fountains in the park among the trees. Fantastic! And so many nightingales! Isay Ben. has a single room, but Mother has to share the room with 4 ladies. One of them is as big as an elephant and, as Mother says, she snores like a hippopotamus, and when she tosses on her bed, the whole room creaks. I. B.'s room is so small that no space remains for a second bed. The meals are wonderful: once they even had strawberries with cream!!!*

[449] Letter [119].

29.5.

My dear, at that exciting detail I had to stop. I had to count out the readings every 2 minutes and it was impossible to write a letter. Now I'm in a terrible hurry: I have to go to Pavlovsk, Nastya isn't in, and I want to drop this letter on my way. There are many things that I'd like to write about; yesterday I had no time at all. Up to 8 o'clock I had to teach, then I scarcely had time to get home and gulp my dinner, then I dictated until 12. After that I was sleepy and couldn't think about the letter. Only the Foreword remains to be done. I'm all covered with glue and ink, because I have been packing your gifts, but the glue wasn't sticky. What on earth, now my fingers are stuck to the pen — why doesn't it glue where necessary!

I didn't have time to buy the textbooks of Russian, because I'm really giddy with being busy. I'll try and do that tomorrow. You know, when on board the ship I was surprised at the fact that all European physicists know each other very well, write postcards, etc. This feeling of one great family is noticeable at once and I liked it terribly. With our scientists everything is different, which may be explained thus: they are much less numerous, they are crowded together in 2–3 Institutes and always meet face-to-face. Certainly, I can render some aid in this respect, but it is too complicated at such a great distance — even purely "philosophic" problems are hard to touch upon. By the way, what about your Privat-Dozentur*? What do you think on the account of Rockefeller? You know, it would be better to do this in the spring (that is, such is my wish). First, in the spring I'll join you for sure; but in the autumn (supposedly I come to Zurich in October) I'd like to have time for a quiet life and getting accustomed to circumstances, that are new for me — rather than begin with hotels, suitcases, trains, and seasickness. Yes, seriously speaking, I'd like to begin our life together in a routine, usual way — unlike we would have it during the Rockefeller year. Can you understand that, can you?*

Well, on the 27th we went to see the ballet, and a sensational scandal took place — Abbot deserted to Vecheslova's fans,[450] *Yeropkin is gleeful,*

[450] Tatiana Vecheslova (1910–1991), Russian ballet dancer and first Soviet ballerina to tour abroad when she went and Chabukiani went to the US in 1934.

Ambarts holds him off. Vecheslova is really a charming woman and actress, but as a dancer is weaker than Ulanova, she lacks "technique". Everything was very agreeable. After the ballet we took a walk past the New Holland building and along the canals to the Neva River, and then along the riverside. It was amazingly beautiful: the white night, the outlines of edifices before the greenish skies, distant fire-works and the quiet city. Fantastic! On the 2nd we are going to a concert, on the 10th, to the Arts Theater, and on the 13th, to the ballet. Maybe something else shall happen in-between. These days I have to walk quite a lot, as they require buying many things for the laboratory. I go shopping all day long — that isn't so pleasant: it is hot and repairs on the streets started. Sometimes it is more difficult to cross the street, than to cross the Alps. Rudi, darling, how much I wish to meet you, how much I wish to live a quiet life with you, without counting the days. I kiss you affectionately. You'll be reading this letter at night. Good night.

Oh, I'll miss my train!

The weather is nice these days: it is sunny and windy, and I love you so!

G

Came back home but didn't find your letter. Can it be that the one of May 10 got lost, too? If tomorrow they will not bring it, it is lost. Only once it took 4 days for the letter to be delivered; as usual I have them on the 3rd day. Rudi, that is too bad! We are so far apart, and the letters are getting lost! What on earth is going on! What a funny sensation I'm feeling right now: with a chocolate in my mouth and the tears in my throat — a mixture of salt and sweet. I was going to write you in the train, but this bright idea wasn't realised (as usual). I had scarcely prepared a sheet of paper and a pencil, when a woman on the next bench play-acted sunstroke, a heart attack and an epileptic fit — all at once. Luckily there was a minimum of epilepsy; the strongest was the heart attack. The train was overcrowded, people occupied the aisles and a half of the windowpanes wouldn't open. There was no water in the whole train at all; the wash-stands in the toilets were out of action (as it was a suburban train, it ran for 45 minutes without stops) — there was only a water closet. The 10-years-old daughter of the woman was howling, the 70-year-old mother-in-law was howling; I thought of using their tears instead of water — but that was too complicated. I had to obtain water from the toilet, using special techniques to get the pure water; then to drive out the audience that got more numerous, as the crowd from other carriages was attracted by the interesting occurrence, and the oxygen content fell to 60%; then I had to dry the tears of the girl and bring the old woman to her senses; unbutton her dress, take the pulse, pour water, unclench the teeth, hold the woman's head and hands — in one word, the whole course of first aid. I should add, that I helped to carry the stretcher to the hospital (located across the street from the railway-station). As you can see, the entertainment was grandiose! Yes, seriously speaking, she had a strong attack and long fainting fit; and not one doctor around, no medicine, no competent people — only curiosity and "Ah"s. This seemed to end happily: in the hospital she regained consciousness. Never in my life had I seen, so close, a person faint, without having the pulse, complicated by convulsions and the tongue thrust out. In such cases I lose all "sentimentality" and act "abstractly", as if it doesn't happen in reality, but I read about it somewhere. I'm neither frightened, nor disgusted. You see?

I made up my mind to write letters spontaneously, just as I speak, without thinking first. Otherwise, whatever I think of doesn't seem worthwhile writing. I'm alone at home, writing this in the study. There is bright sky seen through the window, smokestacks, many of them, and rosy smoke. I'm feeling very sad because of those letters, but still I'm not in low spirits. The summer days pass much faster, though they are longer. Why that is so, I don't know. Why? Maybe, because everybody is planning life for the summer, and lives a month or two ahead of themselves. Everybody is speaking only of July and August. You know, only now, when my temperature is normal, I recall how ill I was. In March and April I was feverish and got used to that state, but still I was feeling so weak. I wasn't so joyful, do you understand? I'm much more cheerful when I'm well. In Moscow, for example, I was dull and constantly wanted to lie down. You know, I read quite a lot these days; sometimes I have such fits of reading. Mainly memoirs, biographies etc. I have read, among other books, the journals of Tolstoy's wife — very interesting. You become aware of how complicated life in general is (the journals touch upon the intimate life). There is no "writer of genius", however close to reality he may be, who can expose one hundredth of the "complications" that are created by everyday life. That is extremely interesting. Complicated things are always interesting, aren't they? I do so much want to settle down with you and live together — not just spend the honeymoon. I'm afraid, when you come, there will be again suitcases, tickets, days to count, new places and new people to change every half-an-hour. In one word, the honeymoon again — not settled routine. (Won't I have my passport in August!?!) Certainly, love is not the gift of heaven, if it were so, that would be too tedious. To live together requires mastering a special science. The two should "enter each other". Do you know the way they make ground joints? At first the glass squeaks and screeches, and only after many turns the ground end enters properly. Many knots will have to be cut; that forms the main interest in life. I think, if we didn't marry, we should have gone apart and wouldn't keep in touch. Love doesn't endure static life; by March 15 we came to the end of all that was possible in unmarried state. Our dilemma was either to go further or give up, to stay in the same position was boring. Oh, yes, how much philosophy! You can see at once that I first write, then think. It is already 11.30, I have to go and wash my hair; otherwise when Mother comes, she will be angry. Rudi, darling, darling, it is a great pity there is no one to kiss you, to wish you good night. I'd like to write you many sweet things, but have no bent for it. I have no bent for saying these things either, though it isn't worthwhile saying, better put the hand on your head. Now I cannot reach you, you are far away. Good night, my sweet little boy. How nice would it be, if you were by my side. Good night!

G

[121] Genia Peierls to Rudolf Peierls

[Leningrad], 2.6.1931
No. 9

If on your birthday you aren't well-shaven because of me, I beg your pardon, congratulate and kiss you, even unshaved and unwashed. Affectionate kisses! I sent you some gifts on the 29th and hope you'll get them in time. Rudi, how I wanted to pack myself up and mail myself! That would be great, wouldn't it? And you wouldn't need to meet me — they bring the box to you, you only have to unpack it and get me out. I like Fedin:[451] *in this collection "Anna Timofeyevna" is good, and something different. The play is rubbish, you needn't read it. Besides, Fedin is one of our contemporary writers that has simple language and could be easily read. I admire Babel,*[452] *but you will not understand a thing; myself, I don't understand a 1/3 of his words, and it is inconvenient to ask anybody as to their meaning — you may make a person feel awkward. Darling Rudi, I am to blame — I haven't bought those books: I get out early, when bookshops aren't open yet, and get back when they are closed. There is no way to do it. On the 30th we went to Petershof to see our parents (they are already back); I was too short of time to buy books in the morning. But tomorrow everything will be correct. Please apologize to Fanny Moscovskaya for me — I'm really terribly busy these days. It is wonderful in Pavlovsk, the house and the garden are excellent, but I haven't moved there because the repairs aren't yet through. It is necessary to mount holders, supply running water in the rooms, etc., that is to say, to transform a dwelling house into a laboratory. On the 6th some of the instruments will be delivered, and on the 10th–12th I'll move in. That is all very nice in summer, but shall I have to go there in winter? Ah! We had a nice trip to Petershof, though we experienced the "cold wave"; on our way back we get cold, in spite of putting on all sorts of warm things that Mother gave us. In the afternoon it was amazing: transparent air, bright green leaves and clouds on the horizon. The surroundings in Petershof, everywhere, are very beautiful. Oh, Lord, how many things to show you, when you come! What else happened during these 4 days? Dau came to see me, he was as sour as* Sauerkraut. *I decided to take special measures to find him a female of the 1st class. Be so kind, send me as soon as possible, the picture of Bretcher's bride: she is the only lady that Dau classifies as*

[451] K.A.Fedin (1892–1977), Russian writer, journalist and editor, closely associated with the *Serapion Brothers*, a literary group dedicated to individual freedom in creative arts.

[452] Isaak E.Babel (1894–1940) Russian journalist, playwright and short story writer.

the 1st class! Please, send 4 copies: I'll hand them to Abbot, Nina and Ambarts, let them look for the like in trams, theaters, etc., then make her acquaintance and present her to Dau. I cannot think out a different way!!! Last night I had a terrible headache. At $10\frac{1}{2}$ *Abbot and I took a walk in the Summer Gardens. The weather was rather alarming: great clouds in the sky, a strong and warm wind, twinkling stars, a distant thunderstorm. Though it was nice — the Summer Gardens are best on a white night. The ghosts of trees and white statues, stretching out their hands. How nice! We were philosophising — mostly Abbot, while I was silent and thinking. We spoke about "happiness". Oh, "happiness", this is a "great" word. I know a delightful verse by Anna Akhmatova*[453] *(the best poetess — in general, one of the best contemporary poets): "Once I was daring, angry and merry — I was unaware it was happiness". This is very true: you are happy when you aren't aware of it, you shouldn't muse upon happiness, nor* <u>*strive*</u> *for happiness — you should be happy. I'm very happy now, and what about you? You know, when I was last in Petershof, I was once standing in the garden, lilac was in bloom, swallows crossed the sky and someone in the sitting room was playing Presto of the Moon Sonata — I was feeling like a cut across the heart: "and then I was aware it was happiness". To be 22 years of age, in the springtime, missing you so much, and Beethoven's music — that is happiness. Yes, sure I'm happy and sound now, I can't help laughing because my hands and feet can move, the breath is deep, and I'm in love — how good! We possess this joy to be living (I mean, you also). As to Dau, he hasn't got it: he is analysing too much — every step he makes, every bite of food he swallows. He will change later, maybe, but it will be too late, because by then his teeth will have become dull. You see, at 40 one cannot be as merry as at 25. Abbot, too, is a bit too "rational", but he is younger and sometimes he forgets to reflect and lives for "the pleasure of living". You know, the time will come, when you'll be a professor at 45 — then you'll recall the warm nights and quiet streets of Zürichberge, and how much you wanted to see me; then you'll say, "Oh, I was unaware at all, it was happiness!" Maybe still, you are aware?*

[453] Anna A. Akhmatova (1889–1966), widely considered to be the greatest woman poet in Russian literature.

Oh, my poor one, I write a "philosophic" letter on your birthday. Are you bored too much? Have patience to bear it a little longer. Now I'm going to answer your letter. I'm glad you have pondered much, I believe you realise this time, what you speak. I'm not afraid any longer of "frightening things", because I know you much better than before, and I find you different from what I imagined. Excellent — you "have many parameters" and you may show surprises. You wrote, we might have married some time later, than we did, then everything would have been easier. To my mind, you aren't right. You know, marriage means the closest intimacy, and we got to know each other so well in those 6 weeks. If we weren't married, we wouldn't have learned that in 6 months. "Philosophising" is possible only at nighttime, with switched off lights, etc. As I think, "the woman" would have screened me off and become a hindrance. So, all is well that ends well, right? Oh, what a tedious letter, what an awful handwriting! I quite forgot to thank you for the parcel. Everything is ravishing, especially chocolate; Mother and Nastya enjoyed mostly flour, rice and coffee. Rudi, if it isn't too much trouble for you, in spite of duties, please send us some tea: we have consumed the tea from Moscow and now have to do without it. Congratulations again. What would you like me to wish for you? I wish you to grow up and get rid of all complexes. I wish for you to meet me soon. Rudi, I kiss you so tight that our lips flatten. My love for you gets stronger, I'd so much like to say you "good morning, congratulations" in the morning of the 5th. Good-bye, my dear.

Genia

P. S. You are a very funny boy, how could you think I didn't guess that you might be worried with "[?]"? Certainly, I did; I know you so well, that you needn't speak of what you think. (Know)^5! What an awful style! All is in order, no worry.
At the post-office. Oh, what a thunderstorm! I'm wet through: my feet, my hat, everything. Though it is warm!
I have read Nina's postcard. How did she dare? I miss you so greatly; I put on weight because of that. (I always put on weight when I'm bored.) Our folks congratulate you and send their regards — I forget to mention this each time.

[122] Rudolf Peierls to Genia Peierls

[Zurich], 24.6.1931

Mein Liebes!
Du hast wirklich recht, ich sollte Dir mehr auf Deutsch schreiben, zür Übung. Aber nicht nur zu Deiner Übung sondern auch zu meiner, es ist für mich sehr schwer, denn in Gedanken spreche ich immer russisch mit Dir. Es kommt mir ganz komisch vor, Dir deutsch zu schreiben.Wie unsere "offizielle Verkehrssprache" sein wird? Ich glaube, gemischt. Natürlich wirst Du fliessend deutsch sprechenlernen, spätestens in einem halben Jahr. Und wenn wir viel mit anderen Leuten zusammen sind, und mit ihnen natürlich deutsch reden, so wird es schon ganz von selbst herauskommen, dass wir auch miteinander deutsch sprechen. Aber russisch ist auch unentbehrlich: wenn wir etwas zu zu sagen haben, was die anderen nicht verstehen sollen.

Ich bin heute beinahe ebenso nass wie Du neulich, als Du mir den letzten Brief schriebst, aber das ist meine eigene Schuld, ich war ohne Mantel fortgegangen, weil es morgens so schön war.

Ich bin heute in sehr guter Laune, es sieht nämlich so aus, als ob ich jetzt endlich die Lösung dieses Frenkelschen Problems gefunden habe.[454] Sie ist ganz einfach; viel zu einfach, und daher habe ich Angst, dass ich noch einen Fehler dabei gemacht habe.

Ausserdem war ich natürlich schon wieder im Strandbad und bin jetzt schon ganz braun. Das Strandbad ist jetzt in vollem Betrieb, man kann sicher sein, dass man dort Bekannte trifft; *Fania, Motschan* Lilly Fenigstein, Physiker etc. Ausserdem gehe ich jetzt oft noch abends spazieren, es ist sehr nett, wenn man bis 11 Uhr zu Hause gelesen oder gearbeitet hat, und sich dann noch mit anderen Leuten trifft. Neulich lag ich um 11 Uhr schon im Bett, Als plötzlich *Fania* und *Motschan* vor meinem Fenster erschienen und spazieren gehen wollten. Und ich habe mich mutig noch einmal angezogen und bin mitgegangen. Dass man

[454] J. Frenkel had published a paper on the coupling between electrons and lattice vibrations. J. Frenkel, 'On the transformation of light into heat in solids' part I. *Phys. Rev.* **37**, 17–44 (1931), part II, *Phys. Rev.* **37**, 1276–94 (1931).

Dich so oft allein lässt, finde ich eine *disgrace* besonders an Deinem Geburtstag. Wo wirst Du an Deinem Geburtstag sein, *at home or at work? Poor dear, are you struggling already? Perhaps I use too many difficult words. Well, it is better that we write about "business" things in Russian. I hope very much that at the end of July you will have authorization to leave. If that is the case, for various reasons I think it will be best in every way to leave the Caucasus and meet up in Berlin straight away. Firstly, because the more time we have together, the better. This way we will have time to see Berlin, and there will be time to show you all the interesting things and dear people: in particular my parents, Nina and Annie and their husbands and children etc., and then we can be here until the beginning of the semester, and we will have plenty of time to organise everything, and get to know everybody. Secondly, I have asked officially for our affairs to be hurried, and if they do indeed hurry them through and we then don't travel, they may be offended.*

If you still don't have your passport by then, I don't think it will be a problem as long as we have some money. If you still don't have your passport when we return from Teberba then we can nonetheless stay in Leningrad; it might be worth arranging something with Rentgenovsky,[455] *but we don't need to do that now. I doubt whether finding anything free in a hurry will be a problem, but if it is taking a long time, and we need something organised quickly, ask Dau to arrange it from the 10th August. It's always possible to refuse.*

Darling, what a strong husband you have! It's only 11.30 now, and I'm tired already. Well, I'm off to bed. Until tomorrow, darling. Good-night!

25.6.

Dearest, do you remember the letter which we wrote to Fräulein Kobylinski's parents? They found it very funny and have indeed posted the parcel which we will destroy tonight.

[455] Leningrad physicist, who developed a spectrometer named after him.

Recently, Fania, Motchan and I visited a strangely interesting doctor. She's Russian, but has lived here for twenty years or more. She was born in a small town near Khersop and only began studying when she was eighteen. Then, after a year and a half, she finished school and began to study medicine. But as she is a Communist, she was in prison rather frequently, and somehow finally emigrated. Here she worked hard to become a doctor, and lived in those interesting Russian circles with Lenin, Lunacharsky, and the rest; and all the while yearned for Russia. Then the Revolution came and the rest of them returned. But by then she couldn't. She had a son who was at school here, and a job here, and most importantly a Swiss husband. Her attitude to her husband is very strange. He is an intelligent man, it's true, but very different to her. A typical Swiss petit bourgeois. She frankly admits that, outside the bedroom, she doesn't want anything to do with him. Not merely that: he is rarely home, and when she has guests she arranges that he won't be around that evening etc. We only met him accidentally. I don't like this attitude at all, but apart from that she is a terribly interesting woman: very capable, and very lively. She has a sanatorium for nervous illnesses and has previously worked in a lunatic asylum. She told us some horrible stories from there; not about the patients, but about the doctors. It seems that there are still a great deal more unenlightened attitudes in Switzerland than I had realised. I don't know how things like this are still possible: she had a girl there, whom it was rather difficult to educate. Her father was an alcoholic, and had never paid much attention to her education; she had lived in Italy, but still couldn't speak either Italian or German very well. Anyway, Frau Doktor *tried her hardest, and after two years the girl was fit to go to school. She went, but the teacher didn't like the way she held her pen. He had this special theory, which he considered necessary, and didn't allow pens to be held any other way. Of course, she didn't know how to do this, and on her second day there, he rapped her hand with a ruler so hard that after a week her fingers were still hurting. She ran home crying.* Frau Doktor, *naturally, was concerned, and telephoned a member of the school committee who she knew. He replied "it's not my area". They sent the girl to another school, of course. But what about all the other children who were still at that school?*

When I speak to someone who works in a profession where he can really help others, and is in a position to do something about all this terrible behaviour, I always feel ashamed that I belong to such an "abstract" profession. Still, I shouldn't complain, everyone must do what he is best at.

Anyway, I hope that soon we will be able to go and see this Frau Doktor together!

All my love, darling, a thousand kisses.

Your Rudi

P.S. I posted the book of Burns to Nina today.

[123] Genia Peierls to Rudolf Peierls

[Leningrad], 25.6.1931

Dear Rudi,
Why wasn't there a letter from you today? To tell you the truth, I'm not taking it so much to heart, though it is the 5th day: I'm so busy that no time left for emotions. Terrible!

Mein Kleiner, *this letter is going to be unpleasant, at least partially. The point is the book.*[456] *Today I'm going to return it, at the end of prolonged term; and they treated this awful delay rather mildly. Anyway, they have too many books stored and this one would lie for another week. Rudi, when I wrote it down for the first time and started to make corrections, I blotted it here and there. I had to copy it once more, but the style remained rather bad. I had to take a lot of trouble over it, terribly hard. You wrote it in a great hurry, and I was afraid to change anything in your text lest I made it worse. I may have distorted it in some places, but not much. Abbot has read one half of it — everything was all right; today I'm going to ask him about some unclear items; besides, there must be proofreading. So everything is more or less all right*

[456] Genia was translating a book written by Rudolf Peierls into Russian. She was concerned about her style and dissatisfied with the translation.

with it. But just look at the style! Each phrase is as heavy as 1000 kilograms. Rudi, aren't you angry? If you are, then pardon me and forget it. Really, I did what I could. If only you could imagine how difficult it was! And don't feel upset; it isn't worthwhile getting hurt because of my style. Won't you? I can say, I didn't realize being alive those 10 days. The Institute, the lessons of German, the lessons with these boys — and then this book. Ouch!! I like to be occupied, but that was too much; even in my nightdreams I saw urgent business, I was to hurry somewhere, I was too late. I became thinner, acquired my bygone slenderness. In one of my dreams, Rudi, I saw you were meeting me at the station in Berlin, etc. That was exciting. I haven't had any entertainments these weeks. Only once with Dau and others we went to see a tr-r-ragic play. The impression was very strange. First, they were shooting all the while (the action takes place during the Civil War in Siberia). Shooting was from all sorts of guns, even the cannon. All that was good, in places, but rather boring in general. I and Ambarts left after the 3rd act; we went to the refreshment room to drink some tea and play with a kitten. Then everybody, of course, said Act 4 was the best. Today we are going to the theater where you saw "The Nose". We shall see "Belle Helène"*. They say, though, that the staging was remade, even the plot and the time of action. That must be strange. Today Dau and Kibel went to Petershof to see the young girl with whom* Jonny *fell in love long ago. She stays there in the Scientists' Rest Home. I didn't go because of the book, bad weather, washing my hair, but the main reason is: tomorrow I'm going to Pavlovsk by car, delivering sensitive instruments (galvanometers, electrometers, microscopes, etc.) that aren't to be carried by train. Yesterday all the tables and chairs, etc. were hauled at last to the new place. I adore moving to another place, and I nearly fell from the top of the wagon, when I loaded the last box. Everything turned out all right. I wonder how we shall drive tomorrow. If you don't get my letters for a long time, it may mean that the car broke down and I perished in a pool between Leningrad and Pavlovsk. Will you weep for me? Will you grieve for your* gutes Weib*? Rudi, I forgot to mention: Leonid Isaakovich Mandelshtam came to see us. He asked about the reprint that you had promised to send him. I'll take one from Dau and send it to Moscow, but please send him one too as if on your own initiative,* *otherwise I'll feel uncomfortable. They were so nice to us in Moscow. Agreed?*

Today before going to the theater I'll undergo a medical check, which is necessary to get a holiday assignment to Teberda.[457] *I registered for August 10th to 25th. I think we can stay here till September or even later, if we want. My dear, it is time for me to go to the post-office, then wash my hair, etc. So many things to do, my head may burst of them all. I used to live a decent life (whatever that means), but now ...*

Rudi, I'm reading Schnitzler in German. Yes! And you don't write German letters! Die Geliebte *is rather funny at some places, but with a sentimental tinge. Is that right?*

Rudi, I love you so and I'm eager to see you so much, that I was offended because there was no letter. Myself, I haven't written you for 6 days, but I was so busy, my darling; I went to bed at 2 o'clock, and got up early, sometimes even at 8!
Kisses, kisses, kisses,

G

[124] Rudolf Peierls to Genia Peierls

[Zürich], 29.6.1931

Genia, darling,
why are you worrying about this book so much? I never expected that it would be so much work for you, and of course I thought that you'd finished it a long time ago. And you're still a little worried about it, right? Why didn't you say anything earlier? My silly girl, it really isn't important, and it won't make any difference to this bloody book if the style isn't completely as it ought to be. Anyway, I'm sure that the style is perfect if you've been working on it all this time; you always notice the tiniest stylistic mistakes, which others miss. And besides, you probably have a very good feel for style. Please, don't think about it any more. Thank you once more. If I knew in the first place how

[457] Village in the Caucasus.

much work it would be for you, of course I wouldn't have suggested it. Anyway, it's too late to discuss that now. My poor darling; you've had such a horrible time with it all the same.

Things with Mandelshtam really are unpleasant: for all sorts of reasons, I forgot to send him our work,[458] which is awful. Not only that: I haven't kept a duplicate, so I can't send it now, but I'll write to him that I've sent it and obviously it has been lost. I don't like doing these things, but it's better like this.

Clearly the post system isn't working properly again. It seems that my letter of the 22nd (no. 14) has been delayed again!

Your letter arrived quite late too, and I had begun to think about sending a telegraph, but it wasn't the post which was to blame ... Darling, don't think for a minute that I'm offended, of course I know how busy you are; don't worry about writing if it's difficult.

Am 26. war ich bei Hauser's, einem Architekten und seiner Frau, ich glaube ich habe Dir schon von ihnen erzählt. Besonders sie ist sehr nett, unbedingt zweite Klasse. Ausserdem [?] alle Physiker war noch ihre Freundin da, eine Physikerin aus Jena, die ich schon auf einer Tagung in Leipzig kennengelernt habe. (Du hast schon recht, man kennt hier doch fast alle Physiker) sie ist aber nur ein Dreier.

Am 27. (Sonnabend) gingen wir alle segeln. *(... on a Russian boat).* "Alle" heisst *Motschan,* Bretscher's, *Fania,* Stobbe und Frl. Kobylinski. Wir sassen alle im Badeanzug auf dem Boot, und sprangen, wenn es uns zu warm wurde, ins Wasser. Ausserdem war der Wind wundervoll — also alles was man sich nur wünschen kann. Abends war ein grosses Feuerwerk auf dem See, das wir uns vor dem Balkon von Roms aus ansahen. (Rom ist dieser Arzt, den ich durch Hermann Krebs kenne). Diese Roms sind aber doch schrecklich schweizerisch, und wir gingen nach dem Feuerwerk so schnell wir möglich fort. Dann musste man es aber noch ausnützen, dass an diesem Tage die Polizeistunde (*the hour when all the cafes, dining rooms etc. have to close*) nicht um 12 sondern um 4 Uhr war, so dass man abends noch tanzen konnte. Das taten wir

[458] Landau and Peierls' work on uncertainty, see letter [93], note 380. See also Genia's letter [123].

auch, und *Fania* brachte es fertig, mit Stobbe zu tanzen, obwohl dieser sich entsetzlich sträubte und behauptete, er könnte nicht tanzen. Aber er wurde einfach mitgenommen. Gestern abend waren wir bei *Fania's* Schwester; man muss es immer wieder bewundern, wie dieses Mädchen sich hält. Es ist jetzt wieder etwas schlechter mit ihr geworden, sodass im September bestimmt noch eine Operation gemacht wird, und es ist noch gar nicht sicher, ob sie inzwischen nach Hause fahren darf, oder ob die Reise zu gefährlich ist.

Wie Du siehst, war ich in der letzten Zeit sehr unsolide, aber jetzt geht es wieder ordentlich mit der Arbeit los. Ich habe jetzt endlich verstanden, was ich mit der Lichtabsorption der festen Körper machen muss,[459] und es bleibt jetzt nur noch übrig, wirklich zu rechnen ("ixen"). Aber das wird noch viel Zeit brauchen. Und ausserdem gibt es noch eine Menge anderer Sachen.

"Die Geliebte" ist eines der schlechtesten Stücke von Fulda.[460] Ich mag überhaupt seine Stücke nicht, sie sind zwar alle sehr gut gemacht, aber so schrecklich "harmlos" und uninteressant. Er hat vor allem ein grosses Formtalent, und wo es nur auf Form ankommt (z.B. bei Übersetzungen) da ist er einfach nicht zu übertreffen.

Stobbe *is in a good mood today. He's received a place with Lennard-Jones in Bristol.*[461] *Living there will be very good (despite what Dau says), and even the fact that Lennard-Jones is an old fool won't bother him. Because of that, Stobbe won't be at Rentgenovsky's, at least not this year; tomorrow he'll write to Dau about it. I'm very pleased for him: the place at Rentgenovsky's won't be there forever, and he can stay in Bristol for as long as he wants. (And probably will stay as long as he wants, as he finds changing jobs rather difficult — he's not as adaptable as he seems.)*

---

[459]See letter [126] note 476.

[460]Ludwig Fulda, Else Peierls' brother-in-law and playwright.

[461]John Edward Lennard-Jones (1894–1954), studied mathematics at Manchester and Cambridge where he obtained his Ph.D. in 1924. He taught at Bristol (1925–1932) until he took up the Plummer Chair in Theoretical Chemistry which he held until 1953.

Liebes, ich warte jetzt schon jeden Tag auf das Telegramm, dass Du kommst. Ich möchte Dich sehen, Dich küssen können! Es ist jetzt schon spät, und ich will noch arbeiten, also gute Nacht! Bei Euch ist es jetzt schon $\frac{1}{2}$2 und Du schläfst sicher schon und sprichst im Traum *Good night, my dear!*

Until the next time I see you! (Probably in about six weeks, if not before). Kisses,

Rudi

[125] Wolfgang Pauli to Rudolf Peierls

Ann Arbor,[462] 1.7.1931
[1. Brief]

Lieber Herr Peierls!
Dank für Brief und Manuskript.[463] Obwohl ich mit dem Inhalt dieses Manuskriptes einverstanden bin, bin ich doch gegen eine Publikation. Erstens halte ich es für schädlich, wenn die (relativ zu mir) jüngeren Physiker sich an die Größenordnungsphysik gewöhnen. Zweitens ist der Restwiderstand ein Dreckeffekt und im Dreck soll man nicht wühlen. Drittens schwebt die Theorie stark in der Luft, da man nicht genau genug weiß, wieviel bei der gegebenen Temperatur T die additiven Zusatzterme, die zu höheren Potenzen von T/Θ proportional sind, für den elektrischen Widerstand ausmachen. Viertens sollten Sie doch vernünftigere Fragestellungen haben als solche kleine Problemchen; ich finde Sie verzetteln sich in letzter Zeit zu sehr mit Kleinigkeiten. Ich bin nicht so wie die Bolschewisten, die die Kirchen, nachdem sie sie geschlossen haben, auch noch in die Luft knallen. Also zerreiße ich Ihr Manuskript nicht, sondern sende es Ihnen zurück, aber mit dem

[462]Wolfgang Pauli spent much of the spring and summer of 1931 in the United States, initially at a physics meeting in Pasadena and then teaching at the summer school at Ann Arbor. See Enz, *No Time to Be Brief.*, chapter 6.

[463]This refers to a draft of the manuscript 'Zur Frage des elektrischen Widerstandsgesetzes für tiefe Temperaturen', which Peierls — against Pauli's advice — published the following year. *Ann. Phys.* **5**, 154–68 (1932).

ausdrücklichen Rat, * es vorläufig in ihrer Schublade liegend aufzubewahren. Vielleicht schreiben Sie irgendwann einmal ein größeres zusammenhängendes Referat, dann können sie es ja vielleicht versenden. Leider ist das Annalen-Heft mit der Nordheimschen Arbeit noch nicht hier.[464] Falls Nordheim etwas unrichtiges behauptet hat, würde ich Ihnen empfehlen an ihn zu schreiben. Eventuell soll er selbst dann eine Berichtigung zu seiner Arbeit veröffentlichen. Wenn Sie in der Lichtabsorptionsfrage mehr als Größenordnungen herausbringen könnten, wäre es ja sehr schön, aber ich bin sehr skeptisch.[465] Frenkel kommt vielleicht bald nach Zürich oder nach Leipzig, vielleicht können Sie ihn überzeugen.[466]

Das ist ja eine schöne Bieridee, magnetische Wechselfelder zu betrachten, deren Frequenz mit der Dublettaufspaltung übereinstimmt, denn der Zeemaneffekt, der vom Magnetfeld des Lichtes herrührt, ist doch wegen der kleinen Amplituden minimal. Ich glaube nicht, daß sich da irgendetwas wird beobachten lassen. (Über eine Abänderung der Fragestellung siehe später.)

Was die magnetische Seite der Frage betrifft, so muß man schon bei normalem Zeemaneffekt wesentlich die Fälle unterscheiden, wo das Feld seine Richtung beibehält und den, wo es seine Richtung ändert. (Ihr Brief ist so schlampig, daß Sie nicht angeben, welchen Fall Sie betrachten; da Sie aber sagen, das Resultat sei einfach, vermute ich,

---

[464] L. Nordheim, 'Zur Elekronentheorie der Metalle. I and II', *Ann. Phys.* **9**, 607–40, 641–78 (1931). Here Nordheim had suggested an interpolarisation formula for electrical, resistance. It was not derived from theory but was a good empirical approximation of experimental data. See Pauli, *Wissenschaftlicher Briefwechsel*, II, 87, note c. Peierls himself published a paper in 1932 dealing with this question. 'Elektronentheorie der Metalle', *Ergebnisse der exakten Naturwissenschaften* **11**, 264–322 (1932).

[465] This refers to Peierls work for his 'Habilitation'. This was published a year later as 'Zur Theorie der Absorptionsspektren fester Körper', *Ann.d.Phys.* V **13**, 905–52, (1932).

[466] See letter [122]. Peierls regarded Frenkel's methods as too crude to produce a reliable result for the relation between electrons and lattice vibrations. He presented his criticism in a seminar paper at a Theoretical Physics Conference at Kharkov, published as R. Peierls, 'Über die Absorptionsspektren fester Körper', *Phys.Z.Sowjetunion* **1**, 297–8, (1932). A more elaborate discussion is found in his 'Habilitationsschrift', see above note 334.

es handelt sich um feste Richtung des Feldes.) Bei fester Richtung des Feldes ist das Resultat in der Tat beim normalen Zeemaneffekt schon aus der klassischen Oszillationsvorstellung ableitbar (ich weiß nicht, ob es explizit in der Literatur irgendwo zu finden ist, aber es muß bestimmt vielen bekannt sein, nicht nur mir.) Wellenmechanisch kann man am einfachsten mit den in der Güttingerschen Arbeit stehenden Formeln rechnen.[467] Sind $u_k(H, x, y, z)$ die Eigenfunktion, die zum jeweiligen momentanen Wert von H gehören und sind die gesuchten Lösungen der Schrödingergleichung für das Wechselfeld

$$\psi = \sum c_k(t) u_k$$

so gelten für die $c_k(t)$, abgesehen von Vorzeichenfehlern, Faktoren 2, etc. die Gleichungen

$$\dot{c}_k + \frac{i}{\hbar} E_k c_k + \sum_{\substack{l \\ (Q \neq l)}}' \frac{\dot{H}_x (\underline{M_x})_{kl} + \dot{H}_y (\underline{M_y})_{kl} + \dot{H}_z (\underline{M_z})_{kl}}{E_u - E_l} c_l = 0$$

Die $\underline{M_x}, \underline{M_y}, \underline{M_z}$ sind die Matrizen des totalen magnetischen Momentes des Atoms; es ist zu summieren über alle Zustände l des Atoms (bei der betreffenden Feldstärke; die E_l sind von $\underline{H_x}, \underline{H_y}, \underline{H_z}$ und damit implizite auch von t ahängig), die vom Ausgangszustand k verschieden sind. (Nichtdiagonalenelemente der Matrizen).

Beim normalen Zeemaneffekt wird H parallell der festen Richtung z, enthält $\underline{M_z}$ keine Nichtdiagonalelemente, die Summe $\sum_l$ verschwindet, und die Lösung wird

$$c_k = \text{const.} e^{-\frac{i}{\hbar} \int^t E_k dt} = \text{const.} e^{-\frac{i}{\hbar} g \mu_0 m \int H_z(t) dt}.$$

Die Lichtabsorption oder -emission folgt durch Bilden der elektrischen Momentmatrix und Fourierzerlegung nach der Zeit. Wenn $H_z(t) =$

[467]Paul Güttinger (1908–55) was a student of Pauli's working on atoms in alternating magnetic fields. P. Güttinger, 'Das Verhalten von Atomen im magnetischen Drehfeld', *Z. Phys.* **73**, 169–84 (1931). This paper contains the formula used by Pauli.

$A \sin 2\pi\nu t$, dann kommt man auf Besselfunktionen. Das Vernachlässigen der $\sum_l$ ist für H parallell z auch bei Dubletts solange erlaubt, als ν klein gegen die Dublettaufspaltung (und Larmorfrequenz klein gegen Dublettaufspaltung). Für den von Ihnen betrachteten Fall des Dubletts $\Delta\nu$ und Wechselfrequenz $\nu \sim \Delta\nu$ habe ich die Gleichungen für $c_{j,m}$ und $c_{j',m}$ (j und j' die beiden j-Werte des Dubletts) hingeschrieben und glaube kaum, daß sie allgemein integrierbar sind, da die Verschiedenheit der g-Werte bei den beiden Dublettniveaus komplizierte Terme bedingt. [Die (in Bezug auf j) nichtdiagonalen Matrixelemente von $\underline{M_z}$ sind natürlich gleich denen des totalen Spinmomentes S_z.] es gibt ein Experiment von Fermi und Rasetti H_i [Z.Phys.33, 246, (1925)][468] über die Depolarisation der Resonanzstrahlung in magnetischen Wechselfeldern. Das ist eine viel vernünftigere Fragestellung als der Zeemaneffekt in Wechselfeldern, da diese Depolarisation gerade bei Feldern von einigen Gauss eintritt und Wechselfelder so kleiner Feldstärken leichter herstellbar sind. Wenn es Ihnen Spaß macht, könnten Sie folgende Frage untersuchen. Sie haben ein Hyperfeinstrukturmultiplett $\Delta\nu$ und untersuchen die Polarisation der Resonanzstrahlung in (eventuell auch der Richtung nach) wechselnden Magnetfeldern als Funktion und dessen Frequenz ν. Was passiert mit dieser Polarisation, wenn $\nu \sim \Delta\nu$ wird? Vielleicht (wenn die Resultate nicht zu kompliziert werden) könnten derartige Effekte dazu benützt werden, um unaufgelöste Hyperfeinstrukturen nachzuweisen.

Nun haben Sie also außer Kritik auch noch eine positive Fragestellung. Ich hoffe, Ihre Physik wird besser, ich bin wenig zufrieden mit ihr.

Bitte sagen Sie Güttinger, er soll mir möglichst bald Korrekturen seiner Arbeit über den Stern-Versuch hierhersenden![469]

Trotz Gelegenheit zum Schwimmen leide ich sehr unter der großen Hitze hier. Unter der "Trockenheit" leide ich aber gar nicht, da Laporte

[468] E.Fermi and F. Rasetti, 'Über den Einfluß eines wechselnden magnetischen Feldes auf die Polarisation der Resonanzstrahlung', *Z.Phys.* **33**, 246–50 (1925).

[469] Otto Stern (1888–1969), professor of physical chemistry at Hamburg. He had carried out experiments relating to Güttinger's diploma thesis. See note 481.

und Uhlenbeck[470] ausgezeichnet mit Alkohol versorgt sind (man merkt die Nähe der kanadischen Grenze). Physik (und Physiker) gibt es hier sehr viel, aber ich finde sie zu formal. In Chicago, Detroit und New York werde ich sicher noch viel amüsantes erleben. Bisher hatte ich wenig Zeit. Wäre gerne länger in Kaliforniern geblieben. Grüße Sie, Scherrer und das ganze Institut.

von W. Pauli

*Ich habe keine legale Handhabe, Sie zu zwingen, diesen Rat zu befolgen. Auch bin ich kein Dorfmann, der Ihnen die Assistentenstelle kündigt, wenn Sie ihn nicht befolgen.

[126] Wolfgang Pauli to Rudolf Peierls

Ann Arbor, 1.7.1931[471]

Lieber Herr Peierls!
Bei der Beantwortung Ihres Briefes vom 30.VI. möchte ich mit dem Ende anfangen,[472] der Frage der diamagnetischen Drehung in Wechselfeldern. Zunächst war meine seinerzeitige Bemerkung zu Bretscher[473] die, daß rein klassisch auch im veränderlichen Magnetfeld die Präzessionsbewegung in jedem Augenblick die Periode $\sigma = \frac{e\mathfrak{H}}{4\pi mc}$ hat (σ jetzt zeitlich veränderlich). Das ist ein allgemeines Theorem. Wenn ich mich auf die Felder fester Richtung ($\mathfrak{Z}$) der harmonischen Oszillator

[470]George Eugene Uhlenbeck (1900–1988), studied at Leiden, Rome and Copenhagen before completing his doctorate under Ehrenfest in Leiden in 1927, then took up an appointment at Michigan. In 1935 he accepted a Chair in Utrecht and throughout his carreer he worked for periods in the US and in the Netherlands.

[471]The letter is misdated, as the letter from Peierls which it refers to was sent on 30 June. Furthermore, the accident described in the letter, according to other sources occurred on 14 July 1931. See Pauli, *Briefwechsel*, II, p. 90, note a.

[472]Letter could not be located.

[473]Egon Bretscher, at the time assistant in Zurich, was working on this problem. See E. Bretscher and W. Deck, 'Magnetrotation bei nichtadiabatischer Änderung des Magnetfeldes" *Helv.Phys.Acta* **6**, 229–31 (1933).

beschränke, so lauten die Bewegungsgleichungen für $x_1 = x + iy$ und $x_2 = x - iy$

$$\ddot{x}_1 + \omega_0^2 x_1 - 2i\sigma\dot{x}_1 - i\dot{\sigma}x_1 = \frac{(-e)}{m}E_1$$

$$\ddot{x}_2 + \omega_0^2 x_2 + 2i\sigma\dot{x}_2 + i\dot{\sigma}x_2 = \frac{(-e)}{m}E_2$$

(E_x, E_y äußere Feldstärke, $E_{1,2} = E_x \pm iE_y$.)

Die Terme mit $\dot{\sigma} = \frac{d\sigma}{dt}$ kommen von der Induktionskraft $\mathfrak{E} = -\frac{1}{c}\mathfrak{A}$ Feldes. Wenn man, wie immer, Terme $\sim \sigma^2$ vernachlässigt, ist die Lösung für die freie Schwingung ($E_1 = E_2 = 0$ gesetzt):

$$x_1 = Q_1 e^{i(\omega_0 t + \omega)}, \qquad x_2 = Q_2 e^{i(\omega_0 t - \omega)}$$

Abkürzung $w = \int_0^t \sigma dt$. Dies ist die Aussage des erwähnten Theorems für unseren Spezialfall.

Um aber die Drehung der Polarisationsebene zu bekommen, muß man die erzwungene Schwingung mit $E_{1,2} = E_{1,2}^0 e^{i\omega t}$ untersuchen, d.h. die Lösung der inhomogenen Gleichung (1). Diese lassen sich für allgemeines $w(t)$ angeben.

$$x_1 = (\frac{-\epsilon}{m}E_1^0) \cdot \frac{e^{iw}}{2i\omega_0}[e^{i\omega_0 t}\int^t dt' e^{i(\omega-\omega_0)t'-iw} - e^{-i\omega_0 t}\int^t dt' e^{i(\omega+\omega_0)t'-iw}]$$

$$x_2 = (\frac{-\epsilon}{m}E_2^0) \text{ [} w \text{ wird durch } -w \text{ ersetzt.]}$$

Es scheint richtig, daß das Resultat für $w = \sigma_0 \cdot \frac{\sin \nu t}{\nu}$ und $\nu \gg \sigma_0 (\sigma = \sigma_0 \cos \nu t)$ anders wird als für $\nu \ll \sigma_0$. Man kann, wenn man will, in diesem Fall $e^{iw} = e^{i\sigma_0 \frac{\sin \nu t}{\nu}}$ in Fourierreihen zerlegen, was auf Besslfunktionen von $\frac{\sigma_0}{\nu}$ führt und dieses in (2) nachher einsetzen. Sie können das noch weiter durchdiskutieren, wenn Sie wollen.

Nun zur Frage der Übertragung des Resultates in die Quantentheorie. Da Sie die Arbeit von Blaton in diesem Zusammanhang erwähnen[474] (bei Ihnen muß man das wichtigste immer indirekt aus irgendwelchen Andeutungen entnehmen: Zeichen für große Zappligkeit Ihrerseits) vermute ich, daß Sie Diracs Strahlungstheorie (Quantelung des magnetischen Wechselfeldes der Frequenz ν) treiben wollen. Obwohl es wahrscheinlich ist, daß dies zu richtigen Resultaten führt, ist mir in diesem Fall die Benützung dieser Methode sehr unsympathisch. Denn es ist ja in Wirklichkeit nicht so, daß die Anzahl der Quanten $h\nu$ im Anfangszustand durch Beobachtungen (Observations) fixiert wird, und es scheint mir deshalb dem physikalischen Sachverhalt näher zu kommen, das Wechselfeld als klassisches Feld mit gegebener Zeitveränderlichkeit einzuführen (vgl. meinen letzten Brief). Man hat einfach die Matrixelemente R_{nm} des Atomsystems als Funktion der Zeit unter Einfluß a) des magnetischen Wechselfeldes , b) des Feldes der Lichtquelle zu ermitteln. Ich vermute, die klassischen Gleichungen für den Oszillator werden dann im wesentlichen auch für diese Matrixelemente gelten.

Über die Ungleichungen für die Feldstärken in der Arbeit von Ihnen und Landau habe ich inzwischen noch nachgedacht.[475] Ich halte es für sehr wahrscheinlich, daß sie für makroskopische Probekörper mit Ladung $e > \sqrt{\hbar c}$ falsch sind.

Nun über die Lichtabsorption im festen Körper. Es ist sehr merkwürdig, daß Sie die Breite der Linie berechnen können, ohne die angeregten stationären Zustände explizite zu kennen.[476] Wenn es richtig ist, ist es ein ganz lustiger Trick. Schreiben Sie doch die ganze Sache zusammen und schicken Sie sie mir, bitte. Wenn die Resultate herauskommen, die sich an der Erfahrung prüfen lassen, bin ich im Prinzip damit einverstanden, daß die Arbeit als Habilitationsschrift verwendet wird. (Dabei wird man auf die Polemik gegen Frenkel ein bißchen aufpassen müssen. Könnten Sie nicht versuchen, ihn zu erreichen und ihn zu überzeugen? Er muß in Europa sein.) In diesem Fall müssen Sie in dem Widerstreit zwischen Ihrem Habilitationsehrgeiz und Ihrer Publikationswut also dem ersteren das Übergewicht geben.

---

[474] J. Blaton, 'Gibt es eine Doppelstreuung von Lichtquanten?', *Z.Phys.* **69**, 835–49 (1931).
[475] See note 375.
[476] See letter [125].

Übrigens wird es bei den Professuren eine kleine Verschiebung geben. Landé[477] wird definitiv in Amerika bleiben, so daß die Tübinger Professur frei wird. Es besteht, wie ich höre, eine gewisse Absicht, Unsöld dahin zu berufen, zugleich wegen der dortigen Astronomie. Dann würde also (meine frühere) Hamburger Assistentenstelle wieder ledig ...[478]

Dummerweise bin ich neulich (in etwas angeheitertem Zustand) so ungünstig über die Treppe gefallen, daß ich mir die Schulter gebrochen habe und nun im Bett liegen muß, bis die Knochen wieder ganz sind — sehr langweilig.[479]

Grüße

Pauli

Wenn Sie in Leipzig sind, grüßen Sie Heisenberg, Bloch und die anderen sehr von mir. Was gibt es Neues?

[127] Wolfgang Pauli to Rudolf Peierls

[Ann Arbor], 3.7.1931

Lieber Peierls!

Heute habe ich es schon wieder auf Ihre Physik abgesehen und zwar diesmal auf die Arbeit von Ihnen und Landau. Als ich hier im Kolloquium über diese vortragen sollte, merkte ich, daß Ihre Begründung der Ungleichungen für die Feldstärken nicht richtig ist. (Vielleicht sind die Ungleichungen selbst aber richtig, und Sie können die Ableitungen verbessern.) Es ist offenbar unrichtig, daß die ausgestrahlte Energie $\frac{e^2}{c^3}\frac{(v'-v)^2}{\Delta t}$ eine <u>unbestimmte</u> Energieänderung ist. (Achten Sie auch darauf, daß ja die ganze Betrachtung nur für <u>makroskopische</u> Probekörper

[477] Alfred Landé (1888–1976), studied physics at Marburg, Munich and Göttingen. After an initial appointment in Frankfurt, he became professor at Tübingen. In 1931 he decided to extend shorter visits to the US and to take up a professorship at Ohio State University where he remained until his retirement in 1960.

[478] Peierls applied for the assistantship at Hamburg, but decided not to return to Germany in 1933 after Hitler's ascent to power.

[479] See Enz, *No Time to Be Brief*, pp. 223–4.

mit Ladung e, für die $\frac{e^2}{hc} > 1$, in Frage steht.) Es mag sein, daß die ausgesstrahlte Energie auch eine Unbestimmtheit enthält, wegen der Unsicherheit im zeitlichen Verlauf, aber in erster Näherung ist die ausgestrahlte Energie sicher eine bestimmte Änderung. Daher ist auch die Gleichung

$$(v - v')\Delta P > \frac{e^2}{c^3}\frac{(v - v')^2}{\Delta t}$$

sicher nicht richtig als Unsicherheitsrelation. Das geht ja schon daraus hervor, daß sie h nicht enthält und, wenn richtig, eine prinzipielle Unbestimmtheit des Impulses geladener Teilchen in der klassischen Theorie postulieren würde!

Schreiben Sie mir also, bitte, Ihre Meinung über diese Sache. In diesem Falle bin ich übrigens für publizieren. Sie müssen entweder (zusammen mit Landau) eine verbesserte Herleitung der Ungleichungen für die Feldstärken publizieren oder Sie müssen öffentlich bekennen, daß diese Ungleichungen (für makroskopische Probekörper großer Ladung) bisher nicht begründet.[480]

(Die §§ über die Ortsmessung von Lichtquanten und Elektronen sind übrigens so schlecht geschrieben, daß man kaum die Argumente verstehen kann. Auch scheint mir die Frage nach der Ausmeßbarkeit der Diracschen Dichtefunktion in seinem stationären Zustand — wo man sich bei den Experimenten Zeit lassen kann — viel interessanter und wichtiger als die in Ihrer Arbeit diskutierten Fragen. Sie sollten das noch näher untersuchen statt Größenordnungsphysik zu treiben.)

Die Arbeit von Güttinger habe ich inzwischen bekommen.[481] Er soll die Kleinsche Arbeit, Zeitschrift für Physik, 41, 407, 1927 für seine Dissertation durchstudieren, ebenso die neue Arbeit von Blaton über Doppelstreuung.[482]

Heute ist es nicht so heiß und ich fühle mich hier im allgemeinen sehr wohl. Nur die Physik ist hier sehr unbefriedigend, weil zu formal.

---

[480] Although Peierls accepts his mentor's criticism (see letter [132]), he does not take his advice with regards publishing the proposed qualification.

[481] See letter [125].

[482] O. Klein, 'Elektrodynamik und Wellenmechanik vom Standpunkt des Korrespondenzprinzips', *Z.Phys.* , **41**, 407–42 (1927) and Blaton, see letter [125], note 474.

Grüße an Sie und das ganze Institut. Richten Sie bitte Scherrer auch aus, daß Laporte und Oppenheimer[483] als Ersatz für Mr. Wellmann sehr zu brauchen sind. Wissen Sie kein vernünftiges physikalishes Problem für mich? Ich würde im September gerne in New York wohnen, am Tag ixen und nachts bummeln.

Ihr W. Pauli

[128] Rudolf Peierls to Genia Peierls

[Zurich], 4.7.1931

My darling!
I really should be working at the moment, but I received your lovely letter yesterday[484] *and wanted to write to you. Dearest Genia, I miss you terribly. Every day I await a telegram. But I've noticed that already I find it difficult to imagine you arriving here. I think about it all the time and think about all the details. But the more I think about it, the less I can believe that it isn't just a fantasy, that it will really happen; I don't know when, but the time will certainly come (and perhaps even soon, perhaps there is a telegram at home now!)*

Thank you for the letter, I am so pleased that you're not worrying about that book any more; I'm absolutely sure the style is all fine. Good news about the money, but even better that you've received a place at Teberda. Let's hope that you won't have to make use of it.

When my work here is finished, I'll go to Leipzig, around the 16th July. There my address will be: Physikalisches Institut, Linnestr. 5, Leipzig C1.

In der letzten Woche gab es sehr viel zu tun. Einen oder zwei Tage habe ich nur an der Maschine gesessen und Briefe geschrieben, z.B. einen

[483]Otto Laporte (1901–1972), had moved to the University of Michigan where he became professor of physics. Robert Oppenheimer, another of Pauli's former students was also at Ann Arbor in the summer of 1931 which must have triggered Pauli's remark.

[484]Letter [123].

langen Brief an Pauli,[485] in dem ich ihm über meine Arbeit schrieb, die jetzt fast fertig ist (genauer gesagt, sie fängt jetzt erst an, denn jetzt kommen die numerischen Rechnungen) und ihm mitteilte, dass ich mich seit dieser Arbeit im Winter habilitieren will (*like* Privatdozent). Ich glaube, er wird einverstanden sein.[486]

Am 1. erschien hier im Institut die Physikerin aus Jena (weisst Du, die ich beim Architekten getroffen habe) und es zeigte sich, dass sie mit einem jungen Theoretiker in Jena befreundet ist (oder mehr), und der eben erst seinen Doktor gemacht hat, und keine Stelle bekommt. Sie fragte mich, was man da tun kann, und ich bemühte mich, ihr einige Adressen zu geben. Aber was kann man da sonst so machen. Sie hat auch eben ihren Doktor gemacht und hat auch keine Stelle. Dann blieb sie gleich bei uns zum Tee, (Wir haben bei uns im Institut jeden Mittwoch vor dem Kolloquium einen Tee) und als Scherrer hereinkam, sah er sie neben mir sitzen und sagte: "Aha, Frau Peierls ...?" Er sagte das sehr nett und liebenswürdig, sodass Du also damit rechnen kannst, dass er Dich nett empfangen wird. Übrigens wurde auch *Fania* schon einmal mit Dir verwechselt. Wir sind hier nicht weniger als in Leningrad allgemeines Gesprächsthema und wenn ich irgendwo irgendjemand kennenlerne, so weiss er gleich, wer ich bin und sagt: "Ach so, Sie sind der ..." Du hast Dir jedenfalls hier dadurch dass Du so lange auf Dich warten lässt, einen guten Auftritt gesichert.

Gestern war ich wieder mit dieser Physikerin und dem Architekten zusammen, und wir haben drei Stunden lang jüdische Witze erzählt.

Scherrer ist mit mir sehr zufrieden. Vor einem Jahr hat ein Jugoslawe hier eine Doktorarbeit gemacht, ganz gute Experimente, aber er hat die Arbeit schrecklich geschrieben und nicht so, wie Scherrer wollte. Besonders ist es hier nötig, eine allgemeine theoretische Einleitung zu einer Doktorarbeit zu schreiben, um zu zeigen, dass man alles verstanden hat. Dieser Mann ist jetzt schon fort, und kann auch nicht herkommen, und Scherrer wollte daher, dass ich diese Arbeit schreibe. Aber das habe ich abgelehnt, denn man kann doch nicht für

[485] This letter is referred to in Pauli's reply of 1. July 1931. See letter [125], note 342.

[486] See letter [124].

jemand anders eine Examensarbeit schreiben. Scherrer sah ein, dass ich recht habe, aber er war sehr unglücklich über die ganze Geschichte; schliesslich habe ich dann doch noch ausgedacht, wie man es machen muss: ich werde diesem Mann einen Brief schreiben, und ihm ungefähr (d.h. natürlich sehr genau) sagen, wie er die Arbeit schreiben soll, und dann wird er sie schreiben und ich kann sie dann noch einmal korrigieren. Es ist praktisch natürlich dasselbe, als wenn ich gleich die Arbeit schreiben würde, aber in dieser Form kann man es machen.

Fania should be leaving on the 6th, but she's going alone, as her sister has got worse: she has an operation in September and traveling by train would be dangerous for her. Because she can't manage by herself, until then she will be staying in a refuge for the blind. It is delightful to see how she is always happy, always busy, despite everything.

Together with the parcel, Frl. Kobylinski's parents posted a very funny letter, saying that they want to know how much she's eating, and that we should send them a photo of her. So we did indeed take a photo of her and today in the Institute we will develop it through special cylindrical glasses so that she looks really thin. I notice that I haven't started working yet; instead I'm spending the whole time writing a personal letter — shame!
Goodbye darling, a thousand kisses!

Your Rudi

[129] Rudolf Peierls to Genia Peierls

[Zurich], 7.7.1931

Genia, dearest,
I am currently sitting and listening to Rasetti[487] *giving a lecture. He is a decent chap, but his lecture is strangely boring. He is talking about the*

[487]Franco Dino Rasetti (1901–2001), Italian physicist who, together with Fermi, discovered key processes leading to nuclear fission. He later refused to join the Manhattan project on moral grounds.

effects of Raman's Theory[488] *— why do experimentalists always want to construct theories?*

I am cursed; a letter probably arrived today, and I couldn't wait for the post because I had to go to the dentist, which was fine incidentally. But the landlady cannot receive registered letters for guests, so your letter will only get to me tomorrow. And still no telegram to say when you will be here! I miss you so much my dear!

My life has been fairly uninteresting again recently. On Sunday (the 5th), I want dancing again with Fania, Motchan, and the new assistant from the physi-chemical institute, Kostüm. *He's "very German", but apart from that he is very nice and easy to talk to. We went there to organise a goodbye celebration for Fania, though she hasn't left yet because things have got worse with her sister. She needs to lie down all the time these days, as any movement could be dangerous for her eyes. In theory, the surgeon should operate straight away, but he has to go away for two months. It is generally a very dangerous operation; there is a 50–50 probability that it will go badly and she will become very ill. He will probably be able to operate in September, and Fania will be with her sister all the time, and not have a day to herself.*

I was talking to her yesterday, and she said that she had thought there was something wrong with her sister before, but only accepted the fact when she was forced to. I told her that she has no right to hold her back; her sister isn't so weak that she might do something stupid when agitated, and if everything finally becomes too much for her, then still no-one has any right to decide things for her. Fania said that I was right, but all the same, she can't bear the thought that her sister is ill, and that she is helpless. It is understandable, but also all theoretical, as eventually her sister will not be able to take many decisions herself.

I completely forgot to write to you, because I received a very funny book in order to write a review. Zwei Dialoge über Raum und Zeit. *It's by Jaffé (a theorist in Gießen) in the style of the ancient Greek philoso-*

[488] Rasetti had published among others F. Rasetti, 'Incoherent Scattered Radiation in Diatomic Molecules', *Phys. Rev.* **34**, 367–371 (1929), R.G. Dickinson, R.T. Dillon, and F. Rasetti, 'Raman Spectra of Polyatomic Gases', *Phys. Rev.* **34**, 582–589 (1929).

phers.[489] *I wrote a light-hearted review (for* Zeitschrift für physikalische Chemie) *and now I'm terribly curious as to whether Simon will print it or not.*

Aha, now Rasetti has begun to talk about intelligent things and I need to listen. Bye for now, dear wife!

8.7.1931

Last night after the lecture we went to Wentzel's, and it does indeed seem that Rasetti is a good fellow; now I understand why Bethe liked him when he was in Rome.

I came home late after that; I saw that your letter hadn't arrived and was sure that I would receive one in the morning. So I didn't set my alarm clock, and thought: "Let Genia wake me up!" How wonderful you are! Thank you for the letter, dearest. I'm always interested to read about your work, but I still don't understand what problems you are working on. Write and tell me, I'm fascinated.

But it's oddly interesting that you were thinking about this! I think a lot about how we could live together, and I had also thought of a few possibilities, but it's not worth talking about them now.

But there are also many problems, about which we will have to talk "late at night". Firstly, what to do if we have different tastes in people. Of course, we already know to some extent that there won't be big differences. It probably won't be the case that one of us always wants to see someone who the other can't stand. But it may be, of course, that one of our friends is more interested in one of us than the other. Without any difficulties it's easy, of course, to be "polite", but it's hard to say whether this will be the case, or whether it will sometimes be better for one of us to go alone to see some people, who the other isn't that keen on? Anyway, this is all terribly theoretical, most likely these questions won't arise, because I think you'll like all my friends here: Motchan primarily, the Bretschers, Fania (if she's still here when you arrive), Hauser (the architect who I told you about) and many others.

[489] George Jaffé, *Zwei Dialoge über Raum und Zeit*, Leipzig: Akademische Verlagsgesellschaft, 1931.

And you'll get to know plenty of people without me anyway, as I'll be at the Institute all day.

The Bretschers are getting married on the 21st. I'm very pleased about that; it's such a pleasure to see them together! He was recently talking to me about room furnishings etc. and asked my opinion, which he said was the advice of an old woman! Terribly funny.

I've just received a letter from Simon accepting my review. It will be terribly amusing: I'm very rude about philosophers. Simon sends you his regards, by the way.

Darling, I'd love to write a whole lot more, but I don't have much time (after lunch there is a colloquium, and after that a lecture on the radio)[490] *and I want to get this letter sent today.*

Kissing your mouth, your eyes, your hair ... everywhere! Goodbye, darling!

Your Rudi

P.S. On the 12th I'll write to you from Leipzig

[130] Rudolf Peierls to Genia Peierls

[Zurich], 11.7.1931

My next letter will probably be from Leipzig!

Genia,
dearest, of course, you're not an idiot — but you are a little crazy all the same! I understand perfectly how something can be so interesting that one forgets everything else, even if this leads to all sorts of problems. I do it too, so I must be a little crazy too!

Incidentally, we think here that you really need special shoes to play tennis, and I wouldn't recommend playing in normal shoes. But that should become obvious if you try to play in ordinary shoes, no?

There's a saying here: Wer den Schaden hat, braucht für den Spott nicht zu sorgen. *But your loss in this case isn't too great: uncomfortable nights, it's true, are never pleasant, but if it means that you can lie on*

[490] See letter [130], note 494.

the balcony in the sun during the daytime, there's no call to feel sorry for you! But all the same, I hope that in the meantime your nights have returned to normal, and all of Genia is healthy to the point of frivolity!

Darling! I love it so much that you "chatter" in your letters; all the same we should both remember to talk about how much we miss each other. How could it be otherwise? It's ten weeks since I left; it's not been too bad, especially as before we were together night and day. The time has gone quickly, and the remaining four weeks will pass quickly. All the same, it's an awfully long time!

But I probably won't be "chattering" too much today, as I'm strangely tired. I wanted to write to you earlier but I didn't because I was very moody and would have written a very gloomy letter. But I'm feeling better now. I had received a letter from Ehrenfest,[491] *in which he asked me about the evidence in my work on the Hall-effect, but when I went to explain, I found that some of it was actually wrong. At first, it appeared that the whole theory is flawed, which would have been unspeakably horrible. This was my first work and is still one of my favourites. Eventually I found that the evidence displays a few mistakes, but the results are correct, and I can collectively prove this result even in a special case, which means that qualitatively (and probably even collectively) it will always be correct. That there is a mistake in the evidence is unfortunate, but not that surprising.*

Then I received a letter from Pauli, in which he was oddly rude and said that he doesn't like my physics work, etc.[492] *Of course, I'm quite accustomed to him being rude, but it's not very nice all the same.*

I also received a letter from Frenkel, who doesn't want to believe that his work is incorrect.[493] *But if he is having trouble believing this, it will be terribly difficult to prove all sorts of little things to him. There*

[491]The letter does not survive. For details of the work referred to see letter [17], note 48. Initiated by Ehrenfest's comments, Peierls revises some arguments of his 1929 paper on the Hall-effect and publishes these revisions as R. Peierls, 'Zur Frage des elektrischen Widerstandsgesetzes für tiefe Temperaturen', *Ann. Phys.* **12**, 154–68 (1932).

[492]Letter [125].

[493]The letter does not survive. See also letter [125], note 466.

will have to be a long discussion. I will probably see him in Leipzig or Berlin.

On the 8th my lecture was on the radio, as I expected, but it was very amusing to read the lecture with no-one in the room, just a microphone, and I did find myself grinning at the microphone a couple of times.[494]

But you know, it was only then that I realised how important these things are to me. I don't think it's ambition, but public speaking is important to me. I still remember the feeling when I first read a paper in a seminar, and first gave a lecture. These things are quite commonplace for me now, interesting and enjoyable nonetheless, but this is something different. Naturally, I'll become accustomed to it after a while, but then I'll look for something else.

I don't know, of course, whether anyone will listen; yes, I know that at least one person will listen to the end: the radio station official who is obliged to.

Tomorrow a big group of us are going sailing for the last time; on the evening of the 15th I'll leave for Stuttgart, where I'll stay one day with Ewald, Barbara Zarniko and Ruhemann, and on the 17th I'll go on to Leipzig. It will be very interesting there: Heisenberg and Bloch, and loads of other physicists. And among them, who? Guth, of course! Curses!

Yesterday I went to the Strandbad and saw Lilly Fenigstein. I find her less and less interesting, and I think she likes me less and less.

Today I was at the Strandbad, again (you see how dull my life is?) with Fania, Motchan and Kostüm *(the new physical chemistry assistant) and so I'm exhausted now. I'm writing really slowly and after each sentence my eyes close.*

It would be better, I think, to go to bed now, even without you. And how many more times without you? Thinking about you, and kissing you.

Goodnight, my love!

Your Rudi

[494] Rudolf Peierls had recorded a short lecture for the Swiss radio which was broadcast on 8th July 1931.

12.7.

It's frustrating that you're not here: the seven of us have just been out sailing; it was sunny and hot, everyone was in their bathing costumes and just jumped in the water if they were overheating. We took lunch with us, and were on the lake among the mountains until 4 o'clock.

Motchan, Stobbe (who everyone calls "Be-cause"), Kostüm, Fania and a friend, and Mott-Smith[495] *were there. He's the American physicist I think I told you about — very quiet, terribly modest and unbelievably polite. But now I know him better. In the boat everyone wanted to steer, of course, but whenever anyone asked him whether he wanted a turn, he replied "No thanks, I'm a bit tired today, and I'm not that keen on it anyway", and no-one could persuade him otherwise. After a while I simply said: "Please take the rudder, I can't steer any more, and I should be preparing lunch", and everything was fine. I hope that this excessive politenesses won't creep into our circle of friends, that one won't start deceiving another in order to get what one wants. But that won't happen, will it?*

Yes, you're absolutely right that kissing someone's hand is unnecessary and merely a habit, like taking someone's hat etc. But all these customs together create a style — they don't exist for any particular reason, solely because nowadays one is used to them. The style to which hand kissing belongs isn't one I'm terribly fond of. I don't detest it as much as Dau does, of course, but it's just not 'me'. I don't have a personal style as such, but if I did it certainly wouldn't be that. What about you? I don't think that's really your style either, though oddly it seems to appeal to you sometimes, and you're more than willing to play along with it. But I don't think it's really 'you'.

We wrote a letter to you together in the boat, but it got lost somewhere and we were too lazy to rewrite it. Anyway, it was mainly along the lines that your future husband has excellent domestic abilities!

I've got a terribly busy day tomorrow. After lunch I'm seeing Professor Stodola and we'll no doubt be discussing pointless questions. He,

[495]Harold Meade Mott-Smith, graduate student from the University of Illinois Urbana Champaign.

of course, believes that there is no such thing, and thinks that philosophy and questions about "the existence of reality" etc. have some value. Nonetheless, it should be amusing.

I want to go and look for rooms for October now. It probably won't be possible to arrange anything yet, but I'd like to have a look around to see what we'll be able to get. Of course, there are plenty of rooms, but some are dear; others are in a terrible part of town, miles from anywhere and needing a tram journey to reach, which is dull; others will be horribly furnished, old and dirty. But I hope I should be able to find something nice, and we should be able to have two rooms, or one that we can convert into a bedroom at night.

[...][496] *This doctor hasn't said all this to us, of course, but to Fania; still, it's quite clear that she doesn't consider men very intelligent! What else appears clear to me is that she is worried others will find out how dumb her man is, so she talks all the time to prevent him from getting a word in edgeways; in reality, she would be better off not thinking about it!*
See you soon!

Your Rudi

[131] Genia Peierls to Rudolf Peierls

Pavlovsk, 14.7.1931

Well, my darling, you are terribly absent-minded. Why didn't you send your letter of the 4th by airmail? It was delivered on the 12th, together with your last letter, and I nearly got crazy during 4 days, and even my day-off was already lost because of worry. Could it be that you had to mail such a lot of letters that you lost your head and mixed everything: which letter was to be sent where?

Rudi, I haven't written to you for 5 days, maybe more, but the last three days I ran off my feet in connection with various affairs and troubles. These do not concern me, but my friends and relatives. So, the friend of my aunt (Mother's sister) went to the country, as a doctor, for

[496] Line missing in manuscript.

2 months and caught scarlet fever. She is very aged and she has very weak heart: can you imagine how worried the aunt is! Yesterday I came back at 9 p.m. from Pavlovsk and almost at once we set off for her place to keep her company. When she is alone, she loses her head because of worry, and she also has very weak heart and she works a lot. I[say] B[enedictovich]'s brother has been ill for some days. The diagnosis was pneumonia; he lives alone, so everybody had to go to his place, bringing some food, medicines, etc.

My parents are unable to leave tomorrow or the next day, as the pay wasn't provided in time. They'll have to stay in the over-heated flat for several days. Some days ago they visited one well-known doctor and professor, who declared the following: 1) both of them are terrible neurasthenics (that I had known before); 2) Mother has an awful blood pressure, about 200, while the normal 160/80; I.B.'s diagnosis - the initial stage of sclerosis and the heart valves are weak. One may say, "no end of pleasurable things". And so on. Oh, my poor boy, this is not a letter but a medicine file.

Rudi, you are an impossible optimist, if you think that I won't go to Caucasus. If we leave together, that would be very nice. My dear, Eric was in Moscow and I asked him to call on the Central Executive Committee. He did, and then wired me the following: they don't know of anything, the file is in the Foreign Department, they advise to apply for urgency. So write this application once more (it is better if you write yourself) and send it as a registered air letter to the Foreign Department of the Executive Committee. Agreed?

And so you enjoy life in Leipzig? You'll visit Berlin before your departure to Russia, and where should the visa be sent? Ioffe isn't here, hence a terrible mishmash, everybody mixes and forgets the papers. At present everything seems to be done properly. Dau and Abbott are very busy with it and keep visiting various red tape officials busy too. I hope everything will be well. If you meet your father, tell him about our circumstances, he may give a useful piece of advice. Unless it is unpleasant for you. As to the money: from the local publishing house they wrote to the Moscow centre, and there is no certainty yet.

I haven't yet got an accommodation for Teberda and it seems I won't get it, because they arrange it for a whole month only, from the 1st to 1st.

This is very inconvenient, if you arrive on the 10th; then staying there for a whole month is of no use. We'll spend there some 10 days, and then walk to Sukhumi to bathe and eat peaches. Right? Don't forget to bring trousers for me, and rubber pillows. A Rucksack *can be borrowed from Eric, so you may take only one for yourself. Oh, you are in Leipzig already; so, if you forgot to take trousers for me, there is nothing to be done. Didn't you forget it? Rudi, my golden one, besides this letter I'm going to write 4 – 5 more. Not so bad, but still there remain 26 days! What a nuisance to think you could be here in 4 days! And it is so nice here, especially in Pavlovsk: "The lime-trees blossom out in woods, and birds are singing in their branches." That is a line from a poem of one of my favourite poets. Isn't it beautiful?*

I'm sitting on the haycock unter den Linden *and eating strawberries that grow in my garden, everything is nice — that is Moslem Paradise, and I feel like a jobless houri. Unfortunately, I can enjoy these intervals for 15 to 30 minutes only; in this Moslem Paradise on a branch there hangs a watch that I have to look at with each berry, and then go up to the first floor, look into the tube and write 4 figures with this red ink. This bright colour of my letter isn't too irritating for you? You ask me what sort of work I do. Well, there are radioactive substances in the ground and, consequently, there is an emanation in the air. It is rather intensive, but the question is, what quantity and how it changes depending on the place, etc. About all this very little is known. There were special methods of measurement, rather prolonged and inexact. We develop better methods, one of them seems to be found: it is accurate,* "absolute"*. The instruments aren't yet ready, and they will be rather expensive. Now I have one of them on trial. It isn't that good, but portable, convenient and rapid. For the last 5 days it was capricious, today it is in full order. I use petroleum as a source of* e-radioactivity *— this is only the beginning, in 10 days I'll start serious work. My other task (you may remember I told you this in winter) is to establish the dependence between indications of ionization devices and the material they are made of (for γ-rays). It is strange, that lead shows twice as much ionization compared to aluminium, while for a long time they spoke of γ- and cosmic rays intensity, proceeding from the indications of such instruments.*

Oh, how late it is! Rudi, $\frac{3}{4}$ of my day off was lost because of you. We went to Pulkovo[497] *to play tennis. Dau said I have a great aptitude, I think it is for flattery. Even Abbot played with great excitement. There was a good boy with us, a radiologist; he stays in the same flat as Dau. We all went to Ambarts' place to arrange a drinking bout and eat ice cream. We were met there as though we returned from another planet, because Ambarts had promised to come back at 8 o'clock and didn't tell us that. We returned at 11 a.m.; this morning we had planned to go on a rowing boat but, as it was cloudy, we abandoned this idea. Ambarts' mother thought we had drowned and remained at the bottom of Finnish Gulf. She wasn't prudent enough and phoned to my Mother. You may imagine her state — ah! ah! Then we had a very long walk, it was so beautiful all around and no people, only rare optimistic anglers. It was 4 a.m.*

I saw a sound film, and became firmly convinced that Dau is sentimental. During the show one can hear marvelous croaking of frogs and chirping of a nightingale. One of the actors is excellent and there are some exciting moments. The most touching are homeless children — the hooligans and the pickpockets — transformed by the end of the film into conscientious and law-abiding people (the children were 15-year-old). The film is not higher, than 3rd class. Tomorrow our "Great Armada" is going to cross the Gulf from the Islands to Lakhta.[498] *I hope the weather won't interfere.*

July 15

The weather was all right; it was Mother who interfered. When she heard I was going to be the Admiral of the Great Armada, she got scared and said she wouldn't be quiet for a minute, especially in the company of such "navigators", as are Dau and Abbot. [...][499]

---

[497] Village near Leningrad.
[498] Village on the northern coast of the Finnish Gulf.
[499] Two pages missing in manuscript.

July 16

Kleiner, *I had no chance to mail the letter yesterday, as we took a walk up to the tram stop — the distance of some 15 km — and returned only by 9 in the evening, while the last mail is collected at 8. I'm suntanned and no longer look disgustingly greenish, as I used to in winter. Why does a tanned person appear to be dressed? That is true, while a white-skinned person seems undressed even wearing a bathing suit. Oh, all this is nonsense, and for me it is urgent to go to the Division of Scientists. They phoned me yesterday and said that I received an accommodation at the sanatorium in Teberda. I'm going to reject it, because it means I must go there on the 28th and spend there a whole month. I'm almost sure we'll get places in the Scientists' Holiday Home. It is much better, as we'll be staying there for only 2 weeks and then we'll march to the sea and thus get a 75% discount on the railway tickets. Oh, Rudi, I have already described all this once.*

My dear, yesterday I had a letter from you, when I returned home;[500] *I even said "Ah!" being pleasurably surprised. Then I was very sorry for you: such a lot of "professional" troubles all at once. Errors — all these are trifles; every living man makes errors, but Pauli is really unpleasant (unpleasant for me), I haven't got used to him scolding you, and I'm very angry. Well, to hell with him!*

I send my regards to Frenkel; tell him that here the gossipers made him marry a nun and don't expect him to return alone, but accompanied by a Creole beauty. By the way, Sarra Isak[ovna] feels rather unwell now, and Yakov Ilyich is a real swine: he abandoned her and the two children for 10 months, and goes "for entertainment" to California and around Europe. Of course, you shouldn't tell him this (including Sarra's indisposition, too).

I'm worried with your visa, because neither Dau, nor Abbot tell me anything for certain. Seems to be all right, but it would be stupid if, as a result of some trifles, we meet several days later. Rudi, my dear, now I must be going. Excuse me for this utterly indecipherable letter. You'll get it on the 18th; that means in Leipzig you will remain without letters

[500]Letter [130].

for one day only, and you'll hardly notice it being under the impressions of the arrival.

I miss you terribly and can hardly imagine, when you are to come. I'm tired of waiting for letters, writing letters; I'm not bored only with reading them. Please, write; and though you are diligent and write me often enough, sometimes you send letters by ordinary mail — not by Luft *— and I begin to worry that you have drowned, together with [?] and the boat. Can one drown in Leipzig?*

I've always known of your outstanding aptitude in doing household chores; I'll tell you frankly, and you should bear it in mind, that I'll soon get spoilt, if you indulge me when I come to Zurich. You'll have to tolerate for some time until I cope with the duties of a housewife. Don't pity me and order me about, otherwise I won't be able to learn: I'm a terrible lazybones. That is the kind of wife you have — outspoken. I'm moved to tears by my "disadvantageous" frankness!

The shoes are a failure. Mother bought them in my absence. Here you are! I kiss you affectionately, I long to see you, I'm going crazy ...

[132] Rudolf Peierls to Genia Peierls

Stuttgart, 16.7.1931
On the train

Genia,
darling, how wonderful — I've left Zurich at last! Still not on my way to see you, it's true, but closer all the same. It's an interesting day today — four months since we were married! When I left Zurich, I looked around me at every street, every house, every person, and thought: when we see each other again it will be with Genia! It's true, isn't it?!

Everybody was terribly sweet, and it was particularly funny at the train station because all my friends were there, all with presents for the journey: books, fruit, etc. and I had to eat as much as possible so that I could fit the rest in my hands. Recently I've been thinking what a nice group they all are, and I'll be very interested to know whether you like them all.

Over the last few days I've been manically busy, of course, especially looking at rooms etc. Looking around accommodation is an exhausting business: especially in Zurich, where you have to keep going up and down hills. But it was very interesting to see how so many people live. I disliked immensely most of the rooms I looked at, but I've found two or three possible places. They were far more expensive than I had expected, but not impossibly so. The majority of rooms I saw are occupied until past October, but I just wanted to get an impression of the general situation, in order to make it easier to find something nearer the time.

Darling, how funny, I also found some old letters when I was packing up my stuff, and I found it hilarious too, when I looked at the letters in English. Every other word is "you"! It's impossible to imagine that we really did communicate like that once.

Fania's sister was operated on yesterday. The professor was terribly worried, because it's a very dangerous operation. He said to her himself that it's va banque *and if she was scared, she could still say no at the last minute. During the operation he couldn't examine the area where he was operating: he had to find the painful places beforehand, and then try to disturb precisely only those places. It's not clear yet whether the operation was successful or not, but he seemed to think it was. She is now lying in complete darkness; she can't move and can't rotate her eyeballs at all for a week. Naturally, Fania is completely worn out.*

When I said goodbye to Scherrer today, he said that if I return without you in October, he will assume that you don't exist, and that I only invented you to scare Pauli!

My old landlady was there too, she was very sweet, and said that her little niece in particular will be very pleased when I return.

I received a terribly funny letter from Eva Gaebeler. She wrote before that she always takes her holidays in Switzerland, and will be passing through Zurich. She wanted to go to the mountains with Schneider, the architect who we met in Arosa — he's a pretty decent chap, but they didn't come after all. She sounded rather upset in her letter: apparently it suddenly emerged that Schneider is married, and his wife had written to her parents, who are very European, and there was an enormous scandal. She's obviously now in love with Schneider. Terribly funny woman, I only just realised how serious her history with Bübchen *is. The problem with her is simply that she finds it impossible to get angry.*

Incidentally, I keep meaning to ask you: are you happy with the fact that I am corresponding with Eva and will probably meet up with her? You may be completely at ease that you have no reason to be jealous, as you know. But maybe you're not happy about it because of the episode in Arosa. If that's the case, just let me know — the correspondence isn't important to me, and I don't want the smallest complication to come between us because of it. If you would like to, you can probably see her in Berlin, and I'm sure you'll like her, but I understand perfectly if you don't want to.

Ach Genia, everything is so good here, why aren't you here too?! In Zurich it was pouring with rain, but here (I'm in South Germany already) it's sunny, and there are huge green forests, and mountains. The train is going slowly, oh so slowly.

Here (and in Zurich) there is an unusual degree of agitation because of the economic situation in Germany.[501] *You have probably read about it in the newspapers, and nobody has the faintest idea what will happen. Hmmm, I prefer thinking about physics. And, indeed, I need to, because I received another letter from Pauli in which he is cursing again.*[502] *But this time, I think he has a point, namely in relation to Dau's work, and I have an equation which hasn't been sufficiently proven and now I should think up some better proof, especially as I want to give a paper about it in Stuttgart tomorrow.*

Do you remember, I wrote to you recently about the special-focus photos we did for Eva Kobylinski's parents? They received them fine, and today we received a letter saying that they are delighted with them.[503]

[501]The Great Depression reached even more dramatic heights, when the financial markets collapsed. The crisis led to a reconsideration of Germany's financial obligations, and in particular her reparations obligations.

[502]Letter [127].

[503]Letter [128].

Stuttgart, 16.7.[1931], morning

This city is reasonably clean, with nice houses. Last night I went drinking with Ruhemann and Ehrenberg[504] *again (he is crippled, but only very slightly). They drink wine like water here, very good, light wine! I should go to the Institute now. Goodbye, my darling. A thousand kisses!*

Your Rudi

[133] Genia Peierls to Rudolf Peierls

Leningrad, 19.7.1931

My dear,
I have just received your letter,[505] *and it is so pleasant to answer without delay — just like we are talking. Is that right? I'm dissolutely staying indoors; though it is the 19th today, I'm not going to Pavlovsk, because there are lots of things to do in the city. I don't do them either, as I'm quite unwilling; so you will soon get this reply. I stayed in Pavlovsk for two days and even spent the night here. In one of the three rooms assigned to the chief, a bedstand was brought. I slept marvelously under the analytical balance. The bed stays there only temporarily, as the whole house is in repairs — so there is nothing to do. I slept very soundly: some of the last few days I went to bed at 2 o'clock, and got up earlier than 9. Oh, what a terrible dream I have had: as though in the morning a lot of people entered into the adjacent big room, separated from my room by the glazed door; so I didn't know how to get out of bed. I must have talked in my sleep, but there was nobody around to laugh at me.*

A wonderful watchwoman takes care of me: makes tea, boils milk and eggs, etc. What I appreciate greatly is that she does all this of her own good will. Rudi, there is a cat I fell in love with. A tiny kitten, all black except paws and a tie, green-eyed, he is very tender and lively. He looks straight into the eyes, meowing, asking to be taken on your arms. Or, which is less pleasant, climbs up the leg and arm. I'm all

---

[504] W. Ehrenberg, physicist working with H. Mark on Röntgen spectra.
[505] Letter [132].

scratchy as a result of his acrobatic tricks. Isay Benedictovich left for his mother and sister, who occupy a cottage near Kharkov. He'll stay there for 10 days and then Nina will join him, and they will go to Gagry until September. It is complicated to deal with Mother, for she doesn't want to go anywhere, she says the best holiday for her is when we all leave, and she remains with Nastya only. She says she relaxed a little during last 2 days, when Isay was away. This is nonsense, of course, as Mother has to ascend our staircase, and it is hot here, etc. Indeed, she cannot go to Gagry with Nina and Isay, as over there the weather is too hot, which is forbidden for Mother with her weak heart. It makes no sense either for all of them to go anywhere, because Mother will start at once to force Nina to drink milk, liquidate the bedbugs that will bite Isay, worry they wouldn't drown while swimming or eat too much fruit, etc. It is impossible for Mother to live a quiet life. When her family members are away, she finds some other people to worry about. Last year, for example, in Kislovodsk she nursed a 4-year-old girl, while the mother of the girl was flirting with everybody in Kislovodsk, etc. That is why she has such a bad heart and bad nerves. I'll try and get accommodation for her in Kislovodsk for September (she needs mineral water baths). In the meantime, as I hope, Mother will go to a village near Pskov in the company of some friends. But it is problematic now, that Isay is away. Mother had promised him she would go to the country, and he left calmed. Yesterday she started to say she doesn't know, if she would go. Oh, Rudi, for you it may seem insignificant, but I'm really worried about all this! If only we succeeded in getting the accommodation in Kislovodsk! Right now she shouldn't go there because of the large crowds, the queues in canteens, noise, etc. — she will suffer with all that.

Yesterday I dropped in at the Scientists' Club. I learned that we may possibly obtain that lodge for tourists in Teberda — it is an excellent place; everything will be really great. Rudi, my dear, yesterday Abbot came to see me and made me so downcast, that I was about to wire to you. The matter is the visa: this idiotic red tape office in the Roentgen Institute acts as 3 horses pulling in 3 directions, while all the aids of Ioffe (Rubanovsky, and oth.) are now in Moscow. So Abbot says he doesn't know what to do, he is in deep despair, and he isn't to blame, because whomever he addressed, whatever office he entered, he met Kuka everywhere, who bustles about for two weeks without result. And if even Kuka cannot obtain what he needs, there is really nothing to do. Luckily Dau phoned me and said that he, by chance, came across Rubanovsky (the aid of Ioffe), brought him to the office, where the latter started a frightening row; they promised to write to Moscow at once and send the visa by wire to Berlin as soon as they get it from Moscow. Dau will go there every day and quarrel — to be on the safe side; still I'm not quite sure. So, if you meet Frenkel, please tell him everything and ask him if it is possible for him to do something in the Embassy in Berlin. As you have already been to Leningrad twice, that should not be too complicated.

Because I permanently go to Pavlovsk, I don't go out, though now on tour there is a Japanese conductor and a Japanese singer. On the 22nd the concert is on again, maybe I'll go to listen. The programme is Tchaikovsky and Japanese composers. Interesting? Rudi, is Pauli really scared by your marriage so much, that Scherrer thinks it is a fancied marriage, especially to frighten him? A very witty remark. The affair with Eva is tragicomic and resembles a humoristic novel. I wonder, what did she answer her parents? And the parents, who start a row — that is very funny. Schneider's wife is a scum, and he isn't any better: why didn't he say at once that he was married? He distracted the "innocent"" girl, however, who distracted whom — is a question. Writing complaints to the parents, blaming the girl who takes away her husband — that is too stupid. There cannot be a worse position than Schneider's. Both women must have scolded him fiercely. Altogether it is really a humoristic story, and if Eva possesses but a drop of irony, she will soon be comforted and merry. Are you able to laugh at yourself? I am, and that is very important and sometimes funny. You know quite well I'm not jealous at all; so you can, without any pricks of conscience, correspond and meet whomever you like — that wouldn't be unpleasant for me. (I'll be glad to make Eve's acquaintance — she seems to be an agreeable girl, but how couldn't she detect at once that the man is married? Even if he is without a wedding ring?) But I wouldn't like to be dumb and deaf as well. You see? When I feel some day that I have lost "the first place", even for a moment, I'll leave at once, without saying goodbye, maybe. Anyway, I'm not going to write to the girl's parents, and you'll never fall in into some sort of the "Schneider-situation" —

of this you may be quite sure! If I were Schneider, I'd abandon them both and run to the end of the world. And what about you? By the way, you aren't jealous either, are you? Or just a little bit? If you are, you must tell me at once. It is terribly pity when a man is upset for nothing. Rudi, haven't they sent you the address of that American lady in Zurich? You know, in the last letter I was surprised at that "kissing the hand". You wrote that it isn't my style, unless I play it for fun. That is correct, but it doesn't happen nowadays. Some 5 years ago I played it rather often. When did you notice it? I don't remember that I played it in your presence. You are very attentive, mein Gemahl, *bravo!*

Ah, 6 pages are already written, and I didn't do anything! Shame on me! Shame on me! Shame on me!

Rudi, I forgot to write you about my director — you may remember, he is the 2nd class, even better — he left for $1\frac{1}{2}$ *month on board the ice-breaker with Nobile and other explorers. They got to 82° of latitude. Great, isn't it? I don't envy him: to spend 6 weeks on board the steamer with winds and storms around. How awful! Still it is a lot of fun, if you don't suffer from seasickness. One of our professors is leaving for the North Pole flying the Zeppelin. He is married and has a small kid. I'll never allow you to go on such trips, bear that in mind, though I don't like to be a hindrance. This would be a limit, I'd go crazy. I don't believe they'll reach the goal; they might fall down. Even if they do reach, then that will be nothing else, but a sport record, frankly speaking. Am I right? Oh, Lord, I cannot stop writing! I kiss you affectionately, my dear. How painful, you are in Leipzig and not here!*
Good-bye.

G

Oh, I forgot the most important thing. Rudi, that is great! Enough to drive one crazy! Now I can understand why Alfred and you married Russian women. You aren't the first Peierls in Russia. Yesterday I discovered by sheer chance the Schwiegers *ancestor, and as early as in the 17th century! Fine? In 1604 when the False Demetrius was on the Russian throne, the Augsburg merchant* German Payerls *came here together with his companion* Natan *(*Natan *seems to be a Jewish surname, and* Payerls *was obviously a Jew too). Spelling* ay *instead of* ei *has no impor-*

tance. He left the memoirs of Russia of that time, entitled Rayse *etc., instead of* Reise. *Haven't you any relatives in Augsburg? This merchant possibly had several "adventures" in this savage country, and you inherited sympathy towards Russian women. It cannot be proved, of course.* Payerls *is almost certainly your ancestor, this is a rare surname, isn't it?*

After all these genealogical digressions I kiss you again — the last kiss in this letter. G

[134] Rudolf Peierls to Genia Peierls

Leipzig, 19.7.1931

Genia,
my dear, here's some good news: the letter which you sent on the 16th arrived here on the morning of the 18th. I had already begun to worry so it was a very pleasant surprise when it arrived so quickly. With the letter from the 4th it went thus: I was very busy, and Fania went to the post office as usual, so she took my letter with her, but I had put a gold card on it: 'Luftpost'. Then she returned and said that the letter was paid wrongly, so I thought that they won't send it by airmail, but I still wanted the letter to be sent. Then the next day a post office official came by, apologised that his colleague had made a mistake and charged too little, then took some more money and said that in the mean time the letter had, of course, been sent by airmail. It's a colossal disgrace, if only I were in Zurich I would cause a big scandal, but writing a letter about it hardly seems worth it.

The economic situation in Germany at the moment is even worse than you have been reading in the newspapers, and they are now passing new laws which are absolutely horrific. Firstly: anyone who wishes to travel abroad has to pay 100 marks for a passport. I don't know whether this will apply to me even though I have been invited by the scientific institute. In any case, I will ask Dau to send me an official invitation if possible. Then they banned the sending of money abroad, and my father can't send anything for us to Zurich if we run short: of course I spoke to my father a while ago and he did everything he could to move things on.

My dearest, it seems you are somewhat upset that I haven't been to visit you yet? Surely you know that it is impossible to work if there is no-one around to discuss physics with, and without work I can't be your husband, right? I have no "professional" reasons for passing through Berlin, but it would be a very bad idea not to visit my parents. If I am confident that I will be able to come through Berlin on the way back, then it will be less of an issue. But perhaps if you receive your visa late, then it would be better if I can be in Russia for as long as possible. You understand, don't you? And you're not offended?

In Stuttgart it was very nice but quite busy. I gave a report in the morning and after lunch. Ewald was very nice, and there were lots of other interesting people; Barbara Zarniko wasn't there, she will be leaving Berlin for Stuttgart soon, so I won't see her this time. I arrived here on the 17th, I traveled all night and was quite tired when I arrived. The institute here is very nice: plenty of theorists: Heisenberg, Bloch, Santer, Weisskopf, Houston, Waller, Beck, Placzek, unfortunately Guth as well;[506] *and anyway I can't discuss all of them in this letter. But the town itself is a disgusting horror! Not merely that they all speak Saxon here, all the people look like, as someone recently said: the French people who invented the word "bocke" probably lived in Leipzig. It's true that everybody who one meets on the street has something unpleasant about them, everything you look at has something unpleasant about it, the houses are old and dirty and the flat smells of dust, that particular smell which I have only encountered in Leipzig. Last time I was in Leipzig I stayed in a work flat, which was nicer, but this time it wasn't worth looking for a room and I'm staying in a pension with most of the other theorists — it's a fairly unpleasant place.*

[506] Viktor Weisskopf (1908–2002), whom Peierls met for the first time at Leipzig in the summer of 1931, remained a life-long friend. He would succeed Peierls and Hendrik Casimir as Pauli's assistant in 1933. Similarly, Georg Placzek (1905–1955) would remain closely associated with Peierls and collaborate closely in the subsequent decades.

My darling, you're absolutely right that a sun-tanned person seems "clothed"; I am already almost completely "clothed", and only my stomach is still completely white.

Genia, dearest, I really can't wait to see you, but we have about another three weeks to wait! Kissing you. Goodnight!

20.7.1931

The new pictures look far better. In my opinion "Leibphotograph" is not enough for Erika and I recommend "Oberleibphotograph"!

Nina[507] *said that you haven't replied to her letter: she's not offended (she rarely writes herself), but it would be better to answer her letter, I think.*

My parents are at Annie's today, then meeting the Briegers: my father's sister and her husband. They are practically the only relatives of mine who don't like me, and with whom I am in contact only because they are relatives. But it's not a big problem because they are rarely here. The worst thing about them is that she has never worked, and he retired very early (at 40, I believe). Of course they are at home all day, bored and cursing at each other, and they have been living like this for more than 30 years, quarreling all the time. Their daughter and her husband are similarly argumentative, even in front of the children (who are delightful).

If Dau is going to be staying with us, he and I should try not to discuss physics apart from when you are asleep etc. But, to be honest, I am far less bothered by the physics issue than that he will be with us all the time. In Leningrad I expect everyone will be madly busy, and when we are in Teberd it will probably be the first time when we will be able to have some peaceful time together. It will be very tiresome if we're never allowed any time to ourselves (I'd be very surprised if we are allocated a room together). Dau is a very dear chap, but being married to him wouldn't be much fun.

[507] Nina Peierls, Peierls' sister-in-law.

I must go to the post office to make sure this letter is posted today. Although it will arrive after the telegram!
Many kisses, darling,

Yours, Rudi

[135] Rudolf Peierls to Genia Peierls

[Leipzig], 21.7.1931

Darling,
It will already be your birthday by the time you receive this letter, and so I must wish you many happy returns, although it seems very funny to do so when "we're not allowed to talk about birthdays"!

I wish I could be there on the morning of the 25th to kiss you, hold you close for hours, and say, "A new year of your life is beginning, may it be wonderful". May it be a wonderful year, and may you have many new and interesting experiences, and be thoroughly happy. You see, Genia, I always try to wish for nice things for you, but these things are really just what make me happy. Because, if you are seeing so many new things, it means that you will be with me. And if you are feeling happy, it makes me so happy too. So perhaps it would be better to say: "A new year of <u>*our*</u> *life is beginning, I hope that it will bring us happiness!"*

It has worked out nicely for you that the 25th falls on a day off so you can celebrate in peace and without everyday problems. It will be sad, of course, if I. B. and Nina aren't there, but your mother probably will be and that will help.

It's a great shame, but I'm afraid I have to tell you that your "genealogical theory", though very witty, is completely wrong. The truth is that the Payerles are a Bavarian family, and it is quite a common name. But "Peierls" comes from "Peirels" (in Hebrew perel *means "pearl", and* peirel *is the plural); my grandfather simply changed his name to Peierls, because he preferred it. Because of that, we know everyone in the world called Peierls: my father and his two sons, his brother and nephew in America, and his cousin in Breslau, who also took this surname. It's a shame, because your theory was a good one.*

When I met Sack for the first time he didn't say anything in particular. Then I was told that when he received the card, he didn't believe it and thought it was a joke; it was only when he heard from Bloch (via Pauli) that he believed it, and was quite astonished. He sends his regards to you. It seems that he is still a little offended, because you haven't replied to his letter with the pictures.

There are a quite unbelievable number of theorists here and, though it's pleasant, it's almost impossible to speak with Heisenberg as he is busy the whole time. It's now half past one in the morning; we've been playing a huge ping-pong tournament in the institute with everyone here, apart from Guth of course, to whom nobody really talks much. It's quite awful that Guth has turned up here. One has to feel quite sorry for him, but that's not right, because if he wasn't here, we would all be worse off. But I really love ping-pong. It's a good thing that we don't have a table in Zurich, otherwise I wouldn't do anything else. Today I presented a paper at the seminar about the work which I finished in Leningrad, and it seems that it wasn't quite as poor and boring as I had thought.[508] *Tomorrow I will probably speak again about the absorption of light in solid bodies. I have a few more things which I should speak to Heisenberg and Bloch about, but when that's done, I can quietly leave for Berlin, because if I can leave here a little earlier I can come to see you sooner. But I haven't told anybody yet.*

The mood here is far better than in Zurich. The subjects are less interesting, but there are plenty of great people, who are all working on their particular theme which they are interested in. It's possible here to work on something even if it's not important, as long as it's still interesting and not too trivial. It's also very useful to remember how many of my colleagues here are working on things about which I understand nothing. I had begun to forget already.

But it's now two o'clock, and I have to sleep. Goodnight my darling, you were asleep long ago, but I'm kissing you while you sleep.
Until I see you — very soon!

Your Rudi

[508] This paper was based on Peierls' earlier work on the Hall-Effekt. See Rudolf Peierls, interview with Charles Weiner, 11.–13.8.1969, Niels Bohr Library, AIP. It was published in 1932; see letter [130], note 491.

23.7.1931 morning

I held on to this letter so that it will arrive on the 25th, I hope I haven't made an error of calculation. Yesterday I spoke to Bloch for a long while, he's a very nice chap. I spoke in the seminar again etc. Oh yes, I forgot to tell you: I received a letter from a doctor in southern Germany, saying that he had heard my paper on the radio[509] *and found it very interesting, but unfortunately didn't have time to take notes. He would like a transcript, because it would be particularly useful for some courses he gives for charitable sisters! Wonderful, isn't it? I send my love and kiss your lips once more.*
Love to all!

Rudi

[136] Genia Peierls to Rudolf Peierls

In the movie-theatre, 23.7.1931

No, you aren't nice. Certainly, I understand quite well: you are in Leipzig, you are in Berlin, and I'm not a bit hurt with your being late. But such a short letter is a swinishness;[510] *not a word about Zack — shame on you, shame on you!*

I have the news about Teberda: you and I got accommodations and discount on the railway ticket for me. Magnificent! But I'm terribly worried about the visa. Two days ago the letter was sent to Moscow, and in some 4 days they are to wire; but it is reasonable to have less hope, as long as the business depends on those red tape offices. Rudi, couldn't you try and speak with the officials in the Berlin Embassy and arrange, to be on the safe side, another visa from Germany? Tell them about our situation; maybe the visa will sooner be got from Berlin? Anyway, it is better to have two visas in due time, instead of none.

I'm an awful pessimist in such matters and I always wish to have both "the belt and the suspenders" to support. I spoke to Sarra Isaak[ovna]

[509] See letter [130], note 494.
[510] Letter [134].

— she suggested to tell Kuka, that Yakov Ilyich asked him to accelerate this process; but I think we'd better not address Kuka, despite of his performance in things like these he is better than Dau. In Ukrainian they say, "don't touch the trouble" — i.e. we may raise an additional trouble. Kuka is unpredictable, out of malice he may even slacken the pace. Am I right?

In two days Dau will come to the Institute and will send you the invitation sponsored by the Institute. Now the business matters seem to be finished; as to the other ones, I have neither time nor wish to write about them. A short reply to my 8-page letter: aren't you ashamed? You have been to two cities and written only 3 pages! When you receive this letter, your wife becomes a 23-year-old matron. I feel this burden of years, and what about you? Well, so be it, I continue: yesterday I was nearly killed by a falling tree. There was a terrible thunderstorm and a hurricane. In Pavlovsk 100-year-old trees fell down like sticks. Horror! These lightnings! And the thunder produced the incessant tremble of windowpanes! I have never seen such a furious thunderstorm in my life. Even in Gagry it wasn't so intensive. One house was destroyed completely: a pine-tree fell on it. Luckily there was nobody in, and the instruments remained intact by a miracle; the fall of the roof spared them. A tree of enormous size fell at a little distance from me!

You would have cried if I was killed and left you to be a true widower, not a grass-widower!

Well, the film by Charlie Chaplin now begins.

I kiss you affectionately, though you are a pig.

July 24

I'm at the post-office now, and I'm in a hurry: I have to call on the Gosizdat[511] *to correct the mistaken number of my bank account, then I'm going to Pavlovsk. Yesterday I wrote you a fierce letter, today I don't feel so much hurt. Nina leaves on the 27th. I'm afraid, Mother still will be here, when you come. The thing is, a lady, the friend of my aunt (I have written you about her scarlet fever) is now in a bad state. The aunt is terribly worried, so Mother doesn't want to leave her now.*

---

[511] State publishing house. Genia had been working on a translation which was published by Gosizdat.

Well, my dear, goodbye. I hope, till 10th–12th. Yes?

To be serious, I'm not at all hurt by Leipzig and Berlin, but I feel anguish because they exist, these Leipzig and Berlin, and you are not yet here.

Charlie is a humbug, in general, though there were several funny episodes. The film is very old, judging by the pre-war fashion, seen in it.

I kiss you and pull your ears. Don't be a muff and write a long letter. I feel bored!

Don't buy any trousers. Where is you tiger-striped suit you wore in Kislovodsk? It could fit me. This morning I had a bright idea. We shall leave the suitcase with galoshes, dresses, etc. in Moscow, because we'll need it only in Moscow. I'm afraid we'll have to spend a long time there, on our way back home. The Central Executive Committee, etc. We'll take only Rücksack's *and blankets with us.*

G

[137] Rudolf Peierls to Genia Peierls

Leipzig, 26.7.1931

Darling, it's a long time since I've received a letter from you! I wasn't in the Institute yesterday and I was frustrated all day, because I thought there was a letter there. Today I went in, although it's Sunday and it's strictly forbidden to go to the Institute, and there was no letter. But maybe it's my fault and you think I'm in Berlin already. No, I leave for Berlin on the 28th (Reichskanzlerplatz 10, Berlin-Charlottenburg). True, I received your last letter only five days ago; please forgive me if I'm impatient to hear from you again already!

Yesterday we all went to Düben, a little place between Berlin and Leipzig, to meet the Berlin physicists: Becker, Schottky[512] *and others,*

[512]Hans Becker, collaborator of Walter Bothe's; Walter Schottky (1886–1976), had studied physics in Berlin and received his Ph.D. in 1912 with a thesis on special theory of relativity. He became professor of physics in Rostock before moving back to industrial research at Siemens where he focused on research into semiconductors.

and discuss ferromagnetism. We all went by bicycle and covered 60 kilometres in all. We went to a school there, where the teacher was very proud to have such famous people visit, and gave us a classroom for our debating. Then we all bathed and continued to argue about ferromagnetism and semi-conductors in our bathing costumes. The lake there had very dirty water, in fact the water was actually blacker than the Saxon coffee which we then went off to drink.

I'm staying so long here because there are still so many people I want to speak to. Not just with Devus etc. about organising publicity, but principally with the publishers and so on. There are many possibilities here for finding paid work, far more than in Zurich, so I still need to think hard about it. I'll also have much of the same to do when I arrive in Berlin, with Becker et al, and all the physicists. It's an unpleasant business, but a necessary one.

On the evening of the 23rd we went out in Sack's car. He has a splendid little motor, and drives it "like an adult". He sends his love to you, and Nina in particular. It's time to go to the baths in order to lose all the dirt from the lake. Until tomorrow, dearest! Many kisses. (Not too many more times on paper, I hope!)

Evening

It was a very good trip to the baths, I saw Heisenberg there and he told me that Brillouin[513] *has written a new edition of his book,*[514] *in which he is very rude about me.*[515] *Tomorrow I should read it and find out whether it's true or not. In the evening I went to a birthday party for the wife of the local doctor, and yet again caused a sensation. They still knew absolutely nothing and it caused a great effect when I*

---

[513]Léon Brillouin, (1889-1969), professor of physics at the Sorbonne (1928) and subsequently at the College de France (1932-1949).

[514]L. Brillouin, *The Quantenstatistik*, Berlin: Springer, 1931.

[515]In his paper 'Zur Theorie der elektrischen und thermischen Leitfähigkeit von Metallen' *Ann. Phys.* **4**, 121–148 (1930), Peierls criticises one aspect of Bloch's theory of electrical conduction and presents a modified version leading to different results. He discusses the one-dimensional case of a weak periodical potential which was later extended by Brioullin to three dimensions leading to the 'Brioullin zones'.

"accidentally" mentioned that I'm married. They all send their best wishes and in particular they asked me to tell you that they really liked your photo.

27.7. morning

My love, yesterday my eyes closed, I was simply too tired to write, but this morning your letter arrived. Thank you! I was so pleased to read it, although you sound so unhappy about the short letter.[516] *Do you know why I always write such short letters at the moment? It always seems as if "I'm already sitting on the train" and Leipzig and Berlin are just big stations en route to you, and therefore it's scarcely worth writing letters. This isn't the case of course, we still have about two weeks left to wait, but I'm not thinking about that for now.*

Half of me is already with you in Leningrad and so everything I do here seems unimportant.

And Genia, what wonderful things we will do there! We can stroll among the falling trees. Maybe you already do regularly? Just think! If a tree fell on your head I'd have a headless wife, a quite horrendous thought![517]

So, Nina is leaving today and you will be left alone anxiously at Maria Abramovna's, because if Nina goes out without her coat, nobody will notice. But will Maria Abramovna really stay indoors all summer? She will be running around manically all day as usual, never stopping to relax.

On the business side of things: I was thinking that we could go to Moscow with our suitcases and travel on from there with only rucksacks. But wouldn't it be better to stop and relax for a little while in Moscow? It would be a great shame to lose even one day in Teberd, but if it means hurrying the passport business along, that's probably more important, right?

---

[516] Letter [136].

[517] See letter [136] where Genia relates how she almost became a victim of a falling tree.

You're absolutely right that it will be better to do everything without Kuka,[518] *although I don't think that he will do us any harm. Incidentally, the day after tomorrow I will go to the consulate in Berlin and I'll also speak to Frenkel when he is in Berlin (29th or 30th).*

I have learned a lot here. There is a wonderfully pleasant atmosphere at Heisenberg's. Similar to how things used to be at Sommerfeld's: he has a "school" where people work on all sorts of interesting topics, and there is a tremendous sense of optimism around the place. Heisenberg himself is terribly energetic and does all kinds of sports: ping-pong, tennis, skiing, mountaineering, cycling. All these things he does, not like a professor, but like a young boy; and, of course, he does them all very well. For example, when we play ping-pong, nobody can beat him; when we go cycling he arrives back an hour before everyone else, and so on. He also plays the piano very well and it's common for people to meet him in society and hear how he plays the piano, then hear from me that he is my professor, and ask in astonishment: "what, he understands physics as well?"

But it is terrible my dear, I've noticed that have begun to forget all my Russian. Nobody here speaks Russian — it's time to leave.

...

Here comes Placzek, a very decent chap. He has been everywhere: born in Czechoslovakia, lived a long time in Vienna, then in Holland (I stayed with him when I went there with Pauli), and he has spent several periods in France, Spain, etc. He understands an awful lot, and is good with both practical experiments and theories, but is also strangely lazy. He is very capable of laughing at himself, but he does it so often that I suspect he has very little faith in himself. Perhaps it is simply because his father was very rich, that he is always so self-deprecating. At any rate, he's a good fellow to discuss all sorts of things with.

Teller[519] *is coming any moment with a copy of Brillouin's book and I want to get to the post office before he arrives. So goodbye, my 23-*

[518] Refers to attempts to get Peierls' paperwork for his Russian travels organised. See letters [134, 136].

[519] Edward Teller (1908–2003), obtained his Ph.D. from Leipzig in 1930; in 1935 he took up a post in Washington; played a prominent role at Los Alamos and later in the development of the H-bomb.

year-old wife, I'll be with you on the 10th or 11th, as long as my visa arrives in time.
I just can't wait to see you!
Thousands of kisses,

Your Rudi

P.S. If you are counting the pages, don't forget that my pages are much bigger than yours! True, those old Chaplin films are boring now; he has changed a lot.

[138] Genia Peierls to Rudi Peierls

[Leningrad], 26.7.1931

Dear Rudi,
darling, darling, darling; last time I sent you a quarrelsome letter, but I miss you greatly — so greatly, that it is impossible to wait any longer. Lieber, *thank you for the wire, thank you for the gifts and the letter. I got the wire while still in bed, and I thought how wonderful it would have been, if you were by my side. But the wire, dated by 23 h 40 m, represented 0.25 of the person. The letter was delivered only today, but it is better that way, because after so many pleasant things, that occurred the day before, I was feeling rather lonely — and here you are, your letter. On board the train I was thinking about the letter: would I receive it or not? And I did have it, a very interesting one. The doctor is a nice man; had you thought that you would ever give lectures for the hospital nurses? I regret about* Payerls*: such a bright idea it was. Sack has no observation — that exceeds all the limits, how stupid of him! Didn't he believe the postcard?* Sport und Jagd *is great, especially some dogs. And the genial ending with pigeons on St. Marco. By the way, there is a photograph of Ioffe with his wife, when being in Venice; they feed the pigeons, and a dove sits on the Academician's bold head. How touching! I enjoyed Cranach, I like him immensely, but don't know many of his paintings: I had seen only 3 ones — terribly great, especially women, aren't they? Not women, but girls, with thin arms, slender shoulders and slanting, amazed eyes. Wonderful women, naive and wise; his landscapes are surprisingly poetic. By the way, the*

men aren't good at all, they look a bit stupid and weak-willed — except portraits, Luther is magnificent. Each painter is recognizable by the particular "type" of people, landscapes, sometimes even animals in their pictures (remember Vrubel[520] *or Renoire). I think it was difficult to find models in the olden days, so the painters portrayed mostly their wives and children as Madonnas, Dianas, and the Bible heroines. Am I right? Cranach women are all alike: luckily, he had a charming wife. Then came early Renaiscance: for the first time after the Dark ages undressed women could appear on canvas. The Germans showed Eve at the moment when she was driven from Paradise as a fat-bellied woman with bendy legs — the consequence of wearing heavy garments and staying in airless stuffy rooms. In Italy morals and manner s were more liberal, the women were more approachable; that is why the Italians chose beautiful models, "suitable" for the subject matter. I am reading the book by Domel, it is very interesting, though there are words that I can't understand. But it is easy reading, and I have read 45 pages on board the train.*

July 27

Rudi, I had to interrupt writing because we are having celebrations and Mother is making jam. (You may imagine how enthusiastic I am.) Yesterday all folks were busy cutting off the "tails" of the blackcurrants. I had to participate too. Today they are making jam and my duty is to lick up spoons and decide whether the jam is ready. How delicious! And the froth is so delicious! Have you ever tasted it? This is the foam that appears on the surface of the jam while it is boiling. The froth is to be taken off and given to those, who like it, me including. In the evening we went to see off the Gurviches, and Nina left too. Everybody is leaving for summer!

On my birthday many pleasant things showed up. I received a lot of congratulations by mail and by wire. Also a lot of flowers and sweets. The flowers are everywhere in the flat. The beautiful roses, it is a marvel. In the evening we (i.e. Abbot, Dau, etc.) went to Pavlovsk to

[520]Mikhail Vrubel (1856–1910), Russian painter.

attend the gypsy music concert. From the train window we saw Zeppelin 27, which flew further to the Arctic yesterday. We saw it from afar, and what crowds were in the city — oh, my Lord! Crowds were in all the streets and everybody was looking up into the sky. For fun I started "panic" and cried from the tram platform that the Zeppelin was called off for today or that it had flown in another place. It was awfully funny to watch these dumbfounded faces and speed further on. The concert was interesting and so peculiar; I wasn't acquainted with Gypsy style: they sang choruses and solos, and the dances were really great. At first they sang their original Gypsy camp songs. You know, the music is quite different from European style. When listening, you begin to see the steppe with evening stars and hear the creaking of caravans and rustling of the high grass. It was such a pity you were away. Then they sang "European" pieces, but in these alien gypsy voices, a bit low and velvet, soft contraltos — great! In the intervals we walked in the park, now there are wonderful warm nights in Pavlovsk, so much like those described in the romantic verse of the early 19th century. Mist, silver lakes, lime-trees in bloom, you'd wish to see fairies and mermaids.

Dau has just phoned: the invitation for you was sent out by the Luftpost. *But Dau is absent-minded, he knew you were in Leipzig, but gave them your address in Zurich. All the same you'll get it about the 30th. Oh, how I wish you already had the visa and I could be sure you come before the 10th or the 12th! By the way, Dau may be here in August: his mother is going to come here and he puts off his vacation. Abbot doesn't know where "to lay his head" as he couldn't get accommodation at the Scientists' Rest Home near Moscow. He may go to the Caucasian coast: to Sochi, or Gagry, or Afon. And it is very nice in Teberda: over there near the Sukhumi Road there is a canyon which is much better than the Daryal Canyon; there is a glacier and green, green lakes. Then we can go to Sukhumi to see the monkeys (I'm crazy about them) and bathe at sunset, when the sun sinks to the sea like a big orange and the golden road crosses the sea. Splendid, isn't it? We shall stay in Sukhumi for a day or two, and then we'll go to Gagry for 5–6 days. We'll stay in a hotel, there are wonderful hotels — no other place like Gagry. There are palms, fountains, and high trees in blossom; the climate isn't so damp as in Batumi. The Sukhumi Road is 135 km long, some 20 km can be passed riding horses and 40 km by bus (the less interesting places). That leaves 75 km which we shall walk in 2–3 days. There are no people around and the paths are so narrow that pack-horses can hardly walk along. We shall hire a horse in order not to carry our things, including provisions. A company of 8–10 people will gather (the usual number for these marches), so the horse will not be too expensive. The fellow travelers can be found at the starting place. We shall spend some 10–12 days in Teberda, ride horseback etc.; all the surroundings are marvelous there in 30 km around. We shall stay at the tourists' lodge for scientists. It may well be they haven't single or double bedrooms; are you afraid of the dormitory? You sleep so soundly, that even if all your four room-mates snore, you will not wake up. Is that right? I have developed in detail our plans for this summer, haven't I? The programme seems ideal: sea and mountains, wilderness and culture, ice and palm-trees — the dream!*

Don't forget to bring a good supply of film: there will be lots of scenes to take pictures of. Rudi, there remain only 2 weeks to wait, if only they don't detain your visa, you don't break your leg, etc. The most difficult thing now is to wait these 10 days. The next four will be easier. I'll know that you are on your way or, at least, have packed your suitcase. You haven't forgotten me? When you get this letter, it will be 3 months since we are parted. But now "the style" of longing for you is quite different than it was in winter. I cannot tell which is worse: now or then — but in winter, December 1st was the doomsday. Later on was full of uncertainty, and I could only think of meeting you at the station. Now, on the contrary, I think more of the future. It will be different, of course, from what I imagine, and still I'm feeling a sense of relief. The wish to see you this time is stronger, as if I left a hand or foot somewhere in the tram — it is so uncomfortable without you.

I'm sending you the photos taken yesterday. You can see that I procured some trousers. Eric gave me his old ones; he wore them when he was still quite thin. I'm going to dye them black, though they are linen, and for the walks on ice something warmer is needed. Is that right? I'm absolutely incompetent in mountain hiking, so you will have to inspect my things and tell me what to take, and what to leave. The photos are "the wonder of technique" - they are printed from the damp negatives at 2 a.m., in secret from Mother. She thought I was asleep while I was inspiring Eric for this deed. The negatives are good enough, the prints might be made better. So you may tear up the pictures, later we can make better ones. I just wanted to send them to you as soon as possible — that is why they are so bad (damp negatives, what on earth!). There is also a nice group taken on my birthday, I'll send it in my next letter.

If I begin a new page, I'll never finish it! So I kiss you affectionately, kiss you without glasses — it is easier.

G

All send their regards.

[139] Rudolf Peierls to Genia Peierls

Berlin, 30.7.1931

Genia, dearest,
I received another letter from you today.[521] *It was wonderful to read it in bed this morning. It's also quite amazing that it has arrived already, even though it came via Leipzig.*

Now, at home, it seems that Leningrad is so far away again and that these ten days that I still have to wait to see you will pass very, very slowly. Therefore you probably won't have to worry about receiving short letters. My last day in Leipzig was very busy again: I had a read through Brillouin's book. He is obviously furious with me — on every other page he drops a hint of some kind: "Peierls has already seen these results, but clearly doesn't understand what they signify", "Peierls has written this formula, but he doesn't know what problems it may encounter through thoughtless application", etc. Besides this, in one place he finds an error in my work. He really "doesn't understand what it signifies" himself and makes a mistake of his own, but I now ought to write a new article to say that the resistance of crystals at low temperatures is proportional

[521] Letter [138].

not to T4 but to T5. (I'll need to change it in the book too, but we'll have time to do this at the proof-readers'.)

Fortunately I have now noticed how it is possible to correct the mistake: I really just have to write an article. As well as that, I received a letter from Pauli, which concurs that I will mich habilitieren *in winter but I'll have to write about my work and send it to him.*[522] *I very probably won't have time here so I'll take the typewriter with me and write to him in Leningrad when we return there, if we don't have to leave straight away. Pauli also wrote that "with the assistance of Bacchus" he recently fell down some stairs and broke his shoulder, so he is confined to his bed and is bored stiff.*

But I wanted to tell you about Leipzig. I had a lot of discussions with Teller; he is a Hungarian, a Jew, and has some unpleasant characteristics, but he is clever and very knowledgeable.[523] *On the evening of the 27th I went to the Mathematics Institute, where I heard a very dull paper being given and saw van der Waerden,*[524] *who was recently appointed professor there. He is a very decent chap. I got on very well with his assistant — Deuring ("Bübchen")*[525] *— and we chatted for a long time. I found it oddly interesting to "analyse the situation" and tried to find out whether he is in love with Eva or not. He didn't want to talk about it at all, however, and I didn't manage to discover anything.*

How funny, I completely forgot to mention that Houston (the American physicist) was staying in the same pension as me with his wife (a

---

[522] Letter [126].

[523] Teller's qualities as a theoretical physicist were admired by his colleagues, but he made many enemies throughout his career, most notably when he chose to testify against Robert Oppenheimer during hearings in 1954 during which Oppenheimer's loyalty was scrutinized.

[524] Bartel Leendert van der Waerden (1903–1996), studied mathematics in the Netherlands, at Göttingen and Hamburg. In 1931, based on lectures by Emil Artin and Emmy Noether, he wrote a book, *Modern Algebra*, that revolutionized twentieth-century algebra.

[525] Max Deuring (1907–1984), studied mathematics under Emmy Noether at Göttingen, where he received his Ph.D. in 1930; he then moved to Leipzig and proceeded to teach in Jena, Posen, Marburg, Hamburg and since 1950 in Göttingen.

very ordinary and dull American woman) and Wick,[526] *who is a very intelligent and pleasant Italian. Mrs Houston saw my ring and, when I left, asked Wick whether I was married. He replied: "no, he's far too clever!" He's right in essence, isn't he?*

On the 28th Bloch gave another paper, and then I came here. My parents are very sweet and, as far as I can see, aren't offended. My father seems younger than before. He is working more again, and this is good for him: he's the kind of man who can't live without work. Of course it is especially so now, when everything in Germany is in such a mess, nobody knows whether they will have enough money to buy food tomorrow, and everyone is in a state of agitation about everything. Alfred is very tired; he has been working an awful lot. Now Nina is at the seaside, he is on his own and works all day, until midnight, and gets up again at 6 in the morning. Annie is very lively, busy all the time with all sorts of things to do for her relatives, but always seems to have time and plays a lot of tennis, she seems very well etc. Hermann and the children I haven't seen yet.

Last night I went to a colloquium at the university and saw all almost all the physicists: Houtermans, Simon, London, Rupp, Lane, Schrödinger and many more. Tomorrow I should be giving a paper in a theory seminar, the day after tomorrow I'll go to Meissner's place in the Reichsanstalt.[527] *Today I was at Berliner's and he promised me that I would have an article about conductivity in* Ergebnisse der exakten Naturwissenchaften Band 1932.[528] *In the morning I went to the consulate and my visa wasn't there, of course; they told me that it's pointless to order a visa there, as it will take at least four weeks. Perhaps Yakov Il'ich can do something to help, though I haven't seen him here yet.*

[526]Gian Carlo Wick (1909–1992) a native Italian studied at Turin and later worked at Leipzig and Göttingen before returning to Italy, eventually working under Fermi, before becoming professor at Palermo and then Padova.

[527]Walter Meissner was studying, at the Physikalisch Technische Reichsanstalt, the disappearance of ohmic resistance shown by many metals around the temperatures of the order of that of liquid helium.

[528]R. Peierls, 'Elektronentheorie der Metalle', *Ergebnisse der Exakten Naturwissenschaften* **11**, 264–322 (1932).

Thank you for the pictures. It was lovely to see some different photos of your face; I don't think you've changed at all and haven't got much thinner from boredom! I'm almost offended. The sweater you're wearing doesn't suit you though — it looks like you're about to go and play football!

The programme is remarkable; you are a superb organiser. I know that it will be almost impossible to get a separate compartment only for us, and it seems stupid that we will have to sleep apart, but for a short time only we'll have to cope. I have no idea how I'm going to get on top of all the work I haven't done yet. It will be amusing anyway.

The Graf Zeppelin is flying over our house now. It was a very familiar sight when I was in Zurich, but yesterday it probably went over Leningrad. I need to finish the letter now; I promised I'd visit Juliusburger this evening (you remember the young guy and his wife who were having problems with his parents?).
Goodbye until tomorrow dear, a thousand kisses!

3[1].7

I had a lovely time at Juliusburger's. His young wife is expecting a child, and it seems that all has not been entirely well: she had to lie down for a long while etc. but is clearly better now. She is strangely weak, at least second class (are you jealous?). It's good to have friends like these, with whom one gets on so well; it will be wonderful to visit them together. The same goes for the Bretschers.

With his parents all is now officially in order, which is wonderful, but besides that it's obviously quite a tough situation.

They have an adorable little cat! It was 'exploring' the room and suddenly started leaping and bounding around the room, but then it fell off a chair, got frightened, and had to tentatively tiptoe around the chair to find out why it was so evil!

Igor Evgenich Tamm called me this morning and I'm going to see him this evening, which I'm very pleased about. I'll speak to him about my visa.

I received a postcard from Bübi, which asked whether we will be in Odessa at any point — I suppose not, though. It was a very beautiful postcard of the Caucasus. Incidentally, I forgot to mention: Ivanenko keeps asking when I will come to Khar'kov. On our return journey we could just as easily travel via Khar'kov; it would be easy to stay there for a day or so and I could give a paper to earn a little money. (I don't expect it would be a huge amount.) But we can discuss all this when I arrive.

I also received a very gloomy letter from Fania. Her sister is still confined to her bed, and they don't know how the operation went yet as she is still not allowed to open her eyes; they are all hoping, of course, for the best. But the economic crisis has affected Fania's father, like everyone else; she is no longer receiving any money from him and has to earn everything for herself — the future looks like it will be very hard for her.

It will be time for dinner soon, and then I should head straight into town — I know I haven't explained our situation here: we live in the West of the city, about half an hour on the underground from the centre, in a big square; from my window I can see the square, and every ten seconds a car goes past. The large radio tower is near our house, and there is a projector on top which revolves at night. From our house a large, wide road runs right the way into town, like Nevsky Prospect, but much longer. My father used to walk along there, and now when I sit on the underground, I think about the long, beautiful street above my head.

How I wish I could show you around all these places. But you know, my darling, it's very possible that in just six weeks or so I'll be able to show you around Berlin, isn't it?

I'll be leaving for town in about five minutes, and I'll post this letter when I get there.

See you really soon.
A thousand kisses for your beautiful lips and eyes!

Your Rudi

P.S. Was this letter long enough?

[140] Genia Peierls to Rudolf Peierls

[Leningrad], 31.7.[1931]

My dear boy,
some 2 hours ago it might have been a rather jolly letter, but at this moment, at least, it is going to be melancholy. I'm hot, I can't breathe, I want to bathe and loaf around. And after the train, the trams and a 5-hour-long looking into the tube I'm so uptight, that there is no bathing for me. All this is, certainly, caprice, because I'm in a comparatively good situation this summer; so in an hour everything will be all right. The next page will be written in a different mood. I have to write in a depressed state of mind, to mail the letter in time. Yesterday Abbot formulated a theoretic foundation for an experimental low: "The less you do, the less time you have." In fact, yesterday was my day off and I haven't written to you, I don't even know why. Abbot explains the situation in this way: to write a letter is more pleasant than to do some work; therefore if one has a dilemma of doing something or writing a letter, one naturally starts to write a letter. To do nothing or entertain oneself is more pleasant than to write letters; that is why one rarely writes on a day off. Yes, actually it is nonsense, to write letters after a continuous conversation with you, that is going on in my mind. And then I have to write only 10^{-10} *of what I have had in my thoughts for 3 days. Unfortunately, telepathy hasn't yet reached the technical achievements that are necessary, and there are no gadgets, when for 10 rubles a month I could read your thoughts in Berlin.*

Rudi, what are you doing now, on the 31st at 8 o'clock of our standard time?

My dear, I have just figured out that, if the visa and everything is a success, then my next letter will be the last one. It is so nice — too good to be true. Just think, one more letter from me, 2 letters from you, then a telegram, the railway-station ... May there emerge something to detain you? Oh, I will cry if you don't arrive on the 11th.

I forgot to thank you for the books that arrived on the 29th, though they had been mailed at the same time as the first ones. Laws of the Post are certainly statistical, or even submit to the Heisenberg principle:

if the day of sending off is known, one cannot define exactly the day of delivery; and vice versa. I'm glad that soon we'll cease to scold the post service!

You know, a whole eternity had passed since I slept in my own bed, or rather on my own sofa. I spend the nights either in Pavlovsk or with the aunt, whose maid went on a holiday. She stays alone in her flat — that is not pleasant at all, having a weak heart; in addition, her acquaintance is still seriously ill. In general, old relatives and acquaintances form immense sick quarters. Do you remember the doctor, to whom we "paid a visit" after some of our trips? In winter they excised him a tumor, and now he has metastases in bones (the recurrent cancer); he is dying, but he doesn't know it is cancer, he thinks he feels bad with some unknown disease and hopes to get well. Why, he is a doctor! It means, he doesn't want to know. The "habit of being alive" is very strong in people, especially in aged people. That is strange. What do you think? And what about us: at 70 we, presumably, will be afraid of death and grip at that max. 0.30 of life, remaining for a 70-year-old person. Or shall we be different?

Well, all this is a rather tedious philosophy and I'm passing on to more cheerful things in the next three pages, and then I'm going to see off Natasha who is leaving for Moscow. Her family departed 4 days ago, and she was under our wardship (we fed her, etc.). So I have to go, willy-nilly; there is no pleasure at all in overcrowded trams, on the platform, etc. But ... cousinage oblige*!*

Rudi, darling, yesterday I had a great time! I and Abbot went to the final tennis match Moscow — Leningrad. I felt my heart burst with excitement — and how beautiful it was! Especially male doubles*. For women there is no use to play, because they are worse than men. They cannot properly return the ball, and in general seeing them in one team with men is utterly unpleasant. The most decisive game couldn't go on because of darkness. Today they were going to continue, but I couldn't go; it was so disappointing for me, I nearly cried. Another sensational thing is the "situation" of Dau with a female who belongs to the 92nd class, according to Abbot. Dau is of opinion she is the 2nd class. In any case she is rather vulgar-looking, with brightly rouged lips and yellow face because of freckles. We weren't able to reveal her other features. She is* *married, and Dau is inclined to consider this situation asymmetric or zero. The husband is a terrible bore and Dau already became a pain in the neck to him, so the husband hates to see him. Dau and Abbot nearly drowned the innocent lady in the Neva River. In general, Dau tries his best. That is amusing, but he looks disgusting, is depressed and the "situation" (though Dau rejects this term) isn't jolly at all. You know, he may go to Teberda with us and later, when we set off for Sukhumi, he will go to Gagry. Are you glad? But I proclaim* Pauli Verbot *on* theoretische Physic *there, otherwise I'll fall asleep and roll down from a mountain. Will you protest? Dau gives his consent. Tomorrow Abbot is leaving for the Caucasus — to Sochi and Sukhumi; then to Odessa. Here everybody either departs or arrives: the summer condition. It is unbelievable that in 2 weeks I may be on board the train with you. Nina and I[say] B[enedictovich] must be in Gagry today. Nina sent a funny letter written in verse and mailed somewhere on the way.*

Oh, how late it is!

My dear, here we have a term "the suburban husband": a man, whose family stays in a country cottage, and he goes there from the city, loaded with some errands. This "suburban husband" is a personage of humorous writings, including Sholom-Aleichem's ones. Will you feel hurt, if for one time I'll ask you to be the "suburban", or better to say, "foreign" husband? Won't you? Well then, if it isn't difficult, troublesome, etc. (I try to be courteous as yet), please bring along:

1. *Neurosmon* (schwach) *in tabl, N50 (Fabr. Promonta Hamburg), two packs*

2. *Piramidon: the more, the better — Mother eats it like bread*

3. *If it isn't too much trouble, please buy a gray felt hat No. 56*

4. *Bring some Vaseline, etc. for the mountains. You may remember my state the last year?*
   Well, you know that yourself. That is all for the time being. If I recall something else, I'll add it in the next letter.

5. *Bring yourself, first of all! Please! That will be for me personally — not for Mother. I don't need anything else.*

My dear, dear, I kiss you affectionately, wringing your neck! My regards for everybody, particularly to Nina.

G

P. S. I enclose two good photo prints by the request of Eric. He was indignant when he learned that I had sent you bad ones, and now he demands rehabilitation of his aptitude in photography. As a fee, he asks to bring any desensitizer. Anyway, he wrote you something himself.

[141] Rudolf Peierls to Genia Peierls

Berlin, 3.8.1931

Genia, my darling,
I have very little time today, so I hope you won't be too upset if this letter isn't very long; I'm mainly writing to you because this letter may be the last to which you can reply, though if I receive any more from you it will be a delight. Of course I'll take the thermals and whatever things we'll need in the Caucasus, and perhaps we'll need a few things which I'll have to buy (for you or your mother etc.) here. If that's the case, write and let me know straight away.

I'm still not sure when I'll be able to leave here, because I don't have a visa yet (Dau's invitation hasn't arrived either, incidentally, but that isn't too important). If the visa arrives in time (I'm still optimistic!) I'll arrive in Leningrad in the evening on the 10th or 11th. I'd prefer the 10th, but the 8th is a Saturday, and my parents will be offended if I leave on the Saturday evening and I'm not with them for the Sunday.

Here's the business side of things — I'm desperately missing you, but I still have many things to do before the journey. In particular, I have to quickly finish the new work on conductivity:[529] *when I first arrive in Leningrad I probably won't have much time, and I can't wait another month. But it's more complicated than I first imagined, and I now have a horrible load of mathematics to do.*

[529] See letter [137], note 515.

On the 31st I presented a paper at the university. Schrödinger was there, and I got the impression that he is much cleverer and livelier than I had thought. He's not at all conservative, and understands everything very quickly.

Besides him, London was there,[530] *less* verbonzt *than before; Elsasser,*[531] *who sends his best wishes to you; Tamm*[532] *and Yacov Il'ich,*[533] *who was talking a load of rubbish. I spoke with him for a long while about his work, but of course we didn't agree. Yesterday he came to ours' for lunch and told many good Hebrew anecdotes. Yesterday evening he left.*

On the morning of the 1st I went to Meissner's in the Reichsanstalt, he showed me everything (apparatus for fluids etc.) and seemed very pleased to see me. I visited Edgar (my American cousin) for lunch — he is living in Hamburg for a short while — and met his wife, whom he married for money. She is the daughter of a doctor and not a particularly nice woman.

In the evening I went to Juliusburger's place again, and Boehm was there too. He's a very decent mathematician with a Russian wife. (It seems to be a professional illness!)

Yesterday I saw Nina and Vera. They seemed very well (they've been at the seaside) and very happy, but they had other guests at the time, so I didn't get the chance to chat to them much.

---

[530] Fritz London (1900–1954), studied at Bonn, Frankfurt, Göttingen, Munich (where he obtained his Ph.D. in 1921), and Paris. With Walter Heitler he devised (1927) the first quantum mechanical treatment of the hydrogen molecule. He later went to Oxford and again to Paris before taking up a professorship of theoretical chemistry at Duke University.

[531] Walter Maurice Elsasser (1904–1991) studied at Heidelberg, Munich and Göttingen where he obtained his doctorate in 1927. After appointments in (1928–1930) and Frankfurt University (1930–1933) he became guest lecturer at the Sorbonne (1933–1936) before emigrating to the US.

[532] Igor Y. Tamm (1895–1971), Russian physicist and close collaborator of L. Mandelstam, Tamm taught at the Moscow State University (1924–1937) and was in charge of the theoretical division of the Lebedev Institute of Physics of the Russian Academy of Science from 1934.

[533] Y.I. Frenkel.

Houtermans told me a little piece of gossip: Jonny arrived here on the 1st but flew out to Moscow in the evening, probably to travel on to Khar'kov.

Today the Houtermans are coming here, and I am fascinated to see whether they will be very 'bohemian' when their parents visit.

Oh yes, and I received a card saying that Barbara Zarniko and Ruhemann have got married in Stuttgart. She's done her exams already, by the way.

Incidentally, I still haven't told you that I have to get up at 8 am every day here! Is this healthy? I have breakfast with my father, who leaves for work at 9.30, and he likes to eat breakfast slowly and peacefully.

As I said at the beginning of this letter, I need to stop here so that this letter makes the airmail today.

See you in just 8 days!

Quick kisses, but no less lovingly for that.

Your Rudi

[142] Rudolf Peierls to Genia Peierls

Berlin, 6.8.1931, midnight

My dear,
it is terribly hot here, even though it's the middle of the night, the window is open, and I'm sitting in bed completely naked to write to you. Your picture is in front of me (you know the "obizyana" in the style of Rembrandt, it's always on my table) though I often feel that it's indecent to sit unclothed in front of such a beautiful image. I hope you won't be offended.

It's a disgrace, it really is: my visa still hasn't arrived. At the consulate they told me that I have to send a telegraph to the Rentgenovsky Institute. If only Frenkel would arrive soon; Igor Evgenich[534] *also promised me that he would phone Narkomindel*[535] *when he arrives*

[534] Refers to Igor Evgenich Tamm.
[535] NARCOMINDEL: People's Commissariat of Foreign Affairs of the Soviet Union.

in Moscow. I'm not sure how much sense this makes. I'm so tired of waiting!

I should reply to you today because your letter of the 3rd arrived here on the 4th![536] *But when the second letter arrived today, I was terribly worried because I was still in bed (the post arrives here at 7.30) and I didn't have my glasses on. I could only read the first sentence without my spectacles and I believed it literally. Of course I was very concerned, and only when I got my glasses from the other end of the room could I read further, and I saw that it wasn't as strange as it first appeared. All the same, my darling, you shouldn't do these things to me!*

It's quite funny: on the 31st at 5 o'clock Russian time I was sitting in the university arguing with Frenkel.

Everything is in order with Brillouin. He wrote me a very polite letter — nobody does courtesy like the French.[537]

Today I visited Nina. She is always exhausted these days because her housemaid is ill and she has to do everything for herself; her flat is quite large and she regularly has guests etc. but she's not very strong. Vera is adorable. First of all she asked me: "who do you love most in all of Germany?" and, when I hesitated, added: "what about the whole world? I know: Genia". When I agreed, she continued: "do you know who I love most of all? Firstly papa (Alfred) and then mama and you" and after a brief pause "... and Genia". "Why Genia", I asked, "you've never met her; how can you love her?". "Oh, I only said that so that she wouldn't be sad".

But she is a very highly-strung child, and is hard work for Nina.

Du hast aber recht, ich muss Dir ein bisschen auf deutsch schreiben, sonst vergisst Du alles.

Heute morgen traf ich Käthe Sperling (weisst Du, die Mathematikerin, von der ich in Leningrad einen komischen, aber doch melancholischen Brief bekam).[538] Es war ganz nett, sich mit ihr zu unterhalten,

[536] Letter has not survived.

[537] See letter [137], notes 513–514.

[538] Käthe and her sister Lotte, a physics student, were part of the circle of student friends during Peierls' Berlin student days. Käthe later married the physicist Werner Fenchel.

aber viel interessantes hat sie nicht erzählt. Dann war ich bei Wertheim, um einzukaufen, und das ist wirklich sehr bequem. Was Du für Augen machen wirst, wenn wir zum erstem Mal in einem grossen Warenhaus (*department store*) herumlaufen werden! Das bequemste ist, dass man ein Buch bekommen kann, in das überall die Sachen eingeschrieben werden, die man kaufen will, wenn man fertig ist, gibt man das Buch ab, bezahlt, und am nächsten Tag bekommt man alle Sachen ins Haus.

Gestern war ich bei Schottky, der bei Siemens arbeitet, er hat ein sehr schönes Zimmer in der 12. Etage des grossen Siemens Hauses, und aus seinem Fenster hat man einen wundervollen Blick über Berlin. Vorgestern war ich mit Eva zusammen. Ihr Roman ist noch viel verrückter und komplizierter als ich glaubte, indem z.B. nicht Frau Schneider an ihre Eltern geschrieben hatte, sondern eine eifersüchtige Freundin von Schneider in Zürich. Aber die Geschichte ist zu lang für einen Brief, man muss sie erzählen. Jedenfalls hat sie mir ausführlich ihr Herz ausgeschüttet (*this is a very nice German expression!*) Überhaupt ist sie ein sehr komisches Mädchen. Am selben Nachmittag war ich bei Annie, und war wieder begeistert von den Kindern. Der Kleine fängt jetzt schon an, ein ganz venünftiges Gesicht zu machen, und er hat schon viel gelernt, z.B. spucken (*to spit*). Ilse, seine Schwester, fühlt sich schon sehr verantwortlich und passt auf ihn auf. Annie hat ein reizendes Verhältnis zu ihren Kindern, ein viel netteres als Nina, aber dafür sind sie bei ihr auch der Hauptinhalt des Lebens.

You are probably tired of German now, and I am tired because it's 1 o'clock in the morning. A thousand kisses, dearest.

Maybe this will be my last letter! Maybe not just the last for now, but also the last for a long time! Sleep well!

Your Rudi

7.8.1931

No visa today either; they are so stupid. In all probability it won't be here tomorrow, and the day after is Sunday, so I'm afraid I probably won't be able to leave before the 10th. Hopefully no later! I've been working all day and my head is "spinning with conductivity". It's all very difficult and I'm not sure how I'm going to get everything done before I leave.

[143] Rudolf Peierls to Genia Peierls

Berlin, 9.8.[1931]

The devil take them all! The visa didn't arrive yesterday, of course, and so now, as "my" train is already in Poland, I am sitting at home and still don't know when I can leave. I miss you so terribly, my darling, I can't wait to see you, but you can't come here and I can't get there! If only I knew when my visa would arrive, but the waiting ... I know the telephone number of the consulate from memory now! Every day I phone them twice to find out whether my visa has arrived.

No, dearest, I'm not worried at all about the Houtermans living in such a Bohemian manner, just that their parents will find out. The parents (hers in particular) won't be pleased at all if they do discover how they are living.

Nothing new to report of late; last night I went to the cabaret with my parents: it was very amusing. They are having guests around tonight, but I feel strangely uninterested. At the moment I can only think about the visa and the journey, I just want to leave, to leave ...

All the same, it's possible that this letter will arrive there later than me! Look at what an utterly daft and boring letter I'm writing, it would be better if I go to bed now and see what the situation is tomorrow.

My father is an awful pessimist: he obviously doesn't think the visa will arrive in the near future. He hasn't said anything to me (which is a shame) but Else has told me what he thinks — she's not all that tactful. She said to Annie today, "We'll be leaving the day after tomorrow, and Rudi might be stuck here for a long while yet, so look after him"! I just hope father isn't right!

Anyway, thinking about it does no good. Goodnight, my darling. More kisses for your beautiful face!

10.8.

Thank you for your letter. I went to the bureau again today, and it seems that I don't have to send the form off to Moscow, but I can fill it in here. They assure me that I can leave on the 14th or 15th.

What medicine does Sarra Isaakovna need? Last time I only brought Ovarin, but that will be no good for tonsillitis, which is a different matter altogether.

I don't have any other news; I'm starting work again, which is probably stupid, but I'm going to anyway.

A thousand kisses! Until I see you again... soon!

Your Rudi

[144] Wolfgang Pauli to Rudolf Peierls

[New York], 29.9.[1931]

Lieber Peierls,

Für Ihren letzten Brief vom 4.VIII.[539] noch besten Dank. Hoffentlich haben Sie eine anständige Klärung der Frage der Ungleichungen für die Feldstärken, wenn ich nach Zürich zurückkomme.[540] Übrigens möchte ich die Frage der Ortsmessung im relativistschen Gebiet noch genau mit Ihnen durchdiskutieren, vielleicht kann man weiterkommen, wenn man diese Begriffe klarer faßt. So wie es in der Arbeit von Ihnen und Landau steht, ist es mir nicht klar genug.

In der Frage der Strahlung im zeitlich wechselnden Magnetfeld haben Sie viel zu viel Respekt vor der Linienbreite. Es ist immer ganz leicht diese zu berücksichtigen. Ferner ist der Polarisationsgrad der Resonanzstrahlung eine ganz alte Methode, um Hyperfeinstrukturen nachzuweisen und nach meiner Meinung ist darüber gar nichts mehr zu rechnen, weil alles schon bekannt ist. Daß die klassischen Ansätze mit gedämpften virtuellen Oszillatoren richtig sind, muß und kann man allgemein zeigen, nicht bei jedem Beispiel extra. Aber diese Fragen sind nicht sehr interessant.

---

[539] Could not be located.

[540] See letter [126].

Als rein mathematische Kraftleistung habe ich sehr bewundert die Arbeit von Bethe,[541] in der er das richtig macht, was Ludloff[542] falsch gemacht hat. Vielleicht kann man jetzt endlich doch eine anständige Theorie des Curiepunktes ferromagnetischer Substanzen machen. Über Halbleiter soll man nicht arbeiten, das ist eine Schweinerei, wer weiß ob es überhaup Halbleiter gibt. Ihre Arbeit über die Lichtabsorption in festen Körpern will ich genau studieren, wenn ich in Zürich bin. Hoffentlich haben Sie sich inzwischen mit Frenkel darüber geeinigt.[543] Am 1. Oktober haben wir ein Magnetpol*-Neutronen-Kolloquium in Princeton, in dem Dirac[544] und ich die Hauptrollen spielen werden (a first national attraction). Vom rein logischen Standpunkt aus scheinen mir die Magnet-Pole befriedigender als die magnetschen Dipole aber man kann dann die falsche Statistik der N-Kerne nicht damit verstehen. Es ist auch durchaus möglich, daß es weder Magnetpole noch Dipol-Neutronen gibt.

Bitte sorgen Sie in Zürich dafür, daß meine Kollegs angekündigt werden und zwar sollen beide (Kurs- und Spezialkolleg) am Dienstag, den 20. Oktober beginnen. Ich werde wohl erst am 19. Oktober im Institut auftauchen und will dann auch den Antrag wegen Ihres Gehaltes schreiben.

Amerika gefällt mir gut. Am 2. Oktober fahre ich von hier direkt nach Neapel, wo ich am 11. eintreffe. Mein Arm ist längst wieder heil. Viele Grüße

Ihr W. Pauli

*Sie haben wohl Diracs Arbeit im September-Heft der Proceedings of the Royal Society gelesen.

[541] H. Bethe, 'Zur Theorie der Metalle. Eigenwerte und Eigenfunktionen der linearen Atomkette', *Z. Phys.* **71**, 205–26 (1931)

[542] H. Ludloff, 'Zur Frage der Nullpunktsentropie des festen Körpers vom Standpunkt der Quantenstatistik, I, II, und III' *Z. Phys.* **68**, 433–92 (1931).

[543] See letters [122] and [125].

[544] P.A.M. Dirac, 'Quantised Singularities in Electromagnetic Fields', *Proc. Roy. Soc.* **133**, 60–72, (1931).

[145] Hans Bethe to Rudolf Peierls

München, 12.11.[19]31

Lieber Peierls,
Man soll doch immer alle Annahmen, die man macht, in eine Arbeit schreiben, und nicht, wie ich diesmal, die Hälfte weglassen.

Ich habe nämlich stillschweigend angenommen, dass die Formel (5) nur dann gilt, wenn,

$$m_1 < m_2 < m_1 + N \tag{I}$$

ist, d.h. wenn der Abstand der beiden Rechtsspins kleiner als die Länge der Periode ist. Eine solche einschränkende Bestimmung zu treffen, ist offenbar notwendig, sonst entwickelt man sich mit der Periodizitätsbedingun in Widersprüche – Kostproben dafür haben Sie mir ja genug gegeben. Macht man die Einschränkung (I), so ist alles in Butter: Für $a(m_1, m_2 + N)$ gilt dann die Formel (5) nicht mehr, weil $m_2 + N > m_1 + N$, und es bleibt ausser $a(m_1, m_2)$ selbst nur $a(m_2, m_1 + N)$ als einziges a übrig, dessen Argumente die Bedingung I erfüllen. Damit kommt man zwangsläufig auf meine Randbedingung (10).

Wenn meine Periodizitätsbedingung (10) erfüllt ist, gilt offenbar auch die Gl[eichung] (4b) auch dann, wenn $m_2 - m_1 = N - 1$, d.h. falls der zweite Spin unserer Periode neben dem ersten Spin der nächsten Periode der Kette liegt. Die Gültigkeit von (4b) in diesem Fall zu verlangen, erscheint doch notwendig, gleichgültig ob man einen geschlossenen Ring von Atomen hat oder eine Kette mit aufgezwungener Periodizitätsbedingung – warum Sie einen Unterschied zwischen diesen beiden Fällen konstruieren, habe ich nicht recht verstanden.

Die Einschränkung (I) für die m's zu treffen, scheint mir nicht nur notwendig, sondern auch vernünftig: Man darf an jeder beliebigen Stelle der Ketten anfangen, die Rechtsspins aufzuzählen, muss sie dann aber in der Reihenfolge aufzählen, wie sie aufeinanderfolgen, und darf nicht zwischendurch einen auslassen. Allerdings muss ich nach alledem zugeben, dass meine Grenzbedingung (3) irreführend oder sogar direct verkehrt ist, man darf tatsächlich nur fordern

$$a(m_1 m_2 \ldots m_r) = a(m_2 \ldots m_r m_1 + N)$$

wenigstens, wenn man will, dass beide a's durch die geschlossene Formel (5) bezw. (25) gegeben sind. Ich glaube nicht, dass Ihnen meine Randbedingung (10) jetzt noch so künstlich erscheint. Ausserdem ist es durchaus notwendig, eine scharfe Randbedingung zu stellen und nicht zu sagen, dass es auf die Abzählung erst in zweiter Linie ankommt: Wenn man z.B. den tiefsten Eigenwert im nicht ferromagnetischen Fall $\mathfrak{J}$ ausrechnet (der ja einem Gesamtspin Null entspricht), so gehen die Phasen ϕ_{ik}, die ja eben durch die Periodizitätsbedingung hereinkommen, ganz wesentlich in das Resultat ein. Sie bewirken nämlich, dass die Wellenzahlen f_i der Spinwellen sich alle um $f_i = \pi$ herum anhäufen, und da eine Spinwelle im Fall $\mathfrak{J} < 0$ umso weniger Energie hat, je näher ihre Wellenzahl an π liegt ($E = \mathfrak{J}\ (1 - \cos f_i)$) so wird dadurch die Energie wesentlich herabgesetzt. Es kleben also nicht nur die Spins zusammen, sondern auch die Wellenzahlen der Spinwellen! (Übrigens ist der tiefste Eigenwert für $\mathfrak{J} < 0$

$$e = E_0 + N\mathfrak{J}\,(\lg 4 - 1) = E_0 + 0{,}386N\mathfrak{J} \tag{II}$$

Slater fand $e = E_0 + 0{,}293N\mathfrak{J}$, also erheblich zuviel. (II) ist exakt bis auf Grössen der Ordnung $\mathfrak{J}$.)

Noch ein paar Kleinigkeiten zu Ihren "Angriffen" gegen meine Periodizitätsbedingung: Die komplexen Wellen befriedigen selbstverständlich auch die Periodizitätsbedingung. Denn so ist ja gerade das $\varphi = \psi + i\chi$ nach (16) gewält. Die a's sehen etwa so aus (siehe (23))

$$a(m, m_2) = e^{\frac{1}{2}ik(m_1+m_2)} \cdot \cos(\frac{1}{2}N - (m_2 - m_1)) \tag{III}$$

Die Amplitude a fällt also exponentiell ab, wenn m_2 sich von m_1 entfernt, erreicht ein Minimum für $m_2 - m_1 = \frac{1}{2}N$ und steigt dann wieder wenn m_2 sich dem ersten Spin der nächsten Periode der Kette $m_1 + N$, nähert. Für $m_2 = m_1 + N - 1$ bekommt man genau dieselbe Amplitude wie für $m_2 = m_1 + 1$. (III) gilt übrigens nur für gerade k, für ungerade tritt ein hyperbolischer Sinus an die Stelle des Cosinus. Man kann das alles leicht nachrechnen, wenn man (16) in (5) einsetzt.

Warum ich in (11) $\lambda_1 = \lambda_2$ ausschliesse? Weil sich (11), (8) dann überhaupt nicht lösen lässt. Wenn $f_1 = f_2$, so ist unweigerlich $\phi = \pm\pi$, d.h. sowohl für $\lambda_1 = \lambda_2 + 1$ wie $\lambda_1 = \lambda_2 - 1$ führen zu

$f_1 = f_2$. Ist $f_1 > f_2$ so muss $\lambda_1 > \lambda_2 + 1$ sein, und für $f_1 < f_2$ muss $\lambda_1 < \lambda_2 - 1$ sein (ausgenommen für die ganz kleinen $f < \frac{1}{\sqrt{N}}$, für die die komplexen Lösungen in reelle degenerieren). Jedenfalls gibt es überhaupt keine reellen Werte $f_1 f_2$, für die $\lambda_1 = \lambda_2$ werden kann — bei den Komplexlösungen kommt dann ja (für gerade k) $\lambda_1 = \lambda_2$ vor. Schliesslich noch die Feststellung, dass man die komplexen Lösungen bei der Abzählung mitnehmen muss — ich bin sogar nur dadurch darauf gekommen, dass die reellen Lösungen nicht die nötige Anzahl gaben. Für Gesamtspin Null ($r = \frac{N}{2}$) bleibt sogar nur eine einzige Lösung mit reellen Wellenzahlen, wegen des Verbots, dass nie zwei Wellenzahlen gleich sein dürfen (diese Lösung führt bei $\mathfrak{J} < 0$ auf den tiefsten Eigenwert), praktisch alle Lösungen sind also komplex. Ihren Einwand, meine komplexen Lösungen befriedigten die Perioditätsbedingung (10) nicht, habe ich ja schon oben besprochen. Ihre Umschreibung meiner Rechnungen gefällt mir sehr gut. Besonders erkennt man viel besser als bei mir, warum es bei gegebenem k nur einen diskreten Eigenwert gibt und warum $f_1 = f_2$ auszuschliessen ist — dass das eben an der "Randbedingung" für $l = m_2 - m_1 = 0$ liegt und nicht etwa an der Perioditätsbedingung, wie es bei mir aussehen könnte. "Aneinenderkleben" tun die Spins übrigens natürlich nur bei der komplexen Lösung, im allgemeinen (bei den reellen Wellen) ist sogar eine grössere Entfernung wahrscheinlich, weil im Mittel über alle reellen und komplexen Lösungen jede Entfernung zwischen m_1 und m_2 gleich wahrscheinlich sein muss.

Ihre Ersatzperiodizitätsbedingung $u(l_0) = 0$ geben Sie vielleicht mir zuliebe auf, weil Sie sich von der mathematischen Eindeutigkeit meiner Bedingung überzeugt haben. Wenn nicht, stets gern zu weiteren Diensten.

Ob man mit Ihrer Schrödingergleichung beim dreidimensionalen Fall etwas anfangen kann, weiss ich nicht. Die grosse Schweinerei ist nämlich, dass (schon bei zwei Dimensionen) sich die Definition der a's für den Fall zweier Rechtsspins am gleichen Atom nicht mehr willkürfrei machen lässt: Definiert man $a(m_1 m_2)$ einmal so, dass die Gleichung, die (4b) entspricht, befriedigt ist, wenn m_2 rechts neben m_1 liegt, das andere mal für m_2 über m_1, so widersprechen sich die beiden Definitionen. Dann kriegen Sie auch keine Grenzbedingung für Ihre Schrödingerfunktion mehr. Aber ich muss mir die Sache nochmal überlegen.

Hoffentlich rechnen Sie bei der Lichtabsorption in festen Körpern nicht zuviele Sachen, die ich in Rom gerechnet habe. Aber ich habe mich ja immer auf die Atome und Ionen mit unabgeschlossenen Schalen kapriziert, sodass wir uns wohl nicht ins Gehege kommen. Natürlich würde es mich interessieren, was an nicht-trivialen Dingen herauskommt. Ich schwitze an einem Handbuchartikel über H und He und komme garnicht vorwärts,[545] teils aus eigener Insuffizienz, teils weil mich alle fünf Minuten jemand stört. Ausserdem merke ich jetzt nach meiner "Weltreise"[546] erst richtig, dass hier im Institut meist nur Mathematik gemacht wird und keine Physik.

Kann denn irgend jemand von uns, d.h. von Ihnen "selbst wirtschaften"? Ist es möglich, in St. Antonien Lebensmittel zu kriegen, oder steht eine Hungersnot zu befürchten? Glauben Sie dass das billiger ist als eine Pension? Kurz gesagt: An sich gerne, aber ich mag mich noch nicht binden. Ich habe auch nebenbei noch ein zweites "Projekt" Sachs laufen, dessen Entwicklung ich noch abwarten muss. Ihre Dispositionen sind ja wohl auch unabhängig von der genauen Anzahl der Beteiligten. Es wäre aber nett, uns auf die Weise endlich wiederzusehen. Grüssen Sie Ihre Frau, die Sie doch überraschend schnell herausgekriegt haben, Ihren Chef und sich selbst von Ihrem

H[ans] Bethe

[545]Initiated by his teacher Sommerfeld, Hans Bethe was writing the core of a substantial review article to be published under joint authorship of Sommerfeld and himself. Sommerfeld, A. and H. Bethe, 'Electron Theory of Metals', *Handbuch der Physik* **24**, 2, (1933).

[546]Refers to his year abroad as part of his Rockefeller fellowship, when he spent six months in Cambridge and six months with Fermi in Rome.

3. Zurich — Rome — Cambridge — Manchester — Birmingham:
The Journey Towards Academic Success

When Rudolf and Genia settled into their married life at Zurich they knew that this could only be a temporary arrangement. It was known that Pauli liked to change his assistants frequently. His first assistant, Ralph Kronig, had stayed for two semesters, Felix Bloch only for one, and Peierls knew that it was time for him to look elsewhere for the following academic year. His earlier plans to join Franz Simon's institute in Berlin had not materialised, as the latter had taken up a Chair in Breslau in 1931. Generally, job prospects for your theoreticians in Germany were far from rosy. Max Born, in a letter to Arnold Sommerfeld contemplated that all the positions were filled and that there would be no new ones in the foreseeable future. On the other hand, there was an unusually large number of excellent young people looking for long-term employment, among others 'Wigner, London, Heitler, Bethe, Nordheim, Peierls, ... Weisskopf, Stobbe, Delbrück and Maria Goeppert.'[547] Peierls' close friend from the Munich years, Hans Bethe, spent the academic year 1931/32 on a Rockefeller fellowship in Cambridge and Rome, and in view of the difficult job market for young academics, Peierls considered this as an attractive option. With Pauli's encouragement, he applied for a fellowship,[548] and subsequently, he spent the winter semester 1932/33 in Rome working with Fermi, and the summer semester 1933 in Cambridge. Prior to these work-related travels, however, the spring and summer of 1932 saw Rudi and Genia enjoy other trips, including a short visit to Paris, funded by the generous wedding

[547]Max Born to Arnold Sommerfeld, 13.11.1930, Eckert, *Sommerfeld Briefwechsel*, II, 320.

[548]Letter [146].

gift of a relative, and a journey to Russia which allowed Genia to visit her family and Rudi to meet and work with his Russian colleagues again.

In Rome, Peierls renewed his acquaintance with Fermi and many other Italian colleagues, and just like Bethe a couple of years before him,[549] he was deeply impressed with Fermi's institute, his abilities as a researcher and academic tutor.[550] In the aftermath of the discovery of the neutron by Chadwick earlier in the year, Fermi's group started thinking about more experimental work, as Fermi himself believed at that point that progress in nuclear physics would require a strong emphasis in this area. However, much of Rudolf Peierls' work was unaffected by this new direction in Fermi's work. He joined a project with focus on the numerical solution of differential equations, needed for work with atomic wave functions. But his main research interest still concerned the theory of electrons in metals, and in particular diamagnetic metals. Landau had developed a theory of diamagnetism of free electrons,[551] but it was not clear how this theory needed to be modified in view of the presence of atoms in metals. While in Rome, Peierls completed two papers which discussed the general state of the electron band deriving a general expression for weak fields[552] and considering strong fields and low temperatures.[553] With this work, Peierls could explain the mysterious magnetic properties of bismuth, which showed much greater diamagnetism than any other substance, and, as W. J. de Haas[554] and his students von Alphen had discovered, the diamagnetic susceptibility of bismuth at low temperatures showed periodic variations as a function of the magnetic field.[555]

---

[549]Letter [113].

[550]Peierls, *Bird of Passage*, p. 86.

[551]L. Landau, 'Paramagnetism of metals', *Z. Phys.*, **64**, 629 (1930).

[552]R. Peierls, 'On the Theory of Diamagnetism of Conduction Electrons', *Z. Phys.*, **80**, 763–91 (1933).

[553]R. Peierls, 'On the Theory of Diamagnetism of Conduction Electrons II, Strong Magnetic Fields.', *Z. Phys.*, **81**, 186–94 (1933).

[554]Wander Johannes de Haas (1878–1960) studied physics in Leiden and worked as Kamerlingh Onnes' assistant (1905–11) obtaining his Ph.D. from Leiden in 1912. After working in Berlin, Delft and Groningen he returned to Leiden in 1924.

[555]This famous de Haas-van Alphen effect was later investigated further by David Shoenberg in Cambridge and became an important tool for exploring the states of electrons in metals.

Early during his time in Rome, Rudolf Peierls exchanged some letters with Paul Ehrenfest.[556] The latter had found the discovery of quantum mechanics deeply unsatisfactory. He saw it as a mathematical apparatus built on Hilbert's basis, with operators too abstract for physical meaning. As he put it in a letter to Bohr in 1931, he felt as though he had lost contact with theoretical physics, he felt incompetent to have even a modest grasp of the discipline.[557] In 1932, Ehrenfest published his 'Erkundigungsfragen',[558] his last substantial contribution to physics before his suicide in September 1933. In this paper, he asked 'someone who really understood the subject' and who was capable of explaining it to older physicists in a comprehensible form to explain some of the theory that Ehrenfest felt to have lost touch with. Implicitly this was aimed at Pauli, but Peierls, too, wrote a lengthy reply[559] in which he addressed Ehrenfest's queries. Ehrenfest replied to both Pauli[560] and Peierls,[561] urging them to publish these letters to him in an attempt to make quantum mechanics more comprehensible to his students and other physicists. Peierls did not do so, as he knew that his mentor Pauli intended to publish his letter,[562] but he wrote another elaborate reply to Ehrenfest addressing further outstanding questions.[563]

In the spring of 1933, it was again time for Rudolf and Genia to move, this time to Cambridge where they were planning to spend the second half of Rudolf's fellowship. In order to fulfil one of the conditions for the award of the Rockefeller fellowship, namely a position to

---

[556]Paul Ehrenfest (1880–1933) had studied in Vienna and Göttingen under Ludwig Boltzmann, Felix Klein and David Hilbert. After receiving his Ph.D. from Vienna in 1904 he taught at Göttingen and Leningrad before taking up the Chair for theoretical physics at Leiden.

[557]Letter cited in A. Pais, *Niels Bohr's Times: In Physics, Philosophy and Polity*, Oxford: Oxford University Press, 1991, p. 409.

[558]P. Ehrenfest, 'Einige die Quantenmechanik betreffende Erkundigungsfragen', *Z. Phys.* **78**, 555–59 (1932).

[559]Letter [155].

[560]Ehrenfest to Pauli, 31.10.1932, Pauli, *Wissenschaftlicher Briefwechsel*, II, pp. 135–37.

[561]Letter [156].

[562]See letter [154].

[563]Letter [157].

return to at the end of the award period, Wolfgang Pauli had undertaken an insurance to take Peierls on again, on the understanding that the latter would not take up this offer.[564] Therefore, Rudolf was glad to receive an offer from Hamburg, to take up the assistantship at Otto Stern's institute, a position that Wolfgang Pauli had held some years previously. However, living under Mussolini's regime in Italy had given Rudolf and Genia a taster of fascism, and of course Genia had already had experience of living under Stalin's totalitarian regime, albeit at a time when its worst excesses had not become apparent. Hence, both were sceptical about the wisdom of returning to Germany at a time when National Socialism was becoming an ever stronger force. In June 1932, von Papen had replaced Brüning as German Chancellor, only to be replaced himself by Schleicher in December 1932. All this convinced Peierls of the political volatility which made it risky to take up an appointment in Germany. When the official invitation for the Hamburg post arrived, early in 1933, Peierls turned it down.[565] This decision proved right in view of the fact that soon afterwards Hitler was elected Chancellor and National Socialist rule established firmly within a short period of time.

At that time, Cambridge was at the centre of exciting discoveries in physics. 1932, rightly or wrongly referred to as the annus mirabilis in nuclear physics, saw the discovery of the neutron by James Chadwick, the identification of the positron by John Anderson and the use of a particle accelerator, by John Cockroft and Ernest Walton, to perform the first artificial disintegration of the atomic nucleus. At the same time, Blackett was developing the Wilson cloud chamber into a more sophisticated instrument in the enterprise of studying nuclei and cosmic radiation. Peierls joined the Cavendish in this nuclear heyday, and he tried to learn as much as possible about this new exciting discipline. But his research was still concerned with two other subjects that he had previously worked on, the electron theory of metals and relativistic field theory, including Dirac's hole theory.

---

[564]See letter [159], note 636.

[565]Peierls, *Bird of Passage*, p. 90.

In a letter to Pauli in July 1933[566] Peierls seemed to have modified his highly critical view of Dirac's theory as expressed in his paper with Landau in the spring of 1931,[567] probably as a result of working more closely with Dirac in Cambridge at that time. The remarks have to be seen in the context of a wider debate about Dirac's hole theory. In the run-up to the 7th Solvay Conference to take place in Brussels in October 1933, which was to be devoted to the structure and properties of the atomic nuclei, Pauli, Heisenberg and Peierls, and later Pauli's new assistant Weisskopf, engaged in an animated scientific exchange, a review of mathematically equivalent formulations of the theory. But all of these led to the same difficulties, among others they did not solve the infinite self energy problem.[568] In his letter to Pauli, Peierls comes to Dirac's defence by pointing out that the difficulties of Dirac's theory and the divergences of electrodynamics had to be solved together and that Dirac's theory could not be criticised, when the fundamental problem had to be seen as lying with electrodynamics.

Many of the letters exchanged between Peierls and his friends and colleagues in 1933 dealt with both, physics and physicists, as the developments in Germany had dramatically altered the academic landscape and the already difficult job prospects for young scientists. This was particularly evident in the correspondence between Peierls and Bethe in the summer of 1933.[569] Both were looking for employment, and given the similarities of their background and the proximity of their research interests, it was hardly suriprising to see them competing for the same jobs. In the event, both succeeded in obtaining positions in Manchester, Bethe as a university-funded lecturer and Peierls as a fellow sponsored by the Academic Assistance Council, an organisation, funded by voluntary donations and founded to support refugees with grants until they could secure employment in the United Kingdom or elsewhere.

---

[566]Letter [163].

[567]R. Peierls and L. Landau, 'Quantenelektrodynamik im Konfigurationsraum', *Z. Phys.* **62**, 188–200 (1930).

[568]See in particular Pauli to Heisenberg, 25.7.1933, Pauli, *Wissenschaftlicher Briefwechsel*, II, pp. 208–209.

[569]Letters [160–162].

Family correspondence in the summer of 1933 was dominated firstly by the birth of Rudi and Genia's first child, their daughter Gaby, in August 1933; and secondly by the deteriorating situation in Germany which affected more and more Jewish friends and acquaintances of the Peierls'.[570] Heinrich Peierls, despite his Jewish origins, was not yet suffering financial hardship himself. He still offered to support his children within the limits regulating financial transfers to foreign countries,[571] and Heinrich and Else Peierls were still able to go abroad themselves.[572] Rudolf Peierls' one-year position at Manchester was much appreciated at the time but gave no medium- or long-term security. In January 1934, therefore, he was among a number of physicist friends who applied for a position at Chandrasekhara Raman's new physics institute in Bengalore.[573] None of Peierls' young colleagues obtained the position, it went to Max Born, a more senior colleague. But the episode illustrates that, as European and American positions were increasingly competitive, young physicists, especially German refugees, began to look further afield.

The time in Manchester was among the most prolific period of Peierls' career in publication terms. He still worked on aspects of the electron theory of metals publishing a significant number of papers.[574] But in addition to this, he continued his work on Dirac's hole theory, which also resulted in a publication.[575] The paper is clearly result of Peierls extensive discussions with Dirac, intially at Cambridge and later in correspondence.[576] These communications led Peierls to the conclu-

---

[570]Letters [164], [166–168], [170–171].

[571]Letters [166], [175], [177], [184].

[572]In August 1933, Heinrich and Else Peierls spent several weeks in Switzerland, in March 1934 they traveled to Italy.

[573]See letters [171–172].

[574]R. Peierls, 'Bemerkungen zur Theorie der Metalle', *Z. Phys.* **88**, 786–91 (1934); R. Peierls, 'Electric conductivity at low temperatures', *Phys. Z. Sowjetunion* **516–19** (1934); R. Peierls, 'Über die statistischen Grundlagen der Elektronentheorie der Metalle', *Helv. Phys. Acta* **7**, Supp. 2, 24–30 (1934); and R. Peierls, 'Quelques Propriétés des corps solides', *Ann. Inst. Henri Poincaré*, **5**, 177–222.

[575]R. Peierls, 'The vacuum in Dirac's theory of the positive electron', *Proc. Roy. Soc. Lond. A* **146**, 420–41.

[576]Letters [169], [176], [178], [180].

sion that the theory as it stood was not complete, because it made use of infinite quantities which were inadmissible in physical equations. Therefore, he believed that the underlying equations had to be modified. He took up the problems encountered in the Dirac Theory in a later paper, in which he reacted to a prediction of Brinkman who had forecast in Dirac's theory an infinite dispersion of light by a vacuum.[577] Peierls, again supporting Dirac's theory as such, showed that the dispersion of light did not present any new difficulties for the theory which were not already present from the start, as in the well-known infinities in electrodynamics.[578]

The Manchester period also marked the beginning of Peierls' more intense interest in nuclear physics. It was a time of fruitful collaboration between Bethe and Peierls, and in February 1934, together with Hans Bethe, Rudolf Peierls published his first paper dealing with something other than solid state or quantum dynamics.[579] quickly followed by several other joint papers on nuclear physics questions.[580]

And also those papers which Peierls published as sole author, bore the mark of Hans Bethe's influence. In another solid state paper,[581] which appeared to have been inspired by Bragg who had formulated the problem of variation of degree of order in an alloy with temperature,[582] Peierls treated this problem mathematically, using a technique which Bethe had devised[583] to deal with a similar crystal problem. None of

[577]H.C. Brinkman, *Physica* **1**, 93 (1933).

[578]R. Peierls, 'Die Dirac'sche Löchertheorie und die Lichtgeschwindigkeit im Vakuum', *Physica* **2**, 399–402 (1935).

[579]R.E. Peierls and H.A. Bethe, 'The "Neutrino"', *Nature*, **133**, 532–33 (1934).

[580]H.A. Bethe and R.E. Peierls, 'The Neutrino', *Nature* **133**, 689–690 (1934); H.A. Bethe and R.E. Peierls, 'Photoelectric Disintegration of the Diplon', *Proc. Int. Conf. Phys. London, 1934; vol. 1, Nucl. Phys.*, 93–94; H.A. Bethe and R.E. Peierls, 'Quantum Theory of the Diplon', *Proc. Roy. Soc. Lond. A* **148**, 146–56 (1935); 'The scattering of neutrons by protons', *Proc. Roy. Soc. Lond. A* **149**, 176–83 (1935).

[581]R. Peierls, 'Statistical Theory of superlattices with unequal concentrations of components', *Proc. Roy. Soc. Lond. A* **154**, 207–22.

[582]W.L. Bragg and E.J. Williams, 'The Effect of Thermal Agitation on Atomic Arrangement in Alloys', *Proc. Roy. Soc. A* **145**, 699 (1934); W.L. Bragg and E.J. Williams, 'The Effect of Thermal Agitation on Atomic Arrangement in Alloys. II', *Proc. Roy. Soc. A* **151**, 231 (1935).

[583]H.A. Bethe, 'On the Annihilation Radiation of Positrons', *Proc. Roy. Soc. A* **150**, 552 (1935).

this is reflected in the Peierls correspondence, as Bethe and Peierls were not only employed in the same institution, but also shared a house while they were at Manchester. As Bethe put it in a letter to his former teacher Sommerfeld, this arrangement was not beneficial for English language acquisition, but very much so for Bethe's general well being and their physics.[584]

While Manchester was an exceptionally successful period in Rudolf Peierls' academic career, both Rudi and Genia had serious concerns about their families. The situation in Russia and in Germany deteriorated, and Genia's sister and parents were exiled in 1935. The reasons were never officially communicated, but as one of Genia's cousins had been implicated in the murder of the head of the Leningrad Secret Police in 1918, the family had always had to expect some repercussions in a regime that was altogether unpredictable.

In the spring of 1935, Rudolf Peierls was offered a position (which he took up later in the year) at the Cambridge Mond Laboratory, the laboratory for magnetism and low temperature physics which had been built for P. Kapitza. After Kapitza's detention in Russia in 1934 which prevented him from returning to Cambridge,[585] the Royal Society was persuaded by Rutherford to use the earmarked unclaimed salary for the establishment of two fellowships, one of which was offered to Peierls, and the other one to J.F. Allen, a Canadian experimentalist.

The two years at Cambridge were, again, a productive period for Rudolf Peierls, partly inspired by old and new colleagues at Cambridge, partly continuing earlier collaborations. Peierls outstanding collaborative abilities, his qualities as a 'scientific tennis player' came to the fore. Among the papers based on his Cambridge contacts was one on supraconductors,[586] which showed traces of Peierls' collaboration with David Shoenberg[587] and a statistical mechanics paper,[588] which was inspired

[584]Bethe to Sommerfeld, 23.12.1933; Eckert, *Sommerfeld Briefe*, II, p. 396.

[585]See letter [193], note 763.

[586]R.E. Peierls, 'Magnetic Transition Curves of Supraconductors', *Proc. Soc. Roy. Lond. A* **155**, 613–28 (1936).

[587]R.E. Peierls, interview with Lillian Hoddeson and Paul Hoch, 13.5. and July 1981, Niels Bohr Library, AIP.

[588]R.E. Peierls, 'Statistical Theory of Adsorption with Interaction between the Absorbed Atoms', *Proc. Camb. Phil. Soc.* **32**, 471–81 (1936).

by earlier work of Ralph Fowler.[589] This paper was another example of work that took its impulse from someone else, here Fowler, and utilised methods first developed by yet another colleague, here Hans Bethe,[590] to solve a physical problem mathematically, or here statistically.

Bethe's move to Cornell in 1935[591] had not put a stop to the collaboration of Bethe and Peierls. Transatlantic and inter-European correspondence during Bethe's summer visit of 1936 demonstrate that the intellectual exchange continued at high frequency. Many of the letters were concerned with nuclear physics, resulted in several publications[592] and fed into many others.[593] Bethe had by then become editor of the American journal *Physical Review* which had advanced to being one of the key journals in the field, particularly for American contributions.

Another paper originating at Cambridge in which Peierls acknowledged Bethe's contribution was his now famous paper on the Ising model.[594] The advent of quantum theory had sparked renewed interest in Ising's 1925 model for ferromagnetism.[595] Heisenberg had replaced Ising's model with one based on exchange forces[596] and Bloch had extended the theory.[597] Ising's model had solved the problem only for the one-dimensional case, and Peierls expanded the model to two dimensions by giving an elementary proof that in two dimensions, the

---

[589]R.H. Fowler, 'Adsorption Isotherms — Critical Conditions', *Proc. Camb. Phil. Soc.* **32** 144–151 (1936).

[590]H.A. Bethe, 'Statistical Theory of Superlattices', *Proc. Roy. Soc. Lond. A* **150**, 552–575 (1935).

[591]In 1935 Hans Bethe was appointed assistant professor at Cornell University.

[592]H.A. Bethe, 'Nuclear Radius and Many-Body Problem', *Phys. Rev.* **50**, 977–79 (1936); H.A. Bethe and M.E. Rose, 'Nuclear Spins and Magnetic Moments in the Hartree Model', *Phys. Rev.* **51**, 205–13, and H.A. Bethe and G. Placzek, 'Resonance Effects in Nuclear Processes', *Phys. Rev.* **51**, 450–84 (1937).

[593]R.E. Peierls, 'Interpretation of Shankland's experiment', *Nature* **137**, 904–5 (1936); R.E. Peierls and P.L. Kapur, 'The dispersion formula for nuclear reactions', *Proc. Roy. Soc. A* **166**, 277–95 (1938).

[594]R.E. Peierls, 'On Ising's Model of Ferromagnetism', *Proc. Camb. Phil. Soc.* **32**, 477–81 (1936).

[595]Ising, 'Report on the Theory of Ferromagnetism', *Z. Phys.* **31**, 253–258 (1925).

[596]W. Heisenberg, 'Zur Theorie des Ferromagnetismus', *Z. Phys.* **49**, 619–636 (1928).

[597]F. Bloch, *Handbuch der Radiologie* VI, 2 (Leizpig), p. 375.

Ising model showed ferromagnetism and concluded that the same held *a fortiori* also for the three-dimensional model.[598]

Throughout his time at Cambridge, Rudolf Peierls also continued his correspondence with Niels Bohr[599] and his regular visits to Bohr's institute at Copenhagen which would eventually lead to more intense collaboration leading to the most-cited unpublished research paper, the Bohr-Placzek-Peierls paper.[600]

Almost as important as the impressive research record for his future scientific career, however, was the intellectual exchange and networking that took place in Cambridge. In a letter to Peierls, Hans Bethe commented on the 'disintegration of the Cavendish[601] Chadwick had moved to Liverpool, Blackett had gone to London, Kapitza had not returned from Russia, and effectively had provided the means for Peierls' appointment at Cambridge in the first place. In 1937, again, it was the move away from the Cavendish of another eminent physicist, Mark Oliphant, that opened new possibilities for Rudolf Peierls. Oliphant had been appointed to the chair of physics at Birmingham in 1936 but had delayed his move until October 1937. In the spring of that year he asked Peierls whether he would be interested in joining him as professor of mathematical physics, a chair not yet established. Peierls' appointment to this chair in 1937 was his first permanent job, and it would indeed be a long-term appointment, as he stayed at Birmingham, apart from war-time interruptions for more than a quarter of a century.

---

[598]See note 575.
[599]Letters [191], [197–199].
[600]See below, pp. 522–23.
[601]Letter [190].

[146] Rudolf Peierls to Arnold Sommerfeld

Zürich, 29.1.1932

Sehr geehrter Herr Geheimrat!
Ich komme heute mit einem grossen Anliegen zu Ihnen: Im Einverständnis mit Pauli habe ich für den Herbst dieses Jahres ein Rockefeller-Stipendium beantragt, mit dem ich ein Semester in Rom und ein Semester in Cambridge arbeiten möchte. Ich möchte Sie daher bitten, ein Empfehlungsschreiben für mich an die Rockefeller Foundation (20 rue de la Baume, Paris) zu schreiben. Nach ihren Vorschriften möchte die Foundation "special reference to the candidate's character and personality, his scholarship and capacity for original work, his relative standing among the other men of his institution, and his standing compared with any past or present fellows of this Foundation who may be known to the recommenders".

Dr. Hanson von der Foundation machte ich darauf aufmerksam, dass ich mit Schwierigkeiten zu rechnen hätte, weil ich eigentlich noch zu jung für ein Stipendium sei. Es wäre besonders liebenswürdig, wenn Sie bei der Abfassung des Schreibens diese Tatsache im Auge behalten wollten.

Bereits in Voraus herzlichen Dank für Ihre Bemühung. Mit besten Grüssen Ihr Ihnen ergebener

Rudolf Peierls

[147] Heinrich Peierls to Rudolf Peierls

Berlin, 17.3.1932

Lieber Rudi.
Ich hoffe Euch wohl zurückgekehrt und danke herzlich für Deinen Brief vom 7. d[iesen] M[onats] aus Zürich sowie Eure Karte aus Paris. Wir würden uns freuen, näheres über Euern offenbar sehr genussreichen Aufenthalt zu hören.[602]

---

[602]Genia and Rudi had been given some money as a wedding gift, which they decided to spend on a trip to Paris in the spring of 1932. See Peierls, *Bird of Passage*, p. 75.

Was den Inhalt Deines Schreibens vom 7. d[iesen] M[onats] betrifft, so werde ich mich demnächst hier als Prophet niederlassen. Ich kann nämlich behaupten, dass ich ganz genau gewusst habe, sogar wörtlich, was Du mir auf meine Einwendungen bezüglich Russland antworten würdest.[603] Ich habe diese Antwort im voraus gewusst und ebenso weiss ich, dass Du auf mein Urteil keinen grossen Wert legen wirst. Ich muss es aber nach jeder Richtung aufrecht erhalten; diese Offenheit bin ich Dir schuldig.

Bitte teile uns rechtzeitig mit, wann Ihr in Berlin sein wollt. Viele Grüsse, auch an Genia

Dein Vater
Viele Grüße Else

[148] Rudolf Peierls to Genia Peierls

Moscow, 2.4.1932

My dear Genia,
I feel very strange, I'm unaccustomed to travel without you and I feel as if I had no limbs. Now that I can write, it means I only have no feet. When are you going to come? I'm now at the Central Telegraph in waiting for the phone connection with Leningrad. I'm going to phone you tomorrow because today I couldn't find out anything definite about your visa in Narcomindel. Only tomorrow I'll have an appointment in the department that is competent in your case. That is why I'll phone you tomorrow, and as to Leningrad, I'm going to do it today to be able to tell you how they are doing.

[603] Rudolf and Genia were planning another trip to Russia in the spring of 1932. Heinrich Peierls advised against this, for two reasons. He was concerned about the political situaion both in Germany and in Russia; and he thought it irresponsible, in the prevailing economic climate, to undertake such an expensive journey without pressing need to do so. See corrrespondence *Peierls Papers*, Supp., A. 103. As evident from the letters below, letters [149–153]. Rudi and Genia did go, despite Heinrich Peierls' advice to the contrary.

My trip was very good, my fellow travelers were some Polish Jews: they were met at various stations by their wives (of course, each one had only one wife) and quite openly on the platform they kissed each other; that was a touching scene. The Germans in the train applauded and arranged bets as to who could kiss best.

A woman, being a 2nd class passenger, went by the 3rd from Wien and with three children. One of them quite charming, 3 years of age. Such a pity you weren't there.

Then, without any troubles, I got a place, the linen, etc. I had supper in the Russian dining car, breakfast too; and in the meantime had a nice sleep. There were many foreigners in the compartment who didn't understand Russian. I had to take on the job of the interpreter — with great pleasure. One of them wasn't met by anybody, so I had to accompany him up to his office. Then I went to the Institute. Everything was all right there, everybody was glad to see me, but they regret you couldn't come. I'm staying with Tamm, who isn't in high spirits now, because his daughter fell ill with malaria that wouldn't cure in spite of quinine injections. Bubi has had influenza; he looks unhealthy and is downcast. Though in general they are doing well.

I continue writing this on the 10th, at Tamm's apartment, as I had no time to write a long letter. Yesterday I went to the visa department and was informed about the routine. At first you produce the inquiry, then they look for the file and only on the second day after this you can have an interview with the official who works on these cases. And so today I called on them, filled in the same questionnaires once again and they promised to give the reply in 4 or 5 days. Then they'll send the decision to Berlin by wire. But they warned me that a refusal might follow, so the probability isn't very high. Still it is higher than 27%, therefore you should stay in Berlin, just as I have told you over the phone.

I regret you spent a whole day waiting for the phone to call. Hadn't we agreed I would call in the evening? It was awfully pleasant to hear your voice. I miss you greatly. How are you doing there? In spite of your low spirits you should entertain yourself. Berlin is an interesting city, and very soon we'll get back to our Swiss neighbourhood in Zurich. How well can you make out the city map, do you feel comfortable at home? Can you visit Nina? I put a lot of questions, but I shall not receive written answers because we'll meet soon either in Leningrad or in Berlin. Tomorrow or the following day I'll go to Leningrad to wait for the answer that is expected by the 15th. If it is positive, I'll give up prolongation and will wait for you in Leningrad: you are to arrive on the 17th or the 18th. But if there is a refusal, I leave for Berlin at once.

Tonight I gave a report for the theorists and tomorrow another report to Schottky is scheduled. They say Ioffe is here and may attend the conference.

Everybody here regrets that you haven't come. They all send greetings to you. Besides the local folks there is Galya who wants to find job here and Tanya Ehrenfest who will have the work for a long time forth. Heitler is going to come soon, which makes the situation rather interesting.

I haven't yet fond out anything about the Ruhemanns,[604] *but I'll do that tomorrow and then wire her if there is a need; otherwise I'll send him a postcard.*

It's now half past 12, and tomorrow at 9 in the morning I'll have to go to the railway station for the ticket; so it's time to go to bed.

Good-bye, my dear, I kiss you affectionately. I have started counting the days till we meet each other. Greetings to everybody, goodnight.
Rudi.

I quite forgot to describe the funniest events. Firstly, the telegram that I sent Tamm from Negoreloye[605] *(To: Moscow, ... str., ..., Tamm. Arrive tomorrow, Peierls) was delivered later than I came. The postman cursed the stingy sender who didn't mention the name of the addressee; the text was: "To: Moscow, ... str., ... There is, arrive tomorrow, pomerlets".*

---

[604]Martin Ruhemann had been invited by Weissberg to take charge of a new low-temperature research facility to develop a nitrogen industry for Narkomtiazprom at Kharkov. Ruhemann and his wife eventually lived through the Ukranian famine of the 1930s and only left in 1938, when his contract was not renewed.

[605]Village near the border, now the Minsk region.

Secondly, Tamm's son, a fresh and flippant fellow, very much like his father, greeted me with "Auf Wiedersehen!" *and then in the next room he said, "You know, a German has come to us; I speak with him in German".*

Today (on the 11th) I was trying to find out about Ruhemann's affairs; no news so far. In Narcomindel they promised to speed up matters, and if any other papers are necessary they write at once to the Consulate in Berlin. I'm going to post a card to Ruhemanns today.

[149] Genia Peierls to Rudolf Peierls

[Leningrad], 29.5.1932

I miss you terribly and love you very much. I see you in my dreams every night, but in the morning I forget everything, all the dreams, and cannot recollect them. How are you doing without me? No letters arrived for only two days, but it seems long, long; and I'm listening whether the postman is coming. The weather has now become quite warm, very warm, it is sunny and windy and the white nights came; but Mother embraces me tight, so I don't go out, though a lot of people visit me. In the evening of the 27th Ambartz with Vera, Abbot and Dumus came to see us. Ambartz has lost weight, become pale and grown wild hair; he is very merry and companionable. His hair is like a crow's nest after a terrible storm and hail. All the hairs point different ways, and are black, of unbelievable length. Vera is a dear, plump and merry girl, but because of an awful cold in the head, unfortunately, she laughs less than usual. She has recently had her hair cut, but Ambartz is against it and demands tresses: she began to grow them obediently. There are obedient wives! Abbot looks gloomy, downcast and depressed, though he assumes an air as if all was well. I was scolding him and at least learned his opinion of those unpleasant events. It turned out that he started to destroy everything, crushed some prestige and provided for Gamov the membership of the Academy — all because Yakov Ilyich put obstacles in Dau's path, discredited him in experimenters' opinion,[606] *All this was*

[606]The episode is recounted by Gorelik and Frenkel, *Matvei Bronstein*, pp. 54–56.

very strange for me, as I had the impression Y. I. always held Dau in high regard, praised him highly and even was afraid of him. Sarah had told me the same. In this turmoil the devil himself would break both his legs, but wouldn't understand anything. At least Abbot's point of view is, in theory, acceptable for me. Dumus has also grown thin; he is nice and merry and doesn't rush at anyone. Jonny flings mud at him, but he accepts this with calm humour. It is pleasant to look at him and listen to him. We were talking all simultaneously and so loudly, that Mother finally had a headache. Yesterday I went to Detskoye to see Olga. There is also fair amount of "tzores"[607] *here. Olga's life in Detskoye is very nice; she even occupies a separate room, has a good meal, the doctors and nurses are fine. But she is bored and, anyway, she is to leave in two weeks as the term comes to an end. It isn't clear what to do next, because the process is still going on with the temperature 37.6 and she increased her weight only by 2 kg in two months. This increase is reflected in her face: she looks much better. Nevertheless her weight (43.5 kg) is a tragic feature. She may not be able to work and will have to apply for occupational disability, which means she would get a small amount of money. So she would have to take on some job. A vicious circle! In Detskoye it is awfully pleasant, everything is bright green; the ponds are like mirrors, beds of flowers — wonderful! It's a pity that you haven't seen these parks and palaces. I would have shown you everything, everything!*

On my way back I dropped in at the Scientists' Rest Home, where Potin and Oksana, Dumus' wife, are staying now. Her life story is rather dramatic: she used to be a stout healthy woman, but after child-birth fell ill with the 2nd stage of tuberculosis and dilation of the heart. Now she got terribly meager, spends her time sitting under the bush, she can neither run, nor walk, misses her daughter (who lives in Leningrad with Oksana's mother — she cannot be taken to this sanatorium, being 6 months old) and gets still thinner with grief. The only good feature is her suntan, as if she had spent a month in the mountains, so she doesn't look too bad. On the train I advocated the good effect of the air draughts

[607] Jewish expression for troubles.

and was nearly eaten up by three sweet ladies; to be exact, such was their wish, but I would get stuck in their throats. I answered all their invectives with a lot of fun: they went silent with amazement. Today I have spent the whole morning fixing the "anti-Semite"', that broke into separate parts, as before. It was a torture that lasted for 2 hours with no good result. I'm ashamed, because it took you 10 minutes to fix it. Disgrace on me, but what can be done?

Bob Kravtzov[608] *came to see me yesterday. I looked at him with some fun because of his lack of scandals. He is as merry as a bird. In the summer he is going to the Pamirs, where he will stay at 4,000 meters height and is going to climb as high as 7,600. He says, though, that perhaps he will not reach this height. I think that after these records it would be funny to climb the hills of Switzerland and we'll go to the seaside. Won't we? Today I plan to kiss with my Bob and uncle Yakov; I didn't let Sergey Ark. To hell with him!*

Here is the most detailed account of my activities. Is it well written? Sometimes I feel that everything was a dream and I saw you only in my dream. Then I get sad, as it happens after a nice dream. Now I'm about to leave for the post-office to post this letter and a congratulation letter to Else and Ostroumova.[609] *I don't know how I should tie it up for it not to break; therefore I'm angry and troubled. Shall I get a letter tomorrow, or at least a postcard? Have you forgotten about me and do you court Mme Solomon, Frau Bretscher or Asya? A wide choice! If you flee with Mme Solomon, wire me: then I will not write in vain! It would be very useful for your education — you'll learn French and in general it's good for your promotion to become* beau fils *of the great man. Therefore you'd better flee with Mme Solomon; all others aren't worth it. I'm a little worried at your words, that "our financial state isn't hopeless". Not hopeless, but bad, isn't it? Very bad? You could in your own manner be mistaken for the better twice, and it is really hopeless. All right, I won't try and guess; when I return we'll think it over.*

[608]T.P. Kravtsov, member of Frenkel's theory team at Leningrad.

[609]Genia sent a birthday letter and a book of the Russian painter Ostroumova to Else Peierls.

I kiss you affectionately, I'm yearning to see and hear you! Well, given our practice, three week is a trifle! How are you feeding yourself without me? Don't worry about money! Greetings to everybody, to all, all. Here everybody send their love and kisses.

Genia

[150] Genia Peierls to Rudolf Peierls

[Leningrad], 1.6.1932

My dear,
my darling, this letter may arrive on the eve of your 25th birthday, so I wish you many happy returns, nice weather (I wish this to myself as well), unexpected fees, the academic fame (I feel flattered to be Frau Professor), and also good business with Rockefeller — as I'm worried about the American grant — I'm afraid he may become bankrupt before we leave for Rome and Cambridge on his money. I'm sending you rather few gifts, but cheer up! I'll bring some remainders with me, and they are very impressive.

I would so much like to kiss you in the morning of the 5th and wake you up, for you to open your right eye and say "m-m-m ..." but, unfortunately, the tele-kiss hasn't yet been invented. It may be even better if I'm away, otherwise your spectacles would be in great danger. Both glasses would be broken and you would become blind. Tomorrow I'm going to send another letter, so you will be heaped up with envelopes and not so lonely. If you miss me still and haven't forgotten about me in the arms of Frau Solomon or Lilly Fenigstein! (R-r-r!)

Now I'm sitting on the sofa and watching the graceful profile of Lucy from Odessa: she saved us in Moscow last year. Yesterday we went to an excellent recital of grand Casadesus.[610] *I had the greatest pleasure, though was a little saddened by your absence. (If you have "forgotten me", you may cross out these lines).*

[610]Robert Casadesus (1901–1972), French-born concert pianist who on his world tour of the early 1930s also played in Leningrad.

Scores of acquaintances rushed at me, everyone said the same thing: that I had become mature — I even began to feel hurt. I went to the concert with Isay, Nina and (our) Bob: Mother had a headache, so her ticket was given to Bob. The Philharmonic has been rebuilt completely; if compared to that of Zürich it is a masterpiece of architecture. It gives you a thrill when you enter (just as when you enter the Notre Dame). Yes, seriously speaking, the Philharmonic became much better and now it has an organ. Casadesus played Beethoven, Schumann, Chopin and Debussy — superbly. He is the best pianist after Schnabel[611] *of those whom I have listened. When we returned home we found Mother, Ernestino and Lucy sitting at the table. Lucy, having taken a bath, was steaming and kept her eyes to 9/10 shut because of tiredness. Oh, yesterday's papers were late and it was while we were at the concert we learned of the foul tricks of Hindenburg.*[612] *How do you like that old scum? (Isay says that* Luder *is the singular, and* Lumpenpack *is the plural). What a swine! And all this at the age of 84 – "the terrible age and terrible hearts!" (it's a useful literary quotation from Pushkin). Only think, 18,000,000 naïve Germans trusted him. I have never liked him. How can it be: a 84-year old General is elected President, there is always the risk he will die or get softening of the brain. To my mind, he entered his second childhood; such playfulness at his age is indecent.*

[handwritten addition from Genia's mother]

My dear Rudi; my best wishes on your birthday, a lot of happiness and good health. Affectionate kisses.

[continued by Genia]

This congratulation was written by Mother who was in a hurry going to the market. There wasn't enough ink to sign, as the "anti-Semite" is out of order. On the 30th Nina and I visited Olya and then met Dau with his sister in the parks. We went for a very long walk. Dau is

---

[611]Artur Schnabel (1882–1951), Austrian-born pianist and music teacher. Between 1925 and 1931 professor of music in Berlin; later emigration to Italy and the US.

[612]Hindenburg had been re-elected as German President in April 1932. On 1st June 1932 he appointed Franz von Papen as Reichskanzler.

melancholic, even deeply melancholic and at first I tried to divert and comfort him; but after that started to scold him for he is a brawler, an impudent and heartless fellow; I called him a crocodile that swallowed a pig. Now I'm terribly sorry for him, I'll phone him straight away and we'll meet again some time later.

Of other entertainments to be mentioned is the ballet on the 10th: "Giselle" and "Carnival" with Vecheslova, but without Ulanova (oh, that's my Jewish luck: the part of Giselle is performed by [?]). Smaller entertainments: visits and visits. Today we are going to see Bob's daughter; tomorrow, somebody else's daughter — oh, no: tomorrow we have Dumus and Ambartz, and only then it will be a daughter or a son or something like that. Nina feels agitation, because yesterday their Institute launched a yacht, a motor-boat and three rowing boats. That's why today we have Northern wind, dust and clouds. Well, my dear, it's time to hurry to the post-office to wrap up the books. Ostroumova[613] *is charming, to my mind; I sent the same to Else.* [614] *I always write full address, but my complaints can be endless. Do you still have your dinner at home? I'm afraid you do it in order to economize and it's a pity. Go to the restaurant with Racuh, he is merry boy. Tell Feenberg*[615] *that I empathize with him and [?]. Why didn't you describe in detail the celebration on the birthday of Mrs. Solomon? Why didn't you send the photos?!! I kiss you affectionately, without taking care of your spectacles. And love you very much,*

Genia

[handwritten addition by Genia's stepfather Isay]
My congratulations! A quarter of a century — that's a serious thing. To live thus far isn't as easy as it seems. Let me remind you in connection with this that Einstein on his 26th year achieved as high position

[613] A.P. Ostroumova-Lebedeva (1871–1955), Russian painter.

[614] See letter [149].

[615] Eugene Feenberg (1906–1977), studied at Austin and gained his Ph.D. from Harvard under E.C. Kemble; then studied in Munich, Zurich, Rome and Leipzig, and Wisconsin where he collaborated with Gregory Breit.

as Newton, and I'd like to advise you to use your age to get the same success. Certainly, I understand how easy it is for me to give a piece of good advice and how difficult it is for you to follow it. You need not be obedient in this case. With love I. M.

[151] Genia Peierls to Rudolf Peierls

[Leningrad], 6.6.1932

My dear, dear, wise, twenty-five-years-old husband,
this morning (at 11 o'clock) your letter was brought to my bed and I'm reading it on all sides and in all directions. You write wonderful letters but why don't you send photos? I can't understand. I sleep now in the study on the sofa, two armchairs and a chair, fenced off by the table and the rocking chair, because Nina and Lucy get up early and they don't want to wake up I. B. for nothing. The window is curtained with Eric's torn oriental robe from Turkestan, which I'd like to take to Zurich. I hope to get it as a gift (isn't it impertinent?). The nights here are now as light as the days, but that doesn't hamper my sleep, so I sleep on and on, and see you in my dreams. Instead of your shoulder I put my head on the arm of the armchair — but I really don't know which is harder! We go to bed, as a rule, late at night, about one o'clock a.m., and in spite of Mother's attempts to do otherwise, we talk in beds in various combinations, so it is noisy in the rooms. Each evening we have guests or I go to visit someone and I curse the good acquaintances and relatives, but there's nothing to be done! One of these days I ate a marvelous pie with cottage cheese and listened to very witty mockery at Minoga. Yesterday all our family went to see Natasha and Anelya. Nina went there straight from her Institute, while the three of us went from home. In this light one never knows what time it is, so all of a sudden we realised that it was 10 o'clock, so we hurried to the tram-stop. Mother has an outstanding way of going by tram: she jumps in while it is still in motion and is ready to jump out at the third stop before the necessary one; she cries, "Be quick! Be quick!" and has a row with drunkards. Such a lively trip! We were coming back quite late — about one o'clock. It was so unbelievably beautiful, that I nearly wept at the

thought that you have never seen Leningrad in a white night. Especially after Switzerland there is an impression of transparency and clarity, every detail of iron-wrought gates and every volute of columns are clearly seen on the background of the green, with rosy edge, sky. The spires are bright with tarnished gold. The Neva is splashing under granite embankments. The lovers are sitting on stone benches and kissing, kissing, without paying any attention to passers-by — and to gray cats, who are surprised at the young people: why are they so diligent?

Oh, my dear, I quite forgot to write that on 5th I got back the wrapping sent to you by post (a book by Ostroumova) with a label: "Additional pay 17 rubles 05 kopeks". It was such a tiresome business posting it that I wasn't even surprised — and didn't pay any money; I'd better bring it to you by myself. It is an unfortunate thing!

Yesterday I was very busy; I visited the officials, as I hadn't known the visa should be registered and hadn't done that up to then. I went to the Hermitage, then to the Intourist office to book the ticket for the ship on the 18th. So I'm going to arrive at Zürich *on the 22nd. I went to Torgsin*[616] *and to Karpovka*[617] *in the evening. I got devilishly tired. Besides I was worried and scared by today's visit to ... the dentist! Are you shocked? I went to my good friend Semyon Moisseyevich who loves me tenderly and he even sighs when switching on the drill machine: he is so sorry for me. He got into my mouth and luckily found only one hole, rather large — as big as* Saint Gottard *tunnel. If there are some more, they are quite small, infinitesimal of 5th order, and they can be filled easy. That is why I'm very cheerful and smile to my sister Nina, who is sitting, her legs crossed, on the windowsill. We are trying to mock Abbot and Dau, who invited me to help them writing a* Revue *for the joined party of two Institutes: physical-chemical and physical-mechanical. We want to represent them as two Ajaxes from "Helene the Beautiful". Yesterday some "significant" events took place. Here came the Americans from the Rockefeller [fund] and suddenly Sarra phoned us (I wasn't in) asking for Abbot, very urgently. Mother started to phone everybody and found him somewhere. As it turned out, he was to call on the American*

[616]Store of goods and curiosities for foreigners.
[617]Small river and its embankment.

in connection with the Rockefeller grant. Abbot hesitated somewhat but went and ... and ... and ... the American wasn't in. Abbot declared proudly he doesn't like caddishness and left. Yakov Ilyich phoned to check the address: it is correct. Then they phoned the American but he had already left. Isn't he an idiot? But now Abbot is glad he isn't obliged to Yakov Ilyich, etc. Oh, the three of them have just gone away. Kit was here, too. They were sitting and laughing. Dau is unshaven, meek and nice, Abbott is thoughtful, Kit enjoys his marital bliss and invites me to look at Ro's iconography — I'll drop in some day. Well, my dear, now I hurry to Sarra and I'm going to drop this letter on the way. I hope you'll get it soon.

In spite of my dissipated social life, I miss you awfully and feel lost, especially in the morning or at night. I kiss you, stroke your head, and take off your glasses. Good-bye, my dear.

G

[152] Genia Peierls to Rudolf Peierls

[Leningrad], 9.6.[19]32

My dear, dear Rudi, I'm now sitting at home at the table. The lamp is lit for the first time, only because it has rained and it is rather "dark", that is, dark for reading and writing without the lamp. I feel very sad as, firstly, I haven't sent the letter today and you'll be disappointed not to find anything in the box and, secondly, my teeth. Yesterday my Herod [the dentist] was drilling my teeth for 45 minutes, he worked with such zeal — I already thought he would drill through my jaw and the drill would get out of chin, but after some 25 minutes he changed the direction from vertical to horizontal. He zoomed and zoomed and then suddenly said, "Well, let's take the cover off the next tooth". When he did that, the next tooth came into view and a hole in it became visible. Danke Schön! *Well, he started to treat that hole and I was shivering with anger and resonance. At last, he finished by putting in a cotton pellet with some disgusting medicine that made me spit for half an hour. And even later I ate the jam without any pleasure. Today when I ate radish*

I felt toothache. There is no pain until I touch these teeth, though, but I'm worried at the thought that the nerve has to be killed — and I suffer awfully! Very unpleasant — all this! Besides I have eaten 15 radishes and a pound of rye bread-and-butter. My belly now is something fantastic. If the process goes on, I'll fly off like a dirigible at night. Yes, there are some pleasant things, too: my recently washed hair, nice little candies brought by Eric, expectance of tomorrow's flight [?], and, as I hope, your letter. Today Nina left her lunch at home, so I had to bring it to her to the Institute and got into heavy rain, hail, lightning and thunder; my mackintosh didn't wet through, but my beret turned into red clay — my feet are still cold now, though they aren't wet. At the Institute I saw a meager but sympathetic-looking rabbit with the profile that was much like yours. They took a blood sample from his ear and he was covered with cotton pieces, looking just like you do after [??]. Yesterday I went to the Scientists' Club and I was amazed at the luxurious sitting rooms, carpets, armchairs, cleanliness and a wonderful view on the Neva. I took a little, but fine snack there and then enjoyed the society of three theorists and $(n+1)^3$ *acquaintances. I met my Gymnasium messmate; as a rule, I meet her once in two years — so she didn't know anything about you and was thrilled mightily. Oh, good Lord, I haven't written to you that I visited Yasha on the 6th. I liked little Vitya who is a rogue and a great gourmand; they also have a nice doggie. There I met Talmud: he had nearly died in Berlin. He had a horrible goiter (*Kropf*) and almost the whole of his thyroid gland had to be ablated; as a result of ether narcosis he caught pneumonia. They even ordered a place in the crematorium for him, but he somehow survived and is now feeling rather well, though he weighed only 49 kg after the operation. Isn't it great? In general, I incessantly listen to someone, see someone, open the door for someone and I begin to confuse people: I address Dau as Abbot, Ambartz as Dau, etc.*

In connection with all these talks of the coming summer I start to deliberate where you and I may go in August. The plans are as follows: the parents spend August in Koktebel (a village in Crimea near Sudak, short of Sevastopol, if you go from Odessa) at a holiday home. As to July, there are no definite plans. Nina wants to spend July in Teberda, in August she comes down to Gagry and stays there with Valya or goes to Odessa. And, my dear, what about us? I have an awful wish to go with you alone, without Bretschers, Solomons, etc. Let us choose some calm place to stay and not travel for the whole month. To my mind, we should settle somewhere, for we'll have to go to the conference in September. We'll go to Berlin and by the 1st of October we'll be in Rome. Without having a rest in August, we'll only get tired and have no pleasure. Is that correct? I yearn for the sea with its tides and ebbs, sands and [?], but without people (I mean without tourists). I look forward to some small hamlet in Brittany with the single inn and many heavy fishing boats, the odour of the sea — a mixture of tar, fish and salty wind. Would you like to go to Brittany? In August they still have nice weather there, and it isn't expensive. The Swiss mountains are unpleasant because everything is terribly expensive there with lots of tourists. Even near this hole of St. Anthönien *there are 3 boarding houses! If you really prefer mountains we may choose some little-known place in Austria or Germany, some quiet village. Over there they at least won't charge too much. Well, all this is* lange Geschichte*; before that I have to go to you and not to die on board the ship of seasickness! Well, good night, my dear, Lucy has just come and must go to bed. I again sleep in (I nearly wrote "the nursery"!) our room, as Nina has an excellent hotel room and may come whenever she wants. Poor Lucy has to get up at 7 every morning! Of course, I don't hear anything and never open my eyes.*

Continued at 13.30, July 10. I have been to the Foreigners' office where I had to queue. In the morning I received your letter, so thick, so nice, so detailed, and with pictures.

My poor dear, I'm so sorry that on your birthday it rained, and your only entertainment was Harry Peel. What a shame! You'd better go to [?]. No other news except your letter. You cried when Feenberg *went away, didn't you? Why can't you do anything concerning Asya and Racuh? And why should the girl marry? Schneider is wretch — meet him just with a cold greeting. Pauli is a dear, but why did he go off his mind again? He had got cool once, hadn't he? Oh, there is news, but don't gossip about it: Gladis, with whom Max was in love in Bristol, has married an Egyptian student. Max wants (maybe not for this reason) to come to Leningrad with 3 British physicists and mathematicians in a car to travel across Russia. To my mind this is sheer nonsense; the distances here are fantastic, compared with the European scale; the landscapes are monotonous, and they never get to the Caucasus. Well, that is all, definitely. I kiss you affectionately, long and tender. I wish you to be in good health and cheerful; I'll come soon. Why do you eat potatoes every day? And oranges that aren't so good now? Mother taught me to fry beef steaks and veal cutlets. I kiss you once more. Thank you for your letters, you are a clear head and I love you immensely. Regards and kisses from everybody. Send my greetings to all, all.*

Genia

[153] Genia Peierls to Rudolf Peierls

[Leningrad], without date
(probably mid-June 1932)

My dear Rudi,
my "dictator", I have just had your letter, so I'm very happy that you have good weather in Zurich; probably, you bathe already; and my next letter will be the last one, and we'll see each other very soon. Well, first of all to business matters. I have been to the Intourist; our old man is no longer there, a young boy is walking around instead, very likeable and very inapt. We extracted all timetables and all tickets, and we found out that now they can provide for only one route: Stettin — Berlin — Frankfurt — Eisenach — Basel — Zürich. I think I'll go this way. As a matter of fact, any other route is impossible, as they can offer tickets to book only for Berlin — Basel, *while other witty combinations like* Nürnberg — Friedrichshafen *are absent. But I wanted to go via this route anyway, and I booked the ticket. Let us agree on this: I leave* Berlin *at 22.16 and get to Zürich at 16.11. Here is the alternative: if the boat is not late and I arrive in* Berlin *at 10 o'clock, I'll have to run in haste with my tongue hanging out, produce a lot of trouble for* Anlauf *and the parents, and board the train at the hottest and the dustiest hour. In* Basel *I'll have to change at 6 in the morning; in addition, there will be only 30 minutes for customs and to go over to another platform —*

you can understand how it is pressing at 6 o'clock. You'll come to meet me early in the morning, without breakfast and badly shaven; you'll be hungry, bristled and angry, and you'll be yawning so wide as to swallow — if not me, then the grammophone. And taking into consideration 836 hours, 8 extra hours compared to those is a trifle. So, to put it in a definite way, I'll arrive on 22nd, at 16.11 at Hauptbahnhof. Is that all right, yes? Oh, I quite forgot, the 22nd is Wednesday, the day of the colloquium, but you can play truant once or, if you don't want to, you can nevertheless be in time having one hour. These are the most important things.

These days we had wonderful weather; Abbot and I took walks in the Summer Garden. Three days ago I walked downtown to buy some gifts. I bought 4 boxes like those you had bought in Moscow (one of them looks just the same, and three other ones are jollier), 2 small chairs with the seat that can be raised and on the inside there is a box, and a round box. I'll fill them with sweets and present them to Ilse and Vera. The round box is for [?], she'll be very glad. I don't know what to do with the fourth one. Shall I give it to Frieda or to the Bretchers? Well, what is the use of question marks, if you cannot answer this letter! I had an amiable letter from Berlin. Else liked Ostroumova very much, and in general a lot of amiable words. They expect I wire them about my arrival, but without knowing the exact time it is stupid to wire, so I'll write a postcard today. I haven't bought a present for Nina, but I think I'll find something — may be some books.

I'm going to have little luggage, so I won't hire any porters if I haven't much money. In Berlin there are 10 to 15 marks that belong to me, so over there I'll be provided for. You can see how clever your wife is and how well she can arrange everything! I know the timetable beforehand! I don't loose luggage tickets! I always know where the tickets are kept! So I can be the dictator on our next trip. And where shall we go? Let's go to Bretagne! Yesterday Natasha, her husband and Leonid Isaakovich came to see us. He is going to the conference in Paris at the end of June; so he will spend July and some days in August abroad, though he doesn't know where exactly — may be in Berlin. Natasha's husband is very likeable, though looking like a young boy: he is 24 but looks like 20–21. He is fair-haired — even his eyebrows are white, red-cheeked, blue-

eyed, bashful and blushful, he laughs nicely. But he is silent like a log, only answers the questions, and never tries to strike up a conversation himself. The two of them look like a flea with a polar bear! Today I'm going to visit Yasha, then there are a lot of small things to be done. Tomorrow I'll bathe, the day after tomorrow get the ticket and good-bye! This time it will be easier to go, as this is physically sensible — to see you! Last time it looked rather senseless: why go when it is so good to stay? Still, I'll feel sad. What did you do on Sunday? How is Jacob [?] ? What does this scum of Solomons think about the car? Oh, again these senseless questions. I haven't seen the doctor because I'm feeling excellent, my temperature is normal, I wouldn't know what to tell him. To complain that I lost 14 kg in 6 months and had 37 for two weeks — that would be silly. Thank you for the picture. I look like a camel in an oasis, while Asya looks like a primeval sculpture with glasses. Still they are neat, these pictures. My dear, why do you eat potatoes every day? This is bad for the stomach and in 8 years you'll look like Pauli: "the front is where the tie is". With your family strain you must not eat many potatoes! Are you feeling tired reading this letter? I have an especially disgusting handwriting today. Pardon me, my dear. In general I'm too lazy to be diligent and too lazy to write a letter that will be delivered two days before my arrival. My affectionate kisses! Soon I'll be kissing you in real earnest (shave yourself!) and I'll break your spectacles. I wonder, what part of me will get to Zürich. All of a sudden the rolling and sea-sickness and [time?]-effect. For 3 days I wouldn't be able to eat or drink, only feel sick. I'd like to try my resistance to rolling, or there in Georgia that was the influence of grapes. And now we have been acquainted for 2 years. 2 years! It seems to be much less, as this half-year overshadowed the whole romance in letters. Is that right? Mein Herr Gemahl, *why didn't you send me any letters in German? Let us speak only German, when we meet we say to each other,* "Guten Tag" *– and so on. Is that all right?* Aufwiedersehen! Ich küsse Dich Tausend Mal. *Everybody send their regards and kisses.*

Genia

[154] Wolfgang Pauli to Rudolf Peierls

Zürich, 24.10.1932

Lieber Peierls!

Dank für Ihren mir viel zu komplizierten Brief.[618] Ich selbst habe Überlegungen in anderer Richtung angestellt betreffend die Feldmessung, vielleicht können Sie nun beides zusammen bringen.[619] Ich betrachte nicht die Feldmessung, sondern die einmalige Impulsmessung während der Zeit t (abgebrochene Wellenzüge). Es interessiert uns der Fall, wo nicht nur gewöhnliche Comptonprozesse mit merklicher Wahrscheinlichkeit auftreten, sondern solche höheren Prozesse, wo etwa (im Mittel) N Lichtquanten beteiligt sind. Dann ist mir ungeheuer plausibel, daß man hier, selbst wenn man Energie und Impuls der gestreuten Quanten genau messen würde, auf die Energiedifferenz des Teilchens vor und nach dem Prozeß nur mit der Genauigkeit

$$\Delta(E - E') > \frac{h}{\Delta t} \cdot N \tag{1}$$

zurückschließen kann. Plausibilitätsbegründung: Infolge des Abhackens der Wellenzüge wissen wir über das Vorhandensein von Quanten $h\nu'$ mit $\nu' < \frac{1}{\Delta t}$ im gestreuten Licht nichts.

Es kann nun passieren, daß N solche Quanten mehr oder weniger gestreut werden. Das ist nicht ganz streng.

Wenn (1) wahr ist, sind wir fertig, dann gilt

$$(v - v')\Delta P > \frac{h}{\Delta t} \cdot N \tag{2}$$

(wenn $v = v' = c$); eben [so] $N \cdot \frac{h}{\Delta t}$ = ausgestr[ahlte] Energie $< \frac{e^2}{c^3} \frac{(v' - v)^2}{\Delta t}$.

[618]Peierls spent the autumn semester 1932 in Rome, working with Fermi as part of a Rockefeller fellowship. Hence he communicated with his mentor Pauli by letter.

[619]W. Pauli, 'Die allgemeinen Prinzipien der Wellenmechanik', in *Handbuch der Physik*, ed. by Geiger and Scheel, 2nd edn., vol. 24, part 1, pp. 257ff. See letter [99].

Das, was ich vermute, aber noch nicht ganz richtig begründen kann, ist also: Selbst wenn ich die einfallenden und gestreuten Photonen genau messe, ist, falls Prozesse mit N Photonen mit merklicher Wahrscheinlichkeit möglich sind, die Ungenauigkeit, mit der aus der Messung die Impulsänderung der Teilchen zurückgeschlossen werden kann, bei gegebenen Δt um den Faktor N heraufgesetzt. Über diese These müssen wir also diskutieren.

Eine zweite Frage, die mir noch nicht ganz klar ist, ist diese: wie können die eventuell in dem zu messenden Feld vorhandenen Photonen, über die ich ja noch gar nichts weiß, von den bei der Impulsmessung einfallenden und gestreuten Photonen unterschieden werden?

Haben Sie Ehrenfest's "Erkundigungsfragen" in der Zeitschrift für Physik gelesen?[620] Ich habe ihm einen mit etlichen Bosheiten versehenen langen Antwortbrief geschrieben.[621] Finden Sie nicht, man sollte auch in der N[ature] darauf antworten? Natürlich könnte ich aber nichts sagen, was Sie und die kompetenten Gelehrten nicht sowieso wissen. Hier sind ein bischen wenig Leute in diesem Semester.

Viele Grüße

Ihr W. Pauli

[155] Rudolf Peierls to Paul Ehrenfest

Rom, 26.10.1932
(carbon copy)

Lieber Herr Ehrenfest!

Ich habe mit grossem Interesse Ihre "Erkundigungsfragen" gelesen,[622] die mir von keinem Standpunkt aus sinnlos erschienen, sondern im Gegenteil sehr interessant. Ich glaube, einen Teil von ihnen ziemlich

---

[620] P. Ehrenfest, 'Einige die Quantenmechanik betreffenden Erkundigungsfragen', *Z. Phys.* **74**, 555–59 (1932).

[621] In fact, Pauli's 'long letter' was only the beginning of an even longer exchange of letters between Pauli and Ehrenfest in the winter of 1932. See Pauli, *Wissenschaftlicher Briefwechsel*, II, letters 294, 295, 296, 299, 300, 301, 302, 303.

[622] See letter [154], note 620.

genau beantworten zu können, ob diese Antworten aber so klar sind, wie Sie es wünschen, werden Sie selbst sehen.

1. Eine axiomatische Klärung dieser Frage hängt natürlich davon ab, was man bereits als bekannt voraussetzen will. Ich möchte mich einmal auf den etwas bequemen Standpunkt stellen, dass man schon weiss, dass die Wellengleichungen linear sind, wie Sie das auch zu tun scheinen

Eine weitere interessante Frage ist dann die Begründung dieses Axioms, die nicht sehr einfach ist, trotz der anscheinenden Trivialität. Man kann zwar viele Gründe für die Linearität angeben, aber keiner läßt eine befriedigend einfache und strenge Ableitung zu.

Setzt man die Linearität schon voraus, so ist es natürlich evident, dass die Wellengleichung nicht eine Gleichung für eine reelle Funktion von erster Ordnung in der Zeit sein kann, denn sonst wäre sie gar nicht vom Typus einer Wellengleichung sondern der Wärmeleitungsgleichung, und es ist sehr leicht, sich genau davon zu überzeugen, dass man auf diese Weise nicht den Anschluss an die klassische Mechanik bekommen kann. (Die Gleichung $\frac{\partial u}{\partial t} = \alpha \frac{\partial a}{\partial x}$, die Wellen als Lösungen hat, weil dort alle Wellen in einer Richtung laufen). Anschaulich liegt dies daran, dass man in der klassischen Mechanik zwei Grössen — Ort und Impuls — zur Beschreibung nötig hat. Aus diesem Grunde muss man nämlich aus der Wellenfunktion zwei Wahrscheinlichkeitsfunktionen, für Ort und Impuls, berechnen können. Gäbe es nur eine reelle Wellenfunktion, so wären diese beiden Wahrscheinlichkeiten stärker gekoppelt, als es die Korrespondenz zulässt.

Also muss die Wellengleichung mindestens aus zwei simultanen reellen Gleichungen erster Ordnung in t bestehen. (Der Fall einer Gleichung zweiter Ordnung lässt sich ja durch eine triviale Substitution hierauf zurückführen).

Nun bleibt aber noch die Frage, warum diese Geichungen gerade besonders bequem werden, wenn man sie in eine komplexe Gleichung zusammenfasst. Diese Frage ist natürlich weniger physikalisch als formal (aber deswegen keineswegs etwa sinnlos) und lässt daher viele Antworten zu. Der wesentlichste Grund scheint mir in folgendem zu liegen. Solange man es mit einem Elektron zu tun hat, macht es keinen merklichen Unterschied für die Bequemlichkeit, mit der man die Gleichung behandeln kann, ob man sie als eine komplexe oder zwei reelle betrachtet. Geht man aber zur Behandlung eines Mehrkörpersystems über, so hat man erstens zu fordern, dass sich die Wahrscheinlichkeiten als Produkte der einzelnen Wahrscheinlichkeiten ergeben (solange die Teilchen ungekoppelt sind). Ausserdem sollen sich die Energien addieren, was man bei Beibehaltung der Forderung $E = h$ für das Gesamtsystem auch so formulieren kann, dass die Frequenz des Gesamtsystems die Summe der Teilchenfrequenzen ist. Diese beiden Forderungen lassen sich in einfacher und übersichtlicher Weise dadurch erfüllen, dass man die Wellenfunktion als Produkt ansetzt, was aber nur im Falle einer komplexen Funktion möglich ist. Spaltet man dagegen in Real- und Imaginärteil, so werden die Zusammenhänge sehr verwickelt. Ich würde also sagen: Die komplexe Schreibweise verdient deshalb den Vorzug, weil sie die Zusammensetzung mehrerer Wellenfunktionen für die einzelnen Teilchen einer Funktion für das Gesamtsystem bequemer erlaubt.

Dass dies richtig ist, sieht man insbesondere aus der Analogie zu den Maxwellschen Gleichungen. An sich ist in diesen die komplexe Schreibweise mit der reellen völlig gleichberechtigt, auch wenn man sie nicht als Differentialgleichungen für die beobachtbaren Feldstärken, sondern als Wellenfunktion eines Lichtquants auffasst. Geht man aber zu mehreren Lichtquanten über, oder zur Wechselwirkung mit der Materie, so ergibt nur die komplexe Schreibweise einfache Formeln.

Dies führt aber eigentlich schon auf die zweite Frage.

B. Wie weit reicht die Analogie zwischen Materie und Licht? Ich glaube, dass die Grenzen dieser Analogie, die Sie anführen, reell sind. Der Wellenfunktion der Materie wird nämlich niemals eine beobachtbare Grösse entsprechen, aus zwei Gründen, die einen fundamentalen Unterschied zwischen Licht und Materie ausmachen.

Erstens ist es wesentlich, dass die materiellen Teilchen der Fermi-, Photonen aber auch Bosestatistik genügen. Dies hat folgende Konsequenz: Auch für Teilchen mit Fermistatistik kann man zwar, wie aus den gelehrten Arbeiten von Jordan, Wigner,[623] etc. folgt, formal einen Operator einer Gesamtwellenfunktion im dreidimensionalen Raum

[623] P. Jordan and E. P. Wigner, 'The Pauli Exclusion Principle', *Z. Physik* **47**, 631–51 (1928).

definieren. Für diese Funktion gelten jedoch andere Vertauschungsrelationen, insbesondere z.B. für $P \neq P'$: $\phi(P)\phi^+(P') + \phi^+\phi(P) = 0$ D.h. auch wenn man zum klassischen Grenzfall übergeht, $(h = 0)$ werden die Wellenfunktionen an verschiedenen Punkten nicht miteinander vertauschbar, entsprechen also keinen klassisch messbaren Grössen. Anschaulich, wenn auch sehr grob, habe ich diesen Sachverhalt einmal so zu formulieren versucht: Damit das Feld beobachtbar wird, müssen sich die Phasen der einzelnen Partikeln irgendwie so koppeln, dass insgesamt eine definierbare Phase entsteht. In der Fermistatistik müssen aber die Teilchen immer verschiedene Zustände haben, und können sich infolgedessen nicht in der richtigen Weise zu einer Welle zusammensetzen. Aber es ist sehr Geschmackssache, wieweit man auf solche sehr groben Veranschaulichungen Wert legt.

Zweitens steht die Messung der Feldstärke in einem ausschliessenden Verhältnis zur Messung der Teilchenzahl (wie man entweder durch die Betrachtung der entsprechenden Operatoren in der Diracschen Strahlungstheorie, oder durch Diskussion der Messmöglichkeiten beweisen kann — deswegen kann übrigens das Feld eines Photons niemals gemessen werden — und man kann die Feldstärke nur mit Hilfe von solchen Prozessen messen, bei denen sich die Protonenzahl mit grosser Wahrscheinlichkeit um mindestens 1 ändert. Da aber solche Prozesse für materielle Teilchen bisher unbekannt sind, so folgt, dass man ihnen kein messbares Feld zuordnen kann. Andererseits kann man auch zeigen, dass für hypothetische materielle Teilchen, die der Bosestatistik genügen würden und für die Prozesse mit Erzeugung und Vernichtung von Teilchen hervorgerufen werden können, eine beobachtbare Wellenfunktion definierbar wäre.

Für die wirklichen materiellen Teilchen ist aber infolge dieser beiden Eigenschaften (von denen allerdings nur die erste bewiesen, die zweite u.U. nur näherungsweise richtig sein kann) die Analogie mit dem Licht durchbrochen.

Umgekehrt fragen Sie nun: Inwieweit lassen sich die Wellenfunktion eines Lichtquants in Analogie zur Materie als Aufenthaltswahrscheinlichkeit interpretieren? Auch hier ist m.E. die Analogie durchbrochen, denn eine solche Interpretation ist auf keine Weise möglich. Physikalisch deswegen nicht, weil es gar keine Messung gibt, die sich sinnvoll als

Ortsmessung eines Photons interpretieren lässt, und die das Photon genauer als bis auf die Wellenlänge (die es schon vorher hatte) festlegt. Das liegt daran, wie Landau und ich in unserer vielbeschimpften Arbeit[624] des näheren diskutiert haben, dass die Messungen an Photonen immer Zeit brauchen, mindestens eine Periode des — ungefähr bekannten Anfangszustandes des Photons, sodass der Zeitpunkt, auf den sich die Messung bezieht, um diesen Betrag unbestimmt und daher der Ort zu einem gegebenen Zeitpunkt um die Wellenlänge unbestimmt ist.

Das heisst also, man kann vom Ort des Lichtquants noch in dem Sinne reden, dass man sagt: In einem der geometrischen Optik genügenden Paket, bei dem also sowohl $\mathfrak{F}$ wir $\mathfrak{G}$ ausserhalb eines bestimmten Bereichs veschwinden, befindet sich das Photon sicher im Innern dieses Bereiches. Dagegen ist es nicht möglich, ein Experiment anzugeben, das einem bestimmt, wo sich nun das Photon innerhalb des Pakets befindet. (Während dies in der Wellenmechanik des Elekrons gerade das Experiment ist, dessen Wahrscheinlichkeit durch $\phi\phi^+$ gemessen wird).

Mathematisch äussert sich dies in den Scheusslichkeiten auf die man im Konfigurationsraum der Lichtquanten stösst (denn wenn der Lichtquantenort keinen Sinn hat, so ist der Konfigurationsraum natürlich keine sehr adäquate Schreibweise) und die — darin haben Sie ganz recht — noch viel schlimmer sind als eine nicht definierte Dichte. (Die Ihnen unangenehme Tatsache, dass die Dichte durch einen Integraloperator definiert wird, lässt sich dadurch umgehen, dass man eine neue Funktion $g = \frac{1}{\sqrt[4]{\delta}} f$ einführt und gg^* als Dichte definiert. Diese Dichte ist definiert, und normiert, aber transformiert sich falsch bei Lorentztransformationen, wie jede andere aus F und G quadratisch gebildete Grösse).

In einem tieferen Sinne ist dies übrigens kein Versagen der Analogie, denn auch bei materiellen Teilchen trifft man auf dieselben

[624]See letter [93], note 380. When the paper was still in draft, Landau and Peierls discussed it with Bohr who strongly objected to the results and later wrote his own theory of the problem with Rosenfeld. Bohr, N. and Rosenfeld, L., 'Zur Frage der Messbarkeit der elektromagnetischen Feldgrössen', *Mat.-fys.Medd.Dan. Vid.Selsk* **12**, 8 (1933). An English version is found in R.S. Cohen and J. Stachel (eds.), *Selected Writings of L. Rosenfeld*, Boston: Reidel, 1979, p. 357.

Schwierigkeiten, sobald man den Ort sehr genau messen will. Nur liegt hier die Grenze der Ortsmessung weiter als für ein Photon, für langsame Elektronen unterscheidet sie sich von der Wellenlänge um einen Faktor v/c, sodass es dort durchaus noch einen Sinn hat, von einer Ortsmessung innerhalb der Wellenlänge zu reden; für sehr schnell bewegte Elektronen wird aber auch hier die Grenze mit der Wellenlänge identisch. Der einzige Unterschied besteht also darin, dass man für materielle Teilchen eine nichtrelativistische Näherung machen kann, während Photonen sich eben immer mit Lichtgeschwindigkeit bewegen.

Man kann die Grenzen der Analogie also auf folgendes Schema bringen:

Grenzfall:	Elektronen:	Photonen
Grosse Teilchenzahl	—	Feld messbar
Banddimension gross wegen Wellenlänge	Bahnbegriff	Bahnbegriff
$v \ll c$	Ort oder Impuls messbar	—
$v \sim c$, kräftefrei	Nur Impuls messbar	Nur Impuls messbar

Das einzige, was unter allen Umständen sinnvoll bleibt, ist also der Impuls eines freien Teilchens, und hieraus bezieht sich die Analogie allein in Strenge. Oder, wie dies Pauli einmal formulierte: Die Analogie wird vollständig durch die gleichmässige Gültigkeit der Relationen

$$E = h\nu\,, \qquad p = \frac{h}{\lambda}$$

für Photon und Materieteilchen zum Ausdruck gebracht.

Zu Ihrer Bemerkung 3 ist noch zu sagen, dass die Arbeiten über Zuordnung von Maxwellschen und Diracschen Gleichungen fast ausnahmslos von vorneherein dadurch als Pathologie charakterisiert werden, dass sie eine lineare Zuordnung zwischen Vektoren und Spinoren einführen.

Für die Fragen unter C fühle ich mich zu wenig kompetent. Ich wäre Ihnen sehr dankbar, wenn Sie mir mitteilen, wie weit Ihnen diese Antworten verständlich, wie weit befriedigend erscheinen und an welchen der Kürze halber mit "Es ist leicht zu sehen..." verschlampten Stellen Sie noch detailliertere Auskunft haben wollen.

Wenn ich weiss, auf welche Punkte in dieser Diskussion Sie besonderen Wert legen, habe ich übrigens die Absicht, diese Bemerkungen in etwas ausgefeilterer Form an die Zeitschrift zu schicken, sofern dort nicht inzwischen schon von berufenerer Seite eine Antwort veröffentlicht wird.

Mit vielen Grüssen
Ihr Ihnen ergebener

[R.E. Peierls]

[156] Paul Ehrenfest to Rudolf Peierls

[without place], 31.10.1932

Lieber Herr Peierls!
Heute morgens empfing ich zu meiner grossen Freude Ihren freundlichen, interessanten Brief.[625] Ich beeile mich, darauf mit einigen Bemerkungen zu reagieren.

Vor allem: ich glaube, dass es sehr nützlich sein wird, wenn Sie (und hoffentlich auch noch andere, die ueber diese Dinge nachgedacht haben) kurz und uebersichtlich Ihre Bemerkungen publizieren wollten. Denn ich weiss aus laengerer Erfahrung, dass meine verschiedenen Bekannten durchaus nicht identisch ueber diese (didactisch sehr wichtigen!!) Fragen denken. Und oft sind verhaeltnismaessig feine Differenzen in der Auffassung hier BESONDERS LEHRREICH! — Und es wird mir ein Stein vom Herzen genommen, wenn ich meine Studenten auf solche KURZE, OFFENHERZIGE Publicationen werde verweisen koennen. Und ich HOFFE gerade auf feine Differenzen. Denn nichts ermutigt die Studenten mehr, als dann durch eigenes Urteil zwischen diesen Differenzen waehlen zu koennen. Das IST doch so!

Und nun hintereinander Einzelbemerkungen zu den aufeinanderfolgenden Darlegungen Ihres Briefes.

---

[625] Letter [155].

1. Geben Sie doch wenigstens einige aphoristische Bemerkungen betreffs der LINEARITAET der Fundamentalgleichungen. Lassen Sie dann den Versuch einer STRENGEN Begruendung ruhig beiseite.

2. "Anschaulich liegt dies darin - - - als es die Correspondenz zulaesst." Diese Andeutungen waren fuer mich ALL zu kurz. Vielleicht wuerde ich sie verstehen, wenn dieser Passus dreimal laenger waere. Ich kann natuerlich herumraten, was Sie so etwa meinen koennten. Aber das ist ein Unangenehmes, aergerliches Geschaeft. Seien Sie also in der Publication ausfuehrlicher und senden Sie mir, bitte, eine Copie des Manuscriptes.

Auch ist mir traurig, dass Sie (und wahrscheinlich werden das alle ausser Schroedinger und de Broglie and Einstein thun!) "natuerlich" die "Wellen" schon als "Wahrscheinlichkeitswellen" ansehen. Nun in Gottesnamen werde ich nun meinen Studenten einblaeuen: Es ist UNANSTAENDIG — oder was noch viel viel schrecklicher klingt: es ist wissenschaftlich GESCHMACKLOS heutzutage noch zu fragen: warum unerbittlich "Wahrscheinlichkeits-Wellen"? (Also DARAUF brauchen Sie nicht weiter einzugehen). Ich kenne die Argumente.

3. Hat man schon, etwa auf Grund Ihrer Argumente zugegeben, dass zwei Scalare notwending sind, so FOLGT daraus doch noch nicht, dass sie in eine complexe Loesung EINER Gleichung zusammengefasst werden koennen. Sondern das erfordert doch eine bestimmte Einschraenkung in der Structur der Simultangleichungen, denen die zwei reellen Gleichungen genuegen!! (Also z.B. in den Maxwell-Gleichungen kommt die feine Nuance des Vorzeichenunterschiedes der beiden Systeme in Betracht). Dieses Element haette mich noch interessiert, nicht aber das, was Sie ueber die grosse formale Bequemlichkeit der complexen Zusammenfassung sagen.

4. Ich bedauere, dass ich nirgendwo eine von der Schroedingergleichung und der Transformationstheorie moeglichst unabhaengige (!) und umgekeht an Bohrs Correspondenztheorie anschliessende Besprechung der Rolle von i in der Heisenberg-Bornschen Matrizentheorie finden kann. Ich weiss, wie sehr das DIDACTISCH wichtig waere fuer meine Studenten!!!! (Was mit darueber meine Bekannten hochnaesig sagten, waren lauter verschiedene brummige Bemerkungen, von denen ich fast nichts klar verstehen konnte. Ich weiss also nicht, ob einer davon es selber klar versteht.

5. In dem, was Sie ueber die Messung der Feldstaerken sagten, war mir Vieles neu und sehr lehrreich. Ganz besonders war mir unbekannt, dass hier der Gegensatz zwischen Fermi- und Bosestatistik eine wesentliche Rolle spielt. Und ebenso, dass die Moeglichkeit der Vernichtung eines materiellen Theilchens eine Voraussetzung fuer die Messbarkeit des Feldes der Gesamtwellenfunktion waere. Diese und aehnliche doch im Grunde GANZ einfache Bemerkungen sind fuer Leute wie ich einer bin in der Litteratur (selbst in den herablassenden "zusammenfassenden Berichten") einfach schon voellig unzugaenglich gemacht worden. Die Misthaufen von Gelehrtheit, die darueber ausgeschuettet sind (das Charakteristikum von Stagnationsperioden!!!!) — Also KURZE EHRLICHE (!!!) Formulierung solcher Dinge werde ich als Segen für meine Studenten ansehen. Besonders falls sie UEBERDIES auch noch RICHTIGE Folgerungen aus den Grundauffassungen sein sollten. Sind sie falsch, nun dann muss freundschaftlich-ehrlich darueber discutiert werden. Wenn man mich nur schimpfen laesst, wird sich schon klaeren, was.[626]
Also besten Dank, [???]

Paul Ehrenfest

[157] Rudolf Peierls to Paul Ehrenfest

Rom, 10.11.1932
(carbon copy)

Lieber Herr Ehrenfest!
Vielen Dank für Ihren Brief, zu dessen Beantwortung ich wegen einer kleinen Erkrankung leider erst heute komme. Ich habe mich bemüht, die einzelnen Punkte, die Sie angeben, noch ein bisschen klarer zu überlegen, und möchte der Reihe nach auf sie eingehen, wobei ich zunächst nur die Linearitätsfrage übergehen will.

---

[626]The last few lines of the letter have been torn off and make the ending illegible.

2. Mit der anschaulichen Begründung meinte ich, dass es eine andere Ableitung für die Notwendigkeit von mindestens zwei reellen Funktionen gibt, die physikalisch noch durchsichtiger ist, die aber eine logische Lücke enthält, die sich nur durch ungeheuer gelehrte Betrachtungen überbrücken liesse, sodass ich diese Ueberlegung nicht für die Ableitung wählte.

Man ist nämlich versucht, folgendermassen zu schliessen: Aus der Wellenfunktion muss ja die Dichte und die Impulsverteilung folgen. Wäre nun die Wellenfunktion ein reeller Skalar, so würde sie bereits durch die Dichte im wesentlichen — d.h. bis auf gewisse endliche Mehrdeutigkeiten — bestimmt sein. Die Impulsverteilung wäre also schon durch die Dichte bestimmt. Soweit ist die Ueberlegung richtig. Jetzt möchte man aber weiter schliessen: Das geht nicht, denn ich weiss doch aus der Korrepondenz, dass man z.B. zwei Wellenpakete bauen kann, die zu Anfang am selben Ort sind und dieselbe Ausdehnung haben, also ungefähr die gleiche Dichte, andererseits aber ganz anders laufen, weil sie zu verschiedenen mittleren Impulsen gehören.

Aber hier liegt ein Haken verborgen, denn aus der Korrespondenz, d.h. aus makroskopischen Messungen, kann man natürlich nur die "Grobstruktur" der Dichte kennen, nicht aber, ob sie nicht noch feine Schwankungen auf Gebieten von der Dimension einer Wellenlänge hat. Diese Freiheit könnte aber schon ausreichen, um die Verschiedenheit von zwei derartigen Wellenpaketen zu erklären.

In Wirklichkeit kann man sich in einem Spezialfall davon überzeugen, dass sie hierzu nicht ausreicht. Man denke etwa daran, dass man das Quadrat des Realteils der Schrödingerschen Funktion als Dichte erklärt. Dann kann man wirklich aus der Dichte diesen Realteil (bis auf ein uninteressantes Vorzeichen) ermitteln, und aus einer Fourierzerlegung die einzelnen Werte des Impuls voneinander trennen; man kann jedoch nicht die Komponenten, die zu einem bestimmten Impuls gehören, von denen mit dem entgegengesetzen Impuls — d.h. die nach rechts laufenden von den nach links laufenden Wellen —trennen. Hierzu braucht man vielmehr unbedingt den Imaginärteil der Wellenfunktion (oder die zeitliche Ableitung des Realteils).

Dieses Beispiel scheint mir einerseits zu zeigen, dass die Zahl 2 der Wellenfunktion wirklich in einem Zusammenhang mit der Zahl 2 der in der klassischen Theorie notwendigen Vektoren steht. Aber andererseits zeigt es auch, dass es sehr kompliziert sein dürfte, nicht nur diesen Spezialfall auszuschliessen, sondern ein strenges Argument gegen die Möglichkeit zu finden, dass feine Schwingungen der Dichte die Impulsverteilung bestimmen.

3. Hier war wirklich noch eine Lücke in meinen Ueberlegungen, die sich aber ausfüllen lässt, mindestens für den Fall freier Elektronen. Man braucht hierzu folgende Forderungen: Es soll eine Dichte existieren, die positiv definit ist; für diese Dichte soll ein Erhaltungssatz gelten; der Raum soll isotrop und homogen sein, d.h. mit $u(x,t)v(x,t)$ sind auch $u(x-at)$ und $v(x-at,t)$ und $u(-x,t)$ Lösungen. Nimmt man diese, wohl recht plausiblen Forderungen zu dem früheren Resultat hinzu, dass es sich um zwei lineare Gleichungen für zwei relle Skalare handelt, so sieht man schon, dass die Dichte quadratisch sein muss (weil sonst kein Erhaltungssatz gelten könnte) so dass man sie ohne die Linearität der Gleichungen zu zerstören, auf die Form u^2+v^2 bringen kann.

Dann haben die Gleichungen die folgende Form:

$$u'(\nu) = \int A(\nu,\nu')\mu(\nu')d\tau' + \int B(\nu,\nu')\omega(\omega,\nu')d\tau'$$

$$v'(\nu) = \int C \qquad\qquad\qquad D$$

$A-D$ sind hier beliebige Funktionen, lässt man für die auch singuläre Funktionen zu, so hat man hier wirklich die allgemeinste Form eines solchen Gleichungssystems vor sich. Wegen der Voraussetzung über die Homogenität müssen $A-D$ nur Funktionen des Vektors $\nu+\nu'$ sein, wegen der Spiegelinvarianz folgt noch $A(\nu-\nu') = A(-\nu+\nu')$ etc. Schreibt man sich nun den Ausdruck für die zeitliche Aenderung von $\int(u^2+v^2)d\tau$ auf, der — bis auf Oberflächenglieder — verschwinden soll, so findet man:

$$\frac{d}{dt}\int(u^2+v^2)d\tau = 2\{\int u(v)A(v-v')U(v')d\tau dt^{-1} + \int u(v)B(v-v')v(v')d\tau dt$$
$$+\int v(v)(v-v')u(v')d\tau dt^1 + \int v(v)D(v-v')v(v')d\tau dt'\}$$

Damit dieser Ausdruck identisch in u und v verschwindet, muss offenbar $A(\nu) = -A(-\nu) = 0, B = -C, D = 0$ sein, und hieraus folgt, dass sich die Gleichung in eine komplexe zusammenfassen lassen [kann]. Es folgt sogar speziell, dass diese Gleichung i nur in dem Glied mit $\frac{d}{dt}$ enthält.

4. Diese Frage konnte ich nicht beantworten, weil ich mich mit dieser Ableitung nie sehr intensiv befasst hatte. Fermi hat aber hierzu eine sehr hübsche Bemerkung gemacht, die ich mit seiner Erlaubnis hier wiedergebe.

Solange man die Zerlegung nach Partialschwingungen im streng klassischen Modell durchführt, ist es natürlich gleichgültig, of man sie als eine νt und $\cos \nu t$ oder als eine $e^{i\nu t}$ und $\overline{e}^{i}\nu t$ schreibt. Nun will man eine von Ihnen als Amplitude des Quantensprungs nach "oben" die andere als die des Sprungs nach "unten" interpretieren. Dabei lässt sich jedoch die Trennung in sin and cos nicht willkürfrei, d.h. ohne Bezugnahme auf den Anfangspunkt der Zeitkoordinate vornehmen, während die Zerlegung in zwei Exponentialglieder hiervon unabhängig ist. Insbesondere wird diese Frage in der Quantentheorie akut, wo ja diese beiden Partialschwingungen nicht mehr streng dieselbe Frequenz haben.

5. Die fraglichen Beweise sind sehr einfach zu erbringen. Erstens die Unmöglichkeit, im Falle der Fermistatistik Feldgrössen im klassischen Sinn zu bekommen.

Angenommen, es existiert ein derartiges Feld. Man kann es dann nach Partialwellen zerlegen, die den Zuständen eines einzelnen Teilchens entsprechen, und die Energie wird dann die Summe aus den Energien aller einzelnen Partialwellen. Die Energie jeder Partialwelle hat dann (in gewissen Einheiten) ganzzahlige Eigenwerte, da nur eine ganze Zahl von Teilchen in dem betreffenden Zustand sein kann. Jede Eigenschwingung entspricht also formal einem harmonischen Oszillator. Für das Licht ist dies eine ganz alte Weisheit, aber es gilt natürlich ebenso für die Materie.

Ist nun das Feld in klassischem Sinn messbar, so muss das auch für seine Fourierkomponenten gelten. Die Amplitude jeder dieser Partialschwingungen muss also nach Betrag und Phase mit einer vernünftigen Genauigkeit messbar sein. Mit vernünftiger Genauigkeit ist hierbei gemeint, dass die Ungenauigkeit in der Phase gegen [??], die in dem Betrag klein gegen den Wert des Betrages selbst sein soll.

Nun weiss man aber, dass Energie und Phase eines Oszillators nur dann mit einer solchen "vernünftigen Genauigkeit" gleichzeitig messbar sind, wenn die Quantenzahl gross gegen l ist. Mit anderen Worten: ein

Feld kann nur dann im klassischen Sinn messbar sein, wenn in jeder der betrachteten Eigenschwingungen viele Teilchen sitzen.

Im Fall der Bosestatistik führt dies auf die bekannte Aussage, dass die klassische Elektrodynamik den Grenzfall vieler Lichtquanten behandelt.

Im Fall der Fermistatistik folgt aber die Unmöglichkeit der Voraussetzung, da sich dort in jedem Zustand höchstens ein Teilchen befinden kann.

Fragen wir nun weiter nach der gleichzeitigen Bestimmbarkeit von Feld- und Teilchenzahl, so brauchen wir nur wieder die Analogie mit dem harmonischen Oszillator heranzuziehen. Wir interessieren uns jetzt nicht für die Messungen im klassischen Sinn, sondern überhaupt für eine Messung irgendeiner Feldgrösse z.B. nur von $\mathfrak{F}$ oder $\mathfrak{H}$ (bezw. bei der Materie Φ oder Φ^*) wollen aber gleichzeitig etwas über die Teilchenzahl wissen. Für unseren Partialschwingungsoszillator bedeutet das, dass wir nur seine Koordinate, *oder seine Geschwindigkeit mit "vernünftiger" Genauigkeit messen wollen, und gleichzeitig nach der Quantenzahl fragen. Aus den Vertauschungsrelationen des Oszillators folgt aber sofort

$$\underline{q}\underline{n} - \underline{n}\underline{q} = \frac{1}{kv}(\underline{q}\underline{E} - \underline{E}\underline{q}) = \underline{q}$$

und folglich

$$\delta q \delta n \sim q; \qquad \frac{\delta q}{q}\delta n \sim 1$$

Will man also q so genau messen dass $\delta q < q$ — und nur dann hat die Messung einen Sinn — so muss n um mindestens l unbestimmt sein. Um also über die Amplitude einer Partialschwingung eine Aussage zu machen, muss die Zahl der Teilchen in dieser Partialschwingung mindestens um l unbestimmt sein. Das gleiche gilt für die Gesamtteilchenzahl, da die Teilchen in anderen Zuständen keine Rolle spielen.

Man kann noch überlegen, dass dasselbe nicht nur für die Messung einer Fourierkomponente der Feldstärke in einem Punkt gilt, doch will ich den Beweis hierfür übergehen.

Schliesslich möchte ich noch bemerken, dass man alle diese Ueberlegungen, statt aus den Vertauschungsrelationen, direkt aus der Diskussion der Messmöglichkeiten herleiten kann, da man ja, (z.B. wie es

Heisenberg in seinem Buche tut[627]) die aus den Vertauschungsrelationen folgenden Unbestimmtheiten auch direkt ableiten kann. Das hat den Vorteil, dass man nicht von der Feldquantelung Gebrauch machen muss (auch nicht in der verdaulicheren Form der Analogie mit dem Oszillator), dafür ist man aber gänzlich auf die Analogie mit dem elektromagnetischen Feld angewiesen, denn für die Materie gibt es eben prinzipiell kein physikalisch mögliches Experiment, das man diskutieren könnte.

6. Ueber den Wurzel-Delta Operator als solchen, d.h. als mathematisches Symbol, kann ich sagen, dass ich mit ihm sehr zufrieden bin, er ist auch bei vielen anderen Gelegenheiten sehr bequem.[628]

Die von Landau und mir mit seiner Hilfe geschriebenen Gleichungen sind aber hässlich, eben wegen der Fernwirkung. Der Grund hierfür ist eben, dass der Lichtquantenort keinen physikalischen Sinn hat, und es daher nicht vernünftig ist, einen Konfigurationsraum einzuführen. Die Gleichungen sind zuweilen bequem, wenn man sie nur in der Näherung der geometrischen Optik braucht, in der die Lichtquanten lokalisierbar sind. Z.B. wenn man Wert darauf legt, die Versuche von Compton-Simon oder Geiger-Bothe etc. mit der Quantenmechanik nachzurechnen.[629] Sie sind ferner zuweilen bequem, wenn es sich um ein einziges Photon im Vakuum handelt, wo es daher nicht darauf ankommt, die Zustände negativer Energie auszuschliessen, und wo man daher die Gleichungen auch ohne Wurzel-Delta schreiben kann.

7. Diese Bemerkung wird sicher in dem Paulischen Handbuchartikel[630] gedruckt stehen. Ich habe mich übrigens nie geniert, sie laut

[627]W. Heisenberg, *The Physical Principles of the Quantum Theory*, Chicago: University of Chicago Press, 1930.

[628]See Correspondence with Heisenberg, Peierls Papers, Ms.Eng.misc.b.208; C.138.

[629]See W. Bothe and H. Geiger, 'Über das Wesen des Comptoneffekts: ein experimenteller Beitrag zur Theorie der Strahlung' *Z. Phys.* **32**, 639–663 (1925); Arthur H. Compton and Alfred W. Simon, 'Measurements of *beta*-Rays Associated with Scattered X-Rays', *Phys. Rev* **25**, 306–313 (1925); Arthur H. Compton and Alfred W. Simon, 'Directed Quanta of Scattered X-Rays' *Phys. Rev.* **26**, 289–99 (1925).

[630]See letter [99].

auszusprechen und sie bei Gelegenheit jedes Referates in den Physikalischen Berichten angebracht, das sich auf eine derartige Arbeit bezog.

Im übrigen möchte ich aber eigentlich darauf verzichten, meine Bemerkungen zu publizieren, denn Pauli hat mir inzwischen von seinem Brief an Sie Kenntnis gegeben,[631] den er publizieren will, und dieser unterscheidet sich von meinem inhaltlich gar nicht und in den Argumenten so wenig, dass es schon ein Missbrauch der Zeitschrift wäre, wegen so kleiner Unterschiede eine zweite Antwort zu drucken.

Zur Linearitätsfrage weiss ich ausserordentlich wenig. Man sieht natürlich sofort, dass man auf ganz scheussliche Verhältnisse kommt, wenn man nur versucht, eine nichtlineare Gleichung aufzuschreiben. Aber Komplikation ist an sich noch kein Grund um eine Möglichkeit auszuschliessen, und andererseits macht eben diese Komplikation es unmöglich zu diskutieren, warum man nun mit einer solchen Gleichung sicher auf Widersprüche mit der Erfahrung kommt.

Für die Linearität der Wellengleichung des freien Teilchens könnte man daran denken, sie direkt aus den Beugungsexperimenten zu erschliessen, die sich ja wohl kaum mit einer anderen Gleichung so gut beschreiben liessen, aber es ist nicht leicht, das zu beweisen.

Für gebundene Teilchen könnte man vermutlich mindestens die Homogenität der Gleichung daraus erschliessen, dass es möglich sein muss, die Lösung mit einem konstanten Faktor zu multiplizieren. Man möchte nämlich gern haben, dass die Wellenfunktion eines Teilchens, von dem man nicht weiss, in welchem von zwei räumlich getrennten Wellenpaketen es sich befindet, an jedem Raumpunkt zeitlich genau so verläuft, als ob nur das dort befindliche Wellenpaket vorhanden wäre. Dann muss wegen der Gesamtnormierung die Dichte in jedem Punkt um einen Faktor 2 kleiner sein. Nur muss man mit einer solchen Ueberlegung sehr vorsichtig sein, denn zu einem Zustand, wo man nicht weiss, ob sich das Elektron in dem einen oder in dem anderen Paket befindet gehört nur dann eine Wellenfunktion (er ist nur dann ein "reiner" Fall), wenn dafür

[631] Letter Wolfgang Pauli to Ehrenfest, 28.10.1932, Pauli, *Wissenschaftlicher Briefwechsel*, II, 124–34. The letter was published in a slightly revised version. W. Pauli, 'Einige die Quantenmachanik betreffende Erkundigungsfragen', *Z. Phys.* **80**, 573–86 (1933).

eine andere Grösse gemessen worden ist. Und diese Messung könnte die Bewegung beeinflussen.

Eine weitere Frage wäre dann, wie man aus der Homogenität auf die Linearität schliesst; auch das ist eine Frage, die ich nicht beantworten kann, von der ich aber glaube, dass sie ohne sehr einschneidende physikalische Postulate erledigt werden kann.

Ich will mir in der nächsten Zeit hierüber noch ein bisschen den Kopf zerbrechen, und ebenso über die Frage, ob man die Möglichkeit, die zwei reellen Gleichungen in eine komplexe zusammenzufassen, nicht auch im Falle von Teilchen in einem Potentialfeld einfach beweisen kann.
Mit herzlichen Grüssen
Ihr Ihnen ergebener

R[udolf] P[eierls]

[158] Heinrich Peierls to Rudolf Peierls

Berlin,[632] 21.4.1933

Lieber Rudi.
Wir haben uns gefreut, aus Eurer Karte aus London zu ersehen, dass Ihr gut in England angekommen seid und hoffen, dass ihr Euch inzwischen in Cambridge gut eingerichtet habt.

Heute bitte ich Euch um folgendes:
Karli Fulda ist von der Justizbehörde beurlaubt worden und wird voraussichtlich — eine bestimmte Entscheidung steht noch nicht fest — seine Ausbildung unterbrechen müssen, sodass er zum Assessorexamen

[632]In early 1933, Reichskanzlerplatz, where Heinrich and Else Peierls lived, was renamed Adolf-Hitler-Platz, and the letterhead in the Peierls correspondence reads accordingly. While few of the letters of the previous two years contain references to the increasingly difficult situation of Jews in Berlin, correspondence from now on is concerned with these issues, especially with attempts of Heinrich Peierls to help friends build an existence outside Germany.

möglicherweise nicht zugelassen wird.[633] Aus diesem Grunde bitte ich Dich festzustellen, ob für junge deutsche Juristen in England eine, wenn auch noch so bescheidene, Existenzmöglichkeit vorliegt. Es braucht auch in diesem Falle vorläufig keine rein juristische Tätigkeit, sondern es könnte auch eine andere Position sein, bei der er seine vorhandenen juristischen und journalistischen Fähigkeiten verwerten kann. Er lässt Dich bitten, einmal hierüber auch die Meinung des jungen Lancaster einzuholen, bei dessen Eltern Du seinerzeit gewohnt hast.[634] Bitte richte die Antwort hierauf so ein, dass ich sie im Original weitergeben kann. Im voraus besten Dank.

Von hier ist nichts Neues zu melden. Die weitere Entwicklung der Dinge wirst Du ja in den Zeitungen verfolgen können.[635]
Euch Beiden herzliche Grüsse

Dein Vater

Meine Lieben!
Ihr habt so viel Schönes gesehen + werdet nun froh sein, ein eigenes Nest zu bekommen, das Ihr uns ja schildern werdet. Fischers + Erich sind auch beurlaubt. Wir sind heute nachmittag z[um] Thee bei Henry Goldmann, der immer s[ehr] anregend ist.
Also alles Gute + viele Grüße

Else

Wäre für Erich Bruck eine Möglichkeit in England?
Wie geht's Hansi (Fischer)?
Das Geschenk für Genia kostete 5 M Zoll, ich sage das nur, damit Ihr für England bescheid wißt + gebt.

---

[633]The civil service was among the first sectors to be affected by the anti-Jewish legislation, and hence Peierls' cousin, Karli Fulda, was prevented from completing his legal training.

[634]See letter [20].

[635]The wording of the letters between Peierls and his parents makes it clear that both sides were aware of censorship, and the correspondence never contains direct references to or criticism of the political situation in Germany.

[159] Wolfgang Pauli to Rudolf Peierls

Zürich, 22.5.[19]33

Lieber Herr Peierls,
Vielen Dank für Ihren Brief vom 19. Ich muß Ihnen allerdings gestehen, daß mir Ihre Weise, Physik zu treiben, wesentlich weniger gut gefällt als die von anderen, die auch für meine Assistentenstelle in Frage kämen.[636] Ihre letzte Arbeit über starke Felder hat sich hinsichtlich Form und Inhalt sehr abstoßend auf mich gewirkt.[637] Und die Mitteilung Ihres Briefes, daß Sie immer noch nicht von der Physik des festen Körpers losgekommen sind, macht mir die Aussicht, Sie wieder als Assistent hier zu haben, auch nicht gerade verlockend.

Trotz der genannten Umstände kann ich es doch – auch abgesehen von der gegenüber der Rockefeller-Stiftung übernommenen Verpflichtung – nicht mit meinem Gewissen vereinbaren, Sie jetzt einfach auf der Straße liegen zu lassen, wenn Sie im Herbst keine Stelle finden.[638]

Ihre Aussichten auf das Lorentz-Stipendium scheinen – nach dem, was ich gehört habe, nicht groß zu sein. Wie groß Ihre Chancen in Belfast sind, kann ich nicht beurteilen, aber ich höre, daß auch Bethe sich um diese Stelle beworben hat. Natürlich können Sie meinen Namen ohne Weiteres als Referenz angeben. Aus Amerika bekam ich Antwort von Herrn Rabi, daß es für Sie zwar schwierig aber vielleicht nicht ganz

[636] Peierls' Rockefeller Fellowship was conditional on his having an academic position to return to after the year abroad. Pauli had undertaken to offer such a post to Peierls on his return from Cambridge, but this had been done with an understanding that Peierls would not take up the appointment, as the three years Peierls had been working with Pauli was a longer than usual collaboration for Pauli who preferred to change his assistants regularly. At the time, he was deliberating between Hans Bethe and Victor Weisskopf. Pauli decided in favour of Weisskopf, largely on the strength of his research interest which was closer to his own likes. See Pauli, *Wissenschaftlicher Briefwechsel*, II, pp. XXI, 161–2.

[637] R. Peierls, 'Zur Theorie des Diamagnetismus von Leitungselektronen. II. Starke Magnetfelder. *Z. Phys.*, **81**, pp. 186-94 (1933).

[638] The employment situation for young physicists had deteriorated signifiantly in the early 1930s, particularly since the economic crisis of 1932. See e.g. Kevles, Daniel J., *The Physicists: The History of a Scientific Coummunity in Modern America*, New York: Knopf, 1977, pp. 278–86; Eckert *Sommerfeld Briefwechsel*, vol. 2, pp. 351–77.

unmöglich wäre, eine Stelle in Amerika zu bekommen.[639] Schreiben Sie auf jeden Fall an Slater (Boston),[640] es sollen dort noch die meisten Chancen sein. Momentan ist Elsasser hier, ich bemühe mich, für ihn eine Stelle in Paris zu finden.[641] Heute tauchte auch Fröhlich auf, ein Opfer des dritten Reiches und der Metallphysik![642]

Was nun meine Assistentenstelle betrifft, so wäre es mir am liebsten, Casimir würde im Herbst noch hier bleiben. Es ist aber nicht wahrscheinlich, daß ich das durchsetzen kann, denn von Ehrenfest kommen dringende Hilferufe, Casimir möge nach Leiden zurückkehren und Casimir ist momentan sehr geneigt, das auch zu tun. Bloch dürfte sehr wahrscheinlich im Herbst sein Rockefeller Stipendium bekommen und für die Zeit nach Ablauf hat Bohr sich verpflichtet, Bloch anzustellen.[643] Güttinger ist endgültig in einer Versicherungsgesellschaft in Basel verschwunden. Ich werde jetzt auf jeden Fall mindestens noch einen Monat mit der Entscheidung über die Besetzung meiner Assistentenstelle warten und hoffe, von Ihnen bis dahin zu hören, ob Sie untergekommen sind.

Für den Fall, daß dies eintritt, denke ich ernstlich an Weißkopf und an Bethe. Bethe allerdings nehme ich nur unter der Bedingung, daß er mit dem Handbuchartikel über Elektronentheorie der Metalle bis zum Herbst fertig ist und der Physik des festen Körpers abschwört (Es ist dies eine Art "Antimodernisten-Eid", den ich einführen will.) Es dürfte

[639]Wolfgang Pauli had written to Rabi on Peierls' behalf to enquire about possibilities in the United States. See letter Pauli to I.I. Rabi, 26.2.1933, Bodleian *Peierls Papers*, Ms.Eng.misc.b.212.

[640]John Clark Slater, (1908–1976) received his Ph.D. from Harvard in 1923; then before moving to Europe for postgraduate work with Bohr and Heisenberg. In 1930 he was appointed head of the physics department at MIT where he stayed until his retirement in 1966.

[641]See letter [141], note 531.

[642]Herbert Fröhlich (1905–1991), after completion of his habilitation in Freiburg, had moved to Leningrad, where he was working on his book on the electron theory of metals. H. Fröhlich, *Elektronentheorie der Metalle*, Berlin: Springer, 1936.

[643]Felix Bloch (1905–1983) initially lectured in Paris at the Institute Henri Poincaré, before he used his Rockefeller fellowship to work with Fermi in Rome in 1934. From there he moved to Stanford. He later worked at Los Alamos. After the war he returned to Stanford, where — with the exception of a short spell at Geneva, as the first director of CERN — he remained until his retirement.

letzteres allerdings nicht so schwer für ihn sein, da er sich immer für Stöße interessiert hat. Immerhin ist von denjenigen, die nicht schon versorgt sind, Weißkopf der einzige, gegen den bei mir gar nichts spricht.

Delbrück[644] ziehe ich jetzt kaum mehr als Kandidat für meine Assistentenstelle in Betracht. Erstens gehört er zu den wenigen jungen Theoretikern, die in Deutschland Unterkunftsmöglichkeit haben, zweitens hat er sich wissenschaftlich nicht günstig entwickelt, hat auch im letzten Jahr nichts geleistet.[645]

Nun von den Fragen des "Physikermarktes" zu den Fragen der Physik. Was die Kernphysik betrifft, so glaube ich wieder sehr an die Erhaltungssätze — nicht nur an die von Energie, Impuls (und selbstverständlich der Ladung), sondern auch an die des Impulsmomentes und den Symmetriecharakter (Bose oder Fermi) des Gesamtsystems. Denn ich bin in meiner früheren Vermutung der Existenz eines "Neutrinos" und seiner Emission beim β-Zerfall durch die Entdeckung des positiven Elektrons sehr bestärkt worden.

Ich habe darüber bewegliche Briefe an Blackett[646] und an Joliot geschrieben.[647] Letzterer versprach, zu untersuchen, ob die künstlichen

---

[644]Max Delbrück (1906–1981) received his Ph.D. from Göttingen, worked in Berlin, Copenhagen and Zurich, later became professor of Medicine. See Ernst Peter Fischer and Carol Lipson, *Thinking about Science: Max Delbrück and the origins of molecular biology*, New York: W.W. Norton, 1988.

[645]Lise Meitner had expressed an interest in taking Delbrück as her assistent. See Letter Pauli to Lise Meitner, 29.5.1932, Pauli, *Wissenschaftlicher Briefwechsel*, II, pp. 113–4. Delbrück had turned to questions of molecular biology, a decision that had led to fewer publications in theoretical physics.

[646]Pauli was interested in P.M.S. Blackett's and Giuseppe Occhialini's work on the positive electron (P.M.S. Blackett and G. Occhialini, 'Some photographs of the tracks of penetrating radiation', *Proc. Roy. Soc.* A **139**, 699–726 (1933)) which he hoped would provide further proof of the existence of the neutrino. See letter Pauli to Blackett, 19.4.1933, Pauli, *Wissenschaftlicher Briefwechsel*, II, letter 307, p. 158.

[647]Bothe's work on beryllium (W. Bothe, 'α-Strahlen, künstliche Kernumwandlung und Anregung, Isotope — Bemerkungen über die Ultra-Korpuskelstrahlung', *Convengo di Fisica Nucleare.* Roma, 1932, pp. 83–106) had spurred Iréne Curie and Fréderic Joliot to new experiments that had led to better understanding of the berrilium rays. I. Curie and F. Joliot, 'Emission de protons de grande vitesse par les substances hydrogénées sous l'influence des rayons γ très pénétrants', *C. R. Acad. Sci. Paris* **194**, 273–75 (1932); I. Curie and F. Joliot, 'Sur la nature des du rayon-

γ-Strahlen des Be Neutrinos enthalten (und zwar wäre das Kriterium der Unterscheidung die Geschwindigkeits- und Richtungsverteilung der Sekundärelektronen, die einerseits γ-Photonen, andererseits Neutrinos auslösen).[648]

Was nun die oben genannten Erhaltungssätze von Impulsmoment und Symmetriecharakter des Gesamtsystems bei Kernprozessen betrifft, so würden sie zunächst erklären, warum ein Wasserstoffatom sich nicht zunächst in ein Neutron verwandeln kann. Denn das H-Atom hat als ganzes Bosestatistik und ganzzahligen Spin (von Proton und Elektron), das Neutron halbzahligen Spin und Fermistatistik. Das Neutron könnte dann nicht aus Proton und Elektron bestehen, wenigstens nicht aus diesen allein.[649]

Ferner ist es wahrscheinlich, daß das positive Elektron ganzzahligen Spin (vielleicht Null) und Bose-Statistik hat. Dies könnte man aus den Erhaltungssätzen (und der empirischen Tatsache, daß die Massenzahl der Kerne ihre Statistik und den Geradheitscharakter des Spins eindeutig bestimmt) allerdings nur dann schließen, wenn man wüßte, daß es bei einem Elementarprozess in ungerader Zahl erzeugt werden kann. Es ware sehr zu wünschen, wenn die Experimentatoren genau feststellen würden, bei welchem Elementarprozess das positive Elektron entsteht.

Die Idee, daß das positive Elektron Bose-Statistik hätte, hat Elsasser zuerst gehabt (wenn auch aus anderen Gründen), und ich habe ihm geraten, darüber eine Note an die Nature zu schicken. (Daß dieser Gesichtspunkt der Diracschen Löchertheorie entgegengesetzt ist, spricht nur für ihn. Denn die Erfahrung zeigt doch, daß positive und negative Elektrizität sich nicht genau gleich verhalten und es ist mir sehr unsympathisch, diese Asymmetrie in den Anfangszustand der Welt zu verlegen wie Dirac es tut.)[650]

---

nement pénétrants excité dans les noyaux légères par les particules α', *C. R. Acad. Sci. Paris* **194**, 1229–31 (1932).

[648]See also I. Curie and F. Joliot, 'Origine des électrons positifs', *C. R. Acad. Sci. Paris* **196**, 1581–83 (1933).

[649]Ettorio Majorana had discussed details about the composition of the neutron in 'E. Majorana, 'Über die Kerntheorie', *Z. Phys.* **82**, 137–45 (1933).

[650]P.A.M. Dirac, 'A Theory of Electrons and Protons', *Proc. Roy. Soc. A* **126**, 360–365 (1930).

Wenn ich also glaube, daß Bohr mit seiner Bekämpfung des Energiesatzes[651] in der Kernphysik ganz unrecht hat,* so glaube ich andererseits, daß er in der Frage der Feldstärkemessung unbedingt Recht hat. Ich glaube dies nicht aus allgemeinem Bohr-Autoritätsglauben (der war bei mir nie vohanden), und nicht einmal nur deswegen, weil ich in seinem Beweis keinen Fehler gefunden habe. Sondern ich glaube es, weil umgekehrt kein einziger stichhaltiger Grund für die Ungleichungen der Feldstärken in der Arbeit von Ihnen und Landau besteht. (Die Diskussion einer speziellen Meßanordnung ist nie ein Grund!)[652] Wenn eine Messungenauigkeitsungleichung besteht, die nicht in den Grundlagen des theoretischen Formalismus als Vertauschungsrelation direct vorkommt, so muß eine solche Relation einen angebbaren physikalischen Grund haben. Bei den Relationen über die Ungenauigkeit der Impuls- und Ortsmessung von Teilchen in Ihrer Arbeit ist dieser Grund das tatsächliche Fehlen von Zuständen negativer Energie (welcher Tatsache die Theorie widerspricht). Aber bei der Feldstärkemessung spielen die Zustände negativer Energie keine Rolle und für eine aus den Vertauschungs-Relationen nicht folgende Ungleichung, die nicht die Ladung, Masse oder Dimension der Probekörper enthält, fehlt jeder Grund. (Die scheinbaren Paradoxien über die "Nullpunktschwankung" der Feldstärke wurden von Bohr geklärt.)

Andererseits kann natürlich etwas passieren, wenn der Raum-Zeit-bereich über den bei der Feldstärkebestimmung zu mitteln ist, Lineardimensionen der Größenordnung h/mc (bzw. H/mc^2) bekommt (m=Masse des Probekörpers) bzw. allgemeiner, wenn die atomistische Konstitution der Probekörper eine Rolle spielt.

*Er wollte übrigens, als ich ihn im März sah, von dem positiven Elektron nichts wissen und meinte, daß Blackett eben "pathologische Photographien" gemacht habe.

[651]See Pais, *Niels Bohr's Times*, pp. 312–13.
[652]See Letter [155], note 624.

Das meint auch Bohr. (Die versprochenen Korrekturen der Arbeit von Rosenfeld und mir sind übrigens bis heute noch nicht gekommen.)[653]

Was meine eigene Physik betrifft, so ist zunächst das Streuproblem bei kurzwelligen γ-Strahlen dank einer Arbeit von Casimir gelöst.[654] Die Heisenbergsche Vermutung, daß in gewissen Grenzfällen die Klein-Nishina-Formel gelte, ist damit endgültig widerlegt. Ich selbst schreibe eine längere zusammenfassende Arbeit über 5-dimensionale Physik, deren erster Teil fertig ist und deren zweiter Teil bald fertig wird.[655]

Dann möchte ich mir die Wellenfunktion in mehreren Zeiten von Dirac, Fock, Podolsky, die mich sehr interessiert, hinsichtlich ihrer physikalischen und korrespondenzmäßigen Bedeutung noch näher ansehen.[656]

Ich habe mich gefreut, daß aus Diracs Quatsch vom vorigen Jahr nun doch etwas Vernünftiges herausgekommen ist.[657]

Beste Grüße an Sie und alle Cambridger Physiker (besonders an Weißkopf; bitte zeigen Sie ihm auch diesen Brief).

Ihr W. Pauli

---

[653]No joint paper was published.

[654]See W. Pauli, 'Über die Intensität der Streustrahlung freier Elektronen', *Helv. Phys. Acta*, **6**, 279–86 (1933), based on work by Casimir and Pauli. See also H. Casimir, Über die Intensität der Streustrahlung gebundener Elektronen', *Helv. Phys. Acta* **6**, 287–304 (1933).

[655]W. Pauli, 'Über die Formulierung der Naturgesetze mit fünf homogenen Koordinaten. Teil I (klassische Theorie)', *Ann. Phys.* **18**, 305–336 (1933); W. Pauli, 'Über die Formulierung der Naturgesetze mit fünf homogenen Koordinaten. Teil II (Die Diracschen Gleichungen für die Materiewellen)', *Ann. Phys.* **18**, 337–372 (1933).

[656]P.A.M. Dirac, V.A. Fock, B. Podolsky, 'On Quantum Electrodynamics', *Phys. Z. der Sowjetunion* **2**, 468–79 (1932).

[657]Pauli refers to Dirac's criticism of the Heisenberg–Pauli theory in P.A.M. Dirac, 'Relativistic Quantum Mechanics', *Proc. Roy. Soc.* A **136** 453–64 (1932). See letter Pauli to Dirac, 11.9.1932, Pauli, *Wissenschaftlicher Briefwechsel*, II, pp. 115–21.

[160] Rudolf Peierls to Hans Bethe

Cambridge, 14.6.[19]33

Lieber Bethe!
Ich habe Ihnen immer noch nicht auf Ihren Brief geantwortet, hauptsächlich in der Annahme, Ihnen demnächst irgendwelche offiziösen Mitteilungen in der lecturer-Angelegenheit machen zu können.[658] Aber die Entscheidung hierin geht erstens beliebig adiabatisch vor sich und zweitens ist hier nichts herauszubringen, sodass Sie vermutlich mehr darüber wissen als ich. Dem Vernehmen nach soll "einer von Ihnen" eine Stelle bekommen haben, aber ob Sie das sind, oder ob das eine lectureship ist oder eine sehr schofel bezahlte assistant l[ectureship] weiss ich nicht. Ich stehe jedenfalls nicht in der engeren Wahl.

Was haben Sie sonst fuer Ideen? Ich habe ziemliche Aussichten auf ein Stipendium von Langevin,[659] und mache Ihnen ausserdem Konkurrenz in Belfast.[660] In letzterem scheinen allerdings die Aussichten fuer Sie und mich ziemlich infinitesimal zu sein, da es sich um eine Stelle mit viel Verwaltungsarbeit zu handeln scheint, sodass man wohl vermutlich einen der hiesigen Leute hinsetzen wird. Ich liebäugele auch ein bisschen mit Manchester, wo bei Bragg[661] sogar vermutlich zwei Stellen frei sind — wenn nämlich Williams[662] sein Rockefeller bekommt. Bragg scheint auch Sie nach hier umlaufenden Gerüchten sehr stark in Betracht zu ziehen.

---

[658] For details about the Cambridge lectureship see letter [161].

[659] Paul Langevin (1872–1946) had been director of the École municipale de Physique et Chimie in Paris since 1929.

[660] See letter [159].

[661] William Lawrence Bragg (1890–1971) studied mathematics at Adelaide before completing a physics degree at Cambridge. In 1919 be was appointed Langworthy Professor of Physics at Manchester University. In 1938 he became Cavendish Professor of Experimental Physics at Cambrigde.

[662] Evan James Williams (1903–1945), at the time reader in physics at Manchester, was appointed professor of Physics at Aberystwyth in 1939 where, interrupted by his service for the ment of Naval Operational Research, he stayed until his premature death in 1945.

Ich habe jetzt meine Metalle ein bisschen aufgesteckt und ixe mit den Dirac-Löchern=Blackett-Elektronen herum. Vielleicht hat diese Diracsche Theorie doch in irgendeiner Näherung einen vernünftigen Sinn.[663]

Sonst gibt es von hier nicht viel Neues, denn dass es in Cambridge sehr nett ist, dass Wilson noch immer nicht seinen Fehler einsieht,[664] dass Frau Blackett allen Leuten erfrischende Frechheiten sagt, und dass Fowler jedes zweite Kollegium vergisst, dürfte wohl für Sie nicht neu sein.

Mit besten Grüssen Ihr

R. Peierls

Gruss auch von Genia

[161] Rudolf Peierls to Hans Bethe

Cambridge, 28.6.[19]33

Lieber Bethe!

Vielen Dank für Ihre prompte Antwort. Was den Lorentzfonds betrifft, so habe ich von Fokker[665] eine Absage bekommen, das sie Ihre Mittel jetzt für dringendere Aufgaben (d.h. offenbar für kurzfristige Einladungen) brauchen und ich ja ausserdem immer nach Zürich zurückkönnte (Ob das richtig ist, hängt allerdings von der Definition des Wortes "können" ab).[666] Ich weiss übrigens nicht, ob ich Ihnen geschrieben habe, dass für mich merkliche Aussichten auf ein Pariser Stipendium[667] zu bestehen scheinen.

---

[663] See letter [159].

[664] Alan Herries Wilson (1906–1976) had criticised Peierls' version of his doctoral thesis (See Letter [36], note 164. and attempted to prove that he was wrong. In response, Peierls refuted Wilson point by point in his paper R. Peierls. 'Zur Theorie der Metalle', *Z. Phys.* **81**, 697–99 (1933).

[665] Adriaan Fokker was professor of physics in Leiden between 1928 and 1955. See also letter [44], note 186.

[666] Letter Hans Bethe to R. Peierls, 22.5.1933, *Peierls Papers*, Ms.Eng.misc.b202, C.15.

[667] One of Peierls' options of a fellowship at Langevin's institute; see letter [160].

Bezüglich Cambridge hat sich die Situation insofern wieder geändert als plötzlich wieder zwei Stellen frei sind. Erstens hat man Bloch eine Stelle mit 120-200£ angeboten, die er wohl mit grosser Sicherheit ablehnen dürfte, da er nicht auf das Rockefeller Stipendium verzichten wird.[668] Ausserdem hat man eine zweite Stelle — ich glaubte fälschlich es wäre dieselbe — Born angeboten, der inzwischen auch abgelehnt hat.[669] Für mindestens eine von diesen beiden Stellen haben Sie erhebliche Chancen, als erster berufen zu werden, was nämlich erstens aus Aeusserungen des — allerdings nicht immer korrekt informierten — Mott[670] hervorgeht und mir auch a priori plausibel erscheint. Allerdings kann man bei den hiesigen Bonzen nie wissen; ich war z.B. sehr erstaunt, dass man Bloch vor Ihnen berufen hat, zumal man sie hier sehr zu schätzen scheint. Haben Sie sich etwa hier mit jemand verkracht? Das ärgerlichste ist nun, dass das Komitee, das die Stellen neu besetzen soll, von der Leistung Bloch und Born zu berufen, so erschöpft war, dass es schleunigst in die Ferien ging, und vor Oktober nicht wieder zusammentritt. Trotz dieser unbegrenzten Adiabasie sollte man ja nun eigentlich schon vorher unoffiziell feststellen können, wer eingeladen wird, denn da es ziemlich fest zu stehen scheint dass man Physiker berufen wird, so wird es wohl ganz von Fowler abhängen. Der ist aber so ein Geheimniskrämer, dass ich nicht weiss, ob man vor Oktober irgendwas erfahren wird.

Mott liebäugelt übrigens mit einem Tauschhandel. Wie Sie wissen, hat er Stobbe gegen Heitler eingetauscht, um den ersteren loszuwerden, und scheint dies nun schrecklich zu bereuen, da ja Stobbe jetzt auch ohne Tausch beliebig viele Stellen bekommen kann. Er wird nun weiter tauschen, und jedenfalls wenn Sie in Manchester sind, wird er sicher

---

[668]Bloch had not rejected the offer. See letter [162].

[669]Max Born (1882–1970), studied at Breslau, Heidelberg, Zurich, Göttingen, where he obtained his doctorate in 1907. He later tought at Göttingen, Berlin, Frankfurt and again as professor in Göttingen. Being forced to flee Germany in 1933 Born initially went to India and in 1936 accepted the Tait professorship in Applied Mathematics at Edinburgh.

[670]Nevill Mott (1905–1996), received his Ph.D. from Cambridge in 1926, and taught at Manchester, Cambridge and Bristol before returning to Cambridge as Cavendish professor of Physics in 1954, where he stayed until his retirement in 1971.

nach einem halben Jahr versuchen, Heitler gegen Sie einzutauschen. Allerdings weiss ich nicht, ob Bragg sich dies gefallen lässt, und ob Sie das Risiko eingehen werden, ev[entuell] nach einem halben Jahr gegen Heisenberg eingetauscht zu werden. All dies ist übrigens nicht offiziell.

Die zweite Stelle in Manchester hat sich inzwischen auch auf Null reduziert, d.h. wird mit einem Engländer besetzt werden.[671]

Wissen Sie übrigens dass auch die Royal Society Stipendien zu vergeben hat? Über die allerdings noch keine näheren Einzelheiten bekannt sind. Wenn Sie sich dafür interessieren, so schicken Sie doch Ihre Personalien an das Academic Assistance Council,[672] Burlington House, Piccadily, London.

Wir haben mit Ihnen herzliches Mitleid, nachdem uns Mott erzählt hat, wie Sie in Manchester gehungert haben. Genia macht die ganze Zeit Pläne, wie man es einrichten kann, dass Sie bei uns wohnen und bei uns anständig zu essen kriegen (in Quantitäten wie etwa in St. Antönien), aber das geht natürlich nur dann, wenn wir in einer Stadt wohnen, wofür die Wahrscheinlichkeit sehr gering ist.

Sehen werden wir Sie ja vermutlich auf dem Weg nach Manchester, denn wir bleiben bis Anfang Oktober hier und Sie kommen doch wohl sicher über Cambridge, aber dann wird Genia keine Zeit haben, Sie zu füttern da sie dann damit beschäftigt sein dürfte, einen Mitbürger zu füttern, der 27 Jahre und zwei Monate jünger ist als Sie.

Sonst nichts Neues, ich ixe Diraclöcher.

Beste Grüsse

Ihr

R. Peierls

---

[671]See Rudolf Peierls, interview by Charles Weiner, 11.–13.8.1963, Niels Bohr Library, AIP. Bethe was appointed to a lectureship at Bragg's institute at Manchester, but it became clear that it would not have been acceptable to appoint another German to a second lectureship, should this become available. Peierls was then appointed at Manchester to a fellowship funded by the Academic Assistance Council.

[672]The Academic Assistance Council was formed in May 1933 as a British response to help central European scholars persecuted by the Nazis. It provided, among others, scholarships to distinguished academics.

[162] Rudolf Peierls to Hans Bethe

Cambridge, 6.7.[19]33

Lieber Bethe!
Der Mensch denkt und das Faculty Board of mathematics lenkt. Und daraus ergeben sich dann alle möglichen Falschmeldungen. Z.B. stimmt es nicht, dass B[loch?] abgeleht hat, im Gegenteil, er kommt bestimmt her, (nur weiss ich nicht, wie weit das offiziell ist). Trotzdem gibt es ausserdem noch zwei Stellen, von denen eine wohl ein hiesiger bekommen wird, während die andere mit grosser Wahrscheinlichkeit — nach vertraulicher Aussage von Newman[673] — einer der drei Nordheim,[674] Sie oder ich bekommen werden. Da die Entscheidung hierüber nicht vor September zu erwarten ist, so muss man anfangen, Eventualprogramme zu machen. Können Sie jetzt schon feststellen, was Sie machen würden, wenn Sie hier eingeladen würden, ob Sie annehmen oder nach Manchester gehen würden? Das ist mir deswegen von Interesse, weil ich im ersteren Fall zu Bragg gehen würde. Hängt diese Frage mit der Höhe des hier gezahlten Gehalts empfindlich zusammen?
Mit besten Grüssen
Ihr

R. Peierls

Warum sitzen Sie eigentlich in München und kommen nicht her? Ist es nicht in dem Münchener Institut jetzt sehr langweilig?

[673]Max Newman (1897–1984), mathematician at the time lecturer of mathematics at Cambridge.

[674]Lothar W. Nordheim (1899–), studied at Hamburg, Munich, and Göttingen, where he obtained his doctorate under Max Born. In 1927 he took up a post at the Cambridge Cavendish Laboratory, and in 1835 he moved to the US, taking up positions at Purdue University (1935), Duke University (1937) and at the John Hopkins Laboratory (1956).

[163] Rudolf Peierls to Wolfgang Pauli

Cambridge, 17.7.1933
(carbon copy)[675]

Lieber Herr Pauli!
Ich bin nicht ganz so pessimistisch, wie Sie, bezüglich der Frage, wieweit das Diracsche Löchermodell so formuliert werden kann, daß es eindeutige Antworten liefert.[676] Es scheint mir nähmlich ganz ähnlich zu liegen, wie bei der Quanteneletrodynamik, insofern, als man in erster Näherung bezüglich der Wechselwirkungen zu vernünftigen Resultaten kommt, während höhere Näherungen divergieren. Das gilt im Fall des Diracmodells, wie ich näher ausführen werde, auch besonders bezüglich der Wechselwirkung mit einem Potentialfeld. In der Quantenelektrodynamik ist man ja bisher immer mit dem Prinzip durchgekommen, daß diejenigen Effekte, für die man nicht divergierende Resultate bekommt, auch der Wirklichkeit entsprechen. Ich würde daher vermuten, daß es vernünftig ist, das Gleiche in dem Löchermodell zu versuchen. Dann wäre einer von den gesichertsten Effekten sicher die Erzeugung von Elek-

---

[675]This letter appears to be based on a letter drafted by Pauli for Heisenberg. The handwritten mathematical formula are Pauli's but the letter is signed by Peierls, which suggests that Pauli had sent a draft to Peierls for approval. (See comment to letter in Pauli, *Wissenschaftlicher Briefwechsel*, II, p. 200.) A typewritten copy of the letter was then sent on to Heisenberg. See letter Pauli to Heisenberg, 22.7.1933, Pauli, *Wissenschaftlicher Briefwechsel*, vol. 2, p. 208.

[676]The letter is a reply to a query by Pauli, who had been thinking much about the Dirac equation and quantum electrodynamics since 1932. (See Enz, *No Time to be Brief*, pp. 253, 256, 295–309). In anticipation of the Solvay Conference in October 1933, Heisenberg, Pauli, Peierls and later Weisskopf exchanged a large number of letters about the subject. Pauli himself approached the president of the Solvay committee Paul Langevin to invite Dirac to give a paper on his hole theory. (See Pauli, *Wissemschaftlicher Briefwechsel*, II, p. 203, note 1). Furthermore, he suggested that Heisenberg and Peierls could contribute. (Letter Pauli to Heisenberg, 19.7.1933, Ibid., pp. 201–203). The assumption that this letter was part of the fruitful collaboration between Bethe and Peierls during their Manchester period, however, is not correct. Peierls did not move to Manchester until the autumn, and the correspondence between Bethe and Peierls in the summer of 1933 only contains vague references to Peierls' calculations.

tronenpaaren durch Licht,[677] weil der Formalismus der Berechnung so eng mit dem des Photoeffekts verknüpft ist, daß unmöglich das eine falsch und das andere richtig sein kann, sofern man natürlich überhaupt die ganze Diracsche Idee akzeptiert. (D.h. die Hypothese, daß es Zerstrahlungsprozesse gibt, und daß ihre Existenz mit der Nicht-Existenz der Zustände negativer Energie zusammenhängt.)

Sofern man sich auf solche Effekte erster Näherung beschränkt, läßt sich die Theorie so formulieren, daß keine Unendlichkeiten auftreten. (Denn die Selbstenergie ist ja ein Effekt zweiter Näherung.) Das kann man etwa in der Weise tun, wie wir es uns überlegt haben.

Auf diese Weise bekommt man natürlich keinen fertigen Formalismus, sondern nur ein heuristisches Rechenschema, denn ein anständiger Formalismus muß in jeder Näherung anwendbar sein. In höherer Näherung passieren aber allerhand Schweinereien.

Die erste Frage, die Sie in Ihrem Brief erwähnen, ist allerdings keine ernsthafte Schwierigkeit, denn man soll ja gar nicht die unendliche Coulombsche Wechselwirkungsenergie der Elektronen in negativen Zuständen wegstreichen, sondern man soll die Gleichungen so schreiben, daß diese unendliche Ladungsdichte überhaupt kein elektrisches Feld erzeugt. Daß das ohne Widerspruch möglich ist, kann man grob anschaulich so sagen, daß man einfach eine ebensogroße unendlich positive Ladungsdichte hinzufügt. Zur Beruhigung ängstlicher Gemüter kann man diese sogar als die Ladungsdichte der — zweifellos auch existierenden — Protonen in negativen Zuständen interpretieren. Daher glaube ich, daß im kräftefreien Fall alles in Ordnung ist. Man muß die Wechselwirkung zwischen den Teilchen natürlich insofern in Rechnung stellen, als man etwa bei dem Prozeß der Erzeugung von einem Teilchenpaar die Wechselwirkung dieser beiden Teilchen berücksichtigen muß. Aber das macht nur eine vermutlich kleine und jedenfalls endliche Korrektur.

---

[677] Hans Bethe and W. Heitler were concerned with similar questions at the time. See H. Bethe and W. Heitler, 'On the stopping of Fast Particles and on the Creation of Positive Electrons' *Proc. Roy. Soc. A* **146**, 83–112 (1934). See also similar results in J.R. Oppenheimer and M.S. Plesset, 'On the Production of the positive Electron', *Phys. Rev.* **44**, 53–55 (1933).

Das Delbrücksche Argument mit der Änderung der Teilchendichte durch die Wirkung eines Potentialsfeldes ist nicht neu,[678] Fermi fand es gleich, als er von der Theorie hörte.[679] Wenn man den Effekt in dem Diracschen Bild mit den unendlichen vielen negativen Elektronen überlegen will, so bekommt man kein eindeutiges Resultat. Man kann nämlich wenn man zunächst in quasi-klassischer Näherung bleibt, auf zwei Weisen verfahren: Entweder berechnet man den Unterschied der Dichte, die von den Elektronen in negativen Zuständen herrühren, an einem bestimmten Raumpunkt, indem man zunächst die von Elektronen aus einem bestimmten Energieintervall $E, E + dE$ herrührende Dichte an dem Punkt mit oder ohne Potential vergleicht, und diese Differenz über E integriert. In diesem Fall bekommt man einen unendlichen Effekt, und zwar im Sinne einer Verstärkung des Potentialfeldes. Denn wenn man z.B. das Feld einer positiven Ladung annimmt, so wirkt es ja auf die Elektronen in negativen Zuständen abstoßend, die zu jedem Elektron gehörende Dichte wird also durch das Potentialfeld verringert, man bekommt also eine neue positive Ladung, die noch dazu, wie man leicht nachrechnen kann, an jedem Punkt unendlich wird.

Die zweite Möglichkeit ist, daß man die an einem bestimmten Punkt x von Elektronen mit der Energie E herrührende Dichte vergleicht mit der Dichte, die im Fall des Potentialfeldes von Elektronen mit der Energie $E - V(x)$ — also mit der gleichen kinetischen Energie — am gleichen Punkt erzeugt wird. Diese Differenz verschwindet aber in der hier betrachteten Näherung, da zu den beiden Intervallen dasselbe Volumenelement im Phasenraum gehört, und wenn man sie über E integriert, so bleibt sie Null. Bei dieser Betrachtungsweise bekommt man also in der quasi-klassischen Näherung keine Änderung der Ladungsverteilung, man könnte also nur noch Quanteneffekte erwarten.[680]

[678]M. Delbrück, 'Zusatz bei der Korrektur', p. 144 appendix to L. Meitner and H. Kösters, 'Über die Streuung kurzwelliger γ-Strahlen', *Z. Phys.* **84**, 137–44 (1933).

[679]E. Fermi and G. Uhlenbeck, 'On the Recombination of Electrons and Positrons', *Phys. Rev.* **44** 510–11 (1933).

[680]For further elaboration of the argument see R. Peierls, 'The Vacuum in Dirac's Theory of the Positive Electron', *Proc. Roy. Soc. A* **146**, 420–41 (1934).

Diese Zweideutigkeit muß natürlich in jeder Theorie behoben werden, in der keine expliziten Unendlichkeiten auftreten. Man kann nun zum Glück allgemein sagen, was aus einer solchen Theorie herauskommen muß, ohne sie noch explizit aufzustellen. Dabei ergibt sich, daß die zweite Betrachtungsweise die richtige ist. Man kann sich das folgendermaßen überlegen: Betrachten wir einen Fall, in dem das Potential als kleine Störung behandelt werden kann. Dann ist also ohne Potential keine Ladung vorhanden und beim Einschalten des Potentials V entsteht eine Ladungsdichte ρ. Da das Raumintegral dieser Dichte verschwindet, ist die nächste einfache Größe, die man berechnen wird $\int \rho V d\tau$. Diese Größe ist aber nichts anderes als die Eigenwertstörung zweiter Näherung des Gesamtsystems, und von der weiß man, daß sie negativ sein muß, weil das System in einem relativ tiefen Zustand ist. (Es gibt natürlich noch tiefere Zustände, die man so beschreiben kann, daß man jedes negative Elektron in der Energieskala ein bißchen nach unten verschiebt, aber diese können natürlich nicht mit Hilfe nicht verschwindender Matrixelemente erreicht werden.) Sofern es also eine widerspruchsfreie Schreibweise der Theorie ohne explizite Singularitäten gibt, die der üblichen Transformationstheorie genügt (oder sich mindestens aus einem Variationsprinzip ableiten läßt) so muß $\int \rho V d\tau$ negativ sein. Das heißt aber, die Ladung hat das umgekehrte Vorzeichen wie das Potentialfeld und hat die Tendenz, es zu kompensieren. Das mit Hilfe der ersten Überlegung der vorherigen Seite gewonnene Resultat kann also aus einer anständigen Theorie nicht herauskommen.

Man kann dieses Argument präzisieren: Falls das Potential eine kleine Störung ist, so bekommt man in verständlicher Bezeichnungsweise für die Eigenfunktion erster Näherung jedes einzelnen Elektrons in einem negativen Zustand i:

$$u_i + \sum_k \frac{V_{ik}}{E_i - E_k} u_k$$

Dabei durchläuft k sowohl negative wie positive Zustände. Die gesamte Änderung der Dichteverteilung wird also:

$$\rho = \sum_{-i} \sum_{+k} \frac{1}{E_i - E_k} [u_i^* u_k V_{ik} + u_i^* u_k V_{ki}] + \sum_{-i} \sum_{-k} \text{(formal gleicher Term)}$$

wobei getrennt der erste Term die Teilsumme bezeichnet, in der k positive Zustände durchläuft und der zweite die mit negativen. i durchläuft natürlich immer nur negative Zustände, da nur diese besetzt sind. In der zweiten Summe sind nun je zwei Terme entgegengesetzt gleich, da der Summand beim Vertauschen von i und k antisymmetrisch ist. Eine vernünftige Theorie muß nun eben so eingerichtet sein, daß diese mehrdeutige Summe nicht auftritt, denn allen diesen Sprüngen der einzelnen Elektronen von einem negativen Zustand zum anderen entspricht kein möglicher und nach dem Pauliprinzip erlaubter Übergang des Gesamtsystems, und wenn die Quantenmechanik richtig ist, so muß jede Änderung des Erwartungswertes jeder physikalischen Größe mit den möglichen Übergängen des Systems verknüpft sein.

Dagegen bleibt nun die erste Summe stehen. Sie läßt sich auf keine Weise weginterpretieren, da sie mit Übergängen verknüpft ist, die in der Theorie unbedingt vorkommen müssen, nämlich mit der Erzeugung von Teilchenpaaren. Sie liefert einen Beitrag, der immer in der Richtung der Kompensation des Potentialfeldes geht, denn der Zähler wird, wenn man noch mit V multipliziert, und über den Raum integriert, positiv definit, während der Nenner negativ ist $(E_i < 0, E_k > 0)$. Leider divergiert nun diese Summe.

Genauer betrachtet liegen die Verhältnisse so: Fragt man nach der ξ-ten Fourierkomponente von ρ, so läßt sich allgemein zeigen, daß sie in dieser Näherung proportional zur ξ-ten Fourierkomponente des Potentials ist:

$$\rho(\xi) = -B(\xi) \cdot V(\xi).$$

B ist hierbei ein bestimmtes Integral über die Eigenfunktion der positiven und negativen Zustände, das logarithmisch bei den Zuständen unendlich höher bzw. unendlich tiefer Energie divergiert, für jeden Wert von ξ.

Wir haben andererseits die Potentialgleichung, die in Fourierzerlegung die Form bekommt:

$$\xi^2 V(\xi) = 4\pi[\rho(\xi) + \rho_0(\xi)].$$

wobei ρ_0 die "äußere", das Potentialfeld erzeugende Ladung ist. Zusammen bekommt man also:

$$V(\xi) = \frac{4\pi\rho(\xi)}{\xi^2 + 4\pi B(\xi)}.$$

Wenn also $B(\xi)$ für alle Werte von ξ unendlich wird, so verschwindet V identisch an jedem Raumpunkt.

Dieses Resultat bedeutet natürlich eine erhebliche Schwierigkeit für die Diracsche Theorie. Es ist von derselben Art wie die Selbstenergie, aber wohl nicht schlimmer, denn zur Beseitigung der Selbstenergie muß man sicher etwas machen was darauf hinausläuft, daß die Matrixelemente die zu den ganz kurzen Wellen gehören, keine felderzeugende Wirkung mehr haben. Dann wird aber auch die Größe B endlich, und da sie nur logarithmisch divergent war, sicher hinreichend klein, um nicht mit der Erfahrung in Widerspruch zu kommen.

Die Divergenz tritt hier allerdings schon an einer früheren Stelle auf, also die Selbstenergie, nämlich ohne die Feldquantelung; ich glaube aber nicht, daß das ein prinzipieller Unterschied ist.

Man könnte noch fragen, wie weit dieses Resultat daran hängt, daß man sich auf die erste Näherung bezüglich des Potentials beschränkt hat. An sich wird zwar diese Annahme dadurch gerechtfertigt, daß sich zum Schluß das Potential zu Null ergibt, was sicher klein ist, aber man könnte unter Umständen wegen der Singularität des Coulombpotentials Bedenken haben. Ich habe mir das nicht anständig überlegt, glaube aber, daß dort nicht die Rettung zu suchen ist.

Zusammenfassend möchte ich also sagen, daß diese Schwierigkeit eine anständige Umschreibung des Diracschen Modells im Augenblick unmöglich macht, sondern daß diese Schwierigkeit zur zusammen mit der Selbstenergiefrage zu lösen ist. Andererseits bedeutet dies aber keinen ernsthaften Mangel der Diracschen Idee, denn sie dürfte dieselbe Reichweite haben, wie sie irgendeine Theorie haben kann, die die Selbstenergie nicht beseitigt.

Mit besten Grüßen
Ihr

R. Peierls

[164] Heinrich Peierls to Rudolf Peierls

Berlin, 18.7.1933

Meine Lieben.
Vielen Dank für Eure auf der Fahrt und in Cambridge geschriebenen Zeilen vom 13. d[iesen] M[onats]. Zunächst freue ich mich sehr über die Ehrung, die Dir, lieber Rudi, durch die Einladung nach Brüssel zuteil geworden ist[681] und hoffe, dass die Tagung für Dich interessant sein wird. Wann wird diese Versammlung stattfinden? Auch dass Du die Möglichkeit hast, nach Manchester zu gehen, freut mich, obwohl die Stadt selbst nicht sehr reizvoll ist.[682] Ich kann natürlich nicht von hier aus beurteilen, ob es richtig ist, den Ruf anzunehmen oder etwas anderes abzuwarten. Du wirst das von dort aus besser beurteilen können.

Ueber Fischers machen wir uns andauernd grosse Sorgen, da es sehr schwer wird, für Armin irgendetwas zu finden. Umso bedauerlicher wäre es, wenn Hans sich dort nicht halten kann.[683] Uebrigens soll nach meinen Informationen das Tageblatt, das hier eine besondere G.m.b.H. bildet, die Gehälter weiterzahlen und nicht in die Zahlungsschwierigkeiten von Rudolf Mosse einbezogen sein.[684] Ganz Positives hierüber konnte ich aber nicht feststellen; jedenfalls sind wir Euch für jede Freundlichkeit, die Ihr Hans Fischer erweist, dankbar.

Karli Fulda hat trotz aller Bemühungen und guter Empfehlungen in Paris nichts gefunden. Er will in einiger Zeit die Idee einer weiteren juristischen Ausbildng in Genf, die sich ihm wahrscheinlich bietet, verfolgen.

Wir freuen uns zu hören, dass es Dr. Thorner, wenn auch mit grossen Anstrengungen, zu gelingen scheint, sich in England eine Position zu

---

[681] Rudolf Peierls had been invited to take part in the 1933 Solvay Conference.

[682] See letter [161], note 671.

[683] The Fischer family were friends of Heinrich and Else Peierls. Father Armin had lost his job early after the establishment of the NS regime, and their son Hans tried to secure a position in England.

[684] The *Berliner Tageblatt*, founded by Rudolf Mosse, as one of the leading liberal newspapers in Germany, was one of the early targets of Nazi political harrassment, and the Mosse family left Germany the day after Hitler's appointment as Chancellor.

schaffen. Da seine Frau auch sehr tüchtig ist, so ist zu hoffen, dass sie durchkommen.

Unsere Hausnummer ist in der Tat plötzlich von 10 auf 5 abgeändert worden. Es schadet aber nichts, wenn noch Briefe mit der Hausnummer 10 unterwegs sind.
Viele herzliche Grüsse

Euer Vater

Meine Lieben!
Hocherfreut hat uns Rudis Stellungs- + Stipendiumsangebot + d[ie] Einladung z[um] Congress. Es ist s[ehr] schön, wie sein Weg harmonisch weiter aufwärts geht. Hoffentlich glückt Genia auch alles und Ihr seid fein dran. Nehmt Euch nur des Hansi an so oft Ihr könnt, er kennt nur Sorgen, schon als Schulbub. Weiter alles Gute + viele Grüße

Eure Else

auch hier kühler regnerischer Sommer.

[165] Rudolf Peierls to Hans Bethe

Cambridge, 24.7.[19]33

Lieber Bethe!
Das Laufende ist nicht viel weiter gelaufen, ich möchte Sie aber trotzdem darauf halten. Bragg war jetzt hier ich hatte ihn vorher nur flüchtig gesehen und der reizende Eindruck, den er schon damals auf mich gemacht hat, ist so vestärkt worden, dass das fast alle Unannehmlichkeiten der Stadt Manchester aufwiegen dürfte. Ich bin daher fast ganz sicher entschlossen, das Angebot von Bragg anzunehmen, es sei denn dass sich Fowler doch noch in den nächsten Tagen entschliesst, mir etwas hier anzubieten, wofür die Wahrscheinlichkeit nicht höher als 5,73% anzusetzen ist. In diesem Fall würden wir in Manchester zusammen sein, was natürlich sehr erfreulich wäre. Armstrongs Taktlosigkeit nimmt natürlich hier niemand ernst, zumal er ja in seinem Brief auch die Tatsachen entstellt hatte, eine Richtigstellung ist bisher nur in der Zeitung veröffentlicht worden. Im übrigen stimmt man aber insofern mit ihm überein, als die allgemeine Praxis die ist, uns keine

etatmässigen Stellen zu geben, sondern nur Stipendien oder ad hoc geschaffene Stellen. Ob das noch in einem oder zwei Jahren genau so sein wird, kann man natürlich nicht sagen. Vielleicht hat man sich bis dahin mit unserer Existenz schon abgefunden.

Noch eine technische Frage: Wir werden uns in Manchester natürlich eine Wohnung nehmen, und es würde ein relativ geringer Aufwand an Mühe sein, zu dieser Wohnung noch ein Zimmer mehr zu nehmen und das Ihnen zu geben. Da Sie das englische Essen nicht lieben, und wie Mott erzählte, sogar in Manchester besonders zu leiden hatten, so könnte ich mir denken, dass Ihnen diese Lösung angenehm wäre, zumal Sie dadurch in den Stand gesetzt würden, "langsam und reichlich" zu essen! Was die Kostenfrage betrifft, so würden Sie natürlich nur zu bezahlen haben, was Sie uns kosten, was wohl sicher weniger ist als in jeder Pension. Da wir schon sehr bald nach Manchester fahren wollen, um die Wohnungsmöglichkeiten zu studieren, wäre es mir lieb, wenn Sie bald antworten würden, wie Sie sich zu diesem Vorschlag stellen würden — natürlich vorbehaltlich der Entscheidung, ob Sie und ob wir überhaupt in Manchester sein werden.[685]

Die positiven Elektronen werden sicher durch Zusammenstoss des Lichtquants mit einem Kern erzeugt, also einem Vorgang, der sich in dem Diracschen Modell durch Photoeffekt aus dem negativen ins positive Energiespektrum beschreiben lässt. Die Wahrscheinlichkeit hierfür habe ich noch immer nicht fertig geixt. Oppenheimer hat eine Formel hierfür in der letzten Phys.Rev. veröffentlicht,[686] die, soweit ich sehen kann, für Oppenheimer ausserordentlich korrekt ist, es sind nämlich anscheinend nur die Zahlfaktoren falsch, jedenfalls für $E \ll mc^2$. Im entgegengesetzten Fall kinetischer Energien scheint mir seine Formel überhaupt verdächtig, ohne dass ich allerdings beweisen kann, dass sie falsch ist.

---

[685] Bethe did indeed live with the Peierls family in Manchester. As he wrote to Arnold Sommerfeld at the end of the year, this arrangement had 'disadvantages for the English' of the two physicists, but 'great advantages for their general well being and for their physics'. See Eckert, *Sommerfeld Briefwechsel*, vol. 2, p. 196.

[686] J.R. Oppenheimer and M.S. Plesset, 'On the Production of the positive Electron', *Phys. Rev.* **44**, 53–55 (1933).

Ueber die Frage, ob man Ihnen hier noch eine Stelle anbieten wird, und wenn was für eine, ist mir nichts Neues bekannt. Mit besten Grüssen Ihr

R. Peierls

Genia lässt grüssen.

[166] Else Peierls to Rudolf Peierls

St. Moritz, 10.8.[19]33

Lieber Rudel + Genia!
Euer lieber ausführlicher Brief war wirklich eine große Freude für uns, der ich gleich nach Erhalt Ausdruck geben möchte. Zunächst freut uns sehr, daß Ihr nun ein nettes Heim gefunden + Euch einrichten werdet.[687] Zufällig reisten wir im Zug mit einem früherem Collegen-Freund von mir, Prof. Robert Heger, der von der Wiener Hofoper nun a[n] d[ie] Staatsoper nach Berlin engagiert ist + mir sagte, daß er im Okt[ober] od[er] Nov[ember] in Manchester Concerte dirigiert. Merkt Euch das mal vor da müsst Ihr unbedingt hin.

Du bist wirklich ein lieber Kerl, Rudi, daß Du Dir um uns alle Gedanken machst.[688] Wenn es auch nicht schön in Deutschland ist und wir hier d[ie] Freiheit wunderbar empfinden (ich setzte diese Reise auch durch), so ist's persönlich für uns in Berlin erträglich + Vater immer noch so angesehen, + d[er] Berater v[on] Vielen. Außerdem ließe er sich nicht gut verpflanzen, + fühlte sich unfrei im Ausland. Wir haben bis 1. April unsere schöne Wohnung (sehr ermäßigt) + bis dahin wird sich vieles klären. Ob und wie wir uns verkleinern, ob Vater wie jetzt

---

[687] Rudi and Genia Peierls had moved to Menchester in the autumn of 1933.

[688] This letter written from Switzerland and others written from Switzerland in the summer of 1933 and not reproduced here (*Peierls Papers*, Supp. A 104) were among the few letters in which the family could assume not to be censored. Therefore they are franker in explanations concerning individuals, and also about the difficulties for Jewish family and friends in Germany.

noch Bergmann beibehält oder was wir sonst tun. Im Ausland leben geht der Gelder wegen, die man nicht angreifen darf, + d[er] Pension wegen nicht. Wie Ihr mich kennt, bin ich so beweglich, wie es velangt wird, + würde mich in alles schicken. Durch Vaters Einfluß und Diplomatie sind die Kinder in ihren Stellungen + ich meine die Initiative müßte von Alfred selbst ausgehen. Alfred hat sie nicht und Nina noch weniger. Sonst käme in erster Linie Milano in Betracht. Sorgen machen uns am meisten (da Nina und Vera arisch sind) Annies Kinder, aber auch da hat's noch vier Jahre Zeit. Ehe Ilse dann in die Höhere Schule käme, müßte sie ins Ausland. Hermann ist ja sehr geschäftstüchtig + schmiedet allerlei Pläne + knüpft ausländische Beziehungen. Hier kann er ja leider auch nicht weiterkommen. Wer kann prophezeien, was wird. Man muß warten und sich abfinden. Mit Vielem! Wenn Vater öfter mal verreist, was wir tun, wie Ihr seht, geht's ganz gut. Schrecklich sind die Verhaftungen Unschuldiger. Jedenfalls möchte man so gerne Vielen helfen + kann nicht. So war es nett, uns ins Bild zu bringen. Der arme Hansi hat nun seine Entlassung + Fischers machen uns weiter Sorgen. Am 1. Okt[ober] geben sie ihre Wohnung auf. Auch Riele schreibt sehr unglücklich. I[n] d[er] Provinz ist es viel schlimmer u[nd] sie wird auf der Straße angepöbelt. Wie geht's in Rußland, D[einen] Eltern, l[iebe] Genia, wenn man darüber liest, ist es auch schwer. Jedenfalls hast Du ein glückliches Temperament + bist energisch + so wirst Du schon mit allem fertig werden. So schließe ich mit den herzlichsten Wünschen für das Ereignis.[689] Telegr[aphiert] uns gleich.
Seid beide herzl[ich] gegrüßt v[on] Eurer Else
Entschuldigt die Feder ist schlecht.

Meine Lieben.
Else hat Euch so ausführlich berichtet, daß wenig hinzuzufügen ist. Auch ich habe mich über Eure ausführlichen Nachrichten sehr gefreut. Aber ich möchte nicht, daß Ihr übertrieben spart. Benützt doch ruhig auch das Geld, das Ihr Euch in der Schweiz erspart habt. Wenn dies trotzdem nicht reicht, so kann ich Euch in denjenigen Monaten,

[689] Genia was expecting her first child.

während der wir nicht selbst im Ausland sind, die gesetzlich gestatteten 200M schicken! Bitte m[ir] dann Meldung zu machen. Auf Antrag werde ich vielleicht auch etwas mehr von der Devisenstelle freibekommen. Übrigens: ich habe, bevor ich Berlin verließ mein Testament neu redigiert, wovon man ja nicht stirbt. Bei dieser Gelegenheit habe ich jetzt den Passus gestrichen, von dem ich mit Dir einmal in Berlin sprach, daß Dein Erbteil zunächst von dem Testments-Vollstrecker verwaltet wird, und Du nur die Zinsen erhälst. Bei einem 26-jährigen und Familien-Vater in spe schien mir das nicht mehr erforderlich zu sein. Allerdings weiß ich vorläufig nicht, wie Ihr [?] halber das Geld aus Deutschland herausbekommt. Daraufhin werde ich mir jedenfalls so viel Zeit wie möglich nehmen und mit meinem Ableben zu warten suchen, bis sich die Verhältnisse geklärt haben. Ferner habe ich, da die anderen Kinder nun [??] erhalten haben es Dir l[ieber] Rudi, zum Ausgleich ein Voraus-Vermächtnis in angemessener Höhe eingeräumt.

Für Deine Bemühungen wegen Dr. Samson vielen Dank. Mir liegen Fischers, Karli Fulda, Erich Bruck + Hermann Hamburger + Hans Fischer sehr stark auf der Seele. Wir haben es hier weiter sehr angenehm. Briegers kommen in einigen Tagen.

Wenn zu Hause nicht besonderes anfällt, so bleiben wir bis zum 27. hier und fahren direkt nach Hause wo wir also am 28. eintreffen wollen. Dies für Briefe und etwaige Depeschen, denen wir hoffnungsvoll entgegensehen.

Eure Dispositionen betr[effend] Manchester scheinen mir durchaus richtig zu sein. Hans Juliusberger kann erst im Herbst übersehen, ob sich sein Geschäft halten kann. Alfred hofft, seine Englandreise im Herbst machen zu können und dabei werdet Ihr alles besprechen können. Ich halte seine hiesige Stellung jetzt nicht mehr für gefährdet.
Viele Grüße

Euer Vater

[167] Heinrich Peierls to Rudolf Peierls

Berlin, 16.9.1933

Meine Lieben.
Euer Schreiben vom 12. d[iesen] M[onats] haben wir erhalten. Ich hoffe, dass Du, lieber Rudi, inzwischen aus Kopenhagen gut zurückgekehrt bist und Deine Familie wohl angetroffen hast. War die Konferenz interessant und nicht zu anstrengend?[690]

Gern haben wir davon Notiz genommen, dass die Namen Eurer Tochter Gaby Ellen lauten,[691] wobei der erste Name der Rufname ist.

Fräulein Dr. Ullmann ist nicht mehr bei uns. Sie ist vom Dienst beurlaubt und wird voraussichtlich damit einverstanden sein müssen, wieder als Studienrätin in eine kleine Stadt zu gehen. Wenn Du ihr eine Zeile schreiben willst, so ist die Adresse bis auf weiteres: Stettin, Gesenius-Wegener Oberlyzeum, Friedrich-Karl-Str. 40–42.

Dass Ihr nicht an Onkel Siegfried geschrieben habt, ist mir nicht angenehm. Ich habe es ihm vor einigen Tagen direkt mitgeteilt, und er wird Euch wohl auch gratulieren. Jedenfalls bitte ich, ihm dann sofort zu antworten. Seine Adresse für die nächsten Wochen ist: Aix-les-Bains, Hotel Splendide.

Von Hermann und Rita Weigert liegen bereits Nachrichten aus Kapstadt vor, die vorläufig noch recht wenig ermutigend klingen.[692] Es lässt sich aber nach der kürze der Zeit noch nichts Bestimmtes sagen. Ihre Adresse ist vorläufig: Kapstadt, Hotel Royal.
Viele Grüsse Euch Dreien

Dein Vater

---

[690]This refers to the physics conference which took place in Copenhagen between 14 and 20 September 1933. This was the last conference attended by Ehrenfest who committed suicide five days after the end of the meeting.

[691]The first daughter of Rudolf and Genia Peierls was born in August 1933.

[692]The Weigerts, a brother of Rudolf Peierls' mother and his wife, had emigrated to South Africa.

[168] Heinrich Peierls to Rudolf Peierls

Berlin, 17.11.1933

Lieber Rudi.
Endlich erhielten wir Eure Nachrichten vom 13. d[iesen] M[onats]. Es muss also tatsächlich eine Mitteilung von Euch verloren gegangen sein, denn wir haben über zwei Wochen lang nichts von Euch gehört. Jedenfalls freue ich mich, dass Ihr wieder alle gesund seid und hoffe, dass es so bleiben wird. Bitte vergiss nicht, mir gelegentlich den für Dr. Kurt Weigert bestimmten Brief bezüglich seines Sohnes zu schicken.

Der englische Nebel ist auch mir in unangenehmer Erinnerung geblieben. Ich war an einem nebligen Sonntag in London zum Lunch eingeladen und war erst nach langer Verspätung imstande, die Wohnung, in die ich bestellt war, richtig aufzufinden. Auch wir haben von Hermann und Rita verhältnismässig gute Nachrichten. Es darf wohl angenommen werden, dass sie sich, wenn auch langsam, einrichten.

Wir wollen Weihnachten zu Hause bleiben. Dagegen habe ich Krebses und Alfreds eingeladen, zu unserer Vertretung am 24. Dezember in Meran zu sein. Da es gerade um die Weihnachts- und Neujahrszeit ist, wird sich dies gut ermöglichen lassen.
Sonst nichts Neues. Viele Grüsse

Euer Vater

Wenn Ihr diese Zeilen erhaltet, begeht Gaby ihren Vierteljahresgeburtstag. Herzliche Glückwünsche!

D[ein] V[ater]

Gratuliere ebenfalls.
Liebe Leute!
Wie fein, daß Gaby so abgehärtet ist. Ich habe auch wieder meine Erkältung weg (Kehlkopfkatarrh/zu dumm). Habt Ihr nichts von einem Concert des Kapellmeisters Robert Heger gehört; er hatte in vielen Städten Englands große Erfolge + muß auch in Manchester gewesen

sein; ich schrieb Euch schon seinerzeit davon.[693] Hermann Schey bleibt nun hier + wird auch im Culturbund jüdischer Leiter + Mitwirkende, Zuhörer nur Juden) singen. Sollen glänzende Kräfte haben. Wir sahen einen Film mit bildschönen Menschen, spielt in Amerika: "Seitenwege des Lebens".[694] Seht Ihr dort Kino's? Es gießt. Alles Gute!
GrußKuss

E[lse]

Armin Fischer fährt für 3 Wochen nach England.
Wie geht's Genia's Angehörigen?

[169] Rudolf Peierls to Paul Dirac

Manchester, 8.12.1933

Dear Dirac,
I had some difficulties in understanding your letter,[695] but I believe that I now see what you mean.

Although you do not state your form of the matrix $\mathfrak{R}_a$ I think it very likely that your method is equivalent to mine.

One can express my method also in the form of a matrix (although, as you say, it is not easy to get the physical results directly from this form). My method is identical with yours, if one chooses (apart from factors)

$$(xt/\mathfrak{R}_a/x't') = (xt/\mathfrak{R}_o/x't') \times e^{\frac{ie}{\hbar c}\int_{xt}^{x't'} \mathfrak{A} d\xi} \qquad (*)$$

where $\mathfrak{R}_o$ means the corresponding matrix for free electrons and the integral is to be extended along the staight line joining the world point xt with $x't'$. while $\mathfrak{A}$ denotes the component of the vector potential parallel to this line. (*) is obviously relativistically invariant, and gauge invariant (one sees, by the way, in this form, that any method is relativistically invariant without making any assumption even for strong fields which I did not know before).

---

[693]Letter [166].
[694]John M. Stahl's 1932 melodram, *Back Street,* had been dubbed for German audience in 1933.
[695]Letter could not be located.

As $*$ is very simple, too, I expect it to be identic[al] with yours. It would be very nice, if you could write me whether this is true.

As to the equation for $\mathfrak{R}_b$, I cannot quite understand what you mean, because it is not possible that equations of the form

$$-\frac{\hbar}{i}\frac{\delta\mathfrak{R}_b}{\delta t} = X\mathfrak{R}_b \qquad \frac{\hbar}{i}\frac{\delta}{\delta t'}\mathfrak{R}_b = \overline{X}\mathfrak{R}_b \tag{+}$$

can hold, where X is not the operator occuring in the equ[ations] of motion. If the equations of motion are

$$TR = 0 \qquad ; \qquad R\overline{T} = 0$$

one gets immediately an equation for $\mathfrak{R}_b$:

$$T\mathfrak{R}_b = -T\mathfrak{R}_a \qquad ; \qquad \mathfrak{R}_b\overline{T} = -\mathfrak{R}_a\overline{T}$$

Now either $\mathfrak{R}_a$ satisfies the equ[ations] of motion; then the right hand sides will vanish and $\mathfrak{R}_b$ is a solution of these equations as well; or the right hand sides do not vanish; but then the equ[ations] for $\mathfrak{R}_b$ become non-homogeneous and therefore are not of the type (+). If the latter is the case if I am right in assuming that your $\mathfrak{R}_a$ is equal to ($*$). Then my calculations show that $\mathfrak{R}_b$ is not regular at the point $xt = x't'$ but retains a logarithmic singularity.

If this is true it would confirm me in the view that the physical conclusions of my last letter are correct. If not, I would very much like to see in what respect your $\mathfrak{R}_a$ differs from ($*$)
Yours sincerely

R. Peierls

[170] Heinrich Peierls to Rudolf Peierls

Berlin, 16.1.1934

Meine Lieben.
Für Eure Zeilen vom 6. d[iesen] M[onats] vielen Dank. Wir haben uns sehr gefreut, dass Ihr angenehm in das neue Jahr hereingekommen seid.

Bei uns war es, wie wir Euch ja inzwischen gschrieben haben, gleichfalls der Fall. Euer Familienbild hat uns sehr erfreut. Gaby sieht darauf ja schon sehr erwachsen und verständig aus.

Einen Besuch in England werden wir in absehbarer Zeit nicht machen können, da nach der jetzigen Devisen-Ordnung nur RM 200,- pro Person mitgenommen werden können und dies nur einen Aufenthalt von wenigen Tagen bedeuten würde, der die weite Reise nicht lohnt.

Bei dieser Gelegenheit bitte ich Dich, lieber Rudi, mir mitzuteilen, in welcher ungefähren Zeit Deine Abmeldung von Deutschland nach Zürich erfolgt ist. Es muss ungefähr der Termin gewesen sein, zu dem Du Dein Studium in Zürich begonnen hast. Du wirst aber eher als ich den Zeitpunkt feststellen können. Wir haben damals noch in der Roonstrasse gewohnt. Ich bitte Dich um gelegentlichen Bescheid.
Viele Grüsse

Dein Vater

[171] Heinrich Peierls to Rudolf Peierls

Berlin, 22.1.1934

Lieber Rudi.
Unsere Briefe müssen sich gekreuzt haben, da Du in Deinem Schreiben vom 19. d[iesen] M[onats] mitteilst, Ihr hättet lange nichts von uns gehört. Inzwischen werdet Ihr wohl unsere letzten Nachrichten erhalten haben.

Es tut uns schrecklich leid, dass Genia, von deren Eltern wir übrigens einen sehr liebenswürdigen Brief erhalten haben, nicht wohl war. Ich habe den Eindruck, dass sie sich etwas überanstrengt hat und wiederhole unsere Bereitwilligkeit, Euch einen Zuschuss zu geben, wenn dies zweckmässig erscheint. Ich bitte, ungeniert hiervon Gebrauch zu machen. Bitte teile uns baldigst mit, ob sich Genias Befinden inzwischen gebessert hat, was wir zuversichtlich hoffen.

Dass Gaby sich zufriedenstellend entwickelt, freut uns sehr und wir bedauern nur sehr, noch nicht persönlich ihre Bekanntschaft gemacht zu haben.

Die indische Angelegenheit interessiert uns natürlich sehr, aber wir werden wunschgemäss nicht hierüber sprechen.[696] Ein Rat lässt sich von hieraus nicht erteilen, zumal wir nicht wissen, welche Chancen Du in Stamm-England hast. Einerseits wäre ein Fortkommen in Europa sicher viel angenehmer für Dich, andererseits gebe ich zu, dass die Chancen sich stark verschlechtert haben. Was Indien selbst betrifft, so würde unser grosses Bedauern, Euch so weit weg zu wissen, für Eure Entschliessungen nicht in die Waagschale geworfen werden dürfen. Aber auch abgesehen hiervon, weiss ich nicht, ob Du recht daran tätest, es anzunehmen, wenn sie Dir angeboten werden sollte. Kann man sich übrigens im vorliegenden Falle bewerben und schliesslich, wenn man akzeptiert wird, immer noch zurücktreten? Jedenfalls will eine derartige Sache sehr reiflich überlegt sein, denn ich sehe in allen ähnlich gelagerten Fällen, dass es doch nicht so einfach für Europäer ist, sich in überseeischen Ländern, noch dazu Indien, einzuleben. Aber letzten Endes müsst Ihr die Entscheidung selbst treffen. In beiden Fällen, ob Du in England bleibst oder nach Indien gehst, bitte ich, mir gelegentlich zu schreiben, wie Deine Naturalisierungs-Aussichten stehen.

Von hier aus ist nichts Neues zu berichten; es geht uns allen gut. Mit herzlichen Grüssen und vielen guten Wünschen für Genia

Dein Vater

Lieber Rudi!
Ich kann nur sagen, dass ich wieder einmal ganz meiner Meinung mit Vater bin. Meinem Gefühl nach wird ja gerade Asien der Schauplatz schwerster Crisen + Kämpfe für lange Zeit werden + so schön Indien sein muß, (es war immer mein höchster Wunsch, es kennen zu lernen) so glaube ich für Dich und Deine Familie wäre England schon besser. Dies nur gefühlsmässig. Kennst Du keinen Inder? Gehörige Auskünfte wirst Du ja einziehen. Es tut mir sehr leid, daß Genia sich überanstrengt hat; sie macht's wie ich öfter "plus ultra"! Bitte Genia, nimm Dir auf unsere Kosten ein richtiges Mädchen + rackere Dich nicht unnütz selbst ab.

[696] Peierls was thinking of applying for a position at Raman's newly opened institute in Bengalore. See letter [173].

Das Leben giebt uns ausserdem genug zu tun auf.
Über den lieben Brief D[eines] Vaters und D[einer] Mutter habe ich mich sehr gefreut. Wir erwarten mit Spannung weiteres von Euch Dreien. I wish you a speedy recovery.
Love,
Yours sincerely

Else

[172] Niels Bohr to Rudolf Peierls

Copenhagen, 27.1.1934

Lieber Peierls,
Gleich beim Empfang Ihres Briefes habe ich ein paar Linien geschrieben, die ich einschliesse, und die ich hoffe für Ihre Bewerbung in Bengalore behilflich sein wird.[697] Übrigens werde ich sehr bald an Raman in einer anderen Angelegenheit schreiben und werde dann auch Ihre Bewerbung auf das wärmste empfehlen. Sobald ich näher über die von Ihnen erwähnten Fragen der Diracschen Theorie[n] nachgedacht habe,[698] und wir sie im Institut besprochen haben, sollen Sie wieder von uns hören. Heute sende ich [mit] derselben Post ein Exemplar der Arbeit von Rosenfeld und mir, in deren letzten Fassung wir hoffen die Problemen etwas klarer ausgedrückt zu haben.[699] Übrigens bin ich eben beschäftigt mit

---

[697]See letter [171].

[698]The early part of 1934 saw a lively discussion of Dirac's hole theory across the centres of theoretical physics. Heisenberg and Pauli, in particular, communicated about this, and after exchanging their views at the 7th Solvay Congress in Brussels in October 1933, they exchanged more than 50 letters in 1934 most of which concerned Dirac's theories. See Pauli, *Wissenschaftlicher Briefwechsel*, II, pp. 243–75. See also letters [169], [176] and [178].

[699]This refers to their paper on measurability of electromagnetic field quantities. The paper was first submitted in December 1932, but according to Rosenfeld it took more than a year and fourteen drafts until the paper reached its final form. See L. Rosenfeld, " in W. Pauli (ed.), *Niels Bohr and the Development of Physics*, New York: McGraw Hill, 1955; L. Rosenfeld (ed.), *Niels Bohr, his life and work as seen by his friends and colleagues*, Amsterdam: North Holland, 1967, p. 125.

einer kleinen Artikel über die Grenzen des Korrespondenzargumentes in der Elektronentheorie, die ich hoffe, bald abzuschliessen.
Mit vielen freundlichen Grüssen an Sie und Ihre Frau und alle gemeinsame Freunde in Manchester.
Ihr ergebener

Niels Bohr

[handwritten addition]
P.S. Ich sehe dass ich nicht an Ihre Frage über eventuelle Ferienbesuche in Copenhagen geantwortet habe, dass solche Besuche für uns immer eine Freude sein sollen, brauche ich aber auch kaum zu sagen.

[173] Rudolf Peierls Arnold Sommerfeld

Manchester, 28.1.1934

Sehr geehrter Herr Geheimrat,
Ich komme heute mit einer Bitte zu Ihnen: Ich habe die Absicht, mich um eine Stelle in Ramans neuem Institut in Bengalore zu bewerben, die in der "Nature" vom 6. Januar ausgeschrieben ist. Ich wäre Ihnen nun ausserordentlich dankbar, wenn Sie für mich eine Empfehlung schreiben könnten, da dies sicher meine Chancen für die Stelle wesentlich erhöhen würde.

Falls Sie diese Freundlichkeit haben wollen, so möchte ich es Ihrem Ermessen überlassen, ob Sie mir diesen Brief schicken wollen, oder direkt an Raman schreiben, aber vielleicht ist das letztere das einfachste. Ramans Adresse ist: The Indian Institute of Science, Bengalore, Indien. (Flugpostbriefe gehen eine Woche schneller als gewöhnliche.)

Ich bin hier in der Gesellschaft von Bragg und Bethe ausserordentlich zufrieden, es ist eben nur schade, dass so wenig Aussichten bestehen, sich hier für dauernd festzusetzen.

Mit nochmals herzlichen Dank und in der Hoffnung Ihnen mit diesem Anliegen nicht zu viel Mühe zu machen,
mit besten Grüssen, auch von Bethe
Ihr Ihnen ergebener

R. Peierls

[174] Arnold Sommerfeld to Rudolf Peierls

Munich, 2.2.1934

Lieber Peierls!
Beifolgender Schrieb ist mit Luftpost abgegangen.[700] Ich habe inzwischen erfahren, dass Hopf's[701] Bewerbung keine Concurrenz für Sie bedeutet, da es sich bei ihm um eine Professur für angewandte Physik handelt. Ich wollte aber die Absendung nicht verzögern. Bangalore ist sehr schön und das Institut von Raman ist das beste in Indien. Sagen Sie bitte Bethe, dass ich die Separata des Handb[uch] Art[ikels][702] versenden will u. diesbez[üglich] sehnsüchtig auf seine Antwort warte, auch wegen der anderen Punkte, die mir am Herzen liegen.

Sie verstehen, dass ich Hopf in seiner prekären Lage in erster Linie dienlich sein wollte;[703] ich freue mich jetzt dass mein warmes Eintreten für Hopf Sie nicht schädigen wird.
Alles Gute!

A. Sommerfeld

---

[700] Copy of Sommerfeld's reference in Eckert, *Sommerfeld Briefwechsel*, vol. 2, pp. 406–407.

[701] Ludwig Hopf (1884–1939), former sudent of Sommerfeld's.

[702] See letter [145], note 545.

[703] Sommerfeld had supported Hopf's application for a post at the same institute in Bangalore, as Hopf was in a more precarious financial situation than Peierls.

[175] Heinrich Peierls to Rudolf Peierls

Berlin, 7.2.1934

Meine Lieben.
Ich habe Euch für zwei Briefe zu danken, ebenso für Eure Glückwünsche zu meinem Geburtstag, der programmgemäss verlaufen ist. Besonderen Dank sende ich Euch auch für das offenbar sehr interessante Buch, das pünktlich hier eintraf. Ich werde mit der Lektüre demnächst schon beginnen können.

Eine Frage hast Du, lieber Rudi, übersehen, mir zu beantworten. Bitte hole dies umgehend nach.

Ich wollte wissen, wann Du seinerzeit von München nach Zürich abgemeldet worden bist. Es handelt sich um folgendes: Es ist jetzt, bei dem augenblicklichen Stand der Devisengesetzgebung möglich, einen Antrag zu stellen, wonach ich Dir anstatt laufender kleiner Ueberweisungen einmalig eine grössere Summe zukommen lassen kann. Hierzu ist folgendes erforderlich:

Erstens ist die Beibringung einer Kopie Deiner Abmeldung auch München nach der Schweiz notwendig. Ich erweitere also mein Ersuchen in der Weise, dass ich bitte, zu sehen, Dir diese Abmeldung schleunigst zu beschaffen und sie mir weiterzugeben.

Zweitens würde ich für eine gewisse Zeit Deinen Pass brauchen, aber ich möchte erst wissen, wie lange er noch läuft.

Drittens brauche ich eine behördliche Bescheinigung darüber, dass Du Dich augenblicklich in Manchester aufhälst.

Da ich sehen möchte, die Angelegenheit noch vor unserer Abreise, die wahrscheinlich in etwa 14 Tagen erfolgt, in die Wege zu leiten, so wäre ich Dir jedenfalls für umgehenden Bescheid dankbar.

Was die Sache Indien betrifft, so wollen wir erst einmal den Erfolg Deiner Bewerbung abwarten. Im übrigen gebe ich zu, dass — falls Du gewählt wirst — die Sache auch ihre Lichtseiten hat und überlasse alles weitere, wie ich schon mehrfach erwähnt habe, Euch selbst.
Viele herzliche Grüsse

Euer Vater

Meine Lieben!
Wir feierten gestern im Familienkreise nachmittags mit den Enkelkindern + abends kamen dazu d[ie] Kinder, Gretel, Uli + seine Großmama. Im Theater nichts Aufregendes, so gehen wir als Nachfeier in die "Fledermaus" heute mit Besuch. Das Wetter ist wechselnd + macht müde. Ich habe einen reizenden kl[einen] Teckel, der s[ehr] wild + verspielt ist. Wie geht's Gaby + Genia? Alles Gute Euch Dreien + viele Grüße

E[lse]

[176] Rudolf Peierls to Paul Dirac

Manchester, 8.2.1934

Dear Dirac,
Many thanks for your manuscript[704] and the letter. I cannot agree with you that it is not easy to find other expressions for $\mathcal{R}_b$ which have the same properties of being finite giving conservation of charge, having right invariance and vanishing for vanishing field (and, of course, not giving the same value for $H\mathcal{R}_a$). Instead of choosing $\mathcal{R}_a$ and $\mathcal{R}_b$, one may write

$$\mathcal{R} = (\mathcal{R}_a - S) + (\mathcal{R}_b + S)$$

Let us consider first the quantities

$$S_\mu(x't') = \begin{cases} i\sum_{spin} S(x't';x't') & \mu = 4 \quad (x_4 = ict) \\ \sum_{spin} \alpha_\mu S(x't';x't') & \mu = 1,2,3 \end{cases}$$

which are important for the production of field. They must satisfy the following conditions:

1. $\delta\mu$ form a four vector
2. $\delta\mu$ finite everywhere

[704] Dirac had submitted his long-awaited manuscript about his hole theories in early February 1934. P.A.M. Dirac, 'Discussion of the Infinite Distribution of Electrons in the Theory of the Positron', *Proc. Cambr. Phil. Soc.* **30**, 150–163 (1934).

3. $\sum_\mu \frac{\partial S_\mu}{\partial x_\mu} = 0$
4. $S_\mu = 0$ for $F_{ik} = 0$ (F = field tensor)

In three dimensions one can construct vectors satisfying cond[ition] 3 by the operation rot In analogy we construct the following operation:

$$\mathcal{E}_{iklm} = \begin{cases} 0 & \text{if two of the indices are equal} \\ +1 & \text{if iklm is an even permutation of 1234} \\ -1 & \text{'' '' '' odd '' '' ''} \end{cases}$$

$\mathcal{E}$ forms a pseudo-tensor, i.e. behaves like a tensor under Lorentz-transformation with the determinant $+1$, but changes sign if the determinant is -1. If therefore, we form the expression

$$S_\mu = \sum_{klm=1}^{4} \mathcal{E}_{\mu klm} \frac{\partial}{\gamma x k} A_{em} \qquad (*)$$

where A is a pseudo tensor, it is obvious that conditions 1 to 3 are satisfied. The simplest expressions are obtained by choosing for A the "dual" field tensor F (which is obtained by F_{ik} by interchanging electric and magnetic field)

In this case

$$S_\mu = a \sum_\nu \frac{\partial F_{\mu\nu}}{\partial x_\nu}$$

with an arbitrary constant a. This means that S is proportional to the total current density, i.e. means an additional constant polarisibility. The next simple is

$$A_{em} = \sum_{i,k} F_{ik} \overline{F}_{ik} F_{em}$$

which leads to

$$S_\mu = ib \sum_i \frac{\partial}{\partial x_i} (E_i(\vec{E}\vec{H}))$$

$$S_l = b \frac{d}{dt} \{E_l(\vec{E}\vec{H})\} - b\, rot_l \{\vec{H}\ (\vec{E}\vec{H})\}$$

(with arbitrary b).

In this case the additional current density would not only depend on the "external" current density. You see that an infinite number of such expressions may be obtained by introducing higher powers of the field tensor. A further modification may be obtained by introducing differential operations of any order, i.e. by applying the symbol $c^{\frac{1}{2}}\frac{\partial^2}{\partial t^2} - \nabla^2$ to any of the above expressions. Finally one has to see whether to each of these functions S_μ corresponds a matrix

$$S(x^{'}t^{'}, x^{''}t^{''})$$

Once $S_\mu(x^{'}t^{'}, xt)$ is given this means that $S(x^{'}t^{'}, xt)$ is partly defined the freedom one has in choosing diagonal elements (diagonal with respect to xt) consists in adding any function whose diagonal spin sum vanishes, also after multiplication with an α. One can add any tensor function of the co-ordinates, multiplied $\alpha_i\alpha_k - \alpha_k\alpha_i$, by $\alpha_i\alpha_k\alpha_l$ or by a suitable combination containing $\alpha_1\alpha_2\alpha_3 \cdot \beta$.

If all spacially diagonal elements are given, we are left with the problem of finding a matrix with given diagonal elements which is everywhere finite and vanishes if the diagonal elements vanish. Furthermore, it must be defined in a relativistically invariant way, and (as you emphasize) must not satisfy the equation of motion.

There are many ways of constructing such matrices, the simplest being the following.

1. Put $S(x^{'}t^{'}x^{''}t^{''})$ equal to the arithmetic mean of the diagonal values at $x^{'}t^{'}$ and $x^{''}t^{''}$.

2. Put $S(x^{'}t^{'}x^{''}t^{''})$ equal to the diagonal element at a point half way between $x^{'}t^{'}$ and $x^{''}t^{''}$.

More complicated solutions may be obtained by introducing functions of the distance between $x^{'}t^{'}$ and $x^{''}t^{''}$ and of the field. You will agree that there is such a variety of different solutions of our conditions for $\mathfrak{R}_b$ that it is hardly possible to say it is very difficult to find a different solution. There seems to be, however, one more restricting condition which seems to have escaped your attention:

Let us restrict ourselves, as before, to Hartree's approximation and let me choose three dimensional equations which are mathematically simpler if there is no question of invariance. The one has to find a matrix $R_b(x', x'', t)$ so that

$$-i\hbar R_b' = HR_b - R_bH + D$$

$$divE(xt) = \sum_{spin} 4\pi eR_b(x, x, t)\ divH = 0$$

$$\frac{1}{c}\frac{d}{dt}E(xt) - rotH(xt) = \sum_{spin} 4\pi e\alpha R_b(x, x, t)$$

Here

$$H = i\hbar c\alpha_k \frac{\partial}{\partial x_k} + \beta mc^2 + eV - \frac{e}{c}\alpha_k \mathfrak{A}_k(xt)$$

and

$$D = HR_a - R_aH + i\hbar R_a$$

is a given matrix that depends only upon the field. These equations are compatible if R_b satisfies the conservation of charge. In order that they may be quantized, however, we need also a conservation of energy and momentum.

Let us consider the energy. One might expect in analogy to the usual theory, that the energy has the form:

$$E_0 = \sigma(HR_b) + \frac{1}{8\pi}S(\vec{E}^2 + \vec{H}^2)d\tau$$

where the symbol σ means: diagonal sum with respect to all variables. From this point of view one would subject the theory to the condition that E_0 must be a constant of motion. That obviously would be a very narrow restriction and would probably have no freedom whatsoever in R_a if there is any possible solution at all. I believed this to be a correct point of view, but in the meantime I have seen that it is not a very natural one. For if we make a theory in which the vacuum has a polarisability, this obviously means that we modify the electrodynamics, and if we do so there is no reason why the energy expression should not be changed. It is essential, however, that there is some energy expression, i.e. that there is a function E, such that

$$\frac{d}{dt}(E_0 + R_1) = 0$$

We calculate $\frac{d}{dt}E_0$

$$\frac{d}{dtE_0} = \frac{i}{\hbar}\sigma(\dot{H}R_b - R_b\dot{H}) = \frac{i}{\hbar}\sigma HD + \frac{1}{4\pi}\int(\vec{E}\ \dot{E} + \vec{H}\ \dot{H})d\tau \qquad (+)$$

$\dot{H}$ in the first term means the differentiation of H with regard to t as far as it occurs explicitly. This gives a term $e\sigma(\vec{E}\ \alpha R_b)$ if, for simplicity (and without loss of generality) we assume the scalar potential to be zero. This term cancels with the corresponding expression arising from the last term in (+) (in agreement with the fact that $\frac{d}{dt}E_0 = 0$ if $D = 0$, i.e. if there is no negative electron distribution.) The second term remains only and gives $-\frac{dE_0}{dt} = \frac{dE_1}{dt} = -\frac{i}{\hbar}\sigma HD$

D must be subject to the condition that this equation has a non-trivial solution for E_1 (i.e. where E_1 is not equal to $-E_0$).

According to your assumption D depends upon the field only, and as E_1 is the only function containing essentially R_b and having the property that its derivative depends upon the field only (because the current and change of density can be expressed by the field with the aid of the field equations) E_1 will have to be a function of the field only.

I am not going to investigate which matrices D fulfil this condition, but I shall only show that this really means a restriction as to the possible choice of D, but that this restriction is not strong enough to make D unambiguous. (if there is any solution at all). For this purpose I assume that a certain R_a fulfils the requirements and I replace it by $R_a - S$ as above. Then

$$D' = D - (HS - SH) - i\hbar\dot{S}$$

Therefore, the new D will again give conservation of energy if

$$\sigma H\dot{S}$$

is a final derivative of a function of the field variables which does not contain t explicitly.

As

$$\frac{d}{dt}\sigma(HS) = \sigma(H\dot{S} + \dot{H}S)$$

obviously has this property, this is equivalent with the condition that

$$\sigma H\dot{S} = e\int d\tau \sum_{k=1}^{3} E_k S_k$$

does so (S_k has the same meaning as in the beginning.)

An equivalent condition might be put in the form the conservation of momentum but the above condition together with relativistic invariance, is sufficient. Now, if you insert for S the second of the possibilities given in the beginning, you will see that the resulting expression cannot be written as a final derivative of a function not containing it explicitly.

This shows that not every R_a that satisfies your condition is possible physically. It is not proved that your particular R_a is one of the possible ones.

On the other hand, the condition of energy conservation is not sufficient to make R_a unambiguous. This is clear because one can chose such matrices S that $e\int d\tau \sum_{k=1}^{3} E_k S_k$ is a final derivative.

At first this property is common to all those S which have vanishing diagonal current density (They may, however, lead to different results for some physical questions).

One of the non-trivial solutions is the first expression given above, corresponding to a constant additional polarisability. There are many even less trivial solutions, because any field theory in which Maxwell's equations are modified without destroying their invariance defines a polarisability which has the required properties. (One example is Born's theory. I do not, however, believe that the very special modification he proposes, has any physical reality.)

In conclusion, I think one must admit the following points:

1. There are a great many matrices R_a satisfying your conditions, if there is one.

2. If one postulates conservation of energy, there is a further restriction in the possible choice of R_a already in the Hartree approximation. It is not proved that your particular R_a satisfies this condition.

Piz da Daint, Engadin, 1930. Gamow "working".

With Pauli.

With Genia Peierls and Placzek.

Genia and Nina Kannegiser with Landau.

Peierls family with Hans Bethe.

With Pauli.

Arnold Sommerfeld, 1927.

With Genia.

Landau, mid-1930s.

With Placzek.

Skiing with Klaus Fuchs, second from right.

With colleagues at the Mond Laboratory, Cambridge.

In his office.

Leipzig 1931. With Heisenberg, Gentile, Placzek, Wick, F. Bloch, Weisskopf and Sauter.

With Genia, New York, 1943.

With colleagues in Russia, 1931.

1946.

3. There are, however, still many different matrices R_a fulfilling all requirements. A further restriction upon them might arise from the proper theory of interaction, but this point has not yet been investigated.

4. If there is any freedom left, the actual choice would have to be left to experimental evidence. One might guess the result, if one chooses the mathematically simplest of all admissible solutions, but I am neither convinced that your function is admissible, nor that it is the simplest.

With best regards
Yours sincerely,

R.Peierls

[177] Heinrich Peierls to Rudolf Peierls

Berlin, 14.2.1934

Lieber Rudi.
Deine Mitteilungen vom 9. d[iesen] M[onats] habe ich mit Dank erhalten. Sowie die Kopie Deiner Abmeldung aus Leipzig gekommen ist, bitte ich, sie mir zuzuschicken. Dabei macht es nichts, wenn wir schon verreist sind, da ich hier dafür gesorgt habe, dass diese Sachen ordnungsgemäss erledigt werden.

Ebenso bitte ich, die Bescheinigung von Manchester baldigst einzusenden.

Der Hautzweck meines heutigen Schreibens ist, Dich zu bitten, Deinen Pass umgehend per Einschreiben an mich zu senden. Du bekommst ihn nach wenigen Tagen ebenfalls unter Einschreiben zurück. Die Absendung des Passes möchte ich deswegen beeilt haben, weil ich ihn selbst wieder an Dich zurück schicken will. Es geschieht nichts weiter mit dem Pass, als dass ein Stempel darauf gedrückt wird.

Ich schicke Dir per Drucksache drei Formulare, bei denen Du in blanco Deine Unterschrift geben und sie mir auch baldigst wieder zurückschicken willst. Du brauchst nur zwei Formulare zu schicken, das dritte ist nur beigefügt für den Fall, dass eines der Formulare unbrauchbar wird. Ich nehme an, dass Du von Deinen Schweizer Ersparnissen her noch ein Guthaben in England oder Schweiz hast. Wie hoch ist dieses?

Ferner bitte ich Dich, baldigst eine Vollmacht folgendes Inhalts zu schicken:

Hierdurch bevollmächtige ich Herrn Willy Ritter, Berlin, Alexanderplatz 1, mich bei der Auswanderungsberatungsstelle und Devisenbewirtschaftungsstelle in Berlin zu vertreten.

Es scheint, dass ein Brief von uns verloren gegangen ist, in dem wir geschrieben haben, dass unser Hund Charlie mit einer Spritze ins jenseits befördert werden musste, weil er eine organische Erkrankung hatte, die nicht heilbar war.

Wir hatten Dir ferner mitgeteilt, dass wir am 24. d[iesen] M[onats] nach Rom fahren wollen, dort etwa eine Woche bleiben und dann via Neapel und Sorrent für einige Zeit wieder nach Capri gehen wollen. Es kommt mir sehr merkwürdig vor, dass wir erst nach Rom gehen, nachdem Ihr schon weggezogen seid. Aber bedauerlicherweise liess es sich nicht anders machen. Vom 25. Februar bis mindestens 4. März würden wir also in Rom, Eden-Hotel, erreichbar sein.

Erich und Riele sind noch in Nürnberg, Bayernstr. 154.[705] Wahrscheinlich werden sie nach Frankfurt/Main übersiedeln, da sie dort bei ihrer Mutter besser und billiger leben können.

Mit der Bitte um postwendenden Bescheid bin ich mit herzlichen Grüssen für Euch drei, auch von Else,

Dein Vater

P.S. Wenn Du mir den Pass sofort nach Erhalt dieses Schreibens schickst, worum ich bitte, so schreibe mir dann gleichzeitig, falls Du

[705]See letter [166].

— wie ich annehme — die Abmeldung aus Leipzig noch nicht erhalten hast, wenigstens den ungefähren Termin, an dem Du von Leipzig nach Zürich übersiedelt bist.

D[ein] V[ater]

[178] Rudolf Peierls to Paul Dirac

Manchester, 17.2.1934

Dear Dirac,
There is a technical question about which I would like to ask your advice. As you will have seen from my last letter,[706] my point of view in the polarisability question differs from yours and it might be worth while to publish my calculations as well. I would like to do that in the Proceedings of the Royal Society.[707] My difficulty is this: as this paper would be in some way a criticism of your point of view, I could imagine that you would not like to communicate such a paper to the Royal Society, because this seems to imply more or less that you agree with the content. On the other hand it would look strange if such a paper is forwarded through somebody else. What is therefore your opinion as to what I ought to do?

There is one more point that occured to me when I read your paper again. The matrix R_b, even apart from satisfying its equation of motion cannot be any arbitrary matrix but must be in agreement with the Pauli principle. This, in general, is very complicated. Don't you think that if one makes mathematical simplicity an argument for the choice between various ways of defining R_b one ought to arrange at least that also Pauli's principle takes a comparatively simple form?
Yours sincerely,

R. Peierls

---

[706]Letter [176].

[707]Later in the year Peierls published 'The vacuum in Dirac's theory of the positive electron', *Proc. Roy. Soc. Lond. A* **146**, 420–441 (1934). It had, indeed, been communicated by Dirac.

We consider a Lorentz transformation which changes

$$\begin{aligned} x &\to x + vt \\ t &\to t - vt \end{aligned} \tag{1}$$

(i is taken to be unity)

If we consider a function of the co-ordinates and compare the value at a point xt with the value at a point with the same co-ordinates in the new system the change is

$$F(x,t) \to F(xt) - v(x\frac{\partial}{\partial t} - t\frac{\partial}{\partial x})F(x,t)$$

Therefore a ψ function is transformed:

$$\psi(xt) \to \psi(xt) + \frac{\psi}{2}\alpha\psi - v(x\frac{\partial}{\partial t} - t\frac{\partial}{\partial x})\psi$$

an expression of the form

$$(xt/R/x't') = \sum \psi(xt)\psi^*(x't')$$

therefore transforms into

$$(xt/R/x't') + \frac{\psi}{2}[\alpha(xt/R/x't') + (xt/R/x't'/\alpha]$$

$$- v[x\frac{\partial}{\partial t} + x'\frac{\partial}{\partial t'} - t\frac{\partial}{\partial x}](xt/R/x't') \tag{2}$$

If we have a finite number of occupied states, densities + current are to be calculated from:

$$\begin{aligned} \rho &= S(xt/R/xt) \\ j &= s\alpha(xt/R/xt) \end{aligned} \tag{3}$$

where S means diagonal sum with respect to the spin variables.

By inserting (2) into (3) one easily deduces

$$\begin{aligned} \rho &\to \rho - vj - v(x\frac{\partial}{\partial t} - t\frac{\partial}{\partial x})\rho \\ j &\to j + v\rho - v(x\frac{\partial}{\partial t} - t\frac{\partial}{\partial x})j \end{aligned} \tag{4}$$

which expresses the fact that ρ , j is a four vector.

After the re-interpretation one has to insert $R \to R_0$ instead of R into (3). One must therefore know the transformation properties of R_0. R_0 is defined from the fact that in every system of co-ordinates the $(q, q^{'})$ matrix element of R_0 is given by

$$(xt/R_0/x^{'}t) = \sum_s u_s(xt)u_s(xt^{'}) \tag{5}$$

where u_s are the eigenfunctions of the operator $E = \alpha(p - \frac{e}{c}\mathfrak{A}) + pmc$ at the time t and the sum is to be extended over the negative values of this operator.

If therefore R_0 is known in one system, it may be constructed in the next one in the following way. The potential in the new co-ordinate system will be

$$\mathfrak{A} \to \mathfrak{A} + v\mathfrak{A}_0 v[x\frac{\partial}{\partial t} - t\frac{\partial}{\partial x}]\mathfrak{A} \tag{6}$$

($\mathfrak{A}_0$ denotes the scalar potential.)

We may here regard the terms with v as small perturbation and the new eigenfunction will be:

$$u_s \to u_s + v\sum_r \frac{u_r}{E_s - E_r}(r/\mathfrak{A}_0 - (x\frac{\partial}{\partial t} - t\frac{\partial}{\partial x})\mathfrak{A}/s)$$

(the last bracket denoting the matrix elements of the perturbation function.)

Therefore, the change in $(xt/R_0/x't)$ becomes

$$(xt/R_0/x't)$$

$$\to (xt/R_0/x't) + v\sum_r u_r(x)\frac{(r/\mathfrak{A}_0 - (x\frac{\partial}{dt} - t\frac{\partial}{\partial x})\mathfrak{A}/s)}{E_s - E_r}u_s^*(x')$$

$$+ v\sum_r u_s(x)\frac{(s/\mathfrak{A}_0 - (x\frac{\partial}{\partial t} - t\frac{\partial}{\partial x})\mathfrak{A}/r)}{E_s - E_r}u_r^*(x') \tag{7}$$

Therefore the change in $(xt/R - R_0/xt)$ becomes:

$$(xt/R - R_0/xt)$$

$$\to (xt/R - R_0/xt) + \frac{v}{2}(\alpha(xt/R/xt) + (xt/R/xt)\alpha)$$

$$- v(x\frac{\partial}{\partial t} - t\frac{\partial}{\partial x})(xt/R/xt) + v\sum_r u_r(x)\frac{(r/\mathfrak{A}_0 - (x\frac{\partial}{\partial t} - t\frac{\partial}{\partial x})\mathfrak{A}/s)}{E_s - E_r}u_s^*(x)$$

$$+ v\sum_r u_s(x)\frac{(s/\mathfrak{A}_0 - (x\frac{\partial}{\partial t} - t\frac{\partial}{\partial x})\mathfrak{A}/r)}{E_s - E_r}u_r^*(x)$$

I.e. the change in ρ becomes:

$$\rho(x) \to \rho(x) - vj(x) + vS[\alpha(xt/R_0/xt)] - v(x\frac{\partial}{\partial t} - t\frac{\partial}{\partial x})\rho(x)$$

$$+ v\int (x\frac{\partial}{\partial t} - t\frac{\partial}{\partial x})(xt/R_0/xt)$$

$$+ v\sum_r \int u_r(x)\frac{(r/\mathfrak{A}_0 - (x\frac{\partial}{\partial t} - t\frac{\partial}{\partial x})\mathfrak{A}/s)}{E_s - E_r}u_s^*(x)$$

$$+ v\sum_r Su_s(x)\frac{(s/\mathfrak{A}_0 - (x\frac{\partial}{\partial t} - t\frac{\partial}{\partial x})\mathfrak{A}/r)}{E_s - E_r}u_r^*(x)$$

Comparing this with (4) we see that the theory can only be invariant if the diagonal sum of the following expression vanishes:

$$\alpha(xt/R_0/xt) + (x\frac{\partial}{\partial t} - t\frac{\partial}{\partial x})(xt/R_0/xt)$$

$$+ \sum_{rs} u_r(x)\frac{(r/\mathfrak{A}_0 - (x\frac{\partial}{\partial t} - t\frac{\partial}{\partial x})\mathfrak{A}/s)}{E_s - E_r}u_s^*(x)$$

$$+ \sum_{rs} u_s(x)\frac{(s/\mathfrak{A}_0 - (x\frac{\partial}{\partial t} - t\frac{\partial}{\partial x})\mathfrak{A}/r)}{E_s - E_r}u_r^*(x) \tag{8}$$

Here s takes all the negative Eigenvalues of E, r all the positive.

Now the expression cannot vanish, as $\mathfrak{A}_0$ is contained only in the last two terms and as one can easily see that the terms containing $\mathfrak{A}_0$ do not vanish because

$$\sum u_r u_s^* \frac{(r/\mathfrak{A}_0/s)}{E_s - E_r} + \sum u_s u_r^* \frac{(s/\mathfrak{A}_0/r)}{E_s - E_r}$$

is just the induced polarisation of the vacuum by an electric field, which we calculated before and which does not vanish.

The theory therefore is not invariant.

[179] Heinrich Peierls to Rudolf Peierls

Neapel, 6.3.1934

Meine Lieben.

Wir bedauern sehr, aus Deinem hier pünktlich angekommenen Nachrichten vom 2. zu ersehen, daß Du l[iebe] Genia, große Unannehmlichkeiten durchzumachen hattest.[708] Wir freuen uns aber zu hören, daß Alles wieder in Ordnung ist, + hoffen, daß es so bleiben wird. Nun zu Deiner Angelegenheit, l[ieber] Rudi. Die Angabe "um sich eine Existenz zu gründen" hat bei Bethe genügt, da er ja beim Fortgehen nicht wissen

[708]Genia had had an operation.

konnte, wo + wie er landen würde. Bei Dir wird nach Details gefragt, aber ich werde schon sehen, wie man die Antwort giebt. Zunächst warte ich also Deine definitive Antwort ab, was ich ja sowieso tun muß bis ich nach Hause zurückkomme. Übrigens nehme ich an, daß Euer Conto infolge Eurer aufgetretenen größeren Ausgaben, stark reduziert worden ist. Ich kann wohl die Hälfte, also 45£ angeben, womit ich Dich einverstanden hoffe. Vielleicht ist auch diese Summe noch zu hoch? Jedenfalls bitte ich um weitere Nachricht, sowie Du mit den deutschen Bekannten gesprochen hast. Selbstverständlich würde die Überweisung für mich keine finanzielle Belastung bedeuten. Bei näherer Überlegung glaube auch ich, daß der Paß-Vermerk für Dich keinerlei Unannehmlichkeit bedeutet, aber ich wollte Deiner Zustimmung sicher sein. Die Angelegenheit kann vor Ostern nicht erledigt werden, da wir erst ungefähr zu dieser Zeit nach Hause kommen. Leider kann ich nicht nach England kommen, zumal ich im Sommer nochmals verreisen will. Eher könnte vielleicht Else für einige Tage herüberkommen. Wäre es aber nicht einfacher, wenn Ihr im Anschluß an Eure Pariser Reise für einige Tage nach Berlin kämet? Wir sind um diese Zeit sicher zu Hause und dann könnte, wenn Du außer dem Paße auch die neulich Dir wieder zurückgeschickte Aufenthaltsbescheinigung mitbrächtest, die Sache am besten erledigt werden. (wenn sie überhaupt genehmigt wird, was man im Gegensatz zu früher jetzt nie weiß). Vielleicht könnt Ihr trotzdem die Reise nach Russland, wenn hierdurch der Zeit-Verlust nicht zu groß wird, für nächstes Jahr aufheben. Wir sind wieder mit unserer Reise sehr zufrieden; übermorgen fahren wir nach Capri. Hotel Quizizona, wo wir, wenn wir Euch nichts anderses melden, mindestens 14 Tage bleiben wollen.
Viele Grüße

Euer Vater

Meine Lieben!
Hoffentlich erholt sich Genia bald ganz ohne Complicationen. Wir haben so viel Kunst studiert, daß es für Vati Zeit wird, sich in Capri zu erholen, wo wir 14 Tage bis 3 Wochen bleiben wollen, um dann direkt nach Hause zu fahren.

Kommt Ihr zu uns, ja? Ev[entuell] hoffe ich im Juni 1 Woche nach England zu kommen. Vati will so große Reisen nicht mehr machen. Wir müssen fort, daher Schluß + 1000 Grüße Eure

Else

[180] Rudolf Peierls to Paul Dirac

Manchester, 20.3.1934

Dear Dirac,
I am sorry I did not notice at once the remarks on the cover of the Proceedings.[709] I am now enclosing a new drawing, which I hope will do and the three copies of the summary.

I had missed the point that in you paper R depends only on the field in the neighbourhood and that only a limited number of powers of the field strength occur. Therefore, the argument in my last letter[710] about what happens when a pair was created was wrong.

If one makes the postulate that the expressions must be of a certain order in the field strength and its derivatives one reduces the possible choice of S in $R = (R_a - S) + (R_b + S)$ to a finite number of expressions. But there is still quite an appreciable freedom left, which, for instance, maybe made us of in such a way that the vacuum polarisation is brought to zero, while, as far as I can see it does not vanish with your choice of R_a.

But it is no great use to make guesses about this and I now believe that it is not so difficult to solve your equations for R_a (or rather for HR_a) explicitly and then it will be easier to see what happens.
With best wishes.
Yours sincerely,

R. Peierls

[709] See letter [176], note 704.
[710] Letter [178].

[181] Carl-Friedrich v. Weizsäcker[711] to Rudolf Peierls

Leipzig, 3.7.1934

Lieber Herr Peierls,
Heisenberg bittet mich, Ihre Anfrage betr[effend] Fermis β-Theorie zu beantworten, da ich mich selbst in der letzten Zeit ein wenig damit beschäftigt habe. Verzeihen Sie die Verzögerung, an der zum Teil die Ablenkung von der Physik durch die politische Spannung schuld war. Über Leitfähigkeit im Mangetfeld wird Ihnen Herr Titeica[712] noch genaueres schreiben oder hat es vielleicht schon getan (Er sagt mir gerade: noch nicht).

Das Austauschintegral, das aus Fermis Theorie folgt, ist semikonvergent. Heisenberg hat seinerzeit folgende Integrationsweise gewählt (unter Weglassung der Spinorenprodukte):

$$\int dy \int dy' \frac{e^{\frac{i}{\hbar}(y+y')w}}{p+p'} = \lim_{\alpha\to 0} \int dy \int dy' \frac{e^{\frac{i}{\hbar}(y+y')w-\alpha(p+p')}}{p+p'}$$

$$= \lim_{\alpha\to 0} \int_\alpha^\infty d\alpha \int dy \int dy' e^{\frac{i}{\hbar}(y+y')w-\alpha(p+p')}$$

$$= \lim_{\alpha\to 0} \int_\alpha^\infty d\alpha [\int dy\ e^{\frac{i}{\hbar}yw-\alpha p}]^2$$

Dies gibt etwas Endliches und aus Dimensionsgründen natürlich etwas in der Form $\frac{1}{r^5}$. Ob ein Resultat, das so von der Integrationsreihenfolge abhängt, Vertrauen verdient, ist freilich unklar. Aber jedenfalls sieht man, dass die Rechnungsweise nicht zwangsläufig zu Unfug führt.

Empirisch ist die so berechnete Austauschkraft ja nur um den Faktor 10^5 bis 10^{10} zu klein. D.h. die Energie wird $\sim mc^2$ erst für einen Radius $r_o \stackrel{\sim}{<} \frac{h}{Mc}$. (Änderung von v_o um den Faktor 10 ändert ja

[711] Carl-Friedrich v. Weizsäcker (1912–), studied physics, astronomy and mathematics at Berlin, Göttingen and Leipzig. He became Heisenberg's assistant at Leipzig and later joined Heisenberg in his work on the German nuclear project.

[712] Serban Titeica, (1908–1985) studied in Leipzig for a Ph.D. which he completed in 1934 ('On the behaviour of electric resistance of metals in magnetic fields', Leipzig, 1934).

die Energie v bei einem festen Abstand um 10^5.) Hier weiss Heisenberg bloss die Ausrede, dass in der Quantenelektrodynamik bei Berechnung der Wechselwirkung zweier Elektronen als zweiter Näherung der (virtuellen) Emission von Lichtquanten ebenfalls zunächst etwas um 10^6 zu Kleines herauskommt, nämlich nur die magnetische Wechselwirkung. Um die elektrische Wechselwirkung zu erhalten, braucht man die Longitudinalwellen, denen keine mögliche Lichtquantenemission entspricht. Ähnlich, meint er, könnte auch das Fermi-Feld noch Glieder enthalten, die zur β-Emission nichts beitragen, wohl aber zur Austauschkraft.

Ein wenig gefreut hat mich in diesem Zusammenhang das folgende, an sich ebenso unsaubere Resultat. Die Austauschwirkung lässt sich ja wie bei den Molekülen in einer gewissen Näherung einfach als Zusatzenergie in der Hamiltonfunktion der schweren Teilchen alleine darstellen. Ich habe nun nachgesehen, ob ich nicht in der Dirac-Gleichung für ein Proton ein Zusatzglied bekommen kann, das einem anomalen magnetischen Moment entspricht. Van der Waerden hat uns neulich in Heisenbergs Seminar vorgerechnet, dass man aus Invarianzgründen zur Diracgleichung gerade nur ein Glied der Form const. $\gamma_i \gamma_k F_{ik}$ hinzufügen kann. Das bedeutet gerade: mechanisches Moment $1/2\hbar$, magnetisches Moment $\frac{e\hbar}{Mc}$ + const, sowie eine (wesentlich kleinere) zusätzliche Wechselwirkung mit elektrischen Feldern.

Lassen Sie nun ausser der Fermischen Störungsenergie noch ein äusseres Feld als Störung wirken und betrachten folgenden Prozess dritter Näherung: 1. Das Proton emittiert ein Positron und ein Neutrino (oder Antineutrino). 2. Das Positron springt unter Einfluss des äusseren Feldes in einen anderen Zustand. 3. Positron und Neutrino verschwinden wieder. Wir berechnen also ausser der Selbstenergie im Vakuum, die, wie Sie auch schreiben, divergiert und daher weggelassen werden muss, die Änderung der Selbstenergie durch ein äusseres Feld. Diese divergiert nun leider auch. D.h. das Proton bekommt nach Fermi zu allen anderen Unendlichkeiten noch ein unendliches magnetisches Moment. Man kann es aber endlich machen, wenn man — beispielsweise — einen "Fermischen Protonenradius" R einführt durch die Bedingung: Fermische Selbstenergie = Mc^2, D.h. $\frac{g^2}{\hbar c R^5} = Mc^2$. Ausgerechnet habe ich die ganzen Spinorenprodukte der dritten Näherung bisher nicht; aus

Dimensionsgründen möchte ich aber folgendes über die Grössenordnung schliessen. Die Störungsenergie dritter Näherung muss die Form haben $f \cdot \frac{g^2}{\hbar c R^5} = f \cdot Mc^2$. f ist dimensionslos und muss den Faktor e sowie, nach van der Waerden, H (Magnetfeld) enthalten. R, h, und c dürfen vorkommen (die Ruhemasse des Elektrons natürlich nicht, da der Hauptbeitrag von sehr $(mc^2 < E < \frac{\hbar c}{R})$ hohen Elektronenenergien beruht). Demnach muss sein $f = \frac{eHR^2}{\hbar c}$. Die gewöhnliche Energie im Magnetfeld wäre $\mu H = \frac{e\hbar H}{Mc}$. Das Verhältnis beider ist also die Grösse des zusätzlichen Magnetons in Einheiten des aus der Diracschen Theorie folgenden. Es ist $= \frac{R^2}{\hbar/Mc)^2}$. Wie gross ist nun R? Nach Fermis Zahlwert für g liegt R etwa in der Mitte zwischen $\frac{\hbar}{Mc}$ und $\frac{e^2}{Mc^2}$. Dann wird das Zusatzmagneton nur $\sim i/137$ des ursprünglichen. Gibt man dagegen g die Grössenordnung, die notwendig ist, um Heisenbergs empirische Austauschkräfte darzustellen, (d.h. $\frac{g^2}{\hbar c r^5} \sim mc^2$ für $r \sim \frac{e^2}{mc^2}$), so wird $R \sim \frac{h}{Mc}$ und das Verhältnis der Magnetonen von der Grössenordnung eins.

Bemerkt sei noch, dass das Neutron das Zusatzmagneton mit umgekehrtem Vorzeichen haben müsste, aber natürlich kein Diracsches Magneton, und folglich $\mu_{\text{Proton}} + \mu_{\text{Neutron}} = \frac{e\hbar}{Mc}$, was zu Sterns neuesten Messungen am H_2 passen könnte.[713]

Wenn ich versuche, mir Ihre Reaktion auf diesen Brief auszumalen, komme ich etwa auf die Formulierung: Glaubst Du etwa ich soll genesen in diesem Wust von Schlamperei? Etwas Besseres wird man wohl nicht machen können, ehe man weiss, wie man Selbstenergie etc. zu vermeiden hat. Immerhin sieht es nicht unmöglich aus, dass die Kernkräfte und das magn[etische] Mom[ent] des Protons mit Fermis Ansätzen zusammenhängen.

Mit besten Grüssen auch von Heisenberg,
Ihr

C.F. Weizsäcker

[713]Otto Stern had measured the magnetic moment of the proton by using a molecular beam and found that it was actually about $2\frac{1}{2}$ times the theoretical value.

[182] Heinrich Peierls to Rudolf Peierls

Berlin, 19.7.1934

Lieber Rudi.
Endlich kann ich Dir die erfreuliche Mitteilung machen, dass die Devisenstelle Deinen Antrag, allerdings nur mit einer ermässigten Summe von RM 5.000.-, definitiv genehmigt hat. Jetzt muss aber die Reichsbank die dafür nötigen Devisen zuteilen, und es ist anzunehmen, dass die Ueberweisung nicht auf einmal sondern in Raten erfolgt. Für mögliche Beschleunigung wird Alfred, der alle Vorgänge besitzt, sorgen.

Deinen Pass und die Aufenthaltsbescheinigung sende ich Dir anbei zurück, nachdem der Vermerk eingetragen ist.

Nachträglich bestätige ich noch den Eingang Deines letzten Schreibens in der Pass-Angelegenheit. Die darin angegebenen Daten waren für die Bewilligung Deines Antrags sehr wertvoll.

Wir fahren heute nachmittag nach St. Moritz - Dorf, Hotel Schweizerhof, wo uns bis auf weiteres Nachrichten treffen.
Herzliche Grüsse Euch Dreien auch von Else

Dein Vater

[183] Heinrich Peierls to Rudolf Peierls

Berlin, 21.9.1934

Lieber Rudi.
Ich habe Dir noch den Empfang Deines Schreibens vom 14. d[iesen] M[onats] aus Leningrad zu bestätigen und hoffe, dass Du bei Erhalt dieses Schreibens nach guter Reise wohlbehalten eingetroffen bist.[714]

[714]Rudolf and Genia had visited Genia's family in Leningrad, and while Genia stayed with her parents Rudolf travelled to the Caucasus with Lev Landau. Peierls, *Bird of Passage*, p. 109.

Ihr habt unsere Depesche ganz richtig aufgefaßt. Uns war die Idee gekommen, dass sich Genia und Gaby vielleicht lieber einmal eine Woche in Berlin ausruhen. Wir müssen aber nach reiflicher Ueberlegung Euern Ausführungen zustimmen, und ich habe bereits Genia vor einigen Tagen geschrieben, dass es bei unserer Verabredung am 1. und 2. Oktober in Hamburg bleibt. Uns selbst bereitet das keinerlei Unannehmlichkeiten, nach Hamburg zu fahren und ich schrieb Genia bereits, dass wir entweder selbst am Hafen sein werden oder unseren Wagen, dessen Chauffeur und Nummer wir ihr angaben, hinschicken. Wir freuen uns sehr, dass es beiden jetzt wieder gut geht.
Viele herzliche Grüsse

Dein Vater

Lieber Rudi!
Hoffentlich klappt unser rendez-vous diesmal, es ist nur sehr schade, daß wir Dich nicht zu sehen bekommen. Große Pläne im voraus kann man ja jetzt nicht machen + mein Traum mal n[ach] England + zu Euch zu fahren wird sich wohl nie realisieren. Hoffentlich hattest Du eine gute Reise es ist ja noch immer herrliches Wetter. Fuldas kommen am Sonntag zurück, meine arme Schwester leider gar nicht erholt + mit dem Herzen nicht in Ordnung, da ihr die Höhenluft nie bekommt!!—

Alfred hat sich endlich entschlossen, eine kl[eine] Auto-Erholungsreise m[it] Nina zu machen. (Weimar Eisenach) Krebse alle 4 munter. Alles Gute f[ür] D[eine] Arbeit u[nd] viele liebe Grüße
v[on] D[einer]

Else

[184] Heinrich Peierls to Rudolf Peierls

Berlin, 25.9.1934

Lieber Rudi.
Die Reichsbank hat nunmehr von dem durch die Devisenstelle bewilligten Betrag den Gegenwert von RM 200.- an Dich freigegeben, den

Du nächster Tage durch Deine Bank erhalten wirst. Hoffentlich werden in absehbarer Zeit weitere Summen genehmigt. Du wirst Dich leider jetzt auf eine sehr lange Zwischenpause einrichten müssen und es bleibt nichts übrig als dass Du Dich nach Möglichkeit einschränkst.
Von hier nichts Neues. Viele Grüsse

Dein Vater

L[ieber] R[udi]
Hoffentlich klappte Deine Rückreise und Du bist zur rechten Zeit daheim angelangt.

Wir freun uns sehr Gaby kennen zu lernen. Hoffentlich haben sie eine gute Reise. Wirst Du die Vorträge der jungen Madame Curie miterleben?[715] Fabelhaft.
Viel Liebes

Else

[185] Heinrich Peierls to Rudolf Peierls

Berlin, 5.10.1934

Lieber Rudi.
Deine Mitteilungen nach Hamburg haben wir mit grossem Interesse gelesen und freuen uns, dass Du rechtzeitig angelangt bist.

Wir haben einige sehr angenehme Tage in Hamburg verlebt und uns mit Genia und Gaby ausserordentlich gefreut. Unser langes Warten ist reichlich belohnt worden, dadurch, dass wir Eure Tochter kennengelernt haben. Wir sind entzückt davon, wie liebenswürdig und angeregt

[715] Rudolf Peierls attended a conference in Geneva. Peierls, *Bird of Passage*, p. 110.

sie sich verhalten hat, trotzdem sie doch in eine ganz neue Umgebung gekommen ist. Wir sind sehr stolz auf unsere Enkeltochter und hoffen, die Bekanntschaft mit ihr recht bald erneuern zu können. Wir hoffen ferner, dass beide Damen inzwischen wohlbehalten in England angekommen sind und wären erfreut, dies recht bald bestätigt zu erhalten.
Viele herzliche Grüsse für Euch Drei

Dein Vater

Lieber Rudi!
Gaby war ganz goldig und so voller Leben, Humor + Lachen daß es eine Freude ist. Sie schlief überhaupt am ganzen Tag nur 1 Stunde — Genia hingegen ist etwas übermüdet + blaß, ich hoffe, sie hat keinen Sturm gehabt + die lange anstrengende Reise endlich hinter sich. Alles Beste für Euch alle Drei + Hans[716] + ev[entuelle] Mieter + viele Grüße
Eure

Else

[186] Heinz Rudolph to Rudolf Peierls

Berlin, 10.3.1935

Lieber Rudi!
Wieder einmal jährt sich der Tag unseres Abiturientenexamens. In üblicher Weise werden wir zu diesem Anlass bei Jux (Westfälischer Hof) zusammenkommen. Ich bin beauftragt worden, Dich zu diesem Treffen gleich allen anderen einzuladen. 21.III.35. Freilich wirst Du kaum (der Entfernung wegen) dieser herzlichen Einladung Folge leisten können. Es wäre deshalb angebracht, dass Du in bekannter Weise in einem Brief alles das niederlegst, was uns an Deinen letzten Lebensereignissen interessieren könnte. Ich nehme an, dass das eine ganze Menge sein wird. Wenn Du also diesem Vorschlag folgen würdest, wären wir Dir sehr

[716]Hans Bethe was lodging with Genia and Rudolf Peierls in Manchester.

dankbar. Es ist übrigens in diesem Jahr das zehnte Mal, dass wir den Tag unseres Abi durch eine Zusammenkunft begehen.

Von mir ist nichts besonderes mitzuteilen. Es geht alles seinen Weg. Mit herzlichen Grüssen

Deine Klassenkameraden und Dein Heinz

[187] Wolfgang Pauli to Rudolf Peierls

[Zurich], 13.6.1935

Lieber Herr Peierls!
Da Sie mir Ihre etwas komische Note über die Neutrinotheorie des Lichtes,[717] geschickt haben, möchte ich Ihnen doch kurz schreiben, was ich selbst zu dieser Theorie (nicht etwa: zu Ihrer Note!) überlegt habe und meine.

1. Ich habe niemals die beiden ersten Paragraphen der Jordanschen Arbeit[718] gelesen, sondern habe mit der Lektüre immer gleich bei §3 angefangen. An Jordans absurder "Ramaneffekt"-Idee bin ich also völlig unschuldig. Es interessierte mich aber die Fragestellung, ob und wie man aus $a(\mu)$ und $c(\mu)$, die J[ordan]'s Gl[eichung] (18) genügen, wirklich Wellen $b(\nu)$ bekommen kann, die (17) erfüllen. Es war mir nämlich schon lange vor der Jordanschen Arbeit das triviale Argument bekannt, daß man mit kräftefreien Neutrinos (ohne Wechselwirkung), die das Ausschliessungsprinzip erfüllen, nie mehr als ein Paar (also ein Photon) auf eine Eigenschwingung des Hohlraums setzen kann, daher die Paare dann sicher nicht die Bose-Statistik erfüllen. Es muß also jeder Ansatz, der zu (17) führt — entgegen J[ordan]s eigener Meinung — doch etwas einer Wechselwirkung der Neutrinos irgendwie

[717]H. Bethe and R. Peierls, 'The "Neutrino", *Nature* **133**, 532, 689-90 (1934). For comments see also letter Heisenberg to Peierls, 28.1.1935, *Peierls Papers*, Ms. Eng. misc. b208, C.138.
[718]P. Jordan, 'Zur Neutrinotheorie des Lichts', *Z. Phys.* **93**, 464–72 (1935).

physikalisch äquivalentes einführen (wozu auch Kronigs Feststellungen passen![719]). Das sieht man auch in der Tat, wenn man von den $a(\nu)$, $c(\nu)$, $b(\nu)$ zu den entsprechenden $a(x)$, $c(x)$, $b(x)$ übergeht.

2. Was aber die spezielle J[ordan]sche Formel (19) betrifft, so ist die Herleitung von (17), die J[ordan] gibt, mathematisch nicht einwandfrei, da das Resultat eigentlich zunächst $\infty - \infty$ ist und einem Ausdruck diesen Typus dann von Jordan willkürlich ein spezieller Wert erteilt wird. Weisskopf[720] u[nd] ich haben deshalb versucht, in (19) zunächst eine abschneidende Gewichtsfunktion $g(\alpha)$ einzuführen und zum Schluß den $\lim g \to 1$ zu machen. Dann kam aber nicht das J[ordan]sche Resultat (17) heraus, sondern vielmehr Vertauschbarkeit aller $b(\nu)$. Deshalb glaube ich, daß J[ordan]s Ansatz (19) Quatsch ist.

3. Ich habe dann noch de Broglies u[nd] Wentzels Arbeiten studiert[721] und diskutiert und ebenfalls einen sehr negativen Eindruck bekommen. Entweder es kommt Unsinn heraus oder man verbietet es irgendwie, einzelne Neutrinos zuzulassen, betrachtet nur Paare und schreibt hiermit die bisherige Elektrodynamik in unübersichtlicher und komplizierter Weise um. (Wozu war denn dann die ganze Mühe?)

Zusammenfassend möchte ich also meiner großen Betrübnis über den gegenwärtigen Stand der sogenannten 'Neutrinotheorie des Lichts' Ausdruck geben, möchte aber gerne bemerken, daß mich Ihre Note nicht noch betrübter gemacht hat!

---

[719] R. de L. Kronig, 'Zur Neutrinotheorie des Lichtes I, II, III', *Physica* **2**, 491–98, 854–60, 968–80.

[720] Weisskopf was Pauli's assistant between 1933/4 and 1935/6.

[721] L. de Broglie, 'Sur une analogie entre l'électron de Dirac et l'onde électromagnetique' *Comptes Rendus Acad. Sci.* **195**, 536–37 (1932), L. de Broglie, 'Remarque sur le moment magnétique et le moment de rotation de l'électron, *Comptes Rendus Acad. Sci.* **195**, 577–78 (1932); L. de Broglie, 'Sur la nature du photon' *Comptes Rendus Acad. Sci.* **198**, 135–38 (1934); L. de Broglie, *L'électron magnétique*, Paris, 1934; G. Wentzel, 'Zur Frage der Äquivalenz von Lichtquanten und Korpuskelpaaren', *Z. Phys.* **92**, 337–58 (1934).

Den Streit über die ersten zwei Paragraphen der J[ordan]schen Arbeit Ihnen und Jordan als Privatvergnügen überlassend, grüßt Sie herzlich

Ihr W. Pauli

[188] Heinrich Peierls to Rudolf and Alfred Peierls[722]

Meran, 23.4.1936

Durchlag für Rudis
Meine Lieben!
Für die Briefe von Nina und Vera wird Else noch selbst danken. Heute möchte ich Euch sagen, dass Euer gemeinsames Protokoll, für das wir gleichfalls herzlich danken, uns grosses Vergnügen bereitet hat.

Zu den einzelnen Punkten möchte ich, obwohl ja noch genügend Zeit ist, heute schon folgendes erwähnen: Die Nachricht der Amerikaner, ob sie tatsächlich am 8. Juni in London sein können, steht noch aus, aber ich hoffe, sie nächster Tage zu erhalten.

Unter dieser Voraussetzung würden wir tatsächlich sehen, schon am 2. Juni abends abzufahren. Wenn ich in meinem letzten Schreiben die Fahrt über Hoek von Holland erwähnte, die uns an sich sehr unbequem liegt, da wir eine überaus kurze Nachtfahrt auf See mitmachen müssten, so geschah dies, weil ich annahm, dass man von Harwich eine bessere Verbindung nach Cambridge hätte.[723] Da dies nach Eurem Protokoll offenbar nicht der Fall ist, ziehen wir natürlich die uns viel besser liegende Verbindung Ostende-Dover vor und würden dann in demselben Zuge fahren, den Nina benützt hat. Wir würden also unser Billet für die Hinreise, wie für die Rückreise über Ostende-Dover nehmen. Ich nehme an, dass der durchgehende Wagen in Ostende bis zum Pier fährt oder muss man sich erst einen Wagen bis zur Schiffshaltestelle nehmen?

---

[722]Alfred, Nina and Vera Peierls had emigrated to England in 1935, and since then Heinrich Peierls occasionally sent joint letters to the two brothers.

[723]Rudolf and Genia Peierls had moved to Cambridge the previous autumn when Peierls had been offered a research fellowship at the Mond Laboratory.

Ebenso möchte ich wissen, ob wir uns schon in Berlin ein durchgehendes Billet bis Cambridge und ein weiteres Billet von Cambridge nach London lösen sollen. Ebenso ob ev[entuell] ein grosser Koffer direkt bis Cambridge aufgegeben werden kann, oder ob es richtiger ist, Handgepäck mitzunehmen + wie viel Stück sind ratsam? Schliesslich bitte ich noch, mir die genaue Verbindung Dover-Cambridge anzugeben, die ich aus meinem deutschen Reichskursbuch nicht ersehen kann.

Ich brauche nicht nochmals zu erwähnen, dass wir uns auf das Wiedersehen mit Euch riesig freuen. Alfred und Rudi übernehmen wohl die Beantwortung meiner Anfragen.

Ihr könnt uns noch hierher schreiben, da wir Meran erst am 4. Mai verlassen wollen. Sämtlichen Unterzeichnern des Protokolls und den noch unmündigen kleinen Cambridgern viele herzliche Grüsse auch von Briegers

Euer Vater

My dear Rudi + Genia
In a hurry I'll write you only a few words, to let you know how glad I am to see you in a short time. Your invitation is charming. I am looking forward with great pleasure. I am very much interested in the colleges + everything you will show me. Good night. Much love,

Else

[189] Else Peierls to Rudolf Peierls

Berlin, 2.7.1936

My dear Genia + Rudi, Gabylein + Ronnylein.
Father and I thank you very much for your long and interesting letter. I think it's wonderful that Heysenberg found a way to clear problems + that you learned a lot by all these discourses. I wonder whether Genia got the parcel from Jennie[724] from London with the bag etc. Jennie + Siegfried will arrive in: Carlsbad. Hotel Pupp. Tschechoslowakei on

[724] Jennie Peierls, Heinrich Peierls' sister-in-law.

the 7th of July. Briegers will meet them there + have yet arrived at Carlsbad.

The hot in Berlin was awfully one couldn't sleep. Last night we went to Glienicke Savarin[725] to get some fresh air. It was lovely. We have not been there for month + month. Did you get our post card? Fuldas are in Italia, Carezza presso Bolzano. Villa Laurin. Will you please write a card to Ludi's birthday on the 15th. Sometimes we have rains and thunderstorms as even now so that we feel very refreshed + much better. We cannot go to St. Moritz. It's a pity. So we don't know were we may have our holidays. I am very sorry to tell you that father became ill in the sleeping car.[726] The agitation of the last years, the fear of english journey, the exertions in London all this came together to give him a little nervous shock. I brought him home in our car and laid him on the sofa. The doctor was sent for + ordered some medicaments and rest. He is much better now, but very irritable + often very weak. We live very quite in our nice home and sometimes we have a friend + yesterday we enjoyed our Glienicke evening. I did not tell Alfred and Nina because they are nervous themselves. Don't worry about it + don't mention it when you write again. I hope we can leave soon, perhaps to San Marino + I hope it will do him good.[727] We are very glad to know how you live now in your charming house + garden with the sweet children. Many kisses from Misses grandmother + granddad to Gaby + Ronny. With much love from both of us
ever yours

Elsie

Many kind regards to "Olive"[728] please.

---

[725] Café in Berlin Glienicke.

[726] Heinrich and Else Peierls had visited Rudolf and Alfred Peierls and their families in England in June 1936, and they met up with American relatives in London on that occasion.

[727] Travelling outside Germany became increasingly difficult, because foreign currency had to be applied for and was often refused.

[728] The household help of Rudi and Genia.

[190] Hans Bethe to Rudolf Peierls

RMS Queen Mary, 1.8.1936

Lieber Peierls,
ein neuer Monat auf einem neuen Schiff – das ist eine vorzügliche Gelegenheit für gute Vorsätze. Immerhin, da Du netterweise den Deckel des Phys.Rev. als einen jedesmaligen Gruss von mir betrachtest,[729] fühle ich mein Gewissen etwas weniger schlagen (schlägt das Gewissen?) und das erleichtert den Anfang.

Also zunächst: schönen Dank für Deine Standpauke. Ich habe sie glaube ich sehr nötig, weil leider die Amerikaner zu höflich sind, mir welche zu halten. Die nichtamerikanischen Physiker in Amerika sind meist Ungarn, wie Teller und Wigner, und da ist auch nicht viel zu erwarten.

Sachlich bin ich nicht ganz Deiner Meinung. In anderen Worten, es steht mit mir schon ebensoschlimm wie mit Bieberbach:[730] ich glaube sogar, was ich sage.

Zunächst einmal im Prinzip: Ich halte es für äusserst nützlich, wenn gerechnet wird. Nur auf diese Weise kann man doch irgendetwas über die Kernkräfte herausbekommen. Aus dem hohlen Bauch kann man doch gewiss nichts darüber aussagen. Und da die Amerikaner Geduld haben, soll man sie doch dazu ausnutzen.

Eine andere Frage ist, wieweit die bisherigen Rechnungen anständig sind. Da gibt es natürlich so'ne und so'ne. Die anständigsten Rechnungen sind von Present,[731] allerdings nur über das H^3. Er nimmt eine recht gute Eigenfunktion (9 Konstanten, glaube ich) und Variationsprinzip. Resultat: selbst mit Wignerkräften kriegt man nur 5 MV Bindungsenergie bei einer Reichweite von 2.10^{-13} cm, und glaube ich 6 MV bei

[729]Hans Bethe was editor of the journal *Phys. Rev.*, hence his name appeared on the cover.

[730]Ludwig Bieberbach (1886–1992), at the time professor of geometry at Berlin and outspoken Nazi. Founder, in 1936, of the journal *Deutsche Mathematik.* Bethe's comment about 'being as bad as Bieberbach' referred to some recent elaborate calculations, not to his political attitudes.

[731]R.D.Present, 'Must Neutron-Neutron Forces exist in H^3?' *Phys. Rev.* **50**, 635–42 (1936).

1.10^{-13}. Vergleiche das mit 4.4 MV, der doppelten Bindungsenergie des Deuterons — soviel würde man ja selbst bei unendlicher Reichweite der Kräfte bekommen — und 8.4 MV, der beobachteten Bindungsenergie. Der Unterschied gegen Thomas ist einfach, dass Thomas nur zeigt, dass im Grenzfall unendlich kleiner Reichweite die Bindungsenergie von H^3 unendlich wird. Aber das geschieht sehr langsam und sagt gar nichts über die Bindungsenergie bei endlicher Reichweite.[732] Andererseits geben natürlich Rechnungen über das α-Teilchen eine untere Grenze für die Reichweite, und wenn bei dieser unteren Grenze das H^3 zu kleine Bindungsenergie hat, so zeigt das eben die Existenz von Neutron-Neutron Kräften. Thomas Rechnungen zeigen bloss, dass man aus dem H^3 allein nichts schliessen kann (was trotzdem immer wieder versucht wird).

Nun das α-Teilchen: es scheint, dass Inglis[733] eine anständige Rechnung hat, mit Variationsprinzip und etwa 7 verfügbaren Konstanten. Er hat sich zwar einmal dabei ziemlich verrechnet, aber wenn er das das zweitemal nicht wieder getan hat, so gibt seine Rechnung fast identisch dasselbe wie das korrigierte equivalent two-body problem. Korrigiert das heisst, dass Feenberg eine Störungsrechnung gemacht hat, in der er das equivalent two-body problem rechtfertigt und korrigiert.[734] Soviel ich mich erinnere, kommt seine Sache auf eine Näherung von grossen Reichweiten heraus: Als nullte Näherung kann man etwa den Eigenwert betrachten, den man mit der Eigenfunktion $e^{-\mu\left(r_{12}^2+r_{13}^2+r_{14}^2+r_{23}^2+r_{24}^2+r_{34}^2\right)}$ bekommt. Die Korrektur dieses Eigenwertes wird in erster Näherung richtig durch die equiv[alent] two-body Methode gegeben, in zweiter

---

[732]L.H. Thomas, 'The Interaction Between a Neutron and a Proton and the Structure of H^3', *Phys. Rev.* **47**, 903–909 (1935).

[733]David Rittenhouse Inglis (1905–) studied at Amhurst College and Ann Arbor, then moved to Ohio State, Pittsburgh, Princeton and John Hopkins. He published a number of papers on the alpha particle, e.g. D.R. Inglis, 'Spin-Orbit Coupling in Nuclei' *Phys. Rev.* **50**, 783, (1936); S. Dancoff and D.R. Inglis, 'On the Thomas Precession of Accelerated Axes' *Phys. Rev.* **50**, 784 (1936); D.R. Inglis, 'Spin-Orbit Coupling in the Alpha-Model of Light Nuclei', *Phys. Rev. 50* 1175–80 (1939); D.R. Inglis, 'The Alpha Model of Nuclear Structure, and Nuclear Moments' *Phys. Rev.* **60**, 837–51 (1941).

[734]E. Feenberg, 'On Relativity Corrections in the Theory of the Deuteron', *Phys. Rev.* **50**, 674 (1936).

Näherung kommt ein positiver Beitrag dazu (dies gibt die korrigierte two-body Methode), der bei $2 \cdot 10^{-13}$cm Reichweite etwa 2 MV beträgt. Die nächste Näherung gibt, wie üblich, wieder eine Verminderung des Eigenwertes.

Ich glaube also, das Feenbergs Methode lange nicht so schlecht ist wie einem danach werden kann. Allerdings ist dies eine Rechtfertigung a posteriori, und Du hast völlig recht, wenn Du sagst, dass ich das alles nicht wissen konnte, als ich meinen Artikel schrieb. Damals wusste ich nur von Presents Rechnungen und war von ihnen ziemlich überrascht. Aber es war damals schon klar, dass man ungeheuer kurze Reichweite braucht, um das H^3 ohne Neutron-Neutron Kräfte zu erklären, und selbst wenn Feenbergs Rechnungen für das α-Teilchen sehr verkehrt gewesen wären: diese ganz kurzen Reichweiten waren jedenfalls unmöglich. Ausserdem war es nach Presents Rechnungen ziemlich wahrscheinlich, dass Feenbergs Methode auch für das α-Teilchen recht gut sein würde.

Natürlich müssen die Rechnungen noch verbessert werden. Das wird auch geschehen. Ich habe einen Dollar zur Beförderung des Fortschritts der Wissenschaft eingesetzt, indem ich mit Feenberg gewettet habe, dass seine equivalent two-body Methode zu wenig Bindungsenergie für das α-Teilchen gibt. Er sagt in seiner letzten Arbeit, zuviel. Zu entscheiden durch eine anständige Variationsrechnung, die natürlich er machen muss.

Immerhin hat es wohl jetzt schon einen Sinn zu sagen, dass die Bindungsenergien von H^3 und He^4 mit den Beobachtungen über H^2, die Neutron-Proton und die Proton-Proton Streuung, innerhalb der Fehlergrenzen, übereinstimmen. Das ist sehr wichtig, denn es zeigt, dass die Kernkräfte sich als Summe von Kräften zwischen Teilchenpaaren darstellen lassen.

Breit[735] hat ein interessantes Resultat aus der Proton-Proton-Streuung abgeleitet:[736] and Es scheint, dass für antiparallellen Spin

[735]Gregory Breit (1899–1981) studied at John Hopkins where he received his Ph.D. in 1921, in Leiden and at Harvard, before moving to Washington in 1924, New York University (1929–1934), the University of Wisconsin (1934–47) and Yale (1947–1968).

[736]G. Breit, E.U. Condon and R.D. Present 'Theory of scattering of protons by protons', *Phys. Rev.* **50**, 825–45 (1936).

die Kräfte Proton-Proton und Proton-Neutron bis auf die Coulombkraft identisch sind. Was das bedeutet, weiss man natürlich nicht, aber die Übereinstimmung ist überraschend gut. Es ware interessant, herauszufinden, was die Kräfte zwischen gleichartigen Teilchen bei parallelen Spins tun: Die X-Meister, wie Feenberg & Co. versuchen, das Li^6 zu diesem Zweck zu rechnen.

Ich habe vesucht, mit der Hartreemethode (siehe §36 meines Review-Artikels)[737] den Spin der leichten Kerne fest zustellen. Für Li^7 kriegt man ziemlich automatisch einen $^2P_{3/2}$-Zustand als Grundzustand, und ein magnetisches Moment 3.1 was natürlich viel zu gut mit den beobachteten 3.3 übereinstimmt. Der grösste Teil des Moments kommt daher, dass einfach die Neutronenspins antiparallel sind sodass der ganze Spin-anteil des Kernmoments vom Proton herkommt. Zu den 2,85 "Magnetinos" (Autor: Condon[738]) vom Protonenmoment kommt dann noch ein Beitrag vom Bahnmoment.

Viel mehr als ich haben Wigner und Feenberg über die Kerne von He bis O^{16} gerechnet; tiefste Terme, Abstände zwischen Nachbarniveaus etc. Das Quantitative ist wohl noch sehr vorläufig.

Bohrs Papier über die Kerne gefällt mir vortrefflich. Es ist bezeichnend, dass es am 29. Februar erschienen ist.[739] Es ist erstaunlich klar und sagt genau, was nötig ist. (Vergl. dagegen Bohrs letzten Nature letter über die Energieerhaltung.[740]) Man soll aber doch Breit und Wigner nicht unterschätzen.[741] Es ist zwar miserabel geschrieben, aber sie haben doch alles richtig erkannt und ausserdem eine quantitative Formel gegeben, wo Bohr rein qualitativ bleibt.

---

[737]See letter [145].

[738]Edward Uhler Condon (1902–1974), studied at Göttingen, Munich, Berkeley, lectured at Columbia, Princeton and Minnesota before becoming associate director of research at Westinghouse and serving on the National Defense Research Committee during the Second World War.

[739]Niels Bohr had given a lecture to the Danish Academy on 24 January 1936 on neutron capture and nuclear constitution which was summarised in N. Bohr, 'Om Atomkernernes Egenskaber og Opbygning', *Dansk. Vid. Selsk. Virksomh. 6/35–5/36*, 39 (1936).

[740]N. Bohr, 'Conservation Laws in Quantum Theory', *Nature* **138**, 25–26 (1936).

[741]G. Breit and E. Wigner, 'Capture of Slow Neutrons', *Phys. Rev.* **49**, 519–531 (1936).

Bohrs Papier hat mich zu einer Rechnung über die Anzahl der Niveaus veranlasst. Das Modell ist beliebig schlampig: Fermistatistik, Potentialloch, keine Wechselwirkung zwischen den Teilen. Die Anzahl der Energieniveaus ist dann im wesentlichen die Entropie des Fermigases. Berücksichtigt ist, dass nur Zustände mit bestimmtem Drehimpuls für Prozesse wie Neutroneneinfangung in Frage kommen. Das Resultat unterscheidet sich von Oppenheimers (Seattle Meeting der Amer. Phys. Soc.[742]) dadurch, dass es im Rahmen des Modells richtig gerechnet ist; es sieht infolgedessen sehr anders aus. Ich würde gerne eine vernünftige Methode zur statistischen Abschätzung der Matrixelemente für Neutroneneneinfangung und γ-Emission haben, meistens kriegt man ja dafür zuviel heraus.

Nebenbei: Du bist doch von Deinen magnetischen Rechnungen her Expert in Fermistatistik mit allen möglichen komplizierten Wechselwirkungen. Weisst Du nicht eine Methode für die Kernprobleme?

Shanklands Experiment[743] hat mich eigentlich nicht sehr aufgeregt, selbst ehe es als falsch nachgewiesen war.[744] Ich habe vorgezogen, es zunächst nicht zu glauben a) weil nicht sein kann, was nicht sein darf, b) weil mir das Experiment nicht so sehr sauber vorkam, c) weil es soviel andere Dinge, auch mit hochenergetischen Lichtquanten, gibt, bei denen Energieerhaltung gilt. Z.B. Paarerzeugung. Was hält Dirac davon? Ich habe wirklich nicht verstanden, warum Dirac so begeistert davon war. Er schien schon vor 1 1/2 Jahren die Nichterhaltung der Energie zu lieben, und fand die damals noch hohe Masse des Be^9 wunderschön.[745] Was ist mit ihm passiert? Die Neuigkeiten über die Disintegration des Cavendish sind sehr sehr interessant. Besonders dass

[742]'Minutes of the Seattle Meeting' *Phys. Rev.* **50** 384–93 (1936).

[743]Robert S. Shankland, 'Scattering of Gamma Rays', *Phys. Rev.* **50**, 571 (1936), Robert S. Shankland, 'An Apparent Failure of the Photon Theory of Scattering', *Phys. Rev.* **49**, 8–13 (1936); R.S. Shankland, *Nature* **137**, 241 (1936).

[744]Rudolf Peierls had published a paper about the experiments. R.E. Peierls, 'Interpretation of Shankland's experiment', *Nature* **137**, 904–5 (1936).

[745]P.A.M. Dirac, 'Does Conservation of Energy Hold in Atomic Processes?', *Nature* 298–99 (1936).

Oliphant fortgeht.[746] Ist das diesmal wirklich um wiederzukommen? Hast Du übrigens inzwischen einen besseren (in beiden Bedeutungen) Eindruck von ihm bekommen? Bei uns gibt es von Personalnachrichten nur, dass Wigner nach Wisconin berufen ist, an die selbe Universität wo Breit ist. Er scheint gehen zu wollen, aber Princeton hat bisher noch nichts unternommen. Also vielleicht wird er doch in P[rinceton] bleiben.

In Cornell kommen allmählich einige Theoretiker. Ich habe einen research assistant, Rose,[747] früher bei Uhlenbeck, sehr gut und schrecklich fleissig, bezahlt von der American Philosphical Society. Er soll offiziell die Rechnungen über die Vielfachstreuung von Elektronen fertigmachen, unoffiziell rechnet er Kernmomente. Ausserdem kommt nächstes Jahr Konopinski[748] als National Research Fellow.

Du erziehst also den Bhabba.[749] Ich dachte er wäre ganz von der Sorte Hulme-Masse, die in Bezug auf Rechnungen tausendmal schlimmer sind als die Amerikaner.

Was macht die family? Placzek erzählte Wunderdinge von Gabys Intelligenz. Und Ronald Frank muss jetzt auch in dem Alter sein, wo man von seiner Intelligenz reden kann.

Ich habe weiterhin keinerlei family. Immerhin habe ich meinen car der Bestimmung jeden amerikanischen cars zugeführt. Er ist dafür sehr praktisch und auch sonst sehr schön. Wie sehr ich amerikanisiert bin, siehst Du daraus dass ich in einem Jahr 7000 Meilen gefahren bin, obwohl ich ursprünglich geschworen hatte, den Nepomuk nur zu Ausflügen in die Umgebung Ithacas zu benutzen.

---

[746] In 1935 Chadwick left the Cavendish to accept a Chair at Liverpool. Oliphant replaced him as Assistant Director of Research. Blackett moved to London, Kapitza did not return from Russia, and in 1936 Oliphant left to become Poynting Professor of Physics at Birmingham.

[747] Morris Edgar Rose (1911–), joined Hans Bethe as a research assistant before moving on to Yale, Oak Ridge and in the 1960s to the University of Virginia.

[748] Emil Jan Konopinski(1911–1990), graduated from the University of Michigan in 1933, later worked at Los Alamos and co-patented the device that triggered the first Hydrogen bomb.

[749] Homi K. Bhabha (1909–1966), studied mechanical engineering, maths and physics at Cambridge where he worked with Dirac. He did research at the Cavendish before joining Raman's institute at Bangalore in 1940. In 1945 he set up the Tata Institute of Fundamental Research in Bombay.

Zurzeit fahre ich also zu meiner Mutter, und mit ihr nach Seelisberg, Switzerland. Ich glaube meine Mutter schrieb, wir wohnen im Grand Hotel dortselbst. Sie ist momentan sehr aktiv, also sehr gesund.[750] Viele herzliche Grüsse and Genia und Dich selbst Dein

Hans Bethe

Wiedereröffnung des Briefes nach Lektüre Deiniges:
Man soll sich doch wohl über Goldhabers Experiment[751] nicht aufregen (siehe oben Shankland). Das Resultat ist höchstwahrscheinlich falsch; Tuve und Hafstad[752] haben angeblich das Experiment mit besserer Technik nachgemacht und finden den theoretischen Wert. Der Haupteinwand gegen Goldhabers Anordnung scheint mir, dass beim Durchgang durch den Paraffinabsorber viele von den durchgehenden Neutronen verlangsamt werden (kleine Ablenkungen der Doppelstösse) und dadurch grössere Wahrscheinlichkeit haben, gezählt zu werden. Der Effekt ist von derselben Grössenordnung wie die Absorption. Ausserdem ist natürlich die theoretische cross section jetzt gar nicht mehr so gross, nachdem Fermi den Faktor 4 von der chemischen Bindung entdeckt hat;[753] seine letzten Messungen entsprechen einem virtuellen

[750]Bethe's mother, a talented musician, had suffered from a hearing impairment as a result of contracting influenza. This had lasting physical and psychological effects. See letter [10].

[751]Maurice Goldhaber (1911–), completed his Ph.D. in Cambridge in 1936, In 1938 he took up a post at the University of Illinois before moving to Brookhaven in 1950 where he stayed until 1973, Since 1961 he has also been Adjunct Professor at Stony Brook. M. Goldhaber, 'Scattering of Neutrons by Protons', *Nature* **137**, 824–25 (1930).

[752]Tuve, Hafstad and Heydenberg had carried out a series of experments on proton-proton scattering and on neutron-proton scattering. M.A. Tuve, N.P. Heydenburg and L.R. Hafstad, 'The Scattering of Protons by Protons', *Phys. Rev.* **49**, 402 (1936); M.A. Tuve and L.R. Hafstad, 'The Scattering of Neutrons by Protons', *Phys. Rev.* **50**, 490–91 (1936); M.A. Tuve, N.P. Heydenburg and L.R. Hafstad, 'The Scattering of Protons by Protons', *Phys. Rev.* **50**, 239–46 (1936); M.A. Tuve, N.P. Heydenburg and L.R. Hafstad, 'The Scattering of Protons by Protons', *Phys. Rev.* **56**, 1078–91 (1939).

[753]E. Amaldi and E. Fermi, 'On the Absorbtion and the Diffusion of Slow Neutrons', *Phys. Rev.* **50**, 899–928 (1936).

Niveau bei 120 kV, was den Wirkungsquerschnitt bei 200 kV auf etwa 2/3 des früheren Wertes reduziert.

[191] Rudolf Peierls to Niels Bohr

Cambridge, 17.8.1936
(carbon copy)

Lieber Herr Bohr,
Ich möchte mir erlauben, Sie mit einer Frage zu belästigen, die sich auf Ihre Kerntheorie bezieht. Ich habe nämlich in der letzten Zeit eine Rechnung über die Dichte der angeregten Energieniveaus von schweren Kernen gemacht, die in Ihre Theorie eingeht. Diese Rechnung war eigentlich nur zu meiner eigenen Information gemeint, aber inzwischen habe ich gefunden, dass verschiedene Leute über diese Frage ziemlich falsche Ansichten haben, und ich überlege mir, ob es sich nicht lohnt, die Rechnung zu veröffentlichen. Nun weiss ich aber nicht — und konnte aus Kalckars und Placzeks Vortrag in Kopenhagen nicht entnehmen, — ob Sie nicht schon eine entsprechende Rechnung gemacht haben, oder ob Sie die Absicht haben, sie im weiteren Verlauf Ihrer Arbeit zu machen. Ich wäre Ihnen sehr dankbar, wenn Sie mich das wissen lassen wollten, denn in diesem Falle würde ich natürlich meine Ueberlegungen nicht zu veröffentlichen brauchen.

Es handelt sich um folgende Ueberlegung: Die Anzahl der Zustände pro Energieintervall ist viel weniger empfindlich gegen die speziellen Annahmen über das Modell, da störende Kräfte, auch wenn sie gross genug sind, um die dynamischen Verhältnisse ganz zu verändern, im allgemeinen nur eine schwache Wirkung auf die grobe Verteilung des Energieniveaus haben. Infolgedessen scheint es vernünftig, die grobe Verteilung der Eigenwerte so abzuschätzen, dass man sich die Neutronen und Protonen als unabhängige Teilchen in einem Potentialfeld denkt. Wie ich zeigen werde, ist es kein Widerspruch, anzunehmen, dass dies die richtige grobe Verteilung der Energiewerte gibt, und dass Wechselwirkungskräfte zwischen den Teilchen existieren, die stark genug sind, um die Voraussetzungen Ihrer Kerntheorie zu befriedigen.

Das statistische Problem ist dann natürlich einfach das der Bestimmung der Entropie eines Fermigases, und die Anzahl der Niveaus des Kerns zwischen U und UdU ist im wesentlichen gegeben durch

$$e^{S/k}$$

wenn S die Entropie eines Gases von Neutronen und Protonen in dem betreffenden Potentialfeld, und der Energie U, und k die Boltzmannsche Konstante ist. Vor diesem Ausdruck steht noch ein Faktor, der nicht von der typischen thermodynamischen Behandlung her geläufig ist, der sich aber leicht bestimmen läßt.

Man muss ferner berücksichtigen, dass man sich ja nicht für die Anzahl aller Niveaus interessiert, sondern nur derer mit dem richtigen Drehimpuls, sagen wir z.B. mit $j = 0$. Berücksichtigt man dies, so bekommt man für die Anzahl der Zustände zwischen U und UdU

... [754]

Hierin ist U_o der Grundzustand, dz_1/d_1 ist die Dichte von Zuständen für ein Proton in dem benachbarten Potentialfeld pro Energieintervall, und zwar in der Nähe der höchsten besetzten Niveaus, dz_2/d_2 ist die entsprechende Größe für die Neutronen, m ist der quadratische Mittelwert einer Komponente des Drehimpulses (Spin und Bahn) eines Teilchens, wobei der Mittelwert über die Teilchen in den höchsten besetzten Zuständen zu nehmen ist. Mit den vernünftigen Werten [...][755]$=(0,2MV)^{-1}$ $U - U_o = 10MV, m = 10$, bekommt man einen mittleren Abstand der Niveaus des Gesamtkernes von ca. 14 volt. Dies beweist natürlich nicht viel, denn der Exponent ist hier bereits so gross, dass eine kleine Veränderung in dz/d ausreicht, um jeden Wert für den Abstand zu bekommen. Es zeigt aber jedenfalls, dass die Entropie bereits gross ist und es jedenfalls zulässig ist eine hohe Dichte der Niveaus zu postulieren. Man kann versuchen, umgekehrt den auf diesem Wege bestimmten Wert für dz/d mit anderen Daten zu vergleichen, und im Rahmen dieses Modells ist der konsequenteste Weg wohl der, dass man

[754] Handwritten calculations missing in carbon copy.
[755] Handwritten formula missing in carbon copy.

von der Differenz zwischen Anzahl der Protonen und Neutronen in stabilen Isotopen ausgeht und fordert, dass die bei der Verwandlung eines Protons in ein Neutron freiwerdende elektrostatische Energie gerade durch den Energieunterschied des höchsten besetzten Protonen- und Neutronenniveaus aufgehoben werden soll. Dies führt für mittelschwere Elemente wieder zu einem Abstand der Niveaus von einzelnen Teilchen von ca. $0,2MV$, in Einklang mit dem oben angenommenen Wert.

Ein direkter Vergleich mit experimentellen Daten sollte möglich sein (1) durch die Abhängigkeit von U, d.h. im Wesentlichen den Faktor exp((const.U − U_o) wenn es möglich wäre, aus den Messungen an schnellen Neutronen die Anzahl der virtuellen Niveaus abzuleiten; (2) durch die Abhängigkeit von dz/d indem man feststellt, wo im periodischen System die Existenz von virtuellen Niveaus im Gebiete langsamer Neutronen anfängt. Hierzu wäre aber eine eingehendere statistische Untersuchung des Vorkommens von Resonanzniveaus nötig, zu der das experimentelle Material nicht ausreicht.

Die wichtigste Frage ist nun natürlich, wie weit das betrachtete Modell einen Sinn hat. Hierzu kann man sich überlegen, dass wenn man zu einem statistischen System das im thermischen Gleichgewicht bei der Temperatur kT ist, irgendwelche störende Wechselwirkungen hinzufügt, deren Grössenordnung durch eine Energie W gegeben wird, so treten in allen Gleichgewichtsfunktionen nur Korrektionen auf, die von der Grösse (W/kT) abhängen (im allgemeinen tritt sogar erst deren Quadrat auf[)]. Nun ist in dem hier betrachteten Fall die "Temperatur" von der Grössenordnung $1MV$, also erheblich grösser als die Abstände benachbarter Niveaus eines Teilchens. Es ist daher möglich, solche Wechselwirkungskräfte zwischen den Teilchen anzunehmen, dass sie einerseits klein gegen "kT" sind und daher die grobe Verteilung der Energieniveaus des Gesamtsystems wenig stören, andererseits aber grösser sind, als die Energiedifferenzen der Niveaus von individuellen Teilchen sodass hinreichende Kopplung zwischen den Teilchen vorhanden ist, um Ihre Theorie anwenden zu können.

Man hat allerdings zu berücksichtigen, dass solche Kopplungen den Grundzustand verschieben können, und dass man dann die richtige Verteilung der Engergiewerte, aber in einer falschen Skala bekommen würde.

Wenn man aber von diesem letzten Umstand absieht, der keine sehr grosse Aenderung bedingt, so kann also die Formel einen vernünftigen Sinn haben, wenn nämlich die Korrelation zwischen den einzelnen Teilchen im Kern nicht zu gross ist. Ob das der Wirklichkeit entspricht, ist natürlich eine andere Frage. Um einen Ueberblick zu bekommen, kann man sich den entgegengesetzten Grenzfall überlegen, in dem die Korrelationen zwischen den Teilchen stark sind. Dann würde das Bild des Kerns etwa analog zu einem grossen Molekül sein, und die angeregten Zustände wären durch die Normalschwingungen zu beschreiben. Da aber der Kern viel mehr Teilchen enthält als die schwersten Moleküle, so kann man zur Statistik der Normalschwingungen die Debyesche Theorie verwenden und kann so eine Formel für die Verteilung der angeregten Zustände finden, die der oben angegebenen für unabhängige Teilchen sehr anlog ist, nur steht im Exponenen $(\text{const.})^{3/2}(U - U_o)^{3/2}$ statt $U - U_o$. Hier geht in die Formel natürlich eine Grösse von der Dimension einer "elastischen Konstanten des Kerns" ein. Der [???] Neutron etwa 50 volt zu machen, muss man diese Konstante so wählen, dass das kleinste Schwingungsquant im Kern bei einigen Hundertausend Volt liegt. Es scheint daher, dass auch dieses Modell nicht mit Ihrer Theorie im Widerspruch wäre und dass die Dichte der virtuellen Niveaus ein zu unspezifischer Zug des Modells ist um sogar eine Entscheidung zwischen diesen zwei extrem entgegengesetzten Fällen zu erlauben. Wenn es dagegen möglich wäre, eine empirische Abschätzung für die Veränderlichkeit der Dichte mit der Energie zu bekommen, so sollten die beiden Modelle schon zu unterscheiden sein.

Das zuletzt genannte Modell mit Normalschwingungen scheint natürlich übertrieben, denn es ist wohl sicher, dass die Teilchen im Kern nicht um feste Gleichgewichtslagen schwingen, sondern dass sicher ihre Nullpunktsenergie grösser ist als die Potentialschwelle zwischen verschiedenen Gleichgewichtslagen. Dieser Fall ist mathematisch schwer zu übersehen, aber zum Glück können wir hier, wenigstens qualitativ, die Natur für uns die Gleichungen lösen lassen, denn flüssiges Helium am absoluten Nullpunkt ist offenbar von dieser Art (d.h. die Atome können ihre Gleichgewichtslagen wegen der Nullpunktsenergie verlassen). Und dort weiss man, dass die spezifische Wärme sehr ähnlich verläuft wie die eines festen Körpers.

Ich glaube, dass ein solches Modell analog einem Stück Kristall oder einem Tropfen Helium im absoluten Nullpunkt eine bessere Näherung an den wirklichen Kern darstellt als das Modell unabhängiger Teilchen und möchte gerne diese[r] Möglichkeit etwas weiter nachgehen. Für dieses Modell spricht, dass die Existenz von starken Quadrupollinien in den Kernspektren sich in diesem Modell sehr zwanglos erklärt, denn bei den langwelligsten Schwingungen sollten jedenfalls die Bewegungen aller Teilchen ähnlich sein, und daher der elektrische Schwerpunkt angenähert mit dem Massenschwerpunkt zusammenfallen, was man auch immer über die Verteilung der Ladung annimmt. (Das Argument von Gamow und Bloch in der Phys. Rev. vom 1. August[756] auf Grund der Heisenbergschen Austauschkräfte scheint mir auf einem Fehlschluss zu beruhen.)

Wenn man dies Modell des fest-flüssigen Kerns erstnimmt, so entsteht auch die Frage der Rotationsniveaus die ja bei einem starren Kern nur einen Abstand von wenigen 1000 Volt haben sollen. (für Atomgewicht 200). Wenn die Teilchen ihre Plätze tauschen können, so wird die zur Rotation erforderliche Energie erhöht. Wieviel sie in konkreten Fällen beträgt, konnte ich bisher nur unter sehr vereinfachten Bedingungen abschätzen.

Ich möchte aber nicht weiter in Einzelheiten gehen, denn dieser Brief ist schon sehr viel länger geworden als ich eigentlich wollte und ich habe schon zu grosse Anforderungen an Ihre Geduld gestellt. Ich wäre Ihnen also schrecklich dankbar, wenn Sie mich bei Gelegenheit wissen lassen würden, ob diese Ueberlegungen Ihnen bekannt sind, oder ob sie zu Ihrem Programm für die nähere Zukunft gehören.

Nochmals vielen Dank für die anregenden Tage in Kopenhagen. Mit vielen Grüssen,
Ihr

[R.E. Peierls]

---

[756] F. Bloch and G. Gamow, 'On the Probability of γ-Ray Emission', *Phys. Rev.* **50**, 260 (1936).

[192] Hans Bethe to Rudolf Peierls

Selisberg, 19.8.1936

Die quantenhafte Ausstrahlung wird kontinuierlich — wenn die Frequenz unseres Briefwechsels sich weiter so vermehrt, konvergiert es gegen einen Häufungspunkt in ein paar Wochen.

Also zuerst meine humble apologies für den nichtgemachten Besuch.[757] Aber so schlimm wie Du Dir das denkst, war es nicht. Ich fuhr nicht über Southampton, sondern verliess die Queen in Cherbourg. Sie machte sich besonders gut vom Tender aus. Auch bei der Rückreise wird wohl diesmal nichts mit einem Besuch in England werden: wegen summer school teaching in Ann Arbor ist es diesmal sehr knapp mit der Zeit in Europa. Ich würde sonst gern das hochherrschaftliche Haus sehen, in dem sich sogar Deine Eltern wohlfühlen, und zum Tee kommen, zu dem auch Männer zugelassen sind: Aber das nächste mal - bis dahin darfst Du vielleicht jeden Tag im St. Johns College auf eigene Kosten dinieren.

Der Hauptgrund dieses Schreibens (Handschreibens, würde Placzek sagen) sind natürlich die Kerne. Ich hätte mir eigentlich denken können, dass Du die Niveaudichte auch ausgerechnet hast. Fermi hat es nebenbei auch getan, und Oppenheimer verkehrt. Ich hätte gern unsere Weisheit gepoolt, aber das geht leider nicht mehr, denn meine Arbeit ist schon vor vier Tagen im Phys. Rev. erschienen (so jedenfalls schrieb man mir).[758] Sie enthält nur das freie-Teilchen-Modell, und dies mit einer nicht sehr eleganten Ableitung - ich habe zu viel gerechnet, weil ich es zu speziell machte.

Das andere Modell — fester Körper - leuchtet mir eigentlich sehr ein. Ich hatte schon auch daran gedacht, ungefähr 3 Tage ehe Dein Brief kam. Dein Argument mit der spezifischen Wärme des He scheint mir ziemlich beweisend; und die Argumente gegen das freie-Teilchen-Modell sind natürlich recht schlimm.

---

[757]Bethe had intended to visit the Peierls family in England but had to change his travel plans going via France to Switzerland instead.

[758]H.A. Bethe, 'An Attempt to Calculate the Number of Energy Levels of a Heavy Nucleus', *Phys. Rev.* **50**, 332–341 (1936).

Ich habe versucht, etwas über dies Modell zu rechnen, ohne grossen Erfolg. Zunächst finde ich, dass der Exponent wie $U^{3/4}$ geht (siehe Anlage). Du schriebst $U^{3/2}$. Auch so dürfte der Unterschied gegen $U^{1/2}$ beobachtbar sein. Dann enthält der Exponent einen Faktor $A^{1/4}$ (A = Atomgewicht) gegen $A^{1/2}$ beim freien-Teilchen-Modell; beides unter der üblichen Voraussetzung Kernvolumen proportional Teilchenzahl. D.h. dass die Niveaudichte bei schweren Kernen nicht so rasch steigt, was mir im Hinblick auf die experimentellen Daten nicht unsympatisch ist. Weiter habe ich probiert, die Schallgeschwindigkeit auszurechnen, indem ich sie auf die Kompressibilität zurückführe, die man recht gut aus einer Majorana-Heisenberg-Weizsäcker-Rechnung ableiten kann. Dann kriegt man einen viel zu grossen Niveau-Abstand, nämlich 3MVC(!) bei $A = 100$ und $U = 8MV$. Aber ist natürlich die Annahme, dass die Kompressibilität massgebend ist, verkehrt, M[it] a[nderen] W[orten], die transversalen Schwingungen müssen sich ganz anders verhalten als die longitudinalen, für die die Kompressibilität sicher richtig ist. Das ist kein Wunder, denn schliesslich ist der Kern bestenfalls eine Flüssigkeit und kein fester Körper. Und da ist die Frequenz der Transversalwellen sicher sehr viel kleiner, besonders derjenigen langer Wellenlänge die ja für spezifische Wärme bei niedrigen Temperaturen wesentlich sind. Abschätzen kann man das ja wohl nicht.

Von den empirischen Daten her kann man versuchen, den Elastizitätsmodul für die Transversalschwingungen abzuschätzen: Der Abstand Δ zwischen benachbarten Niveaus ist wohl etwa 100 Volt; da nach der Rechnung $\Delta \approx 30 \cdot 10^6 e^{-5}$ Volts, so muss $S = \log 300000 = 12.6$ sein, d[ann] i[st] $12.6/2.3 = 5.5$ mal so gross wie der Wert den ich aus der Kompressibilität bekam. Da S proportional $E_0^{-3/4}$, so muss E_0 etwa kleiner sein. E_0 ist aber proportional u, also proportional der Wurzel aus dem Elastizitätsmodell. Letzteres muss also für Transversalschwingungen etwa 100 mal kleiner sein als für longitudinale.

Wegen der Wichtigkeit der Transversalschwingungen und des Unterschiedes zwischen Flüssigkeit und Festkörper glaube ich auch nicht sehr an Deine Rotation des Kerns als Ganzes. Er würde sich dadurch doch stark deformieren. Was hälst Du von folgender, sehr klassischer Überlegung zur Bestimmung der Rotationsniveaus: Man bestimme die Form und Energie des rotierenden Kerns als F[unktion] der Winkelgeschwindigkeit ω. Das ist wahrscheinlich einfach eine Frage der Oberflächenspannung. Wenn $E = \frac{1}{2}\Theta\omega^2$, dann ist Θ das "effektive Trägheitsmoment", das sicher viel kleiner ist als das wahre. Der Abstand der Rotationsniveaus wird entsprechend grösser.

Sollen wir diese Dinge über das Tropfenmodell zusammen publizieren? Die Idee ist ja allerdings im wesentlichen von Dir, also ist dieser Vorschlag meinerseits eigentlich eine Unverschämtheit. Ich glaube jedenfalls, es würde dem Fortschritt der Wissenschaft förderlich sein, dies der Oeffentlichkeit zu übergeben.

Da wir dich beim Publizieren sind: Was macht Dein Buch über die Quantentheorie?[759] Ich habe allen Leuten in USA so davon vorgeschwärmt, dass mich jetzt jeder fragt: Wann kommt endlich? So frage ich auch — ich würde es gern meinen Studenten in die Hand drücken.

Mit der Energie-Reichweite-Beziehung war ich natürlich ein Schwein. Ich habe die Daten in Baden-Baden und werde hoffentlich daran denken. Ausserdem werden bei Breit Messungen gemacht, bei kleinen Protonengeschwindigkeiten bis etwa 2MV.

Viele herzliche Grüsse an Dich und Genia

Dein Hans Bethe

Meine Mutter lässt auch grüssen

(Anhang)

1. Exponent in der Niveaudichte und Energie $\mathfrak{U}$ des System
   Annahme: $\mathfrak{U} = \alpha\tau^n$ ($\tau = kT, \alpha =$ fester Koeffizient)
   Dann ist

$$\underline{S} = \int \frac{c_v}{kT} dT = \int \frac{d\mathfrak{U}}{d\tau}\frac{d\tau}{\tau} = \frac{\alpha n}{n-1}\tau^{n-1}$$

[759]The book was finally published in 1955. R.E. Peierls, *Quantum Theory of Solids* Oxford: OUP, 1955.

$\tau = (\mathfrak{U}/\alpha)^{1/n}$, also

$$\underline{S} = \frac{n}{n-1}\alpha\left(\frac{\mathfrak{U}}{\alpha}\right)^{\frac{n-1}{n}} = \underline{\frac{n}{n-1}\alpha^{\frac{1}{n}}\mathfrak{U}^{\frac{n-1}{n}}}$$

Beim Schwingungsmodell ist n=4, also $S = \frac{4}{3}\alpha^{1/4}\mathfrak{U}^{3/4}$.
Wenn die Energie pro Volumenteinheit nur von der Dichte der Teilchen abhängt, und diese Dichte bei allen Teilchen konstant ist, so ist α proportional der Anzahl der Teilchen A, also $S \sim A^{1/4}\mathfrak{U}^{3/4}$.

2. Explizite Formel für die Niveaudichte $\rho(\mathfrak{U})$
   F ist definiert durch

$$e^{-F(\beta)} = \sum_i e^{-\beta Ei} = \int \rho(\mathcal{U}')d\mathcal{U}'e^{-\beta\mathcal{U}'}\beta = 1/??$$

Der Integrand hat ein scharfes Maximum bei $\mathcal{U}' = \mathfrak{U}$, wo $\mathfrak{U} = \frac{dF}{d\beta}$ die mittlere Energie des Systems bei einer Temperatur $1/\beta$ ist. Wir setzen $\rho(\mathfrak{U})e^{-\beta\mathfrak{U}} = \lambda(\beta)e^{-F(\beta)}$ wo λ eine langsam veränderliche F. Dann ist genau

$$\int \rho(\mathfrak{U}')d\mathfrak{U}'e^{-\beta\mathfrak{U}'} = \lambda(\beta)\int e^{-F(\beta')+\beta'\mathcal{U}'-\beta\mathfrak{U}'}d\mathcal{U}'$$

wo β' die zu $\mathfrak{U}'$ gehörige Temperatur. Folglich

$$\lambda(\beta)\int e^{F(\beta)-F(\beta')+(\beta'-\beta)\mathfrak{U}'}d\mathfrak{U}' = 1$$

Der Exponent ist

$$(\beta-\beta')\left(\frac{dF}{d\beta}\right)_{\beta'} + \frac{1}{2}(\beta-\beta')^2\frac{d^2F}{d\beta^2} + (\beta-\beta')\mathfrak{U} = \frac{1}{2}(\beta'-\beta)^2\frac{d\mathfrak{U}}{d\beta}$$

da $\mathfrak{U} = \frac{dF}{d\beta}$. Also

$$\frac{1}{\lambda(\beta)} = \int e^{\frac{1}{2}(\beta'-\beta)^2\frac{d\mathfrak{U}}{d\beta}}d\mathcal{U}' = \int e^{\frac{1}{2}(\mathfrak{U}'-\mathfrak{U})^2/\frac{d\mathfrak{U}}{d\beta}}d\mathfrak{U}'\sqrt{2\pi|\frac{d\mathfrak{U}}{d\beta}|}$$

($\frac{\mathfrak{U}}{d\beta}$ ist immer negativ). Demnach Abstand benachbarter Niveaus:

$$\Delta(\mathfrak{U}) = \frac{1}{\rho(\mathfrak{U})} = \frac{1}{\lambda(\mathfrak{U})} e^{F-\beta\mathfrak{U}} = \sqrt{2}\pi \left|\frac{d\mathfrak{U}}{d\beta}\right| e^{-s}$$

wo

$$\frac{dS}{d\beta} = -\frac{dF}{d\beta} + \mathfrak{U} + \beta\frac{d\mathfrak{U}}{d\beta} = \beta\frac{d\mathfrak{U}}{d\beta}, \quad \frac{dS}{d\tau} = \frac{1}{\tau}\frac{d\mathfrak{U}}{d\tau}$$

Wenn $\mathfrak{U} = \alpha\tau^n$, so ist

$$\Delta = \sqrt{2\pi n \alpha}^{-\frac{1}{2n}} \mathfrak{U}^{\frac{n+1}{2n}} e^{-\frac{n}{n+1}\alpha^{1/n}\mathfrak{U}^{n-1/n}}$$

3. Schallgeschwindigkeit nach Kompressibilität
Wenn Kernvolumen $= \Omega = \frac{a\pi}{3}R^3$ und Schallgeschwindgkeit $= u$ für Longitudinal- und Transversalwellen, so ist die Anzahl der Schwingungen pro $d\omega$

$$\mathfrak{N}(\omega)d\omega = 3\frac{\Omega}{(2\pi)^3}4\pi\omega^2 d\omega = \frac{2}{\pi}\frac{R^3}{u^3}\omega^3 d\omega$$

die Energie

$$\mathfrak{U} = \frac{2}{\pi}\left(\frac{R}{u\hbar}\right)^3 \int \frac{(\hbar\omega)^3 d(\hbar\omega)}{e^{\hbar\omega/\tau} - 1} = \frac{2}{\pi}\left(\frac{R}{\hbar u}\right)^3 \tau^4 \int_0 x^3 dx (e^{-x} + e^{-2x} + \ldots)$$

bei tiefer Temperatur: $\int = 6(1 + \frac{1}{2^4} + \frac{1}{3^4} + \ldots) = \frac{6\pi^2}{90} = \frac{\pi^2}{15}$;
$\alpha = \frac{2\pi}{15}\left(\frac{R}{\hbar u}\right)^3 A$ wo $R^3 = r_0^3 A$, r_0 unabhängig von A.
Nun ist $u^2 = \frac{\rho^2}{M}\frac{d^2E}{d\rho^2}$, wo M die totale Kernmasse, E die totale Kernenergie und ρ die Dichte der Kernmaterie. Da $\rho \sim R^{-3}$, so ist auch $u^2 = \frac{1}{9M}R^2\frac{d^2E}{dR^2}$.
Die Energie als Funktion von $x = const/R$ ist nach Formel (159, 159s) ein meinem Kernartikel[760]

$$\frac{E*}{AT_o} = \frac{3}{10}x^2 - \frac{B/T_0}{\sqrt{x}}(2x^{-3} - 3x^{-1} + (x^{-1} - 2x^{-3})e^{-x^2} + \sqrt{\pi}\Phi(x))$$

[760]Letter [147].

$T_0 = \frac{\hbar^2}{Ma^2}$. Zweimal differenziert gibt das:

$$x^2 \frac{d^2E}{dx^2} \left(R^2 \frac{d^2E}{dR^2} \right)$$

$$= 6\mathrm{AT}_0 \left[\frac{1}{10} x^2 + \frac{B}{\sqrt{\pi}\mathrm{T}_0} x^{-3} (x^2 - 4 + (x^4 + 3x^e + 4)e^{-x^2}) \right]$$

Mit den Werten in (164) etc. ($T_0 = \frac{70}{9.6} = 7.3$ MV, $\frac{B}{T_0} = 9.6$, $x = 2.7$) gibt das

$$R^2 \frac{d^2E}{dx^2} = 6 \cdot 7.3 \cdot 1.65A = 72A \text{ MV}$$

$$u^2 = \frac{8MV}{M_0} (M_0 = \text{Protonenmasse})$$

$$\sqrt{\left(\frac{\hbar u}{r_0} \right)^2} = \sqrt{\frac{\hbar^2}{M_0^2} \cdot 8MV} = 13.3MV$$

$\alpha \approx (0.75 \frac{r_0}{\hbar\omega})^3 A = \frac{A}{E_0^3}$, wo $E_0 = \frac{13.3MV}{0.75} = 18MV$. Folglich

$$S = \frac{4}{3} \alpha^{1/4} \mathfrak{U}^{3/4} = \frac{4}{3} A^{1/4} \left(\frac{\mathfrak{U}}{E_0} \right)^{3/4} = 2.3 \text{ für} A = 100. \mathfrak{U} = 8MV$$

$$\sqrt{2\pi n} \alpha^{-\frac{1}{2n}} \mathfrak{U}^{\frac{n+1}{2n}} = \sqrt{8\pi} A^{-1/8} E_0^{3/8} \mathfrak{U}^{5/8} = 5 \cdot \frac{1}{1.8} \cdot 18^{3/8} \cdot 8^{5/8} = 30MV$$

$$\underline{\Delta} = 30 \cdot e^{-2.3} = \underline{3MV}.$$

[193] Rudolf Peierls to Hans Bethe

Cambridge, 25.8.1936

Lieber Bethe,

Unsere Korrespondenz nimmt in der Tat ein atemberaubendes Tempo an![761]

Die Rechnungen, die Du schreibst, sind im grossen und ganzen dieselben, die ich gemacht habe. (Der Exponent 3/2 in der Entropie war natürlich ein misprint, denn die Rechnung ist so einfach, dass selbst ich dabei den Exponenten nicht falsch machen kann). Der Faktor in der Entropie ist dagegen bei Dir falsch, denn ich möchte eigentlich dafür sein, dass $\sum_1^\infty n^{-4} = \pi^4/90$, nicht, wie Du schreibst $\pi^2/90$ ist. Also sogar der grosse Bethe macht Rechenfehler? Teller hatte übrigens grossen Erfolg in Kopenhagen mit der Bemerkung, dass eine Tabelle in Deinem Handbuchartikel[762] zwei numerische Rechenfehler enthält und dass Gamow dies entdeckt hat. Ist das Amerika oder das Auto?

Deine Abschätzung der Schallgeschwindigkeit aus der Kompressibilität scheint mir aus zwei Gründen inkonsequent: 1. weil, wie Du selbst sagst, die Transversalwellen vermutlich viel kleinere Frequenzen haben, und 2. weil es nicht erlaubt ist, in diesem Modell mit der Heisenbergschen Methode zu rechnen, die doch fast unabhängige Teilchen vorausgesetzt, während andererseits wir doch jetzt auf dem Standpunkt stehen, dass das Modell mit den beinahe unabhängigen Teilchen nicht einmal die richtige spezifische Wärme gibt! Es scheint mir daher notwendig, den Faktor α in dem Gesetz $E = \alpha(kT)^4$ offen zu lassen. Uebrigens soll man wohl auch den Faktor 3 von der Polarisation auf jeden Fall durch einen Faktor 2 ersetzen, da eben praktisch nur die transversalen Schwingungen etwas ausmachen. Ich würde versuchen, umgekehrt das α so zu bestimmen, dass die Sache vernünftig wird. Dazu muss die Entropie offenbar von der Grössenordnung 10 werden (im Gegensatz zum Fall der Fermistatistik, wo wegen der vielen Nebenbedingungen wie Konstanz der Neutronenanzahl und Protone-

---

[761] Reference to Bethe's comment in letter [192].

[762] See letter [145].

nanzahl und des Impulsmoments, der Faktor vor dem e^s viel kleiner wird). Hieraus folgt:

$$\alpha \sim 6(MV)^{-3}$$

Andererseits hat α die folgende Beziehung zur Grundgeschwindigkeit des Kerns:

$$\epsilon = \gamma \left(\frac{\pi^5}{45\alpha} \right)^{1/3}$$

Hierin ist ϵ die kleinste Anregungsenergie, γ ist ein numerischer Faktor, der für eine elastische Kugel sehr nahe an 1 ist, und es ist vorausgesetzt, dass die Kugel nahezu inkompressibel ist (nur bei Transversalwellen). Folglich müsste dann ϵ von der Grössenordnung 1 MV sein, was wohl zu hoch.

Jedenfalls weiss man von den radioaktiven Kernen, dass dort die Anregungsenergie höchstens 100 000 Volt ist, auch dazu muss man schon alle die vorhandenen Niveaus von 4.10^4 Volt auf Feinstruktur oder auf Rotation schieben, was ich nur ungern tun würde. Nun kann man sich allerdings darauf ausreden, dass man ja wenig über die Resonanzniveaus von so schweren Kernen weiss, und es wäre ja denkbar, dass für die radioaktiven Kerne wirklich der Exponent beträchtlich grösser, d.h. die virtuellen Niveaus bereits selbst für thermische Neutronen praktisch kontinuierlich sind. Aber dazu müsste die Schallgeschwindigkeit mit wachsendem Atomgewicht ziemlich rasch abfallen, was zu ad hoc gemacht ist.

Nun scheint es mir auch nicht sicher, dass es richtig ist, die Debyesche Formel so Ernst zu nehmen. Denn wir wissen ja, dass der Kern nicht ein Kristall mit Gleichgewichtslagen ist, sondern eher eine Flüssigkeit wie flüssiges He. Je mehr ich mir darüber den Kopf zerbreche, umsomehr scheint es mir, dass dieses ein neuer Aggregatzustand mit definierten Eigenschaften sein muss. Denn flüssiges He ist offenbar makroskopisch eine Flüssigkeit, jedenfalls bei nicht zu tiefen Temperaturen, andererseits muss bei sehr tiefen Temperaturen seine Entropie verschwinden. Hieraus folgt unmittelbar, dass bereits bei einer endlichen Temperatur sehr kurze Transversalwellen nicht mehr die Frequenz Null haben können (obwohl lange das offenbar noch tun), denn wenn man z.B. annimmt, dass Wellen von der Länge L noch keine Frequenz haben, so hat ein Tropfen He von der linearen Dimension $D(n/L)^3$ Freiheitsgrade ohne potentielle Energie und daher eine abschätzbare Entropie (obwohl es mir noch nicht gelungen ist, diese Abschätzung vernünftig durchzuführen, man muss das natürlich quanteln, und dabei gibt es unangenehme Unklarheiten). Aus einer — bisher noch nicht vorhandenen — anständigen thermischen Untersuchung von He bei extrem tiefen Temperaturen könnte man also eine Abschätzung für L bekommen. Ferner erwarte ich auch Aufschluss von Untersuchungen an Ultraschallwellen in flüssigem He, die jetzt hier gerade angefangen werden.[763]

Es ist jedenfalls wohl sicher, dass sehr kurze Transversalwellen eine endliche Frequenz haben werden, während lange das nicht haben. Das kann man sich auf zwei Weisen einfach vorstellen: Entweder nimmt man an, dass die Substanz sich für lange Wellen wie eine Flüssigkeit, für kurze wie ein Kristall benimmt, wobei die kritische Wellenlänge dann von der Temperatur abhängen müsste. Oder aber, was mir seit heute früh sympathischer scheint, man nimmt an, dass die Frequenz in diesem komischen Aggregatzustand proportional zum Quadrat des Wellenvektors ist.

(Eine dritte Möglichkeit wäre, dass die Elastizitätsgrenze von der Wellenlänge abhängt, und bei langen Wellen bereits bei der Amplitude Null liegt. Dies ist natürlich schlampig ausgedrückt, denn wenn die Gleichungen nicht mehr linear sind, so darf man überhaupt nicht mehr von einzelnen Wellen reden.)
Nimmt man die Proportionalität zwischen ω und k^2 an, so wird die Entropie $U^{3/5}$ statt $U^{3/4}$, was natürlich den Faktor nicht mehr stark ändert, aber die Anregungsenergie des Kerns wird viel kleiner.

Für das flüssige He ergibt sich, dass U prop $T^{5/2}$ wird, was allerdings ein bisschen zu schwach scheint. Es liegt aber immerhin innerhalb der Messgenauigkeit, es scheint nur sicher zu sein, dass die spezifische Wärme rascher als proportional mit T verschwindet.

---

[763]In 1933 the Mond Laboratory had opened in Cambridge and in 1934, largely spurred by Kapitza's efforts, the first liquid helium had been produced there in 1934. After Kapitza's detention in Russia, John Cockroft and Ernest Rutherford supervised work at the Cambridge facilities and managed to make some significant contributions to low temperature physics until the war interrupted work.

Es müsste eigentlich möglich sein, diese Frage theoretisch zu entscheiden, aber jedenfalls etwas mehr herauszubekommen darüber, wie ein Modell mit Gleichgewichtslagen aber mit Nullpunktsenergie > Potentialschwelle sich benehmen sollte. Hierüber verstehe ich jedoch vorläufig noch immer Bahnhof.

Mit Deinem Vorschlag, die Rotationsniveaus klassisch abzuschätzen, bin ich eigentlich gar nicht einverstanden, und zwar aus folgendem Grunde: was man klassisch untersuchen kann, sind im wesentlichen zwei Dinge: Erstens gibt es eine Zentrifugalkraft, unter der der Kern deformiert wird. Diese Zentrifugalkraft ist aber erstens sehr klein, weil die Winkelgeschwindigkeit bei einem schweren Kern sehr klein ist, und zweitens bewirkt sie eine Erhöhung des Drehmoments (d.h. Erhöhung der Rotationsenergie bei fester Winkelgeschwindigkeit) und daher eine Verkleinerung der Rotationsenergie bei festem Drehimpuls. Zweitens kann klassisch z.B. im Fall einer Flüssigkeit ein Teil des Kerns unabhängig vom Rest rotieren, und hierzu würde natürlich ein kleineres Trägheitsmoment gehören. Damit beweist man aber nur, dass es auch Rotationszustände mit höherer Energie gibt — woran man natürlich sowieso nicht zweifelt aber klassisch gibt es jedenfalls sicher auch die Rotationsform, wo der Kern als Ganzes rotiert, bis auf eine von der Zentrifugalkraft herrührenden Deformation.

Dagegen gibt es einen quantenmechanischen Effekt, der im Prinzip die Rotationsenergie sehr wohl erhöhen kann, der aber klassisch nicht zu behandeln ist. Man kann sich von der Existenz am Besten etwa so überzeugen: Denken wir uns n Teilchen mit beliebiger Wechselwirkung, die aber, statt umeinander herumlaufen zu können wie im wirklichen Kern, an einen Kreisring gebunden sind, so dass man nur ein n-dimensionales Problem hat (so wird man die Eulerschen Winkel los) und bezeichnen wir die Lagen der Teilchen durch N Winkel. Die Energie hängt dann von deren Differenzen ab (natürlich mod 2π) und die Randbedingung lautet, dass die Wellenfunction ungeändert bleiben soll, wenn man irgendeinen dieser Winkel um 2 vermehrt. Man kann nun leicht das gesamte Impulsmoment das natürlich eine Integrationskonstante ist, abseparieren, indem man in der Eigenfunktion den Faktor $e^{2\pi i(\phi_1+\phi_2+\ldots\phi_n)m/N}$ abspaltet. Der Rest der Wellenfunktion enthält dann noch die $N-1$ Differenzen, und genügt dann einer Wellengleichung, die bis auf den Term $m^2\hbar^2/2NMR^2$ in der Energie von m unabhängig ist. Dagegen hat sich nun die Randbedingung geändert und verlangt jetzt, dass bei Änderung eines Winkels um 2π die Eigenfunktion sich mit $e^{2\pi i m/N}$ multiplizieren soll. Diese Veränderung der Randbedingung bewirkt sicher eine Erhöhung der Energie, wieviel hängt natürlich von den Kräften ab. Für $N = 2$ ist die Sache natürlich trivial, man kann dort die Erhöhung leicht für die Fälle anschreiben, wo das W.K.B. erlaubt ist und zwar entweder mit überall reeller oder streckenweise komplexer Wellenzahl. Im ersten Fall wird die Energieerhöhung $\frac{1}{2}dE/dn$ ($n =$ Quantenzahl der Relativbewegung) d.h. die Rotationsenergie wird gleich der inneren Änderungsenergie. Im anderen Grenzfall dagegen, wenn es ein grosses Gebiet mit V.E. gibt, so ist die Erhöhung proportional $\exp(-\int pdq)$ über diesen Bereich also sehr klein. Der Übergang zwischen diesen beiden Grenzfällen wird hier gegeben durch die Bedingung dass das eine Teilchen relativ zum anderen gerade eine merkliche Wahrscheinlichkeit bekommt, einmal ganz herumzulaufen. Es fragt sich man, wie dies auf N Teilchen zu veallgemeinern ist, ob die Bedingung dort lautet, dass ein Teilchen relativ zu allen andern gerade eine merkliche Wahrscheinlichkeit haben soll, einmal um den ganzen Ring herumzulaufen (was natürlich nicht der Fall ist) oder ob die Wahrscheinlichkeit nur beträchtlich sein soll, dass die Teilchen ihre Reihenfolge wechseln (was sicher der Fall ist). Diese Frage muss entscheidbar sein, aber ich sehe noch keinen Weg dazu.

Was die Frage der Publikation betrifft, so habe ich den Eindruck, dass wir im Augenblick auf ziemlich verschiedenen Pfaden wandeln. Ausserdem aber scheint mir die ganze Sache noch nicht publikationsreif, bevor die diversen ungelösten Fragen (elastische Eigenschaften der "Heliumphase", Grösse der Rotationsquanten, etc.) etwas weiter geklärt sind. Ich würde daher vorschlagen, dass man dies abwartet; wenn es dann soweit ist, können wir ja immer noch eine Pfütze (ich meine pool) machen. Oder bestehst Du auf einer Publikation in amerikanischem Tempo?

Ich würde übrigens auch auf jeden Fall warten, bis ich von Bohr gehört habe, ob er das alles schon weiss, bezw. machen will, denn ich habe keine Lust, in unfertige Bohrsche Arbeiten hineinzupfuschen.[764] Mein Buch hatte ich seit Oktober nicht mehr angerührt. Dein gutes Zureden wie das von Mott haben aber jetzt den Erfolg gehabt, dass ich es wieder ausgegraben habe, und dass mir die Stelle wegen deren ich schon die Hoffnung verlor, eigentlich gar nicht mehr so schlecht vorkommt. Vielleicht wird doch noch etwas daraus.

Thomas hat mir jetzt eine Methode zur Behandlung von H^3 erzählt, die wirklich wunderhübsch ist und für ein gegebenes Potentialfeld mit beliebiger Genauigkeit durchgeführt werden kann (jede weitere Näherung dauert etwa zwei Tage, Konvergenz wie eine geometrische Reihe mit einem Faktor von ca. $\frac{1}{2}$).

Resultate damit hat er noch keine. Sein früheres Resultat, wonach Presents Methode sehr schlecht ist, hat sich übrigens als Rechenfehler herausgestellt. Er glaubt aber immer noch an die Richtigkeit seiner Behauptung.[765] Bloch betreibt Gamowsche Physik, die dann in der Physical Rev. in der Gamowsprache veröffentlicht wird[766] (wieso lässt Du als Redakteur das eigentlich zu?
Nu aber Schluss! Beste Grüsse

Dein Peierls

Hier sind übrigens viele Leute "settled". Born hat die Darwinsche Professur in Edinburgh,[767] Berg eine Stelle bei Kodak in London, Ehrenberg scheint bei HMV sehr zufrieden und hat eine kleines Auto.

---

[764]See letter [191].

[765]See letter [190].

[766]Felix Bloch had been working on magnetic scattering of neutrons and on γ-ray emission. He had published several papers one of which with Gamow. F. Bloch and G. Gamow, 'On the Probability of γ-Ray Emission', *Phys.Rev.* **50**, 260 (1936); F. Bloch, 'On the Continuous γ-Radiation Accompanying the β-Decay', *Phys. Rev.* **50** (1936).

[767]Max Born (1882–1970) had left his Chair at Göttingen in 1933, and after short spells at Cambridge and Bangalore he became Tait professor in Edinburgh in succession to Charles Galton Darwin.

[handwritten addition by Genia Pierels]
Ach wie schade dass Du nicht zum Tee kommst. I am sooo sorry! Aber, I hope, nächstes Jahr. Ja übrigens bist Du noch nicht berühmt genug in U.S.A um mich (das heißt Rudi) einzuladen? Wenn nicht, hast Du zwei Jahre verloren. Hast Du den Gamow gesehen? Family blüht und gedeiht und Gaby ist so hübsch geworden (von wo kommt das) dass sie wird für Reklame photographiert. Ronny ist ein judisches Affe, aber für Junge das macht nichts. Ich habe eine amerikanische Party gehabt und 20 Amerikaner haben gefressen wie 200 Krokodile! Warum warst Du so verhungert? Viele Grüsse auch an Deine Mutter.

Genia

[194] Hans Bethe to Rudolf Peierls

Seelisberg, 27.8.1936

Lieber Peierls,
Fortsetzung über den Fortschritt der Wissenschaft:

1. Ich habe versucht, die Kernniveaus mit dem Tropfenmodell abzuschätzen unter der Annahme, dass nur die Oberflächenspannung für die Transversalschwingungen eine Rolle spielt. Die Oberflächenenergie nimmt man natürlich proportional zur Oberfläche an. Dann spielt bei einer angegebenen Oberfläche nur die Verschiebung der Teilchen senkrecht zur Oberfläche eine Rolle. Wenn die Oberfläche in der xy-Ebene liegt und u die Verschiebung ist, so ist die relative Änderung der Oberfläche

$$\delta\sigma = \frac{\int dxdy \left(\sqrt{1 + \left(\frac{\partial u_z}{\partial x}\right)^2 + \left(\frac{\partial u_z}{\partial y}\right)^2} - 1 \right)}{\int dxdy}$$

$$= \frac{1}{2} \frac{\int d\sigma \left[\left(\frac{\partial u_n}{\partial x}\right)^2 + \left(\frac{\partial u_n}{\partial y}\right)^2 \right]}{\int d\sigma}$$

Wenn nun $u = \underset{\sim}{a} \sin \underset{\sim}{k} \cdot \underset{\sim}{r}$ so ist $\delta\sigma = \frac{1}{2} a_n^2 k_\|^2 \overline{\cos^2 k \cdot r}$, wo $k_\|$ die Komponente von k in der Oberfläche. Bei einer Kugel muss man natürlich über die Oberfläche mitteln. Nehmen wir $\underset{\sim}{k}$ parallel zu z, $\underset{\sim}{n}$ parallel zu x, so ist in Polarloordinaten

$$a_n^2 k_\|^2 = a^2 k^2 \sin^4 \vartheta \cos^2 \varphi$$

gemittelt $= a^2 k^2/10$.

Also die Änderung der Oberflächenenergie$= \frac{1}{20} S_o a^2 k^2 \overline{\cos^2 kr}$ wo S_o die Oberflächenenergie ohne Schwingung ist. Wenn M die Gesamtmasse des Kerns ist, so ist dann sie Schallgeschwindigkeit $u = \sqrt{S_o/10M}$.

S_o kann man aus den empirischen Kernenergien abschätzen (à la Weizsäcker)[768]; man bekommt dann mit den neuen Kernradien (siehe unten) $S_o = 10A^{2/3}M$, wo $A =$ Atomgewicht. Wenn man dann die Formel für die Totalenergie des Kerns schreibt:

$$\mathfrak{U} = \frac{4\pi}{45} \left(\frac{R}{\hbar u}\right)^3 \tau^4 = \frac{\tau^4}{E_o'^3} A$$

so ist

$$E_o' = \left(\frac{45}{4\pi}\right)^{1/3} \cdot \sqrt{\frac{\hbar^2}{M_o r_o^2 A} \frac{S_0}{10}} = \left(\frac{45}{4\pi}\right)^{1/3} \sqrt{\frac{\hbar^2}{M_o r_o^2} \cdot \frac{10MV}{10}} A^{-1/6}$$

$$= E_o A^{-1/6}$$

($M_o =$ Protonenmasse; $r_o = 2 \cdot 10^{-13}$cm entsprechend den Bemerkungen unter 2) Also $\frac{\hbar^2}{M_o r_o^2} \approx 10MV$, $E_o = 4.8MV$

$\mathfrak{U} = \frac{\tau^4}{E_o^3} A^{3/2}$; $\alpha = \frac{A^{3/2}}{E_o^3}$ (siehe vorheriger Brief)[769]

---

[768]In 1935, von Weizsäcker had formulated a semi-empirical approach explaining the difference between the mass of a nucleus and the mass of its constituent protons and neutrons the binding energy, as being the result of energies associated with the interactions of the nucleus. Bethe and R. Bracher simplified Weizsäcker's formulation in 1936 and E. Wigner extended to formula in 1937.

[769]Letter [192].

$$\tau = (\mathfrak{U}/\alpha)^{1/4} = \mathfrak{U}^{1/4}E_o^{3/4}A^{-3/8}$$

$$S = \frac{4}{3}\frac{\mathfrak{U}}{\tau} = \frac{4}{3}\left(\frac{\mathfrak{U}}{E_o}\right)^{3/4} A^{3/8}$$

$$\Delta = \sqrt{8\pi}\sqrt{\mathfrak{U}\tau}e^{-5} = \sqrt{8\pi}\mathfrak{U}^{5/8}E_o^{3/8}A^{-\frac{3}{16}}e^{-\frac{4}{3}\cdot 5.61\cdot 1.67^{3/4}} = \text{ca.}200V_0$$

Das sieht eigentlich recht vernünftig aus!

2. Kernradius: Man hat die Radien der radioaktiven Kerne bisher immer mit der Einkörpertheorie ausgerechnet und etwa $9 \cdot 10^{-13}$cm gefunden. Jetzt muss man natürlich auch hier die Vielkörpertheorie anwenden. Das heisst, dass es schon unwahrscheinlich ist, dass das α-Teilchen überhaupt gebildet wird, und dass daher der Potentialberg viel niedriger werden muss. Die reziproke Lebensdauer ist $\lambda = \gamma\alpha e^{\frac{-G}{\hbar}}$, wo γ_α die "α-Breite" des Kernzustandes wäre, wenn es keine Barriere gäbe. e^{-G} ist die Durchlässigkeit des Potentialbergs. Die Hauptfrage ist also die nach der Grösse von γ_α. Die einzige Evidenz, die man über Niveaubreiten hat, ist die von Neutronenexperimenten. Dort findet man etwa 0.005 Volt. Die Neutronenbreite enthält einen Faktor. $v \sim \frac{1}{\bar{\lambda}}$, wo $\bar{\lambda}$ die Wellenlänge ist. Als "normale" Neutronenbreite kann man etwa die Breite bezeichnen, die man bei einer Wellenlänge $\bar{\lambda}$ von der Grössenordnung des Kernradius enthält, d.h. bei einer Neutronenenergie von vielleicht 500 000 Volt. Dann wird die Breite von der Grössenordnung $\sqrt{500000/5} \cdot 0.0005 \sim 1Volt$. Über α-Breiten weiss man natürlich nichts, ich habe einfach auch $\gamma_\alpha = 1$ Volt angenommen. Dann kann man G, und daraus die Radien der radioaktiven Kerne berechnen. Die meisten liegen zwischen 12.5 und 13.5$\cdot 10^{-13}$ cm. Nimmt man $R = r_o A^{1/3}$ an, so wird r_o etwas mehr als $2 \cdot 10^{-13}$cm. Konservativerweise, und der Einfachheit halber, habe ich $r_o = 2 \cdot 10^{-13}$cm genommen. Wenn man das in die Formel für die Niveaudichte mit freien Teilchen einsetzt, kriegt man viel zu kleine Abstände der Niveaus; Bruchteile eines Volt schon bei $A = 100$ und $\mathfrak{U}$=8 MV.

Es erhebt sich die technische Frage der Abfassung unseres Meisterwerkes, die womöglich vor meiner Abreise nach USA beendet sein sollte. Wollen wir die Schreibarbeit verteilen? Vorschläge erbeten![770]

Seit Sonntag ist Plazcek (im Dublettzustand) hier. Wir haben gestern eine Tour gemacht, wobei er hauptsächlich seinen Puls fühlte. Viele schöne Grüsse

Dein Hans Bethe

[195] Rudolf Peierls to Hans Bethe

Cambridge, 30.8.36

Lieber Hans,
Ich bin ja ganz einig, aber, aber, muss ich sagen. Oder in Variation des neuesten Kopenhagener Schlagers: Erst kommt das Denken, dann das Integral![771]

Oder mit noch anderen Worten: Wir sind in der Tendenz natürlich sehr einer Meinung, nicht aber in den Ansätzen, was zunächst die Kernradien betrifft, so habe ich so etwas ähnliches auch gerne immer haben wollen, war aber immer zu faul, es mir zu überlegen. Du hast natürlich völlig recht damit, dass die Kernradien in Wirklichkeit grösser sind als aus der Gamowschen Rechnung folgt, und natürlich wird Dir auch gern geglaubt, dass sie dann ein bisschen grösser als 10^{-12} werden. Nicht geglaubt wird dagegen die Behauptung, dass die α-Breiten von derselben Grösse wir die Neutron-Breiten werden, denn ich kann auf Wunsch ebenso gute Argumente produzieren, dass sie grösser wie dass sie kleiner sein sollten (Man könnte natürlich nach dem Rezept des "equivalent two-body-problems" hieraus schliessen, dass die Gleichsetzung eine ausgezeichnete Näherung sein muss. Es ist natürlich richtig, dass der

[770] No joint paper was published.

[771] 'Adaptation' of the German saying: 'Erst kommt das Fressen, dann kommt die Moral.'

Radius von dieser Breite ziemlich schwach abhängt, für einen Fehler in γ ist der zugehörige Fehler in r:

$$\frac{\delta r}{r} = \frac{1}{2}\frac{\lambda\!\!\!^{-}_0}{r}\frac{\delta\gamma}{\gamma}$$

wobei $\lambda\!\!\!^{-}$ die Wellenlänge eines α-Teilchens mit der Energie $4Ze^2/r$ ist. Numerisch also ungefähr $\frac{\delta r}{r} \sim 0.06\frac{\delta\gamma}{\gamma}$.

Damit Du also berechtigt bist, den Kernradius mit einer Genauigkeit von 5% hinzuschreiben, wie Du das tust, müsste γ bis auf einen Faktor 2 bekannt sein, was es offenbar nicht ist. Um r bis auf 20% zu bekommen, (und weniger hat offenbar keinen Sinn) kann ich Dir nur einen Fehler von einem Faktor 15 in der Breite erlauben. Du musst mich aber noch davon überzeugen, dass die Breite nicht ebensogut 12 statt 1 volt sein kann.

Es hilft auch nichts, wenn man einen Vergleich aller α-Strahle[n] zu Hilfe nimmt. Denn wenn man in allen γ um denselben Faktor falsch macht, das bei allen denselben Fehler in r und man bekommt noch immer vernünftige Verhältnisse der Radien. (Dagegen stimmte eben das Gamowsche Bild so schön). Ich bin übrigens auch noch nicht so restlos davon überzeugt, dass die Neutronenbreite prop. v geht, d.h. dass die anderen Faktoren, die ausser dem v eingehen, für langsame Neutronen konstant werden. Es klingt natürlich plausibel.

Ich möchte auch gegen die Behauptung protestieren, dass man aus diesen Radien nur das Modell mit freien Teilchen widerlegen kann. Erstens ist nämlich ein Faktor 100 in der Niveaudichte noch nicht so sehr schlimm, denn er bedeutet nur, dass die Entropie um 5, d.h. um 25% falsch ist.

Nun wird zwar Dein Wert für den Radius schon nicht um 25% falsch sein, wohl aber die Annahme eines rechteckigen Potentialtropfens. Wenn Du stattdessen annimmst, dass die Breite des Potentialtropfens mit wachsender Höhe zunimmt, so misst die Zerfallskonstante vermutlich die Breite am oberen Rand, (d.h. am Rand der Fermi-Verteilung) während in die Entropie ein Mittelwert eingeht.

Selbst wenn sich aber aus diesem Modell eine zu grosse Niveaudichte ergibt, so ist das nicht notwendig ein Widerspruch, denn es bedeutet nur dass ein Teil der Zustände nicht stark genug mit dem Kontinuum kombiniert, um als virtuelle Niveaus beobachtbar zu werden. Gerade im Fall fast freier Teilchen wäre das ganz vernünftig. Nehmen wir an, dass die Kopplung der Teilchen durch einen Faktor s gemessen wird, so wird die Wahrscheinlichkeit der Anregung eines Niveaus bei dem n Teilchen nicht mehr im Grundzustand sind, den Faktor s^n enthalten. Durch geeignete Wahl von s kann man dann einen beliebigen Bruchteil der Niveaus ausschalten, denn für $s \rightarrow 0$ kommt man offenbar auf Deine alte Theorie zurück. Man braucht s nicht sehr von 1 verschieden anzunehmen, um den Schaden zu beheben.

Ich glaube natürlich nicht an das freie-Teilchen-Modell, aber ich glaube nicht, dass man es so widerlegen kann. Nun zum Tropfenmodell. Ein Teil meiner Bedenken stehen schon in meinem letzten Brief, der sich offenbar mit Deinem gekreuzt hat.[772] Mein Haupteinwand gegen Deine Rechnung ist folgende: Nimmt man das Tropfenmodell mit Oberflächenspannung Ernst, so hat es offenbar eine von Null verschiedene Nullpunktsentropie, denn es gibt eine Menge von Deformationen, die weder das Volumen, noch die Oberfläche ändern, und daher zur Frequenz Null gehören. Beispiel: man denke sich die Kugel in eine Reihe konzentrischer Kugelschalen beliebiger Dicke eingeteilt und verdrehe jeder dieser Schalen in beliebiger Weise gegen ihre Nachbarn. Nun wirst Du natürlich sagen, dass es in der Quantentheorie anders ist und zwar aus den verschiedenen Gründen. Erstens muss in der Quantentheorie diese Rotation der Kugelschalen mit einer minimalen Rotationsgeschwindigkeit vor sich gehen, und dadurch entsteht eine Zentrifugalkraft, die dann die Oberfläche doch deformiert. Dies ist zweifellos richtig, aber die hierzugehörende Frequenz ist offenbar viel kleiner als Du sie abschätzt, denn wenn man die Oberflächenspannung gegen unendlich gehen lässt, so geht die auf diese Art von Rotation wirkende Kraft keineswegs gegen unendlich sondern sie verwandelt sich nur in eine Rotation ohne Deformation. Andererseits kannst Du sagen, dass die Bose- oder Fermistatistik schon einen Teil der Entartungen aufhebt, sodass im Gegensatz zum klassischen Fall durch Angabe der

[772] Letter [194].

äusseren Form der Zustand des Systems bereits mehr oder weniger eindeutig angegeben ist. So etwas ist möglicherweise richtig. Ich habe das aber noch nicht zu Ende denken können, aber wenn es richtig ist, so ist offenbar die Hydrodynamik nicht mehr anwendbar, und der Zusammenhang zwischen Schallgeschwindigkeit und Entropie geht verloren. Mir scheint, es gibt zwei mögliche Standpunkte:

A: Der Kern ist eine Flüssigkeit bis auf die Oberflächenenergie. Mit anderen Worten, wenn man sich einen makroskopischen Tropfen aus Kernmaterie bauen würde, so würde er wirklich eine endliche Nullpunktsentropie haben. (Genauer gesagt, die Temperatur, bei der die Entropie zu verschwinden beginnt, nimmt mit wachsenden Dimensionen ab. In diesem Fall muss man sich überlegen, welches die Kräfte sind, die dann den Grundzustand einfach machen und welches diejenigen Bewegungen sind, die zur kleinsten Anregungsenergie gehören. Dass es auch Bewegungen mit grösserer Anregungsenergie gibt, nützt gar nichts, wir wissen ja ohnehin schon dass es Longitudinalwellen mit sehr hoher Frequenz gibt. Daher erscheint mir der Ansatz einer ebenen Welle mit festem Wellenvektor ebenso falsch als wenn man eine Welle mit beliebiger Polarisation ansetzen würde (also keine Transversalität voraussetzen) und dann über die Polarisationsrichtung mitteln [würde]. So bekommt man natürlich selbst in einer idealen Flüssigkeit 1/3 der Longitudinalgeschwindigkeit als mittlere Schallgeschwindigkeit, was offenbar Unsinn ist.

Der zweite mögliche Standpunkt ist B: dass die Entropie bereits im makroskopischen Fall gegen Null geht. Dann bekommt man den in meinem letzten Brief vorgeschlagenen neuen Aggregatzustand. Mir ist diese Idee deswegen sympathischer, weil sie offenbar bei Helium realisiert ist und wahrscheinlich doch die Bindungskräfte dieselbe qualitative Charakteristik haben. In diesem Falle haben dann die Transversalschwingungen eine endliche definierte Frequenz, und die Berechnung der Entropie bleibt, ungefähr jedenfalls, zulässig. Aber da man weiss, dass die transversalen Schwingungen jedenfalls für unendlich lange Wellen nicht existieren, so muss man dann feststellen, was dort passiert, und das können eben wohl die drei Möglichkeiten sein, dass die Schallgeschwindigkeit mit zunehmender Wellenlänge entweder kontinuierlich oder diskontinuierlich gegen Null geht, oder dass die elastische Grenze für wachsende Wellenlänge gegen Null geht. Je mehr ich es mir überlege, scheint eigentlich der letzte Fall, jedenfalls für flüssiges Helium der wahrscheinlichste. Aber natürlich ist er der unangenehmste zum Rechnen. Immerhin würde er wohl heissen, dass im Kern, wo ja lange Wellen ohnehin nicht in Frage kommen, vermutlich die Näherung vom festen Körper her ganz anständig sein sollte.

Es tut mir leid, dass ich das Meisterwerk durch so viel Pedanterie aufhalte, aber ich glaube, man sollte sich das alles anständig überlegen. Vielleicht publizierst Du mal zunächst die Sache mit den Kernradien, die ja hiermit nicht direct was zu tun hat? Grüsse Placzek. Hat das Dublett (JJ) oder Russell-Saunders-Kopplung? Ich möchte von ihm gerne wissen, wie weit die im Werden begriffene Arbeit von Bohr auch solche Modellfragen behandelt. Ausserdem ist er herzlich eingeladen, uns hier zu besuchen; ich empfehle die Route über Cambridge als sehr praktischen Weg von Seelisberg nach Kopenhagen! Wann schiffst Du Dich wieder ein?
Grüsse

[Rudi]

[196] Hans Bethe to Rudolf Peierls

Seelisberg, 4.9.36

Lieber Peierls,
Bohr würde sagen: Wir sind uns einiger, als Du denkst. So Ernst habe ich die Integrale vor dem Denken nicht genommen, und ich bin ganz Deiner Meinung, dass man zur Zeit das flüssige Helium noch nicht genügend versteht, um die Kerne anständig zu behandeln. Die Absicht, bei dem vorgeschlagenen Meisterwerk war mehr destruktiv als konstruktiv: Ich fühle mich nämlich etwas schuldig wegen der Publikation des Fermi-Thomas-Modells. Die Amerikaner werden das glauben, genau wie sie vorher die Hartree-Näherung für Kerne zu wörtlich geglaubt haben. Und da wollte ich ihnen mitteilen, dass ich selbst nicht daran glaube. Aber wenn Du meinst, dass das zu destruktiv für ein Meisterwerk ist —

und ausserdem sehe ich ein, dass ich für eine destruktive Arbeit etwas viel integriere. Das ist Amerika.

Das Einzige, was ich ernst meine, sind die Kernradien. Die sind doch wirklich besser als Du schriebst. Nach Deiner Formel ist

$$\frac{\delta R}{R} = \frac{\lambda\!\!\!^{-}_0}{2R}\frac{\delta\gamma}{\gamma}$$

$\lambda\!\!\!^{-}$ ist die Wellenlange eines α-Teilchen der Energie $V_0 = \frac{2Ze^2}{R}$
Mit $R = 12.5 \cdot 10^{-13} = 4.5\frac{e^2}{mc^2}$ ist $V_0 = \frac{2 \cdot 90}{4.5}mc^2 = 20MV$, also $\lambda\!\!\!^{-}_0 = \frac{\hbar}{\sqrt{2MV_0}} = \frac{4.5 \cdot 10^{-13}}{\sqrt{4 \cdot 20}} = 0.5 \cdot 10^{-13}\text{cm}$. Demnach $\frac{\delta R}{R} = \frac{0.5}{2 \cdot 2.5}\frac{\delta\gamma}{\gamma} = 0.020\frac{\delta\gamma}{\gamma}$ (nicht $0.06\frac{\delta\gamma}{\gamma}$).

Eine Änderung von R um 5% entspricht also einer Änderung von $\log\gamma$ um einen Faktor von 12. Deshalb kann man wohl sagen, dass R auf 10% genau ist, denn Du wirst mir wohl zugeben, dass γ wahrscheinlich zwischen $\frac{1}{100}$ und 100 Volt liegt.

Bezüglich der Neutronenbreite bin ich doch von dem Faktor v sehr überzeugt: Das Matrixelement, das die Neutronenbreite bestimmt, enthält die Eigenfunktion des compound-Kerns und des Neutrons. Für den compound Kern ist es sicher ganz gleich, ob er mit $8MV$, mit 10 oder auch mit $15MV$ angeregt ist, weil auf jeden Fall die Anregungsenergie minimal ist gegen die totale Bindungsenergie. Die Neutronenfunktion enthält einen Normierungsfaktor, der eben das v-Gesetz gibt, und eine Ortsabhängigkeit. Letztere kann aber sicher nichts ausmachen, solange die Neutronenwellenlänge grösser als der Kernradius ist. Es sieht mir jetzt eigentlich plausible aus, dass die kritische Wellenlänge nicht der Kernradius ist, sondern etwa die Reichweite der Kernkräfte — oder, was dasselbe ist, der Abstand benachbarter Teilchen im Kern. Das würde die Abschätzung des γ um etwa einen Faktor von 5 erhöhen und R um ein paar % verkleinern.

Mit der Bemerkung, dass alle α-Strahlen dasselbe R geben, wollte ich nicht die Richtigkeit meines $\gamma's$ beweisen. Immerhin ist es nicht ganz trivial: Dadurch dass R soviel grösser wird, wird der Potentialberg niedriger. Die abgekürzte Gamowsche Formel für den Exponenten im Durchlässigkeitskoeffizienten, $G = \frac{\pi\ Ze^2}{\hbar v} - \frac{2e}{\hbar}\sqrt{2zZMR}$ gilt ja nur wenn $E \ll \frac{zZe^2}{R}$. Wenn die Energie vegleichbar ist mit der Höhe des Potentialberges, kann G nicht mehr geschrieben werden als ein Term, der nur von der Energie abhängt minus einem zweiten Term, der nur vom Radius abhängt, sondern es kommt noch etwas dazu, was von beiden Grössen abhängt. Es ist deshalb nicht trivial, dass das Geiger-Nuttall-Gesetz noch genügend genau erfüllt ist.

Die neuen Kernradien sind mir hauptsächlich noch deshalb sympathisch, weil sie die absolute Ausbeute der Deuteronen-Reaktionen in schweren Elementen verständlich machen. Mit den alten Radien kam die Ausbeute immer zu klein heraus — mit der Bohrschen Theorie sowohl wie mit der uralten Gamowschen. Über diese Sachen will ich einen letter to the editor (also an mich selbst)[773] schreiben, es ist ja ganz unabhängig von den Fragen der Niveaudichte.

Die Geschichte mit der Oberflächenspannung hat Teller, der drei Tage hier war, viel gescheiter verstanden als ich sie meinte: Er sagte, dass die Oberflächenspannung ein ganz gutes Mass für die Scherungskräfte sein könnte, da man ja bei der Scherung zwei benachbarte Schichten von der günstigeren Position in eine weniger günstige bringt, also einen Teil einer freien Oberfläche erzeugt. Allerdings macht das meine Integrale nicht besser. (Und wahrscheinlich ist es für die Transversalschwingung etwa dieselbe Schallgeschwindigkeit wie für die longitudinalen.) Dein Einwand gegen die Oberflächenenergie, dass die Entropie bei einer mit wachsender Grösse abnehmenden Temperatur verschwindet, ist natürlich richtig. Ich bin auch ganz davon überzeugt, dass es noch andere Kräfte gibt, die sich der Scherung widersetzen. Es wäre aber an sich denkbar, dass diese Kräfte nur bei sehr grossen Tropfen wesentlich werden, und dass bei kleinen Tropfen, wie wir sie in Kernen haben, die Oberflächenspannung den Hauptbeitrag liefert.

Mit Teller sprach ich auch über die Frage der Rotation des ganzen Kerns. Er hatte sich das schon früher überlegt, und es schien uns herauszukommen, dass auf jeden Fall die Rotationsniveaus von derselben Grössenordnung sind wie bei einzelnen Teilchen, und zwar wegen der Symmetrie der Rotationseigenfunktion. Bei n festgebundenen Teilchen mit Bosestatistik in einem Ring gibt es nur jedes n-te Rotationsniveau,

[773] Bethe was editor of *Physical Review*.

und das hebt das grösste Trägheitsmoment auf. Teller wird nach Cambridge kommen — er ist in London — und kann es Dir viel besser erzählen als ich.

Placzek, der gestern vom Dublett- in den Singlett-Zustand übergegangen ist, sagte, dass man sich um Bohr nicht kümmern dürfte — sonst müsste man einfach die Kernphysik im augenblicklichen Zustand abbrechen. Die Arbeit Bohr & Kalckar wird nach Placzeks Meinung nie erscheinen.[774] Bis zu quantitativen Dingen ist man überhaupt nicht vorgedrungen; anscheinend ist die Hauptsorge eine möglichst komplizierte Ableitung der Dispersionsformel, und es scheint gelungen zu sein, die Breit-Wignersche Ableitung an Komplikation zu übertreffen.[775] Von dieser Seite ist also kaum etwas zu erwarten.

Trotzdem hast Du mich ziemlich überzeugt, dass man das flüssige Helium erst anständiger überlegen soll, bevor man etwas sagt. Nebenbei, die meisten Probleme der Transversalwellen gelten wohl auch für gewöhnliche Flüssigkeiten in derselben Weise. Placzek * hat einmal abgeschätzt, dass Transversalwellen mit einer Frequenz grösser als $\rho u^2/\eta$ sich im wesentlichen wie longitudinale verhalten sollen, wo $\eta =$ Viskosität, $\rho =$ Dichte und $u =$ Schallgeschwindigkeit (hochfrequenter) Wellen. Viel nutzt das nicht, weil die Temperaturabhängigkeit von η wieder problematisch ist.
Sollen wir also die Sache vertagen?
Viele schöne Grüsse and Dich und Genia

Dein Bethe

*Es ist mir bisher nicht gelungen, Placzek davon zu überzeugen, dass der nächste Weg von Seelisberg nach Kopenhagen über Cambridge führt.

---

[774]This presumably refers to a paper eventually published in 1937. N. Bohr and F. Kalckar, 'On the Transmutation of Atomic Nuclei by Impact of Material Particles. I. General Theoretical Remarks', *Mat.-Fys. Medd. Dan. Vidensk. Selsk.* **14**, 10 (1937). See below letter [199].

[775]See letter Hans Bethe to Rudolf Peierls, 1.9.1936, *Peierls Papers*, Ms. Eng. misc. b202 C15.

[197] Niels Bohr to Rudolf Peierls

Copenhagen, 9.9.1936

Lieber Peierls,

Ich bedauere, dass ich Ihrem freundlichen und interessanten Brief[776] nicht früher beantworten konnte, weil Kalckar und ich erst vor kurzem von einer skandinavischen Naturforschertagung in Finland zurückkamen.[777] Inzwischen haben Sie sicherlich die Artikel von Bethe in Physical Review[778] gesehen, von der wir schon vor einigen Monaten durch Placzek hörten, aber erst jetzt haben näher kennengelernt. Bethe's Überlegungen sind wohl im Prinzip ähnlich wie die, die Sie angestellt haben, und die mit seinen Rechnungen weniger eng verknüpfte Diskussion der empirischen Daten über gerade und ungerade Kernen u.s.w. scheint uns durchaus vernünftig.

Die Abschätzung, die Bethe und Sie unter Annahmen der unabhängigen Bewegung der Kernteilchen gemacht haben, ist ja sehr interessant, aber die Begründung, die Bethe für die Übereinstimmung mit den empirischen Befunden gegeben hat, scheint uns wenig durchsichtig zu sein. Ich war daher interessiert zu sehen, dass Sie meine Überzeugung zu teilen scheinen, dass eine mehr tiefgehende Analyse des Niveauschemas nur durch ein näheres Studium der in meiner Nature Artikel[779] als kollective bezeichnete Bewegungsart der Kernsubstanz möglich sein dürfte.

Mit einem einfachen Vergleich der Schwingungen eines festen Körpers muss man aber schon deshalb vorsichtig sein, weil, wie ich bei der Konferenz betonte, die Amplituden der Kernbewegungen schon in

[776]Letter [191].

[777]Nordic Scientists' Meeting in Helsinki, 11–15.8.1936. Here Bohr had given a lecture that summarised his work on nuclear physics. See N. Bohr, 'Atomkernernes Egenskaber', *Nordiska (19. skansinaviska) naturforskamötet i Helsingfors den 11–15 augusti 1936*, Helsinki, 1936, pp. 73–81. See also *Fys. Tidskr.* **34**, 186–94 (1936).

[778]H.A. Bethe, 'Theory of Disintegration of Nuclei by Neutrons', *Phys. Rev.* **47**, 447–59 (1935); H.A. Bethe, 'An Attempt to Calculate the Number of Energy Levels of a Heavy Nucleus', *Phys. Rev.* **50**, 332–41 (1936).

[779]See letter [190], note 740.

den niedrigsten Niveauen von derselben Grössenordnung wie die Kerndimensionen sind. Auch haben wir die Konsequenz dieser Auffassung für die γ-Strahl-Erscheinungen näher untersucht und haben nicht nur in qualitativer Hinsicht (Quadrupolstrahlung), sondern auch in quantitativer Weise eine befriedigende Übereinstimmung mit der durch Neutroneneinfang erhaltenen Abschätzung erreicht. Mit Ihrer Kritik der Bemerkungen von Bloch und Gamow,[780] sind wir insbesonderen einig; sie dürfte, wir wir auch hier im Institut diskutiert haben auf einer ganz unzulässigen Deutung der Formalismus der Austauschkräfte beruhen.

Wir werden uns bestreben, unsere Abhandlung[781] in kurzer Zeit fertig zu bringen und sollen Ihnen möglichst bald eine Kopie des Manuskriptes zugehen lassen. Falls Sie vorher beabsichtigen, einige von Ihren Überlegungen zu publizieren, soll ich dankbar sein, wenn Sie uns freundlichst näheres darüber mitteilen wollen.
Mit herzlichen Grüssen von uns allen.
Ihr

Niels Bohr

[198] Rudolf Peierls to Niels Bohr

Cambridge, 11.10.1936

Lieber Herr Bohr,
Ich habe Ihnen noch sehr für Ihren freundlichen und ausführlichen Brief zu danken,[782] mit dessen Inhalt ich sehr einverstanden bin. Es ist auch richtig, dass die Rechnung, die ich im letzten Teil meines Briefes erwähnte, bis auf unwesentliche Verallgemeinerungen mir der Betheschen übereinstimmt. Ich habe die Bethesche Arbeit erst sehr spät zu sehen bekommen, da hier die Bibliotheken in den Ferien geschlossen waren.

---

[780] F. Bloch and G. Gamow, 'On the Probability of γ-Ray Emission', *Phys. Rev.* **50**, 260 (1936).
[781] See letter [196].
[782] Letter [197].

Inzwischen habe ich versucht, die Eigenschaften des "Aggregatzustandes" näher zu untersuchen, in dem sich die Teilchen in den Kernen befinden. Wie Sie sagen und wie ich auch wohl in meinem letzten Brief betonte,[783] ist dieser Aggregatzustand weder kristallin (wie es eine Rechnung mit Normalkoordinaten voraussetzen würde) noch gasförmig, wie Bethe annimmt. Man darf wohl aber trotzdem sagen, dass es ein Aggregatzustand im gewöhnlichen Sinne ist, d.h. dass solche Eigenschaften wie Energie und Entropie bei fester Temperatur proportional zur Grösse des Kernes sind.

Für die Dichte der angeregten Zustände erhält man dann, in derselben Weise wie Bethe

$$f \cdot e^{\frac{n}{n-1}}(AM)^{1/n}E^{\frac{n-1}{n}}$$

worin f eine langsam veränderliche Funktion, E die Anregungsenergie, M die Anzahl der Teilchen im Kern und A und n die Konstante und der Exponent in dem Gesetz

$$C = nAT^{n-1}$$

für die spezifische Wärme als Fuktion der Temperatur sind.

Es ist mir bisher nicht gelungen, aus rein theoretischen Argumenten etwas über die "spezifische Wärme" des Kerns zu schliessen, selbst wenn man bestimmte Annahmen über die Kräfte macht. Aber ich glaube dass die Analogie mit dem flüssigen Helium hier sehr nützlich sein kann. Denn der Grund weshalb Helium auch am absoluten Nullpunkt nicht kristallisiert ist ja, genau wie beim Kern, die Nullpunktsamplitude. Nun ist zwar das Problem des flüssigen Heliums bisher theoretisch auch nicht behandelt worden, aber wir haben für dieses experimentelle Daten zur Verfügung. Aus diesen folgt, dass der Exponent n sicher grösser als zwei und sehr wahrscheinlich gleich vier ist. Ich hoffe, dass bald bessere Messungen vorliegen werden, die diese Frage definitiv entscheiden.

Auch dann bleibt aber natürlich der Koeffizient A unbestimmt, und zu dessen Berechnung hilft auch die Analogie mit dem Helium nichts. Man kann jedoch eine Abschätzung für A bekommen, wenn

[783] Letter [191].

man die Energie berechnet, unterhalb der gerade ein Aggregatzustand liegt. Die Energie kann man dann mit den ersten Anregungszuständen der radioaktiven Kerne vergleichen und sollte so grössenordnungsmässig richtige Werte erhalten.

Ich wollte gern in einer kurzen Note, wahrscheinlich gemeinsam mit Bethe, der sehr ähnliche Überlegungen gemacht hat,[784] diese Punkte erwähnen, insbesondere die Analogie mit dem flüssigen Helium und die Schlüsse, die daraus über die Abhängigkeit von der Anzahl der Teilchen im Kern folgen. Es scheint mir, dass dies nur lose mit dem zusammenhängt, was in Ihrer Arbeit stehen wird, und dass es daher lohnt, schon jetzt eine solche kurze Note zu schreiben, umsomehr als es ratsam ist, vor der zu wörtlichen Anwendung der Betheschen Resultate zu warnen. Ich möchte diese Note aber zurückhalten, falls Sie der Meinung sind, dass diese Punkte in Ihrer Arbeit zur Genüge erledigt werden. In diesem Falle würde ich Sie bitten, mir bald Nachricht zu geben.
Mit herzlichen Grüssen
Ihr Ihnen ergebener

[Rudolf Peierls]

[199] Niels Bohr to Rudolf Peierls

Copenhagen, 17.10.1936

Lieber Peierls,
ich war sehr interessiert, in Ihren freundlichen Brief vom 11ten Okt[ober][785] und glaube, dass unsere Auffassungen über den Kernzustand weitgehend übereinstimmen. Ich hoffe aber, im Laufe von etwa einer Woche Ihnen einen kleinen Aufsatz schicken zu können, den ich seit längerer Zeit in Vorbereitung habe, und in welchem ich meine allgemeinen Ansichten über die Kernprobleme näher präzisiere. Wenn Sie ihn bekommen haben, würde ich sehr froh sein Ihre Meinung darüber

[784] See letter [194].
[785] Letter [198].

zu hören; und vielleicht wäre es auch zweckmässig mit der Stellungnahme zu der Frage der Veröffentlichung Ihrer Ausführungen bis dahin zu warten.[786]

Übrigens haben Kalckar und ich eben einen kleinen Artikel über die Zertrümmerung des Aluminiums geschrieben,[787] worin wir verschiedene, für die Kernreaktion charakteristische Züge näher diskutieren, und von dem ich eine Kopie gleichzeitig mit dem erwähnten Aufsatz nach Cambridge schicken werde.

Mit herzlichen Grüssen von uns allen,
Ihr

Niels Bohr

[200] Heinrich Peierls to Rudolf Peierls

Berlin, 1.12.1936

Meine Lieben.
Wir haben uns gestern abend telefonisch mit Alfred unterhalten und ausgezeichnete Verständigung gehabt, so daß wir uns auch demnächst mal mit Euch unterhalten können. Wegen geeigneter Zeit müßt Ihr uns Vorschläge machen. Vielleicht kommt Weihnachten in Frage, wenn wir wissen wann und wo wir Euch erreichen können.

Hoffentlich ist das neblige Wetter inzwischen vorüber.

Genia wünschen wir für ihre Pariser Reise das allerbeste. Sie soll sich aber wirklich dabei etwas ausruhen und nicht zuviel mitmachen.

Daß uns Deine Aussichten, lieber Rudi, von denen Du schreibst, auf das lebhafteste interessieren, haben wir schon geschrieben.[788] Ist nicht aber in Birmingham die Möglichkeit in einem erstklassigen Laboratorium zu arbeiten, geringer als in Cambridge? Jedenfalls wollen wir

---

[786]Peierls did not receive the promised paper. See letter [206].

[787]N. Bohr and F. Kalckar, 'On the Disintegration of Aluminium by α-Rays'. The paper was never published. A copy of the manuscript has been reprinted in Bohr, *Collected Works*, vol. 9, pp. 183ff.

[788]Peierls was applying for a Chair in Mathematical Physics at Birmingham University.

hoffen, dass sich Deine Aussichten verwirklichen. Von der "Episode"[789] waren auch wir restlos begeistert.
Euch vieren herzliche Grüße

Euer Vater

Meine Lieben,
hat Rudi die Drucksachen Z[eitungs] Notiz über das hies[ige] Laboratorium f[ür] Physik erhalten? Ist es eben nicht etwas ungemütlich in Paris? Wie lange bleibst Du? Gute Erholung, viele Grüße, Kuss d[en] Kinderlein

Else, Seppl, Hansl

[201] Nevill Mott to Rudolf Peierls

Bristol, 4.12.1936

Dear Peierls,
I am awfully glad that you are interested in semi-conductors etc. and I would love to co-operate.

I agree that your first possibility seems a bit unlikely. The calculation I made (Mott and Jones[790]end of chapter II) was for a degenerate electron gas screening a positive charge. Your electron gas is not degenerate, so the calculation is the same as in the Debye-Hückel theory isn't it?

As regards your second suggestion, surely the forces between trapped electrons will be e^2/kr^2 where k is the electric constant, and between free (mobile) electrons e^2/k_0r^2 where k_0 is the part of the dielectric constant which depends on the polarisation of the ions, but not on their displacement. For alkali halides $k_0 \simeq 2$; I have no idea what it is for copper oxide.

[789] 1935 Austrian production by Walter Reisch featuring Paula Wessely.

[790] N.F. Mott and H. Jones, *The Theory of the Properties of Metals and Alloys*, Oxford: Oxford University Press, 1936.

I am not quite sure that I understand your second suggestion — if I do, I don't agree. I agree with what you say about metals, but I do not see that it has any connection with insulators. I suppose one can assume that at $T = 0$ the (impure) copper oxide is an insulator, in the theoretical sense. In that case, surely the energy of the conduction level is the lowest energy which an electron can have if it is introduced into the metal from outside? What one has to consider is whether this level would be changed much my impurities. I should not think it would.

I cannot see that one renders an electron free by talking it to the next impurity atom. The insulator problem is a real ,,one electron" problem — unlike the much more difficult metal problem.

A point in all this theory, which you probably know, is this. In the equation $\sigma = Ae^{-b/T}$, $2kb$ is not the energy required to lift the electron out of impurity (i.e. the threshold for photoconductivity) but this energy minus the energy given up by the ions when they move into their new positions after the electron has gone. For these very polar media the two are comparable.

Here is another suggestion. What part do F centres play? By an F centre I mean an electron which has become trapped by polarising the medium around it, i.e. by moving the ions from their equilibrium positions. For such a polar medium, these F centres must exist. An electron, moreover, can be trapped in this way at any lattice point.

We thus have the following possible levels for an electron, in ascending energy:

1. electrons trapped at an impurity; if c is the concentration of such impurities, the number of places is cN. (N = number of crystals).

2. electrons in F centres; (N places)

3. electrons in conduction level; (statistical weight $\sim N$).

There will thus be more F centres than conduction electrons; and at high temperatures perhaps most of the impurities would dissociate into F centres (cf. Rev.Mod.Phys. vol 8, p. 309, for work on KBr;[791] a

[791] A.L.Hughes, 'Photoconductivity in Crystals', *Rev.Mod.Phys.* **8**, 294–315 (1936).

disassociation of 10% is reached at 600°C) In that case, b would fall with increasing temperature, and would become equal to the F centre — conduction level energy difference.

This does not explain your effect, though it might help with the screening hypothesis. You could imagine that nearly all the impurities had given up their electrons to F centres, so that the mobile F centres and immobile positively charged impurities would form a real Debye-Hückel gas, and the screening would not depend on the much fewer conduction electrons. The screening ought to become bigger with increasing concentration, but as a field between charges is only e^2/kr^2, I should not think that one could get more than $\frac{1}{5}e.v.$ How much is the observed change?

I should love to discuss all this. I suppose there is no chance of seeing you here? I do not expect I can go to Cambridge until next term.

Could you let me have the experimental figures (i.e. the decrease in b) or the references to the papers?

I had meant to write about thin films. I think Fuchs can get your answer by a more elementary method.[792] Should we (Fuchs and I) in any case write it up for Proc. Camb. Phil. Soc. under the head: Fuchs, Mott, Peierls (alphabetical order) — rather a galaxy of talent for so small a theory?[793]
Yours sincerely,

N.F. Mott

[792] Klaus Fuchs (1911–1976), at the time Mott's doctoral student, later joined Peierls at Birmingham where they became close friends. Klaus Fuchs was among the British contingent working at Los Alamos, and after the war became head of the theoretical division of Harwell, the British Atomic Energy Research Establishment. In 1950 he was arrested, charged with violation of the Official Secrets Act for passing on atomic secrets to the USSR and sentenced to 14 years imprisonment.

[793] Fuchs and Mott had been working on a theory of conductivity of thin metals which Fuchs published as sole author two years later; Fuchs, K., 'The Conductivity of Thin Metallic Films According to the Electron Theory of Metals', *Proc. Camb. Phil. Soc.* **34**, 100–108 (1938).

[202] Rudolf Peierls to Nevil Mott

[Cambridge], 9.12.1936
(carbon copy)

Dear Mott,
I can see no reason why my name should be added to Fuchs' paper: all I did was to obtain the same result by a more complicated method.

About the semi-conductors: I do not, of course, mean to intrude into your field of work by taking up semi-conductors in general, but all I meant we could do together is to clear up this one point on which I saw a possible explanation which would be rather different from the current picture. This explanation may, of course, be wrong, but I do not think it is evidently wrong.

I am afraid I have not explained it at all well in my last letter. What I mean is this:

1. Suppose first of all a Cu_2O crystal with no impurities to be an absolute insulator, and suppose we put one excess electron into it. This will then move in the lowest empty band, and if its kinetic energy is not too high, we shall have
$$E = \frac{1}{2m}p^2$$
where m is an effective mass and $p = h$ times wave vector. The zero level of the energy is assumed to coincide with the bottom of the band.

2. Suppose now we put the whole crystal into the constant potential V. Then clearly
$$E = \frac{1}{2m}p^2 V$$

3. Next let the potential V vary gradually throughout the crystal, but sufficiently slowly. Then evidently
$$E = \frac{1}{2m}p^2 V(xyz)$$

Hence the motion of the electron in the combined field of the external potential V and the periodic potential of the lattice can, if it is near the bottom of the empty band, be described as a motion of a particle of mass m (effective mass) in a potential field $V(xyz)$.

This will apply if (a) the variations of the potential V remain less than the width of the band and if (b) V varies so slowly that its variation over an atom may be neglected.

4. Now, if the potential V is due to an unscreened ion in the crystal such that

$$V = e^2/r$$

the conditions (a) and (b) are satisfied, except in a region of a few Angström size around $r = 0$. On the other hand, the Schrödinger equation for a"potential hole" has an infinite number of discrete levels, if the potential drops like l/r at large distance, however the potential may behave at small distances. From this it follows that in our case there must be an infinite number of discrete levels in which the electron moves in the Cu_2O lattice (the probability of it being inside the central ion will practically vanish, or, at any rate, be much smaller than in the neighbouring atoms) but circles round the ion without having enough energy to escape to infinity.

5. Besides these orbits there is, of course, the state in which the electron is attached to the ion, in the usual sense. Of the position of this level we know nothing a priori, and there are two possibilities:

Either the attached level is very much lower than the lowest discrete level in the lattice. Let us call them A and D respectively. Then, in order to make the electron available for conduction, it has first to be lifted from A to D. In addition we must raise it from there to the level $B = 0$ which is defined as the energy of an electron moving at large distance from [?], and with vanishing wave vector.

The second possibility is that A is higher than D. In this case the electron will [?] at $T = 0$, certainly be in one of the discrete orbits in the lattice and the effect of temperature will merely consist in gradually increasing the radius.

I want to assume that either this lattice condition is realized or, at any rate that $D - A$ is small compared with D.

Then the main work that has to be done is D, and in order to obtain reasonable values for the temperature dependence, D should be of the order of $\frac{1}{2}$ volt. This would require the average radius of the lowest orbit to be about 15Å which seems quite reasonable, if anything you would expect it to be less, but I have not allowed for the dielectric constant. This shows, at any rate, that D is by no means negligible and that, unless $D - A$ is very large, it will be the dominant term. This seems satisfactory in view of the otherwise astonishing fact that one has never found an insulator in which impurities produce a conductivity with small but constant number of conduction electrons.

If you grant the assumption that A is, in the special case of Cu_2O, either above, or very close to, D, then at low temperatures all impurity electrons are in such discrete orbit surrounding their respective ions, and with increasing temperature, only the average radii will increase. For low concentrations one can only get conduction from those electrons whose energy has actually become positive (infinite radius of orbit). But at higher impurity contents it will be sufficient if they largely overlap, because we have now again a problem like that of a metal (or rather liquid) in which we have to count as "atoms" only the impurities, while the lattice only contributes by altering the effective mass. The "atomic radius" is then the average radius of the orbits, and if these overlap, we have a case which is almost entirely analogous to, say, liquid Hg.

I am sorry that my last letter was so muddled. Do you now agree to the possibility of such an effect?

The experimental evidence is, of course, rather vague, like everything in semi-conductors. There is a remark in a report by Gudden & Schottky, Phys.Z. vol. 36, p. 719,[794] bottom where they also give some

[794]B. Gudden and W. Schottky, 'Electronen- und Ionenleitung fester Körper', *Phys. Z.* **36**, 717–21 (1935).

reference to other work, and there is a paper by Meyer, Phys. Z. vol. 36, p. 751 (in particular fig. 2)[795] Schottky was much more definite than that I think his evidence lies mainly in unpublished measurements by Waibel.[796]

I would certainly very much like to come to Bristol and have a talk about all that but my programme for the next time is not quite settled yet. How long will you be away over Christmas?
Yours sincerely,

R.E.Peierls

[203] Heinrich Peierls to Rudolf Peierls

Berlin, 14.12.1936

Lieber Rudi,
herzlichen Dank für Deine uns natürlich wieder sehr interessierenden Nachrichten vom 10. d[iesen] M[onats]. Wir wünschen herzlich, daß sich Deine Hoffnungen erfüllen, und bitten wieder um Bericht, sobald die Angelegenheit in ein weiteres Stadium getreten ist.[797]

Die Krebse sind gestern fahrplanmäßig abgefahren, und wir hoffen, daß sie gut in New York ankommen, wo sie schon am 24. bei Onkel Siegfried eingeladen sind.

Ich nehme übrigens in Aussicht, von jetzt an alle 8 bis 10 Tage einen Bericht über alles Wissenswerte abzufassen, der mit Durchschlägen geschrieben wird und sowohl nach New York als auch nach Cambridge und London geht, damit man nicht alle Mitteilungen dreimal zu machen hat.

---

[795] Wilfried Meyer, 'Beitrag zur Elektrizitätsleitung in Halbleiterwerkstoffen', *Phys. Z.* **36**, 750–54 (1935).

[796] F. Waibel and Schottky had published their results jointly, too. W. Schottky and F. Waibel, 'Electron Conductivity of cuprous oxide', *Phys. Z.* **34**, 858–64 (1933) and **36**, 912–14 (1935).

[797] See letter [200], note 788.

Wir hoffen, daß Du von Genia aus Paris gute Nachrichten hast und daß sie sich erholt. Wann erwartest Du Ihre Rückkehr?
Herzliche Grüße Dir und den Kindern

Dein Vater

Lieber Strohwitwer,
es freut mich sehr, daß Olive so nett für Euch sorgt. Gaby ist ja zu drollig — wir haben arbeitsreiche Wochen hinter uns + sind des Sturmes wegen nun in Sorge um die Krebse. Die Kinderchen wohnten die letzte Woche hier + erholten sich gut + waren sehr vergnügt. Deine Zukunftsaussichten sind sehr beglückend. Wie lange bleibt Genia fort? Grüß Olive schön, küss die Kinderchen

Dir alles Gute D[eine] Else

[204] Max Born to Rudolf Peierls

Edinburgh, 16.12.1936

Lieber Peierls,
Es ist eigentlich schändlich, daß ich Ihnen und Ihrer lieben Frau noch nicht geschrieben habe. Wir waren halt sehr in Anspruch genommen mit Einrichten, Eingewöhnen, neuen Pflichten und Geselligkeit.[798] Ich meine, schließlich hätten Sie auch mal schreiben können. Wir sind sehr zufrieden hier. Unser Haus ist behaglich, viel schöner als das in Cambridge. Die Stadt hat gerade die richtige Größe: sie bietet allerlei Großstadtgenüsse, gute Theater und Concerte, und ist doch nicht allzu ausgedehnt, sodaß man leicht in Freie kommt. In 15 Minuten sind wir zu Fuß am Arthur's Seat, in 10 Minuten am Blackford Hill, wo man schön spazieren kann. Aber die meiste Freude haben wir von Samstags-Ausflügen in the Pentland Hills, die schöne, einsame Wanderungen erlauben. Die Kollegen hier sind sehr nett und liebenswürdig. Prof. Orr

[798]After short periods in India and Cambridge, Born had taken up the Tait Professorship in Edinburgh.

und seine Frau, die Sie kennen, haben wir mehrmals getroffen. Was nun die Arbeit betrifft, so haben mir die Vorlesungen ziemlich viel Mühe gemacht. Mein Assistent, R. Schlapp, ist ein ganz famoser Mensch, der mir alles langweilige Zeug abnimmt. Auch kann er viel Physik und ist nur zu bescheiden es zu zeigen.

Mit Barkla[799] kann man wissenschaftlich nicht anfangen. Er ist aber ein reizender Kerl, auch seine Frau ist sehr nett. Dymond[800] sehe ich oft. Er studiert jetzt kosmische Strahlen, da er im nächsten Semester auf eine Spedition nach Grönland geht. Er hat ein Colloquium etwa alle 14 Tage, das meist gut ist. Mit Whittaker[801] bin ich viel zusammen; er interessiert sich für Physik, aber nur für die formale Seite. Er gibt eine einstündige Vorlesung, in der er zuletzt sehr ausführlich den Inhalt des Eddingtonschen Buches[802] dargestellt hat — völlig kritiklos, als ob das Physik sei. Mathematik scheint es übrigens auch nicht zu sein. Was nun meine eigene Arbeit betrifft, so bin ich erst in den letzten Wochen dazu gekommen, etwas zu machen, und zwar ist es, glaube ich, ganz gut gegangen. Ich habe versucht, eine rationelle Theorie des β-Verfalls zu machen unter Berücksichtigung des Punktes, der bei Fermi unbeachtet bleibt, mir aber das Wesentliche scheint, nämlich die Änderung der Ruhmasse der Partikel. Das geht natürlich nicht im Rahmen der üblichen Quantenmechanik. Aber die notwendige Änderung ist relativ gering. Ich führe keineswegs, was man denken könnte, einen Operator für die Ruhmasse ein, sondern einen für die Ladung (Eigenwert 0,1; -1 ist nicht nötig, da die Löcher dafür sorgen). Ich bemühe den ganzen Apparat, den ich in den letzten Jahren dafür gesammelt habe, z.B. die in Proc. Cambr. Phil. Soc. veröffentlichte "Linearspannung" meiner Feldgleichungen;[803] allerdings in gänzlich anderer Weise. Es scheint mir alles äußerst natürlich — jetzt nachdem es ziemlich fertig ist. "Fertig" bedeutet eine Reihe von Feldgleichungen, die vernünftig

[799]Charles Glover Barkla (1877–1944) at the time professor of Natural Philosophy.
[800]Edmund Gilbert Dymond (1900–1952), geophysicist.
[801]Edmund Taylor Whittaker (1873–1956), professor of mathematics.
[802]Arthur Stanley Eddington, *New Pathways of Science*, Cambridge University Press, 1935.
[803]Born probably refers to M.Born, 'On the quantum theory of the electromagnetic field', *Proc. Roy. Soc.* **A143**, 410–437 (1934).

aussehen. Das Prinzip, aus dem sie entspringen, ist eine Verallgemeinerung von Dirac's Wellengleichung. Sie bestehen aus zwei Gruppen von Gleichungen: die eine Gruppe ist ein gekoppeltes System für zwei Wellenfunktionen (mit je 4 Spin-Energie-Komponenten). Die Kopplungsterme sind teils elektrodynamisch, teils "Fermisch"; letztere enthalten gerade zwei Ableitungen, was mit Heisenberg's Schätzungen und Uhlenberg's Resultaten zu stimmen scheint.* Die andere Gruppe sind verallgemeinerte Maxwell-Gleichungen $\sum_l \frac{\partial p_{kl}}{\partial x_l} = s_k$. Die s_k sind die üblichen Stromkomponenten. Die p_{kl} sind vom Typus $(\varphi\gamma_k\gamma_l\varphi)$, wie zu erwarten. Sie sind aber nicht die einzigen Feldgrößen; es gibt noch eine zweite Art Feldgrößen $f_{kl} = \frac{\partial \phi_l}{\partial x_k} - \frac{\partial \phi_k}{\partial x_l}$, wie in meiner nicht-linearen Theorie. Der Unterschied ist nur der, daß nicht die p_{kl}-Funktionen der f_{kl} sind, sondern erst durch einen Integrationsprozess, bei dem die Teilchenbewegung eingeht, aus ihnen gewonnen werden können. Für langsam veränderliche Felder aber ergeben sich wieder gerade die Formeln meiner alten Theorie, mit der soviel bezweifelten Quadratwurzel! Das ist alles nicht konstruiert, sondern folgt aus dem höchst natürlichen Ansatz für den Ladungs- (oder Kernladungs-) Operator. Wirklich integriert habe ich allerdings noch nichts. Mein Glauben an die Sache beruht daher nur auf der "Schönheit" der Formeln. Die Mathematik ist nicht ganz einfach und ziemlich ungewohnt. Es gibt Andeutungen für eine große Menge neuer Zusammenhänge. Na von diesem Bericht werden Sie nicht viel haben. Schade, daß ich nicht bei Inhen hereinkommen kann und es Ihnen erzählen kann, wie früher. Lassen Sie nun auch mal hören, wie es bei Ihnen geht, im Hause, im Cavendish und sonst. Mit herzlichen Grüßen an Ihre Frau, die Kinder und gemeinsamen Freunde, auch von meiner Frau und den Kindern,

Ihr

M.Born

*Konstanten kommen nur 137 und 1845 vor, keine neue für Fermi's Kräfte; sie ist von der richtigen Ordnung.

[205] Max Born to Rudolf and Genia Peierls

Edinburgh, 14.1.1937

Liebe Peierls, Mann und Frau,
Ich komme heute mit 2 Fragen. Erst eine für den Mann: Wir leiden hier dauernd an Erkältungen (meine Frau außerdem Grippe mit Bronchitis, und Irene hatte auch eine Grippe). Darum möchte ich eine Ultraviolett-Lampe anschaffen; der Mangel an Sonnenstrahlung ist hier sehr fühlbar, und die Ärzte erklären die vielen Erkältungen der Zugewanderten durch Ausbleiben von Vitaminbildung in der Haut wegen zu geringer Strahlung. Ich habe Pohl in Göttingen angefragt,[804] was man machen sollte, da er, wir ich weiß seit langem solche Lampen benützt. Er hat mir einen Katalog der "Hanauer Höhensonne" geschickt, (Quarz-Quecksilber-Lampe), daneben aber einen anderen von Osram-Vitalux, und dazu geschrieben, daß er für seine Kinder die letztere Lampe benützt. Es ist eine überhitzte Wolfram-Lampe, kostet mit Gestell nur 65 Mark und hat den Vorteil neben Ultraviolett bei 300mμ eine Menge Wärmestrahlung zu geben, so daß man sich nicht während der Bestrahlung erkälten kann. Soviel ich weiß sind die Haupt-Hg-Linien kurzwelliger, bei 260mμ. Was ich wissen möchte, ist das: Genügt der Bereich bis 300mμ, kontinuierliches Spektrum relativ geringer Intensität, zur Verhütung von Erkältungskrankheiten (d.h. zur Bildung der betr. Vitamine?) oder ist der kurzwellige Bereich der Hg-Lampe nötig? Ein hiesiger Chemiker (Deutscher), Prof. Schönberg, war der Ansicht, daß der kurzwellige Bereich gefährlich und unnötig sei. Die Hg-Lampe ist übrigens auch wesentlich teurer. Vielleicht wissen Sie etwas darüber, wenn nicht, so kennen Sie vielleicht einen Spezialisten, den ich um Auskunft fragen könnte.

Zweite Frage an Sie beide, hauptsächlich aber die Dame. Wir sprachen einmal davon, daß Sie Irene für einige Zeit ins Haus nehmen wollten. Jetzt möchte ich wissen, ob das wirklich ginge. Sie muß nämlich mit dem Heiraten noch mindestens bis Sommer 1938 warten. Hier ist sie

[804] Robert Wichard Pohl (1884–1976), professor of experimental physics in Göttingen since 1916.

in einem Kindergarten für Kinder von Arbeitslosen 3 Tage in der Woche tätig. Die anderen Tage hilft sie im Haus, Kochen Aufräumen etc. und macht ihre eigene Kleidung. Sie ist jetzt sehr willig und nett, viel leichter zu handhaben als in Cambridge. Aber wir haben das Gefühl, daß sie in unserem Haus nicht genug lernt, um später den eigenen Haushalt zu führen, besonders im Kochen. Auch ist es immer noch schwer, sie zum Früh-Aufstehen (d.h. etwa 8 Uhr) zu bringen. Sie selbst wünscht sich einmal, vor ihrer Ehe aus unserem Haus zu kommen und unter Leitung einer anderen Hausfrau zu arbeiten. Cambridge hat für sie die Anziehung, daß ihr Verlobter nicht allzu weit entfernt ist, so daß sie sich ab und zu sehen könnten. (Ich bin dagegen, daß es allzu oft ist, höchstens einmal im Monat), und daß sie in Cambridge viele Freunde und Bekannte hat. Ihr Haus hat die spezielle Auszeichnung, daß sie Sie beide kennt und gern hat. Zu fremden Leuten zu gehen, ist ihr unheimlich. Ich dachte, sie könnte vielleicht nach Ostern für ein paar Monate fort gehen. Würde das Ihnen passen? Sie müssen ganz offen schreiben, wenn's Ihnen nicht passt. Ich wollte nur nicht anderswo anfragen, ehe ich bei Ihnen gefragt habe. Nächsten Winter wollte ich Irene gern in Paris unterbringen, damit sie gut französisch lernt.
Meine Frau liegt jetzt die dritte Woche im Bett, es geht ihr mrklich besser. Die Jungens und ich sind wohl. Mit den herzlichsten Grüßen von uns allen
Ihr

M. Born

[206] Rudolf Peierls to Hans Bethe

Cambridge, 13.2.[19]37
(carbon copy)

Lieber Bethe,
Diesmal bin ich derjenige welcher, und noch dazu ohne jeden Vorwand mit dem Titelblatt, denn auf dem Umschlag der Camb. Phil. Soc. steht nur der Name von meinem Freund A.H. Wilson, und Du bist berechtigt, das nicht als Gruss anzuerkennen. Als mildernden Umstand muss ich

immerhin plädieren, dass ich immer hoffte, jeweils etwas besseres über die angeregten Kerne und flüssiges He herauszubekommen, aber dann wird es immer Essig (flüssiger).

Was ich bisher weiss, geht nicht weit über das hinaus, worüber wir uns schon im Sommer einigten. So lohnt sich vielleicht jetzt, die Situation zusammenzufassen:

Wenn man ausser der Energie noch r Integrale der Bewegungsgleichungen $F_1 \ldots F_r$ konstant hält, (Protonen- und Neutronenzahl, Spin), so wird die Anzahl der Niveaus pro $dEdF_1 \ldots dF_r$:

$$\sqrt{(2\pi)^{r+1} D_{r+1}} e^s \tag{1}$$

wo s die Entropie ist und $D_r + 1$ eine Determinante von

$$d_{i,k} = \frac{\delta^2 S}{\delta F_i \delta F_k} \qquad i, k = 0, 1, \ldots r, \qquad F_0 \equiv E$$

Dies dürfte Dir keine Neuigkeit sein. Man kann so die Formeln Deiner Arbeit ein bisschen hübscher und symmetrischer ableiten, aber das ist natürlich egal.

Im Falle von $r = 0$ bekommt man natürlich anstatt der Determinante einfach d^2S/dE^2, und der Korrektionsfaktor um den sich der Ausdruck (1) hiervon unterscheidet, geht natürlich wie eine Potenz der Teilchenzahl A. Der Fehler, den man begeht, wenn man diese Korrektur weglässt, ist also ein Term der Entropie der Grössenordnung $\log A$.

Dies bedeutet nun aber, dass eine makroskopische Analogie diesen Faktor nie ergeben kann. (Der Unterschied ist natürlich der zwischen mikrokanonischer und kanonischer Verteilung bezüglich $F_1 \ldots F_r$). Ferner zeigt dieses Argument aber auch dass es ungerechtfertigt ist, diese Feinheiten mitzunehmen, ohne Oberflächeneffekte in Betracht zu ziehen, denn wenn Oberflächenterme vernachlässigbar sind, so sind es die logarithmischen Terme erst recht.

Solche Oberflächeneffekte sind natürlich in Deiner Formel für das Fermigas vernachlässigt, insofern nämlich als Du die Anzahl der Niveaus für das Einzelteilchen proportional zum Volumen annimmst. Das könnte man im Prinzip zwar leicht richtigstellen, sobald man das Potentialfeld kennen würde. Da man es aber nicht mit hinreichender Genauigkeit kennt (bezw. da man es vermutlich gar nicht mit derartiger Genauigkeit definieren kann) so hilft das nicht viel.

Hieraus würde ich schliessen, dass es ohne ein vollständiges mikroskopisches Modell unmöglich ist, die Oberflächeneffekte zu berücksichtigen, und einem daher auch die logarithmischen Terme, d.h. die $F's$ in (1) schnuppe sein können. Dabei begeht man natürlich im Absolutwert der Dichte einen Fehler von mehreren Zehnerpotenzen, und muss sich mit einer ganz groben Annäherung begnügen.

Wenn man das tut, so hat man

$$N = \sqrt{2\pi \frac{\delta^2 S}{\delta E^2}} e^S$$

und braucht nur noch mit der Funktion $S(E)$ oder $E(T)$ zu spielen. Da die Temperaturen tief sind, so ist es eine gute Annäherung, diese Funktion durch eine Potenz darzustellen

$$E = A.B.T^n$$

Dies entspricht der Annahme völliger Entartung in Deinem Fall und bedeutet, obwohl der Fehler im Exponenten so klein ist, vermutlich auch wieder eine beträchtliche Unsicherheit im Resultat. Ferner ist der Faktor A schon so geschrieben, dass bei Vernachlässigung von Oberflächeneffekten B unabhängig von A wird. Wenn man das annimmt, so bleibt nur noch n und der mehr oder weniger universelle Faktor B verfügbar.

Für n ist es wohl vernünftig, anzunehmen, dass es mit dem entsprechenden Exponenten für flüssiges He identisch ist, und man könnte denken, den letzteren experimentell zu bestimmen. Im Mond[805] hier sind derartige Experimente geplant, sie sind aber nicht leicht, da man wesentlich unter 1° heruntergehen muss, um das asymptotische Gesetz zu finden. Die Leidener Messungen der spezifischen Wärme gehen etwa bis auf 1° und wenn man dort C/T^3 gegen T aufträgt, so bekommt man eine gerade Linie gerade durch den Nullpunkt. Nimmt man diese ernst, so bekommt man $n = 5$, was wohl Blödsinn ist, denn dann sollte

[805] Mond Laboratory, Cambridge. See letter [194].

das flüssige He schwerer zu Schwingungen anregbar sein als ein fester Körper! Es ist natürlich leicht möglich, dass die Kurve, statt durch den Nullpunkt zu gehen, schliesslich mal umbiegt und horizontal wird, wie das für $n = 4$ erforderlich wäre, aber um weniger als 4 zu bekommen, müsste sie schliesslich gegen unendlich gehen, und das ist ein bisschen viel verlangt. Infolgedessen scheint mir vorläufig $n = 4$ der plausibelste.

Die beste Abschätzung für B scheint mir zu sein, dass für die mittlere Anregungsenergie der radioaktiven Elemente und für $A = 200$, S von der Grössenordnung 1 werden muss. Nimmt man für diese Anregungsenergie 0.2MV, so erhält man für $A = 100$ und $E = 8MV$ einen Niveauabstand von $100V$. Das stimmt natürlich viel zu schön, ich glaube man muss mindestens zwei Zehnerpotenzen in jeder Richtung offenlassen aber jedenfalls ist die Sache im Rahmen dieser groben Ueberlegungen in Ordnung. Besser kann man es, glaube ich, nicht bekommen, ohne ein anständiges Modell für die Bewegung des Kerns zu haben und das wäre natürlich das eigentlich interessante, aber dafür habe ich nichts herausgebracht. Das einzige, was sicher ist, ist dieses: Die schlampige Ueberlegung, die ich damals mit Longitudinal- und Transversalwellen machte, lässt sich nicht präzisieren, denn wenn die Substanz kein Kristall ist, so lässt sich ihre Bewegung halt überhaupt nicht mehr streng in Wellen zerlegen, und für die langen Wellen für die das noch geht benimmt sie sich halt wie eine ideale Flüssigkeit.

Der einzige Schritt vorwärts, der aber auch nicht viel zu helfen scheint, liegt in der folgenden Feststellung: Man kann kaum hoffen, die angeregten Zustände besser behandeln zu können als den Grundzustand. Für diesen kommt man aber natürlich nur mit einem Variationsprinzip weiter. Es ist daher ein gewisser Trost, dass es auch für die Zustandssumme ein Variationsprinzip gibt. Es gilt nämlich folgendes: es sei H eine Hamiltonfunktion, und ich bilde ihre Diagonalwerte H_{ii} für irgendein beliebig gewähltes Orthogonalsystem. Wenn ich dann so tue als ob dies die richtigen Eigenwerte wären, und daher die Funktion bilde

$$F(T) = \sum_i e^{-H_{ii}/kT}$$

so ist diese immer <u>kleiner</u> (d.h. die zugehörige freie Energie immer grösser) als die wahre Zustandssumme. Wenn man ein vollständiges Orthogonalsystem mit freien Parametern nimmt (also etwa die Eigenfunktion eines lösbaren Problems mit Parametern) (im Potential), so kann man vermutlich durch Variation der Parameter der wirklichen freien Energie recht nahe kommen. Ich habe aber bisher für diesen Fall noch kein geeignetes Funktionensystem finden können. Aber vielleicht fällt Die hierzu etwas Kluges ein.

Das ist natürlich alles herzlich dünn, und ich weiss nicht, ob es sich lohnt, darüber überhaupt etwas zu veröffentlichen, aber vielleicht lohnt es sich immerhin einen Brief an die Phys.Rev. im Anschluss an Deine Arbeit. Die Frage wird noch dadurch kompliziert, dass eine schüchterne Anfrage bei Bohr zu dem Ergebnis geführt hat, dass er sehr bittet, mit einer evtl. Publikation über dieses Thema zu warten, da er sehr ähnliche Dinge macht und er würde mir das Manuskript "nächste Woche" schicken.[806] Da das aber schon sechs Monate her ist, und ich ausserdem davon überzeugt bin, dass er sich für diese Punkte nicht interessiert hat, so braucht man sich, glaube ich, nicht mehr gebunden zu fühlen — vorausgesetzt natürlich, dass es sich überhaupt lohnt, etwas zu schreiben. Was findest Du?

Die Schwierigkeit ist natürlich wieder, dass der Kern, so wie flüssiges He, so wie ein Supraleiter, eben eine Art von Ordnung hat, und man nicht weiterkommt, ohne ein Bild davon zu haben, worin diese Ordnung besteht, und wie sie zu beschreiben ist. Natürlich soll man das nicht so machen wie Herr Slater.[807]

Anlässlich der oben erwähnten Variationsmethode für die Zustandssumme habe ich mir übrigens folgendes überlegt: man kann natürlich für die Wechselwirkung der Metallelektronen etwas ähnliches probieren. Dazu muss man natürlich in nullter Näherung die Zustandssumme ausrechnen, die sich ergibt, wenn man die Diagonalele-

---

[806]See Letter [199].

[807]John Clarke Slater (1900–1976) received his Ph.D. from Harvard in 1923, then he worked with Niels Bohr and Werner Heisenberg. In 1930 he became head of physics at MIT, where, among others, he worked on the electromagnetic theory of microwaves. Slater had just published a paper with some tentative suggestions about the links between the superconducting states of metallic electrons and the perturbation theory of Bloch's theory. J.C.Slater, 'The Nature of the Superconducting State', *Phys. Rev.* **51**, 195–202 (1937).

mente der Gesamtenergie (einschliesslich Wechselwirkung) für die Eigenfunctionen freier Elektronen ohne Wechselwirkung ausrechnet. Dabei bekommt man etwas sehr lustiges. Bei $T = 0$ kommt natürlich die übliche Verteilung der Elektronen heraus, bis auf einen konstanten Zusatzterm in der Energie. Die spezifische Wärme enthält aber, ausser dem üblichen Term mit T, dessen Koeffizient etwas modifiziert wird, noch einen Term mit $T. \log T$, der natürlich bei tiefen Temperaturen überwiegt. Was das physikalisch heisst, ist mir noch nicht ganz klar. Vermutlich gar nichts, denn obwohl die so berechnete Zustandssumme sicher kleiner ist als die wirkliche, hat die spezifische Wärme, die sich ja durch Differentiation ergibt, keine Extremaleigenschaft, und dieser komische Term kann also leicht bei besserer Annäherung verschwinden. Immerhin zeigt dieses Argument doch, dass die spezifische Wärme in Wirklichkeit nicht proportional zu T ist, und ich bemühe mich ein vernünftiges Funktionssystem mit Parametern zu finden, mit dem man weiter variieren kann.

Sonst habe ich mir den Kopf hauptsächlich über β-Theorie zerbrochen. Ich habe jetzt einen ganz vernünftigen Doktoranden[808] den ich alle möglichen Ausdrücke vom Uhlenbeck-Konopinski-Typus habe ausrechnen lassen. Es gibt natürlich, bereits mit den ersten Ableitungen, ziemlich viel, und insbesondere kann man noch neue Ausdrücke gegenüber den in Deinem Artikel angegebenen finden, indem man zwei Deiner Ausdrücke linear kombiniert. Das ist weniger pedantisch als es auf den ersten Blick aussieht, denn wenn man z.B. die vier Wellenfunktionen anders paart, etwa:

$(\psi^*_{\text{Neutron}} \cdot \psi_{\text{Neutrino}})(\psi^*_{\text{Proton}} \cdot \psi_{\text{Positron}})$ so erhält man einen Ausdruck, der sich als Linearkombination von Ausdrücken der von Dir oder U-K angegebenen Form schreiben lässt. Solche Ausdrücke braucht man insbesondere, um in zweiter Näherung eine Majoranakraft zu bekommen.

Die neue Freiheit, die man damit bekommt, kann man insbesondere dazu benützen, um das β-Spektrum am unteren Ende ($E \ll mc^2$) beliebig zu verändern. In ähnlicher Weise kann man auch das obere

[808]Fred Hoyle, (1915–2001) studied at Cambridge where he received his Ph.D. in 193 After wartime work on the rader he returned to Cambridge and became Plumian Professor of Astronomy in 1958 from which he resigned in 1972.

Ende beliebig anpassen, wenn die Neutrinomasse von Null verschieden ist, was man ja annehmen muss, wenn man die Messungen von Lyman glaubt.[809]

Dieser Doktorand, Hoyle, hat sogar alle Ausdrücke zweiter Ordnung ausgeixt, einschliesslich Linearkombinationen, und kunstvoll derartige Ausdrücke konstruiert, die sich von U-K innerhalb der Messgenauigkeit nicht unterscheiden. Aber sie erscheinen mir etwas zu kunstvoll*, um glaubhaft zu sein.

Wir zerbrechen uns jetzt den Kopf darüber, wie man die ganze Theorie logisch in Ordnung bringen kann, denn wie Du wohl weisst, ist ja die U-K Theorie logisch nicht in Ordnung, da die Hamiltonfunktion sich nicht mit den Vertauschungsrelationen verträgt. Das muss immer passieren, wenn die Hamiltonfunktion, deren Zweck doch nur darin besteht, die zeitlichen Ableitungen von Grössen zu bestimmen, selbst wieder zeitliche Ableitungen enthält. Der plausibelste Ausweg schien mir zunächst der zu sein, dass man dem Neutrino eine Wellengleichung zweiter Ordnung [...][810] Dann ist nämlich die zeitliche Ableitung der Wellenfunktion eine Grösse die zu jeder Zeit noch beliebig vorgegeben werden kann, und die daher auch ungestraft in der Hamiltonfunktion vorkommen darf. Das würde physikalisch bedeuten, dass es einen Spin 3/2 hat, und dass sein Ort ebenso unbeobachtbar ist, wie z.B. der eines Lichtquants. Beides würde mich nicht stören. Aber die Sache scheint an den Zuständen negativer Energie zu scheitern, denn um diese nach dem Diracschen Rezept wegzubringen, braucht man einen positiv definierten Ausdruck für die Dichte, den es nicht gibt, und um das Pauli-Weisskopfsche Rezept anzuwenden, braucht man eine positive Energiedichte, die es im Falle von halbzahligem Spin gleichfalls nicht gibt. Im Augenblick bemühen wir uns, die Sache auf konservativere Weise in Ordnung zu bringen.

*Wir haben eben einen sehr glaubhaften Ausdruck dieser Art gefunden, d.h. eine Summe von zwei Gliedern, mit Koeffizienten 1.

---

[809]Elisabeth Reed Lyman, Berkeley, Urbana.

[810]Missing in manuscript.

Ein anderer ganz anständiger Student arbeitet über das Oppenheimer-Philips Problem.[811] Ich glaube gar nicht, dass dieses Problem als Zweikörperproblem behandelt werden kann, denn wahrscheinlich spielen dort doch Resonanzen eine Rolle. Das Oppenheimersche Argument, dass die kontinuierliche Anregungsfunktion Resonanzen ausschliesst, ist natürlich Unsinn, sobald es sich nicht um reine Capture handelt, sondern das Proton wegläuft. Dann kann sich das Proton eben immer gerade eine solche kinetische Energie aussuchen, dass der Rest in einem Resonanzniveau zurückbleibt. Aber unabhängig davon hat es Interesse, das Zweikörperproblem zu studieren, da das einer der wenigen Fälle ist, in denen man W.K.B.[812] in mehr als einer Dimension anzuwenden wünscht. Streng geht das anscheinend nicht, aber man kann offenbar die Abschätzung machen: Offenbar ist ψ von der Form $Ae^{-S/h}$, wo S der Hamilton-Jakobigleichung genügt, aber im allgemeinen komplex ist. Mit $S = a + ib$ folgt dann, dass grad a auf grad b orthogonal ist und dass $(\text{grad}\, a)^2$ grösser ist als $2m(V - E)$. Um nun den Wert von S an einem Punkt x zu bekommen, kann ich den Gradienten von einem klassischen Umkehrpunkt, an dem ψ von der Grössenordnung 1 und daher S klein ist, bis zu x integrieren. Dabei müsste ich als Integrationsweg natürlich eine Fallinie von a wählen. Wenn ich aber das Integral von grad a über alle möglichen Wege betrachte, und mir den kleinsten Wert heraussuche (was mit Hilfe von Variationsrechung leicht geht) so bekomme ich wieder etwas zu kleines. Es folgt also, dass $a(x)$ sicher grösser ist als der kleinste Wert des Integrals von $2m(V - E)$ über beliebige Wege vom Umkehrpunkt bis zu x. Im Falle von Oppenheimer-Philips stellt sich heraus, dass die von O.-Ph. berechnete Zertrümmerungswahrscheinlichkeit bereits grösser ist als die so gewonnene obere Grenze, was natürlich an ihrer adiabatischen Näherung liegt. Wir versuchen jetzt, die Methode zu verbessern, um eine nähere Abschätzung zu bekommen.

Nebenbei spiele ich noch ein bisschen mit Metallen, Halbleitern und tiefen Temperaturen, wo es alle möglichen spassigen Dinge gibt.

---

[811]Bethe published a paper on the Oppenheimer-Philips problem the following year. H. Bethe, 'The Oppenheimer-Phillips Process', *Phys. Rev.* **53**, 39–50 (1938).

[812]Wentzel-Kramers-Brillouin Approximation.

Auch sonst ist es in Cambridge noch recht lebendig. Im Mond arbeiten zehn Experimentatoren, teilweise recht hübsche Sache. Der neueste "thrill" ist, dass die Wärmeleitfähigkeit von flüssigem Helium II ca. 2000 Watt/cm.grad ist also noch etwa zehnmal grösser als Keesom behauptet.[813] Das ist doch schon eine erschütternde Grössenordnung! Ausserdem werden wir hier wahrscheinlich bald den Rekord für tiefe Temperaturen haben was natürlich zu nichts führt ausser Reklame.

Das neue Hochspannungslaboratorium ist gebaut und sieht sehr imposant aus (von aussen Kino, von innen Kathedrale) und die erste Stufe mit 1,2 MV funktioniert bereits. Auch ein Cyclotron ist im Bau. Dee und Genossen messen die Winkelverteilung für Proton-Neutron-Streuung sehr sauber, zunächst bei 2 MV. Die Resultate stimmen sehr schlecht mit Gans & Harkins.[814] Apropos, wie gefällt Dir die Theorie von Fisk & Morse[815] die so schön mit Gans & Harkins stimmt, wenn man nur annimmt, dass die Neutronen Rn+Be alle eine Energie von 25 MV haben?

In den letzten beiden Termen hatten wir sogar ein Kolloquium über Kerntheorie, (zu dem Fowler meistens zu kommen vergass) und zu Fowlers grossem Bedauern fangen die meisten Doktoranden an, sich nicht mehr an so blödsinnige Probleme setzen zu lassen, mit denen er sie gewöhnlich beschäftigt.

Ich habe jetzt auch mehr mit Studenten zu tun, Kursvorlesung über Kerne, College supervision (d.h. je eine Wochenstunde Privatunterhaltung mit undergraduates, was hier obligatorisch und sehr nützlich ist) und für all das kriegt man extra gezahlt. Das einzige, was in diesem Winter ein bisschen fehlt, ist Diskussionen mit "Erwachsenen", denn Dirac ist durch seine Heirat zu sehr abgelenkt,[816] Darwin hat als Master über Christs zu viel anderes zu tun, Bhabha und Hulme sind fort.

---

[813]Willem H. Keesom (1876–1956), student of Heike Kamerlingh Onnes; in 1926 Keesom developed a method to solidify helium.

[814]William D. Harkins, David M. Gans, 'Inelastic Collisions with Changes of Mass and the Problem of Nuclear Disintegration with Capture or Non-Capture of a Neutron or Another Nuclear Projectile', *Phys. Rev.* **46**, 397–404 (1934).

[815]J.B. Fisk, Philip M. Morse, 'The Elastic Scattering of Neutrons by Protons', *Phys. Rev.* **51**, 54–55 (1937).

[816]Paul Dirac had married Eugène Wigner's younger sister Margit Balasz (Manci) earlier in the year.

Jetzt sind allerdings meine Studenten schon bald so weit, dass man sich mit ihnen unterhalten kann, aber das dauert noch eine Weile.

Sonst gibt es nette Leute hier natürlich wie immer im Ueberfluss und man hat viel zu wenig Zeit, um alle zu sehen. Wenn nur die politische Lage nicht wäre. Aber wegen der hast Du doch wahrscheinlich Recht, dass Du nach Amerika gegangen bist, denn sehr lange kann das ja hier nicht mehr gut gehen, und dann nützt einem die ganze Prosperity nichts.

Apropos Prosperity, weisst Du in den USA eine Stelle für einen ganz ordentlichen, aber nicht genialem, sehr geduldigen und auf dem Gebiet der Fluoreszenz allwissenden Experimentator namens Hirnschlaff? Er hat sich hier mit einigen Leuten verkracht was wohl zu 25% seine Schuld war und wird durch ca. monatlich verlängerte Stipendien am Leben gehalten. Er würde vermutlich jede Stelle annehmen, und wenn es ein bisschen Aussicht hat, würde er hier wohl genug Geld zusammenkratzen um auf eine Fahndungsreise hinüberzufahren, auch ohne eine Stelle in der Tasche zu haben.

Kommst Du diesen Sommer in das senile Europa, solange es noch nicht in die Luft gegangen ist?
Mit besten Grüssen

Dein R. Peierls

[207] Rudolf Peierls to Paul Dirac

Portscatho, Cornwall, 2.7.1937

Dear Dirac,
You see we are taking our vengeance by having an address which is almost as hard to remember as yours [on holiday in Cornwall]. I have heard from Stobbe[817] who is now in Bristol (I suppose this was really

[817]Martin Stobbe (1903–1944). Martin Stobbe had resigned from his post in Göttingen in protest of the Nazis, spent some time in England and later Oslo where he died in 1944. See Alan D. Beyerchen, *Scientists under Hitler*, Yale University Press, 1977, p. 30.

the origin of the rumour your heard that he would come to Cambridge). He would, however, also like to spend some time in Cambridge, and therefore, if you still wish to allow him to use your room while you are away, that would be very helpful to him. Charlesworth said it was most unlikely that the College could do anything as regards dinners &c., but Stobbe has many friends in Cambridge with whom he could have most of his meals. The other, more difficult, question is how to find a job for him, but unfortunately I do not see any prospects. Neither does Lennard-Jones[818] who knows Stobbe well and to whom I talked about him.

We arranged for Ridgeaons to come and cut our lawn once in three weeks (although they tried to convince us it was much less work to cut it every week).

We are now having a very pleasant holiday here and hope you have a nice time.

With kindest regards to you and Mrs Dirac.

Yours sincerely,

R. Peierls

[handwritten addition from Genia]

Yours house is getting on very lovely, they chopped down off all the ivy. The turnips and carrots were growing marvellously.

Here everything is very nice, but a little bit too cold for my liking. But we manage to survive and fish and the cliffs are lovely. Many pleasures in Moscow.

Sincerely yours,

E. Peierls

[818] John Edward Lennard Jones (1894–1954), Plummer Professor of Theoretical Chemistry at Cambridge 1932–1953.

[208] Heinrich Peierls to Rudolf Peierls

Berlin, 12.8.1937

Meine Lieben.
aus Eurer gemeinsamen Karte haben wir zu unserer großen Freude ersehen, daß Ihr zusammen mit Alfred angenehme Stunden verbracht habt.

Auf Dein Schreiben, lieber Rudi, aus Birmingham erwidere ich, daß wir an unserer Reiseroute festhalten wollen. Selbst wenn wir einige Tage länger in Amerika bleiben, würde dies an unseren Dispositionen nicht viel ändern können, da es erstens mit dem Abfahrtermin der Dampfer nicht richtig klappt, und es zweitens auch zweifelhaft wäre, ob Ihr in so wenigen Tagen mit Euren Renovierungsarbeiten fertig sein würdet. Wir machen doch in England Station, um Euch Alle einschließlich der Kinder wiederzusehen. Weder auf einen Aufenthalt in London noch auf eine Besichtigung von Birmingham würden wir besonderen Wert legen. Bezüglich des letzteren würde uns natürlich die Lage, in der Euer Haus sich dort befindet, interessieren und das können wir uns auch während der Reparaturen ansehen. Überdies will ich vor Ende Oktober wieder in Berlin sein.

Wir halten also daran fest, daß wir am 4. September mit der "Europa" nach Amerika fahren. Wir würden dann am 9. Oktober mit der "Bremen" New York verlassen und am 14. Oktober etwa in Southampton ankommen, von wo wir uns zunächst nach Beckenham begeben. Am 21. Oktober würden wir von Southampton wieder mit der "Europa" nach Bremerhaven abfahren. Bitte besprich die Angelegenheit weiter mit Alfred, dem ich eine Kopie dieses Schreibens sende.

Für Eure Reise nach Dänemark und ev[entuelle] Weiterfahrt nach Rußland wünschen wir Euch das Beste.[819] Ihr tätet wohl aber gut

[819] Rudi and Genia travelled to Copenhagen for a conference on nuclear physics in September 1937 and then to the 2nd All Union Conference on nuclear physics in Moscow (20–27.9.1937).

dieses Mal nicht wieder wie das letzte Mal die offenbar schlecht funktionierende Dampferlinie zu benutzen.
Euch Vieren herzliche Grüße

Euer Vater + Else

4. 1937–1939: A Provincial Chair and Non-Provincial Science

When Rudolf Peierls moved to Birmingham in 1937, it was the first time he took up a position at a university which was not considered to be at the heart of theoretical physics. Berlin, Munich, Leipzig, Zurich, Rome, Cambridge, and also Manchester with Rutherford's legacy and Bragg's and later Blackett's presence could be regarded as such. Birmingham could not look back on a strong history of theoretical physics research comparable with any of these places, and in fact the chair of applied mathematics was only being created at the time of Peierls' arrival. Initially he was the only theoretician among the physicists. Few people would have anticipated that within just over a decade this department of applied mathematics would become one of the foremost centres of theoretical physics teaching and research not just in England, but in Europe and arguably across the globe. This was largely due to the effort of its first professor: Rudolf Peierls.

The move to Birmingham brought many other changes. For the first time Peierls had substantial teaching and administrative responsibilities, but this did not distract the young professor from pursuing an intensive programme of research, too. In 1936, Bohr had developed the concept of the 'compound nucleus', which had had an immediate impact on the understanding of the structure of the nucleus. His ideas instigated a range of new research enterprises in nuclear physics, and Peierls played an increasingly active role in this research field. In September 1937, just before moving to Birmingham, he was among many colleagues taking part in the Copenhagen Conference on Nuclear Physics, from where he moved on to Moscow to attend the second All Union Conference on Nuclear Physics which was held at the end of the same month.[820] His meeting there with Pauli gave rise to an exchange of letters with his former tutor. Pauli was not too keen on some of the new ideas formulated by Bohr, many of which were based largely on approximations; he would often refer to these as 'schmutzige Physik' — dirty physics. Nevertheless, he contributed to the conferences and the discussions.[821] Peierls' letters to Pauli in the autumn of 1937 largely refer to the discussions at the two conferences.[822] But their correspondence also took up the discussion of β-decay, on which Peierls had written up a paper for a Russian journal[823] and the emission of light quanta which both Peierls and Pauli had been working on.[824] One of Peierls' students, Fred Hoyle, at the time was considering some aspects of the theory of β-decay. He had been working with Peierls at Cambridge, and when Peierls moved to Birmingham continued his work at Cambridge under the supervision Maurice Pryce. But Hoyle also occasionally met Peierls to seek advice. Fermi's famous 1934 paper[825] had turned Pauli's neutrino idea into a quantitative theory, and this was used as a base for the interpretation of a large number of experimental results. Under Peierls' guidance, Hoyle considered the quantisation of wave equations for particles with higher spin. At the same time, Pauli's assistant Markus Fierz was working on a related topic which was later published as his habilitation.[826] Throughout the summer of

[820] Genia had initially planned to accompany her husband on this trip, but as her parents were still exiled (her sister Nina had been allowed to return to Leningrad), it was thought unwise for her to travel to Russia at that time.

[821] At Moscow, he had given two papers which papers were translated into Russian for publication. W. Pauli, 'Some Basic Remarks about the Theory of β-Decay', *Bull. Acad. Sci. URSS*, Série phys., 148–52 (1938) and W. Pauli, 'The conservation laws in relativity and atomic physics', *Modern Problems of Physical Chemistry and Chemical Technology* **2**, 21–35 (1938).

[822] Letters [209–210], [214].

[823] R.E. Peierls, 'The present state of the theory of β-decay', *Izv. Acad. Nauk* SSSR, 583–606 (1937).

[824] See Pauli's letters to Weisskopf, Pauli, *Wissenschaftlicher Briefwechsel*, II, 527–33, particularly 531–3.

[825] E. Fermi, 'A theory of β-rays' *Z. Phys.* **88**, 161–77 (1934).

[826] M. Fierz, 'Über die relativistische Theorie kräftefreier Teilchen mit beliebigem Spin', *Helv. Phys. Acta* **12**, 3–37 (1939).

1938, Peierls and Pauli tried to clarify some of the key questions in order to give appropriate guidance to their tutees.[827]

Another project in which Peierls joined forces with an advanced student originated from another one of Bohr's many findings. Bohr had shown that in a collision between two nuclei of which at least one was heavy, an unstable compound was formed temporarily and that the lifetime of such a compound, measured on a nuclear scale, was usually very large. The well-defined energy levels of these compound nuclei, the so-called 'resonance levels', their position and the properties of the metastable states belonging to them determined the cross sections for all nuclear reactions. After Breit and Wigner had given a formula for the probability of a nuclear reaction in terms of the virtual state, taking account one state only,[828] Bethe and Placzek generalised the formula to take account of all possible virtual states.[829] Together with his student, P.L. Kapur, who had followed him from Cambridge to Birmingham, Peierls took the applicability of this so-called dispersion formula further by cutting out some of the assumptions made by Bethe and Placzek. The former showed that the dispersion formula held rigorously as long as the widths of the virtual levels which contributed to the expression were smaller than the distances between levels.[830]

Rudolf Peierls' interest in nuclear physics is evident in the fact that he paid several visits to Niels Bohr in Copenhagen between 1937 and 1939, and discussion topics at the time were invariably linked to nuclear physics. It also finds expression in an animated exchange of letters involving Peierls, Bohr and George Placzek, who was working at Bohr's institute for much of the time in question.[831] Again, it was the

[827]Letters [219], [232], [235], [237–39], [246].

[828]G. Breit and E. Wigner, 'Capture of Slow Neutrons' *Phys. Rev.* **49**, 519–31 (1937).

[829]H.A. Bethe and G. Placzek, 'Resonance Effects in Nuclear Processes', *Phys. Rev.* **51**, 450–84 (1937); H.A. Bethe, 'Nuclear Physics B, Nuclear Dynamics. Theoretical', *Rev. Mod. Phys.* **9**, 69–244 (1937).

[830]R.E. Peierls and P.K. Kapur, 'The dispersion formula for nuclear reactions', *Proc. Roy. Soc. London* A **166**, 277–95 (1938).

[831]In the summer of 1938, with the situation in Europe, and in particular in his home Czechia, deteriorating, Placzek first spent some time in France before emigrating to the United States.

compound nucleus, which occupied the three physicists. In the collision of a slow neutron with a nucleus, the resulting compound system has resonance levels which can be narrower than their spacing. At higher neutron energies, the width of the resonance increases and spacing decreases, leading to an overlap. In this region of overlapping resonances, two different ideas of expressing Bohr's compound nucleus formation, the Breit-Wigner formula on the one hand and detailed balancing on the other, gave conflicting answers. During various conversations and by exchanging views in writing,[832] Bohr, Peierls and Placzek eventually arrived at an understanding of the problem and its solution. By the summer of 1938 the challenge of solving the physics problem had been superseded by the challenge of writing up the solution. Bohr was notorious for labouring over the formulation of his research results, often leading to unwelcome and unnecessary delays of their publication. In that respect the fate of the Bohr-Placzek-Peierls calculations was not unusual. As the correspondence between the three scientists between the summer of 1938 and the spring of 1939 shows, the logistic difficulties of communicating among three people in three different countries was one problem, the lack of agreement over the nature of the publication was another.[833] In contrast to Bohr's intention of presenting a broad qualitative argument, Placzek and Peierls wanted a more mathematically rigorous exposition. As the formulation of the more extensive publication proved so difficult, it was agreed to send a short note to *Nature* initially, to be followed by the more detailed paper containing all the derivations at a later stage. When Bohr came to Birmingham to receive an honorary degree in June/July 1939, this note was written and appeared shortly afterwards.[834] The full paper was not finalised by the beginning of the war and work had to be abandoned until after the war. However, as a result of the numerous drafts that had been prepared and circulated to others for comment, the results of the proposed publication were already widely known and were being used in the scientific

[832]Letters [219], [221], [225], [228], [233–34], [236], [240].

[833]Letters [245], [264], [272], [288–89], [299].

[834]N. Bohr, R. Peierls and G. Placzek, 'Nuclear Reactions in the Continuous Energy Region', *Nature* **144**, 200–201 (1939).

literature. Not surprisingly, therefore, the paper has gained considerable fame as being the most frequently cited unpublished paper![835]

In early 1939, Rudolf Peierls had the opportunity to revisit some of his earlier work on the de Haas-van Alphen effect completed in Rome in the early 1930s.[836] David Shoenberg, whom Peierls had met in Cambridge and who eventually moved into the house that the Peierls had occupied there, had visited Kapitza in his newly-built laboratory in Moscow in early 1937, and he began to study in some more detail the oscillatory variation of magnetic susceptibility with magnetic-field strength at low temperature, first observed in bismuth in 1930. He discovered that many other metals showed the same kind of oscillations as bismuth, but they were only observable at lower temperatures and with stronger magnetic fields. Shoenberg was in possession of some of Landau's notes which provided the theoretical background to his experiments and measurements. At the time when Shoenberg was publishing his results, in early 1939, Lev Landau was under arrest in Russia, and therefore Shoenberg asked Rudolf Peierls to help him make sense of Landau's calculations.[837] The letters exchanged between Peierls and David Shoenberg, among others, deal with the question of how to utilize Landau's results without jeopardizing his position in Russia. Shoenberg deleted all references to Landau's theory when publishing his results in a Russian journal but was able to include a full account of it in the version of the paper that was published by the Royal Society.[838]

Lev Landau was one of many scientists directly suffering when the political and ideological climate changed across Europe as a result of the establishment of dictatorships in Italy, Germany and Russia. The distinct nature of those totalitarian regimes meant that people were affected differently in different countries. In Italy, where fascism

[835] Several drafts of the paper survive and some more or less complete versions have been published. See R.E. Peierls (ed.), *Niels Bohr. Collected Works*, Vol. 9. Nuclear Physics (1929–1952), Amsterdam: North-Holland, 1986, p. 49, note 86 and p. 50, notes 87–89.

[836] See above p. 351.

[837] Letters [281], [283].

[838] D. Shoenberg, 'Magnetic properties of bismuth. III. Further measurements of the de Haas-van Alphen effect', *Proc. Roy. Soc.* A **170**, 341–64 (1939).

manifested itself generally in a less oppressive form, many did not suffer persecution as such. But they were dissatisfied with the government and the conditions in Italy and some tried to leave. Others, especially scientists of Jewish origin, such as Racah were forced to emigrate. Much more brutal and uncompromising in the execution of its racially motivated policies was Nazi Germany. Legislation and policies drove into emigration a huge proportion of the country's most talented scientists, old and young. In Stalinist Russia, ideologically motivated terror also targeted many outstanding academics by means of arrest, prison, forced labour or even execution. The 'Abbot', Matvei Bronstein, was among those who perished in the terror wave of the late 1930s. Landau, Houtermans, Weissberg, Fock and many others were arrested, too. Western scientists wanted to act in support of the victims of Soviet terror, just as they were keen to support those colleagues who had to flee National Socialism. But few were in a position to do so effectively. Much of the correspondence between Bohr and Peierls in late 1938 concerns the question of how best to act in support of Landau who had been arrested in April 1938. Peierls, who not only had first-hand knowledge of Russia as a result of his frequent visits, but also had insightful counsel in his Russian-born wife Genia, was frequently asked for advice on how to deal with Soviet authorities. Niels Bohr, unsure about his attempts to help secure Landau's release from prison, turned to his former student for such advice, and indeed many letters exchanged between Bohr and Peierls circled around the logistics of approaching the Soviet authorities in order to ascertain Landau's fate and positively influence it.[839]

While the difficulties of his Russian friends and colleagues were of some concern to Rudolf Peierls, even more prominent in the correspondence was the fate of his German colleagues and of course of his own family. By 1938, both of his siblings had left Germany to settle in the US and England respectively. But his father and stepmother found it difficult to face emigration, although the conditions for Jews worsened by the day in Germany. In August 1938, Hans Bethe traveled to Germany and on that occasion paid a visit to Rudolf Peierls' family. In

---

[839] See letters [240], [242–44].

a letter to his friend he reported back about the living conditions and about the reluctance of the Peierls parents to make the move away from Germany.[840] Just a few days later, Peierls received a letter written by his stepmother from Switzerland. Freed from German censorship, Else Peierls writes about their dilemma of finding life in Germany increasingly unbearable, while at the same time not wanting to have to rely on the support of relatives, if they chose to emigrate.[841] The situation of the Peierls family was in many respects typical of Jewish middle-class families at the time. Frequently, the younger generation saw the signs of the time early; they had less to leave behind and were still in a position to start afresh and build a new life outside Germany. In contrast, the older generation, often more reluctant to face change had a home, friends and material goods to leave behind with less of a chance of starting a new life. By the time, many older people were ready to face emigration, it had become too difficult to carry it through, because of the increasingly stringent regulations about transfer abroad of possessions and money. Rudolf's immediate response to Else's letter was a reply urging them to depart from Germany as early as possible.[842] His parents may still have been wavering, but for Rudolf it was evident that the time would soon come when either Germany would no longer allow them to leave or the potential recipient countries would no longer allow them to immigrate.[843] In the last quarter of 1938 and the early part of 1939, he continued to impress on them the need to arrange for emigration and to agree to settling in Birmingham. Only a few letters of the rather voluminous correspondence are included here.[844] Finally, in early 1939, Heinrich and Else Peierls got the necessary permissions from Germany and the UK and emigrated to England. In 1940, they continued their journey to the United States, where they settled in New Jersey near Rudolf Peierls' sister Annie and Heinrich's brother Siegfried.

---

[840] Letter [249].
[841] Letter [251].
[842] Letter [252].
[843] See letters [256], [263].
[844] Letters [263], [276], [280].

Some of the letters exchanged between Rudolf Peierls and colleagues of his generation, in particular with Hans Bethe and George Placzek, give a vivid insight into the precarious situation of many young (and older) scientists who were forced out of Germany and Central Europe by the policies of the totalitarian regimes on the Continent. Many were competing on an increasingly tight job market; the number of foreign born applicants for positions in the UK and the US created more and more resentment among British and American colleagues; and the appointment of foreigners at times was not considered for fear of alienating the academic competition at home.[845]

By the outbreak of the war in 1939 Rudolf Peierls had lived outside Germany for about a decade, and he had spent more than half of this time in England. He felt gratitude towards the country that had provided him with a safe home at the time when his birth country had failed to do so, and by the late 1930s it was clear that Rudolf and Genia Peierls had no intention of returning to Germany. In May 1938, therefore, Rudolf Peierls applied for naturalisation, but he also felt the need actively to end his association with the country that Germany had become under Hitler, rather than terminating it by default once the process of his naturalization had been completed. In July 1939, he enquired at the German Embassy in London, whether he could renounce his German nationality.[846] As the advice received by the Embassy made clear, this was technically impossible, as some of the documents needed for an application to end his status as member of the German nation were also required for his naturalisation, which of course took precedence.

Peierls, who had been a staunch opponent of appeasement[847] was convinced that Hitler could only be stopped by a determined policy

[845]The problem was acute for Peierls and Bethe, when they were looking for employment themselves in 1933. After Bethe had secured his appointment in Manchester, it was thought to be unacceptable for another German to be appointed to a second vacant position. Eventually Peierls was appointed, but on the bases of a grant provided by the Academic Assistance Council, and not the university. In 1938/9 similar though not identical questions of appointability were among those discussed in letters concerning Placzek's chances of securing a position in either England or the US. See letters [253], [259], [264].

[846]Letter [291].

[847]See comments in letters [267], [275], [293].

which, in the last resort, could not rule out war. In a letter to Hans Bethe in August 1939,[848] he remarked that the choice for Europe was between War and Appeasement. The former would be bitter, but the latter would be a disaster, as only war would give Europe a chance and hope to rid itself from its dictators. Had Europe opted for appeasement, Rudolf Peierls would have tried to emigrate to America, and he asked Hans Bethe for support, should that scenario arise. Once, however, war was on the horizon, he attempted to ensure that he could play an active part to help his chosen country of residence, and on the day before the German invasion of Poland, he applied to the Home Office to speed up his application for naturalisation.[849] With the outbreak of the war, these 'routine' enquiries were of course put on the back burner, and technically still an enemy alien Rudolf Peierls found himself excluded from many lines of work which he would have regarded as useful and productive in a country at war. Therefore, in mid-September, he approached Edward Appleton, then Secretary of the Department of Scientific and Industrial Research, for advice on how to gain a status that would give him early permission to serve the United Kingdom during a time of national emergency.[850] Little did Peierls know that his contribution, more than that of most, would later have a decisive impact on a significant aspect of war-related research and weapons development in the UK and subsequently in the United States.

[848]Letter [293].
[849]Letter [294].
[850]See letters [298], [302].

[209] Rudolf Peierls to Wolfgang Pauli

Birmingham, 12.10.1937
(carbon copy)

Lieber Herr Pauli!
Nach näherer Untersuchung Ihres komischen Faktors in der Ausstrahlungsformel (oder genauer gesagt in der Streuformel) scheint es mir, daß er doch reell ist, und weder auf der Vernachlässigung der Retardierung, noch der relativistischen Effekte beruht.[851] Man kann nämlich die Existenz eines solchen Faktors fast ohne jede Rechnung einsehen.

Um das zu zeigen, will ich den Fall der Bornschen Näherung betrachten, d.h. ich will annehmen, daß das streuende Kraftfeld schwach ist, während aber über die Wechselwirkung zwischen den Teilchen und dem Strahlungsfeld nichts vorausgesetzt werden soll. Unter Vernachlässigung des streuenden Feldes kann man die Zustände dann wie bei Ihnen durch den Impuls p des Teilchens und die Lichtquantenzahlen N beschreiben. Und zwar will ich annehmen, daß ich die strengen Lösungen des Problems hätte, d.h. ohne Vernachlässigung von Retardierung oder Relativität.

Wenn man nun das streuende Potentialfeld einführt, so geben dessen verschiedene Fourierkomponenten zu verschiedenen Übergängen Anlaß. Da sie unabhängig voneinander sind, darf man sie als getrennt betrachten. Es handelt sich also [darum], die Matrixelemente von einer einzigen Fourierkomponente, z.B. e^{ikx}, zu finden. Die Matrixelemente lassen sich in der Diracschen Beziehungsweise so aufschreiben:

$$(p, Ne^{ikx}p', N').$$

Nun ist das Absolutquadrat des Operators e^{ikx} offenbar gleich Eins, und das bedeutet für die Matrix, daß

$$(p, N, e^{ikx}p'N')^2 = 1.$$

[851] W. Pauli and M. Fierz, 'Zur Theorie der Emission langwelliger Lichtquanten', *Nuovo Cimento* **15**, 167–88 (1938).

Also: die Quadratsumme der Matrixelemente vom Anfangszustand zu einem beliebigen Endzustand hat denselben Wert wie ohne Strahlung.

Nun ist aber das Quadrat des Matrixelements nicht mit der Übergangswahrscheinlichkeit identisch. Um diese zu bekommen, hat man noch die Dichte der Endzustände pro Energieinterval einzuführen. Für diejenigen Endzustände, in denen $N' = 0$, in denen also kein freies Lichtquant auftritt, ist dieser Faktor derselbe wie ohne Strahlung, für die anderen Zustände ist er aber kleiner. Insbesondere für die Zustände, in denen ein Lichtquant vorhanden ist, dessen Energie größer ist als $(p^2 - (p - hk)^2)/2\ m$, wird dieser Faktor Null. Die Endzustände geben also keinen Beitrag zur Übergangswahrscheinlichkeit; aber da sie einen Beitrag zu der obigen Summe geben, so bedeutet dies, daß die übrigen Matrixelemente und daher auch die Übergangswahrscheinlichkeit entsprechend kleiner wird. Also muß die Gesamtwahrscheinlichkeit irgendeiner Ablenkung (mit oder ohne Ausstrahlung) kleiner werden als wenn man die Wechselwirkung mit der Strahlung vernachlässigt.

Diese Überlegung ist natürlich im wesentlichen mit dem äquivalent, was Sie gerechnet haben und was Sie in Kopenhagen hätten erzählen sollen,[852] aber so zeigt sich etwas mehr, nähmlich daß das Resultat nicht von den Vernachlässigungen herkommen kann. Denn wenn die Berücksichtigung der Retardierung das Resultat ändern sollte, so müsste dies nach dem obigen bedeuten, daß alle diejenigen Matrixelemente, $(p, Ne^{ikx}p', N')$ verschwinden, in denen die Energie $N'h$ größer ist, als die maximal bei einer Ablenkung um k ausgestrahlte Energie. Mit anderen Worten, die Übergänge, für die der Energiesatz nicht erfüllt ist, sollten dann doppelt verboten sein, erstens wegen dem Energiesatz, und zweitens soll außerdem noch das zugehörige Matrixelement verschwinden. Es ist evident, daß dies bei Berücksichtigung der Retardierung sicher nicht herauskommt.

[852]Both Peierls and Pauli had attended a nuclear physics conference in Copenhagen in September. W. Pauli to Niels Bohr, 31.8.1937, Pauli, *Wissenschaftlicher Briefwechsel* II, pp. 534–5. See also notes to Pauli's talk at Copenhagen, 'Zur Theorie der Emission langwelliger Lichtquanten', *ibid.*, pp. 531–2.

Eine andere Frage ist natürlich, ob dieses Resultat etwas mit der Wirklichkeit zu tun hat, und es wäre lustig, wenn sich diese Frage experimentell entscheiden ließe. Dieser Faktor würde eine Korrektion von etwa 6% in der Rutherford-Streuung schneller Elektronen unter großen Winkeln bedeuten (wenn man bei e^2/m^2 abschneidet), und das liegt nicht unbedingt innerhalb der Meßfehler. Bedauerlicherweise gibt es gerade bei der Rutherfordstreuung schneller Elektronen, abgesehen von Abschirmungsfragen, noch alle möglichen anderen Komplikationen die von der Löchertheorie herrühren, sodaß man ein Experiment nicht so leicht interpretieren könnte.

Es tut mir leid, mich in Moskau nicht von Ihnen verabschieden zu können[853] und insbesondere Ihrer Frau für ihre Freundlichkeit mit dem Kleid zu danken, aber ich konnte Sie abends nicht mehr finden.
Mit bestem Gruß
Ihr

[Peierls]

[210] Wolfgang Pauli to Rudolf Peierls

Zürich, 15.10.1937

Lieber Herr Peierls!
Ich habe Ihren Brief erhalten und möchte dazu Folgendes sagen: So sehr mir das Resultat Ihrer Überlegungen sympathisch ist, scheint mir diese doch in zweierlei Hinsicht unvollständig zu sein.

1. Ihre Behauptung, man könne die strengen stationären Zustände des Systems bei Abwesenheit äußerer Kräfte, aber bei Berücksichtigung von Relativität und insbesondere der Retardierung, immer noch durch Größen p und N charakterisieren (die dann als Teilchenimpuls und als freie Lichtquantenzahlen zu

[853]Between 20 and 26 September 1937, the second All-Union conference on nuclear physics took place in Moscow. Pauli and Peierls (the former accompanied by his wife) attended this Copenhagen meeting.

deuten wären), scheint mir durchaus nicht so trivial zu sein. Z.B. kann man dann wohl kaum mehr erwarten, daß die Energiewerte des Hohlraums in Strenge von der Form $E = \frac{p^2}{2m} + \sum N_\nu h\nu$ sein werden. Ich habe noch gar keine bestimmte Vorstellung darüber, wie diese strengen stationären Zustände aussehen werden und wie groß ihre Dichte pro Energieintervall sein wird.

2. Selbst die fraglichen Behauptungen zugestanden, liefert Ihre Überlegung zu wenig: nämlich nur, daß das Strahlungsfeld die Gesamtwahrscheinlichkeit irgendeiner Ablenkung (gegenüber der strahlungslosen Theorie) vermindert. Ihre Überlegung zeigt aber nicht, ob und wie dieser hinzutretende Faktor (< 1) vom Radius a des geladenen Körpers abhängt. Und meine Gegner nehmen nur daran Anstoß, daß dieser Faktor (auch im Fall $\omega_0 < \omega_1$, d.h. $\frac{1}{2}\frac{mV_0^2}{\hbar} < 2\pi\frac{c}{a}$) von a abhängt. Mit Hilfe Ihrer Überlegung läßt sich z.B. nicht die Möglichkeit ausschließen, daß — bei strenger Berücksichtigung der Retardierung — sich herausstellen könnte, daß in meinem Korrektionsfaktor $\left(\frac{E}{\hbar\omega_1}\right)^{\frac{e^2}{\hbar c}\frac{(\vec{p}-\vec{p}\,')^2}{c^2}}$ die Größe $\hbar\omega_1$ durch m_0c^2 zu ersetzen wäre.

Ich selbst glaube allerdings gar nicht an diese Möglichkeit, denn es scheint mir plausibel, daß die Streuformel für kleine Werte von $\frac{\hbar\omega_1}{m_0c^2}$ eine gewöhnliche Potenzreihenentwicklung nach diesem Parameter gestatten muß, deren 'nulltes' Glied durch meine Formel richtig wiedergegeben wird.

Es würde mich also interessieren, ob sich Ihre Überlegung hinsichtlich der Einwände 1. und 2. verfeinern läßt.

Bloch ist inzwischen hier aufgetaucht[854] (er wird den ganzen Winter hier sein) und ich bemühe mich, ihn davon zu überzeugen, daß seine Behandlung der Retardierung unrichtig ist.

[854]Felix Bloch, who had left Germany after Hitler's ascent to power, had accepted a post at Stanford. He spent 1937/8 in Europe and frequently visitd his teacher Pauli in Zurich.

Für gelegentliche Mitteilung über die bei Ihnen gemachte Arbeit über die Diracgleichung für Teilchen mit beliebigem Spin[855] wäre ich wegen meines Diplomanden dankbar.[856]
Mit besten Grüßen
Ihr

W. Pauli

[211] Rudolf Peierls to Niels Bohr

Birmingham, 28.10.1937
(carbon copy)

Lieber Herr Bohr,
Zunächst möchte ich Ihnen noch, mit grosser Verspätung recht herzlich für die Tage in Kopenhagen danken.[857] Diese Zusammankünfte in Kopenhagen bedeuten für mich, wie sicher auch für die anderen, die Auffrischung und Anregung, von der man das ganze folgende Jahr zehren kann. Ich wollte eigentlich schon früher schreiben, aber ich war krank, als ich aus Russland zurückkam und hatte dann mit der Übersiedlung und den neuen Pflichten in Birmingham alle Hände voll zu tun.

Placzek hat Ihnen hoffentlich berichtet, dass die Diskussion mit Pauli wenigstens das eine Resultat ergeben hat, dass kein logischer Widerspruch zwischen dem Resultat von Pauli-Fierz und dem Argument von Williams besteht.[858] Hiervon unabhängig ist aber

[855]Pauli and Fierz were working on particles of arbitrary spin. See, e.g., W. Pauli and M. Fierz, 'On Relativistic Wave Equations for Particles with Arbitrary Spin in an Electromagnetic Field', *Proc. Roy. Soc.* **A173**, 211–32 (1939). A shorter paper based on a talk at the Swiss Physical Society had been published in German in *Helv. Phys. Acta* **12**, 297–300.

[856]The candidate is Pauli's subsequent assistant Josef Maria Jauch.

[857]Bohr had hosted the nuclear physics meeting in Copenhagen in September 1937. See letter [209].

[858]E.J. Williams, 'General survey of theory and experiment for high energy electrons', in E. Bretscher (ed.), *Kernphysik*, Berlin: Springer, 1936, pp. 139–41.

natürlich die Frage, ob das Paulische Resultat etwas mit der Wirklichkeit zu tun hat. Ueber diesen letzteren Punkt habe ich inzwischen noch ein bisschen nachgedacht, und bin zur Ueberzeugung gekommen (auf Grund einer Ueberlegung, die ich weiter unten wiedergeben möchte) dass das Paulische Resultat, sicher nicht auf den mathematischen Vereinfachungen der Paulischen Rechnung (Vernachlässigung von Retardierung und Relativität) beruht. Ferner scheint es mir dass dieses Resultat nur in sehr geringem Masse von den speziellen Eigenschaften der jeweiligen Strahlungstheorie Gebrauch macht; ich würde es daher für wahrscheinlich halten, dass es auch aus einer verbesserten Theorie folgen würde. Es würde mich sehr interessieren, ob Sie glauben, dass Gründe dafür vorhanden sind, dass dieses Resultat nicht der Wirklichkeit entsprechen kann, denn wenn man die Forderung stellen könnte, dass eine Theorie dieses Resultat nicht liefern soll, so würde dies eine sehr einschränkende Bedingung sein, und würde vielleicht ein Fingerzeig geben, in welcher Richtung man die Theorie abändern soll.

Ich darf vielleicht die Tatsachen noch einmal zusammenfassen: Das Pauli-Fierzsche Resultat lässt sich folgendermassen formulieren. Man denke sich ein geladenes ausgedehntes Teilchen von der Ladung e, Masse m und Radius a, und es sei $a \gg e^2/mc^2$ und $a \gg \hbar/mc$ ferner sei $e^2/\hbar c \ll 1$. Dieses Teilchen, welches natürlich dauernd in Wechselwirkung mit dem Strahlungsfeld steht werde nun an einem äusseren Potentialfeld $V(x, y, z)$ gestreut. Dieses Feld sei so schwach, dass es genügt, es in erster Näherung zu behandeln. Dann ist die Wirkung aller Fourierkomponenten

$$V(\vec{k}) = \int e^{\vec{k}\,\vec{x}}\, 1/(xyz) d\tau$$

additiv. Nimmt man $V(xyz)$ als zentralsymmetrisch an, was bequem ist, so hängt $V(\vec{k})$ nur von $k = \mid \vec{k} \mid$ ab. Wir wollen nun den Effekt aller Fourierkomponenten des Potentialfeldes mit festem k (d.h. fester Wellenlänge, aber beliebiger Richtung), anschauen. Diese ergeben, wenn man den Impuls der ausgestrahlten Lichtquanten vernachlässigt, Streuprozesse, in denen die vektorielle Impulsänderung des Teilchens den Betrag $\hbar k$ hat. Mit diesen Bezeichnungen kann man zunächst nach der Wahrscheinlichkeit fragen, dass, unter dem Einfluß der so ausgesuchten Fourierkomponenten des Potentials überhaupt eine Ablenkung des Teilchens stattfindet (unabhängig davon, ob und wieviele Lichtquanten ausgestrahlt werden). Diese Wahrscheinlichkeit will ich $w_e(k)$ nennen.

Man kann diese Wahrscheinlichkeit vergleichen mit $w_o(k)$ d.h. mit der entsprechenden Wahrscheinlichkeit für ein ungeladenes Teilchen. Die letztere läßt sich natürlich einfach aus der Bornschen Formel entnehmen. Das Resultat der Pauli-Fierzschen Rechnung ist nun, dass W_e <u>kleiner</u> ist als $w_o(k)$ und zwar um einen Faktor

$$\left(\frac{\omega_k}{\omega_a}\right)^{e^2/\hbar c\, k^2 (\frac{\hbar}{mc})^2} \qquad *$$

der allerdings sehr nahe an Eins ist. Hierin ist $\omega_a = \frac{2\pi e}{a}$ und $\hbar\omega_k$ ist der maximale Energieverlust, der durch die von uns ausgesuchten Fourierkomponenten den das Teilchen erleiden kann wenn der Betrag der Impulsänderung $\hbar k$ ist. Falls k klein gegenüber dem Anfangspunkt p_0 des Teilchens ist, ist

$$\omega_k = p_0 k/m$$

Also: Die Wahrscheinlichkeit eines Streuprozesses (unabhängig von der damit verbundenen Ausstrahlung) ist um den Faktor (*) kleiner als für ein ungeladenes Teilchen.

Man kann nun weiter folgende Fragen stellen:
<u>Erstens</u>: Angenommen das Teilchen werde gestreut, was ist dann die mittlere Anzahl der ausgestrahlten Lichtquanten mit Frequenzen zwischen ω und $\omega + d\omega$? <u>Antwort</u> nach Pauli-Fierz: Unter den zu Anfang genannten Voraussetzungen ist diese mittlere Anzahl <u>gleich</u> der in naiver Weise mit der Sommerfeldschen Formel berechneten. Die gesamte Anzahl ausgestrahlter Lichtquanten wird daher natürlich um den Faktor (*) kleiner als bei Sommerfeld.

<u>Zweitens</u>: Was ist die Wahrscheinlichkeit dafür, dass, wiederum vorausgesetzt, dass das Teilchen abgelenkt wird, der Energieverlust zwischen E und $E + dE$ liegt. Williams hat gezeigt dass man diese zweite Frage beantworten kann, sobald man die Antwort auf die erste weiss.

In der Tat stimmt, im Gegensatz zu dem, was wir in Kopenhagen der Diskussion zugrunde legten, die Antwort von Pauli und Fierz auf diese Frage genau mit der Williamschen Ueberlegung überein. Die Wahrscheinlichkeit dafür, dass das Teilchen eine Impulsänderung vom Betrage [...][859] erfährt und dass dabei der Energieverlust einen bestimmten Betrag hat, ist dann natürlich wiederum um den Faktor (*) als man sie aus dem Williamschen Argument bekommt, wenn man die Streuformel für ein ungeladenes Teilchen plus die Sommerfeldsche Formel für die Streuintensität bei gegebener Ablenkung zugrundelegt.

All dies scheint im Widerspruch mit dem zu sein, was wir alle in Kopenhagen aus dem Paulischen Bericht zu verstehen glaubten, nämlich dass es sich hier um einen Effekt handeln sollte, der besonders bei langen Lichtwellen (und bei kleinen Energieverlusten) eine erhebliche Rolle spielt, während es sich nun herausstellt, dass einfach die Wahrscheinlichkeit jedes Prozesses um denselben Faktor (*) der sehr nahe an Eins liegt, vermindert wird. Die Aufklärung dieses Missverständnisses ergibt sich daraus, dass wir damals eine merkwürdige Fragestellung diskutierten: es verhält sich nämlich so:

Nach der "naiven" Sommerfeldschen Rechnung ist die mittlere Anzahl der ausgestrahlten Quanten mit Energien zwischen ϵ und $\epsilon + \frac{d\epsilon}{\epsilon}$ pro einfallendes Teilchen gleich

$$A\frac{d\epsilon}{\epsilon}$$

solange man diese als einzeln ausgestrahlt betrachten darf, folgt hieraus, dass die Wahrscheinlichkeit eines Energieverlustes zwischen E und EdE auch gleich

$$w_{\text{naiv}} = AdE/E$$

ist. Wie Williams gezeigt hat, ist diese Formel [...][860] für kleine E unrichtig, weil dann immer mehrere Quanten mit vergleichbarer Energie herauskommen. Wenn man das Williamsche Argument quantitativ durchführt, so bekommt man

$$w_{\text{Williams}} = (\frac{E}{\hbar\omega_k})^{\frac{e^2}{\hbar c}} \frac{k^2\hbar^2}{m^2c^2} A \frac{dE}{E}$$

[859] Missing in manuscript.
[860] Missing in manuscript.

Dies ist bei hohen Energien praktisch gleich w_{naiv}, und wird insbesondere diesem exakt gleich für die größtmögliche ausgestrahlte Energie

$$\hbar\omega_k$$

Die Energie, bei der der Korrekturfaktor z.B. gerade $\frac{1}{2}$ beträgt, (und das ist die Energie, nach der wir in der Diskussion fragten) ist dann, aus der Bedingung

$$w_{\text{Williams}} = \frac{1}{2} w_{\text{naiv}}$$

gleich

$$\hbar\omega_k (\frac{1}{2})^{\frac{\hbar^2}{e^2} \cdot (\frac{mc}{\hbar k})^2}$$

wie Williams auch schon in Kopenhagen angab.

Berücksichtigt man nun nach Pauli, dass die Streuwahrscheinlichkeit um den Faktor (*) kleiner wird so wird die Wahrscheinlichkeit jedes Energieverlustes natürlich auch um diesen Faktor kleiner, d.h.

$$w_{\text{Pauli}} = (\frac{E}{\hbar\omega_k})^{\frac{e^2}{\hbar c}(\frac{k\hbar}{mc})^2} (\frac{\omega_k}{\omega_a})^{\frac{e^2}{\hbar c}(\frac{k\hbar}{mc})^2} A \frac{dE}{E} = (\frac{E}{\hbar\omega_a})^{\frac{e^2}{\hbar c}(\frac{k\hbar}{mc})^2} A \frac{dE}{E}$$

Für hohe Energien ist dies nun wieder <u>ungefähr</u> gleich w_{naiv} aber selbst an der Grenze [...][861] nicht genau. Vielmehr ist es an der Grenze gleich

$$w_{\text{grenz}} = (\frac{\omega_k}{\omega_a})^{\frac{e^2}{\hbar c}(\frac{k\hbar}{mc})^2} \cdot A \frac{dE}{E}$$

Fragen wir nun, bei welcher Energie die Formel auf die Hälfte des Grenzgesetzes heruntergegangen ist, so kommen wir natürlich auf dieselbe Energie wie oben. <u>Aber</u> fragen wir statt dessen bei welcher Energie

$$w_{\text{Pauli}} = \frac{1}{2} w_{\text{naiv}}$$

ist, so ist diese Bedingung <u>ein bisschen</u> verschieden von der Bedingung

$$w_{\text{Williams}} = \frac{1}{2} w_{\text{naiv}}$$

[861] Missing in manuscript.

oder

$$w_{\text{Pauli}} = \frac{1}{2} w_{\text{grenz}}$$

wegen der konstanten Faktoren. Diese Faktoren sind zwar sehr nahe an Eins, aber da die Energie in diese Gleichungen in einer so schwachen Potenz eingeht, so macht schon eine kleine Veränderung sehr viel in der Energie aus. In der Tat ist die Energie, bei der $w_{\text{Pauli}} = \frac{1}{2} w_{\text{naiv}}$ ist

$$\hbar\omega_a \frac{1}{2}^{\frac{\hbar c}{e^2}\left(\frac{mc}{\hbar k}\right)^2}$$

hängt also vom Teilchenradius ab. Das bedeutet aber natürlich nicht wie wir damals glaubten, dass die Form der Energieverlustkurve vom Teilchenradius abhängt, sondern nur, dass wir zwei Dinge verglichen haben, die sich durch einen von a abhängigen Faktor unterscheiden.

Ich hoffe, Sie werden mir verzeihen, dass ich alle diese elementaren Dinge in solcher Ausführlichkeit aufschreibe, aber die Diskussion in Kopenhagen war durch die verschiedenen Missverständnisse so verwikkelt geworden, dass es mir ratsam schien, den Sachverhalt noch einmal ganz explizit festzustellen.

Es stellt sich somit heraus, dass die Williamssche Ueberlegung nicht im Widerspruch damit ist, dass das Resultat der jetzigen Strahlungstheorie von der Teilchengrösse abhängt, wenn auch nur in einer schwachen Potenz. Es schien mir nun von Interesse zu versuchen, ob man dies nicht auch ganz ohne Rechnung einsehen kann, damit man besser versteht, auf welche Züge der Strahlungstheorie es hierbei ankommt. Das kann man in der Tat, und zwar auf folgende Weise:

Man nehme an, man kenne die exakten Lösungen der Strahlungstheorie für ein freies Teilchen, ohne äußere Kräfte. (Retardierung und Relativität sollen also auch berücksichtigt sein). Wenn die Theorie dann überhaupt vernünftig ist, so müssen die sich ergebenden Zustände sich folgendermassen gruppieren lassen: Es gibt erstens Zustände, in denen die mittlere Feldenergie minus Nullpunktsenergie mit dem Abstand vom Ort des Teilchens rascher als mit r^{-2} abfällt. Diese Zustände lassen sich durch den Gesamtimpuls charakterisieren, und die Energie ist für kleine

Impulse $E = p^2/2m + \text{const.}$ Ob hierbei die Masse m rein mechanischer oder teilweise elektromagnetischer Natur ist, ist hierbei belanglos. Ferner muss es Zustände geben, in denen ausser dem Teilchen und seinem Eigenfeld noch N_1 "freie" Lichtquanten im Zustand 1 und N_2 solcher im Zustand 2 u.s.w. vorhanden sind. Die Energie ist dann

$$p^2/2m + N_1\hbar\omega_1 + N_2\hbar\omega_2 \cdots$$

Der Zustand ohne freie Quanten ist natürlich ein Spezialfall hiervon für $N_1 = N_2 = N_3 = \cdots = 0$;

Betrachten wir nun wieder die Streuung eines solchen Teilchens durch ein schwaches Potentialfeld und beschränken wir uns auf die Streuung, die von einer einzigen Fourierreihe verursacht wird. Für diese Streuung kommt es offenbar auf das Absolutquadrat des Matrizelementes von e^{ikx} d.h.

$$|(p_0, 0, 0 \ldots |e^{ikx}|p', N_1, N_2, \ldots)|^2$$

an. Von diesen Matrixelementen können wir die Summe der Absolutquadrate, summiert über alle denkbaren Endzustände, ohne weiteres angeben, denn diese Summe ist nach den Regeln der Matrixmultiplikation gleich dem Diagonalelement von $|e^{ikx}|^2$ d.h. gleich 1:

$$\sum_{p' N_1 N_2 \ldots} |(p_0, 0, 0, 0|e^{ikx}|p', N_1, N_2, \ldots)|^2 = 1$$

Im Falle eines ungeladenen Teilchens ist dieser Satz trivial, denn dann besteht die Summe nur aus dem einen Glied, für das auch auf der rechten Seite sämtliche N verschwinden und von diesem einen Glied sieht man leicht direkt, dass es gleich eins ist.

Aus diesem Summensatz folgt also zunächst, dass durch das Vorhandensein der Uebergänge mit Ausstrahlung die Wahrscheinlichkeit eines strahlungslosen Prozesses reduziert wird (wie die Paulische Rechnung zeigt, und wie man auch leicht aus der Summenregel ablesen kann, wird diese Wahrscheinlichkeit sogar Null.)

Man könnte aber vermuten, dass die Gesamtwahrscheinlichkeit für irgendeinen Uebergang durch diese Aenderung nicht betroffen wird.

Das wäre dann der Fall, wenn die Wahrscheinlichkeit eines Ueberganges direkt gleich dem Quadrat des Matrixelementes wäre. In Wirklichkeit aber enthält aber der allgemeine quantenmechanische Ausdruck für die Wahrscheinlichkeit eines Ueberganges ausser dem Quadrat des Matrizelementes noch einen Faktor, der im wesentlichen die Anzahl der möglichen Endzustände pro Energieintervall darstellt. Dieser Faktor wird um so kleiner, je grösser die Energie der ausgestrahlten Lichtquanten $N_1\hbar\omega_1 + N_2\hbar\omega_2 + \cdots$ ist. Insbesondere wenn die Energie der Lichtquanten den größtmöglichen Energieverlust des Teilchens übertrifft, so ist dieser Faktor Null. Solche virtuellen Uebergänge geben also keinen Beitrag zu der Streuwahrscheinlichkeit, da sie aber einen Beitrag zu der Quadratsumme der Matrixelemente geben, so vermindert ihr Vorhandensein die Größe der Matrixelemente mit geringer Ausstrahlung und daher die Streuwahrscheinlichkeit. Man sieht hieraus also: Die Streuwahrscheinlichkeit ist umso kleiner, je mehr virtuelle Uebergänge es gibt, die energetisch verboten sind. Die Grenze der Uebergänge, die energetisch möglich sind, wird durch $E < \hbar\omega_k$ festgelegt; die virtuellen Uebergänge sind möglich, sofern kein Lichtquant ausgestrahlt wird, dessen Wellenlänge kleiner als a ist, d.h. dessen Frequenz ω_a überschreitet. Nun sind aber Prozesse mit vielen harten Lichtquanten selten, d.h. die meisten virtuellen Uebergänge entsprechen Strahlungsenergien von weniger als $\hbar\omega_a$. Die virtuellen Uebergänge, die keinen Beitrag zur wirklichen Strahlung ergeben, und deren Existenz die Streuwahrscheinlichkeit verkleinert, sind also gerade die Uebergänge, in denen die Strahlungsenergie zwischen $\hbar\omega_k$ und $\hbar\omega_a$ liegt. Demnach hängt also die Voraussage der Theorie für die Streuung von dem Radius des Teilchens ab.

Ich wäre sehr dankbar, gelegentlich von Ihnen zu hören, was Sie über diese Frage denken, insbesondere wieweit man dieses Resultat glauben soll. (Wenn es wahr ist, ist es eigentlich nicht sehr interessant, nur wenn es einen Fehler der jetzigen Theorie darstellt, so wird es aufregend.)
Nochmals mit vielem Dank und herzlichen Grüssen
Ihr

[Rudolf Peierls]

[212] G. Placzek to Rudolf Peierls

Copenhagen, 6.11.1937

Lieber Peierls,
Beiliegend einige Bilder zum ewigen Gedächtnis der Kopenhagener Konferenz.[862] Sie und Gemahlin sind zwar weit weniger photogen als Pauli, Heitler und Rosenfeld, aber immerhin noch besser als die durchschnittlichen dänischen Filmstars.

Neues ist hier ausserdem nichts, die n-te Korrektur der Bohrschen Arbeit wird jeweils morgen fertig, aber trotzdem sieht es jetzt beinahe so aus, als ob Bethe mit seinem Zitat 1939 Unrecht behalten sollte.[863]

In Paris war natürlich auch nichts los. Hingegen ist Beck sehr beunruhigt, weil Sie ihm noch nicht geschrieben haben wie Sie ihm die Tischdecke berechnen, die übrigens seiner Schwiegermutter ausnehmend gefallen hat.

Ich versuche, zur Einleitung des Winter[s] die zahlreichen neuen radioaktiven Isotopen und Isomeren zu verstehen, die allvierzehntäglich von den amerikanischen Cyclotronen im Phys.Rev. ausgespuckt werden, vorläufig mit geringem Erfolg. Schreiben Sie bitte, ob Sie was Neues über das Rotieren oder Nichtrotieren der Kerne wissen, etc. Über den Pauli-Williams-effect hat sich Bohr weitgehend beruhigt.

Ich hoffe, Sie sind in der Zwischenzeit von Ihrer Krankheit genesen und versehen bereits musterhaft Ihre neuen Amtspflichten.[864]
Mit den herzlichsten Grüssen Ihnen und Ihrer Gemahlin,
Ihr

G. Placzek

---

[862]Nuclear physics conference in September 1937. See letter [211].
[863]In the last part of their review article on nuclear physics, Hans Bethe and M.S. Livingston had cited Bohr's joint paper with Kalckar as published in 1939. (See *Rev. Mod. Phys.* **9**, 382 (1937)).
[864]Rudolf Peierls had taken up his professorship in applied mathematics at Birmingham University in the autumn of 1937.

[213] Rudolf Peierls to R.H. Fowler

Birmingham 25.11.1937

Dear Professor Fowler,
May I trouble you with a question about a problem in statistical mechanics? The chemists here are interested in very long chains of molecules, each chain containing some hundreds of equal links. The link in their case is itself a complicated molecule of the size of a benzene ring or so. They have so far taken it for granted that, given the way of preparation, the molecular weight of such a substance is something quite definite. It occured to me that one would have to expect rather large fluctuations in the length.

That leads one to the problem of working out the probability of a chain having a given length assuming one knows the binding energy between successive links and making simple assumptions about the internal p.f. of a chain (e.g. all vibrations classical or all frozen in). This is, of course, a straightforward application of the law of mass action and the result is (to me rather unexpectedly) that the fluctuations are very large indeed, for example the probability of a chain only being half as long as the average size is $2^{-3/2}$ if all vibrations are frozen in and only slightly lower if they are classical.

Do you know whether this problem has ever been worked out? It is very elementary, but it would seem worthwhile to record the answer somewhere if that has not been done. I think if this has been done by anyone anywhere you would be likely to know.
Yours sincerely,

R. Peierls

[214] Wolfgang Pauli to Rudolf Peierls

Zürich, 25.11.1937

Lieber Herr Peierls!
Ich möchte heute nochmals auf das zur Diskussion stehende Strahlungsproblem zurückkommen — und zwar nicht so sehr, um nochmals über Ihre letzten Briefe zu diskutieren, sondern um Ihre Meinung über die folgenden beiden Punkte zu hören. (Ich schreibe nämlich für die Italiener die Arbeit jetzt ausführlich zusammen und möchte doch möglichst sicher sein, nichts Unrichtiges zu sagen).[865]

1. Bloch* hat mir ungeheuer die Hölle heiß gemacht wegen der Gültigkeitsgrenzen der Bornschen Näherung. Er vermutete insbesondere, daß diese beim Strahlungsproblem wesentlich vom Verhältnis der klassisch-mechanischen Stoßzeit l/v (l=Dimension des Kraftfelds) zur "Lichtzeit" $1/\omega_1$ ($\omega_1 = 2\pi c/a$ = Abschneidefrequenz, a = Dimension des geladenen Körpers) abhängen. Nach seiner Meinung sollte auch bei schwachem Kraftfeld die Bornsche Näherung (wenigstens bei Oszillatoren mit großem ω) nur richtig sein bei kurzer Stoßzeit, d.h. wenn $l/v \ll 1/\omega_1$.
Ich kann trotz eifrigem Suchen nach einem solchen Effekt bisher nichts derartiges finden. Im Gegenteil stellte sich immer heraus, daß die klassische Stoßzeit nichts mit den Gültigkeitsbedingungen der Bornschen Näherung zu tun zu haben scheint. Diese scheinen mir vielmehr beim Strahlenproblem die selben zu sein wie beim strahlungslosen Problem, und zwar was immer der Wert von e^2/hc sein möge. Sind Sie auch dieser Meinung? Wir haben z.B. beim Kraftfeld $V(x) = \frac{\hbar^2}{2m} A \frac{e^{-\alpha r}}{r}$ (Dimension von $A = cm^{-1}$) die nächste Näherung der Störungstheorie nachgerechnet (und

*Er ist momentan in Wien.

[865] Pauli had given a talk in October 1937 during the celebrations commemorating the 200th birthday of Luigi Galvani. This was later published as W. Pauli and M. Fierz, 'Zur Theorie der Emission langwelliger Lichtquanten', *Nuovo Cimento* **15**, 167–88 (1938).

zwar in beiden Grenzfällen $\frac{e^2}{hc} \cdot \frac{v_0^2}{c^2} \ll 1$ und $\frac{e^2}{hc} \cdot \frac{v_0^2}{c^2} \gg 1$; für Werte dazwischen ist es mathematisch etwas kompliziert) und fanden, daß die Bedingug

$$A \ll \alpha$$

in diesem Fall vollkommen hinreichend ist, um die relative Kleinheit dieser nächsten Näherung zu garantieren. Irgendeine "Stoßzeit" kommt dabei gar nicht vor.

2. Ich kann Ihre frühere (mündliche) Frage, was bei Bornscher Näherung für das Kraftfeld, aber $\frac{e^2}{hc}\frac{v_0^2}{c^2} \gg 1$ herauskomme, sehr einfach beantworten. {Dabei ist zu beachten, daß aus dieser Ungleichung zusammen mit der oft erwähnten Voraussetzung $\frac{e^2\omega_1}{mc^3} \ll 1$ stets folgt $\omega_1 \ll \omega_0$ $(= \frac{1}{2}mv_0^2/\hbar)$, d.h. die Abschneidefrequenz liegt weit vor der Kante.} Es folgt nun, daß bei gegebener Winkelablenkung des Teilchens dessen Energieverlust stets

$$E = \frac{2}{3}\frac{e^2}{c^3}(v_0 - v_0')^2\frac{\omega_1}{\pi} \sim \frac{2}{3}\frac{e^2}{c^3}v_0^2 2(1 - \cos\vartheta)\frac{\omega_1}{\pi}$$

(ϑ = Ablenkungswinkel des Teilchens)
beträgt, während das emittierte Spektrum nur durch den Abschneidefaktor $G(\omega)$ von ω abhängt und durch

$$\text{const.}\frac{2}{3}\frac{e^2}{c^3}v_0^2 2(1 - \cos\vartheta)G(\omega)\frac{1}{\pi}d\omega$$

gegeben ist. (Die Wahrscheinlichkeit, daß überhaupt die vorgegebene Winkelablenkung eintritt, ist dieselbe wie in der strahlungslosen Bornschen Näherung.)
Die „anschauliche" Bedeutung des Resultats ist die, daß dieses elektrodynamische Verhalten des geladenen Körpers genau dasselbe ist wie in der klassischen Elektrodynamik, wenn man eine plötzliche Änderung der Geschwindigkeitsrichtung des Körpers von v_0 nach v_0' annimmt – d.h. daß diese Änderung (für den Körper als Ganzes) in einer Zeit erfolgt, die kurz ist gegen $1/\omega_1$. Nun, hier haben wir keine klassisch mechanische Bewegung, höchstens eine zeitliche Änderung von Materiewellenpaketen

> bei der Streuung. Aber es ist doch bemerkenswert, daß in diesem Falle der Bornschen Näherung das Resultat an keine Annahme über die Stoßzeit bzw. die "Plötzlichkeit" der Ablenkung gebunden zu sein scheint. Ich kann jedenfalls mathematisch nicht sehen, wo hier eine solche Annahme hineinkommen soll. (Jedenfalls gibt es keine klassisch-mechanische Stoßzeit.) Die Frage hängt natürlich sehr eng zusammen mit der früheren nach den Gültigkeitsbedingungen der Bornschen Näherung beim Strahlungsproblem. Können Sie eine "anschauliche elementare Überlegung" finden, bei der diese Gültigkeitsbedingungen sowie das hier geschilderte Resultat deutlich werden — zur Beschwichtigung der Gemüter?

Zu Ihren früheren Briefen will ich nur kurz bemerken, daß ich bei Mitnahme der Retardierung eine allgemeine Charakterisierung der stationären Zustände des gel[adenen] Körpers + Strahlungsfeld (bei Abwesenheit von äußeren Kräften) durch Teilchenimpuls p und Anzahlen n von freien Lichtquanten nicht für möglich halte. Man kann hier höchstens noch bei gegebenem Totalimpuls des Systems (Teilchen und Strahlung) den Zustand kleinster Energie als den ohne "freie" Strahlung definieren. Deshalb glaube ich doch nicht, daß die Überlegungen Ihrer Briefe wirklich allgemeiner sind als meine. Andererseits bin ich aber (sogar Bloch ist es) ziemlich beruhigt darüber, daß die Retardierung unter den von mir gemachten Voraussetzungen nichts schaden kann.

In der Hoffnung, daß die ganzen Diskussionen über das Strahlungsproblem** bald vollständig erledigt sein werden.
grüßt Sie bestens
Ihr

W. Pauli

**Bohr äußerte in Bologna darüber verschiedene Meinungen, mit denen ich nicht einverstanden bin.[866]

[866] Bohr and Pauli had met in Bologna in October 1937, and had discussed the question of diffraction of light. See Pauli's and Fierz' publication, *ibid.*

[215] R.H. Fowler to Rudolf Peierls

Cambridge, 1.12.1937

Dear Peierls,
What you say about the probability of long chain molecules having a given length is extremely interesting, and I have no doubt correct. I am quite sure that noone realizes that the fluctuations are so very large, theoretically at least. I do not know any published treatment of the problem, but it is certainly worth while recording. As it is such an important point, I think it ought to go to the Royal Society, and I should be very glad to send it there if you like when it is ready.

I saw Kapur this morning, somewhat worried about his lack of a problem in nuclear physics. I told him — I hope you will agree that this was sound — that I thought he had done very well so far, and that it was really quite reasonable that he should be expected to look round for himself a bit in the near future, there being no danger that he would not have a perfectly adequate degree thesis ready at the right time. I therefore encouraged him to read for himself and to try to think of a problem of his own, and I made the suggestion that he should look at Weisskopf's paper on nuclear processes regarded as evaporation phenomena,[867] to see whether he could not generalize those calculations. That is a problem anyhow on which I might be able to help him. It will certainly do him no harm to look into this, but if you think there is no chance whatever of any further calculations being possible on Weisskopf's lines, you might drop me a hint some time. I have only just skimmed through the paper, and have not yet had time to study it myself. I hope to do so in the Christmas vacation.
Yours very sincerely,

R.H. Fowler

[867] V. Weisskopf, 'Statistics and Nuclear Reactions', *Phys. Rev.* **52**, 295–303 (1937).

[216] Wolfgang Heitler to Rudolf Peierls

Bristol, 7.12.1937

Lieber Peierls,
Ich habe aus Ihrem Brief[868] nicht ganz verstanden, welche Art von Übergängen und welche Wechselwirkung Sie annehmen. Verstehe ich richtig, so soll der Drehimpuls des emittierten Teilchens durch einen Bahn-rückstoß-drehimpuls des Protons kompensiert werden? Das ist aber gar nicht unsere Meinung. Das magnetische Moment soll ja dieselbe Richtung wie der Spin haben. Es muß also der Drehimpuls des emittierten Teilchens mit der Spinrichtung des Protons gekoppelt sein. Die fraglichen Übergänge sind vielmehr z.B.

$$P(m_\rho = +\frac{1}{2}) \to N(m_\rho = -\frac{1}{2}) + Y(m = +1)$$

Y = schweres Elektron, m = z-Komponente vom Bahnimpuls, m_ρ = z-Komponente vom Spin. Bei einem Übergang $S_{+\frac{1}{2}} \to P_{+\frac{1}{2}}(?)$ wäre ja der Spin ganz unbeteiligt, und so etwas kann kein magnetisches Moment geben. Die Wechselwirkung, die die obigen Übergänge liefert ist, für eine ruhendes Proton: $(\sigma, \mathrm{rot}\,\psi)_{(0)\to\text{d.h. am Ort des Protons}})$ wo σ = Pauli-Spinvektor, ψ das vektorielle Wellenfeld des schweren Elektrons. Die Wechselwirkung ist aus relativistischen Gründen die einzige, die (bei einem vektoriellen Wellenfeld) in — für das Proton nicht-relativistischer Näherung — (das schwere El[ekron] kann rel[ativistisch] sein) den Spin enthält. Sie enthält aber die Bahn des Protons nicht. Wir rechnen außerdem den Operator des magn[etischen] Moments aus, d.h. nehmen das Proton als punktförmig und ruhend an. Ich sehe deshalb auch nicht, wie hier irgendwelche Landéfaktoren auftreten können. Sie könnten insofern Recht haben, als es fraglich ist, ob die gemessenen Werte von μ_P und μ_N nicht durch solche Landéfaktoren verfälscht sind. Da aber z.B. das Deuteron ein S-Zustand ist (so wird jedenfalls allgemein behauptet, ich weiß nicht, ob mit vollen Recht), so scheint uns das unwahrscheinlich.

---

[868] Preceding letter from Peierls could not be located.

Wir haben jetzt die Kernkräfte und den Absolutwert des magnetischen Moments ausgerechnet, es kommt alles schön heraus. Einige Konstante können auf verschiedene unabhängige Weisen durch die Experimente festgelegt werden, und die stimmen größenordnungsmäßig überein. Das ist übrigens nur so, wenn man ein vektorielles Wellenfeld für das schwere Elektron annimmt.[869]
Mit besten Grüßen
Ihr

W. Heitler

[217] Wolfgang Pauli to Rudolf Peierls

Zurich, 7.12.1937
(postcard)

Lieber Herr Peierls!
Ich kann mich diesmal sehr kurz fassen, da ich im wesentlichen mit allem in Ihrem Brief Gesagten einverstanden bin.[870] Der Deutlichkeit halber möchte ich nochmals betonen: Bornsche Näherung und große Ladung gibt ein Spektrum wie es klassisch bei plötzlicher Geschwindigkeitsänderung folgen würde. Klassisch läßt sich ja die Lichtemission nur berechnen, wenn der zeitliche Verlauf der Bewegung bekannt ist. Hier gibt es aber keine klassisch-mechanische Bahn. Für die Mitteilung über die Diracgleichungen vielen Dank.[871]

Bloch ist wieder zurück u[nd] hat mir schließlich doch recht gegeben! Über die Diffusionsgleichung im Schwerfeld sollte die Literatur wohl

[869]See W. Heitler, 'Remarks on Nuclear Disintegrations by Cosmic Rays', *Phys. Rev.* **54**, 873–76 (1938).
[870]Letter could not be located.
[871]Pauli later published his results in Pauli, W., 'On Asymptotic Series for Functions in the Theory of Diffraction of Light', *Phys. Rev.* **54**, 924–31 (1938).

ziemlich vollständig in Frank-Mises[872] (vormals 'Weber-Riemann') zu finden sein, insbesondere Bd.II Kap. XIV. (Vgl. speziell §2, N° 5.)
Mit vielen Grüßen
Ihr

W. Pauli

[218] Wolfgang Heitler to Rudolf Peierls

Bristol, 13.12.1937

Lieber Peierls,
Mit Ihren Landéfaktoren bin ich jetzt einverstanden, nachdem ich verstanden habe, was Sie darunter meinen. Die Sache ist sogar noch etwas komplizierter, da es noch eine spin-unabhängige Wechselwirkung gibt, die auch Übergänge nach m=0 ohne Spinänderung des schweren Partikels verursacht. Nennt man die Verweilzeit in all den Übergängen β, dann lauten die Gleichungen

$$\mu_P = 1 - \alpha + \alpha \frac{M}{m} - \beta$$

$$\mu_N = \alpha + \alpha \frac{M}{m} - \beta$$

In London hatte ich β weggelassen, Sie hatten $\beta = \frac{1}{2}\alpha$ angenommen. In Wirklichkeit macht es so gut wie gar nichts aus, (da $\alpha \sim \beta$). Weniger einverstanden bin ich dagegen mit Spin 1. Ich habe Ihnen keinen solchen Spin verschwiegen. Die Sache ist fast genau wie beim Licht (wie es da ist, können Sie z.B. in meiner Arbeit in den Cambr. Proc. sehen.)[873] Es ist

---

[872] Weber H., Rieman, B., *Die partiellen Differentialgleichungen der mathematischen Physik nach Riemanns Vorlesungen*, 6th edn., 2 vols, Brauschweig: Vieweg, 1919; Frank, P., Mises, R. von, *Differential- und Integralgleichungen der Physik*, 2nd ed. (1935). The two separate works were later combined in Frank, P., Mises R. von, Weber, H. and Riemann, B., *Die Differential- und Integralgleichungen der Mechanik und Physik*, 2nd ed., 2 vols., New York, 1943.
[873] W. Heitler, *Camb. Phil. Soc.* **32**, 112 (1936).

natürlich Definitions- und Geschmackssache, ob man z.B. den Drehimpuls einer zirkular-polar. Zylinderwelle Spin nennen soll. Jedenfalls ist es so, daß es Zustände mit $l = 0, 1, 2, \ldots$ gibt, mit den zugehörigen m, sogar zwei solche Serien mit entsprechender elektrisch-magn[etischer] Multipolstr[uktur], und alle haben verschiedene Bahnfunktionen. Die einzige — wenn Sie wollen Annahme — die wir gemacht haben, ist dass das magnet[ische] Moment = mechan[isches] Moment ist, d.h. $mH\mu$ die magn[etische] Wechselwirkung (m=mechanisch[es] Moment.) Das hängt natürlich von den Feldgleichungen im äußeren Feld ab, die man wählt. Wir haben als einfachstes solche angenommen, die die obige Eigenschaft haben. Natürlich kann man das auch anders machen und dem Ding z.B. ein klassisches handfestes Moment geben, wahrscheinlich kann man es auch so machen, daß irgendein Quantenspin herauskommt, obwohl das Teilchen sich im klassischen Grenzfall wie ein momentloses Teilchen benimmt (wie beim Elektron). Das schlimmste, was passieren kann, ist, daß $mH\mu$ sich von dem angenommenen Wert um einen Faktor der Größenordnung 1 unterscheidet, denn dann ist eben unsere Masse um diesen Faktor falsch. Das ist sie eigentlich sowieso, aus vielen anderen Gründen — z.B. deshalb weil die berechnete Masse eigentlich ein gewisser Mittelwert der relativistischen Masse ist, von dem man auch nur sagen kann, daß er von der Ruhemasse sich nur um einen Faktor 1–2 unterscheidet. Ich hoffe, daß wir jetzt über diese Sache einig sind.
Mit besten Grüssen
Ihr

W. Heitler

[219] G. Placzek to Rudolf Peierls

Copenhagen, 27.1.1938

Lieber Peierls,
Verzeihen Sie, wenn ich Ihnen erst jetzt schreibe, ich bin eben von einem prolongierten Weihnachtsurlaub zurückgekehrt. Vielen Dank für die Arbeit,[874] welche mich aus mehreren Gründen besonders interessiert.

---

[874]See letter [221], note 882.

a) hoffe ich darin Erleichterung für mein schlechtes Gewissen zu finden und b) sind wir jetzt hier sehr an den Fragestellungen interessiert, die sich ergeben, wenn die Niveaus überlappen. Kalckar,[875] der versucht hat etwas zu machen, ist, wie Sie vielleicht schon gehört haben, vor einigen Wochen ganz plötzlich an einer Gehirnblutung gestorben.

Bitte geben Sie mir etwas Zeit, die Sachen gründlich zu lesen. Wenn ich damit fertig bin, schreibe ich Ihnen ausführlich und sende das Manuskript an Bethe.

Sonstige Neuigkeiten sind hier keine. Nach Birmingham hoffe ich bestimmt zu kommen and freue mich sehr darauf. Momentan kann ich nicht, weil Bohr verwaist ist, aber im frühen Frühling hoffe ich, dass es geht.

Dieser Tage erschien hier Frau Houtermans mit Nachkommenschaft.[876] Den Gatten hat sie an dem Ort, wo sich zurzeit die Mehrzahl der russischen Fachgenossen befindet, zurückgelassen. Mit ihr selbst ist es noch leidlich abgegangen und durch eine Reihe glücklicher Umstände ist es sogar gelungen, sie safe nach London zu expedieren.

Also ich schreibe wieder. Viele herzliche Grüße Ihnen und Gemahlin
Ihr

G. Placzek

P.S. Eben sagt mir Bohr, dass er einen Brief von Ihnen hat;[877] ich werde dazu sehen, dass Sie bald eine Antwort [??].

---

[875]Fritz Kalckar (1910–1938), Danish pysicist and assistent to Niels Bohr.

[876]Friedrich Georg Houtermans (1903–1966), Dutch-Austrian physicist, obtained his Ph.D. under James Frank in Göttingen in 1927. In 1931 he married the physicist Charlotte Riefenstahl (with Pauli and Peierls as witnesses). After the Nazis seizure of power, Houtermans left Germany, working first in England and, from 1935, in Kharkov, where he was arrested in December 1937 and later imprisoned in Moscow.

[877]Letter could not be located. Reply from Bohr letter [220].

[220] Niels Bohr to Rudolf Peierls

Copenhagen, 1.2.1938

Dear Peierls,
I thank you most heartily for your kind lines about Kalckar's death[878] which in every respect is a great loss to us all, and for your kind offer to help us in the present difficult situation. We have had many discussions how to make the best out of Kalckar's work in which he was making great progress, but unfortunately he has left very little writings. When we have reached at a decision, I shall inform you and shall eventually be very thankful for your valuable assistance and advice. Today I enclose a copy of a note I have just sent to "Nature" about some thoughts regarding the nuclear photoeffect[879] with which I have been occupied for some time, but which I under the present circumstances have first yet succeeded in putting in a satisfactory form. I shall be very glad to hear your opinion about the argument which does not seem to me to be quite natural from the theoretical viewpoint, but also to represent an unforced description of the experimental facts.

With kind regards to your wife and yourself and also to the Oliphants from
Yours

Niels Bohr

P.S. Please give my heartiest thanks to Oliphant from my wife and me for the record of Rutherford's speech in Göttingen which we both treasure very highly.[880]

---

[878]Frity Kalckar had died of a haemorrhage in January 1938.

[879]N. Bohr, 'Nuclear Photo-Effects', *Nature* **141**, 326–27, (1938).

[880]Rutherford had died on 19 October 1937. In 1931 he had been invited to Göttingen as part of the bi-centennial of the Royal Society in Göttingen of which he was a foreign member. He delivered a lecture, chaired by Max Born, on 14.12.1931, which was recorded on celluloid disc which was given to Rutherford as a souvenir. Mark Oliphant, who found the disc after Rutherford's death, had it reproduced and given a copy to Bohr.

[221] Rudolf Peierls to Niels Bohr

Birmingham, 8.2.1938
(carbon copy)

Lieber Herr Bohr,
Vielen Dank für Ihren Brief und die Kopie Ihrer Note,[881] die ich sehr überzeugend fand. Wenn ich Sie richtig verstehe, so kommt jetzt also die Selektivität des Photoeffekts in derselben Weise zustande, wie etwa die Absorptionsbanden eines festen Körpers, der ja auch in weiten Gebieten transparent sein kann, obwohl er in diesen Gebieten zweifellos Energieniveaus besitzt, die eben nur nicht mit seinem Grundzustand kombinieren.

Was die absoluten Grössenordnungen betrifft, so ist es zweifellos, dass Ihr Argument die alte Schwierigkeit über die grossen Wirkungsquerschnitte beseitigt. Ich habe versucht, eine formale Ableitung der Formel zu geben, die Sie angeben, insbesondere um mich davon zu überzeugen, dass für veschwindende Neutronenbreite der (langlebigen) Kernzustände der Photoeffekt auch wirklich verschwindet (was er offenbar tun muss) und um zu sehen, in welcher Weise er für hinreichende Neutronenbreite in Ihre Formel übergeht. Das ist mir bisher noch nicht ganz gelungen, aber ich möchte mir das gerne weiterüberlegen, insbesondere mit Hilfe einer formalen Methode zur Behandlung der langlebigen Zustände, von der Ihnen Placzek vielleicht erzählt hat.[882]

Soweit ich die Frage bis jetzt übersehen kann, erscheint es mir wahrscheinlich, dass die Form der Absorptionslinie nicht genau die einer natürlichen Linienbreite ist. Diese Form folgt nur dann, wenn die Wahrscheinlichkeit der Umwandlung der Anregungsenergie aus dem

---

[881] See letter [220], note 879.

[882] Bohr, Peierls and Placzek continued to devote thought to the problem of nuclear photo-disintegration, and the three communicated at length over a joint paper about this topic. The article, which went down in nuclear history as 'probably the most-cited unpublished paper' was not published at the time. Various drafts survive, one of which is reprinted in Bohr, *Collected Works*, vol. 9. Nuclear Physics (1929–1952), p. 49, note 86 and p. 50, notes 87–89. Other drafts in *Peierls Papers*, Ms. Eng. misc. b203, C. 35.

speziellen mit dem Grundzustand kombinierenden Wellenpaket in langlebige Zustände davon unabhängig ist, wieweit die Umwandlung bereits fortgeschritten ist, (denn nur dann zerfällt der spezielle Zustand exponentiell mit der Zeit). Soweit ich bisher sehen konnte, gibt es keinen Grund, warum das in diesem Fall zutreffen sollte, und wenn man an die Analogie mit einem festen Körper denkt, so hat man ja dort auch Absorptionsbanden deren Form komplizierter ist. Natürlich sind Ihre Resultate bis auf numerische Faktoren von der Form der Linie unabhängig.

Es war mir auch nicht ganz leicht zu sehen woraus Sie schliessen, dass die Breite der Linie ungefähr gleich der Breite der γ-Strahlen ist, aber offenbar haben Sie da noch experimentelles Material benutzt das mir nicht bekannt ist.

Es würde mich sehr interessieren, ob Sie mit meiner Bemerkung über die Linienform einverstanden sind.[883] Sie hat natürlich keine Bedeutung für praktische Resultate, kann aber für die Auswahl einer geeigneten Methode für die mathematische Behandlung von Interesse sein.

Mit nochmals herzlichem Dank und vielen Grüssen, auch von meiner Frau und auch an Frau Bohr,
Ihr

R. Peierls

[222] Heinrich Peierls to Rudolf Peierls

Berlin, 3.3.1938

Meine Lieben,
herzlichen Dank für die Übersendung der erbetenen Bescheinigung.[884] Ob sie mir in absehbarer Zeit etwas nützen wird, steht dahin, nachdem

[883] Bohr received many favourable replies to his above-mentioned *Nature* note, (letter [220], note 879) but Heisenberg, Bloch and Peierls queried some of his assumptions, which led to a slight revision of Bohr's treatment of the topic in two subsequent publications N. Bohr, 'Wirkungsquantum und Atomkern', *Ann. d. Phys.* **32**, 5–19 (1938); N. Bohr, 'Resonance in Nuclear Photo-Effects', *Nature* **141**, 1096–97 (1938).

[884] By 1938, the situation for Heinrich and Else Peierls had worsened significantly. While they had been convinced initially that the privileged position Heinrich Peierls

mein Paß wie ich Euch schon schrieb, auf Inland umgestellt worden ist. Hoffen wir das Beste.[885]

Vielen Dank für Eure Nachrichten über Alfred. Er hat inzwischen auch viel beruhigter geschrieben, und ich hoffe, daß er sich halten kann. Jedenfalls wirst Du, lieber Rudi, wohl mit ihm in Fühlung bleiben, für den Fall, daß es später doch nötig wäre, ihn mit Empfehlungen zu unterstützen. Daß für Nina jetzt ein Psychoanalytiker hinzugezogen wird, ist vielleicht ganz gut; es ist möglich, daß er ihr helfen kann.

Wenn Du Demuth[886] einmal sprichst, so erwidere bitte die Grüße herzlichst. Ich war ihm sehr häufig gefällig, und wenn Du glaubst, daß er später einmal Alfred nützen kann, so bringe ihn doch gelegentlich mit ihm zusammen.

Wir freuen uns, daß Genias Finger inzwischen geheilt sind und daß es Euch gut geht.

Viele Grüße Euch Vier

Euer Vater

[223] Wolfgang Pauli to Rudolf Peierls

Cambridge, 19.3.1938

Lieber Herr Peierls!

Ich möchte heute über eine Frage an Sie schreiben, die im Zusammenhang mit der Arbeit von Hoyle* über die Wellengleichung für Teilchen

*Dirac gab mir Hoyles Manuskript zu lesen. Am meisten interessierte mich eine Behauptung, daß halbzahlige Spins $> 1/2$ zu Widersprüchen führen, wenn man nach Ausschließungsprinzip quantelt. Ich hatte noch nicht die Zeit, diesen Fall zu untersuchen.

---

had enjoyed would spare them from the worst excesses of the National Socialist regime, it became increasingly clear that life would be untenable for all Jews, no matter what their previous positions. In early 1938, Heinrich and Else Peierls had to leave their flat in Berlin. They were hoping to emigrate to England or the United States but did not know whether or when they they would be able to leave Germany.

[885] By early 1938, the passports of most Jews had been altered to 'inland only' documents to stop them from travelling abroad. Else, because of her non-Jewish descendancy was not affected directly by the increasingly restrictive regulations.

[886] Family friend who had settled in England.

mit höherem Spin aufgetaucht ist.[887] Für meine Cambridger Vorlesungen habe ich den Fall ganzzahliger Spins näher untersucht. Hier läßt sich der Spinorkalkül ganz vermeiden und man kann — unabhängig von Dirac's früherer Arbeit[888] — das ganze mit Tensoren geeigneter Symmetrie darstellen. Was die Quantelung betrifft, so verläuft sie bei nicht-verschwindender Ruhemasse etwas anders als bei Hoyle und in der Elektrodynamik, da dann alle Nebenwirkungen als q-Zahlrelationen aufgefaßt werden müssen, (d.h. mit allen Größen vertauschbar sind).

Ebenso wie bei Hoyle kann man bei meiner Schreibweise der Gleichungen leicht sehen, daß die Energiedichte zwar bei Spin 0 oder 1 positiv definit ist, bei ganzen Spins > 1 aber nicht. Andererseits konnte ich leicht zeigen, daß die totale Energie (d.h. die über das Volumen integrierte Energie) bei Abwesenheit von Kräften in allen diesen Fällen positiv ist (wegen der Nebenbedingungen).

Und nun entsteht die physikalische Frage, ob das Postulat einer positiv definiten Energiedichte notwendig ist. Exakter: ob es im Falle von Boseteilchen notwendig ist. Denn man muß beachten, daß in der Löchertheorie keine positiv definite Energiedichte existiert. Dies folgt sehr elementar aus einer Diskussion der Werte von

$$W = i\sum_{\rho} \psi_\rho^* \frac{\partial \psi_\rho}{\partial t} - \psi_\rho \frac{\partial \psi_\rho^*}{\partial t}$$

für Wellenpakete, die Lösungen der kräftefreien Diracschen Gleichungen, die nur negative Frequenzen (positive Energien) enthalten. Während dann W an einer bestimmten Raumstelle sowohl positive wie negative Werte haben kann, ist $\int W dV$ stets positiv. Soviel ich weiß, entstehen daraus keine physikalischen Schwierigkeiten.

Nun möchte ich gerne wissen, ob aus einem analogen Verhalten für Boseteilchen doch physikalische Schwierigkeiten entstehen, worüber ich mir noch nicht klar bin. Es ist möglich, denn hier kann man den Limes

[887] Fred Hoyle was working for his Ph.D. under Peierls, investigating the quantisation of wave equations for particles with higher spin.

[888] P.A.M. Dirac, 'Relativistic Wave Equations', *Proc. Roy. Soc.* **A155**, 447–59 (1936).

zu einer klassischen Feldtheorie machen. Man müßte die möglichen Messungen der Energie in einem Teilvolumen (s[iehe] eine ältere Arbeit von Heisenberg[889]) näher diskutieren. Da diese Frage entscheidend dafür zu sein scheint, ob Elementarteilchen mit ganzzahligem Spin > 1 a priori möglich sind, schien sie mir doch einer näheren Untersuchung wert. In Hoyles Arbeit scheint sie etwas unter den Tisch gefallen zu sein.

Antworten Sie mir, bitte, gelegentlich nach Zürich. Hier sind die Meinungen darüber geteilt. Es ist schade, daß weder Sie noch Hoyle hier waren.[890]

Mit besten Grüßen
Ihr

W. Pauli

P.S. Ich habe vorläufig noch nicht die Absicht, etwas über diesen Gegenstand zu publizieren; vielleicht ist das überflüssig.

Wissen Sie von Blochs Verlobung, die er dann plötzlich wieder gelöst hat, nachdem schon das Aufgebot in der Zeitung gestanden ist? Es tat mir leid für ihn; es war doch eine etwas peinlich Sache, auch für das Mädchen. Nun ist er also unbeweibt wieder in Amerika.

[224] Rudolf Peierls to Wolfgang Pauli

Birmingham, 24.3.1938
(carbon copy)

Lieber Herr Pauli!
Es tat mir leid, daß ich nicht zu Ihren Vorträgen nach Cambridge kommen konnte,[891] aber die letzte Woche im Term gibt es immer viel Betrieb, und diesmal hatte ich noch dazu einige wichtige Sitzungen, von denen ich mich nicht drücken konnte.

---

[889] W. Heisenberg, 'Über die mit der Entstehung von Materie aus Strahlung verknüpften Ladungsschwankungen', *Ber. Sächs. Akad. Wiss.* **86**, 317–22 (1934).

[890] Pauli had been invited by Fowler to give some lectures at the Cavendish Laboratory. Between 15 and 18 March 1938 he gave three lectures on 'Contributions to the Theory of Field Quantization'. See also *Nature* **141**, 486 (1938).

[891] See letter [223].

Hoyle hat kein Argument dafür, daß a priori die Energiedichte positiv definit sein muß. Sein Schluß war vielmehr der folgende: Bei ihm enthält die Energiedichte nur die Wellenfunktion selbst, nicht ihre Ableitungen. Wenn es also kleine Nebenbdingungen geben würde, so wäre die Bedingung, daß die Gesamtenergie positiv definit ist, trivialerweise mit der Bedingung positiver Dichte identisch, da die Werte der Wellenfunktionen an verschiedenen Orten voneinander unabhängig sind. Nun gibt es allerdings die Nebenbedingungen, die Beziehungen zwischen den Werten an verschiedenen Orten bedeuten, aber da sie linear und von der ersten Ordnung in den Ableitungen sind, so kann man sehen, daß sie nicht dazu helfen können, einen nicht positiven definiten Ausdruck in den Wellenfunktionen definit zu machen.

Soweit ich sehe, ist dies kein Widerspruch mit Ihrem Resultat, denn bei Ihnen enthält die Energiedichte auch die Ableitungen, und dann liegt die Sache natürlich anders. Der Unterschied zwischen Ihrer und Hoyles Behandlungsweise liegt nicht so sehr daran, daß Sie die Wellenfunktion als Tensor schreiben — denn jeder Spinor gerader Stufe kann natürlich als Tensor geschrieben werden — sondern daß Sie eine endliche Ruhemasse annehmen, während Hoyle die Masse = 0 setzte. Wenn ich Pryces Bericht über Ihre Rechnung richtig verstanden habe,[892] so lassen sich Ihre Gleichungen nicht auf den Fall $m = 0$ extrapolieren, sondern werden dann singulär. Es besteht dann also kein Widerspruch zwischen den beiden Rechnungen, die sich auf verschiedene Fälle beziehen. Ich wäre Ihnen sehr dankbar, wenn Sie mir dies bestätigen würden.

Eine andere Frage ist es natürlich, ob es einen Grund gibt, Gleichungen mit einer nicht definiten Energiedichte zu verbieten. Hierüber weiß ich im Augenblick nichts, aber ich halte es für möglich, daß es solche Gründe gibt. Es ist zwar richtig, daß die schlimmste Schwierigkeit, die z.B. in der Diracschen Theorie ohne Löcher auftritt, nämlich daß die Eigenwerte keine untere Schranke haben, bereits dadurch ausgeschlossen wird, daß man die Gesamtenergie positiv macht. Aber es kann immer noch passieren, daß die Anzahl der niedrigen Energiewerte zu groß

[892] After graduating in Cambridge in 1933, Maurice Pryce continued research under R. Fowler and Max Born before spending two years at Princeton 1935–37. In 1937 he returned to Cambridge. He had attended the Pauli lectures.

wird, z.B. daß es schon für ein endliches Volumen unendlich viele Eigenwerte gibt, die zu größeren und größeren Schwankungen der Feldgrößen gehören. Ich glaube, daß die Lösung dieser Frage sich aus der Untersuchung der Verteilung der Eigenwerte ergeben muß. Wenn diese vernünftig ist, so sehe ich keinen Grund ein, eine derartige Theorie auszuschließen.

Von dem Blochschen Mißgeschick habe ich schon gehört. Er könnte dem Bethe doch wirklich bessere Sachen nachmachen!
Mit besten Grüßen
Ihr

[Rudolf Peierls]

[225] G. Placzek to Rudolf Peierls

Copenhagen, 4.4.1938

Lieber Peierls,
Vielen Dank für Ihren Brief. Die Arbeit von Kalckar und Oppenheimer habe ich auch niemals verstanden.[893] Da mir das Resultat nicht unplausibel erschien, habe ich's zunächst bis auf Weiteres geglaubt,* vielleicht mit Unrecht. Vor einigen Wochen sandte Weinhaupt an Kalckar eine Ableitung, bei der das Endresultat $\sigma = \lambda^2 \frac{\Gamma^{\text{rad}}_{eo}}{\overline{\Gamma}^{\text{neut}}_{eu}}$ herauskommt ($\overline{\Gamma}^{\text{neut}}_{eu}$ Neutronenbreite für den Übergang in ein einziges Niveau des Restkerns), also um den sehr grossen Faktor $\frac{\overline{\Gamma}^{\text{neut}}_{ee}}{\Gamma_{\text{rot}}}$ mehr als Kalckar-Oppenheimer. Es würde dies ungefähr der alten Formel von Bethe und mir [894] für

*Kalckar hat immer gesagt, in der Arbeit stünde es ganz schlecht, aber er könnte es jetzt viel besser ableiten, und da er das und andere Dinge in einer Dissertation zusammenschrieb, hat sich die Diskussion darüber immer verschoben.

[893] F. Kalckar, J.R. Oppenheimer and R. Serber, 'Note on Resonances in Transmutations of Light Nuclei', *Phys. Rev.* **52**, 279–282 (1937).
[894] H.A. Bethe and G. Placzek, 'Resonance Effects in Nuclear Processes', *Phys. Rev.* **51**, 450–84 (1937).

getrennte Niveaus ($\gamma = \lambda^2 \frac{\Gamma_{eo}^{\rm rad}}{D}$) entsprechen, da ja $\overline{\Gamma}_{eu}^{\rm neut} \cong D$ (vgl. Landau[895]) Abgesehen von einem triviellen Fehler (Annahme dass, vgl. (17) $|\sqrt{\frac{N_e}{Z_e}}e^{i\delta} + 1|^2 = \frac{N_e}{Z_e} + 1$), der die Formeln (19) und (20) unrichtig machen dürfte und einer sehr zweifelhaften Abschätzung von $\overline{\Gamma}_{eu}/D$ auf S. 6, scheint mit die ganze Ableitung willkürlich, da die entscheidenden Annahmen über die Phasen nicht diskutiert wird. Ich sende Ihnen beiliegend die ganze Sache (übrigens will sie Weisskopf gar nicht in dieser Form publizieren) und würde mich sehr a) für Ihre bezügliche Meinung sowie b) Ihre eigenen Resultate, die Sie in Ihrem Brief erwähnen [interessieren].

Inzwischen haben Sie ja die Bohr'sche Note erhalten.[896] Wir haben hier viel darüber gestritten. Die Ziffern über die Breiten sind natürlich ganz aus der Luft gegriffen, aber auf die kommt es für das Prinzipielle ja nicht sehr an. Wichtig, und in der letzten Note nur angedeutet, ist, dass nach Bohr im Gebiet der dichten Niveaus auch bei sehr grosser Kopplung, wenn die Kopplungsbreite schon gross wäre gegen den Abstand, der mit dem Grundzustand zu kombinierenden Niveaus des unverkoppelten Oszillatorsystems, der Querschnitt für Photoeffekt noch immer durch die Kopplungsbreite, und nicht durch die Zerfallsbreite des normalen Compoundzustands bestimmt werden soll, und daher** prinzipiell kein Vergleich mit einer Formel, die mit dem gewöhnlichen Begriff des Compoundzustandes operiert, möglich sein soll.

Zu der ganzen Sache wäre ja manches zu sagen, mich interessiert, was Sie dazu meinen, mittlerweile dürften Sie ja Ihre Antwort an Bohr schon angesandt haben.[897]

In Ihre Arbeit dring ich fleissig weiter ein, wenn Sie aber die Behauptung, dass unter Umständen die Strömungstheorie auch für $\Gamma > D$ gelten soll, zurückziehen, so bin ich sehr einverstanden.

**im Gebiet der überlappenden Compoundniveaus

---

[895]L. Landau, 'Statistical Theory of Nuclei', *Zh. Eksp. Teor. Fiz.* **7**, 371 (1937) [*Phys. Z. Sowjetunion* **11**, 556 (1937)].

[896]N. Bohr, 'Resonance in Nuclear Photo-Effects', *Nature* **141**, 1096–7 (1938).

[897]Letter [221].

Weitere Neuigkeiten, falls diese Sie noch nicht erreicht haben: unsere Volksgenossen haben inzwischen sehr erfolgreich für die Vermehrung der Rasse der "weissen Juden" gesorgt. Heisenberg hat pikanterweise Zwillinge verschiedenen Geschlechts, so dass die schlechten Worte über Paarerzeugung kein Ende nehmen, Weizsäcker hingegen bloss einen Sohn.[898]

Ihren Mitteilungen entsprechend, mit den herzlichsten Grüssen, Ihnen und Gemahlin
Ihr

G. Placzek

[226] Wolfgang Pauli to Rudolf Peierls

Zürich, 18.4.1938

Lieber Herr Peierls!
Ich danke Ihnen für Ihren Brief vom 24. III., bin aber nicht mit den darin enthaltenen Betrachtungen über die Totalenergie einverstanden. Erstens scheint es mir, daß Hoyle's Ausdruck für den Energietensor (wenn ich mich richtig erinnere) mit unserem identisch sein muß, wenn die verwendeten Größen einander passend zugeordnet werden; insbesondere enthält unser Ausdruck für die Energiedichte auch <u>nicht</u> die Ableitungen (s. unten).*
[<u>Erläuterung</u>: wir führen im Falle ganzzahligen Spins s erstens einen symmetrischen Tensor mit s Indices $A_{\underbrace{ik\cdots r}_{\text{Anzahl } s}}$ ein, der der Bedingung genügt, daß dieser die Spur über irgendein Indexpaar 0 ist $\sum_{i=1}^{4} A_{iil\cdots r} = 0$. Man findet, daß dieser $(s+1)^2$ unabhängige Komponenten hat.

*Wünscht man die B zu eliminieren, kommen natürlich die 1. Ableitungen der A vor.

[898] Werner Heisenberg had been criticised by the Nazi authorities for not distancing himself from what was termed 'Jewish' physics, and he was often referred to as a 'white Jew'. In January 1938, his twins, the first of seven children of the Heisenbergs, were born.

Weiter einen Tensor $B_{\underbrace{ik\ldots p}_{\text{Anzahl } s-1}}[qr]$, der in den ersten (s-1) Indizes symmetrisch, im eingeklammerten Indexpaar schief ist und die Bedingungen

$$\sum_I B_{ii\ldots p[qr]} = 0, \quad \sum_q B_{ii\ldots p[qr]} = 0,$$

$$B_{ik\ldots p[qr]} + B_{ik\ldots r[pq]} + B_{ik\ldots q[rp]} = 0$$

erfüllt. Man findet, daß dieser Tensor $2s(s+2)$ unabhängige Komponenten hat. Der Energietensor ist eine quadratische Form der Komponenten von A und B, bzw. wenn A und B nicht als reell angenommen werden, eine Bilinear-form in A^*B^* und AB. Bei expliziter Einführung der A und B kommen die Ableitungen im Energietensor nicht vor.

Auf die Differentialgleichungen bzw. Nebenbedingungen, denen unsere Größen genügen, brauche ich hier nicht einzugehen. Die Beziehung zu den bei Dirac eingeführten Größen[899] ist folgende. Dirac führt zwei Spinoren Ψ_A bzw. Ψ_B ein mit je $(2k+1)2l$ bzw. $(2l+1)2k$ Komponenten. Wir betrachten nun speziell den Fall $k = l - 1/2$ und setzen $2k = s, 2l = s+1$ so daß Diracs $\Psi_A(s+1)^2$, Diracs $\Psi_B s(s+2)$-Komponenten bekommt. Es hängt nun unser A eineindeutig und linear mit Diracs Ψ_A zusammen, während unser B zunächst doppelt soviele Komponenten besitzt wie Diracs Ψ_B und zunächst nur letzteres eindeutig und linear aus erstem folgt. Die genauere Untersuchung zeigt in der Tat, daß nur der zu unserem B gehörige selbst-duale Tensor mit Komponenten $B_{\ldots[12]} + iB_{\ldots[34]}$ etc. eindeutig aus Diracs Ψ_B folgt. Dieser kleine Unterschied tritt aber schon in der Proca-Theorie $(k = 1/2, l = 3/2)$ auf; die Einführung des selbst-dualen Tensors erweist sich aber nicht als zweckmäßig.

Dies alles läßt sich verallgemeinern für einen beliebigen halbzahligen Wert von $k-l$, wobei der Spin nur von $k+l$ abhängt gemäß $s = k+l-1/2$. Auch die Aussagen über Vorzeichen der Gesamtenergie gelten ebenso im allgemeineren Fall.]

---

[899] P.A.M. Dirac, 'Relativistic Wave Equations', *Proc. Roy. Soc.* **A155**, 447–59 (1936).

Für $m = 0$ werden nur die Vertauschungsrelationen singulär, in der c-Zahltheorie lassen sich unsere Formeln ohne weiteres für $m = 0$ extrapolieren und die Frage des Vorzeichens der Totalenergie ist allein eine Frage der c-Zahltheorie. Im Fall $m = 0$ kann man den Energietensor durch die Größen B allein (ohne ihre Ableitungen) ausdrücken (die im Falle des Photons $m = 0, k = 1/2, l = 3/2$ in die Feldstärken übergehen — während die A in die Potentiale übergehen) und man kann die ganze Theorie durch lineare partielle Differentialgleichungen erster Ordnung zwischen den B allein ausdrücken. Die Nebenbedingungen (die linear und von erster Ordnung in den Ableitungen sind) genügen in der Tat, um zu verhindern, daß das Volumenintegral eines nicht positiv-definiten Ausdruckes in den Wellenfunktionen negativ wird!

Nur der Integrand — die Energiedichte — kann in einem gegebenen Raumpunkt negativ werden. Ich bin sicher, daß beim Hoyleschen Ausdruck für die Energie (den ich nicht mehr auswendig weiß) — immer im Fall ganz-zahligen Spins — dasselbe gelten muß. Eine kleine Ergänzung zu meinem letzten Brief[900] für den Sonderfall $m = 0$ muß ich hier allerdings machen: für $m = 0$ kann es eben noch passieren, daß die (niemals negative) Totalenergie verschwindet, ohne daß das Feld $B_{ik\ldots p[qr]}$ verschwindet, während für $m \neq 0$ aus dem Verschwinden der Totalenerie vermöge der Feldgleichungen das Verschwinden aller Komponenten von $A\ldots$ und $B\ldots$ folgt. Es scheint mir vernünftig, für $m = 0$ eine "Eichgruppe" einzuführen, der gegenüber die Totalenergie — nicht die Energiedichte! — invariant bleibt. Dann wären die Felder mit verschwindender Totalenergie wegtransformierbar. Aber dieses Detail ist eine Frage der physikalischen Interpretation und der Fall $m = 0$ gefällt mir physikalisch nicht besonders.

Über die physikalische Frage der Möglichkeit negativer Energiedichten bei Teilchen mit Bosestatistik weiß ich nichts Neues. Dagegen haben Fierz und ich nach meiner Rückkehr nach Zürich[901] mit der Un-

[900]Letter [223]. His comments refer to his and Fierz's work on relativistic theory of particles of arbitrary spin. This was published in 1939. M. Fierz and W. Pauli, 'Über relativistische Feldgleichungen von Teilchen mit beliebigem Spin im elektromagnetischen Feld', *Helv. Phys. Acta* **12**, 297–300 (1939).

[901]After his lectures in Cambridge, Pauli traveled on to Leiden and Utrecht. See correspondence Pauli with Kemmer and Heisenberg, Pauli, *Wissenschaftlicher Briefwechsel*, II, pp. 552–61.

tersuchung des Falles halbzahliger Spins begonnen. Bis jetzt fanden wir für $m = 0$ keine Schwierigkeit für halbzahligen Spin $> 1/2$ nach Ausschließungsprinzip zu quanteln und eine Löchertheorie zu machen. Wenn also Hoyle richtig gerechnet hat, so dürfte sein Mißerfolg hier auf der unglücklichen Annahme $m = 0$ beruhen. Ich will darauf später noch zurückkommen, wenn unsere Rechnungen über diesen Fall etwas weiter gediehen sein werden.

Ich wäre sehr dankbar, wenn wir gelegentlich ein Manuskript oder Korrekturen der Hoyleschen Arbeit bekommen könnten. Ich habe es ja nur einmal flüchtig bei Dirac gelesen.[902]

Mit besten Grüßen

Ihr

W. Pauli

[227] Nevill Mott to Rudolf Peierls

Bristol, 22.4.1938

Dear Peierls,

If you have an electron in an excited state on a polar crystal, have you any idea of the temperature dependence of the probability for time *dt* that it makes a transition to the normal state, giving up the energy etc. as heat?

Has this been worked out at all? I can't find it in Frenkel's papers.[903]

I hope you've had a good vac. The Faraday meeting was quite fun, Hirsch + Pohl again!

Yours sincerely,

N.F. Mott

---

[902] See letter [223], note 887.

[903] Frenkel was the first to discuss the electrolytic conductivity of solid polar salts from a theoretical point of view. Frenkel, *Z. Phys.* **35**, 652 (1926). These were then further developed by Jost *J. Chem. Physics* **1**, 466 (1933); *Z. Phys. Chem.* **169**, A 129 (1934); and *Z. Phys. Chem.* **32**, B1 (1936).

[228] Rudolf Peierls to G. Placzek

Birmingham, 22.4.[19]38
(carbon copy)

Lieber Placzek,
Beiliegend zunächst die Manuskripte von Beck und Mercier, die ich natürlich vergass, Ihnen zu geben. Ferner eine Abschrift einer Karte von Frau Schrö. an Bekannte in England. Dies nur zu Ihrer Information.

Schnaks[904] sprach ich telefonisch, und sie schien sehr enttäuscht, dass Bohr nicht schreiben will. Sie sah natürlich ein, dass ein Brief an den Akademiepräsidenten keinen Sinn hat, aber sie hatte offenbar gehofft, dass er direkt an höhere Instanzen schreiben würde, und ich wusste nicht recht, was ich ihr auf die Frage antworten sollte, warum er das nicht tut.[905]

Ich bekam heute einen Brief von Bethe,[906] in dem er u.a. schreibt: Weisskopf und ich interessieren uns hauptsächlich für den Photoeffekt. Uebrigens behauptet Placzek, Du hättest ihm gesagt, die Arbeit von Kalckar, Oppenheimer und Serber über den Photoeffekt wäre in Ordnung.[907] Das kann doch wohl nicht sein (oder Dein) Ernst sein. Sehr traurig fand ich Bohr's Arbeit in Nature über dieses Thema: sowohl die experimentellen Resultate sind nicht, wie er sagt, als auch die "Erklärung" scheint mir reaktionär und Unsinn.[908] Er nimmt ja jetzt das Kristallmodell ernster als je jemand das Hartreemodell!"

Ich wiederhole dies nur, um zur Aufklärung der Situation beizutragen. Es scheint mir insbesondere zu folgen, dass Bethe und Weisskopf vermutlich wissen, warum die Kalckar-Oppenheimer-Formel falsch ist, und, ungefähr das richtige haben.

Ausserdem teilt Bethe noch mit, dass er am 18. Juli nach Europa fährt und ca. am 18. September zurück. Physikalisch weiss ich

---

[904]Probably Charlotte Houtermans.

[905]Charlotte Houtermans had hoped for Bohr's intervention on behalf of her husband's arrest by Russian police. See letter [219], note 876.

[906]Letter could not be located.

[907]F. Kalckar, R. Oppenheimer and Serber, 'Note on Nuclear Photoeffect at High Energies', *Phys. Rev.* **52**, 273–78 (1937).

[908]N. Bohr, 'Nuclear Photo-Effects', *Nature* **141**, 326–27 (1938).

nichts neues. Bezüglich der Frage der Beiträge von Zuständen mit Energien ausserhalb der Breite bin ich, je mehr ich darüber nachdenke, um so mehr überzeugt, dass sie nichts ausmachen können. Z.B. im Falle der Lichtabsoption makroskopischer Körper, wo man doch die Situation ziemlich gut versteht (oder es jedenfalls glaubt) machen sie nichts. Aber es ist natürlich nicht leicht, das in schlagender Form zu beweisen, ohne eine lange Ixerei über Reihenentwicklungen nach nicht-orthogonalen Funktionssystemen oder so etwas zu machen.

Mit herzlichen Grüssen, auch an Frau Bensch, und nochmals tak för sidst.
Ihr

[R.E. Peierls]

[229] Rudolf Peierls to Wolfgang Pauli

Birmingham, 23.4.1938
(carbon copy)

Lieber Herr Pauli!
Vielen Dank für Ihren Brief.[909] Ich möchte auf die Frage, die Sie diskutieren, etwas später eingehen, wenn ich Zeit gehabt habe, sie mir etwas genauer zu überlegen. Leider existiert im Augenblick von der Hoyleschen Arbeit nur ein Manuskript, aber ich werde sehen, daß er möglichst bald noch eine Kopie machen läßt und Ihnen diese schickt.

Ich war über Ostern in Kopenhagen,[910] wo wir u. a. auf die Ultrarotkatastrophe zu sprechen kamen.[911] Dabei war ich erstaunt zu

---

[909]Letter [226].

[910]In early April 1938, Bohr's new institute was opened and a number of colleagues and collaborators met at Copenhagen for the occasion.

[911]The Infrarotkatastrophe, or Ultrarotkatastrophe, as Pauli referred to it, had been discussed in the latter half of 1937 by Heisenberg and Pauli. See exchange of letters, Pauli, *Wissenschaftlicher Briefwechsel*, II, pp. 439, 443, 518–19, 522–25, 528–535, 538, 540, 549, 552.

sehen, daß alle Leute, insbesondere Bohr sowie Kramers und Kronig,[912] die die Sache ganz kürzlich von Ihnen selbst gehört hatten, sich in einem völligen Mißverständnis über die Sachlage befanden, indem sie nämlich der Meinung waren, daß die Verteilung der Energieverluste bei gegebener Ablenkung vom Elektronenradius abhängig wird, und daß insbesondere der Radius besonderen Einfluß auf die langen Wellen hat. Ich hatte eigentlich die ganze Zeit den Eindruck, daß wir uns in Moskau, und in der darauffolgenden Korrespondenz darauf geeinigt hätten, daß in Wirklichkeit nur die Wahrscheinlichkeit der Ablenkung (also im Falle eines Coulombfeldes die Rutherfordsche Streuformel) um einen Faktor sehr nahe an Eins geändert wird, daß aber die Wahrscheinlichkeit, daß ein Teilchen, von dem bekannt ist, daß es eine bestimmte Ablenkung erfahren hat, einen Energieverlust in einem Intervall dE hat, vom Teilchenradius unabhängig ist. (Diese Wahrscheinlichkeit könnte auf keinen Fall um einen konstanten Faktor geändert werden, da ja ihr Integral über dE Eins ergeben muß.)

Ich schrieb damals an Bohr eine Erklärung der Sachlage, aber leider scheint er damals meinen Brief nicht gelesen zu haben.[913] Ich habe ihn jetzt davon überzeugt, daß die Sache so liegt, wie ich behauptete, und er war sehr erleichtert, und nannte die ganze Geschichte "des Kaisers neue Kleider" indem er und Kramers, und wohl auch andere, sich sehr über ein Problem beunruhigt haben, das gar nicht existiert.

Ich glaube, daß Sie über diese Fakten mit mir einig sind (jedenfalls glaubte ich das aus unserer damaligen Korrespondenz zu entnehmen) und daß Bohr, Kramers und Kronig Sie mißverstanden haben. Um weitere Mißverständnisse zu vermeiden, füge ich jetzt eine Kopie meines damaligen Briefes an Bohr bei. Er enthält u.a. meine Behauptung, daß der Einfluß des Elektronenradius auf die Stoßwahrscheinlichkeit unabhängig von den Einzelheiten der theoretischen Behandlung ist. Ich

---

[912] Hendrik A. Kramers (1894–1952), studied in Leiden, Copenhagen (where he was later assistant to Bohr); became professor in Utrecht (1926) and Ehrenfest's successor in Leiden in 1934 where he stayed until his death in 1952; Ralph de Laer Kronig (1904–1995), received his Ph.D. from Columbia in 1925, did research in Copenhagen, London, and in Zurich as Pauli's assistant, before taking up teaching positions in Utrecht, Groningen and Delft.

[913] Letter [211].

weiß, daß Sie damit nicht ganz einverstanden sind, aber ich würde nun gern wissen, ob Sie mit der Beschreibung der Tatsachen auf Seite 2 bis 6 einverstanden sind.

Wenn ja, so wird Bohr Sie sehr beschimpfen, daß Sie sich unverständlich ausgedrückt haben; wenn nein, so werden Sie weiteres von mir hören.

Mit besten Grüßen

Ihr

[Peierls]

[230] G. Placzek to Rudolf Peierls

Copenhagen, 27.4.1938

Lieber Peierls,

Vielen Dank für Ihren Brief.[914] Die Dokumente habe ich ordnungsgemäss abgeliefert, sodass sie nunmehr in Ruhe verschimmeln können. Die Wissenschaft macht keine besonderen Fortschritte, in der Angelegenheit der Gammas ist vorläufig ein Waffenstillstand geschlossen, ich erwarte inzwischen von Ihnen die Lieferung des in meinem letzten Briefe erbetenen neuen Kriegsmaterials, um damit wieder zum Angriff vorzugehen.[915]

Desweiteren hat Bohr eine Note für Nature über den Photoeffekt verfasst,[916] weil er meint man kann nichts falsches stehenlassen. Es scheint mir aber, dass nach deren Erscheinen überhaupt kein Mensch mehr irgendetwas verstehen würde. Daher habe ich alle Energie für ihre Abtreibung eingesetzt und zunächst einmal eine Verschiebung erreicht; wenn sich diese bis Kaladas Graecas erstrecken lässt, so ist ja alles gut.

Die Betheschen Meinungen über den Photoeffekt wundern mich garnicht, aber davon, dass die Selektivität nicht existiert, und die O.-K.

[914]Letter [228].

[915]Peierls had written another letter on 26.4.1938. Only a carbon copy exists in the Bodleian, but due to the large number of missing handwritten passages its content is unclear. *Peierls Papers*, MS. Eng. misc. b203, C. 30

[916]N. Bohr, 'Nuclear Photo Effects', *Nature* **141**, 326–27 (1938).

Formel falsch ist, wird ja die Weisskopf-Formel nicht richtig. Woher die Idee kommt, dass ich die Gültigkeit der O.-K. Formel für hohe Energien behauptet und Sie dafür zitiert hätte, ist mir völlig schleierhaft.[917] In dem Brief, auf den angespielt wird, habe ich nur behauptet (und dies energisch) dass die Weisskopfsche Formel mit abessynischen Methoden abgeleitet ist und ich sie daher nicht glaube, und Sie diesem Urteil beistimmen.

Die falsche Behauptung, die ich gemacht habe, steht in meinem letzten Brief an Weisskopf, auf den sich Bethe noch nicht beziehen kann und auch dort habe ich nie behauptet, dass die O.K.-formel für Licht niedriger Energien (langer Wellenlängen) gilt, was natürlich an sich schon Blödsinn genug ist. Uebrigens habe ich Weisskopf noch nicht geschrieben, weil Bohr will, dass er erst informiert wird, wenn wir einig sind. Ich werde aber nicht mehr lange warten, weil das ja nicht geht. Jedenfalls schreiben Sie doch bitte unabhängig an Bethe.

Von Frau Houtermans haben Møller,[918] Rosenfeld und ich lange Briefe bekommen. Bohr wird nochmals erwägen, ob er schreiben soll. Von der Entscheidung verständige ich Sie.[919]
Sonst ist zurzeit nichts neues. Herzlichste Grüsse von und zu
Ihr

G. Placzek

P.S. Danke für die Karte von Frau S. Der Inhalt ist ja eindeutig. Die Berliner Reise betraf die Planckfeier.[920]
(P.S.)^2 Manuskript folgt separat.
(P.S.)^3 Bohr grüsst und dankt für Ihren Brief.[921] Die Einteilung passt gut. Er wird noch schreiben.

---

[917] See letter [228].

[918] Christian Møller (1904–1980) studied physics at Copenhagen where he received his doctorate in 1932 and continued to teach until his retirement in 1975.

[919] It appears as though Niels Bohr did not intervene on this occasion. In contrast, he tried to act on Lev Landau's behalf. See letters [247–48].

[920] On 23rd of April 1938 Max Planck celebrated his 80th birthday.

[921] Letter Peierls to Niels Bohr, 22.4.1938, not reproduced here. *Peierls Papers*, Ms. Eng. misc. b203, C. 30.

[231] Rudolf Peierls to Nevill Mott

Birmingham, 27.4.1938
(carbon copy)

Dear Mott,
I am not aware of any calculation of this effect.[922] It is probably in many ways similar to the conversion of optical excitation energy into heat which I discussed in my old Annalen paper.[923] In both cases the conversion must create a large number of phonons simultaneously and can therefore occur with appreciable probability only if the obvious approximations break down. There will therefore be a rather critical dependence on the parameters giving the strength of the coupling between electronic motion and lattice vibrations, in the sense that below a certain value of the parameter the series of approximations will converge and there will practically be no conversion whereas for higher values of the parameter there will be appreciable conversion. This coupling depends on temperature. One could imagine that at low temperatures the parameter is below the critical value, but it reaches it as the temperature increases. In this case, there would be a violent temperature dependence. Otherwise I think the temperature dependence would only be slight. But all these are guesses. I know nothing more definite about the problem.
Yours sincerely,

[R.E. Peierls]

[232] Wolfgang Pauli to Rudolf Peierls

Zürich, 27.4.1938

Lieber Herr Peierls!
In Beantwortung Ihres Briefes vom 23. hoffe ich das Thema "Ultrarotkatastrophe" mit folgenden kurzen Bemerkungen abzuschließen:

---

[922] See letter [227].
[923] R. Peierls, 'Zur Theorie der Absorptionsspektren fester Körper', *Ann. d. Phys.* **13**, 905–52 (1932).

1. In der Arbeit von Fierz und mir[924] wurde die Voraussetzung $e^2/hc \ll 1$ nicht von vorneherein gemacht. Ihr Brief an Bohr[925] beschreibt nur den Teil unserer Resultate, der unter dieser besonderen (etwas banalisierenden) Voraussetzung gilt.

2. Der Radius des geladenen Teilchens geht in das Resultat (für Energieverlust $E \ll h\omega_1$) in Form eines Faktors ω_1^{-c} ein, mit $C = \frac{2}{3\pi}\frac{e^2}{\hbar c}\frac{(\vec{p_0}-\vec{p_0}')^2}{m^2c^2}$. (Hierbei ist übrigens die Voraussetzung der Bornschen Näherung wesentlich, sonst würde ω_1 viel komplizierter eingehen.) Dies ist ein konstanter Faktor bei festem $p_0^2+p_0'^2-2p_0p_0'\cos\theta$, es ist kein konstanter (von $E=\frac{1}{2m}(p_0^2-p_0'^2)$ unabhängiger) Faktor bei gegebenem Ablenkungswinkel θ. Letzteres ist nur annähernd der Fall für Energieverluste E, die klein sind gegen die ursprüngliche Energie $E_0=\frac{p_0^2}{2m}$ des Teilchens.
Unter den spezialisierenden Annahmen $e^2/hc \ll 1$ und $E \ll E_0$ ist allerdings Ihre Bemerkung ganz richtig, daß bei gegebenem Ablenkungswinkel des Teilchens die Verteilug der Energieverluste vom Teilchenradius unabhängig wird und dieser nur die gesamte Wahrscheinlichkeit einer gegebenen Ablenkung um einen Faktor < 1, der für kleines e^2/hc sehr nahe an 1 liegt, vermindert.
Über diesen Punkt haben wir uns in der Tat in Moskau geeinigt und er ist inzwischen nicht anders geworden.
In Cambridge hat mir Pryce eine Transformation angegeben, mit deren Hilfe man den Strahlenimpuls in erster Näherung (in $\hbar\omega_1/mc^2$) noch mitberücksichtigen kann. Etwas physikalisch Interessantes kommt aber dabei nicht heraus.

Über die mit der Hoyleschen Arbeit zusammenhängenden Fragen hoffe ich also noch von Ihnen zu hören.[926]
Mit den besten Grüßen
Ihr

W. Pauli

---

[924] W. Pauli and M. Fierz, 'Zur Theorie der Emission langwelliger Lichquanten' *Nuovo Cimento* **15**, 167–88 (1938).
[925] Letter [211].
[926] See letter [223].

[233] Niels Bohr to Rudolf Peierls

Copenhagen, 7.5.1938

Dear Peierls,
Thanks for your kind letter of April 22. I am very glad that you think you can come in the last week of May which would be very helpful for the completion of the work.

In the last days, Placzek and I have discussed the problem of the general validity of the simple dispersion formula in nuclear problems and have become very optimistic as regards its far reaching validity, independent of the ratio between width and distance of the levels and it appeared to us that this is also in conformity with the results of you and Kapur.[927] So far as we can see, will all the deviations of your formula from the usual dispersion formula depend only on the partial width of the levels; therefore it would seem in the first place that the questions of the validity of the usual dispersion formula does not depend on the ratio of the total width to the level distance on the energy region concerned, as it is already directly indicated by considering, instead of the stationary problem, a time limited problem with an uncertainty of energy larger than the level distance but too small to be of essential influence on the general mechanical properties of the system. The only question would therefore seem to be the problem of the influence of the high energy terms for which the partial width is no longer negligible compared to the level distance.

We must naturally distinguish between the case where $\frac{\gamma_0}{d} \ll 1$(γ_0 partial width, d level distance) in the region of incident energy and the case of so large an incident energy, that γ_0 is no longer negligible compared to the level distance. In the first case, the contribution of the "near level" will be exactly given by the usual dispersion formula but, as follows from your and Kapur's calculations, there may be considerable modifications of the terms representing the contribution of the distant high levels, but the total influence of these latter terms will, in the case concerned, only be of relatively small importance.

---

[927] R. Peierls and P.L. Kapur, 'The Dispersion formula for nuclear reactions', *Proc. Roy. Soc.* **A166**, 277–95 (1938).

At the first moment, it may perhaps seem that here is a problem of convergence, but it is clear that the whole procedure can only include finite energy limits and that the summation has to be broken off at least when the total width becomes of the same order as the excitation energy. As far as we can see, it is even necessary to consider this point in the discussion of the one body problem in your paper, where the remark about the convergence seems to assume that the γ are approximately constant although they will rise suddenly as soon as the energy comes near the top of the potential barrier. For such energies, however, the whole procedure will break down due to the lack of definition of the states concerned. In the case of the many body problem the whole situation is of course very different, because we have a large intermediate region where the nucleus is black before it becomes transparent.

We were not yet quite clear where the summation has to be broken off most rationally, but the exact limits for the summation as well as the modifications of these last terms in the dispersion formula would only seem to be of small importance due to the insensitivity of the logarithmic expression to which the unmodified terms lead. In fact, the estimate of γ_0 from the condition that direct sticking must take place where the wave length is of the order of the average distance δ between the nuclear particles lead to values for the contributions to the cross section for elastic scattering of the order of magnitude δ^2 and their influence should therefore even be small compared to the regular "potential scattering" which, in your treatment, comes in so naturally.

In the case that the incident energy itself comes near the sticking region, the whole problem is entirely modified and difficult to treat, but for increasing energy the sticking lends again of course a new simplicity to the problem. It therefore appears to us that, apart from the transitory region between the proper dispersion problems and the sticking collisions, the treatment stands fairly satisfactory an[d] that especially we have conformity between the perturbation treatment and the purely statistical considerations.

I have written so long about many things which you will surely know at least as well as we, in order to hear whether you in the main lines agree with the way we are looking on the whole situation. As already said, we are at the moment very optimistic and will now in the next days

make an energetic attempt of writing a proper account of it all to have it ready, in any rate in outlines, before you come to Copenhagen. As soon as Placzek knows definitely about the date of his sailing for America[928] and can see, whether there is a chance that we perhaps could all three meet before, he will write to you.
With kindest regards from us all, also to your wife,
Yours,

Niels Bohr

[234] Rudolf Peierls to G. Placzek

Birmingham, 11.5.1938
(carbon copy)

Lieber Placzek,
Ich habe Ihnen auf zwei Briefe zu antworten. Zu dem ersten ist insbesondere zu sagen,[929] dass ich nicht davon überzeugt bin, dass Sie recht haben, Bohr von der Veröffentlichung einer kurzen Bemerkung über den Photoeffekt abzuraten. Es ist doch nicht schön, etwas nicht richtiges stehen zu lassen, insbesondere wenn die Veröffentlichung der ausgeführten Arbeit noch länger dauert (woran wohl kein Zweifel besteht) so wäre es doch wahrscheinlich, — und unerfreulich — wenn der Fehler von jemand anders korrigiert wird.

Bethe habe ich geschrieben, und ihn ganz kurz von unseren Beschlüssen informiert, aber ich habe es klar gemacht, dass diese noch nicht endgültig sind, und dass er die endgültige Weisheit "in due course" zugeschickt bekommen wird.

Was Brief Nummer zwei betrifft,[930] so bekam ich gleichzeitig einen Brief von Bohr,[931] und ich lege eine Kopie meiner Antwort bei.[932] Wie

[928] See letter [236].
[929] Letter [230].
[930] Could not be located.
[931] Letter [233].
[932] Letter Peierls to Niels Bohr, 9.5.1938, *Peierls Papers*, Ms. Eng. misc. b203, C. 30.

Sie sehen, bin ich jetzt einig, dass in gewissen Fällen die gewöhnliche Dispersionsformel noch gelten kann, wenn die Niveaus dicht sind, obwohl es hierzu nicht ausreicht, dass die Partialbreite kleiner als d ist. Nach der Stimmung des Bohrschen Briefes zu urteilen, scheint dies ja auch die andere Frage zu erledigen, da er anscheinend daran glaubt, dass der Einfluss der entfernten Niveaus in diesem Falle keine Rolle spielt, und mit dieser Feststellung — wenn auch nicht mit ihrer Begründung — bin ich einverstanden.

Quantitativ lässt sich auch in diesem Fall sehr wenig sagen. Ihre — wie Sie sagen — schlampige Ableitung scheint mir ziemlich wenig zu beweisen, selbst wenn die gewöhnliche Dispersionsformel gilt, erstens wegen der logarithmischen Divergenz der Summe, und zweitens, weil man ja für die elastische Streuung dann sicher nicht die Summe der Quadrate sondern das Quadrat der Summe zu nehmen hat, und daher, soweit ich sehen kann, die numerischen Faktoren nicht mit dem übereinstimmen, was man aus der statistischen Form zu erwarten hat.

Meine Rechnung über die Zeitabhängigkeit war natürlich kein Beweis dafür, dass alles in Ordnung ist. Die Meinung war vielmehr die, dass man aus dieser Rechnung sehen kann, dass wenn man sogar nur ein einziges Niveau in der Summe berücksichtigt (im Falle dünner Niveaus) man die Rechnung auch so umschreiben kann, dass in der Sprache des Oszillatorniveaus ein Wellenpaket mit bevorzugter Behandlung eines kurzlebigen Anfangszustandes benützt wird. Man kann also aus der Tatsache, dass der Mechanismus solche raschverlaufenden Prozesse enthält, also nicht schliessen, dass man den Beitrag weitentfernter Compoundniveaus mitnehmen muss.

Aber da das Argument sich nur auf den Fall dünner Niveaus bezieht, in denen sowieso alles klar ist, so hilft es nur als Analogie und hat daher bestenfalls psychologischen Wert.

Ich bin inzwischen nicht viel weitergekommen, abgesehen davon, dass es mir gelungen ist, meine komische Identität für den Fall eines leeren Kerns bis zur dritten Näherung zu verifizieren, was jetzt meine Bauchschmerzen über die Vollständigkeitsrelation vollständig kuriert hat.

Ich komme also am 21. abends nach Kopenhagen. Da ich nicht weiss, ob Sie dann noch da sind (was ich hoffe) werde ich direkt an die

Frau Elite schreiben. Bemühen Sie sich also nicht wegen Zimmerbestellung.
Mit herzlichen Grüssen von und an Ihr

[R.E. Peierls]

[235] Rudolf Peierls to Wolfgand Pauli

Birmingham, 14.5.1938
(carbon copy)

Lieber Herr Pauli!
Ich habe jetzt die Frage der Wellengleichung mit Spin größer als 1 mit Hoyle ausführlich diskutiert. Es scheint mir jetzt in der Tat, daß sein Argument, wonach die Nebenbedingungen die Positivität der Gesamtenergie nicht retten können, nicht zwingend war. Um mich aber endgültig davon zu überzeugen, daß das Gegenteil der Fall ist, würde ich gern einen expliziten Ausdruck der Energiedichte in der von Ihnen angegebenen Form sehen, wenn das möglich ist.

Hoyle hat gestern seine Arbeit an Sie abgeschickt, und Sie werden inzwischen sein Argument über den halbzahligen Spin gesehen haben, das mir noch immer richtig zu sein scheint. Insbesondere kommt es bei diesem Argument nicht darauf an, ob die Masse verschwindet oder nicht. Der wesentliche Punkt ist vielmehr dieser: Wenn der Spin größer als 1/2 ist, so kann man noch immer in der üblichen Weise aus der Lagrangefunktion konjugierte Impulse zu den Feldgrößen definieren. Sind Q_i die Feldgrößen, P_i die Impulse, so gilt dann

$$P_i Q_k - Q_k P_i = c_{ik} ,$$

wo c eine (unwesentliche) Konstante ist. Wenn man nun über die zu Q_i konjugiert komplexe Göße Q_i^* betrachtet, so ist sie aber nicht mit P_i identisch. Statt dessen ist Q_i^* eine Linearkombination sämtlicher P. Wenn man daher den Ausdruck

$$Q_i Q_i^* - Q_i^* Q_i$$

bildet, so wird er gleich cf_{ii}, wo f_{ii} ein konstanter Koeffizient ist. Hoyles Argument besteht nun darin, zu zeigen, daß man es nicht so einrichten kann, daß alle f_{ii} (und die entsprechenden Koeffizienten, die man erhält, wenn man statt der Q_i beliebige Linearkombinationen benutzt) sämtlich positiv sind. Ein negatives f_{ii} bedeutet aber einen Widerspruch, da die Diagonalelemente von $QQ^* - Q^*Q$ notwendig positiv sind. Es liegt also nicht daran, daß man keine algebraisch widerspruchsfreien Vertauschungsrelationen anschreiben kann, sondern sie haben keine Lösung, die mit den Realitätsbedingungen im Einklang ist.

Es würde mich sehr interessieren zu hören, ob Sie beweisen können, daß die von Ihnen aufgestellten Vertauschungsrelationen Lösungen haben, und gegen welchen Teil des Hoylschen Beweises Sie einen Einwand haben.

Vielen Dank noch für Ihren Brief über die Ultrarotkatastrophe.[933] Wir sind darüber also jetzt völlig einig, und Kramers, Kronig und Bohr müssen Sie also völlig mißverstanden haben.[934]
Mit besten Grüßen
Ihr

[Peierls]

[236] G. Placzek to Rudolf Peierls

Copenhagen, 16.5.1938

Lieber Peierls,
Vielen Dank für Ihren Brief mitsamt dem Durchschlag,[935] der uns die Möglichkeit gab, Ihr Werk im Doppelexemplar zu studieren wie Weiland der [?]. In der Sache sind wir glaube ich, einig, wenigstens wissen wir zurzeit keine Einwände gegen Ihr Argument über die Gültigkeitsgrenze der gewöhnlichen Dispersionsformel. Hingegen scheint mir, dass Sie mit der Behauptung, daß hier die elastische Streuung mit der in meinem

[933]Letter [232].
[934]See letter [229].
[935]Letter [234].

letzten Brief angeführten Methode falsch heraus kommt, nicht recht zu haben. Denn 1.) die logarithmische Divergenz kann man sich doch wohl auf den Hut stecken. bzw. zur Potentialstreuung rechnen, ganz wie Sie das selbst in Ihrem Brief an Bohr so weise ausführen,* und 2.) die numerischen Faktoren können gerade wichtig werden? Es ist im kontinuierlichen Gebiet

$$\sigma_e = \frac{4\pi}{h^2}(2l+1)(\frac{\gamma_a}{\gamma})^2|\int_{-\infty}^{\infty}\frac{1}{\frac{E}{\gamma/2}-i}\frac{dE}{d}|^2$$

Das Integral ist $i\frac{\pi}{2}\frac{\gamma}{d}$ (was man z.B. auch ohne komplexe Intergation schon sehen kann, wenn man Zähler und Nenner mit $\frac{E}{\gamma/2}+i$ mulitipliziert) und man bekommt also $\sigma_{el^2} = \frac{4\pi}{k^2}(2l+1)(\frac{\frac{\pi}{2}\gamma_a}{d})^2$, genau wie aus statistischer Formel + opt. Argument als untere Grenze folgt. Genauer und mit streng ausgeführten Summen sehen Sie dies am beiliegenden Manuskript, das ich etwas langatmig und unter Weglassung der Ixereien für Bohr aufgeschrieben habe. Der Grund für die langatmige Ableitung der Formel für den Anregungsquerschnitt ist die Vorarbeit, um zu sehen wie der in Ihrer Formel heraus kommt. Wir sehen das zurzeit nicht gut. Wenn man die Ableitung kritiklos verallgemeinert, würde es scheinen, dass in Ihrer Formel nur die N_n dazukommen (die Phase der n ist für die Ableitung unwichtig) und somit jeder Term durch eine Grösse kleiner als 1 zu dividieren wäre, was etwas falsches ergeben würde. Natürlich muss man achtgeben, da die N_n selbst wieder durch die anderen Zustände dargestellt werden, aber statt lange zu reden, frage ich Sie, wie wollen Sie die statistische Anregungsformel aus Ihrer Formel ableiten?

Nun zu den praktischen Dingen. Bohr's Telegramm haben Sie wohl bekommen. Der Grund dazu ist, dass es uns schien, dass es vielleicht schön wäre, wenn wir die Arbeit jetzt alle zusammen fertig machen könnten. Infolgedessen hat Bohr an Weisskopf telegraphiert, ob sich amerikanischerseits alles umarrangieren lässt, wenn ich statt mit Aquitania (ab 25. Mai) mit Queen Mary (ab 1.Juni) komme. Je nach Weisskopf's Antwort ergibt sich folgende Lage.

1. zustimmend. Ich fahre Freitag, den 20. nach Prag, weil ich diese Verabredung nicht mehr rückgängig machen kann, und bin Sonn-

*und für unser Problem einfach von $-\infty$ bis $+\infty$ summieren.

tag, den 22. abends wieder in Kopenhagen, wo ich dann bis am 28. bleibe, worauf ich nach Paris fahre.

2. ablehnend. In diesem Falle muss ich von Prag Samstag abend nach Paris. Bohr würde sich in diesem Falle noch überlegen, ob er Sie bittet, Ihren Besuch zu verschieben, weil er gerade in der Woche auch mit Rosenfeld zu tun hat (was weniger machen würde, wenn Sie und ich zusammen arbeiten könnten im Fall 1). und würde Sie eventuell auch unter Umständen später in England besuchen. Ich würde Sie in dem Fall unter allen Umständen gerne noch sehen, das ginge auf 2 Wegen. Entweder Sie und Ihre Gemahlin haben Lust zu einer Spritztour nach Paris, wo wir bis Mittwoch zusammen sein könnten, oder, wenn Ihnen das nicht passt, kann ich von Prag Samstag abend in London aus den Lüften erscheinen und bis Sonntag abend bleiben, wobei wir, wenn es Ihnen passt, dort alle schwebenden Fragen erledigen können.

Im Falle 1. wird es wahrscheinlich zweckmässig sein, wenn Sie erst Samstag, den 21. abfahren und also Sonntag abend hier ankämen, weil Sie wohl sonst sich zu schnell entschliessen müssen, die Antwort von Weisskopf wird wohl kaum vor Mittwoch da sein. Wir telegraphieren Ihnen dann gleich.
Verzeihen Sie dieses pseudodramatische Durcheinander, aber das kennen Sie ja schon.
Ihr

G. Placzek

[237] Wolfgang Pauli to Rudolf Peierls

Zürich, 21.5.1938
(postcard)

Lieber Herr Peierls!
Vielen Dank für den Brief und Hoyles Manuskript.[936] Die Widerspruchsfreiheit des Falles Spin 3/2 bei nicht verschwindender Ruhemasse

[936]Letter [235].

beruht wesentlich darauf, daß die Vertauschungsrelationen nicht kanonisch sind. Dies ist auch in einer Theorie mit Nebenbedingungen von vorneherein zu erwarten. Herr Fierz wird in einigen Tagen seine ganze (übrigens ziemlich einfache) Rechnung für diesen Fall Ihnen schreiben[937] und Ihnen dann auch Hoyles Manuskript zurückschicken. Übrigens sind bei $m \neq 0$ alle Nebenbedingunen q-Zahlrelationen, d.h. mit allen Zahlen vertauschbar. Dabei ist es sehr vorteilhaft, die Methode von Jordan und mir zur Quantelung zu verwenden[938], da man dann der relativistischen Invarianz von vorneherein sicher ist.

Inzwischen studieren wir noch, ob sich der Fall $m = 0$ (Spin 3/2) in Ordnung bringen läßt (auch für Ausschließungsprinzip). Ich hoffe bald etwas Abschließendes darüber zu wissen. Dieser von Hoyle betrachtete Fall ist insofern singulär, als es (bei gegebener Schwingungs- und Wellenzahl) eine Wellenart gibt, deren Energie (und Impuls) verschwindet. Ich will mir also noch überlegen, was man vernünftigerweise mit deren Quantisierung anfangen kann.

Inzwischen beste Grüße an Sie und Hoyle
Ihr

W. Pauli

[238] W. Pauli to Rudolf Peierls

Zürich, 25.5.1938

Lieber Herr Peierls!

Die beiliegenden Rechnungen von Fierz betreffen erstens die Durchrechnung des Falles Spin 3/2 für $m \neq 0$, zweitens die Durchführung des (schon in einem früheren Brief[939] von mir erwähnten) Programms der Eichgruppe für $m = 0$ und beliebigen Spin — auch ganzen — verallge-

[937] This was published later as M. Fierz, 'Über die relativistische Theorie kräftefreier Teilchen mit beliebigem Spin', *Helv. Phys. Acta* **12**, 3–37 (1939).

[938] For details see P. Jordan and W. Pauli, 'Zur Quantenelektrodynamik ladungsfreier Felder', *Z. Phys.* **47**, 15 (1928)

[939] Letter [226].

meinern), drittens den Nachweis, daß die Totalenergie bei ganzem Spin positiv (bzw. für $m = 0$ höchstens 0 aber nie negativ) ist.[940]

Wenn man die Wellengleichung für $m = 0$ überhaupt physikalisch interpretieren will, scheint es mir unbedingt nötig, die Idee der Eichgruppe, die alle Felder mit Totalenergie (und Impuls) wegtransformierbar macht, einzuführen und nur eichinvariante Vertauschungs-Relationen aufzustellen.* Dann verschwinden alle Vorzeichenschwierigkeiten für Fermistatistik und halbzahligen Spin, auch wenn $m = 0$. Die Degeneration bei hohem Spin ist für $m = 0$ allerdings eine sehr hohe, indem von allen Eigenschwingungen mit gegebenem Faktor $e^{i(kx-\omega t)}$ (oder $e^{i(kx+\omega t)}$) nur je eine einzige von Null verschiedene Energie bekommt.

Für $m \neq 0$ besteht diese Komplikation nicht, es gelten dann auch für alle Größen f die Relationen $\dot{f} = i[\overline{H}, f]$. Die kanonische Form der Vertauschungs-Relation allerdings versagt, aber das ist bei einer Theorie mit Nebenbedingungen nicht anders zu erwarten.

Wie von der Hoyleschen Arbeit etwas gerettet werden kann, sehe ich leider nicht. Die letzten beiden Absätze seines Summary sind das Gegenteil der Wahrheit und auch im zweiten Absatz würde ich "complicated" statt "simplified" sagen. Im Falle [daß] Hoyle die Arbeit noch publizieren sollte, könnte man M. Abraham zitieren, der — in einer Entgegnung auf eine beim Münchener Mathematiker Lindemann gemachte Dissertation — festgestellt hat, daß "die Schuld am Mißlingen dieser Arbeit weniger dem Doktoranden als den unglückseligen Gestirnen zuzuschreiben sei, unter denen die Arbeit das Licht der Welt erblickt hat!"[941]

In diesem Sinne grüßt Sie bestens

Ihr

W. Pauli

*Diese genügen, um die Eigenwerte von Totalenergie- und Impuls festzulegen.

---

[940] M. Fierz and W. Pauli, 'On Relativistic Wave Equations for Particles of Arbitrary Spin in an Electromagnetic Field', *Proc. Roy. Soc.* **A173**, 211–32 (1939).

[941] Pauli had studied mathematics with Ferdinand Lindemann and the latter was one of his Ph.D. examiners.

[239] Rudolf Peierls to Wolfgang Pauli

Birminhgam, 3.6.1938
(carbon copy)

Lieber Herr Pauli!
Vielen Dank für Karte und Brief und für das Manuskript von Fierz.[942] Was zunächst den Fall des ganzzahligen Spins betrifft, so haben Sie offenbar recht, und Hoyles Resultat beruht auf einem Fehlschluß. Ich glaube jetzt auch, daß kein physikalischer Einwand gegen eine nicht positiv-definite Energiedichte besteht.

Im Falle halbzahligen Spins hat Hoyle (auf meine Veranlassung) nur die Möglichkeit betrachtet, daß die Vertauschungsrelation für Funktion der Raumkoordinaten enthält, dagegen nicht ihre Ableitungen. Unter dieser Voraussetzung ist seine Behauptung korrekt. Es scheint mir zunächst selbstverständlich, daß diese Voraussetzung unvermeidlich ist, aber das muß man natürlich beweisen. Es ist richtig, daß die Theorie von Fierz und Ihnen[943] widerspruchsfrei ist, aber damit ist noch nicht bewiesen, daß sie physikalisch möglich ist. Die Frage, die noch zu untersuchen bleibt, scheint mir die folgende: Die Vertauschungsrelationen in der Form, in der Fierz sie aufgeschrieben hat, setzen (analog zu denen von Ihnen und Jordan) voraus, daß die Wellenfunktionen der kräftefreien Wellengleichungen genügen, und schließen daher sowohl Potentialfelder wie Emission oder Absorption von Teilchen aus. Um eine Theorie einschließlich Wechselwirkungen zu haben, muß man Größen zu konstanter Zeit hinschreiben, und deren Vertauschungsrelationen aufstellen. Dabei muß man natürlich alle zeitlichen Ableitungen eliminieren.

Die Frage ist dann, ob sich die so aufgestellten Vertauschungsrelationen bei beliebigen Wechselwirkungen beibehalten lassen, und ob sie auch dann noch immer relativistisch invariant sind. Für Vertauschungsrelationen der kanonischen Form folgt das aus der allgemeinen Theorie von Ihnen und Heisenberg, dagegen muß es für den jetzt in Frage stehenden

[942] See letter [238].
[943] See letter [237], note 937.

Typ von Relationen neu bewiesen werden. Es ist natürlich verdächtig, daß Ihre Relationen die Masse explizit enthalten, denn die Masse ist ja im wesentlichen aus der Hamiltonfunktion definiert; es ist also zu vermuten, daß die Relationen nicht für beliebige Hamiltonfunktionen gelten.

Ich behaupte natürlich nicht, daß die Theorie nicht in Ordnung ist, sondern nur, daß es nicht bewiesen ist, daß sie in Ordnung ist. In jedem Fall haben natürlich Hoyle und ich nichts über diese Fragen bewiesen.

Ich sehe natürlich ein, daß Sie ungern meinen Übergriff in Ihr Spezialgebiet zulassen — es war sonst Ihr Privileg, jeden Ihrer Schüler (mindestens) je eine falsche Arbeit schreiben zu lassen!

Mit besten Grüßen Ihnen und Fierz
Ihr

[Peierls]

[240] Rudolf Peierls to Niels Bohr

Birmingham, 8.6.1938
(carbon copy)

Lieber Herr Bohr,
Ich war vor einigen Tagen in Cambridge und habe dort die Gerüchte über Landau näher verfolgt.[944] Dabei stellte sich heraus, dass von den zwei Quellen, aus denen diese Gerüchte entstanden sind, eine auf einem missverstandenen Satz in einem Brief beruht. Die andere macht einen zuverlässigeren Eindruck. In jedem Fall wird einer der Mitarbeiter in wenigen Tagen in Cambridge erwartet, und ich glaube, es ist daher das beste, vorläufig nichts zu unternehmen, sondern diesen Mann abzuwarten, der sicher genaue Auskunft geben kann.

[944] Lev Landau and his friend Moisey Korets had been arrested in April 1938 among others because of their authorship of an anti-Stalinist leaflet. See Gennady Gorelik, 'Lev Landau, Prosocialist Prisoner of the Soviet State', *Physics Today* (May 1995), 11–15.

Inzwischen ist es mir auch gelungen, die Behauptung zu beweisen, deren Beweis Placzek und mir am letzten Tag so misslungen war. Man kann wirklich in ziemlich einfacher Weise aus der Formel von Kapur und mir einsehen,[945] dass unter Weglassung der Potentialstreuung und der entfernten Niveaus) die gewöhnliche Dispersionsformel für den Gesamtquerschnitt und den elastischen Querschnitt im Mittel das richtige ergibt, sofern nur die Partialbreite die zu dem einfallenden Teilchen gehört, kleiner als der Niveauabstand ist. Ich vermute, dass die zuletzt genannte Voraussetzung auch noch überflüssig ist, aber das habe ich noch nicht bewiesen. Ich möchte den Beweis für diese Behauptung noch etwas vereinfachen, und werde ihn in einigen Tagen anständig aufschreiben und Ihnen schicken.

Was die Frage eines Zusammentreffens Anfang August betrifft, so kann ich mich jetzt ganz danach richten, was Ihnen bequemer ist. Wenn es Ihnen besser passt, dass ich nach Dänemark komme, so könnte ich etwa am 8. August (oder am 6. der 7. ist ein Sonntag) von England fortfahren, und wir könnten dann gemeinsam nach Cambridge fahren. Da die erste offizielle Sitzung der British Ass[ociation] am 17. abends ist, so nehme ich an, dass Sie dann am 15. oder 16. von Kopenhagen fahren würden.

Andererseits würde uns natürlich auch sehr freuen, wenn Sie schon etwas früher nach England kommen würden. In diesem Fall würde wahrscheinlich Birmingham die Möglichkeit geben, am ungestörtesten zu arbeiten. Birmingham ist natürlich keine sehr interessante Stadt, aber es hat doch eine sehr hübsche Umgebung (Warwick, Stratford-on-Avon &c) und wenn vielleicht Frau Bohr in diesem Falle Lust hätte, auch schon etwas früher nach England zu kommen, so würde es meiner Frau eine Freude sein, ihr diesen vielleicht nicht so gut bekannten Teil Englands zu zeigen.

Bitte sagen Sie aber ganz aufrichtig, welche dieser Möglichkeit[en] Ihnen besser passt; unsere Pläne sind jetzt soweit klar, dass ich sehen kann, dass beide Möglichkeiten sich ohne die geringste Schwierigkeit

---

[945] B.L. Kapur and R. Peierls, 'The dispersion formula for nuclear reactions', *Proc. Roy. Soc.* **A166**, 277–95 (1938).

einrichten lassen. Auf jeden Fall freue ich mich schon sehr auf die Gelegenheit, wieder mit Ihnen arbeiten zu dürfen.
Mir herzlichen Grüssen
Ihr

[Rudolf Peierls]

[241] Wolfgang Pauli to Rudolf Peierls

Zürich, 9.6.1938

Lieber Herr Peierls!
Zu Ihrem Brief vom 3. d[iesen Monats] möchte ich folgendes sagen:

1. Der Umstand, daß die Vertauschungs-Relation für $t = t'$ auch die räumlichen Ableitungen der δ-Funktion enthalten, gilt nicht nur für halbzahligen Spin, sondern ebenso für ganzzahligen — und zwar "unvermeidlich"! — (sofern $m \neq 0$). In dieser Hinsicht ist überhaupt kein Unterschied zwischen diesen beiden Fällen. Es ist sogar schon in der Procatheorie der Fall, wo

$$\frac{\partial A_\mu}{\partial x^\mu} = 0\,, \qquad \Box A_\mu = m^2 A_\mu$$

$$[A_\mu(x,t), A_\nu(x',t')] = (\delta_{\mu\nu} - \frac{1}{m^2}\frac{\partial^2}{\partial x_\mu \partial x_\nu}) \cdot D\,.$$

Für $t = t'$ erhalten dann die Vertauschungs-Relationen von A_0 und A_i $(i = 1,2,3)$ die erste räumliche Ableitung. Dieser Umstand ist auch schon in der Arbeiten von Kemmer, Bhabha etc.[946] enthalten und wird nur etwas verdeckt dadurch, daß man dort eben A_0 von der Gültigkeit der kanonischen Vertauschungs-Relationen ausschließt. Für höheren ganzen Spin kann man aber nicht

[946] N. Kemmer, 'Nature of the Nuclear Field', *Nature* **141**, 116 (1938); N. Kemmer, 'Quantum Theory of Einstein-Bose particles and nuclear interaction', *Proc. Roy. Soc.* **A166**, 127–53 (1938); H. Bhabha, 'Nuclear forces, heavy electrons and the β decay', *Nature* **141**, 117–18 (1938); H. Bhabha, 'On the theory of heavy electrons and nuclear forces', *Proc. Roy. Soc.* **A166**, 501–28 (1938).

mehr gewisse Komponenten von $A_{\mu\nu\cdots}$ abtrennen, für welche die kanonischen Vertauschungs-Relationen noch gelten, da durch die Spurbedingung $\sum_\nu A_{\nu\nu\rho\sigma\cdots} = 0$ die Komponenten mit Indizes 0 und die mit Indizes 1, 2, 3 verknüpft werden.

2. Sowohl für halb- wie für ganzzahligen Spin besteht in der Tat das in Ihrem Brief formulierte Problem der Verauschungs-Relationen bei Vorhandensein von äußeren elektromagnetischen Feldern und sonstigen Wechselwirkungen (z.B. mit Protonen). Ich verspreche mir sehr viel von der mehrzeitigen Methode zur Lösung dieser Frage (etwa so wie sie im Diracschen Buch[947] dargestellt ist) — wobei dann die betrachteten Teilchen vom Spin S, das elektromagnetische Feld und eventuell die Protonen oder Neutronen zunächst je eine Zeit für sich bekommen. Dies hat den Vorteil, daß man dann zunächst doch nicht $t = t'$ setzen muß und die relativistische Invarianz der Vertauschungs-Relationen trivial wird. Wenn man nachträglich $t = t'$ setzt und alle Zeiten identifiziert, wird man aber Zusatzglieder in den Vertauschungs-Relationen erhalten. Ich habe eigentlich keine Zweifel, daß es glatt gehen wird, aber weder ich noch Fierz (der jetzt seine schon stark angeschwollenen Rechnungen für seine Publikation zusammenschreibt[948]) haben Lust, weiter darüber zu rechnen. Das wäre ein ganz dankbares Feld für eine Arbeit von Hoyle, wenn er nun etwas rechnen will, was aller Voraussicht nach gehen muß. Es wird sehr ähnlich werden, wie sich das A_0 in der Yukontheorie mit Wechselwirkung verhält. (Die mehrzeitige Methode ist übrigens auch sehr bequem, wenn man in der Yukontheorie die Proton-Neutronkräfte ausrechnen will.)

Mit besten Wünschen für das Gedeihen von Hoyles weiteren Rechnungen und Grüßen an Sie selbst
Ihr

W. Pauli

[947]P.A.M. Dirac, *The Principles of Quantum Mechanics*, Oxford: Clarendon, 1930.
[948]M. Fierz, 'Über die relativistische Theorie kräftefreier Teilchen mit beliebigem Spin', *Helv. Phys. Acta* **12**, 3–37 (1939).

[242] Niels Bohr to Rudolf Peierls

Copenhagen, 27.6.1938

Dear Peierls,

I was of course very interested to learn from your kind letter of June 8 about your progress with the dispersion theory and look forward very much to hear how it was further developed.[949] Due to other obligations, not least a great deal of correspondence in connection with the ever more tragic developments in Germany, I am sorry, however, not yet to have been able to do anything serious with the preparation of our paper with Placzek on the nuclear problems. Nor in the next weeks, where I must complete a report for the Varschaw Congress, shall I be able to proceed with it. It will thus be impossible to complete our paper before the meeting of the B[ritish] A[ssociation] and it would therefore seem to me by far the best plan for our collaboration if you could come to Copenhagen after the meeting instead of before. I do not know what your duties are at that time, but I should of course be extremely pleased if you could manage to stay here for a longer time, so that we together could really get something satisfactory out of our collaboration so well started. I think that we could perhaps finish it in a few weeks, but it would of course be much better if you possibly could stay for a month and even participate in our annual conference which is planned to take place in the beginning of October.

Before the meeting of the B[ritish] A[ssociation] I would then, after finishing the Varschaw report, try to get a little holiday and then to concentrate on my report for the discussion at the B[ritish] A[ssociation] which shall of course be much more general and less technical than the account of our common work. If for the preparation of this last report I should feel that I need a conversation with you, I will try and come over to England a few days before the meeting to see you; but I think that I can manage without it, especially if you would kindly send me an account of what you think the outcome of all our discussions about dispersion problems has been. I should appreciate very much

[949]Letter [240].

indeed if you could make such an account, including the photo-effects, as detailed as possible, as it would serve at the same time as material for the preparation of our paper. There is no hurry about it, as I shall not need such an account before the end of July, but, as I already said, I shall of course welcome soon to learn how your work is proceeding and how the above proposals as regards our collaboration would suit you. I am also very eager to learn whether you have received further information about the fate of our common Russian friends.[950]
With kindest regards and best wishes for the summer, also to your family,
Yours,

Niels Bohr

[243] Rudolf Peierls to Niels Bohr

Birmingham, 28.6.1938
(carbon copy)

Lieber Herr Bohr,
Ich habe jetzt das Gerücht bestätigen können, dass Landau verhaftet ist. Das geschah vor ungefähr einem Monat, und seither hat man nichts mehr von ihm gehört.

Ich habe versprochen, nicht zu sagen, woher ich meine Information habe, und ich möchte Sie auch bitten, wenn Sie irgendwelche Schritte unternehmen wollen, nicht zu erwähnen, woher Sie die Nachricht bekommen haben. Vielleicht ist es am sichersten, um die Leute dort nicht zu kompromittieren, wenn Sie zunächsten einen Brief an Landau schreiben, in dem Sie ihm eine bestimmte Frage stellen, die er sofort beantworten würde, wenn er den Brief erhalten würde, und wenn nach einer gewis-

[950]Between 1936 and 1938, referred to in Russia as "1937" many Russian scientists (and non-Russians working in the Soviet Union) were arrested, some of whom had close links with their Western colleagues, e.g. Houtermans, Landau and Bronstein. See letter [240].

sen Zeit keine Antwort kommt, dann erst offizielle Anfragen abzusenden. Das hängt aber teilweise wohl auch davon ab, ob und in welcher Weise Sie sich für ihn einsetzen wollen oder können.
Mit herzlichen Grüssen
Ihr

[Rudolf Peierls]

[244] Rudolf Peierls to Niels Bohr

Morwenstow, Cornwall, 14.7.1938
(carbon copy)

Lieber Herr Bohr,
Beiliegend schicke ich eine Zusammenstellung sämtlicher Resultate, soweit sie mir wichtig erscheinen.[951] Dabei habe ich nur die wesentlichen Tatsachen aufgeschrieben, und alle mathematischen Ableitungen in einen Anhang verschoben. Dies soll nicht etwa bedeuten, dass ich diese Gruppierung für die spätere Publikation vorschlagen würde, sondern es ist nur der Uebersichtlichkeit halber, und auch weil sich dabei die existierenden Manuskripte fast ungeändert verwenden liessen. Den Anhang selbst werde ich in einigen Tagen schicken: ich wollte den Hauptteil, der wohl für die Zwecke Ihres Vortrages das wesentliche enthält, so wenig wie möglich verzögern. Dabei werden die verschiedenen Teile des Anh[angs], den ich zitiere, folgenden Inhalt haben:
Appendix I: Ableitung der optischen Relation und Diskussion
Appendix II: Auswertung der gewöhnlichen Dispersionsformel für äquidistante Niveaus
Appendix III: Wirkung eines Wellenpaketes
Appendix IV: Gültigkeit der optischen Relation in der verallgemeinerten Dispersionsformel (Kapur-Peierls)

[951] Bohr had asked Peierls for a summary of their joint work on the dispersion formula in preparation for his contribution to the British Association and their meeting in September 1938. See letter [242].

Appendix V: Aequivalenz der gewöhnlichen und verallgemeinerten Dispersionsformel für die gemittelte elastische Amplitude.
Mit Ausnahme von Teil V, der neu ist, entsprechen diese einfach den vorhandenen Manuskripten von Placzek und mir, II und III sind eigentlich jetzt mehr oder weniger überflüssig geworden und es wäre zu überlegen, ob man sie nicht fortan ganz fortlassen soll.

Inzwischen ist mein Programm dadurch etwas elastischer geworden, dass ich mich mit Bethe schon vor der British Ass[ociation] treffen kann. Ich kann daher meine Reise nach Frankreich von Bethe unabhängig machen. Immerhin werde ich wohl spätestens ca. am 13. September dort sein müssen.

Was Landau betrifft, so habe ich mir inzwischen überlegt, dass falls Sie noch nicht einen Brief der Art geschrieben haben, wie ich in meinem letzten Brief erwähnte, der Zweck vielleicht auf die folgende Weise am besten erreicht werden könnte: Sie könnten an Landau schreiben, dass Sie sich für Probleme interessierten, die Sie mit ihm in Moskau diskutieren wollen, und Sie hätten Lust, ihn zu diesem Zweck in Moskau zu besuchen. Dies hat den doppelten Vorteil, dass Sie auf diesen Brief plausiblerweise um sofortige telegrafische Antwort bitten können, und dass es ausserdem einen günstigen Eindruck machen kann. Wenn es nicht zu viel Mühe macht, so wäre vielleicht sogar empfehlenswert, gleichzeitig offiziell um ein Visum zu ersuchen, zu dem Zweck um mit Landau zu diskutieren. Ich weiss natürlich, dass Sie nicht in der Lage sein werden wirklich hinzufahren, aber Sie werden auf diesen Brief keine Antwort von Landau (und vermutlich auch kein Visum) bekommen, und wenn etwa Landau plötzlich doch wieder freigelassen werden sollte, so lässt sich ja leicht ein Grund finden, um die Reise aufzugeben.
Mit herzlichen Grüssen, auch an Frau Bohr,
Ihr

[Rudolf Peierls]

[245] G. Placzek to Rudolf Peierls

Peracaliente, 17.7.[19]38

Lieber Peierls,
Bis jetzt hab ich mich an Ihrem Brief geweidet,[952] jetzt wird es, sagt mir mein Gewissen, doch langsam Zeit zum Schreiben überzugehen. Also: Natürlich ist Ihr Beweis viel schöner als der verschollene, falls dieser jemals existiert haben sollte. (Dies trotz meiner geringen Liebe zur Störungsrechnung!) Es erhebt sich aber nun die Frage nach den Gültigkeitsgrenzen. Selbst wenn man annimmt, dass die Anwendung der Störungsrechnung auf den vorliegenden Fall solange berechtigt wäre, als nur die Partialbreite klein gegen den Abstand ist (was in Anbetracht der anderwärts auftretenden $(\frac{\gamma}{d})^2$ und $\frac{\gamma\gamma_a}{d^2}$ Terme, die sich, wenn ich mich recht erinnere, doch dort nicht durchwegs herausgemittelt haben, noch einer genaueren Untersuchung bedarf. Deswegen würde mich z.B. sehr interessieren, ob man Ihren Beweis auch auf die höheren Näherungen ausdehnen kann und wenn nicht, wie dort die eventuellen Abweichungen aussehen), so ist das praktisch noch nicht sehr viel, z.B. nicht anwendbar im Falle des Photoeffekts (da dort die Partialbreite der austretenden Neutronen vergleichbar mit dem Abstand ist). An geraden Tagen habe ich die Hoffnung, dass die Sache noch gültig ist — z.B. solange nur die Partialbreite des eintretenden Teilchens klein gegen den Abstand ist und die anderen beliebig — an ungeraden Tagen gebe ich die Hoffnung aber wieder auf. Uebrigens meinen Sie wohl sicher etwas ähnliches, ich weiss bloss nicht was, wenn Sie schreiben, dass Sie versuchen, die Sache ohne Voraussetzung ... (nicht ausgefüllt) ... zu beweisen. Jedenfalls sehe ich dem Ergebnis dieses Versuches mit Spannung entgegen und hoffe dass Sie mir bald darüber schreiben werden.

Ueber amerikanische Verhältnisse, Physik wird Ihnen wohl Bethe, der ungefähr gleichzeitig mit diesem Brief auftauchen dürfte, mündlich berichten. Ich bin mit den Weisskopfs bis jetzt ohne besonders dramatische Ereignisse bis hierher in die Wildnis von New Mexico, wo

[952]Letter could not be located.

sich Oppenheimers Ranch befindet, und eben haben Weisskopf und ich die Bekehrung O[ppenheimer]'s zur Wahrheit in Sachen Photoeffekt vollzogen. (Weisskopf hatte übrigens seine abessynische Methode schon lange so modifiziert, dass sie richtige Resultate lieferte, aber legt natürlich nach Studium Ihrer Arbeit auf dieselbe keinen Wert mehr).

Im übrigen fand ich unlängst ein Buch, das jede private amerikanische Reisebeschreibung mit Vorteil ersetzen kann und sogar fast den Besuch dieses Kontinents überflüssig macht: Adnoetschnaja Amerika (little golden America) von Ilf und Petroff, den Autoren von solotoi telonok.[953] Es sei Ihnen hiermit heiss empfohlen.

Haben Sie schon authentische Nachricht über Landaus Schicksal?[954] Bitte schreiben Sie mir c/o Bloch, Stanford University (Cal.)

Ich hoffe, Sie geniessen Ihre Ferien trotz Buch und Bohrschen B.A.-Vortrag und bin mit den herzlichsten Grüssen von und zu Ihr

G. Placzek

P.S. Ich habe mittlerweile vergessen, wie man ableitet, dass $\sum \frac{K_n^{(a)^2}}{N-n} = \sum k_a \frac{\partial W_n}{\partial h_n}$ ist. Können Sie mir das, alles andere als bestimmt voraussetzend schreiben?

[246] Wolfgang Pauli to Rudolf Peierls

Zürich, 18.7.1938

Lieber Herr Peierls!
Heute schreibe ich zunächst wegen einer organisatorischen Frage betreffend eine Unternehmung zu Sommerfelds 70. Geburtstag, der am 5. Dezember dieses Jahres stattfinden wird. In Deutschland will man

[953] Ilya Ilf and Yevgeny Petrov, a Russian team of writers and journalists, best known for their satirical novel *The Twelve Chairs* (1928) had written, in 1936, a book based on their transcontinental automobile trip through the United States.

[954] See letter [240].

ein Heft der Annalen der Physik mit einem Bild von Sommerfeld und Arbeiten seiner Schüler herausgeben. Der allgemeine Druck ist dort aber bereits so weit fortgeschritten, daß die Verleger es nicht mehr wagen, die Arbeiten nichtarischer Autoren zu drucken; überdies fürchtet man, daß dies erneute Zeitungsangriffe gegen Sommerfeld von Seiten der Starkgruppe zur Folge haben würde.[955]

Unter diesen Umständen haben Wentzel und Laporte mir vorgeschlagen, daß die außerhalb Deutschlands wohnenden Schüler Sommerfelds im Heft vom 1. Dezember des Physical Review zusammen einige Arbeiten publizieren sollen, worin auf Sommerfeld's Geburtstag und eventuell auf spezielle Arbeiten von ihm Bezug genommen werden soll (ähnlich wie im Planck-Heft der Physica).[956] Die Redaktion würde Laporte übernehmen (seine Adresse ist bis etwa 1. September c/o d'Hengelière, Würzburg, Eppstr. 87, Später: Ann Arbor (Mich.), the University of Michigan, Physics Dept. Manuskripte sind an ihn zu senden). Er hat sich wegen der Sache bereits mit der Redaktion des Physical Review in Verbindung gesetzt. Die Arbeiten sollen natürlich nicht zu lang sein, wie dies bei einer solchen Gelegenheit üblich ist. Wären Sie in der Lage, eventuell einen Beitrag dazu zu liefern? Es ist geplant, u.a. auch Houston, Eckart, Bethe und Rubinowicz dazu aufzufordern.[957]

Nun noch eine andere Frage ähnlicher Art. Heisenberg hat mir geschrieben und mir dann nochmals durch einen auf der Reise befindlichen Physiker sagen lassen, es würde seine Position in Deutschland sehr stärken (was sie offenbar recht nötig hat), wenn ausländische Physiker,

---

[955]Arnold Sommerfeld had retired from his Chair in Munich in 1935, and the subsequent controversy around his succession led to fierce attacks on Sommerfeld by representatives of the so-called "German Physics". National Socialist policy had resulted in stringent rules for publishers so that the official *Festschrift* was to contain only contributions from Aryan researchers. Therefore, his many pupils abroad initiated a special Sommerfeld edition of *Physical Review.* For details see Sommerfeld, *Briefwechsel*, II, pp. 360–77.

[956]Pauli, *Wissenschaftlicher Briefwechsel*, II, p. 588.

[957]*Physical Review* **54**, issue 11 (1938) contained among others papers by Wentzel, Heitler, Houston, Ewald, Eckart, Bethe, Rubinowicz, Pauling, Laporte, Brillouin, Peierls, Condon, Froehlich, London, Landé and Pauli. Peierls contributed a short paper 'On a Minimum Property of the Free Energy'.

besonders Experimentalphysiker, unpolitische aber sachlich energische Erwiderungen auf den Artikel von Stark in der Nature über "Pragmatism and Dogmatism" schreiben würden.[958] Eine solche Aufgabe ist natürlich weder dankbar noch interessant. Aber glauben Sie, daß Sie vielleicht Oliphant (oder einen anderen englischen Experimentalphysiker) dazu überreden könnten, eben um Heisenberg zu helfen?[959]

Was die Physik betrifft, so hat Fierz nun seine Arbeit über die relativistischen Wellengleichungen für Teilchen mit beliebigem Spin (halb- und ganzzahlig) und ihre Quantisierung bei Abwesenheit von Wechselwirkungen fertig gestellt. Bei Vorhandensein von Kräften ist es nicht die Quantisierung, die mir Sorge macht, aber ich habe gesehen, daß schon in der c-Zahltheorie Besonderheiten auftreten, da die Gleichungen nicht verträglich bleiben, wenn man einfach P_ν durch $P_\nu + \frac{e}{c}\phi_\nu$ ersetzen würde. Wir haben eben jetzt angefangen, diese Frage näher zu untersuchen und werden Ihnen berichten, wenn wir näheres darüber wissen.[960]

Mit besten Grüßen
Ihr

W. Pauli

[958] J. Stark had been asked by *Nature* to summarise his views and those of "German Physics" in an article which was published in April 1938. (J. Stark, 'The Pragmatic and the Dogmatic Spirit in Physics', *Nature* **141**, 770–72 (1938)) The same issue also contained an editorial 'The "Jewish" Spirit in Science', *ibid.*, 778. See also letter Heisenberg to Pauli, 23.5.1938, Pauli, *Wissenschaftlicher Briefwechsel*, II, p. 278.

[959] A.S. Eve wrote an article to that effect in November 1938. A.S. Eve, 'Foundations of Physics', *Nature* **141**, 857–62 (1938).

[960] See M. Fierz and W. Pauli, 'Über relativistische Feldgleichungen von Teilchen mit beliebigem Spin im elektromagnetischen Feld', *Helv. Phys. Acta* **12**, 297–300 (1939).

[247] Niels Bohr to Rudolf Peierls

Copenhagen, 19.7.1938

Dear Peierls.
Many thanks for your letter of 14 July with the general discussion of the dispersion-problems, which I have studied with great interest and pleasure.[961] I also look forward to receive the various appendixes, which will surely be an excellent material for our paper, and as regards its preparation I was of course very pleased to hear that you will be able to stay in Copenhagen somewhat longer.

Immediately after I received your former letter, I sent Landau an invitation to our conference in October, of which I enclose a copy for your information.[962] As I feared I have not yet received any answer, and I have been contemplating to write to the highest authorities in Russia in order to explain, how essential it is for the international scientific cooperation, that scientists are allowed to come to such conferences, and how most extraordinary Landau's scientific merits are. I should like, however, to know what you would think of the consequences of such a step.

From Mrs Houtermans I have received copies of the excellent letters, which Perrin and the Joliots have sent to Russia on behalf of Weissberg and Houtermans,[963] and I have been asked from various sides to join in this action. Just on account of Landau's case, which is very much on my heart, I am reluctant, however to take any step in the other cases which may not be necessary and which may make it more difficult to support him.

With kindest regards and best wishes,
Yours

Niels Bohr

---

[961]Letter [244].

[962]*Peierls Papers*, Ms. Eng. misc. b203, C. 30.

[963]Irène Curie, Frédéric Joliot and Jean Perrin had attempted to secure Friedrich Houtermans' and Alexander Weissberg's release by sending telegrams to Stalin and his public prosecutor Vishinsky in June 1938. See http://mason.gmu.edu/~ikatchal/Krawtchouk.html.

[248] Rudolf Peierls to Niels Bohr

Morwenstow, North Cornwall, 21.7.1938
(carbon copy)

Lieber Herr Bohr!
In diesem Brief schicke ich leider immer noch nicht die fehlenden Rechnungen, sondern muss stattdessen eine meiner Behauptungen zurücknehmen. Es tut mir leid, dass ich in dieser Weise — wie Pauli sagen würde — "Verwirrung stifte" aber ich bin in eine sehr einfache Falle gefallen.

Es handelt sich um die Behauptung, dass die Formeln für den Mittelwert der elastischen Amplitude (die für den Totalquerschnitt massgebend ist) sich nach der strengen und der gewöhnlichen Dispersionsformel übereinstimmend ergeben. Dies ist nicht richtig, jedenfalls nicht in der Allgemeinheit, in der ich es behauptet habe. Statt dessen ist es noch immer richtig, dass der Mittelwert der elastischen Amplitude (und also auch des Totalquerschnitts) unabhängig von der Gesamtbreite der Compoundzustände ist. Mit anderen Worten, wenn man an der Beschaffenheit der Kernoberfläche solche Aenderungen anbringt, dass die Wahrscheinlichkeit aller Zerfallsprozesse reduziert wird, mit Ausnahme der Re-emission des einfachen Teilchens, so ändert sich der mittlere Totalquerschnitt nicht. Also z.B. im Fall des Photoeffekts kann man sich denken, dass man künstlich das Entweichen von Neutronen verhindert, und dabei bleibt der Gesamtquerschnitt unverändert (natürlich gibt es dann keinen Photoeffekt, sondern nur Streuung und Fluoreszenz). Insbesondere ist dann auch der mittlere Gesamtquerschnitt unabhängig davon, ob man sich im Gebiet getrennter oder sich überdeckender Niveaus befindet. Dieses Resultat (dessen Richtigkeit bestehen bleibt) drückt wohl etwas sehr ähnliches aus wie das, was Sie mit dem Argument über kurzfristige Beleuchtung meinten. (Es ist mir allerdings noch nicht gelungen, dieses Resultat in zwangsläufiger Weise unter Benutzung einer solchen Betrachtung mit kurzfristigen Wellenpaketen abzuleiten.)

Auf den ersten Blick sieht es nun so aus, als ob dieses Resultat bereits die Gültigkeit der gewöhnlichen Dispersionsformel einschliesst, denn in dem Fall getrennter Niveaus gilt ja die gewöhnliche Formel, und wir wissen ja aus diesem Resultat, dass der Totalquerschnitt unabhängig davon ist, ob die Niveaus getrennt sind oder nicht. Das ist aber nicht ganz so aus folgendem Grunde:

Wenn die gewöhnliche Dispersionsformel in dem Sinne richtig wäre, wie ich das in dem Manuskript dargestellt habe, so müsste der mittlere Totalquerschnitt durch die Formel

[···][964]

dargestellt werden. Im Falle getrennter Niveaus ist dies natürlich richtig, wenn man die Niveaubreite vergrössert, ohne die Randbedingungen für das entweichende Teilchen a zu ändern, so wird die Formel falsch, und zwar nicht weil sich der Totalquerschnitt ändert, sondern weil sich [···][965] verändert. Die Komplikation entsteht also dadurch, dass die Partialbreite der Compoundzustände z.B. beim Photoeffekt, selbst wenn man den Mechanismus der Ausstrahlung nicht ändert, doch nicht davon abhängt, wie gross die Neutronenbreite ist.

Dies ist an sich nicht sehr beunruhigend, denn die Partialbreite ist gar keine sehr anschauliche Grösse, sofern man sich im Gebiete überdeckender Niveaus befindet. Um sie zu beobachten, muss man den Kern in einen bestimmten Compoundzustand bringen, d.h. in einen Zustand, in dem seine Wellenfunktion eine rein exponentielle Zeitabhängigkeit hat. Das einzig Unangenehme ist, dass nun die Beziehung zu dem statistischen Argument, die mir in sehr befriedigender Weise geklärt schien, jetzt wieder unklar wird, denn in das statistische Argument geht ja wirklich die Partialbreite ein. Der Unterschied zwischen der Aussage der statistischen Formel und der Formel für den Totalquerschnitt muss dann doch wieder darin gemacht werden, dass das statistische Argument den Eindringungsquerschnitt bestimmt, der nicht notwendig mit dem Totalquerschnitt identisch ist.

Was alle praktischen Fragen über Grössenordnungen betrifft, so macht die neue Komplikation nichts aus. Die Terme, die die

[964] Missing in manuscript.
[965] Missing in manuscript.

Abhängigkeit der Partialbreite von der Neutronenbreite darstellen, sind Summen aus sehr vielen Termen mit verschiedenen Vorzeichen, die sich im allgemeinen herausmitteln werden, oder jedenfalls nur kleine Beträge geben. Ich hoffe noch immer beweisen zu können, dass sie sich unter gewissen plausiblen Voraussetzungen wegheben, aber das muss noch gründlicher untersucht werden. Jedenfalls verschwinden sie nicht in der Allgemeinheit, wie mir das schien, und in den Fällen in denen sie nicht verschwinden, muss man natürlich noch einsehen, was ihre physikalische Bedeutung ist.

Ich hoffe sehr, Sie werden mir über die Konfusion nicht böse sein, mir schien die falsche Behauptung sehr plausibel sobald ich das Resultat über die Unabhängigkeit des Totalquerschnitts von der Gesamtbreite gefunden hatte, und ein Rechenfehler half mir dann auch, sie zu beweisen.

Ich hatte zuerst die Absicht, Ihnen zu telegrafieren, dass ich einen Fehler gemacht haben aber ich habe es nicht getan, da das wohl die Konfusion nur erhöhen würde, und besonders weil ich glaube, dass der Irrtum für diejenigen Dinge, auf die es Ihnen für Ihren Cambridger Vortrag ankommt, noch keine grosse Rolle spielt.[966]

Ich werde wieder schreiben, sobald ich mir die neu entstandenen Fragen der physikalischen Bedeutung dieser Abweichungen, und die Beziehung zu der statistischen Formel etwas genauer überlegt habe.
Mit freundlichen Grüssen
Ihr

[Rudolf Peierls]

P.S. Eben kommt Ihr Brief vom 19.VII.[967] Mir scheint, dass Ihr Brief an Landau durchaus das richtige war, und ich glaube auch, dass Sie ganz recht damit haben, nunmehr direkt an die höchsten Instanzen zu schreiben. Ich würde allerdings dies nicht in der Form eines Antrags um Ausreisebewilligung tun, ich glaube nicht, dass das sehr viel helfen

[966] Bohr was preparing a talk for the Annual Meeting of the British Association for the Advancement of Science (BAAS) in Cambridge in September 1939.
[967] Letter [247].

würde. Jetzt nachdem seit Ihrem ersten Brief schon beträchtliche Zeit vergangen ist, würde es mir unbedenklich erscheinen, etwa in dem folgenden Sinne zu schreiben: Sie wären in wissenschaftlichem Kontakt mit Landau gewesen; nun hätte er plötzlich auf einen dringenden Brief nicht geantwortet, und man hätte Ihnen gesagt, der Grund hierfür sei, dass er verhaftet worden ist. Dies gibt Gelegenheit, nicht nur auf Landaus wissenschaftliche Bedeutung und internationalen Ruf hinzuweisen, sondern auch auf seine Bedeutung für die russische theoretische Physik, in der Ausbildung einer guten Schule, u.s.w. Ich weiss natürlich nicht, wieviel Aussicht auf Erfolg ein solcher Schritt haben würde, aber ich glaube, es ist das einzige, was man tun kann, und es kann jedenfalls nichts schaden, weder Landau noch irgendjemand anders.

Es empfiehlt sich wahrscheinlich, zu erwähnen, dass Landau im Institut für physikalische Probleme der Akademie gearbeitet hat, denn sein Name ist sehr häufig, und sonst ist es vielleicht schwierig, ihn zu finden.

Ich glaube auch, dass Sie recht haben, Ihre Hilfe auf die Fälle zu beschränken, in denen sie am wertvollsten ist.

Mit herzlichen Grüssen
Ihr

[Rudolf Peierls]

[249] Hans Bethe to Rudolf Peierls

location unspecified, 22.8.38

Liebe Peierlses,
Ich habe also Eure Eltern (bzw. Schwiegereltern) gesehen. Nicht sehr erfreulich (wie überhaupt die Reise nach Deutschland). Besonders die Mutter ist furchtbar nervös, der Vater nimmt es sehr viel ruhiger. Aber er ist noch nicht entschlossen und findet den Gedanken schrecklich, seinen Kindern "zur Last zu fallen" (oder seinem Bruder). Dabei müssen die Zustände in Berlin noch unerträglicher sein als wir draussen wissen. Die Bitte, Ihr solltet keine Briefe schreiben, ging z.B. zurück auf die persönliche Warnung ob die berechtigt war, weiss ich nicht; es scheint

mir, dass einer den anderen verrückt macht in Berlin. Inzwischen sei das nicht mehr aktuell und Ihr könntet ruhig wieder Briefe schreiben.

Dann gabs grosse Aufregung über eine Büchersendung an Euch. Bücher über 20M Wert dürfen nicht mehr ins Ausland geschenkt werden; rechts und links tuns zwar die Nachbarn noch, aber bei Eurem Papa hat sich die Golddiskontbank darüber aufgeregt. Er nahm sich darauf einen Anwalt (!) und schließlich, nach 4 Wochen, gab sich die Golddiskontbank schliesslich zufrieden. Aber immerhin wars eine grosse Aufregung für ihn.

Dann gibts so hübsche Sachen wie Verbot, den Bank-Safe zu öffnen ausser in Anwesenheit eines Beamten der Zollfahndungsstelle; besondere Genehmigung für jede Sendung ins Ausland, die meist mehrere Monate dauert und dann doch abgelehnt wird; und neuerdings Vorschriften der Steuerbehörde über die Höhe der zulässigen Abhebungen vom Bankkonto (Dies allerdings erst ganz vereinzelt, aber so fängts immer an). Hingehen kann man in Deutschland fast nirgens mehr, denn überall beklagen sich die Hotelgäste, wenn man überhaupt aufgenommen wird. Solche Dinge wie das Waldpark-Sanatorium werden (in diesem Fall: sind) verkauft. Also, es wird so systematisch betrieben, dass man seine Freude daran haben könnte, wenn es einen nichts anginge.

Nun sollte man meinen, dass die Entscheidung über Dableiben oder Fortgehen niemand mehr schwer fallen könnte. Und trotzdem gibts da noch immer Zweifel. Angeblich nicht der Verzicht auf Luxus und Wohlleben, sondern nur der Gedanke, Almosen zu empfangen. Ist auch schwer für jemand, der immer anderen gegeben hat, aber schliesslich ... Und dann hat Dein Vater die Idee, er dürfte das nicht tun, weil er dadurch seiner arischen Frau die schwerverdiente und grosse Pension nimmt. Sie ist übrigens sehr viel mehr entschlossen als er, macht ihn aber durch ihre Nervosität ganz verrückt.

Ich hab versucht, was ich konnte, aber genützt hats glaube ich nicht. Sie sehen nicht ein, dass heutzutage alle normalen Gesichtspunkte bedeutugslos geworden sind. Inzwischen ist also das Affidavit von Amerika gekommen und das hat vielleicht eine gewisse Überzeugungskraft. Aber es warten schon Tausende auf ihr Visum, und die Quoten für 1938 und 39 (!) sind schon vergeben, ausser einem kleinen Rest der für dringende Fälle aufbewahrt wird. Das wissen übrigens Deine Eltern noch nicht. Aber schon die Idee, bis nach Neujahr warten zu müssen, erschien unerträglich. Konsequent oder inkonsequent — na ja. Und unter diesen Umständen dachten sie daran, erst nach England zu gehen und dann später nach Amerika. Ich finde das keine sehr glückliche Idee; alle Komplikationen kommen zweimal, das amerikanische Visum wird von England aus schwieriger zu kriegen sein als von Berlin; der grössere Umzug England–Amerika muss in Pfund bezahlt werden usw. Ich sehe auch nicht ein, warum die plötzliche Ungeduld, nachdem man noch nicht einmal fest entschlossen ist.

Ich habe aber jedenfalls den Auftrag, Euch zu bitten, auch Eurerseits ein Affidavit zu schicken für alle Fälle. Dies soll nur eine Formalität sein und Euch finanziell zu nichts verpflichten — das besorgen die Amerikaner. Es soll auf Universitätsbriefbogen geschrieben sein und sagen, dass Ihr gerne hättet, wenn Eure Eltern nach England kämen, und dass Ihr willens und finanziell imstande seid, sie dort zu ernähren. Auf Englisch natürlich. Im übrigen war man (d.h. nur mütterlicherseits) darüber böse, dass Ihr nie mehr zugeredet hättet, nach England zu kommen (nach dem einen Versuch vor vier Jahren) und man war erstaunt zu hören, dass Ihr sehr für Auswanderung seid usw. Ich hoffe, ich habe getreulich alles ausgerichtet.

Andererseits ist meine Mutter entschlossen, nach Amerika zu gehen. “Ich lasse ja hier nichts zurück”, meinte sie. Und ein Zimmer in meiner Nähe in Amerika würde sie glücklicher machen als ein Haus in Baden-Baden. Wir waren eigentlich beide voneinander überrascht: Ich dachte, ich müsste sie überreden, und sie dachte, ich wäre gegen eine Übersiedlung. Für Mütter sind Visa leichter zu bekommen, wir denken also an einen Umzug nächstes Frühjahr. Mein Vater war sehr überrascht von dem Entschluss und hatte Bedenken, aber meine Stiefmutter möchte am liebsten selbst. Jetzt sind wir also mit Pass in der Schweiz, und es ist wie eine Offenbarung für meine Mutter. Es regent zwar in Strömen, aber Adresse: Seelisberg (Uri), Hotel Sonnenberg. Wir werden bis zu meiner Abreise hier bleiben, und die ist am 16. September mit der Britannic ab Le Havre. Schwierigkeiten habe ich zwar weder beim Rein- noch beim Rausfahren gehabt, aber man sollte nicht übermütig werden.

Meine Geschwister sind sehr gross geworden, und sehr ungezogen. Etwas von Eurer Erziehung täte ihnen gut. Meine Schwester ist lange nicht so gescheit wie Gaby, spricht keine so vollendeten Sätze und kann auch weder rechnen noch schreiben. Dafür kann sie, im Sinne des heutigen Reiches, besser rennen als alle ihre Freunde und besser schwimmen als ihr grosser Bruder (wozu nicht viel gehört). Sie liebt mich sehr und ist sehr zärtlich. Klaus liebt mich auch sehr, aber seine Zärtlichkeit äussert sich hauptsächlich in unvermuteten Rippenstössen.

Seit vorgestern, d.h. seit wir hier sind, denke ich wieder an Physik. Ich dachte, ich könnte beweisen, dass die Unendlichkeiten in der Quantenelektrodynamik nicht eine Folge unzweckmässiger Integration sind, sondern in den Grundformeln drinstecken. Mit Hilfe eines Variationsprinzips. Leider geht das nur für einen tiefsten Eigenwert eines Teilchens. Bei der Nichtlöchertheorie ist aber $E = mc^2$ nicht der tiefste Eigenwert, und bei der Löchertheorie muss man die Differenz zwischen zwei Eigenwerten (nur der "See" und See + Elektron) nehmen, wofür Variationsprinzipe im allgemeinen nicht gelten.

Dank für die Woche in Cornwall. Ich wünschte ich könnte jetzt nochmal hin, denn ich glaube, ich könnte jetzt die quaint old villages and cute ancient churches wieder noch mehr geniessen. Und auch das surfboard riding vielleicht würde ich das schliesslich doch lernen. Aber vor allem Dank für die Aussprache, physikalischer und insbesondere persönlicher Natur. Es hat mir sehr gut getan, und ich wünschte, es wäre weniger weit von Ithaca nach Birmingham.

Natürlich habe ich Eure Adresse doch verloren, ich weiss zwar noch die Nummer 38, aber nicht mehr die Strasse (Calthorpe?). Aber jetzt seid Ihr sowieso beim britischen Esel. Grüsst ihn von mir; kann er mehr als 9 stones 3 tragen?

Schönste Grüsse

Hans

[250] Rudolf Peierls to Hans Bethe

Birmingham, 24.8.[19]38

Lieber Bethe,
Ich komme eben von der Tagung in Cambridge zurück,[968] die sehr interessant war. Insbesondere gab Blackett einen sehr guten Bericht über die Höhenstrahlung, wonach vieles sehr schön in Ordnung zu kommen scheint. Die Versuche von Auger[969] und anderen, wonach die schrägen Strahlen anscheinend rascher als die senkrechten absorbiert werden, (wenn man Höhenabhängigkeit und Winkelabhängigkeit der Koinzidenzien vergleicht) klären sich dadurch auf, dass das Yukon unstabil ist, und daher die Intensität einfach eine Funktion der Zeit ist. Das stimmt alles auch quantitativ. Die Theorie ist von Heisenberg und Euler.[970]

Ich erzählte Oliphant und Cockroft, was Du über das Cyclotron und die Möglichkeit von Fokussierung durch Abweichungen von der Zylindersymmetrie erzählt hast, und sie waren beide sehr interessiert, und wollten gerne wissen, wieweit die verallgemeinerten Rechnungen schon durchgeführt sind, und wie hoch danach die Grenze (im Prinzip) liegen soll. Kannst Du mir das noch gelegentlich schreiben?

Eine weitere Neuigkeit ist, dass Jones, der mein Nachfolger im Mond-Laboratorium war, ein Readership im Imperial College bekommen hat, und Cockroft daher wieder jemand sucht. Ich erwähnte Placzek und das schien ihn sehr zu begeistern, es wurde daher beschlossen an ihn zu telegraphieren, ob er noch frei ist, da ich nicht wusste, wie sich seine Verhandlungen mit Cornell inzwischen entwickelt haben. Ich wollte Dir dies jedenfalls zur Kenntnis geben. Wenn er sich bezüglich Cornell noch nicht definitiv geäussert hat, und wenn dieser neue Vorschlag seine Stellungnahme zu Cornell irgendwie beeinflusst, so wird er sich wohl gleich mit Dir direkt in Verbindung setzen. Ich werde Dich aber jedenfalls

[968] Meeting of the British Association for the Advancement of Science.

[969] Pièrre Auger (1899–1993), considered as the discoverer, in 1938, of giant air showers generated by the interaction of very high energy cosmic rays with the earth's atmosphere. See P. Auger and Thérèse Grivet, 'Analysis of deep rays', *Phys. Mod. Rev.* **11**, 323–54 (1939).

[970] W. Heisenberg and H. Euler, *Z. Phys.* **98**, 714 (1936).

über die weitere Entwicklung in Cambridge auf dem Laufenden halten. Schreib mir doch bitte auch wann, und mit welchem Schiff Du fährst. Mit herzlichen Grüssen von uns allen, auch an Deine Mutter,
Dein

[Rudi]

[251] Else Peierls to Rudolf Peierls

Zürich vom 25.8.[19]38[971]
(carbon copy of letter sent to all three children)

Meine lieben Kinder alle miteinander. Gelt da staunt Ihr! Ich sitze hier mit Blick auf den herrlichen See und habe fuer zwei Tage die Geschwister besucht, die hier wundervoll und gemuetlich aufgehoben sind. Vati ist unter guter Obhut in Baden-Baden, so konnte ich fort. Vor allem muss ich Euch sagen, wie wir immer wieder Eure warme Fuersorge empfinden. Jetzt endlich ist Vati so weit dass er heraus will. Die Sorge, von seinem Bruder abzuhängen, der so viel fuer Euch tut und tun wird, die Abhaengigkeit und das Durchhalten der Vorbereitungen zur Auswanderung, die enorm und seelisch schwer zu ertragende sind, der Glaube an die Pensionsansprueche, das Zureden mancher Leute — kurzum tausend Dinge erschwerten ihm den Entschluss. Dazu ist leider zu sagen, dass er alt und muede und scheu geworden ist, was ja kein Wunder ist. Wir haben keine Hoffnung aufzubauen wie Ihr, sondern muessen annehmen und uns fuegen, und haben dabei die Sorgen ueber Euer Gelingen und Alfreds Existenz! Der einzig Sichere ist Rudi, und der liess uns nun durch einen Freund sagen,[972] wir muessten heraus und zu ihm kommen. Nun habe ich mit Adolf[973] alles durchgesprochen.

---

[971]This letter was written from Zurich, where Else Peierls visited relatives. Heinrich Peierls could no longer travel outside Germany at this stage. This letter was not affected by German censorship and contained more explicit references to personal details and conditions in Germany.

[972]Hans Bethe had visited Else and Heinrich Peierls when they were in Baden-Baden, where the former had visited his mother. See letter [249].

[973]Adolf Brieger, Heinrich Peierls' brother-in-law.

Das amerikanische Affidavit haben wir in Baden-Baden erhalten und auf Rat eines Attachees, mit dem Vater tel[efonierte] eingeschrieben eingesandt und noch keine Antwort. Wir wissen also nicht einmal welche Wartenummer wir bekommen, und es heisst jetzt schon, es dauert bis Maerz, bis alles aufgearbeitet ist. Wie soll man da durchhalten? Siegfried riet, mit Fuldas und Baerwald zu beraten. Letzterer Bruder arbeitet in Evian mit, und er selbst war jetzt in U.S.A. Wir werden am 9. nach Hause reisen, und Fuldas kommen wohl am 19. Sept[ember] und dann wird alles erledigt. Haben wir erst Vatis Pass eingereicht, koennen wir auch ueber die Auswanderung schreiben. Man bekommt ihn nur zur Auswanderung und nur auf beschraenkte Zeit, und kann Pech haben, nicht fertig zu werden. Ueber das Auslandsgeld koennen wir schon seit zwei Jahren nicht mehr verfuegen und muessen auch die Summe, die Ihr damals legal erhieltet nochmals dalassen. Seit Monaten kaempft Vati mit Eingaben, Transfer, es ist nichts bewilligt, obwohl halb zugesagt war.

Wir haengen an keinem Vermoegen oder Besitz. Am liebsten 1 oder 2 Zimmer und Ruhe. Ich koennte alles alleine machen. In New York sich im Luxus gedrueckt fuehlen, dann kaeme nur Montclair in Frage, aber Euch wollen wir nicht zur Last fallen. Glaubst Du, ich koennte irgendwo in Eurer Naehe 2 Zimmer finden und Kuechenbenutzung? Von Moebeln, wenn man sie uns noch laesst, koennte ich dann nur 2 Betten, 1 Couch, 1 runder Salontisch, Stuehle, 1 Schrank, 1 Commode, Teppiche, Silber und Waesche mitnehmen. Wir haben alles natuerlich aufnehmen lassen. Lasst es Euch mal durch den Kopf gehen und schreibt dann einfach "Freunde" von uns kommen jetzt herueber, wir koennen ihnen behilflich sein, 2 Zimmer Bad und Kuechenbenutzung zu verschaffen. Sie bringen auf unseren Rat nur das und das mit. Fluchtartig kann man nicht weg. Jede Stunde aendert sich alles, und man erschrickt den ganzen Tag ueber Briefe, Zeitungen und Bestimmungen. Da ich fuerchte, es dauert zu lange, bitte ich Alfred und Rudi uns eine Einladung fuer dauernd zu senden, so dass wir ev[entuell] via England fortkoennen. Die Auswanderung und der Pass muessen immer via Deutschland und fuer ein bestimmtes Land lauten. Wir konsultierten allerlei Anwaelte. Es ist sowohl fuer England sowie fuer Frankreich-Nizza ebenso schwer. Euer vorgeschlagener Anwalt am Pariser Platz besprach vor

2 Monaten mit uns alles 3/4 St[unde] lang, wusste keinen Rat, sagte nur, es kommen schwere Monate fuer Sie und als Vati bezahlen wollte, sagte er fuer so eine Beratung kann ich nichts nehmen!! Das deuteten wir Euch brieflich an. Was und wieviel Vati als Notpfenning retten kann, ist nicht zu sagen, vielleicht nichts. Nur ist Adolf der Meinung, dass Vati mit mir lieber nach Nizza, wohin sie auch hoffen zu kommen, auswandern soll. Ihr werdet das verstehen aus folgenden Gruenden: das Klima in New York ist schwer zu ertragen fuer aeltere Leute, die Angst vor der Sprache bedrueckt Vati, obwohl so viele Deutsche dort sind, und er will am liebsten jetzt still und bescheiden in der Natur leben. In Nizza ist es viel billiger, so koennte uns der vielbeanspruchte guetige Siegfried eine viel kleinere Summe ueberweisen, und wir koennten behaglich leben. Bekannte findet man und Ansprache, und ich muesste mein Franzoesich fest wiederholen. Vati hat auch grosse Angst vor Krankheit und Operationen etc. kurzum er quaelt sich und mich dauernd. Wenn Euch die Idee einleuchtet, schreibt nur: Es waere schoen wenn Ihr auf aerztl[iches] Attest nach dem Sueden koenntet. (nicht Nizza nennen). Ob es moeglich ist, werde ich im Sept[ember] in Berlin zu erreichen suchen, da auch die Visa fuer franzoesische Plaetze gesperrt sind.

Was alles das fuer mich bedeutet, dass ich meinen geliebten Mann nicht retten konnte, obwohl ich darueber krank geworden bin, kann ich Euch auch mit Traenen nicht ausdruecken. Da ich mich ganz zu den Juden bekannt habe und nicht mehr zurueck will, und das Meer der Leiden und Tragoedien seit Jahren schwer erleide und miterlebe, lasse ich meine arischen Angehoerigen in Verzweiflung zurueck. Was soll aus Lala und Mann werden und meinem Herzensliebling Walter, der sehr an mir haengt, und fuer den ich als Vermaechtnis meiner geliebten Deli-Schwester immer zu sorgen hoffte, und den ich beschuetzen muss — und werde muessen!? Wie soll ich Geld verdienen, um nicht auf Eure Guete auch rechnen zu muessen und Euch zu beanspruchen?

Frau von Buelow besuchte ich, ehe ich abreiste, in der Angst, sie vielleicht nicht mehr wiederzusehen, und sie wollte nur ueber mich und meinen Mann und unser Schicksal wissen. Sie war so unendlich lieb, guetig, warm und verstehend, und sagte mir: “Mir wertvolle, liebe Menschen wie Sie zu verlieren, ist schrecklich, aber Sie muessen fort, auch wenn Sie von einer Sackgasse in die andere laufen und schwere Abhaengigkeit auf sich nehmen muessen. Das Leid von meinen Freunden ist so gross, dass es schon keine Traenen mehr gibt." Nun schreibt sie mir immer die ruehrendsten Briefe, weil ihr unser Gespraech und Schicksal nicht aus dem Kopf geht. Schweren Herzens schreibe ich Euch alles und will Euch keine Sorgen machen, aber einmal muss es doch ausgesprochen werden. Auf Vatis Wunsch tat ich es also und bitte Euch, den Brief abzutippen und an Rudi zu senden, der ihn Alfred weitergeben kann.

Was Alfred betrifft, hoffen wir, dass er sich halten kann. In U.S.A. ist der Kampf zu schwer, und neue Mittel kann niemand geben, da die Auswanderung der Aelteren erforderlich wird. Rudi, der am gesichertsten und hoffentlich bald Englaender ist, muesste dann Alfred ev[entuell] beistehen. So koennten sich beide ueber Wasser halten. Alfred hat meinen Wink nicht verstanden, als ich schrieb, er solle Hansi Fischer anrufen. Hansi hat sich famos benommen und aus eigener Initiative seine Eltern aufgefordert, hinzukommen, obwohl sie hoechstens bei ihm essen und wohnen koennen. Nun werden amerikanische Verwandte Geld dazu geben, und so werden sie mit ihren Eltern leben koennen. Die Sache laeuft, und Fischers werden es wohl noch am ehesten erreichen. H[ansi] F[ischer] weiss immer Bescheid und koennte Alfred erzaehlen. Ausserdem ist Ruth nun eingewandert. Wir bedauern es unendlich, Euch keine Geschenke, nicht mal Buecher senden zu koennen. Unverdienter, unschuldigerweise hatten wir wochenlang die groessten Aufregungen, die jetzt endlich behoben sind.[974] Der Brief ist etwas konfus. Ich werde oefters unterbrochen und bin etwas aufgeregt. Halb Meran ist hier. Keine Ruhe! Schreibt uns durch die Blume, ob Ihr einen Besuch bei Euch oder im Sueden, d.h. Auswanderung U.S.A. oder Nizza fuer besser haltet, und wir werden ja auch bald zur Entscheidung kommen, und dann offen schreiben. Ich sehe Eure Enttaeuschung, wenn wir nicht nach drueben kommen. Vorlaeufig ist alles gesperrt, und wir muessen sehen ev[entuell] einen Preferenzschein Petitionsform 633 Department of Labour von drueben

[974] There had been some legal complications with a book sent by Heinrich Peierls to England. See letter [249].

zu bekommen. Dies riet mir eben Adolf. Jedenfalls ist unser Amerikaaffidavit eingereicht, und wir hoffen, Antwort zu bekommen. Vati fuerchtet auch sein Alter und die Arztkontrolle. Jedenfalls verstehen wir nicht, dass sowohl U.S.A. wie England keine Hilfe schaffen, dass die vorhandenen Affid[avite] und Einreichungen schneller aufgearbeitet werden und Erleichterungen schaffen von den Konsulaten aus, da sie doch menschliches Verstehen fuer die Auswanderer haben?! Nun kommt die ganze Aerzteschaft ans Auswandern. Vatis Darmarzt geht nach London, wo ihn sein Bruder total unterstuetzt, ihm ein Haeuschen mietet. Honigmanns sind in Verzweiflung, Kurt Weigert und tausend andere Sonnenfelds nehmen zwei moeblierte Zimmer, Oscar Fehr weiss nicht, was wird. Ich schreibe noch kurze verschiedene Briefe an Alfred und Rudi, die nun auch wieder vorsichtige Briefe schreiben muessen. Keinen Dank fuer "grosses Geschenk" wie Veralein und Nina fuer das Buch schrieben! Alle Briefe werden geoeffnet. Wir hatten Grund, lieber Kartennachrichten zu verlangen, und Ihr habt den Wink nicht verstanden und schriebt Briefe, was ja alles gut ging. Eben schreibt Vati eine Karte, Du, lieber Hermann seist sehr zufrieden. Unser groesstes Glueck waere Euer Aufstieg. Also weiter alles Beste. Ich lasse Adolf anschreiben. In herzlicher Liebe mit allen guten Wuenschen fuer Euch und die Kleinen draussen in der Freiheit umarmt Euch
Eure traurige

Else

[252] Rudolf Peierls to Else and Heinrich Peierls

Birmingham, 26.8.1938
(carbon copy)

Liebe Eltern,
Ich schicke beiliegend die gewünschte Erklärung, die hoffentlich die richtige Form hat. (Soweit ich weiss, ist eine offizielle Beglaubigung der Unterschrift nicht erforderlich, aber das koennte ich natürlich auch leicht besorgen.)

Wir rechnen sehr ernsthaft darauf, dass Ihr von unserer Einladung Gebrauch macht, und freuen uns schon sehr darauf, Euch bald hier zu sehen. Ich hoffe, Ihr meint die Bemerkung nicht ernst, dass Ihr hier irgendjemand "zur Last fallen " würdet, denn das wäre ebenso kleinlich wie kurzsichtig. Natürlich ist der Komfort, den wir Euch hier bieten können, wesentlich geringer als Ihr es gewohnt seid, aber ich glaube, Ihr solltet Euch nicht zu sehr davor fürchten; es ist doch wesentlich besser als es aus Eurer Perspektive erscheinen mag, und Ihr werdet Euch rasch an die fehlenden Kleinigkeiten gewöhnen.

Es scheint mir aber sehr wichtig, dass Ihr so bald wie möglich kommt, denn der Klimawechsel ist besonders im Winter ungesund. Wenn Ihr kommt, bevor die Kälte und die Nebel anfangen, so wird es Euch wesentlich leichter fallen, Euch an die Verhältnisse hier zu gewöhnen. Ich glaube nicht, dass Euch die Entscheidung jetzt noch sehr schwer fallen sollte, und ich hoffe, Ihr werdet gleich nach Eurer Rückkehr nach Berlin alle für Eure sofortige Uebersiedlung erforderlichen Schritte einleiten. Je früher wir Eure definitive Entscheidung wissen, um so leichter wird es uns sein, Euch hier alles so gemütlich wie möglich einzurichten, und Euch die Eingewöhnung zu erleichtern.

Das englische Visum dürfte, soweit mir bekannt ist, keine Schwierigkeiten machen, aber ich kann es ev[entuell] von hier aus beschleunigen.

Bei uns geht alles gut. Wir sind nach einer anregenden Zeit in Cambridge wieder zu Hause. Ich muss am 1. IX. wieder für zwei Wochen nach Kopenhagen fahren, um meine Arbeit mit Niels Bohr weiterzuführen.

Mit herzlichen Grüssen
Euer

[Rudi]

[253] Rudolf Peierls to G. Placzek

Birmingham, 27.8.1938
(carbon copy)

Lieber Placzek,
Eben kommt Ihr Telegramm.[975] Ich habe dessen Inhalt an Cockroft übermittelt, und es hängt natürlich von ihm ab, ob er noch ein paar Wochen warten will, oder inzwischen schon anderweitige Schritte unternehmen will. Wahrscheinlich haben Sie bis zum Empfang dieses Briefes schon eine telegraphische Mitteilung über diese Frage bekommen.

Zur Geschichte der Angelegenheit: Die Sache ist der, dass Jones, der mein zeitweiliger Nachfolger in Cambridge war, eine Readership in London bekommen hat, und daher zum 1.X. fortgeht. Cockroft schien die Idee, Sie für diese Stelle zu bekommen, sehr zu gefallen, und da bei der Brit[ish] Ass[ociation] gerade Bohr und Bragg in Cambridge waren, so convergierte die Sache unerwartet leicht. Natürlich ist es noch nicht offiziell entschieden, Ihnen die Stelle anzubieten, wie Sie ja wohl aus der Tatsache entnommen haben, dass die Mitteilung von mir und nicht von Cockroft kam, aber ich glaube, das ist rein formal. Allerdings ist Bragg weit weniger zuverlässig als es Rutherford war, aber es erscheint äusserst unwahrscheinlich, dass er noch im letzten Moment ein Bein stellen wird. Es hängt also praktisch alles davon ab, wie eilig es Cockroft hat. Mein Eindruck ist, dass er versuchen wird, auf Sie zu warten, da ihm die möglichen Konkurrenten für die Stelle nicht zu begeisternd erscheinen.

Die Stelle selbst ist, wie Sie wohl wissen, eine rein beratende. Die Idee ist, sich ein bisschen für die Dinge zu interessieren, die die Leute im Mond machen. (Das macht übrigens viel Spaß, denn es sind sehr lustige und meistens unerwartete Resultate über flüssiges Helium u.s.w.) Man braucht natürlich nicht zu jedem Experiment gleich eine Theorie zu produzieren, aber man kann meistens den Leuten etwas über die phenomenologischen oder thermodynamischen Zusammenhänge zwischen ihren

[975] No copy of the telegram could be located, but it must have concerned the Cambridge position about which Peierls, Bethe and Placzek were communicating. See letters [250], [254].

verschiedene Experimenten sagen, erklären, wie weit man Theorie von anderen Leuten über diese Fragen glauben soll, und warum sie falsch sind. Ferner kann man sich mit ihnen über mögliche Messanordnungen und über die Abschätzung von Fehlern, Auswertung von Resultaten usw. unterhalten. Es kommt für diese Dinge gar nicht so sehr darauf an, dass man das alles selber ausixt, sondern man soll nur erklären, wie es gemacht wird, und worauf es ankommt, dann können sie es.

Darüberhinaus hat man (inoffiziell) natürlich Kontakt mit dem Cavendish, und da stelle ich mir vor, dass Sie sehr viele nützliche Anregungen geben können, aber das brauche ich Ihnen nicht zu erklären.

Das Gehalt ist, wie Sie wissen, £400 minus 5% für Superannuation. Aber in Cambridge verdient man à la carte. Wenn Sie eine Fakultät auffordert, eine Vorlesung zu halten, so bekommen Sie für eine dreistündige Vorlesung in einem Term noch £50. (Wenn Sie sich selber anbieten nur 15.) Ferner ist es leicht, Supervisions zu bekommen, wo eine Wochenstunde pro Term £3 – 3 – 0 einbringt (wenn man mehrere Studenten gleichzeitig nimmt, entsprechend mehr) Alle diese Dinge haben aber eine grosse Relaxationszeit. Für Vorlesungen muss man Mitglied der Universität sein, für Supervisions Verbindungen zu einem College haben. Im ersten Winter werden sie daher möglicherweise noch nicht konvergieren.

Dies zu Ihrer Information. Wenn ich von Cockroft noch etwas definitiveres höre, d.h. ob er sich definitiv entschieden hat, zu warten, oder nicht zu warten, so lasse ich es Sie wissen. Ich fahre am 31.8. für zwei Wochen nach Kopenhagen, bin dann wieder hier. (Die Idee ist natürlich, die Arbeit aufzuschreiben ......

Mit herzlichen Grüssen von und an
Ihr

[R.E.Peierls]

29.8.39

Ich habe den Brief noch zurückgehalten, da bis zum Mittwoch sowieso kein anständiges Schiff geht. Inzwischen habe ich von Cockroft noch keine Antwort (er ist gerade in die Ferien gegangen, und das bedeutet wohl eine Verzögerung) dagegen einen Brief von Bethe, den ich über die von hier unternommenen Schritte informiert hatte.[976] Bethe gibt u.a. eine Schilderung der Stelle, die er Ihnen in Cornell angeboten hat, und danach scheint es mir, dass demgegenüber Cambridge doch viele Vorzüge hat. Das Gehalt ist zwar nominell dasselbe, aber wie ich schon sagte, kann man in Cambridge noch mehr verdienen, und ausserdem sind die Lebenskosten in Amerika doch wesentlich höher, besonders wenn man die Reisen nach Europa einrechnet. Dann kommt dazu, dass Sie in Cambridge gewissermassen Hahn im Korbe sind, da es ja sonst keine Theoretiker gibt, die sich für das, was im Mond oder im Cavendish geschieht, interessieren. Ausserdem ist es relativ leicht, Sklaven, (d.h. Studenten) zu bekommen, besonders wenn es Ihnen gelingt, sich mit Fowler gut zu stellen.

Ich habe von Bethes Brief den Eindruck, dass er durchaus Verständnis für die Situation haben würde, wenn Sie den Plan mit dem Spring Term in Cornell aufgeben, insbesondere wenn Cockroft das zur Bedingung macht. Falls das der Fall ist, und falls Paris nichts wird, so würde ich Ihnen daher raten sich ernsthaft zu überlegen, of Sie Cambridge nicht doch gleich für die ganze Zeit annehmen sollen.

Ich sollte übrigens noch etwas über die Dauer der Stelle sagen. Die formelle Anstellung erfolgt immer für ein Jahr, aber das bedeutet natürlich nichts. Das Geld für die Stelle ist mindestens für zwei Jahre vorhanden, und es erscheint mir, dass die Stelle verlängert wird, da das Mond Lab nicht ohne Theoretiker und ohne die parallele experimentelle Stelle, (die jetzt Allen[977] hat) existieren kann. Auf jeden Fall scheint es

[976] Letter [254].

[977] John F. Allen (1908–), a graduate from Toronto had come to Cambridge in 1935 to work with Kapitza. Instead, he continued Kapitza's work in Cambridge after the latter's detention in Russia in 1934. In January 1938, *Nature* contained two separate articles by Kapitza and by Allen and his student Miserer, which mark the discovery of superfluidity.

mir so gut wie sicher, dass Sie, wenn Sie erst einmal in Cambridge sitzen, sich dort auch auf die Dauer halten können (Sie werden sich ja sicher nicht gerade wie Goldhaber benehmen). Aber Sie müssen darauf vorbereitet sein, dass es keinen langfristigen Vertrag, oder auch nur eine bestimmte Zusage gibt (sogar eine Profressur ist im Prinzip auf drei Monate kündbar). Aber wenn Sie den alten Witz mit dem Diplomaten und der Dame kennen, so sind die Engländer durchaus als Damen und nicht als Diplomaten zu verstehen.

Meine Reise nach Kopenhagen ist natürlich schon wieder telegraphisch verschoben (Bethe würde sagen, ich fahre 1939) und die Meinung ist jetzt, dass ich am 7.9. fahren soll. Aber schreiben Sie sicherheitshalber lieber hierher.

Nochmals viele Grüsse
Ihr

[R.E. Peierls]

[254] Hans Bethe to Rudolf Peierls

Seelisberg, 27.8.38

Lieber Peierls,
das ist ja alles sehr sehr interessant.[978]
Vor allem das mit Placzek. Ich bin im Grunde sehr froh und beruhigt darüber. Natürlich hätte ich ihn, von meinem rein persönlichen Standpunkt aus, sehr gern in Cornell gehabt, aber —

Nr.1 er wäre in Cornell eigentlich am ganz falschen Platz, denn <u>ein</u> Theoretiker ist ja durchaus genug für ein Institut. Cambridge dagegen wäre ideal in dieser Beziehung und braucht einen wirklichen Theoretiker ja dringend. Nr.2 brauche ich eigentlich einen Sklaven, und Placzek ebenfalls, und es wären dann in Cornell keine weiteren Stellen für Sklaven vorhanden. Nr.3 wäre es wahrscheinlich schwierig

978 Reply to letter [250].

mit den anderen Leuten in unserem Institut (verbesserte Auflage des Krachs Gibbs-Richtmyer[979]) und man sollte seine politische Situation nicht verschlechtern. Nr.4 würde Placzek sich wahrscheinlich in Europa wohler fühlen als in einem kleinen Städtchen in Amerika. Nr.5 ist zwar das Gehalt meines Wissens dasselbe ($ 2000 und £400), aber in Cambridge kann man eher was extra verdienen, und ausserdem sind dort die Gehälter der anderen auch nicht höher. Nr.6 war Cornell eigentlich nur als Übergang zu einer besseren Stelle in Amerika gedacht, und ob die Amerikaner je genug Verstand haben würden, über Placzeks "unamerikanische" Eigenschaften hinwegzusehen, ist fraglich. Übrigens Mond ist doch für 5 Jahre?

Also, alles in allem fände ich es eigentlich sehr viel besser, wenn P[laczek] nach Cambridge ginge. Das Einzige ist, ob man durch Einfürung eines weiteren Ausländers Ewalds Chancen in Cambridge noch verschlechtert? Ich glaube aber eigentlich Mond und Cavendish, und vor allem Kerne und Kristalle, sind linear unabhängig. Hat eigentlich Bragg beim Mond etwas zu sagen?

Ich schreibe Dir all diese vielen Gründe, die Du alle selber weisst, nicht nur zwecks Verminderung des Verdienstes der Post. Sondern es wäre vielleicht gut, sie Placzek selbst zu schreiben. Nun wäre es nicht sehr nett, wenn ich das täte, sondern sehr viel netter, wenn z.B. Du ihm die Vorzüge Cambridges gegenüber Cornell schilderst. Wäre das möglich? Ich glaube, es würde seine Entscheidung beeinflussen.

Vor etwa einer Woche bekam ich ein Telegramm von Placzek, er käme gerne nach Cornell, aber erst im 2. Semester. Im 1. Semester wollte er auf jeden Fall nach Europa. Ich telegraphierte zurück, das wäre recht, und telegraphierte in gleichem Sinne an Gibbs. Antwort erfolgte darauf noch nicht. Natürlich kann Cornell jederzeit rückgängig gemacht werden, besonders, da ein Anwärter da ist (Rose), der jederzeit auf Abruf zu kommen bereit ist. Rose würde ich übrigens nur für ein weiteres Jahr nehmen, und dann entweder Critchfield (von Teller) oder

[979]R. Clifton Gibbs and Floyd K. Richtmeyer, both professors at Cornell, were engaged in a continual feud, and as Bethe had been recruited by the Head of Department, Gibbs, his standing with Richtmyer was on shaky grounds.

Schwinger (von Rabi), falls Placzek nicht kommt.[980] Infolgedessen ist es möglich, Cornell sogar für Placzek zu reservieren, mindestens bis Februar, falls aus Cambridge nichts wird. Wie weit ist übrigens Cambridge gediehen? Hängt es hauptsächlich von Placzeks Entscheidung ab, oder muss noch jemand gefragt werden?

Bezüglich der anderen Dinge: Die Rechnungen über das nicht-cylindersymmetrische Cyclotron sind bisher nur bis zur der Näherung ausgeführt, in der der Focussierungseffect gerade auftritt (2. Näh[erung]), man kann also die Gültigkeitsgrenze nicht abschätzen. Es geht aber bestimmt solange der Relativitätseffekt noch als "Störung" betrachtet werden kann — ich würde sagen mindestens bis $\frac{1}{10}Mc^2$.

Die Höhenstrahlung ist sehr interessant. Wenn der Zerfall eine reine Frage der Zeit ist, anders als die ursprüngliche Blackett'sche Annahme der 200MV, glaube ich ihm gern. Nur schien es mir, dass die nötigen Zeiten ziemlich gross sind und z.B. nicht mit der extrapolierten β-Theorie stimmen (Rechnungen von Heitler und Nordheim[981]). Ausserdem muss man dann erklären, wie die Barytronen in der Atmosphäre entstehen — dazu gibt's nicht nur keinen Mechanismus, sondern es führt (nach Nordheim) zu einigen direkten Widersprüchen, die ich allerdings vergessen habe. Aber wenn es so ist, dann muss es ja wohl so sein.
Dein

Hans

---

[980]Charles Louis Critchfield (1910–1994) received his Ph.D. from George Washington University, where he studied with Teller and Gamow. He joined the Manhattan Project in 1942/3. After a brief post-war spell at George Washington University he became professor at Minnesota. Julian Seymor Schwinger (1918–1994), received his doctorate from Columbia University in 1939, worked at Berkeley (1939–41) before moving to Purdue (1941–1945) and Harvard (1945–72).

[981]L.W. Nordheim and G. Nordheim, 'On the Production of Heavy Electrons', *Phys. Rev.* **54**, 254–65 (1938); W. Heitler, 'Analysis of Cosmic Rays', *Proc. Roy. Soc.* **A161**, 261 (1937); L.W.Nordheim, 'On the Absorption of Cosmic-Ray Electrons in the Atmosphere', *Phys. Rev.* **51**, 1110 (1937).

[255] Niels Bohr to Rudolf Peierls

Copenhagen, 27.8.[19]38

Dear Peierls,
I hope you have received my telegram and that it will not be inconvenient to you to postpone your visit here a few days. The matter is that Rosenfeld and I, according to the urgent request of the Organization Committee, are in these days working on the completion of my Report to the Warsaw meeting and that it would be most important to have that off my hands before we start energetically on our paper on nuclear problems.

I do not know what time will suit you to come to Copenhagen. In some way it might be best if you could be here the two weeks from September 11 to 25. But if that it too late four your work in Birmingham, it would also be quite convenient for me, if you came a week or so earlier, as Rosenfeld and I will probably need only about a week's time to bring the Report into shape.

I need not say that it was a great pleasure to my wife and me to see you and your wife at the very successful meeting in Cambridge. According to your advice I went to the Russian Embassy in London, where, however, I could not see Maisky[982] but had a conference with his councillor who had a considerable knowledge of scientific workers and among others had been negotiating with Rutherford the transfer of Kapitza's apparatus to Moscow.[983] I explained to him Landau's unique scientific qualifications and my anxiety of having received no answer from him on my invitation to Copenhagen. But, of course, I did not tell that I had underhand information about his arrest in Kapitza's institute. According to the advice of the councillor I have to-day sent an express letter, of which I enclose a copy, to the President of the Academy in Moscow. Of course I am prepared that this letter will have no direct effect, but I think that it may be the best preparation for a subsequent

[982]Ivan M. Maisky, Soviet plenipotentiary in the United Kingdom 1932–1941.

[983]After Kapitza's detention in Russia in 1934, he founded the Russian Institute for Physical Problems, the equipment of which was purchased, with the assistance of Rutherford, from the Mond Laboratory by the Soviet Government.

writing to the higher authorities. For such a next step I shall of course be glad to have your advice when you are here.
With kindest regards and best wishes from us all.
Yours

Niels Bohr

[256] Rudolf Peierls to Siegfried Peierls

Birmingham, 30.8.[19]38
(carbon copy)

Dear Uncle Siegfried
We are getting rather worried about my parents and conditions in Germany. No doubt they have been in touch with you and you know of their plans to emigrate — in fact I have heard that you have been kind enough to arrange for an affidavit for them, but I do not know exactly how much you know (one has to be very careful in letters from Germany) and so I better start from the beginning. I have heard recently from a friend of mine who saw my parents in Baden-Baden, and according to his reports, conditions are even worse than I imagined.[984]

Apparently it is now practically impossible to travel within Germany, as no hotel can afford to take non-Aryan guests. The few Jewish hotels which are left are being forcibly sold. In addition to irrelevant, but annoying regulations (such as assigning special registration numbers to Jewish-owned cars &c.) Jews are not allowed to open their safe without a customs officer being present, and in some cases the income tax office imposes a limit on withdrawals from banking accounts &c. No doubt my father is in no financial trouble, even in spite of all these regulations, but it is mainly their psychological effect which is so terrible. Apparently my father is very depressed, and particularly Else is losing her nerves altogether and is in an almost hysterical condition.

For all these reasons we have welcomed their decision to think seriously about emigration (we only wish they had followed our advice

[984] Letter [249].

to take this decision long before now) particularly because things are rapidly getting worse and one does not know how long it will be possible to get out at all. You have probably heard that in the case of Rothschild the Germans are trying to get a ransom from his relatives in England, and although this is so far an isolated case, it might occur to them that this is a convenient new source of revenue.

Their plan is to emigrate to the U.S.A., as you know, but apparently there is a terrible rush on the US Consulate. There is no chance to get a visa before the spring, and my own information from a man who is trying to take his mother to the USA where he lives, is that already the quota for 1939 is booked up, except for a few permits that are being kept for very urgent cases. My parents feel that it would not be wise to wait all that long, and they therefore think of coming to England first. With Alfred and me being here, there is no trouble in getting them a visa for England.

So we hope to have them here, at least for the time to get their USA visa business settled (there are arguments, even apart from our own personal wishes, to keep them here longer. One of them is the lower cost of living. But we can settle that when they get here.) Naturally they will not be able to live as comfortably here as they were used to, but they will have to get used to that. The worst will probably [be] the English habit of not heating houses and of having steep stairs without lift. As they are evidently going to stay for some time, the ideal solution would be to get a little flat for them (In flats one can easily find more continental conditions) but I am afraid that would be beyond our means. I do not know — and with the present censorship cannot find out from my father directly — what his means will be after he leaves Germany. Of course he does not get a penny out of Germany, but I do not know whether he has any arrangement with you or whether he can expect support from any side for the time he spends here. If he had some £20 a month, say, we could make life for them very much more comfortable.

Please let me know what your views about this question are, as my parents will probably be rather worn out when at last they arrive and I would like to be able to settle matters for them as much as possible and as quickly as possible.

You will of course realize that all this must not be mentioned in direct correspondence with Germany.
With kindest regards from us all, and to all of you
Yours sincerely,

[Rudi]

[257] Rudolf Peierls to Wolfgang Pauli

Birmingham, 31.8.[19]38
(carbon copy)

Lieber Herr Pauli!
Ich schreibe gerade meine Arbeit für das Sommerfeld-Heft.[985] Soweit ich es verstanden habe, wird dies kein ganzes Heft der Physical Review, sondern nur eine Gruppe von Arbeiten. Wird dann jemand eine geeignete Widmung an den Anfang schreiben, oder soll man in jeder einzelnen Arbeit noch etwas über Sommerfeld sagen?

Die Angelegenheit mit dem Starkschen Artikel verfolge ich weiter.[986] Es ist aber nicht ganz einfach, da eine Reihe von Leuten so etwas nicht schreiben wollen. Außerdem ist der Herausgeber der Nature jetzt in den Ferien und schwer erreichbar. Ich hoffe aber noch immer, daß es sich einrichten läßt, und daß ich Chadwick dazu bewegen kann, etwas zu schreiben.
Mit besten Grüßen
Ihr

[Peierls]

---

[985]See letter [246].
[986]*Ibid.*

[258] Rudolf Peierls to Annie Krebs

Birmingham, 2.9.[19]38
(carbon copy)

Liebe Krebse,
Vielen Dank für Euren Brief, und vor allem für die Glückwünsche zu Gabys Geburtstag, die wir hiermit zu Günthers aufs herzlichste erwidern.

Wir machen uns auch grosse Sorgen um die Eltern. Vor ein paar Wochen war ein Freund von uns bei ihnen in Baden-Baden und berichtete über ihre Pläne, über die Ihr vermutlich orientiert seid, und insbesondere über die Zustände in Deutschland.[987] Es scheint dort jetzt sehr unangenehm zu sein, und dauernd schlimmer zu werden. Man kann schon in keinem Hotel mehr wohnen, und die paar jüdischen Sanatorien, die es noch gibt (wie das in dem sie jetzt in Baden-Baden sind), müssen verkauft werden.

Man darf schon nicht mehr seinen Banksafe öffnen, ohne dass ein Devisenbeamter dabei ist, und einige Leute haben schon Verordnungen bekommen, dass sie nur Beträge bis zu einer gewissen Höhe von ihrem Bankkonto abheben dürfen. Mein Freund berichtet, dass Vati die ganzen Dinge eigentlich ziemlich ruhig aufnimmt, dass aber Else schrecklich nervös ist, und ihn ganz verrückt macht. Das stimmt eigentlich ganz gut zusammen mit einem Brief, den Else aus der Schweiz schrieb wo sie zwei Tage bei Briegers zu Besuch war (leider ohne uns dies vorher mitzuteilen oder uns von dort aus anzurufen).[988] Dabei sind sie anscheinend noch jetzt nicht endgültig entschlossen, auszuwandern. (Vati hat offenbar Bedenken, Else die schwerverdiente Pension entgehen zu lassen.) Unsere Meinung ist nach wie vor, dass sie es unbedingt tun sollten, und wir haben ihnen in diesem Sinne geschrieben.

Die Schwierigkeit ist, dass es anscheinend selbst mit Affidavit nicht so leicht ist, nach Amerika zu kommen. Die Quota für 1938 ist jedenfalls schon voll, man sagte mir, dass sogar die für 1939 schon besetzt ist, und

---

[987] Letter [249].
[988] Letter [251].

in jedem Fall bewegen sich die Schätzungen der Zeit, die man braucht, um ein Einwanderungsvisum zu bekommen, zwischen 8 und 16 Monaten. Ich würde es für Wahnsinn halten, solange zu warten, denn niemand weiss, ob man dann überhaupt noch herausgelassen wird, ob man dann überhaupt noch irgendwelche Sachen mitnehmen darf (schon jetzt gibt es anscheinend schwere Einschränkungen) und ob man dann nicht schon für die Gnade, herausgelassen zu werden, in Devisen bezahlen darf.

Anscheinend haben die Eltern auch die Absicht, wenn sie sich überhaupt entschliessen können, zunächst mal nach England zu kommen, um hier das amerikanische Visum abzuwarten. Das würden wir sehr begrüssen. Es ist überhaupt die Frage, ob es für sie nicht richtiger wäre, hier zu bleiben, insbesondere in Anbetracht der Tatsache der höheren Lebenskosten in USA, aber das kann man ja überlegen, wenn sie erst mal draussen sind.

Jedenfalls werden wir wohl zunächst mal die Verantwortung für sie hier haben, wenn sie kommen. Um daher etwas Bewegungsfreiheit zu haben, überweist doch bitte das Geld, das Vati bei Euch für uns deponiert hat.[989]

Vielen Dank für das Bild, Ihr müsst ja wirklich bald anfangen einen Bräutigam für Ilse zu suchen. Auch Günther sieht ungeheuer erwachsen aus. Wir legen jetzt endlich die Bilder unserer Produkte bei.

Wir hatten es auf der Farm in Cornwall sehr schön, die Gegend wo wir waren ist ganz einsam und gar nicht von Touristen überlaufen, und die Küste ist dort sehr wild und schön. An der Ecke läuft die Küste gerade nord-südlich, so dass die Wellen dort direkt von Amerika herüberkommen, und das merkt man. Es ist die richtige Stelle für surf riding, und das haben wir mit grosser Begeisterung gelernt. Die Kinder haben sich auch sehr gut erholt (sofern sie das nötig haben) und die vielen Tiere auf der Farm waren natürlich eine Sensation.

Dann waren wir noch 10 Tage in Cambridge bei einem grossen Kongress. (Wieder mit Kind und Kegel, denn Freunde von uns waren freundlicherweise in Ferien und wir konnten daher in ihrem Haus wohnen.) Das war die British Association for the Advancement of Sci-

[989]Heinrich Peierls had transferred a sum of $5000 for the three children to the United States. This had been deposited with Annie and Herman Krebs.

ence, eine sehr offizielle Geschichte, wo es ausser den offiziellen Sitzungen noch eine Unzahl grösserer Empfänge Garden und Sherry Parties u.s.w. gab.

Jetzt sind wir wieder eine Woche in Birmingham, und Genia ist im Augenblick sehr beschäftigt, da wir eine Kabinettskrise hatten, und unsere "Perle" Olive innerhalb von 24 Stunden herausflog. Wir hatten sie offenbar zu sehr verwöhnt und sie als gleichberechtigt behandelt, und sie fing an, sich immer mehr herauszunehmen. Schon in Cambridge wurde ihr langfristig gekündigt, und bei einer erneuten Frechheit hier flog sie dann prompt ganz heraus. Jetzt versuchen wir, eine Deutsche oder Oesterreicherin zu bekommen, aber es dauert schrecklich lange, bis man dafür die Erlaubnis bekommt, und wir wissen noch nicht, ob wir so lange warten können. Genia hat zwar noch eine Köchin, aber mit dem grossen Haus gibt es ziemlich viel Arbeit, und sie ist dann doch ziemlich angebunden. Ich soll nächste Woche wieder einmal für vierzehn Tage nach Kopenhagen fahren, wenn ich nicht (wie gewöhnlich) wieder im letzten Moment eine Absage bekomme.

Wir freuen uns sehr, dass Ihr so erfreuliche Dinge ueber das Geschäft berichten könnt. Ich kann mir schon vorstellen, dass es eine Weile dauert, bis die Aufträge wachsen, und bis die Kosten der Anlaufzeit wieder hereinkommen. Hoffentlich reicht das Kapital der Firma lange genug bis schliesslich ein vernünftiger Profit erreicht ist.

Euer

[Rudi]

[259] Georg Placzek to Rudolf Peierls

Pasadena, 4.9.[19]38

Lieber Peierls,
Vielen Dank für 2 Briefe,[990] vollständiges Manuskript sowie Telegram.

1. Wissenshaft. Alles sehr interessant, bloss scheint mir, dass Sie den Stein des Weisen gefunden haben (in der Relation [· · ·][991]) und, anstatt sich darüber zu weiden und ihn anzuwenden, sich darüber zu kränken und ihn bezweifeln. Ihr Argument, dass die Gleichung nicht richtig sein kann, weil die Partialbreiten sonst grösser werden als der Abstand, verstehe ich überhaupt nicht, denn ich kann nur sagen 'und wenn schon'. Dass [· · ·][992] stets kleiner als d sein soll, ist doch bloss dann zu fordern, wenn Sie jeweils das k in der Resonanz betrachten und dieses zugleich mit der Energie des Compoundzustandes variieren, d.h. zu höheren Compoundzuständen gehen. In der fraglichen Gleichung wird aber doch immer der gleiche Compoundzustand betrachtet, und bloß das k variiert, wie Sie das doch selber so schön erfunden haben. Und bei dieser Art Variation sehe ich keinerlei Grund ein, warum, wenn Sie mit einem k einstrahlen, dessen Energie weit oberhalb der Resonanzenergie liegt, die Partialbreite dort unten auch nicht grösser werden soll als der Abstand.

   Also sage ich, man nehme nun freudig diesen Stein der Weisen, und berücksichtige ihn bei der Betrachtung des Ausdruckes für die elastische Amplitude. Dann wird eben logischerweise wegen $\frac{\partial^2 W_n}{\partial k_n \partial h_n} : Im(k_n \frac{\partial W_n}{\partial k_n}) = \gamma_a$ und man hat was man will. Weiterhin verschwindet dann auch die Gelehrsamkeit über den Unterschied von statistischer und gewöhnlicher Partialbreite, denn die beiden Grössen sind identisch.
   Also bitte schreiben Sie mir Ihre diesbezügliche Meinung.

---

[990] Only letter of 27.8.1938 could be located.
[991] Missing in manuscript.
[992] Missing in manuscript.

2. Geschäft. Zunächst möchte ich Sie fragen, was mich der seelige Bucharin fragte, als ich ihn einmal sozusagen im Namen der internationalen Wissenschaft bat, sich dafür einzusetzen, dass Landau ab und zu ins Ausland gelassen werde, nämlich: Ist Ihre Demarche offiziell, offiziös, oder inoffiziell?
Weiterhin möchte mich interessieren, wie eigentlich die Lage in Cambridge ist. Ist es so, dass man wider Erwarten doch beschlossen hat, die Kernphysik einer neuen Blüte entgegenzuführen und als eine der Massnahmen zu diesem Zwecke die Mondstelle mit einem Kernphysiker zu besetzen wünscht? Oder aber ist es so, dass man, wie befürchtet, die Kernphysik langsam verdorren lassen will und man mir davon unabhängig, z.B. auf Bohr's Fürsprache hin, die Stelle freundlicherweise als eine Art Sinekure anbieten will?
Meine Lage ist die Folgende: Bethe hat Ihnen ja wohl von der Corneller Stelle erzählt, die er mir angeboten hat.[993] Diese ist ja nicht sehr würdig, aber wenn ich überhaupt in Amerika bleiben wollte, so wäre das, wie mir gesagt wird, ein zu diesem Zwecke ratsamer Anfang.

Bevor Ihr Telegramm kam hatte ich Cornell, zunächst für ein Semester, mit Beginn im Februar, angenommnen.
Mit Paris steht die Sache so, dass ich offiziell nichts gehört habe, obwohl mir Joliot Antwort für Juli versprochen hatte. Privatim hörte ich von Halban, dass die Sache, wegen Nichterscheinens einiger erforderlicher Leute in der zur Entscheidung anberaumten Sitzung im Juli auf Oktober vertagt worden sei.
Nun kann ich die Pariser nicht so vor den Kopf stossen, dass ich, ohne ihre Entscheidung abzuwarten, auch Cambridge annehme (Cornell bin ich nur ein Semester gebunden, das ist für Paris nicht arg). Ich habe sie um äusserste Beschleunigung der Entscheidung gebeten. Frage ist, wie lange kann die Sache in Cambridge warten? Falls Sie mir das nicht schon geschrieben haben, so telegraphieren

[993] Letter [254].

Sie mir es bitte nach Stanford. Meine Koordinaten sind in den nächsten Tagen etwas unbestimmt, Post wird mir von Stanford aus nachgeschickt. Dorthin bitte aber nur Telegramm. Briefadresse c/o Rabi, Physics Dept. Columbia Univ. New York City. Ich denke, ich fahre am 5. Oktober von New York mit der Queen Mary ab, wenn es unbedingt erforderlich ist, so könnte ich maximal 14 Tage früher kommen, in diesem Falle würde ich ebenfalls um umgehende telegraphische Verständigung, natürlich zu meinen Lasten, bitten. Jetzt muss ich schliessen, damit der Brief noch mitgeht. Vielen Dank + viele Gruesse von & zu Ihr

G. Placzek

[260] Jenny Peierls to Rudolf Peierls

Lake Placid, N.Y., 8.9.[19]38

Dear Rudi,
As Siegfried is quite ill in bed, I shall act as his private secretary + try to answer your questions as well as he would have.

We have been in constant communication with your dear parents, direct through Krebses + also through Briegers, whom Else was able to visit in Zurich. It is unbelievable what those poor people are enduring + we only hope that before long, they will be able to emigrate. Naturally, if going to England will hasten matters, for them to get out of Germany, by all means advise them to do so as quickly as possible. Uncle Sieg[fried] will gladly send the 100.00 per month to make them comfortable.

When they are in England it will be ample time to look into getting them a visa for the U.S.A. should they prefer to come over here. Of course, we do not know if it will be difficult to come to America from England, but you will have to look into this yourself.

Uncle Sieg[fried] has been very miserable. He has a bad attack of arthritis and is unable to move his hands or legs, so we are taking him back to N.Y. + right into a hospital. He suffers terribly + naturally is depressed.

I sincerely hope that you, Genia + the children are well. Best regards from both of us,

Jen[ny] Peierls

[261] Rudolf Peierls to Niels Bohr

Birmingham, 24.9.[19]38
(carbon copy)

Lieber Herr Bohr,
Zunächst möchte ich Ihnen noch sehr für die Wochen in Kopenhagen danken, für die freundschaftliche Aufnahme und für alle Anregung, die ich wieder von dort mitbringen durfte.[994] Es tat mir sehr leid, dass ich weglaufen musste, ohne mich von Ihnen verabschieden zu können, und ich war noch bis zum letzten Moment unsicher, ob es nicht richtiger gewesen wäre, auf Sie zu warten, und später mit dem Flugzeug zu fahren. Ich hatte aber dann doch Angst davor, dass ich auf so kurze Frist vielleicht keinen Platz im Flugzeug finden würde, und beschloss daher, doch den gerade noch erreichbaren Zug zu nehmen.

Aus meinem telefonischen Gespräch mit Frau Bohr von der Fähre (das leider durch schlechte Wirkung eines Verstärkers sehr schwer verständlich war) verstand ich, dass Ihr Besuch an jenem Morgen nicht umsonst gewesen war, und dass sich immerhin Aussicht ergeben hatte, den Brief an seinen Bestimmungsort zu befördern. Man kann nur hoffen, dass er auch Erfolg hat.[995]

Ich habe übrigens unterwegs noch sehr über die Frage der Potentialstreuung nachgedacht, und ich glaube jetzt die Schwierigkeit etwas besser formulieren zu können: Wir diskutierten neulich, dass, wenn man in der Nähe der Resonanz sowohl die Termabstände wie die Partialbreiten als annähernd konstant betrachtet, die Dispersionssumme bei

[994] Peierls had finally spent a fortnight in Copenhagen in September 1938 in an attempt to finalise work on the Bohr-Peierls-Placzek paper.

[995] Bohr had delivered a letter to be sent in support of Landau who was still under arrest in Russia.

hohen Energien logarithmisch divergiert, und dass das zu der Vermutung berechtigt, dass der wesentliche Beitrag dieser Summe von den sehr hohen Termen kommt. Nun ist es aber so, dass man für die sehr hohen Terme das Problem als Einkörperproblem behandeln kann; im Einkörperproblem nehmen aber die Termabstände mit wachsender Energie zu, da sich dort die Terme wie const.n^2 verhalten. Dann konvergiert aber die Summe rasch und die weit entfernten Terme geben keinen Beitrag. Aus der Zusammenstellung beider Tatsachen muss man also schliessen, dass in Wirklichkeit der Beitrag aus dem Zwischengebiet kommt, wo zwar die Partialbreiten schon nicht mehr als konstant (d.h. für alle Terme gleich) behandelt werden dürfen, wo aber andererseits das Einkörperproblem noch keine hinreichende Näherung ist. Dies zeigt wohl, dass man nicht erwarten kann, eine vernünftige Abschätzung dieser Terme aus einem stark vereinfachten Modell zu bekommen, und dass man wohl schon recht komplizierte Züge des Modells berücksichtigen muss. (Ich glaube natürlich auch, dass es sich hierbei nicht um eine Frage handelt, für die die Unsicherheit im Kräfteansatz und ähnliche Schwierigkeiten eine wesentliche Rolle spielen, aber man muss sich eben schon ein sehr genaues Bild von dem Mehrkörperproblem machen, um diese Züge einigermassen richtig zu bekommen.) Alle diese rein negativen Bemerkungen sind natürlich nicht als Beitrag zu der Arbeit gemeint, sondern nur in der Hoffnug, dass Ihnen dies meine Behauptung plausibler machen wird, dass die Erledigung der Frage der Potentialstreuung nicht so einfach ist, dass man die jetzige Arbeit so lange zurückhalten sollte.

Nochmals vielen Dank und herzliche Grüsse, auch an Frau Bohr und Ihre Söhne
Ihr

[Rudolf Peierls]

[262] Rudolf Peierls to Jenny Peierls

Birmingham, 24.9.[19]38
(carbon copy)

Dear Aunt Jenny,
Thank you very much for your letter, to which I reply only today because it arrived during my absence for a fortnight, in Copenhagen. We are so sorry to hear of Uncle Siegfried's illness, and I seriously hope that after your return to town, and after he has got into the hands of his own doctors (who, I trust, are better than the average English doctor) he will soon recover, but in the meantime we feel very much sympathy for him, for what must be a very annoying complaint.

We are all the more thankful to him and to you both for replying so quickly in spite of all the other worries and for being so ready to help my parents should the necessity arise. They have now definitely decided to apply for leave to emigrate, and this means in practice that they will come to England first. I do not think that this will very much prejudice their chances of getting a USA visa (even if conditions in the London consulate are bad they are not likely to be worse than in Berlin where I understand the list is closed now for 1 1/2 years ahead)

It is, of course, impossible to get any information at present how the German Ministry of Finance is going to treat their claims, and I still hope that it will not be necessary to make use of your kind offer.

Of course, all these emigration plans will come to nothing if there is war, a possibility which seemed remote during the last week or so, but just now seems to threaten again. The whole position in Europe is so muddled by now that, if it were not for our private worries, one would hardly know what to fear more — that there will be war or that there will be none.

With kindest regards, particularly to Uncle Siegfried for whom we wish a speedy recovery
Yours sincerely,

[Rudi]

[263] Rudolf Peierls to Else and Heinrich Peierls

Birmingham, 24.9.1938
(carbon copy)

Liebe Eltern,
Eben kommt Euer Eilbrief, dessen Inhalt ich auch Alfred mitteilen werde. Wir werden also Euren Antrag auf Einreiseerlaubnis von London aus unterstützen. Allerdings kann man nicht mehr direkt zum Home Office gehen, sondern Anträge dieser Art gehen jetzt alle zunächst über eine Zentralstelle in Woburn House. Ich werde wohl am Montag dorthin fahren um alles zu verabreden. Da die Behandlung in Woburn vermutlich noch einige Tage dauert, so werde ich den Antrag gleich in Gang setzen. Wenn sich dann herausstellt, dass Ihr noch nicht rechtzeitig alles beim Konsulat eingereicht habt, so kann man die Sache dann immer noch aufhalten, bevor sie an das Home Office weitergeht.

Alle weiteren Einzelheiten haben natürlich Zeit, bis Ihr Euer Visum habt. Immerhin möchte ich aber schon jetzt folgendes sagen. Nach gründlicher Überlegung scheint es uns, dass es wahrscheinlich das praktischste ist, wenn Ihr in Birmingham wohnt. Ich weiss natürlich, dass Alfred und Nina Euch auch schrecklich gern bei sich haben würden, aber in Anbetracht von Ninas Gesundheitszustand, der Grösse ihres Hauses, u[nd] s[o] w[eiter] scheint es doch sehr fraglich zu sein, ob das auf die Dauer praktisch wäre. Wenn Ihr in London selbst wohnen würdet, so müsstet Ihr selbst wirtschaften; dazu ist das Leben in London sehr viel teurer als anderswo und die Entfernungen sind so gross, dass niemand Zeit für Verkehr hat. In Birmingham könntet Ihr leicht in unserem Haus wohnen. Dann müsstet Ihr Euch allerdings an die Abwesenheit von Zentralheizung und andere Unbequemlichkeiten gewöhnen, und daher wäre es wohl eine noch bessere Lösung, für Euch eine kleine Wohnung mit Zentralheizung ganz in unserer Nähe zu finden. Vermutlich lässt sich eine finden, die nicht weiter als 5 min. zu Fuss von uns ist. Dort würdet Ihr Lift und Zentralheizung haben, und eine kleine Küche, so dass Ihr nicht verpflichtet seid, zum Essen auszugehen, im allgemeinen würden wir aber darauf rechnen, dass Ihr bei uns esst und unser Haus als Euer eigenes betrachtet. Birmingham bietet natürlich an Theater, Kino und

Konzerten weniger als London, aber das was es bietet, ist immerhin nicht wenig, und es ist soviel billiger als in London, und Verkehrsschwierigkeiten soviel geringer, dass Ihr in praxis wahrscheinlich hier mehr Unterhaltung finden könnt als in London. Zudem gibt es hier viele nette Leute, und ich glaube, es wird Euch durch meine Beziehungen leicht möglich sein, Euch hier einen Kreis von Leuten zu schaffen, die Euch interessieren, und die Deutsch sprechen.

Ich meine natürlich nicht, dass Ihr über diese Fragen jetzt schon irgendwelche Entscheidungen zu treffen braucht, aber es ist Euch vielleicht angenehm zu wissen, was die Möglichkeiten sind, so dass es Euch leichter wird, Euch ein Bild zu machen.

Ich kam gestern abend zurück nach der ruhigsten Ueberfahrt, die ich auf dieser Strecke je gehabt habe, und fand Genia und die Kinder sehr vergnügt und munter. Gaby erregte bei Schulbeginn grosses Aufsehen damit, dass sie in den Ferien fast völlig lesen gelernt hatte. Das ist wegen der Schwierigkeiten, die das Englische beim Lesen bietet, aussergewöhnlich. Auch mit ihrem Rechnen sind die Lehrerinnen sehr zufrieden, aber mit dem Schreiben ist es noch schwach, und im Tanzen kommt sie wieder in die Anfängerklasse. In dieser Beziehung gerät sie also doch nach den Eltern!

Mit herzlichen Grüssen

Euer [Rudi]

[264] G. Placzek to Rudolf Peierls

Paris, 17.10.[1938]

Lieber Peierls,

Vielen Dank für Ihre Karte. Nach längeren Qualen hab ich also nun den Brief an Cockroft geboren. Ich hab ihm also geschrieben, dass ich wegen der positiven Pariser Entscheidung das ganze Jahr besetzt bin, dass ich aber, sofern er Schwierigkeiten mit der sofortigen Besetzung der Stelle hätte und er denken würde, dass ich zur Aushilfe nützlich sein könnte, gerne versuchen würde, hier eine Regelung zu finden, die mir erlauben würde, ihm für einen Teil der nächsten Monate zur Verfügung zu stehen. Hoffentlich wirkt das als Höflichkeit, wie es gemeint ist, denn wenn es dazu kommen würde, weiss ich nicht sicher, ob es sich hier wirklich organisieren liesse. Weitere Höflichkeiten (etwa wie sehr es mich gefreut hätte mit ihm zusammenzuarbeiten und wie sehr ich die Ehre einer Anstellung in Cambridge appreciate etc.) habe ich ihm nicht geschrieben, weil mir das auf englisch schwer fällt. Sofern Sie ihn nächstens treffen, so etwas erzählen wollten, so möchte es schön sein.

Hier ist nichts besonderes los, die allgemeine politische Stimmung können Sie sich denken. Gerne würde ich Sie bald sehen. Ich weiss nicht wie lange es dauern wird bis ich meine Papiere hier in Ordnung habe und wieder beweglich bin, denken Sie nach ob und wann Sie für ein weekend herüberkommen können, was sehr schön sein möchte.

Ich höre dass der Schönberg jetzt in Cambridge sein soll, wissen Sie etwas authentisches über Dau???[996]

Der Wigner verbreitet in Amerika Greuelmärchen über Ihr Funktionensystem, indem er nämlich die Meinung verbreitet, dass es nicht vollständig ist. Es möchte daher schön sein, ob Sie Ihren Beweis für das Gegenteil publizieren möchten. Leider hab ich besagten Wigner nicht selber getroffen.

Die Washingtoner wollen versuchen, Sie für Ihre nächste Konferenz über die tiefen Temperaturen einzuladen, sie wissen aber noch nicht, ob sie das Geld haben. Dieses unter dem tiefen Siegel der Verschwiegenheit.
Viele herzliche Grüsse Ihnen & Gemahlin
Ihr

G.P.m.p.

[996]See letters [240], [243].

[265] Rudolf Peierls to G. Placzek

Birmingham, 22.10.[19]38
(carbon copy)

Lieber Placzek,
Vielen Dank für Ihren Brief. Ich werde es jetzt so einrichten, dass ich jetzt "zufällig" nächste Woche mal in Cambridge bin und dann kann ich erstens in dem von Ihnen gewünschten Sinne Cockroft die Situation erläutern, falls das nötig ist, und ausserdem will ich versuchen, ihm zu raten, dass er Wick nehmen soll, der, glaube ich, ganz gerne kommen würde, und der nun vermutlich der beste für die Stelle ist.

Shoenberg[997] habe ich gesprochen. Ueber Dau hatte er nicht mehr zu berichten, als wir schon wussten (oder jedenfalls befürchteten). In dieselbe Gruppe gehören auch Rumer und Hellman.[998] Hier in England läuft der Zehden herum, der via Berlin hierher vorgedrungen ist, aber seine russische Frau mit Kind in M. zurücklassen musste, und seit Monaten nicht mehr mit ihr korrespondiert. Es ist eine schöne Welt.

Ich würde Sie auch sehr gerne sehen, aber ich weiss noch nicht, wann es mir möglich sein wird, die Spazierfahrt nach Paris zu machen. Von Bohr habe ich seit einiger Zeit nichts mehr gehört. Ende September schrieb er mir noch, dass er einen schon seit längerer Zeit geplanten Brief wegen Dau jetzt in der geeigneten Form abgeschickt hat,[999] aber im übrigen hat er wohl ein schlechtes Gewissen, weil er die Arbeit seit

---

[997]David Shoenberg (1911–2004), Russian-born British physicist and close friend of the Peierls couple. In 1932 Shoenberg had graduated from Cambridge where he remained for the rest of his working life, working on low-temperature physics and, in particular, superconductivity.

[998]Yurij Borisovich Rumer (1901–1985), known in the West as 'George' had worked under Tamm in Moscow; on its foundation, in 1934, he joined the Physics Institute of the Academy of Sciences (FIAN) in Moscow. Later, he was forced to spend many years in exile. Hans A. Hellmann (1903–1938), German-born quantum chemist had emigrated from Nazi Germany in 1934, as his Ukranian wife was Jewish. He worked at the Karpov Institute in Moscow. In March 1938 he was arrested on charges of espionage and shot in May 1938.

[999]Letter Niels Bohr to Peierls, 30.9.1938, *Peierls Papers*, Ms. Eng. misc. b203, C. 30.

meiner Abreise nicht mehr angerührt hat. Ich fange an, die Hoffnung zu verlieren, dass das überhaupt je konvergiert. Haben Sie übrigens inzwischen Ihre Rechnungen mit dem Poyntingvektor gefunden? Ich war noch immer zu faul, das durchzuixen, aber wenn Bohr doch schliesslich Mal zu einem Entschluss kommt, dann möchte es doch ganz schön sein, ob man da die richtigen Faktoren hinschreiben kann.

Ueber die politische Situation schreibe ich schon lieber gar nicht, man ärgert sich doch zu sehr.

Mit herzlichen Grüssen
Ihr

[R.E. Peierls]

Wieso übrigens m.p.?[1000] M.P. verstehe ich, aber m.p.? 'meritierter professor? marchand de physique? miraculous prodigy? meticulously polite? Nein, ich kanns nicht erraten.

[266] Rudolf Peierls to Nevill Mott

Birmingham, 5.11.1938
(carbon copy)

Dear Mott,
I hope you have had a good time in the USA and returned in time before the outbreak of peace.

I have just seen your paper in the *Camb. Phil. Soc.*[1001] and I was very interested because I had considered the same problem from a similar point some time ago. My point was at the time that the double layer should give a simple rectifying effect, because if one writes down the condition for the equilibrium of the space charge it depends on the current that passes through the layer. If the dimensions of the

[1000] Refers to signature of letter [264].
[1001] N.F. Mott, 'The contact between a metal and an insulator or semi-conductor', *Proc. Camb. Phil. Soc.* **34**, 568–572 (1938).

double layer are large as compared with the mean free path (which seems to be the case if one makes reasonable assumptions about the orders of magnitude) one can get the conductivity merely from the electron density at each point, and in this way it is possible to work out the total potential difference as a function of current. At the time I gave up the idea Schottky convinced me that Cu_2O was an excess conductor (because of the Hall effect) and that would lead to the wrong sign for the rectifying effect. Since he discovered that his measurements of the Hall effect were wrong I have always meant to come back to the questions, but never did. The point is now, did you notice this application of the model and are you intending to work it out? If not I think I might dig out my old calculations again.

In the summer I was in Copenhagen again and there met Wick who seemed rather depressed about conditions for research in Italy, and (though presumably he is not in difficulties) seems anxious to leave.[1002] You remember that we discussed this question some time ago and came to the conclusion that, unless there was a job that offered reasonable chances of permanence one could not take the responsibility of inducing him to give up his job in Italy. I gather now that it would be quite feasible for him to get leave of absence for a year or so and it might therefore be possible to get him for a time without putting a moral or formal obligation on anyone. (It is not, of course, certain, that he can get leave before he has actually tried.) I thought I would just let you know about this change in the position, in case you are interested.

Will you come to the Physics Club? I do not know yet whether I manage; unfortunately I had to take a Saturday morning lecture this term, which is very hard to shift.
Yours sincerely,

[R.E. Peierls]

[1002]See letter Cian Carlo Wick to Peierls, 1.11.1938, *Peierls Papers*, Ms. Eng. misc. b216, C. 325.

[267] Rudolf Peierls to Hans Bethe

Birmingham, 5.11.1938
carbon copy

Lieber Hans,
Dies ist im Wesentlichen über Arbeitsprogramme.

Erstens. Du erzähltest mir in Cornwall, dass Deine Ansicht über den β-Zerfall nunmehr die wäre, dass wahrscheinlich der Fermische Ansatz der richtige ist,[1003] und dass, im Falle von erlaubten Übergängen das Auftreten von anscheinenden K.U.-Spektren immer durch mehrere Endzustände (und Strahlen) zu erklären ist.[1004] Diese Idee hatte ich schon vor längerer Zeit einmal mit Hoyle besprochen, aber wir hatten uns dann von den Experimentatoren einschüchtern lassen, die behaupteten, es gäbe Fälle, in denen es bestimmt keine β-Strahlen gibt. Hoyle hat sich weiter für diese Frage interessiert, und es gibt jetzt neue Experimente in Liverpool, die sehr stark für diese Erklärung sprechen. Ich möchte nun gerne wissen, ob Deine Bemerkung damals nur ein Stimmungsausdruck war, oder ob Du diese Frage ernsthaft untersuchst. In letzterem Falle würde ich Hoyle davon abraten, sich damit weiter zu beschäftigen, da Du das natürlich viel besser und schneller machst. Aber falls Du nicht die Absicht hast, es zu tun, so wäre es eigentlich sehr gut, wenn er das mal anständig diskutieren würde. (Eine Schwierigkeit liegt möglicherweise in folgendem: Wenn es auf der ersten verbotenen Sargent-Kurve viele Elemente gibt, die nachweislich keine β-Strahlen haben, und die alle ein U.K. Spektrum zeigen, so müsste dies auf die Verbotenheit zurückgehen. Nun folgt aber aus der Hoyleschen Arbeit,[1005] dass, falls das Elementargesetz das Fermische ist, zwar gewisse verbotene Übergänge ein U.K. Spektrum bekommen sollen, andere aber nicht. Es wird dann (da man ja über die Kernspins im allgemeinen zu wenig weiss) eine Frage der Statistik, ob die Erklärung plausibel ist.

[1003] E. Fermi, 'A theory of β-rays', *Z. Phys.* **88**, 161–177 (1934).
[1004] E. Konopinski and G.E. Uhlenbeck, "On the Fermi-theory of β-radioactivity', *Phys. Rev.* **48**, 7–12 (1935).
[1005] F. Hoyle, 'β-Transitions in a Coulomb field', *Proc. Roy. Soc* **A166**, 249–269 (1938).

Zweitens: (Ich weiss nicht, ob ich das nicht schon mal fragte) Wie steht es mit der Frage über die Elektronenpolarisation, die wir in Sandymouth besprachen. Ich würde an sich dieses Problem gerne Blackman[1006] vorschlagen, der darüber sicher sehr glücklich ist, da er bei G.P. Thomson sitzt und sich dort von Amtes wegen mit Elektronenbeugung befassen soll. Ich glaube, mit ein wenig Aufsicht, wird er das schon fertig bringen. Aber hast Du noch niemand für diese Problem in petto?[1007]

Auf diese beiden Fragen würde ich gerne bald eine Antwort haben. Hier ist inzwischen der Frieden ausgebrochen, wie Liebermann es einmal so schön in einem anderen Zusammenhang ausdrückte. "Soviel kann man gar nicht essen, wie man kotzen möchte!"

Placzek geht nun doch nach Paris (und dann nach Cornell) und Cockroft muss sich jemand anderen suchen. Von meiner Arbeit mit Bohr hoffe ich sehr, dass sie im Jahre 1939 erscheint, aber ich fürchte, es wird mindestens 1940.[1008]

Wie steht es mit der Übersiedelung Deiner Mutter? Meine Eltern haben natürlich so lange gewartet, bis es auch nach England fast unmöglich ist ein Visum zu bekommen, d.h. man bekommt es, aber wegen der Verstopfung des Home Office mit Anträgen von "modernen" Auswanderern aus der Tschechoslovakei und Italien u.s.w. dauert es mindestens drei Monate, bis irgendetwas passiert. Inzwischen scheinen aber die Auswanderungsbedingungen aus Deutschland immer schärfer zu werden. Wir leben in einer amüsanten Welt.
In diesem Sinne herzliche Grüsse
Dein

[Rudi]

[1006] M. Blackman was working with G.P. Thomson on electron diffraction. See G.P. Thomson and M. Blackman, 'Theory of the width of rings formed by electron diffraction', *Proc. Phys. Soc.* **51**, 425–31 (1939).
[1007] Rudolf Peierls, Hans Bethe and Fred Hoyle published the results in 1939. See letter [273].
[1008] Peierls, Placzek and Bohr published a joint paper in 1939, R. Peierls, G. Placzek and Niels Bohr, 'Nuclear reactions in the continuous energy region', *Nature* **144**, 200–201 (1939).

[268] Nevill Mott to Rudolf Peierls

Bristol, [date unspecified]

Dear Peierls,
The λ point is given by

$$N = \int_0^\infty \frac{N(E)dE}{e^{E/kT} - 1}$$

N = number of particles
$N(E)$ = density of levels
$= VNE \times \text{const.}$ (V = volume).
Thus λ-point is

$$\frac{N}{V} = T^{\frac{3}{2}} \int_0^\infty \frac{\sqrt{x}dx}{e^x - 1} \times \text{const.}$$

and does not depend on the volume.
Yours sincerely,

N.F. Mott

[269] Nevill Mott to Rudolf Peierls

Bristol, [date unspecified]

Dear Peierls,
Thanks for yours.[1009]

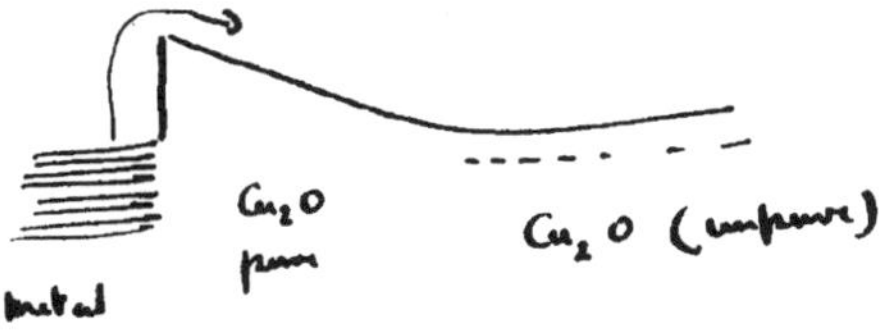

[1009] Letter could not be located.

Since it seems that at room temp. electrons can go over the pot[ential] barrier into ($Cu_2O + O$), it follows that if one had pure Cu_2O (no impurity levels) in contact with Cu, one could get a current from Cu $\rightarrow$ Cu_2O.

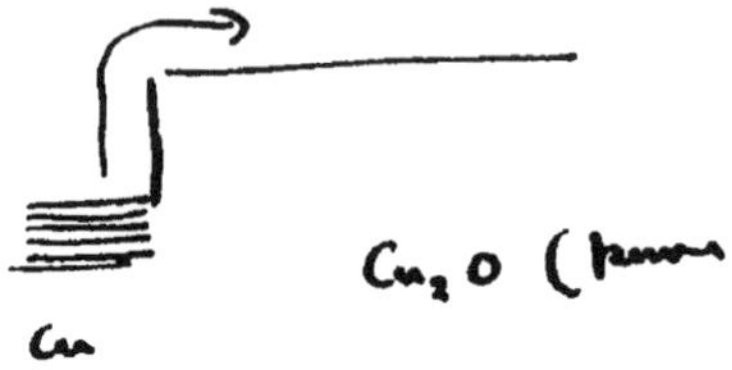

Of course a space charge gets set up — but if one works out the current by equations such as you suggest it seems that currents of the order 10^{-6} amps should be obtainable.

I've been trying to estimate the energies of the lowest empty levels in insulators/~ 0 in alkali-halides, $\sim -3e.V$ in silver halides. If one had a clacium electrode in contact with AgBr, the space charge would be the only thing preventing a current!

Of course, if electrons get trapped in the AgBr, the thing will insulate, but except at v[ery] low temperatures the traps aren't stable.

N.F.M.

[270] Nevill Mott to Rudolf Peierls

Bristol, [date unspecified]

Dear Peierls,
I have just realised that for some unknown reason I was imagining then when you said an excess conductor you meant a hole conductor — which is the origin of the muddle. Sorry.

So, now all is clear. But isn't the point, not that the blocking layer should be large compared with the m[ean] f[ree] p[ath], but that electrons should have to go over the potential barrier instead of through?

Have you seen Hartmann in the recent *Phys. Z.*[1010] who shows that there is no rectifying effect with Cu – Cu_2O if there is no insulating layer?

[1010] W. Hartmann, 'Artificial depletion layer on electronic semi conduction types' *Phys. Z.* **37**, 862–865 (1936).

If electrons can go from Cu into the conduction level of pure Cu_2O at room temperature, one ought to get a small current through pure Cu_2O with Cu electrodes — the current being linked by the space charge?

Do let me know if you are working this out. If not, I'll get a student to do it — but if you do, tant mieux, it will be quicker.
Yours,

N.F.M.

[271] Nevill Mott to Rudolf Peierls

Bristol, 12.11.1938

Dear Peierls,
I think I see what is wrong with Wilson's rectifying theory.[1011]
It all depends on whether the electrons go through the potential barrier separating metal from the semi-conductor, or whether the penetrability is negligible and they have to go over. In the former case the direction of easy flow is, in an excess conductor, from metal to s-c. This is Wilson's theory. But if they have to go over, it seems to me that the direction of the rectification must be reversed. If one tries to make electrons go from metal to s-c, one will get saturation when all the electrons which the metal emits thermally are sucked out through the s.c.; but in the other direction there is no limit to the current that one can get.
Yours,

N.F.M.

[1011] Wilson was among those who, in 1932, developed theories of crystal rectification. A.H. Wilson, 'A note on the theory of rectification' *Proc. Roy. Soc.* **A136**, 487–498 (1932) which he further elaborated in A.H. Wilson, *Theory of Metals*, (1936). Independently, Frenkel and Joffé (*Phys. Z. Sowjet.* **1**, 60 (1932)) and Nordheim, ('Zur Theorie der Detektorwirkung' *Z. Phys.* **75**, 434 (1932)) developed such theories.

[272] Rudolf Peierls to Niels Bohr

Birmingham, 10.12.1938
(carbon copy)

Dear Professor Bohr,
I am now returning the manuscript, of which I have had a copy made. The manuscript is complete, except for pages 13, 14, 20, which are apparently still in Copenhagen. If you have these pages, I would be very grateful if you could let me have them copied or, which would probably be more convenient to you, if you could have them copied in Copenhagen. If you have any difficulties in finding them, please do not trouble to look for them, but let me know: it is probably quite easy for me to reconstruct them from the older manuscript.

During the next few days I shall be fairly free and I shall try to write up a new introduction which would be more in line with your last discussion, and which might serve as a sketch for future modifications. However, I do not want to delay the manuscript until that is done.

I hope you had a decent crossing back. We were all very glad to have you here, and I immensely enjoyed the discussions.[1012] I only regret that so much of your time was wasted on less interesting things — if you come to Birmingham again (as I hope you will come) I shall know better how to avoid this.
With kindest regards, also to Mrs Bohr,
Yours sincerely,

[Rudolf Peierls]

[1012]Niels Bohr had lectured at Edinburgh and subsequently visited Birmingham in order to discuss the joint paper with Peierls and George Placzek.

[273] Rudolf Peierls to Hans Bethe

Birmingham, 10.12.1938
(carbon copy)

Lieber Hans,
Ich kann heute ein Kompliment erwidern, das Du mir vor einiger Zeit gemacht hast, nämlich Dir mitteilen, dass Du soeben eine Note an NATURE gemeinsam mit Hoyle und mir verfasst hast.[1013] Allerdings ist die Situation insofern durchaus anders, als damals die Frage war, ob ich überhaupt etwas dazu beigetragen hatte, während nun umgekehrt die Frage ist, ob Du das nicht alles schon selber weisst, und besser allein publiziertst. Hoyle und ich hatten die Frage schon einmal früher diskutiert, aber wir hatten den Experimentatoren geglaubt, die die Existenz von β-Strahlen leugneten, und ohne Deinen Optimismus in Cornell hätten wir uns die Sache nie richtig überlegt.[1014]

Über die Richtigkeit des Standpunktes kann wohl kein Zweifel mehr sein, und da die Experimentatoren in Cambridge in heller Begeisterung sind, und es gern bald schwarz auf weiss haben wollen, so soll man ihnen den Gefallen wohl tun, und es publizieren. Ich weiss aber nicht, ob nicht die richtigere Form die wäre, dass Du selbst eine Note darüber schreibst oder einen Brief an Dich selbst (als Editor der Phys. Rev.) Dass ich bereits ein Manuskript beilege, soll Deiner Entscheidung in dieser Beziehung in keiner Weise vorweggreifen, sondern nur zur Beschleunigung beitragen. Bitte, teile mir Doch in amerikanischem Tempo telegraphisch mit, ob ja oder nein, und wenn nein, ob Du es selbst schreibst. Wenn Du kleinere Änderungen in der Form wünschst, so kannst Du sie brieflich mitteilen, da wir sie dann noch rechtzeitig für die Korrektur bekommen. Ich schlage daher folgenden Code bei Deiner Antwort vor:

- Accepted = Ihr könnt das Manuskript in der vorliegenden Form absenden, eventuelle kleine Änderungen brieflich.

[1013] R. Peierls, H. Bethe and F. Hoyle, 'Interpretation of β-disintegration', *Nature* **143**, 200-202 (1939).
[1014] See letter [267].

- Amendments = Im Prinzip einverstanden, aber mir gefällt die Formulierung nicht, ich schicke postwendend ein abgeändertes Manuskript.

- Publishing = Ich schreibe selbst eine Note über diese Frage.

Neulich war Bohr auf ein paar Tage hier, und es war sehr interessant, obwohl unsere gemeinsame Arbeit in dieser Zeit nicht wesentlich fortgeschritten ist. In der letzten Zeit habe ich mich ein bisschen mit dem Beweis für die Vollständigkeit des von Kapur und mir verwendeten Funktionensystems beschäftigt,[1015] bei dem es mathematisch recht lustige Dinge gibt.[1016] Es stellt sich nämlich heraus, dass in den Ausnahmefällen, in denen das Quadratintegral einer Eigenfunktion verschwindet, das System in der Tat nicht vollständig ist, aber trotzdem gelten noch alle physikalisch interessanten Formeln, die mit Hilfe der Vollständigkeit abgeleitet sind, wie übrigens aus Stetigkeitsgründen sowieso plausibel ist. Es stellt sich heraus, dass wenn so ein Quadratintegral verschwindet, gerade zwei Eigenfunktionen identisch werden, was bei reeller Randbedingung nicht passieren kann. In dem Augenblick, wo sie gleich sind, hat man natürlich eine Funktion zu wenig zur Vollständigkeit. All das ist natürlich reine Mathematik, aber sehr amüsant.

Meine Eltern hoffe ich, in einigen Wochen hier zu haben. Wie weit ist es mit Deiner Mutter?

Mit herzlichen Grüssen von uns allen
Dein

[Rudi]

[1015]P.L. Kapur and R. Peierls, 'The dispersion formula for nuclear reactions', *Proc. Roy. Soc.* **A166**, 277–95 (1938).

[1016]The Peierls-Kapur paper discussed the response of a nucleus to neutron bombardment in terms of resonance. The paper takes for granted that the resonance states as defined form a complete basis. This is proved by Rudolf Peierls in a paper ten years later. R. Peierls, 'Expansion in terms of sets of functions with complex eigenvalues', *Proc. Camb. Philos. Soc.* **44**, 242–50 (1948).

[274] Hans Bethe to Rudolf Peierls

Ithaca, 12.12.1938

Deine Adresse habe ich wieder verdrängt
38 Calthorpe Road?
Lieber Rudi,
Dein Brief ist jetzt auch schon wieder über einen Monat alt, und Arbeitsprogramme sollte man eigentlich sofort beantworten.

Also erstens: Über die Elektronenpolarisation hat mein Rose viel, und ich glaube Vernünftiges, gerechnet. Das Resultat ist absolut Null. Er hat die Depolarisation durch die Stösse unter kleinen Winkeln berechnet, die zwischen den beiden Ablenkungen um 90° stattfinden. Nun gibt es zwar bei Dymonds Folien etwa 50 solche Stösse, aber die Depolarisation ist doch nur 3%, so dass sodass die theoretische Asymmetrie von 12% auf 11.6 reduziert wird. Berücksichtigt wurden elastische und unelastische Stösse, im ersten Fall durch Einfluss von Austausch mit einem Leuchtelektron entgegengesetzen Spins keine Bornsche Näherung. Er wird jetzt noch versuchen, ob vielleicht die Wechselwirkung mit Leitungselektronen was ausmacht, aber es scheint sehr unwahrscheinlich. Und ich bin überzeugt, dass die Kristallbeugung nichts tut.[1017]

Hingegen habe ich mich nicht mit den β-Spektren befasst, und meine diesbezüglichen Bemerkungen in Cornwall waren nur ein Stimmungsausdruck. Darum möchte es sehr schön sein, ob der Hoyle das anständig diskutieren würde. Im übrigen ist es ja inzwischen einem Californier gelungen, die obere Grenze des β-Spektrums von N^{13} <u>unter</u> die theoretische Grenze (von Kernreaktionen) zu bekommen. Was sind die neuen Experimente in Liverpool?

Wie geht es dem Compound nucleus? Placzek schrieb, traurigerweise "hättest Du doch recht", nämlich damit, dass die schönen Formeln alle nichts nützen.[1018] Ich habe jetzt einen sehr gescheiten Mann,

[1017] Bethe and Rose published the results shortly afterwards. M.E. Rose and H.A. Bethe, 'On the Absence of Polarization in Electron Scattering', *Phys. Rev.* **55**, 277–89 (1939).
[1018] See letter [259].

Groenblom, einen Finnen, der bei Heisenberg doktoriert hat.[1019] Er soll versuchen, eine genügend gute Näherung für schwere Kerne zu kriegen, so dass man wirklich sehen kann, wie weit die "Verschmierung" im Compound nucleus geht. Ob es z.B. wahrscheinlich ist, dass man irgendwelche Quasi-Auswahlregeln kriegt, (ausser den trivialen für Drehim[uls] und Parität), wie sehr Matrix-Elemente schwanken etc. Ich glaube, wir haben eine Methode, die von den Fehlern der Hartree Methode frei ist, indem sie die E[igen]funktion bei kleinem Abstand zweier Teilchen exakt behandelt, und trotzdem hoffentlich noch übersehbar ist.

Persönlich habe ich mich mit den Sternen beschäftigt, bin aber, wie Ehrenberg zu sagen pflegt, jetzt ziemlich damit abgefüttert. Immerhin habe ich herausgefunden, dass praktisch alle Sachen, die die Astrophysiker annehmen, falsch sind. Ausserdem dass kein existierender Stern einen Neutronenkern haben kann, da der Stern dann kleiner sein müsste als die Erde. Ferner, dass es andererseits keine untere Grenze für die Grösse eines Neutronenkerns gibt, wie sie alle Gelehrten anzunehmen belieben. Und weitere derartige destruktive Dinge.[1020]

Wobei mir ein abstract von Houston im neuesten Bulletin of the American Physical Society einfällt, in dem er sich beklagt, dass Dein Erhaltungssatz für $\sum k_{Elektron} = \sum k_{Gitterwellen}$ in der Metalltheorie eine so katastrophale Wirkung auf die Entwicklung der besagten Theorie hatte. Er bringt es fertig, dann Dinge zu schreiben, die sowohl trivial wie falsch sind — das ist immerhin ein Erfolg. (Ich rege mich darüber glaube ich bloss auf, weil einige Leute hier den Houston als Institutsdirektor haben wollen).

Glaubst Du eigentlich auch, wie Møller, dass neutrale Teilchen eine reelle Eigenfunktion haben, d.h. $\psi^* grad\psi$ etc. den elektrischen und nicht den Teilchenstrom bedeutet? Das erschüttert meine tiefsten Überzeugungen — aber es mag ja richtig sein.

[1019] Berndt Olof Grönblom (1913–1941), had studied in Leipzig where he received his doctorate, and after a short spell at Cornell he returned to Heisenberg's institute. When Stalin invaded Finland, Grönblom returned to his home country and died in action in August 1941.

[1020] Bethe had just submitted his first substantial paper on energy production in stars. H.A. Bethe, 'Energy Production in Stars', *Phys. Rev.* **55**, 434–56 (1939).

Speaking of Møller: Wie sollte doch die Retardierung eine Verminderung der Divergenz in der Barytron-Theorie bewirken? Ich vesuchte es Weisskopf zu erklären und brachte es nicht fertig.

Hast Du noch etwas über die Energie des Deuterons mit dem $\sigma_1 \cdot r\sigma_2 \cdot r$ Glied in der Wechselwirkung gerechnet? Ich will nächstens mal publizieren, was ich über diese Probleme gemacht habe und den Optimismus der Yukawisten etwas dämpfen durch Divergenzen.[1021]

Weisskopf und ich haben uns ziemlich viel mit den Divergenzen in der Elekrodynamik beschäftigt. Es scheint, dass die Löchertheorie wirklich viel besser ist als die Dirac-Theorie ohne Löcher, und dass es eigentlich überhaupt keinen Sinn hat, in der letzteren die Selbstenergie zu rechnen. Neuerdings behauptet Weisskopf, er könnte beweisen, dass die Selbstenergie in der Löchertheorie nicht stärker divergiert als $\frac{1}{r}\log r$ (in <u>keiner</u> Näherung!) aber das glaube ich vorerst noch nicht.[1022]

Es gibt schrecklich viele Leute hier zurzeit, und Bacher[1023] spricht nur noch von meiner production line. 5 Doktoranden, 3 zukünftige ditto, besagter Groenbom, Rose, und Ludloff (letzterer kein grosses Vergnügen). Es wird mir etwas zuviel, denn ich komme nicht mehr zum eigenen Arbeiten.

Placzek's Entscheidungen sind mir rätselhaft. Dieser Brief ist nur über Physik, denn wenn man an anderes denkt, wird man wahnsinnig. Meine Mutter ist leider (mit meinem Einverständnis) nach Ausbruch des Friedens wieder nach Baden-Baden zurückgefahren, um die Auswanderung vorzubereiten. Zwischendurch entschied sie sich mal wieder, sie wolle überhaupt in Baden bleiben. Inzwischen ist das trivialerweise vorüber, und ich habe ihr gestern das Affidavit geschickt. Aber —
Und Deine Eltern?

[1021]Bethe published a paper on nuclear forces the following year. H.A. Bethe, 'The Meson Theory of Nuclear Forces', *Phys. Rev.* **55**, 1261–63 (1939).

[1022]Weisskopf published his results in V. Weisskopf, 'On the Self-Energy and the Electromagnetic Field of the Electron', *Phys. Rev.* **56**, 72–85 (1939).

[1023]Robert Fox Bacher (1905–2004), studied at Michigan where he received his doctorate in 1930, after posts at Caltech, MIT, Michigan, and Columbia University, he became professor of physics and director of the Laboratory of Nuclear Studies at Cornell in 1935. Later he joined the Manhattan Project and after the war played a leading role in high energy physics at Caltech.

Solange man nur an die nächste Umgebung denkt, ist es besser. Darum recht schöne Weihnachten und sehr viele Grüsse an Dich und Genia und Gaby und Ronny. Es ist wirklich schade, dass man mit der Queen Elisabeth (meiner) nicht mal über Weekend nach Birmingham fahren kann. So muss man sich mit dem New Hampshire zu Weihnachten begnügen.
Tausend Grüsse

Hans

[275] Rudolf Peierls to Hans Bethe

Birmingham, 25.12.1938
(carbon copy)

Lieber Hans,
Vielen Dank für Deinen Brief,[1024] der sich mit meinem gekreuzt hat.[1025] Ich nehme an, dass Du diesen Brief bereits als vorweggenommene Antwort auf meine Anfrage bezüglich der Publikation betrachtest, und Dir daher das Telegramm sparen kannst, und ich interpretiere das als Einverständnis. Ich schicke daher heute die Note an NATURE ab. Sollte Dir das nicht recht sein, bitte ich um umgehenden Bescheid. Übrigens war Dee etwas böse, dass ich Dir das M[anuskript] geschickt habe, das die Feststellung enthält, dass man jetzt im Cavendish nach den β-Strahlen sucht. Er stellt sich offenbar vor, dass auf die Nachricht hin, dass dies im Cavendish gemacht wird, gleich ganz Amerika auch β-Strahlen suchen, und sie dann durch unzuverlässige Versuche finden werden. Also um ihn zu beruhigen, kannst Du vielleicht von der Information vorsichtigen Gebrauch machen (d.h. nicht von der theoretischen Feststellung, die Dir ja mehr als uns gehört, sondern von den unveröffentlichten Experimenten!)

Was die Polarisation betrifft, so würde ich es gerne sehen, wenn jemand mal anständig über die Frage diskutiert, wie weit man den Effekt

[1024] Letter [274].
[1025] Letter [273].

einzelner Atome einfach suspendieren kann (Denn wenn ich recht verstehe, hat das doch wohl Rose auch gemacht, wenn er auch die Stösse weiterer Atome mit der bereits gestreuten Welle mitnimmt.) Vielleicht ist es trivial, aber dann muss man es auch in ein paar Zeilen anständig begründen können.

Betreffend Möller: Ich habe mich inzwischen auch davon überzeugt, dass es sehr unwahrscheinlich ist, dass die Retardierung in der Spin-Spin-Wechselwirkung der von Neutron-Proton in der Mesotrontheorie etwas retten kann. Möller hat natürlich Recht damit, dass gerade für kleine Abstände die Retardierung wesentlich ist. (Denn damit die vernachlässigbar ist, muss die Präzessionsperiode eines Spins in Felde des anderen lang sein gegen die Zeit, die ein Mesotron braucht, um von einem zum anderen zu laufen, und das ist gerade bei kleinen Abständen nicht der Fall). Andererseits kann man sich leicht überlegen, was passieren würde, wenn man grosse Spins (statt $\frac{1}{2}$) hätte. Dann würde der Zustand tiefster Energie, der der unendlichen Anziehung entspricht, gerade dem entsprechen, dass die Spins parallel zu ihrer Verbindungslinie eingestellt werden. In diesem Falle ist aber das "Feld" in der Richtung des Spins und daher gibt es keine Präzession und also macht die Retardierung nichts aus. Nun könnte es natürlich gerade noch sein, dass die Retardierung für Spin $\frac{1}{2}$ gerade noch die Unendlichkeit schwächer macht, aber das erscheint schon sehr unplausibel. Immerhin ist die Wechselwirkung zweier Spins mit Retardierung, aber ohne ihre Translation in Betracht zu ziehen, ein winziges Problem, und ich möchte gern jemanden das rechnen lassen.

Was die neutralen Mesotronen betrifft, so gefällt mir die Formulierung von Kemmer (Proc. Camb. Phil. Soc.[1026]) sehr gut, weil man dort in so zwangloser Weise die Gleichheit der Kräfte zwischen like und unlike particles bekommt.

Sonst habe ich über die Spindivergenzen nichts weiter gerechnet. Ich habe noch etwas vor; es erscheint mir nicht ausgeschlossen, dass man ohne den Spin-Spin-Term auskommen kann, d.h. ohne den zweiten Wechselwirkungsterm in der Hamiltonfunktion, der für die

[1026]N. Kemmer, 'The charge dependence of nuclear forces', *Proc. Camb. Phil. Soc.* **34**, 354–364 (1938).

ganze Schweinerei verantwortlich ist. Man sagt immer, dass ohne diesen Term man keine Heisenbergkraft bekommen würde, und dass dann der Singlett- und Triplettzustand von ^{2}H dieselbe Energie haben sollte. Das ist aber ein fauler Zauber, denn um dieses Resultat zu bekommen, wird in der Ableitung die Geschwindigkeit der schweren Teilchen vernachlässigt. Nun benutzt man aber für die schweren Teilchen die Diracgleichung und die Geschwindigkeit ist daher immer gleich c (nur ihr Mittelwert ist sehr klein).

Dieses Argument ist also genauso schön als ob man sagen würde, dass die Diracgleichung kann die Feinstruktur der Atome nicht erklären, denn wenn das Elektron sich langsam bewegt, darf man die [...][1027] gleich Null setzen, und dann gibt es keine magnetische Wechselwirkung, also muss man in die Diracgleichung noch die Neutrinoterme hinschreiben.

Nun ist es im Atom allerdings so, dass die Terme, die von der "Zitterbewegung" herrühren, immer um v^2/c^2 kleiner sind als die statischen. Ein entsprechendes Resultat wäre im Kern wohl zu wenig. (Ich bin nicht mal davon überzeugt.) Aber das Eintreten der Geschwindigkeit ist doch dort ziemlich zufällig, in Wirklichkeit steht natürlich da $\ldots/r^3$, und das r^3 hat eben im Atom so eine Grössenordnung, dass das ganze gleich $v^2/c^2 \cdot e^2/r$ wird. Das braucht aber alles im Kern nicht mehr zu stimmen, insbesondere weil ja dort die h/mc für die schweren Teilchen mit dem mittleren Abstand durchaus vergleichbar sind.

Aber all das hat wenig mit dem zu tun, was Du gerechnet hast. Es möchte nur schön sein, ob Du mir eine Kopie schicken könntest, wenn Du darüber etwas dichtest, dann könnte ich Deine Zerschmetterung der Spin-Spin-Wechselwirkung einfach zitieren (falls ich etwas herausbringe) ohne mich darauf näher einzulassen.

Vorgestern kam ein Telegramm von Teller mit (eventueller) Einladung zur Konferenz im Januar.[1028] Mit sehr blutendem Herzen musste ich absagen, denn ich habe hier für den nächsten Term viele Vorlesungen übernommen (11 pro Woche) dass es praktisch unmöglich ist, sie alle zu versäumen, und auch nicht, mich vertreten zu lassen, da ich vorläufig

---

[1027] Missing in manuscript.
[1028] Meeting of the American Physical Society, January 1939.

der einzige Theoretiker in Birmingham bin. Ich möchte gern einen Assistenten haben, und bemühe mich daher, zu beweisen, dass es soviel Arbeit für Theoretiker hier gibt, dass ich sie nicht allein schaffen kann (was übrigens stimmt). Dann wäre es natürlich sehr unpädagogisch, zu versuchen, ob es auch mal vier Wochen ohne mich geht. Die Schiffe gehen nämlich gerade so blöd, dass ich von hier vier Wochen weg sein müsste. (Im Summer Term, der im wesentlichen aus Examen besteht, wäre es natürlich leichter). Aber ärgerlich ist es doch!

Für meine Eltern habe ich das englische Visum bekommen, was ja jetzt gar nicht mehr so einfach ist, und sie bemühen sich jetzt schon seit sechs Wochen, ihre Unbedenklichkeitserklärung u.s.w. zu bekommen. Es ist sehr schwer, denn sie haben natürlich kein Bargeld, um das "Sünegeld" zu bezahlen, und Papiere darf man nicht ohne Genehmigung verkaufen, und die Genehmigung kriegt man nicht. Natürlich hat man das Auto weggenommen, was die Lauferei nicht gerade beschleunigt. Immerhin sind sie noch in ihrer Wohnung, und es sieht doch so aus, als ob sie in absehbarer Zeit alle Papiere zusammenbekommen werden. Es geht ihnen also noch wesentlich besser als den meisten anderen Leuten.

Schrödinger läuft hier in England herum (kriegt wohl eine Stelle in Dublin) und wundert sich sehr, wenn die Leute von ihm verlangen, er sollte doch von seiner Erklärung im Grazer Tageblatt abrücken.[1029] Das müsste doch jeder wissen, dass wenn er so etwas schreibt, es nicht ernst gemeint ist.

Im übrigen kriegt man natürlich massenhaft Briefe von Leuten, die nach England wollen, und den meisten muss man schreiben, dass es so gut wie ausgeschlossen ist. Natürlich gibt es unerwartete Glücksfälle, wie der von Kober,[1030] einem Mathematiker, der mit

---

[1029]Between 1936 and 1938, Schrödinger taught at the University of Graz, where — on the advice of the new Nazi rector of the university — he wrote an open letter to the University Senate that he had 'misjudged up to the last the true will and the true destiny of his country'.

[1030]Hermann Kober (1888–1973) studied at Breslau and Göttingen from where he received his doctorate in 1911. He taught at a Jewish school in Breslau but spent time in Cambridge doing research which helped him later to secure a grant to work in Birmingham in 1939.

Watson[1031] befreundet ist. Er war früher Schullehrer und hat sich 33 zurückgezogen, um privat Mathematik zu betreiben. Jetzt hat sich ein früherer Schüler von ihm gefunden, der bereit ist, ihn und seine Frau vier Jahre lang hier zu erhalten, und bereits das Geld dafür auf den Tisch des Hauses gelegt hat. Kober selbst sitzt noch "im Konzert" wir hoffen aber, dass recht bald herausgelassen wird.

Aber Du hast recht, denken wir lieber an die nächste Umgebung (und auch das mit Auswahl, denn bei uns um die Ecke ist die Villa von Chamberlain). Wir haben eben die obligate Pute gegessen und dabei mit Rührung der Ente gedacht, mit der Genia und Du einen tug-of-war angestellt habt.[1032]

Beste Grüsse von und an
Dein

[Rudi]

[276] Rudolf Peierls to Heinrich Peierls

Birmingham, 27.12.[19]38
carbon copy

Liebe Eltern,
Ich möchte noch einmal wiederholen, dass nicht der geringste Grund zur Beunruhigung besteht. Die einmal erteilte Genehmigung kann nicht wieder zurückgezogen werden, und ausserdem ist auch gar kein Anlass dazu. Der Grund warum überhaupt Schwierigkeiten mit der Erteilung der Visen gemacht werden, ist doch nur die Befürchtung, dass die Einwanderer entweder hier arbeiten wollen, und daher Engländern die Stellen wegnehmen, oder dass sie hier mittellos werden und unterstützt werden müssen. Daher kann es also für die Erteilung eines Visums nur helfen (wenn Ihr es noch nicht bewilligt bekommen hättet),

[1031] George Neville Watson (1886–1965), studied at Trinity College Cambridge, where he became a fellow in 1910. Between 1918 and 1951 he was Mason Professor of Pure Mathematics at Birmingham University.

[1032] Refers to Genia's first attempt at preparing a Christmas dinner while Bethe and the Peierls were living in Manchester.

wenn Ihr Aussichten auf ein USA Visum hättet). Aussderdem schien es mir eigentlich nach einem Brief von Siegfried, dass der Einfluss von George Newgas's Freund dazu dienen sollte, Euch ein englisches Visum zu beschaffen. Ich habe aber den fraglichen Brief von Siegfried an Alfred weitergeschickt, und kann ihn also nicht noch einmal daraufhin ansehen. Gestern konnte ich Alfred nicht erreichen, werde sie aber heute abend anrufen, und noch einmal nach dem Wortlaut dieses Briefes fragen. Mir schient es eigentlich wahrscheinlich, dass Hermann etwas missverstanden hat, und dass es sich um das englische Visum handelt (das Ihr ja inzwischen nicht mehr braucht).

Vielen Dank für Euern Brief, der heute kommt. Wir können über die Wohnung noch keine sehr bestimmten Angaben machen, da wir sehen müssen was wir finden. Die Anzahl der Wohnungen mit Zentralheizung ist nicht sehr gross, weil es zu viel Arbeit macht, sie in Ordnung zu halten. Ich würde eher drei Zimmer raten. Aber es kommt auch etwas auf die Grösse und Lage der Zimmer an. Überflüssige Möbel könnt Ihr immer bei uns unterstellen.

Bis die Wohnung benutzbar ist, können die Lifts, wie ich schon schrieb, bei uns stehen. Wenn Ihr uns die genauen Aussenmasse mitteilt, so können wir schon jetzt sehen, ob sie in die Garage hineingehn, oder im Freien stehen müssen, das macht vielleicht einen Unterschied für die Verpackung. Während dieser Zeit im Hotel zu wohnen, scheint unnötig teuer und ungemütlich. (Ich kenne allerdings das von Preuss empfohlene Hotel nicht und weiss nicht wie teuer und wie anständig es ist.) Ausserdem werdet Ihr gerade in der ersten Zeit, wenn Ihr noch nicht so gut Bescheid wisst, in Kensington recht einsam sein, denn Alfreds können doch nicht so oft in die Stadt kommen, und Eure anderen Bekannten werden wahrscheinlich auch beschäftigt sein. Ich würde glauben, dass es das Beste wäre, (da unser Haus vielleicht wirklich etwas zu kalt wird, besonders wenn das Wetter so ist wie in den letzten Tagen) eine möblierte Wohnung oder ein paar möblierte Zimmer in einem Haus mit Zentralheizung zu finden. Ich glaube, das sollte man hier in der Nähe finden können, und wir werden uns gleich morgen danach umsehen. Wir werden auch gleich an die Suche einer dauernden Wohnung gehen; bisher konnten wir das noch nicht, da wir nicht sicher wussten, was Eure Pläne sind. (Ich glaube eine sehr geeignete Wohnung hier um die Ecke ist uns gerade durch die Lappen gegangen.)

Was Ludwig betrifft, so hatte ich einen Brief von Karli mit Bitte um Rat. Die Lage ist so, dass sie mit Sicherheit nach USA gehen können, sobald Karli dort eingebürgert wird, also in $2\frac{1}{2}$ Jahren. Bis dahin müsste man sie wo anders, z.B. in England, unterbringen und Karli ist bereit, hier eine Summe zu deponieren, die für ihren Lebensunterhalt für diese Zeit ausreichen sollte, und ich nehme an, er würde auch bereit sein, sich zu verpflichen, sie auch weiter hier zu unterhalten, falls irgendeine Verögerung in ihrer Einreise nach USA entsteht. Die Schwierigkeit ist nun, dass seine Garantie für die englischen Behörden keinen Wert hat, da er nicht hier ist, oder jedenfalls nicht hinreicht, wenn sie nicht durch alle möglichen offiziellen Beglaubigungen und andere langwierige Scherereien gestützt wird. Es erschien mir daher das aussichtsreichste zu versuchen, ob nicht einer der englischen Schriftsteller bereit wäre, die Garantie zu unterschreiben. Solange Karli nicht bankrott macht, bedeutet das für den Garantor keinerlei finanzielle Belastung, er übernimmt eben nur die Verantwortung dafür, dass Karli nicht ausrückt. Ich habe, um keine Zeit zu verlieren, gleich auf eigene Faust an einen sehr bekannten englischen Schriftsteller geschrieben, und ihn um Rat gebeten, habe aber noch keine Antwort. Hat Ludwig hier unter den Schriftstellern (oder anderweitig) Bekannte, die ihn persönlich kennen, und von denen es sich vermuten lässt, dass sie sich zu einer solchen Garantie bereit erklären würden?

Die Kälte der letzten Tage war wirklich unerwartet, die meisten Autos sind eingefroren (unseres fing auch schon an) und alle Garagen machen ein Bombengeschäft mit Frostschutzmitteln. In den Häusern ist es natäurlich auch ziemlich kalt, und besonders für Fr[äulein] S[onnenthal], die gerade in diese Kältewelle hineinkam, war das sehr ungewohnt.

Inzwischen weiter viel Erfolg mit Euren Bemühungen, wir können uns vorstellen, wieviel Laufereien und Scherereien notwendig sind, und hoffentlich auf baldiges Wiedersehen. Mit herzlichen Grüssen von uns allen,

Euer

[Rudi]

[277] Siegfried Peierls to Rudolf and Alfred Peierls

New York City, 30.12.1938

Dear Alfred and Rudi
I just received your letters of the 18th inst. and have immediately asked Bankers Trust Co. here to transfer £500.- to the Foreign Department of Barclays in London.

You will see from the enclosed copy of my letter that I have asked them to request Barclays to acknowledge the receipt of this remittance to both of you separately as I do not know whether your father will first stop over in London or go right on to Birmingham. I have sent this large amount although I know that they do not need all of this for the present, but this was the amount that the English consul had mentioned some time ago to your father and I wanted to be sure that your father would not be embarrassed in any way.

Let us all hope that 1939 will bring better days for all of us and find your parents comfortably installed in England.
Lovingly yours,

Uncle Siegfried

[278] Rudolf Peierls to Alfred Peierls

Birmingham, 1.1.[19]39
(carbon copy)

Lieber Alfred,
Zunächst mal Prosit Neumann. Hoffentlich habt Ihr vergnügt gefeiert. Dem alten Jahr wird man ja sicherlich keine Träne nachweinen.

Ich wollte neulich das Gespräch nicht noch ausdehnen, da es erst 6.50 Uhr war. Es fiel mir erst nachher ein, dass wahrscheinlich Deine Uhr vorging. Was die Wohnungsfrage für die Eltern betrifft, so habe ich nach ihren Briefen und Telefongesprächen den Eindruck, dass sie nicht in der Lage sind, irgendwelche Entscheidungen selbst zu treffen. Erstens kennen sie die Verhältnisse nicht (und können sich auch keine genauen Vorstellungen von dem Lebensstandard machen, den sie hier führen werden) und vor allem sind sie durch die vielen Scherereien, die sie jetzt haben, so herunter, dass sie alles nur durcheinander bringen. Wir müssen uns daher selbst überlegen, was für sie richtig ist, und ihnen dazu dann energisch raten, dann werden sie schon einverstanden sein.

Zunächst ist es klar, dass sie unbedingt entweder in unserer oder in Eurer Nähe wohnen müssen und dass dann entweder wir oder Ihr die volle Verantwortung für sie übernehmen müssen. Sie verstehen natürlich nicht zu wirtschaften, und vor allen Dingen, wenn sie irgendwie in Schwierigkeiten kommen (oder nur Schwierigkeiten zu sehen glauben, was sie sicher dauernd tun werden), so muss man ihnen helfen. Besonders wenn Else nicht wohl ist, auch wenn sie nur erkältet ist, oder so etwas, so muss natürlich jemand da sein, der sich um Vati kümmert und für beide sorgt. Aus diesem Grund kommt es nicht in Frage, dass sie etwa in Kensington wohnen, denn ganz abgesehen davon dass das Leben dort viel teurer ist und noch Fahrgelder und solche Sachen dazu kommen, ist das zu weit von Euch, um ständig für sie zu sorgen, falls etwas nicht in Ordnung sein sollte, oder wenn sie etwas brauchen. (oder wenn Ihr es tun wollt, so wird das furchtbar anstrengend.) Ihre Bekannten wohnen ja sehr über London verstreut, und ausserdem würde ich glauben, dass von denen viele ihre frühere Anhänglichkeit vergessen werden, wenn sich die Verhältnisse geändert haben.

Ob sie nun in unserer oder Eurer Nähe wohnen, ist natürlich im Grunde genommen gleich, es spricht aber vieles für Birmingham. Hier würden sie fast immer unser Auto auch tags zur Verfügung haben, da Genia chauffiert und ich auch mit meiner Tageseinteilung nicht so gebunden bin wie Du. Genia kann z[um] B[eispiel] ohne grosse Mühe die Einkäufe für Else miterledigen, was eine grosse Ersparnis bedeutet, da sie auf den Markt fahren kann. Ausserdem kann eben Genia oder ich jederzeit mit dem Wagen bei ihnen sein, sobald etwas Dringendes gebraucht wird. Dann haben sie hier in der unmittelbaren Nähe viele Annehmlichkeiten einer grossen Stadt, während von Euch die Reise nach London doch teuer und für Vati wohl anstrengend wird. Ebenso haben wir hier in der Nähe eine ganze Menge deutscher Bekannter, und Engländer, die Deutsch sprechen, die sie häufiger sehen können als es bei den Londoner Entfernungen möglich wäre. Sie können hier mit den

Kindern spazieren gehen, was sicher für ihre Nerven eine beruhignede Beschäftigung ist.

Ich habe das alles aufgeschrieben, um am Telefon zu sparen, und wenn Ihr Euch die Frage durch den Kopf gehen lasst, so werde ich am Dienstag abend anrufen, sodass wir uns darüber einigen könne, bevor wir den Eltern einen endgültigen Rat geben können.

Wir haben uns hier nach Wohnungen umgesehen, und bisher das beste ist eine sehr gut gebaute Wohnung mit Zentralheizung, einem sehr anständigen Bad, Küche mit Refrigerator u.s.w. zwei sehr grosse Zimmer mit Diele für £120 pro Jahr. Das ist ungefähr 1 1/2 Meilen vom Stadtzentrum und etwa 1 Meile von uns. Das Haus ist sehr "standesgemäss", und ich glaube nicht, dass man in London für dasselbe Geld etwas besseres finden könnte.

Herzlichst Dein

[Rudi]

[279] David Shoenberg to Rudolf Peierls

Cambridge, 18.1.[19]39

Dear Peierls,

Here is the formula in full, though I imagine it is only the isotopic case which is necessary (I am not quite clear how the numerical factors will be altered for the isotropic case). The energy of [?] is

$$E = E(\mu_1) + E(\mu_2) + E(\mu_3)$$

where

$$\mu_1 = \lambda\sqrt{m_1 H_1^2 + m_2 H_2^2 + m_3 H_3^2 + 2m_4 H_2 H_3}$$

$$\mu_2 = \lambda\sqrt{m_1(\frac{H_1 - \sqrt{3}H_2}{2})^2 + m_2(\frac{H_2 + \sqrt{3}H_1}{2})^2 + m_3 H_3^3 - 2m_4(\frac{H_2 + \sqrt{3}H_1}{2})H_3}$$

$$\mu_3 = \lambda\sqrt{m_1(\frac{H_1 + \sqrt{3}H_2}{2})^2 + m_2(\frac{H_2 - \sqrt{3}H_1}{2})^2 + m_3 H_3^2 - 2m_4(\frac{H_2 - \sqrt{3}H_1}{2})H_3}$$

$$\lambda = e^{\hbar}/c\sqrt{m_1(m_2m_3 - m_4^2)}$$

$$E(\mu) = \frac{kT\mu^{3/2}\sqrt{m_1(m_2m_3 - m_4^2)}}{\pi^2\hbar^3}\sum_{\nu=1}^{\infty}\frac{(-1)^r\cos(\frac{\pi r E_0}{\mu} - \frac{\pi}{4})}{r^{3/2}\sinh 2\pi^2 rkT/\mu}$$

$$+ \text{ non-periodic Term}$$

This approximation is valid only if $E_0 \gg kT$ and $E_0 \gg \mu$.

The work of London I mentioned was in the Phys. Rev. about a month ago,[1033] and Allen says it agrees very well with his flow experiments.

I have to give a talk to the Cam. Phys. Soc. (next week) about size effects in superconductors, although I intend to make it mainly experimental, I should like to say something about the theory which I am in an unholy muddle at the moment. So any clarification you can offer would be most welcome just now. The point at issue is, if one assumes any particular law of falling-off of field inside a superconductor, how can the critical field be determined as a function of size. In particular, assuming as London does, that the law is of an exponential type, which is I think equivalent to this equation and curl $j = \lambda H$ (for since curl $H = 4\pi j$ this gives curl curl $H = 4\pi\lambda H$) (is or isn't the size dependence what London or von Laue[1034]) claims it to be. What I am not clear about is whether in von Laue's derivation the relation curl$j = \lambda H$ is used only in determining the field decay or whether it also enters in some other place in a backhand way. If the derivation is correct, then how can one explain the discrepancies between it and the surface energy argument, which leads to a [?] law for large d. $H_C = H_{C_0}(1 + \frac{d_0}{d})$, as against $H_C = H_{C_0}(1 + e^{-d/d_0})$(von Laue) for a plate. (incidentally for a wire there is no discrepancy).

[1033] F. London, 'On the Bose-Einstein Condensation', *Phys. Rev.* **54**, 947–54 (1938). Fritz London had given a proof of the condensation phenomenon of a Bose-Einstein gas and a preliminary outline of its transport properties with a view to its potential bearing on liquid helium discussions. Shoenberg was interested in the properties of superconductors and superfluids.

[1034] Max v. Laue, 'The thermodynamics of superconduction', *Ann. d. Phys.* **32**, 71–84 (1938).

"*Isvestija*" is very full just now of the elections to the Academy of new members and full lists of the candidates in thick black type are being published every day. I notice of the candidates that Kapitza, Fock, Akulov, Arkadiev[1035] are among the candidates for full member, but perhaps significantly, not Tamm. If you are interested I can send you the lists, though I should like to compare them later with the elected candidates. (the elections by secret ballot, are taking place from Nov. till Jan. 26th.) [???]
Yours,

David Shoenberg

[280] Rudolf Peierls to Else and Heinrich Peierls

Birmingham, 23.1.[19]39
(carbon copy)

Liebe Eltern,
Vielen Dank für Euren Brief vom Sonnabend und Eure Karte.[1036] Wie ich schon in meinem letzten Brief sagte, fanden wir Eure Auswahl durchaus richtig und wollten nur für den, (wie gesagt, möglichen) Fall, dass nicht alles hineingeht, einen Rat geben, was von der Liste uns weniger wichtig als der Rest erschien. Goole habe ich nur genannt, weil Ihr das angegeben habt (was mir erstaunlich schien) der normale Hafen für den Verkehr nach Birmingham ist natürlich Hull. Ich weiss natürlich, dass ich die Versicherung von hier aus bezahlen muss, aber ich wusste nicht, ob ich sie von hier aus anmelden soll, oder ob der Spediteur das tut und ich einfach die Rechnung bekomme.

[1035]Vladimir K. Arkadiev and Nikolai Sergevich Akulov (1900–1976) Russian physicists working on magnetism.
[1036]In late 1938 and early 1939, letters between Rudolf Peierls and his parents were exchanged with great frequency, as both sides tried to arrange for the parents' emigration to England. Most of the letters are not included in this collection, as the same topics are revisited. This is one typical example of the communications, demonstrating the obstacles to be overcome in the process of leaving Germany and resettling, either temporarily or permanently, in the UK.

Was Eure sonstigen Bemerkungen über Eure Pläne betrifft, so wissen wir natürlich, dass Ihr in der letzten Zeit viel durchgemacht habt, und dass Ihr mit den Nerven sehr herunter sein werdet, wenn Ihr herkommt, und dass Euch die Eingewöhnung in das neue Land zuerst schwerfallen wird. Gerade deswegen bemühen wir uns doch, alles so einzurichten, dass Ihr es hier so bequem wie möglich habt. Vielleicht ist es wirklich das beste, wenn Ihr zunächst für einige Zeit zur Erholung fahrt, wenn Ihr es einrichten könnt, aber das ändert doch die Pläne auf längere Sicht nicht. Auf längere Zeit in einer Pension zu wohnen, wo man natürlich ganz im englischen Stil lebt, wo es englisches Essen gibt, das Euch sicher nicht schmecken wird, und das Ihr vielleicht nicht vertragt, scheint mir aus eben den Gründen, die Ihr anführt, keine ideale Lösung. Was ich über die Möglichkeit denke, längere Zeit bei Alfreds zu bleiben, brauche ich ja wohl nicht zu wiederholen. Wir hatten uns gedacht, dass eine Wohung, die klein und modern ist, und in der es nur wenig aufzuräumen gibt, (was eine Aufräumfrau leicht in ein oder zwei Stunden am Tage machen kann), das beste wäre. Ihr braucht dort gar keinen Haushalt zu führen, denn es ist so nahe von uns, dass Ihr alle Mahlzeiten bei uns haben könnt (ausser vielleicht dem Frühstück, zu dem es ungemütlich ist erst wegfahren zu müssen und das Ihr daher vermutlich lieber zu Hause habt) und ihr habt dort eine Küche, die Ihr benutzen könnt, wenn Ihr Lust habt, allein zu sein, aber nicht zu benutzen braucht. Was an Einkäufen nötig ist, kann Genia leicht für Euch miterledigen. Wir hatten uns gedacht, dass das nicht zu viel Hausarbeit übriglassen würde und dass das für ein Jahr oder zwei wohl die beste und für Euch angenehmste Lösung ist. Natürlich kann man auch eine Pension, hier oder in London finden, wenn Ihr das vorzieht. Aber Ihr müsst Euch darüber im Klaren sein, dass das Leben in einer Pension, in der Ihr Euch halbwegs wohl fühlt mindestens doppelt so teuer wird, als in einer Wohnung von der Art, wie wir sie im Auge haben (obwohl das hier schon eine Luxuswohnung ist) und Ihr Euch dann eben in anderer Beziehung mehr einschränken müsst.

Die andere Wohnung, die ich in meinem letzten Brief erwähnte, liegt übrigens nicht einsam, es ist eine sehr belebte Wohngegend, nur sind die Verbindungen schlecht. Ich glaube aber auch, dass das nicht in Frage kommt. Es bleibt also vorläufig nur Viceroy Close, wo der Wirt übrigens inzwischen bereit ist, den Vertrag so zu ändern, wie ich es wollte. Aber wenn Ihr Euch noch nicht binden wollt, so werde ich ihm mitteilen müssen, dass er nicht fest auf uns rechnen kann. Es ist immerhin wahrscheinlich, dass die Wohnung noch eine Weile frei bleibt, denn es sind noch zwei von diesem Typ da (davon allerdings eine mit dem Schlafzimmer nach hintern hinaus, wo es eine Garage und daher wohl mehr Lärm gibt) aber sicher weiss man es nicht.

Wir hatten eigentlich gehofft, alles schon so vorbereiten zu können, dass ihr in kurzer Zeit schon Eure Ordnung und Gemütlichkeit haben könnt, aber wenn Ihr lieber alles noch unentschieden lassen wollt, so müsst Ihr das besser beurteilen können.

Die Geschichte mit dem "Brand" bei uns ist übrigens eine Ente, die wahrscheinlich daraus entstanden ist, dass Fr[äulein] Sonnenthal einen Petroleumofen, den sie wegen der Kälte über Nacht im Zimmer hatte, nicht richtig regulierte, und er daher sehr russte. Aber Russ ist noch kein Feuer, und obwohl, Fr[äulein] Sonnenthal, das Zimmer und ihre Sachen nachher sehr gründlich gewaschen werden mussten, ist sonst weiter kein Schaden entstanden.

Wie heißt übrigens der englische Spediteur? Ich konnte den Namen nicht lesen?

Weiter alle guten Wünsche und rascheste Erledigung
Herzlichst Euer

[Rudi]

[281] Rudolf Peierls David Shoenberg

Birmingham, 24.1.[19]39
(carbon copy)

Dear Shoenberg,
I see I shall not be able to write to you more fully for a few days and I want to send at least my comments about the problem of the thin conductor (however incomplete) in time for your lecture.[1037]

[1037] See letter [279].

I do not think that Laue's paper[1038] goes in that respect much beyond the old paper by H. London, except that it is more confused. If you split a thick conductor into two thinner ones, you introduce two more boundary layers which may now transmit lines of force. Thereby you relieve the congestion of lines of force outside the body and thereby admit further lines of force into the body. That amount of energy, taken by itself, would produce just the effect calculated by Pommerantschuk[1039] (or whoever else is the author of the paper you gave me). However, on the London theory there is a further effect which partly compensates the one mentioned. That is the kinetic energy which, on the London theory, has to be attributed to the currents which flow in the surface layer. It turns out that for a slab which is still thicker than the surface this loss of energy almost completely compensates the gain due to the first effect and leaves only the exponential terms found by London. It is quite clear that London's formula is a consistent application of the London theory. I have not been able to see as yet to what extent the exact compensation of the linear term is essential in a theory of the London type, and to what extent it is an accident.

In any case, as London points out, the answer according to which the energy of a number of thinner slabs is, for high enough fields, higher than that of one big slab, is unreasonable because if that were true no superconductor could ever have a sharp magnetic transition. He therefore introduces ad hoc a further surface energy wherever the superconducting and normal phases are in contact (thereby upsetting Laue's thermodynamics). How much of this would also apply to the surface at which the superconductor is in contact with vacuum I do not know. If all of it would apply, one ought, of course, to get a decrease not an increase, of transition field with decreasing thickness. The fact that, for a slab, you get an increase, taken together with the fact that a thick slab has a

[1038]See letter [279], note 1034. Max von Laue also published one paper in collaboration with the brothers Fritz and Heinz London in 1935 (Max von Laue, F. and H. London, 'Zur Theorie der Supraleitung', *Z. Phys.* **96**, 359–364 (1935)), and between 1937 and 1947 he published a further twelve papers and a book on the subject of the theory of superconductivity.

[1039]Issak Yakovlevich Pomeranchuk (1913–1966), Polish-born Russian theoretical physicist.

definite threshold field and does not split (except during the very rapid transition) into alternate layers of superconducting and normal phase, indicates that the surface energy of the outer face is a superconductor. That being so, I do not see how one can apply any simple theory to the problem. I shall write quite soon about the Bi problem,[1040] I believe my trick is the right one, but I have found a snag in justifying one of the necessary steps, and it will take me some time to tidy this up.
Yours sincerely,

[Rudi]

[282] Heinrich Peierls to Rudolf Peierls

Berlin, 9.2.1939

Meine Lieben,
herzlichen Dank für Eure freundliche Glückwünsche zu meinem Geburtstage, den wir ganz still bei Fuldas verlebten.

Leider können wir einen Reisetermin immer noch nicht bestimmen. Wir sind sogar insofern zurückgeworfen, als es vorläufig nicht möglich ist, einen Paß für mich zu erhalten. Else, die sehr beschäftigt ist, wird Euch ev[entuell] rechtzeitig über unsere Dispositionen unterrichten. Voraussichtlich wird unser Lift über Hull gehen.

Besondere Freude habt Ihr uns durch die Uebersendung der Bilder der Kinder gemacht. Wir finden sie beide sehr gewachsen und wohl aussehend.

Bedauerlich ist, daß Genia immer noch nicht in Ordnung ist. Hoffentlich überwindet sie die Krankheit nunmehr schnell völlig.
Viele Grüße allerseits

Euer Vater

[1040] See letter [279].

Meine Lieben!
Ich bin froh, daß es Genia besser geht + hoffe, sie ist bald ganz auf der Höhe. Die Kinderbilder sind ja entzückend. Gaby ist d[ie] reinste Schauspielerin. + der kleine Mönch süß. Wie groß sie schon sind. Wenn wir sie nur bald zu sehen bekämen! Ihr seht wie wir hin- und hergerissen sind. Nun habe ich wieder alles umdisponiert + antworte erst heute auf Rudis ausführlichen Brief, da ich vorhin Rücksprache mit dem Spediteur hatte. Ich nehme nur 1 Lift mit + darum habe ich vieles von d[en] bewilligten Sachen gestrichen. Da wir am 23.+ 24. packen dürfen — aber da wir nicht wissen wann wir reisen können, müssen wir ev[entuell] den Lift nach Hamburg Transit lagern lassen. Sollte es klappen, senden wir ihn nach Hull wie verabredet. Die Versicherung dürfte 20,000 M[ark] ungefähr betragen. Wir folgen Deinem Rat und würden bitten, dann nur gegen Transportrisiko zu versichern. Man schlägt vor: Transport Berlin–England, Bestimmung noch nicht bekannt, in Option mit temporärer Zwischenlagerung im Freihafen in Hamburg. Von da kann es dann ev[entuell] abberufen werden. Die Schränke sind ganz klein, Tischchen ebenso, aber irgendetwas muß man doch haben für Wäsche, Porzellan etc. Wir hängen an nichts, aber 1 Lift wird genommen + gefüllt, da das Leben führen mit eigenen Sachen billiger ist. Vorläufig sind wir mehr als ruhebedürftig, natürl[ich]. Daß man keine Schränke braucht, wußte ich nicht, nur von drüben bei Annie. Es tut mir leid, immer nur Mühe zu machen und zu lavieren, aber das liegt an den Verhältnissen, nicht an uns. Nun lebt wohl. Alles Beste für Euch und die Kleinen, die sich wohl nicht an uns erinnern. Wie oft denke ich an Genias Eltern. Hört sie Gutes über ihre Gesundheit?
Viele Grüße

Eure Else

Nach N[ew] Y[ork] direkt zu schicken, wagen wir nicht, aber unsere Quota Nr. wird im September bearbeitet, acted upon. Fragen kann man nicht, weiß also nicht, ob das heißt, daß man bald das Visum bekommt oder ob es dann noch geprüft wird + lange dauert.

[283] David Shoenberg to Rudolf Peierls

Cambridge, 12.2.1939

Dear Peierls,
Please forgive my bothering you so much about bismuth, but I have just had the proofs of my paper from the R[oyal] S[ociety] and have to decide this question of the appendix.[1041] I saw Fowler about this and told him that you had succeeded in obtaining the derivation, and he left it rather to you to decide on the question of the wisdom (from a political point of view) of adding the appendix (if the R.S. agree, of course). The R.S. seem to be content to leave the matter as it is, but I feel myself that there is every reason (from a scientific point of view) to publish the mathematical derivation, and this appendix method may provide the only convenient means of publication for some time to come. The only question is whether it may not damage anyone at the other end.

Anyway, I enclose a copy of the proofs for your consideration and should be grateful for your advice on the following points.

1. Whether my acknowledgements and references to L[andau] are in a suitable (i.e. not too warm) form. Remember that they were all deleted before I sent the MS to Moscow.

[1041] During his visit to Kapitza's laboratory in Moscow, Shoenberg had started studying in more detail the de Haas-van Alphen effect, an oscillatory variation of a metal's magnetic susceptibility with magnetic-field strength at low temperatures. Peierls had given an explanation of this phenomenon which had been observed in bismuth in 1930. Landau developed a better formulation which facilitated the interpretation of the experimental results, but his arrest made the citing of Landau's work in Russian journals impossible. The exchange of letters between Peierls and David Shoenberg, among others, deals with the question of how to utilize Landau's results without jeopardizing his position in Russia. Shoenberg deleted all references to Landau's theory when publishing his results in a Russian journal but was able to include a full account of it in the version of the paper that was published by the Royal Society. D. Shoenberg, 'Magnetic properties of bismuth. III. Further measurements of the de Haas-van Alphen effect', *Proc. Roy. Soc.* **A170**, 341–64 (1939).

2. Whether the formulae as set out are correct — especially as regards equ[ation] (20) for the number of electrons, in which I think a factor of 2 is in question.

3. About the appendix question. As regards the last point I suggest one of the following three courses

   (a) I return the proofs to the R.S. asking them to inform the referee that a derivation of the formulae has been obtained by you, and asking him whether he would advise publication of this as an appendix — possibly after the paper has appeared.

   (b) I wait with the proofs until you have written out the derivation in a suitably abbreviated form for publication, and only then return the proofs plus your derivation to the R.S., then as (a).

   (c) I let the whole thing drop, and just return the proofs without comment.

In case you should wish to write out the derivation for an appendix, it would be best if you could follow the notation used in the rest of the paper as far as poss[ible] to avoid confusion. As to the method of introducing the appendix, I suggest the following formulation:
"Appendix: Derivation of Landau's formulae.
Landau's formulae have been derived by Prof. Peierls, and he has kindly permitted me to reproduce this derivation here. etc."

This sounds rather lame I know, but I can't think of a better formulation, which will not hint too obviously at the reasons for L[andau] not publishing it himself, or for you not publishing it yourself separately (perhaps you ought even consider that?)

The whole business is very sticky and on the whole I feel inclined to adopt course (c) above, though I feel it will be a bit like compounding a felony.

Did you see that Kapitza and Fock were the only physicists elected finally to the Academy while Leontovitch, Skobeltzina, Alikhanov were elected as corresponding members. Our friend Propper (now known as

Grashchenkov[1042]) was one of the theoretical people elected as corresponding members. I can send you the full lists if you want them.
Kind regards to your family,
Yours,

David Shoenberg

[284] Rudolf Peierls to Siegried Peierls

Birmingham, 19.2.[19]39
(carbon copy)

Dear Uncle Siegfried,
Many thanks for your letter, which Alfred forwarded to me. It is really hard luck for my parents that, in addition to all their other difficulties, there should be this trouble in the US consulate. I rather gather from their letters that they know now what has happened, but they evidently cannot write about these things very openly. At present they are waiting for their passport, and they are rather depressed because apparently they cannot see now how long it will take until they get them and whether they will get them at all. Again they cannot write the reason for which the passport is held up; we believe that it might be in connection with the fact that the second instalment of the "fine" is now payable.

I have never yet written to thank you for the immediate and generous response to Father's request. At the time I was under great pressure of work, and Alfred promised to write and thank you, but I would now like to add my thanks directly.

I hope you and aunt Jenny are quite well; I nearly came to visit you at the end of January, since I was asked to take part in a conference

---

[1042]Nikolai I. Grashchenkov (1901–1965) graduated from Moscow State University in 1926, and later specialised n neurology and neuro surgery. He had visited England and the US between 1935 and 1937 when he had made contacts to Western scientists outside the medical sciences.

in Washington,[1043] but unfortunately I was given too short notice and could not find a substitute to do my work here.
Yours sincerely,

[Rudi]

[285] Rudolf Peierls to Else Peierls

Oslo, 15.4.[19]39
(carbon copy)

Liebe Else,
Ich höre es nicht sehr gerne, dass ich Verpflichtungen übernommen hätte, denen ich nicht nachkommen kann.[1044] Ich habe jetzt die Briefe herausgesucht, und hoffe, dass sich damit die Konfusion klären wird.

Es erschien uns selbstverständlich dass Ihr, falls es sein muss, d.h. falls Onkel Siegfried nicht gern einen Zuschuss gegebenen hätte, bei uns ohne Zuschuss auskommen könntet. Schliesslich erhalten wir ja im Augenblick ein Kind aus Wien, das uns ganz fremd ist, und wegen dem alles so eingerichtet ist, dass wir sie jederzeit loswerden können, da wir darauf rechneten, dass Ihr zu uns kommen würdet. Dabei haben wir natürlich angenommen, dass Ihr von Berlin genügend Kleidung u.s.w. mitgebracht habt, um für einige Zeit keine Anschaffungen zu brauchen. Für grössere unerwartete Ausgaben wie Krankheit u.s.w. gibt es ja noch eine Reserve, die Ihr seinerzeit Hermann für den Fall übergeben habt, dass wir unerwartet grössere Ausgaben haben. Natürlich würde es dann notwendig sein, dass Ihr Euch auf unseren Lebensstandard umstellt, aber wir nehmen an, dass das selbstverständlich wäre, wenn es sein muss.

Es besteht aber glücklicherweise kein Anlass dazu, dass Ihr Euch so einschränkt, da Siegfried so grosszügig ist. Ich glaube man muss

[1043] Washington conference of the American Physical Society. See letter [275].
[1044] In March 1939, Heinrich and Else Peierls had finally been allowed to leave Germany and they first emigrated to England before, at the and of 1939, settling in the US. Prior to their emigration, there had been a discussion among the children about the parents' financial needs and how these could be met given the limited means of all three children.

sich aber überlegen, wie Ihr Euer Leben in England auf die Dauer am besten einrichten könnt, so dass es, ohne Eure Bequemlichkeit und Gemütlichkeit wesentlich einzuschränken, so billig wie möglich wird. Das muss man doch wohl schon mit Rücksicht auf Siegfried tun und auf die Leute, die ganz auf Siegfried's Unterstützung angewiesen sind. Ich weiss nicht, ob ich Dich am Telefon richtig verstand, dass Ihr wenn Ihr aus Bournemouth zurückkommt, eine Pension in London suchen wollt. Ich möchte nicht immer dieselben Bemerkungen wiederholen, aber Ihr wisst, dass unserer Meinung nach London für Euch mindestens doppelt so teuer wird als z.B. Birmingham, und dass Ihr noch dazu ungemütlich und einsam leben werdet. Ich glaube Ihr wisst, dass wir bisher mit unseren Voraussagen recht behalten haben.

Natürlich kommt es in erster Linie auf Eure Wünsche an, und wenn Ihr das Vergnügen, gelegentlich Fischers oder Preussens zu sehen, so hoch bewertet, dass es alle anderen Argumente überwiegt, so gibt das natürlich den Ausschlag. Wir können Euch nur nach besten Kräften beraten, und Ihr müsst selbst entscheiden.

Was die Korrespondenz mit Siegfried betrifft, so habe ich meinen ersten Brief nicht mehr[1045] (oder konnte ihn jedenfalls in der Eile der Abreise gestern nicht finden). Der Inhalt dieses Briefes war folgender. Ich teilte ihm mit, was wir von Bethe über Eure Pläne gehört hatten, da ich nicht wusste, wie weit Ihr mit ihm in Verbindung steht (damals hattet Ihr gerade Bedenken über die Auswanderung zu korrespondieren) und dass Ihr offenbar zuerst für ein bis zwei Jahre nach England kommen müsstet. Wir wollten deshalb versuchen, alles schon so vorzubereiten, dass es Euch hier in England trotz des Klimas u.s.w. so bequem wie möglich gemacht würde, und ich fragte daher an, was Siegfried über Eure finanzielle Lage nach der Auswanderung wüsste. Wenn ich wüsste, dass Ihr mindestens £20 im Monat ausgeben könntet, so könnte ich Eure Wohnung mit Zentralheizung finden (die man einige Zeit im Voraus mieten muss) die wir uns sonst nicht leisten könnten. Ich betonte auch, dass ich nicht in Eurem Auftrage sondern zu meiner eigenen Information anfragte.

[1045] Letter [256].

Danach kam ein recht netter Brief von Jenny[1046] (da Siegfried gerade krank war) die damals gerade Deinen Brief aus Zürich bekommen hatte. Sie schrieb, dass ich selbstverständlich darauf rechnen könnte, dass Ihr einen Zuschuss von Siegfried bekommt. Später, als Siegfried wieder gesund war, schrieb er selbst einen Brief,[1047] den ich beilege. Er kam ungefähr gleichzeitig mit Eurer Bestellung durch Fr[äulein] Sonnenthal, und ich schrieb daher wieder an Siegfried, und lege Euch von diesem Brief eine Kopie bei.[1048]

Ich habe mich sehr gefreut, dass Vati wieder auf den Beinen ist. Natürlich wird er sich noch lange schonen müssen, und wir glauben, dass der Arzt sehr recht hat, dass ihm eine Zeit in Bournemouth oder so sehr gut tun würde. Hoffentlich erweist sich Vera als guter Fremdenführer; im übrigen wäre Genia gern bereit, Dich nächste Woche für einen oder zwei Tage nach Bournemouth zu begleiten, um Dir bei der Suche zu helfen. (Ich kann leider nicht, da meine Vorlesungen wieder anfangen).

Wir sitzen hier in einem sehr netten Häuschen gerade an der See. Augenblicklich regnet es zwar, aber die See ist wunderschön und wild, und wir hatten eine sehr schöne und sonnige Reise. Die Kinder sind in heller Begeisterung, nicht nur über das Meer vor dem Haus sondern auch über die Eisenbahn dahinter.

Mit herzlichen Grüssen von uns allen und an alle

Euer

[Rudi]

[286] Rudolf Peierls to Else and Heinrich Peierls

Birmingham, 6.6.[19]39
(carbon copy)

Liebe Eltern,

Vielen Dank für Eure Glückwünsche. Ich hatte den ganzen gestrigen

---

[1046] Letter [260].

[1047] Letter [277].

[1048] Letter [284].

Tag voll mit Sitzungen u.s.w. aber wir hatten am Sonntag ein bisschen Zeit um herauszukommen.

Es war uns sehr überraschend, von Euch neulich zu hören, dass Ihr nur auf eine Woche oder so kommen wollt. Aus unserer früheren Korrespondenz und unseren Unterhaltungen hatten wir das eigentlich gar nicht verstanden, und wir hatten uns schon ganz danach eingerichtet, hatten schon seit langer Zeit keinen Logierbesuch auf lange Sicht angenommen, das Zimmer im oberen Stock reparieren lassen, und Vorbereitungen getroffen, eine neue Aufräumefrau zu bekommen. Wir hatten uns über Euren Entschluss herzukommen sehr gefreut, nicht nur, weil wir Euch gern für länger hier gehabt hätten, sondern auch aus einer Reihe von anderen Gründen, die ich ja wohl nicht noch einmal zu wiederholen brauche, und hatten darauf gerechnet, dass Ihr für den Rest des Sommers hier sein würdet, ev[entuell] mit Ausnahme der drei Wochen, in denen wir im August vermutlich weg sein werden, und für die sich schon eine Lösung gefunden hätte.

Andererseits kann ich natürlich sehr gut verstehen, dass London für Euch grössere Anziehung bietet als Birmingham. Ich möchte nicht, dass Ihr Euch verpflichtet fühlt, uns zu besuchen. Wir freuen uns natürlich auch über einen kurzen Besuch, aber es fragt sich, ob dafür nicht eine etwas späterer Zeitpunkt geeigneter wäre, da wir in der zweiten Junihälfte nicht sehr viel Zeit haben werden, uns um Euch zu kümmern. Vom 21. bis 28. habe ich jeden Tag ein bis zwei Examenskonferenzen, die gewöhnlich sehr lange dauern, und eine Reihe von gesellschaftlichen Verpflichtungen, denen ich mich nicht entziehen kann. Am 1. Juli ist der Degree Day, an dem diesmal Niels Bohr ein Ehrendoktorat bekommt; Bohr bleibt entweder vor oder nach dieser Zeremonie für ein paar Tage hier, um mit mir zu arbeiten. Wenn Ihr auf längere Zeit hier wäret, und gewissermassen zum Haus gehörig rechnen würdet, so würdet Ihr es natürlich nicht übelnehmen, wenn wir eine Zeitlang mal weniger Zeit für Euch haben würden. Aber wenn es sich um einen so kurzen Besuch handelt, so wäre es doch schade, wenn wir dann so wenig mit Euch zusammen sein können. Wäre es dann nicht besser, es auf Anfang Juli zu verschieben, wenn alle offiziellen Verpflichtungen (mit Ausnahme einer Konferenz in Bristol vom 11.–13. Juli) aufgehört haben. Wir hatten darauf gedrängt, dass Ihr so bald wir möglich herkommt, da wir wollten, dass Ihr Euch so bald wie möglich endgültig hier einrichtet, und aus dem Provisorischen herauskommt, und um Euch die Pensionskosten zu ersparen. Da beides aber sowieso nicht in Frage kommt, so ist es doch wohl das Beste den Zeitpunkt so zu wählen, dass wir am meisten Zeit für Euch haben.

Ich freue es sehr, dass es Vati mit den Zähnen besser geht, und dass er einen vernünftigen Arzt gefunden hat.
Hoffentlich hält die Besserung an.
Mit herzlichen Grüssen von uns allen
Euer

[Rudi]

[287] Niels Bohr to Rudolf Peierls

Copenhagen, 6.6.1939

Dear Peierls,
I thank you for your kind letter and need not to say how happy I am to know — as I first heard from Cockroft a few days ago — that Landau is now back in Kapitza's Institute and has taken up work again.[1049]

I look forward very much to come to Birmingham[1050] and it shall certainly be a great pleasure to me to accept the kind invitation of Mrs. Peierls and you to stay in your home where I spent so delightful days on my last visit.[1051] I have been very thankful in your kind interest in my little article on the human cultures[1052] and I hope that you were fairly satisfied with the final form of the article when it appeared in "Nature". On my next visit I should be very happy if we could work in [for] a few days together on a brief note to "Nature" about the main

[1049] See letters [240], [242–244].
[1050] Niels Bohr was to be awarded an honorary doctorate at Birmingham on 1 July 1939. See letter [286].
[1051] Niels Bohr had stayed with the Peierls family when he visited Birmingham after lecturing at Edinburgh in the autumn of 1938. See letter [272].
[1052] N. Bohr, 'Natural Philosophy and Human Cultures', *Nature* **143**, (1939), 268–72.

results of our common work on the nuclear dispersion theory and, if it would suit you, I shall be glad to come to Birmingham about the 28th of June. Due to the unquiet times and my unavoidable occupation with the fission problem I regret that I have not yet found opportunity to finish our article with Placzek, but, as Placzek suggested, it would be nice if a short account of the result could appear in "Nature" in a near future, and I shall bring with me a draft Placzek and I have written.[1053] When we meet I hope also that we can arrange to work together some later quiet time to finish the greater article to appear in the Proceedings of the Danish Academy. Perhaps you could come to Denmark for this purpose some time in August or September, but about that we can best speak when we meet in Birmingham.[1054]
With kindest regards and best wishes to Mrs. Peierls and yourself from my wife and
Yours,

Niels Bohr

[288] Rudolf Peierls to Niels Bohr

Birmingham, 10.6.1939
(carbon copy)

Dear Professor Bohr,
Thank you very much for your letter, we are so glad you are coming and are looking forward to seeing you here on the 28th. I hope that the present spell of quite un-English weather will last until then, so that you can perhaps see a little of the country round Birmingham, which looks now rather more attractive than when you were here last.

Thank you also for the kind invitation to come to Denmark later in the summer. I would love to come, but there may be some difficulty

[1053] This summary note did, in fact, appear. N. Bohr, R. Peierls, G. Placzek, 'Nuclear Reactions in the Continuous Energy Region', *Nature* **144**, 200–201 (1939).
[1054] Collaboration on the project was interrupted by the war; in 1947, the question was taken up again. See vol. II of the *Peierls Correspondence*, chapter 6.

with my passport just now, since my application for naturalization is still being considered by the Home Office. However, I may perhaps know more about this when you come.

I think it would be an advantage if you had letters which you expect here in Birmingham addressed to my house, 38 Calthorpe Road, Birmingham, 15, as this may avoid occasional delay in forwarding them.

Oliphant tells me that he hopes to have Frisch[1055] here for some weeks in the summer, and we are, of course, extremely glad to have a chance of discussions with him.

The letters to the U.S.A. which Frl. Hellmann sent to me yesterday were immediately re-posted. I could not send them via Liverpool as there will not be a suitable steamer from Liverpool for some time, but I have looked at the list of steamers from Southampton, and made sure that, even if they should miss the mail for which they are marked, no difficulty about the steamer will arise.

With kindest regards,

Yours sincerely,

[Rudolf Peierls]

[289] Rudolf Peierls to Niels Bohr

Birmingham, 16.6.[19]39

Dear Professor Bohr,

Thank you very much for your letter.[1056] The dates which you now

[1055]Robert Otto Frisch (1904–1979), received his doctorate from the University of Vienna in 1926; he continued research in Berlin, Hamburg and Copenhagen before moving to Birmingham in 1939. Here, Frisch and Peierls calculated that the fissioning of uranium had the the fissioning of uranium had the potential to create a chain reaction which could be used to develop an atomic weapon. Frisch later joined the Manhattan Project. After the war he worked for the Atomic Energy research Establishment at Harwell before taking up the Jackson Chair of Physics at Cambridge in 1947.

[1056]Letter Niels Bohr to Rudolf Peierls, 14.6.1939, *Peierls Papers*, Misc. Eng. misc. b203, C. 30.

suggest suit us also very well, and we shall be very glad to see you here on the 25th or 26th. The only disadvantage of the beginning of the week as compared with the end is that by then our university business will not be quite finished. We have a meeting that lasts all morning on Monday, and one for about two hours on Tuesday afternoon. The Monday one is rather important and as Watson will be absent from Birmingham, I have promised to be there and Oliphant will have to attend it also. I may be able to get out of the Tuesday afternoon meeting. There is also another short meeting on Wednesday morning, but that will last only for about an hour and besides, Oliphant will then be free.

Perhaps I better tell you about our programme for the whole week, so that you can decide for yourself in case there are further changes: Besides the meeting I just mentioned I am completely free, except, of course, for the Degree Ceremony on Saturday. That is in the morning, and there is a luncheon after it, to which it is hoped you will come. At this, the Chancellor, Viscount Cecil, is usually in the chair, and I think you will find him a very charming man. In addition, our Vice-Chancellor, Dr. Priestley, is arranging an informal dinner on Friday night, to which he is inviting the honorary graduands (amongst which there is Sir William Bragg). There is no obligation to come to this, and if you come back from London on Friday you may, in fact, not be here in time to come, but you need not decide this at the moment.

You will have had a letter from Oliphant saying that we shall have a joint colloquium, which we regularly hold with Bristol and Oxford, on Thursday at Oxford. According to the arrangements suggested in your letter, you will not be here on that day, but in any case this colloquium will be early in the afternoon so that, if you wished, you would be able to be in London in time for the dinner in the evening.

Lastly, there is a Royal Society discussion on Monday (the 26th) afternoon in London, to which the Royal Society has invited some foreign visitors, chiefly Hahn[1057] from Dahlem, and the subject of which will be

[1057]Otto Hahn (1879–1968) studied chemistry at Marburg and Munich; after receiving his doctorate in 1901 he continued research at Marburg and London before moving to McGill University to work with Rutherford. On his return to Germany in 1907 he moved to Berlin where he remained for most of his research career.

nuclear fission. I was not originally intending to attend it, but I believe Oliphant is going to be there.

I have mentioned all these things for your information, and because I know that you want to see a number of people in this country, you might perhaps find it convenient to go to one or the other of these things and have an opportunity to see several people at a time. I need hardly say that, from our selfish point of view, the fewer of these discussions you choose to attend, and the longer time we shall have you in Birmingham, the better.

If, on account of one of these meetings, or for any other reason, you would like to change the dates, please do not hesitate to say so, we shall be very happy to have you here whenever it is the most convenient to you.

If you leave it at the present arrangement, I would suggest that I come to bring you over from Cambridge by car on the Sunday afternoon. Train services are very bad on Sundays, and by road it is only about three hours' driving through quite pleasant country. I have to see some people in Cambridge anyway so it would not be any trouble.
With kindest regards,
Yours sincerely,

[Rudolf Peierls]

[290] Rudolf Peierls to P.L. Kapur[1058]

Birmingham, 7.7.1939

Dear Kapur,
It is with a very bad conscience that I write to-day's date at the top of

[1058] P.L. Kapur, a student of Peierls' at Cambridge, had followed him to Birmingham, where he worked on the response of a nucleus to neutron bombardment in terms of resonances. The resulting dispersion formula was superceded by an alternative method of Wigner and Eisenbud in E.P. Wigner and L. Eisenbud, 'High Angular Momenta and Long Range Interaction in Resonance Reactions', *Phys. Rev.* **72**, 29–41 (1947).

this letter, because it reminds me how long ago I ought to have written to thank you for your lovely box which you sent us for New Year, and which we appreciated very much. It has been in front of me on my desk ever since, but as you see it has not until to-day produced the reaction it ought to have produced immediately.

The paper with Bohr and Placzek on the properties of nuclear reactions in the continuous energy region is still unfinished, but we have at least sent off a note to NATURE giving them main results,[1059] and we hope to get the paper sent off this summer. I am also preparing a little paper for the London Mathematical Society in which I give a proof for the completeness of the complex set of orthogonal functions which we used in our paper on the dispersion formula; we had then assumed the completeness to be obvious, but apparently several people doubted whether it was correct.[1060]

I hope you are quite settled again in Lahore. With best regards, also from my wife,
Yours sincerely,

R.E. Peierls

[291] Rudolf Peierls to German Embassy

Birmingham, 18.7.1939
(carbon copy)

An die Deutsche Botschaft,
Carlton Terrace,
London.
Ich bitte Sie um Auskunft über die folgende Frage: Ich bin deutscher Staatsangehöriger, Nichtarier, und seit über 6 Jahren in England ansässig. Ich habe einen Antrag auf Naturalisierung gestellt, dessen

[1059] See letter [287].
[1060] The proof was not published until 1948. R.E. Peierls, 'Expansions in terms of sets of functions with complex eigenvalues', *Proc. Camb. Phil. Soc.* **44**, 242–50 (1948).

Erledigung aber noch einige Zeit dauern wird. Ist es technisch möglich, dass ich bereits vor meiner Naturalisierung auf die deutsche Staatsangehörigkeit verzichte, und falls ja, welche Schritte habe ich dazu zu unternehmen?
Hochachtungsvoll

(Professor R. Peierls)

[292] German Embassy to Rudolf Peierls

London, 31.7.1939

In Beantwortung Ihres Schreibens vom 18. d[iesen] M[onats] teile ich Ihnen mit, dass Sie die deutsche Staatsangehörigkeit nicht durch einen blossen Verzicht ablegen können. Es ist Ihnen jedoch anheim gestellt, einen Antrag, (persönliches Erscheinen ist dabei Bedingung) auf Entlassung aus dem deutschen Staatsverband zu stellen. Sie hätten dazu folgende Urkunden vorzulegen:

- Ihre eigene Geburtsurkunde,
- Geburtsurkunde Ihrer Eltern,
- Ev[entuell] Geburtsurkunden Ihrer Ehefrau und Kinder.

Bemerken möchte ich noch, dass bis zum Bescheid der innerdeutschen Behörde in der Angelegenheit etwa 3 Monate vergehen werden.[1061]
Im Auftrage

[name]

[1061] As Rudolf Peierls had to submit some of the required documents mentioned in this letter to the Home Office as part of his naturalisation process, he could not pursue his early release from German nationality. See also letter [298], [302].

[293] Rudolf Peierls to Hans Bethe

Birmingham, 22.8.1939
(carbon copy)

Dear Hans,
Zunächst einmal von uns allen alle guten Wünsche für Deine, wie Rose uns anvertraut hat, so baldbevorstehende Ueberführung in den Dublettzustand. Wir hätten uns auf jeden Fall darüber gefreut, aber seit wir Rose in Cambridge kennengelernt haben, haben wir doppelten Grund, der Unternehmung Glück zu wünschen.[1062]

Die politische Situation in Europa ist ja im Augenblick nicht ohne interessante Seiten. Dass Russland voll Scham, seinen Fehler einsieht, sich selber einzukreisen und nun dem Antikomintern-Pakt beitritt, ist jedenfalls originell. Es gibt jetzt wohl nur zwei Möglichkeiten: Krieg oder Appeasement. Die erste Möglichkeit ist natürlich bitter, aber es ist dann jedenfalls die letzte Chance für Europa, und die letzte Hoffnung, dass die Diktaturen kaputtgehen. Es wird jedenfalls interessant, das mitanzusehen und vielleicht gibt es für uns auch Gelegenheit, in bescheidener Weise mitzuhelfen.

Es ist aber noch immer nicht ausgeschlossen, dass die Sache wie vor einem Jahre in Wohlgefallen und Appeasement endet, und dann hat Freund Adolf Europa in der Tasche. In diesem Falle würde ich auf das energischste versuchen, so schnell wie möglich Europa zu verlassen. Dieser Brief hat teilweise den Zweck, Dich hiervon zu informieren und davon, dass, immer unter der Voraussetzung, dass es ein weiteres Appeasement gibt, ich bereit wäre jede Stelle in U.S. anzunehmen. Mir ist aber klar, dass solche Stellen [nicht] herumlaufen und auf mich warten, und ich würde in diesem Falle Weihnachten (früher wird es wohl nicht gehen) hinüberfahren, um mich selbst zu informieren, und mich, falls möglich, auf einer Konferenz auszustellen. Falls ich dies beschliesse, wovon ich Dich natürlich noch informieren würde, evtl. telegrafisch, wäre ich Dir sehr dankbar, wenn Du mir eine Einladung formaler Art besorgen könntest, zwecks Erleichterung des Visums (Ich würde im übrigen natürlich auf eigene Kosten kommen.)

[1062] Hans Bethe married Rose Ewald in September 1939.

Das ist eine sehr ungewöhnliche Kombination von Gratulation und Geschäft, aber wir leben nun einmal in recht ungewöhnlichen Zeiten.
In Eile sehr herzliche Grüsse,
Dein

[Rudi]

[294] Rudolf Peierls to Home Office

Birmingham, 31.8.1939
(carbon copy)

Dear Sir,
I would be grateful for advice in the following matter: I am an Alien, of German nationality, but have lived in this country for over six years, hold a Professorship at this University, and have applied for naturalisation in May, 1938. I am very anxious to join in some form of ARP work, but I have been given to understand that I cannot be enrolled in any ARP service before my naturalisation is completed.

I am therefore writing now to ask whether there is any way of obtaining a statement to the effect that my application for naturalisation is being considered, or some other statement which might make it possible for me to enrol. After having enjoyed the hospitality of this country for over six years I am disappointed that I should not be allowed to help when there is a need for volunteers, even in capacities that involve no responsibility or confidential information.
Yours faithfully

[R.E. Peierls]

[295] Else Peierls to Genia Peierls

Folkestone, 1.9.[19]39 Vaters Diktat

Lieber Rudi!
Besten Dank für Dein Schreiben von vorgestern. Wir freuen uns, daß die Kinder es so gut getroffen haben + schon an Ort + Stelle sind + hoffen, daß sie sich auch weiter wohl fühlen.[1063] Wir bleiben vorläufig hier. Wir haben es für richtig gehalten, uns für längere Zeit mit Barmitteln zu versehen, da möglicherweise die Guthaben der Ausländer gesperrt werden, mindestens aber mit einem Schließen der Banken für einige Zeit zu rechnen ist.

Da wir in einem Niedrig Parterre Zimmer wohnen + nichts Verschließbares haben, wollen wir nicht das ganze Geld hier aufheben + bitten um umgehende Nachricht, ob Ihr einen Teil für uns aufheben oder noch besser einem zuverlässigen Engländer zum Aufheben übergeben könnt. Wir würden dann später darüber verfügen. Laß uns auch sofort wissen, ob Du in ein Lager kommst, ich nehme an, daß in diesem Fall die Universität Dich reclamieren könnte.
Mit bestem Wunsch für Dich + Genia

Dein Vater

Viele Grüße auch an Marianne. Ist sie verheiratet?

[296] Else Peierls to Rudolf Peierls[1064]

Folkestone, 6.9.[1939]

Dear Rudi,
I was very glad to get your letter (4/9) today. We will see how we are getting on here without your help. Father is well, we both are totally quiet + fatalists. Nevertheless father has become an old man. I am the

---

[1063] Gaby and Ronnie Peierls had been evacuated to the countryside.

[1064] Many of the family letters, especially those sent between Germans within England were now written in English.

nurse. The weather and place here + the parc next to our house are wonderful. The people are not nice at all. A German night nurse, who nursed father only two nights is here with an old lady. She is very nice + loves us and helps me sometimes. Father walks now better, but very slowly. We hope we can stay here this month. It is a pitty that our friends neither Alfred or you can come to pay us a visit. I don't know where my sister is.[1065] We are very glad to hear that Ronnie and Gaby are so happy. Please send Ronnie this card and a present from us and let me know the price. We would like to know very much what you are doing, is the university open? What about Genia and Marianne? + your car? Let us hope the best for the future. With best wishes,
yours

Elsie + father

Alfred was ill but is well again. Nina lost her teeth and seems to be very nervous. They invited us to come, but we don't accept their invitation.

[297] Annie Krebs to Rudolf Peierls

location unspecified, 8.9.[1939]

Mein lieber Rudi.
Ihr koennt Euch garnicht denken, wie man hier, trotzdem man doch eigentlich weit weg vom Schuss ist, um alle zittert, die man in Europa hat, wie die Gedanken um Euch sich immer wieder drehen, besonders da die Nachrichten, die man hier bekommt groesstenteils stark uebertrieben sind, und dann nach einiger Zeit wieder widerrufen werden. So sieht man alle staendig in Lebensgefahr. Else schrieb uns vor einigen Tagen,

[1065] The Fulda's had tried to leave Germany and join their son Karli in the US, but this had proved impossible. Ludwig Fulda, who had found living under National Socialism impossible to cope with, had committed suicide in March 1939. His wife, who like Else was Aryian, left Germany and temporarily went to the US. Her attempt to return to Germany to claim her possessions, failed because of the outbreak of war. See letter [297].

dass Du Dich zum Militaer stellen willst. Ich kann es mir garnicht ausmalen, was das fuer Dich bedeutet, und wie schwer es fuer Genia und die Kinder sein muss. Solltest Du nicht mehr zu Hause sein, so schreibt sie uns vielleicht ein paar Zeilen, was los ist. Durch alle diese Aufregungen haben wir die Geburtstage Eurer Kinder vergessen was wir sehr zu entschuldigen bitten. Ausserdem sehe ich eben, dass ich mich fuer Euern letzten Brief noch nicht bedankt habe, in dem Ihr ausfuehrlichst ueber Vatis Befinden berichtetet. Ich war so froh, ihn auf dem Wege der Besserung zu wissen, und lebe jetzt in der Angst, dass ihm die ganzen Aufregungen wieder sehr schaden koennten. Ich weiss nicht einmal, wo er jetzt ist, da ich annehme, dass er den Hafen Folkestone hat inzwischen verlassen muessen. Die letzten Nachrichten von Else und Vati waren noch vor der Kriegserklaerung, wo sie immer noch hofften, dass das schlimnste verhuetet werden kann.

Lala Fulda, die gerade auf der Heimreise nach Deutschland war, ist garnicht so weit mehr gekommen, sie war gezwungen mit dem gleichen Schiff wieder hierherzukommen, und wir sind froh, dass sie wenigstens sicher ist. Was aus ihrem Haus, ihrem Geld geworden, das wird sie spaeter sehen.

Von meinen Schwiegereltern haben wir seit 2 Wochen keine Nachrichten, haben versucht, ihnen ueber die Schweiz und Holland von uns Briefe zukommen zu lassen. Ob die Juden in Deutschland, wenn sie nicht zum Militaer eingezogen werden koennen noch genug zu essen haben, ist uns nicht bekannt. Natuerlich sieht man immer das schwaerzeste.

Es ist so furchtbar, dass es zu diesem Blutvergiessen kommen musste, und dass dieser Wahnsnnige in Deutschland der Urheber all dieses Unglueckes ist!

Von Alfreds haben wir auch lange nichts gehoert und hoffen nur, dass die Drohungen vom letzten September in der Fabrik nicht wahr gemacht wurden, was seine Stellung anbetrifft.

Wir sitzen die meiste Zeit des Tages am Radio, wo wir direkte Uebertragungen aus Berlin, Paris und London hoeren, aber so viel Widersprechendes, dass man sich kein richtiges Bild machen kann. Privat geht es uns gut. Geschaeftlich ist der Monat August ein ganz guter gewesen, obgleich wir noch immer nicht die Unkosten erreicht haben. Das wird noch immer dauern, aber zumindest wird das noch vorhandene Geld der Firma was arg zu schmelzen anfing etwas geschont, denn es ist sehr zweifelhaft, ob wir neues bekommen. So heisst es weiter sehr sparsam sein, in der Fabrik und auch zu Hause, was uns ja weiter nicht schwer faellt. Unsere Mieterin ist ein angehmer Hausgenosse, und ich empfinde die viele Arbeit im Hause im Moment als besonders wohltuend.

Lasst also bald von Euch wenn es auch nur kurz ist, hoeren und berichtet auch, was Ihr von Alfreds wisst.
Herzlichst Eure

Annie

Meine Lieben,
Dank für Rudis langen Bericht, diese Arbeit, besonders auf der Reise, erkennen wir hoch an!!

Die Quote der Eltern müsste jetzt eigentlich schneller gehen, da doch die Armen in Deutschland nicht raus kommen können. Es ist ganz gut, dass die Eltern als "paradiesisches" Ziel die U.S.A. betrachten, dies gibt ihnen die Kraft und den Glauben.

Geschäftlich sieht es im Augenblick besser aus, und wir kämpfen weiter um den Enderfolg!
Lasst uns nicht ohne Nachricht!
Herzlichst Euer Hermann
Was ist eigentlich mit Nina wirklich los?

[298] Rudolf Peierls to Prof. Appleton[1066]

Birminhgam, 13.9.1939
(carbon copy)

Dear Prof. Appleton,
I am writing to ask for your advice in a difficult situation. I had written

[1066]Edward Victor Appleton (1892–1965), studied physics at Cambridge, became professor in London in 1924 before taking up the Chair of Natural Philosophy at Cambridge in 1936. After the outbreak of the war he became Secretary of the Department of Scientific and Industrial Research.

to Bragg for advice,[1067] and he suggested that you, if anybody, would be able to help me.

The point is that, as you perhaps know, I am still technically, a German subject, and therefore an "Enemy Alien", although I applied for naturalisation in May, 1938. Apart from being subject to police restrictions which do not represent more than a minor nuisance, this means that I shall probably not be allowed to do any work of national importance. At any rate, I have been told that I cannot under any conditions, be enrolled for any aspect of Civil Defence work, for which I wanted to volunteer on a part-time basis.

You will understand that I am extremely anxious to help in some way, for two reasons: After having enjoyed the hospitality of this country for over six years, it is hard not to be allowed to help at a moment when people with technical qualifications are apparently urgently wanted. Secondly, I have some personal experiences of and hence a more intense enmity towards, the present regime in Germany, than most other people in the country.

At present my name has gone forward on the list compiled by the Ministry of Labour, through the University, on the list drawn up by the Royal Society, and on the special list of foreign scientists compiled by the Society for the Protection of Science and Learning. I do not know whether I shall automatically be assigned through some of these agencies, or whether that will be impossible before my case has been considered by one of the tribunals that are to be set up to review the cases of Germans in this country. It will probably take some time until these tribunals are constituted, since people are all busy with more important work, and if it were possible to obtain some preliminary permission for starting some work of national importance before then, I would do all I can do to obtain it.

I understand that Oliphant's department is going to engage in some research work of military importance, and it is likely that this will raise a number of semi-mathematical questions, with which I could be useful, but there may be some other place in which I could be of more use.

[1067]Letter Peierls to W.L. Bragg, 4.9.1939, *Peierls Papers*, Ms. Eng. misc. b203, C. 43.

At the moment I am carrying on normally with my work at the University. But as far as the teaching work goes, I could easily be spared or replaced, or could combine it with other work if the latter can be done at Birmingham; as for pure research, there seems little chance of keeping it going during the period of emergency.

You are possibly in possession of all facts about my qualifications through the Royal Society Questionnaire, but here is the summary of the relevant points as regards my political situation.

I have not lived in Germany since 1929. In March 1933, before the anti-Jewish laws were passed, I declined an offer of a job in Germany, which I had intended to accept, because the Nazi Regime had then just be(en) established and I did not wish to live under it. Since then, I have only once returned to Germany (in 1934) in order to visit my parents; after that I suppose I would not have been able to return even if I wanted to. My parents have now left Germany and I have no near relatives or friends left there. Immediately after my parents had arrived in this country, I attempted to get rid of my German nationality, but found that this was technically impossible while my papers were at the Home Office with my application for naturalisation.[1068] My German passport expired in May, and I have taken no steps to extend it. I have been resident in this country since 1933, and I hold the unconditional permit to stay (which includes an unconditional permit to work) since 1935. My children are British subjects, and my wife, who is Russian by birth, and now German by marriage, has never lived in Germany.

With all this I feel I have a good case, but no-one to present it to. I would be extremely grateful for any advice you could give me as to what steps I could, or should take in order to get early permission to serve this country in some way or another.

Yours sincerely,

[Rudolf Peierls]

[1068] See letters [291–92].

[299] Rudolf Peierls to Niels Bohr

Birmingham, 13.9.1939
(carbon copy)

Dear Professor Bohr,
Here is, at last, the revised manuscript of our paper. I have completely rewritten the introduction on the lines of our discussion. I found that after that the other sections could be considerably shortened, but apart from this they are essentially unchanged. I have omitted from the introduction the part on the limitation of the picture for very high energies, which appears now irrelevant, (but which could also easily be inserted) and I have omitted the quantitative discussion of the photoeffect, since I thought one could not add in this respect to what has been said in the note to NATURE.[1069]

I am sending a copy to Placzek. I hope that you will now find the paper readable, but if you do not, and if it will be difficult to continue the correspondence, I agree, of course, to any alterations which you care to make.
With best wishes and kindest regards to you and Mrs Bohr from all of us,
Yours very sincerely,

[Rudolf Peierls]

[300] Rudolf Peierls to Else Peierls

Birmingham, 15.9.[19]39
(carbon copy)

Dear Else,
Many thanks for your postcard. I am very sorry to hear that you are dissatisfied with your place in Folkestone. What really is the trouble? I can quite imagine that you do not get as much attention or as

[1069]See letter [287], note 1052.

good service, as you are used to, but in present conditions, I am afraid everybody has to bear a certain amount of inconvenience.

I do not know why you wish to go to London. Don't you think that at wartime when everybody who possibly can wants to get away from London, and when even the big government departments move away from London, when the streets are dark and the buses and trams are cut down to a minimum, (so that they will always be crowded, when nobody can use a motor car except for urgent business, and when all places of entertainment have to close, and if they may keep open at all, that in these conditions London will be even more lonely than Folkestone?

In any case, I am afraid that it is just now impossible for us to find you accomodation in London. As you know we have not got our car now (and would in any case not have any petrol for it) and we cannot now get away to look at houses in London.

We could probably find you some suitable place in Birmingham, and, as you know, you are always welcome to stay with us, but as we are without help now, and will both be very busy, the latter would mean that you would have to look after yourself to a certain extent. You could easily find a maid for part of the day. With the restriction on fuel, our house would probably not be any warmer than usually, but we could probably get at least one room warm and comfortable. Alternatively we could probably find a boarding house or family here, but I would be afraid of that having the same defects as your present boarding house (or other ones). If you would like me to look for something, please let me know what actually your trouble is so that I can make sure to find a place at which similar difficulties do not arise.

Do not forget that there are very great advantages in having a doctor in your house (although he is not a registered practitioner) since doctors are now very hard to get hold of, most of them being occupied with military or first aid work.

Please let me know what you have decided. My advice is to stay in Folkestone for the present and realize that if you have difficulties there, you are likely to have discomforts wherever you go as long as we are in wartime. If you prefer to leave Folkestone, come to Birmingham.

I am very glad to hear that Vati is so much better.
Yours sincerely,

[Rudi]

Kind regards from Genia, whom I see now only in the evenings.

[301] Rudolf Peierls to Heinrich and Else Peierls

Birmingham, 6.10.[19]39
(carbon copy)

Dear Parents,
I enclose a letter for the tribunal, which I hope will be of some help. I cannot write a personal letter of this nature on university notepaper, but I do not think that this is very important.

On the whole, the Tribunal will not attach very much weight to my letter, seeing that I am a German myself, and probably the letter would be more useful if I could say that my case had already been cleared by the Birmingham Tribunal. I shall therefore rewrite this letter as soon as I have been before the tribunal, unless I hear from you by then that your case has already come up.

If in Folkestone there exists a Co-ordinating Committee for Refugees, or some similar organisation, they will send a representative to the tribunal. If you have not been in touch with this committee before, it would be advisable to go and see them, so that they will know everything about you and can advise you what further evidence, if any, you ought to prepare.

The epidemic of whooping cough at the school seems to be a rather mild form, and our children are already out of bed. We hope to be able to see them again on Sunday.
With kindest regards
Yours

[Rudi]

[302] Rudolf Peierls to W.L. Bragg

Birmingham, 6.12.1939
(carbon copy)

Dear Prof. Bragg,
After the interest you have taken earlier in my attempt to get permission to do work of national importance, I feel I ought to report to you on the ultimate outcome. As I wrote to you at the time I was passed by the tribunal all right, and that involves, in principle, permission to do war work. However, the Admiralty did not feel, even then, that they wanted me to work with Oliphant's team, and I have not so far found another way of making myself useful except joining the ARS on a part-time basis. This I have done and I shall in due course get a very beautiful uniform, but I still think that I might help the country in a more efficient way than by sleeping in a fire station. So if you can suggest any other way or approach I shall be very grateful for your advice.
Yours sincerely,

[R. Peierls]

5. 1940–45: War

The war had turned Rudolf and Genia Peierls into enemy aliens. Soon tribunals were set up to classify this group of residents.[1070] Being classified as 'category one' meant that the couple in practice had to endure very few restrictions to their every-day life. What concerned Rudolf Peierls more than these relatively insignificant limitations of his private life were those placed on him with regard to civil defence and to his work and research. The former were lifted when his naturalisation was approved in February 1940,[1071] but the latter would continue to present problems throughout the first year of the war.

After the discovery of the first uncharged subatomic particle, the neutron, in 1932, new possibilities of examining the nucleus more closely were opened. Physicists began to discuss whether the neutron could be used to force a nucleus to give up some of its enormous energy by splitting the atomic nucleus and thereby releasing further neutrons. While one of the key figures in nuclear research, Ernest Rutherford, had dismissed such expectations as 'the merest moonshine',[1072] others were less pessimistic about the prospects and saw at least a theoretical possibility of a self-sustaining nuclear chain reaction, as these neutrons could then move on to produce further fission. In 1938, Otto Hahn and Fritz Strassmann, continuing work they had done before in collaboration with Lise Meitner who had to leave Nazi Germany earlier that year because of her Jewish origin, discovered a process that appeared to be nuclear fission. Discussing the results which had been communicated to her by Hahn and Strassmann, Lise Meitner together with her nephew Otto

---

[1070] See letters [298], [301].

[1071] *Peierls Papers*, Supp., A 38.

[1072] E. Rutherford, 'Speech at the British Association for the Advancement of Science', Leicester 11.9.1933, summarised in *Nature* **132**, 432–3 (1933).

Frisch developed a qualitative theoretical explanation of the process.[1073] It was soon understood that the relatively rare uranium isotope U235 was the most promising candidate for fissionable material which would allow a self-sustaining chain reaction. U235, however, was only present in natural uranium to less than 1 per cent, and the separation of this isotope from the more common U238 was a substantial obstacle on the way to producing a chain reaction. Furthermore, it was unclear how much uranium would be needed, if it could be separated at all. It was generally believed at the time that it would be tons rather than kilos.

It was these fundamental questions which two refugee Rudolf Peierls and Austrian-born Otto Frisch, being banned from more important 'war work' tackled when considering the critical mass of uranium. They calculated that a chain reaction was not only theoretically possible but also practically feasible. In a memorandum they suggested that the amount of fissionable material (U235) needed for an atomic bomb based on these principles of a self-sustaining chain reaction was far less than previously assumed and that a sphere of metallic U235 of a radius of about 2.1 cm could be sufficient to be explosive, an amount which corresponded to under 1 kg of U235.[1074] As soon as Frisch and Peierls made this discovery they realised the potential repercussions of their findings. As Frisch wrote to his mother, he felt like a man who 'had caught an elephant in the jungle by the tail and who did not know what to do with it. But he knew it was an elephant.'[1075] Having been told about the contents of the memorandum via Mark Oliphant, Henry Tizard, Chairman of the Committee on the Scientific Survey of Air Defence, and Prof. G.P. Thomson of Imperial College, at the time Chairman of the government committee concerned with the possibility of a nuclear chain reaction set up a government committee, the so-called MAUD Committee, to investigate further the potential implications of the Frisch-Peierls calculations. It met in April 1940 for the first time, but Peierls and Frisch, as only

[1073]Lise Meitner and Otto Frisch, 'Disintegration of Uranium by Neutrons: A New Type of Nuclear Reaction', *Nature* **143**, 239 (1939).
[1074]'The Frisch-Peierls Memorandum of 1940', R.H. Dalitz and R.E. Peierls (eds.), *Selected Scientific Papers of Sir Rudolf Peierls*, London: Imperial College Press, 1997, pp. 277–282.
[1075]Frisch quoted in *On the Right Side of Wrong*, BBC documentary, 17.2.2000.

recently naturalised and enemy alien respectively were excluded from the committee. This exclusion met with misapprehension and caused consternation among the two scientists. As Peierls argued in a letter to the MAUD Committee, while he fully appreciated a policy of secrecy for reasons of national security, he thought that no 'useful purpose [could] be served by trying to keep our own idea a secret from us'.[1076] Indeed, Peierls and Frisch, while not being allowed on the MAUD Committee itself, became members of the technical subcommittee, and remained deeply involved in the developments leading to the production of nuclear weapons. The story has been told in detail elsewhere,[1077] and therefore reference to the history of nuclear weapons will only be made briefly here in order to put the published documents into context.[1078] Correspondence between James Chadwick and Rudolf Peierls throughout 1940 gives some insight into the topics discussed within the technical committee and the direction that nuclear research was taking. For instance, the critical mass was recalculated in the light of experimental data from the US,[1079] and the Frisch-Peierls calculations had to be revised.

But also, at that stage, there were discussions about the nature of the collaboration with the United States. Scientists and politicians differed in their views about the principle and the extent of technology transfer, and a complex interplay between the different parties to the opinion formation and decision making of nuclear policies developed which resulted in a stop-and-go policy which ultimately worked to Britain's disadvantage.[1080] One example of the kind of decisions of significance for the

[1076]Letter [304].
[1077]Still the most comprehensive account is M. Gowing, *Britain and Atomic Energy, 1939–1945*, London: Macmillan, 1964.
[1078]Much of the correspondence concerning nuclear weapons remained classified for several decades after the war. A recent release of documents by the National Archives (NA), however, included the declassification of much of the Tube Alloys material and documents relating to the Atomic Energy research Establishment in Harwell. See National Archives, Public Record Office (PRO): AB1–AB6. Some of this material is included here.
[1079]Letters [305–306].
[1080]S. Lee, 'In no sense vital and actually not even important'? Reality and Perception of Britain's contribution to the development of nuclear weapons', *Contemporary British History*, **20**, 159–185 (2006).

practicalities of Anglo-American co-operation was the discussion about the proposal to send Hans Halban and Leo Kowarski to the US to discuss nuclear research in the autumn of 1940. The two physicists, who had only recently fled to England from France. They knew about certain aspects of the work in the UK, but Peierls regarded it as too risky to see negotiations of such a politically sensitive matter carried out by colleagues with only fragmented knowledge about the UK project.[1081] The theme of coordinating between various departments within the UK but more significantly between the UK and the US was a frequent topic of war-time correspondence between Peierls and Chadwick, but also Akers, Simon, Frisch and others involved in the project in decisive positions. As late as May 1945, shortly before the Trinity tests, views were expressed about the tension in the Anglo-American collaboration.[1082]

After months of research spread across various academic and research institutes, the MAUD Committee, in its final reports in June and July 1941, endorsed the Frisch-Peierls memorandum when concluding that the atomic bomb was feasible, though very costly.[1083] Even before the report, it appeared likely that the United Kingdom would decide to pursue this area of research with some degree of priority, and in the knowledge of the complexity of the mathematical problems to be tackled, Peierls tried to enlist help from fellow theoreticians. Many of them were already engaged in war work of perceived greater urgency. Those least likely to be working in those high-priority fields were foreign-born scientists. Among the best young theoreticians was Klaus Fuchs, a non Jewish German who had to flee Nazi Germany because of his left-wing views. Peierls encouraged Fuchs to join him in his work at Birmingham,[1084] a step with fateful consequences, as Fuchs eventually

---

[1081]Letter [309].
[1082]Letter [378].
[1083]MAUD Committee Report in NA/PRO: AB 1/594. For a reprint of the key parts of the MAUD Committee reports see R.C. Williams and P.L. Cantelon (eds.), *The American Atom. A Documentary History of Nuclear Physics from the Discovery of Fission to the Present. 1939-1984*, Philadelphia: University of Pennsylvania Press, 1984, pp. 16–20.
[1084]Letter [328].

passed on atomic secrets to the Soviet Union.[1085] Initially Peierls was prevented from informing Fuchs about the exact nature of the project, but Fuchs soon received security clearance and was informed fully.

The research, though a scientific enterprise, did not happen in a political vacuum. What spurred the nuclear work in Britain at the time was the knowledge that, if it was theoretically possible and technically feasible to build atomic weapons, Nazi Germany would be capable of staging a nuclear arms project. With Heisenberg, Hahn, von Weizsäcker, Bothe and many others still in Germany, this could be a project with some promise of success. In the summer of 1941 and at regular intervals thereafter, Peierls with the help of Fuchs, studied lecture lists of German universities and publications of German physicists in an attempt to ascertain the activities of those capable of having an impact on nuclear research.[1086] Francis Simon, who was more at home in Physical Chemistry, did the same for this area of research,[1087] and all came to the conclusion that it was unlikely that there was a large-scale effort involving the key scientists in Germany.[1088]

As for the majority of people in England, quite apart from the actual war-work that Rudolf Peierls was doing, his family's life was dominated by everyday concerns of people at war. Rudolf's father and stepmother, in the spring of 1940, emigrated to America,[1089] which relieved him and his brother Alfred from the responsibility of looking after them at a time when travel and communications were becoming increasingly difficult. Their children, Gaby and Ronnie, were initially evacuated from Birmingham to the country. But Rudi and Genia were evidentally concerned about their children's fate, if anything happened to themselves. In the summer of 1940, a German invasion of England was perceived to be a real possibility, and Genia was certain that even if she and her husband survived the bombing and fighting, they would be among the first to be liquidated by an invading German army.[1090] She was not too keen on

---

[1085]See vol. II of the *Peierls Correspondence*, chapters 6 and 7.
[1086]Letters [329], [331], [339].
[1087]Letter [351].
[1088]See reports in NA(PRO): AB3/111.
[1089]Letter [303].
[1090]Letter [307].

the idea of the children being taken into care by their (rich) relatives in America, and therefore Rudi and Genia were pleased that, as part of an arrangement between the universities of Birmingham and Toronto, Gaby and Ronnie were invited to go to Canada for the duration of the war.[1091] While preparing for their children's emigration, Rudi and Genia received a telegram from Hans and Rose Bethe, inviting Gaby and Ronnie to stay with them in the US.[1092] The exchange of letters between Rudi, Genia, Hans and Rose which followed this invitation, is remarkable as a characterisation of the two couples' friendship as well as their way of dealing with a very difficult emotional situation.[1093] In the event, the children did not stay with the Bethe's but joined the Canadian programme, but as a result of the correspondence, the Peierls decided to make Hans a guardian of their children in the case of their coming to harm during the war.[1094]

In early 1941, Peierls engaged in a short exchange of letters with Egon Orowan. While Orowan had been in Birmingham (1937–39), Peierls — on Orowan's suggestion — had written a paper about forces resisting dislocation in a perfect crystal.[1095] The collaboration, if on a much reduced scale, continued after Orowan had moved on to Cambridge in 1939, and the latter asked for advice and comments on some of his latest ideas in the field of plastic gliding, but also some other war-relevant inventions which he was hoping to get patented.[1096] Peierls' reply and the subsequent exchange of letters[1097] are typical examples of the Peierls engaging in a fruitful debate offering constructive criticism even in a research area which was only marginal to his own interests at the time. It also shows Peierls as a theoretician who was always interested in the empirical aspects of the discipline as well as in the practical applications of any theoretical findings.

[1091] Letter [308].
[1092] See letter [310].
[1093] Letters [310], [312–314].
[1094] Letter [313].
[1095] R.E. Peierls, 'The size of a dislocation', *Proc. Roy. Soc.* **52**, 34–37 (1940).
[1096] Letter [319].
[1097] Letters [318–319], [321], [323–324], [326].

After his initial preoccupation with the calculation of the critical mass of U235, Peierls became increasingly involved in the complex problem of isotope separation. In this he collaborated with Simon and his colleagues Kurti and Kuhn in Oxford as well as Dirac at Cambridge and Maurice Pryce,[1098] one of Max Born's students who eventually married one of Born's daughters. The vast majority of papers written by Peierls between 1940 and 1942 discussed the more or less promising avenues on the way to an efficient separation process.[1099] This was reflected among others in his correspondence with James Chadwick. It revealed a sense of urgency in solving the theoretical and experimental problems associated with the various separation methods, and it also reflected the difficulty in obtaining qualified staff to help with the theoretical work. By the autumn of 1941, the nuclear programme, now code-named Tube Alloys, under the Chairmanship of William Akers, had received government backing, and the key technical sub committee including Chadwick, Halban, Peierls and Simon had been set up.

Science and scientific progress are international and scientists in Britain and the United States were convinced about the wisdom of a joint effort. While British scientists were aware of the potential loss of control over the research project, they were not concerned, *per se*, about sacrificing a lead in whichever part of the product development they may have had at this point in time. The task of developing nuclear weapons was enormous, and the (often self-imposed) pressure to succeed and most importantly to stay ahead of the Germans in this scientific race was a powerful catalyst for Anglo-American collaboration

[1098]See letter [336–337].

[1099]E.G. R.E. Peierls, 'Efficiency of isotope separation by thermal diffusion', Directorate Tube Alloys (DTA) rept. MS. 12 (1940) NA/PRO: AB4/837; R.E. Peierls, 'Equilibrium time in a separation plant', (DTA) rept. MS. 12 (1940) NA/PRO: AB4/837; R.E. Peierls, 'Separation of a mixture of three components', (DTA) rept. MS. 20 (1940) NA/PRO: AB4/849; R.E. Peierls, 'Electrolytic method' (DTA) rept. MS. 26 (1940) NA/PRO: AB4/855; R.E. Peierls, 'Isotope separation by means of the thermomolecular effect', (DTA) rept. MS. 32 (1941) NA/PRO: AB4/865; R.E. Peierls, 'Separation by cul-de-sac method', (DTA) rept. MS. 38 (1941) NA/PRO: AB4/872; R.E. Peierls,'Separation by means of evaporation at low pressures', (DTA) rept. MS. 46 (1941) NA/PRO: AB4/880; and numerous papers on the Simon plant.

in the laboratory and beyond. Therefore, the first meeting of the subcommittee was also attended by two American physicists, Harold Urey and G.B. Pegram[1100]

By August 1941 Winston Churchill had approved the British atomic weapons programme. Although close co-operation with the United States was considered beneficial, the independence of a British enterprise was still regarded as an important interest, and in late 1941, when the Americans proposed a joint and largely integrated project, the Churchill government declined and opted for collaboration through exchange of information only. Scientists in general did not share this complacent attitude. In early 1942, a British delegation visited the US to investigate possibilities of further co-operation. They were clearly impressed with what they found. Although there were some complaints that the Americans still had not made the nuclear question a top priority, it had become clear that although the British were ahead in some theoretical work, on the experimental front, the Americans had already far outstripped their British partners. The Japanese attack on Pearl Harbour in December 1941 had accelerated the American programme, and this American effort meant that the position of the two countries relative to each other had reversed in many important parts of the nuclear project by the middle of 1942. The United States had a potential wealth of equipment, industrial financial backing and a large supply of first-class chemists and physicists. In contrast, Britain had a handful of excellent scientists and considerable determination to succeed. Not surprisingly then, in June 1942, when Prime Minister Churchill, talking to President Roosevelt, suggested that Britain and the United States should 'at once pool all our information, work together on equal terms, and share our results if any equally between us', it was the Americans' turn to put a break on all-out nuclear partnership. The British had missed the proverbial bus. They had refused to co-operate on mutually acceptable terms at a time when they were scientifically ahead of the Americans. When they finally realised that the tables had turned, and they had fallen behind the United States in all but some limited theoretical explorations,

[1100]Peierls, *Bird of Passage*, p. 166; see also letters [336–337].

they were not longer able to determine the conditions of the association and had to settle for co-operation largely on American terms.

Rudolf Peierls was among the participants of the British delegation who visited the US in the spring to 1942 to exchange information and investigate collaboration. He used the opportunity of being on the other side of the Atlantic to see his children who were still evacuated in Canada. It also gave him the opportunity to see his relatives in New Jersey, many of whom he had not seen since their emigration to the US, and to catch up with some old friends such as the Bethes.[1101] Among the contacts which Peierls renewed during this visit to America was his acquaintance with Robert Oppenheimer, whom he knew from his Leipzig and Zurich days.[1102] Both respected each other as scientists and during the second half of the war they would have the opportunity to cooperate more closely at Los Alamos.

Throughout much of 1942 and the first half of 1943, Peierls and Chadwick communicated extensively about their theoretical work, but also on organisational matters, both concerning the logistics of theoretical and experimental work in the UK, and about research done in the United States and Canada.[1103] Many of the practical problems were a function of Britain's war-affected economy having to meet conflicting demands of different sections of the war effort.[1104] But some of the problems of co-operation and collaboration were a result of different personal priorities of researchers, political interference in the decision making process which affected scientific progress or sometimes a simple clash of personalities.[1105] By May 1943 Peierls felt the need to air his frustrations about the negative effect of political manoeuvring on the scientific process in a detailed letter to Wallace Akers. Looking back on the short history of nuclear research and collaboration with the US, Peierls concluded that much of this history was characterised by political indecision. He alluded that further delay in decisions about the

---

[1101] Letters [342–343].
[1102] Letter [344].
[1103] See letters [345–353].
[1104] See letter [341].
[1105] Letters [353–354].

general direction of the British nuclear effort, caused by a lack of willingness on the part of the political establishment, would be detrimental and 'great responsibility [was] incurred by doing nothing'.[1106]

There is reason to believe that other scientists shared Peierls misgivings, but Akers' hands were tied. But the political process was beginning to move towards a convergence of views about transatlantic nuclear co-operation, at least to the kind of convergence that would allow meaningful scientific collaboration.

In August 1943, at the Quebec Conference, the official nuclear relationship between Britain and the USA was agreed on for the duration of the war with the American President Roosevelt trading nuclear co-operation with the British for Churchill's agreement for a cross-channel invasion of Europe during 1944. The USA and the UK agreed never to use nuclear weapons against each other, not to use them against other states without each other's consent, and not to pass on information to other countries without each other's consent. Moreover, they undertook to ensure full and effective co-operation to bring the atomic weapons project to fruition. The one provision which demonstrated clearly that the agreement nevertheless was not one between equals was that dealing with the post-war exploitation of the research results. Here, unilateral British concessions on post-war commercial applications amounted to a disclaimer of any interests in industrial and commercial aspects beyond 'what may be considered by the President of the United States to be fair and just and in harmony with the economic welfare of the world.'[1107]

While the Quebec Agreement forced the British Government to make substantial concessions, it enabled their scientists to continue nuclear research at a time when the Tube Alloys project in the UK was under threat. Soon after the signing of the Quebec Agreement, James Chadwick, Francis Simon, Marc Oliphant and Rudolf Peierls arrived in Washington as part of another fact-finding mission about the British role within an Anglo-American project, and by the end of the this visits

[1106] Letter [363].

[1107] 'Agreement Governing Collaboration between the Authorities of the U.S.A. and the U.K. in the Matter of Tube Alloys', *Foreign Relations of the United States*, The First Quebec Conference, Washington, 1970, pp. 1117-9.

in the autumn of 1943, it had become clear that the key figures of the British effort would join their American colleagues in the US.[1108]

Given the nature of the nuclear project, the secrecy surrounding the research and the setup up the Manhattan Project which Peierls joined eventually, it is not surprising that correspondence became much less voluminous between mid-1943 and mid-1945. Genia and Rudolf both went to the US and were eventually reunited with their two children. But also professionally, there was less need to correspond in writing, as the scientists were concentrated in a hand full of research centres with many of the exchanges being made on the ground in personal discussions and seminars. Communications continued between British scientists in America and some of the colleagues who continued research in England, though on a much smaller scale. Peierls, for instance exchanged letters with Kurti and Simon in Oxford, but also with his colleagues in Birmingham. Some of these have been included here, as they give interesting insights into the research which was still being done successfully in the United Kingdom, but also into the secrecy surrounding the work in the US on which Peierls was unable to report in any detail.[1109]

In the latter half of 1944, after Peierls had joined the plant at Los Alamos and thereby gained a clearer picture of the details of the production plant, he wrote to Wallace Akers again. At the time, the UK was contemplating its post-war weapons programme, and Peierls impressed on Akers the need to secure the support of the most knowledgeable personnel in the initial stages of the project when fundamental questions were being tackled. At the time, as Peierls judged, the impact of any individual (and in particular Klaus Fuchs, who he was writing about) at a place like Los Alamos (code-named Y) was far less than it would be in a key position in the British enterprise, if this were to come off.[1110] This letter, is interesting in another aspect, too. Klaus Fuchs, at the time member of the British group working in New York on the gaseous diffusion programme[1111] seemed content on returning to

[1108]Letter [365–366].
[1109]Letters [367–368].
[1110]Letter [370].
[1111]The other members of this group were Peierls, Skyrme, Kurti and Kearton.

the United Kingdom in order to play an active part in whatever kind of atomic research programme the United Kingdom was going to embark on. It is not clear why Fuchs was sent to Los Alamos, but in view of his subsequent passing of information about the Anglo-American nuclear project to the Soviet Union, it was certainly a far-reaching decision.[1112]

By the spring 1945 the Manhattan Project was close to the first test of a nuclear device. It is in this phase when the bulk of the work had been completed, that many tensions between the British and the Americans resurfaced, and that the British began expressing concerns about the flow information and the process of decision making. This is evident in some of the weekly letters sent from Peierls at Los Alamos to the head of the British mission in the US, Chadwick, in Washington.[1113] The exchange is interesting for several reasons. By sending letters beyond the weekly communication which was meant to summarise developments at Los Alamos, Peierls tried to circumvent General Groves, who had demanded to see copies of letters passed between the British at Los Alamos and Washington. Peierls was clearly dissatisfied with arrangements on the ground, in particular with arrangements for the impending Trinity tests. He regarded preparations as unsatisfactory and ultimately as endangering the success of the test. It is unclear whether he expected Chadwick to act upon his complaint, but it is unlikely that the latter would have been in a position to influence events. However, in view of growing political tensions between the United States and the United Kingdom over nuclear issues, the attitudes of scientists on the ground are enlightening.[1114]

[1112]Brian Cathcart, www.nuclearweaponarchive.org.Usa/Med/Fuchs.html.
[1113]Letters [376], [378].
[1114]Letter [378].

[303] Rudolf Peierls to Heinrich and Else Peierls

Birmingham, 20.3.1940
(carbon copy)

Dear Parents
We were so glad to see from your two telegrams that you have safely arrived;[1115] I hope there will soon be a letter to tell us that you were not too tired after the long journey. It must be big relief to know that you have at last arrived at the destination you have chosen, and that there are no more formalities and papers in your way. I hope that when you have had time to recover from the voyage, you will soon successfully solve the last problem of the moment, that of finding and furnishing a flat of your liking and to have again a place which you can look upon as your home.

There is not much to report about us, we are all well. The children are coming home to-morrow for a few day and then go back to school for a fortnight before their vacation starts. The school will stay in the country at any rate for the summer term, i.e. until the long vacation. We have saved up a little petrol to be able to make an excursion with the children over Easter, if the weather is good enough, but we shall not go far.
Give our love to the Krebses and Siegfrieds.
With kindest regards from us all
Yours

[Rudi]

[1115] In the spring of 1940 Heinrich and Else Peierls were finally allowed to enter the US. They settled in Mont Clair, New Jersey, near Heinrich's brother Siegfried and his daughter Annie.

[304] Rudolf Peierls to MAUD Committee[1116]

Birmingham, 22.4.1940
(carbon copy)

Dear Sir,
I understand that there will be another committee meeting to discuss the matters raised by Dr. Frisch and myself in our recent report,[1117] and that we shall not be asked to take part in these discussions. I should not in normal circumstances try to interfere, but in view of the importance of the issue involved I feel I cannot permit myself the luxury of reserve.

It is clear that the matter is of great urgency, at least until it is satisfactorily proved that the construction of a super-bomb is impossible. By excluding us from the discussions the work must necessarily be delayed. We have spent the last weeks in working out the details of the physical problems arising in this connexion, and we would be in a position to answer many of the queries or objections that may be raised and that have apparently been raised by the first meeting.

In addition I would like to point out that Dr. Frisch is one of the discoverers of the phenomenon of nuclear fission and one of the first experts in the field. It is certainly not in the national interest to exclude him from discussions instead of making full use of his experiments.

[1116]Peierls and Otto Frisch had calculated the critical mass of uranium 235 and summarised their findings in a memorandum. 'The Frisch-Peierls Memorandum of 1940', R.H. Dalitz and R.E. Peierls (eds), *Selected Scientific Papers of Sir Rudolf Peierls*, London: Imperial College Press, 1997, 277–282. On being informed about the Frisch-Peierls findings, a government committee, the so-called MAUD Committee, was set up to investigate further the potential implications and Peierls engaged in detailed calculations. It met in April 1940 for the first time, but Peierls and Frisch, as only recently naturalised and enemy alien respectively were excluded from the committee. Peierls did not even know the name of the Chairman of the committee and therefore sent his communications through Oliphant, merely addressing the letter to 'Dear Sir'. See Margaret Gowing, *Britain and Atomic Energy, 1939–1945*, London: Macmillan, 1964, p. 46. Most of these papers, which were initially top secret have now been released and are accessible in the National Archives, Public Record Office, AB4.

[1117]Frisch-Peierls memorandum; see note 1074.

In this connexion the vital importance of absolute secrecy has been mentioned. We are in full agreement with this and you will remember that we underlined this point repeatedly in our first memorandum. But I do not think that any useful purpose can be served by trying to keep our own idea a secret from us. I quite realize that one of the first tasks of your committee had to be to discuss our reliability, and I was not surprised therefore, that the first meeting was held without our being present. But I hope strongly that the committee will not waste any further time on these personal questions, and will come down to business.

It is not a pleasant task to have to write this letter, but at least, if further delay is caused by wasting time in discussing questions to which we know the answers, and if this delay becomes dangerous, I shall feel free from responsibility.

Yours faithfully

[R.E. Peierls]

[305] James Chadwick to Rudolf Peierls

Liverpool, 5.6.1940
(carbon copy)

Dear Peierls,

There is a point in your calculation of the critical size for a bomb of uranium 235 which seems to me very doubtful.[1118] You assume that the cross-section for fission of fast neutrons, say of 1 MV or lower, is about 3.5×10^{24} cm. sq. This does not agree with the experiments of Tuve reported in Bohr and Wheeler's paper in Physical Review, vol. 56, p. 44.[1119] Tuve gives the cross-section for fission for 0.6 MEV neutrons as $.003 \times 10^{24}$ cm. sq. for ordinary uranium.[1120] If your assumption were

[1118]See letter [304], note 1074.

[1119]N. Bohr and J.A. Wheeler, 'The Mechanism of Nuclear Fission', *Phys. Rev.* **56**, 426–50 (1939).

[1120]The figures were based on Tuve's report at the Princeton meeting of the American Physical Society.

correct, the cross-section would be $.25 \times 10^{24}$ The cross section seems therefore to be not more than 1/3 of the value you have taken. These experiments of Tuve have not yet been published as far as I know, but he is a reliable worker and there is no reason to suspect his result. This is a point which needs investigating immediately, and I am taking steps to do so.

Yours sincerely,

[Chadwick]

[306] Rudolf Peierls to James Chadwick

Birmingham, 12.6.1940

Dear Chadwick,

Thank you very much for your letter.[1121] We had not noticed the reference to the Washington experiments, but we had decided that this was precisely the point which should be investigated at once, and Frisch has been measuring the fission cross-section with Rn-Be photo-neutrons. The experiments are nearly complete. If the present results are taken at their face value, they would indicate a cross section of $2.5 \pm 0.5 \times 10^{-24}$. Actually, however, there is a possibility that the shielding against scattered thermal neutrons may have been insufficient in the experiments, and before this point has been fully investigated the results must not be accepted.

Frisch hopes to be able to clear up this point fairly soon and will then report on the experiments in detail. Meanwhile, I think, one should not place too much reliance on Tuve's remark, which apparently referred to somebody else's experiments in Washington which have not since been published.

Yours sincerely,

R. Peierls

[1121] See letter [305].

[307] Genia Peierls to Hans Bethe

Birmingham, 17.6.1940
(carbon copy)

Dear Hans,
Things are moving so quickly now that I don't know where we shall be and what we shall be when you get this letter. At the moment, our children are with their school in the country, and we are both rather in the front line with the A[ir] R[aid] P[recaution] and would be in exposed position during a raid. It is quite possible that we may both get killed and that the children would then be left in a rather awkward situation.

If things should get very bad here I suppose the United States will organize some large scale scheme to get the children removed and if they are really left without us perhaps you could use your influence to get them taken along. If this country wins the war they will, of course, be all right. We have good friends who will be looking after them whatever happens to us. But if we should lose the war, their existence here would become impossible. In that case I imagine the Germans would shoot us, even if nothing happens to us during the air raids. But perhaps the children will be spared.

They are very nice children, very jolly, easy going and bright. They will get all the possible scholarships and so on and be a lot of fun to the people who take them. If they do manage to get taken into some kind of refugee camp in the USA perhaps you could occasionally keep an eye on them and bring them up in the good tradition of theoretical physics.

You know Rudi has a lot of relations in the USA, some of them even are not so bad. But I would much rather that the children would go to people who take them because they want to and not because they are the children of their brother-in-law or so. On no account would I let them feel as "poor relations" and that always happens in these conditions.

I am writing a similar letter to Fankuchen. We are very well, terribly busy and in good spirits. Qui vivra, verra, and if even "ne vivra pas", we had lots of fun and on the whole a long and interesting life. Nothing can be worse than death because you can always have that choice.

I hope to see you once more somewhere and some time.
Love to you all

[Genia]

[308] Rudolf Peierls to Else and Heinrich Peierls

Birmingham, 11.7.[19]40
(carbon copy)

Dear Parents,
Thank you for your two postcards. We are now getting more regular news from Alfred and Nina who seem well.[1122] Alfred writes that he is bathing in the sea. Nina seems to have some quite nice people around her, with whom she can play bridge &c. Their addresses are: Alfred, House 31, Onchan Internment Camp, Douglas, Isle of Man and Nina, Imperial Hotel, Port Erin, Isle of Man. They are allowed to receive letters, which should not be too long. Vera's address is c/o Mrs Ward, 14 Rafford Way, Bromley, Kent.

Yes, we are thinking of sending the children away. They will be included in a scheme by which the Canadian Universities have invited the children of members of staff of English universities. They are ready to go, and it is now merely a question of shipping space. I hope they will leave in a week or two.

I hope you will have a nice time in Lake Placid.
With kindest regards to you and Siegfrieds from all of us,
Yours sincerely

[Rudi]

[1122]As enemy aliens, Alfred and Nina Peierls had been interned after the outbreak of hostilities.

[309] Rudolf Peierls to James Chadwick

Birmingham, 12.7.[19]40

Dear Chadwick,
Oliphant told me about the proposal to send Halban and Kowarski[1123] to the United States, and also told me that the final decision was largely left with you. It is pretty clear that this would mean giving all information away, since it is quite impossible for them to separate completely the question of slow neutrons from that of fast ones, or to forget completely what they have heard in this country. I am all in favour of collaborating with the U.S.A. as much as possible. They obviously have the bigger resources and can get the job done more quickly than anybody else. But if they go over, the matter is entirely out of our hands; I have no doubt that the Americans know the answer first or at any rate will put it into production more rapidly. There would then probably be no point in continuing the work in this country. If it is agreed to hand over the idea it would seem desirable to do that in such a form that a certain amount of control over it can be retained. It would be better then to send over somebody who would know all the details about the proposals discussed in this country and could take part in their development, whereas H[alban] and K[owarski] could pass on the information about this, but would still be more interested in the thermal aspects of the problem.

I also gather that G.P. Thomson is strongly opposed to the transfer to the Ministry of Supply. I suppose that there is nothing that can be done (and that is worth doing) about it at the moment.

I also enclose a list of well-known scientists, who to my personal knowledge have been interned, besides those you mentioned when I was in Liverpool. No doubt you realize that the Society for the Protection

[1123] Hans van Halban (1877–1947) and Lew Kowarski (1907–1979), who, together with Frédéric Joliot, in 1939 had established the certainty of chain reactions, had fled from France to England in 1940. They continued their work in Cambridge.

of Science and Learning are probably in the possession of an almost complete list.
Yours sincerely,

R. Peierls

[310] Rudolf Peierls to Hans and Rose Bethe

Birmingham, 12.7.[19]40
(carbon copy)

My dear Bethes,
I always knew you were friends but your telegram left us quite speechless.[1124] This is really very nice of you. Meanwhile you will have got our cable explaining that, the children have been invited by Toronto University and that we hope, to send them there in a week or two. There is at present a great rush on steamer accommodation, some universities just managed to get their children away before the rush but I hope our party will be able to go quite soon. We thought for a moment whether we should now cancel this and accept your invitation instead, but we decided not to do this at the moment. The children have their passports and permits &c. ready to go to Canada, no visa is required there, they are travelling there as a party with other children (many of whom they know) and also as regards streamer accommodation the official party sent by the university is likely to get better treatment than people who try to book privately. Moreover, if we sent them now, this would probably mean that you would have to rush back from your holidays. We have made sure that the number of children sent is actually a little below that of the number of places offered to us by Toronto, so that there is no question of our children going to the exclusion of anybody else.

The main thing is to get them across to America first, and then there will be enough time to sort out everything. We shall certainly make

[1124]The Bethes had sent a telegram with an offer to provide a home for the Peierls children for the duration of the war.

use of your generous offer to the following extent: For one thing I shall arrange (or try to arrange) that in the event of our being killed Hans will be made guardian of the children. I shall have to see a solicitor about the best form of arranging this without getting into difficulties with any of the laws of the three countries involved. I shall also write to the people in Toronto with whom the children will stay, asking them to communicate with you in the event of our being killed, or if communication with us should become impossible.

Also, if sometime in the future you could go to Toronto and see for yourself how the children are getting on we would be extremely grateful. I believe Toronto is quite near from Rochester, where you will be now and then anyway. I cannot judge whether in the long run it will be better for the children to stay in Toronto or to come and stay with you, as this depends on so many things I cannot judge. I do not know how much they would be in your way, and whether they would fill your house beyond capacity. On the other hand, of course, I do not know, of course, what kind of people they will find in Toronto. Needless to say that we would be very happy to know the children with you. But in all probability they will be in good hands in Toronto.

Incidently among the mothers going to Toronto is Mrs Shapiro[1125] whom, I believe, you have met in Manchester. She went ahead of the rest of the party, because she is again in the same position in which she was when you met her then, and is therefore in a hurry. She knows our children well and will keep an eye on them. We shall ask her to get in touch with you.

Well, in any case, whatever happens, and whatever you ultimately decide is best to do we shall not forget this telegram.
Yours sincerely

[Rudi]

[1125]Isaac Avi Shapiro (1904–2004) and his wife Pauline had first met Rudi and Genia Peierls in Manchester, where Shapiro had taught English between 1933 and 1935, before returning to Birmingham where he spent the rest of his academic career and long retirement.

[311] Rudolf Peierls to Else and Heinrich Peierls

Birmingham, 14.7.1940
(carbon copy)

Dear Parents,
You have probably read in the papers that large numbers of internees have been sent overseas and I think it is possible that Nina and Alfred will also have to go (or, in fact, that they have gone already). As soon as I hear anything definite about their new address, I shall, of course, let you know. If they go to Canada, they will, I suppose, get in touch with you directly.

Meanwhile we are trying to think out what should be done about Vera. I think that if Alfred and Nina have to go overseas, they would feel very much happier if Vera could go, too, as otherwise they would worry too much. As things stand at present, she will not be allowed to go to Canada on any of the children's migration schemes because she is over 16 and also because she is not a British subject. We are just reading in the newspapers that America has abolished the quota for refugee children from the war areas. I do not know yet whether this would include Vera. In any case could you find out as soon as possible whether you can obtain for Vera either an affidavit or an invitation (or guarantee) of the type which are now sent in large numbers to children in Britain. If Uncle Siegfried is not in a position to invite her, perhaps one of his friends would? I have not heard from Alfred whether he would want Vera to go overseas himself, but I imagine that would be his wish, and in any case communications are now so slow that it is best to take all possible steps about an affidavit as quickly as possible, so as to have the papers ready when they are wanted.

Perhaps, in asking this favour for Vera, I should make quite clear the position as regards to our own children. As I wrote you in my last letter,[1126] they are going to Toronto on an invitation sent by members of the staff of Toronto University, and they will stay there for the present. They have also been invited by my friend Bethe, and they will go to

[1126] See letter [308].

Bethe if it is felt that the people they are staying with in Toronto are not suitable or that the children represent a burden to them.[1127] Bethe will take them in any case if the war goes on for very long, or if at the end of it they cannot return to this country. I have arranged to have Bethe appointed executor and guardian of the children, and he is also prepared to take full responsibility for them if for any reason my life insurance cannot be paid out or if the amount cannot be transferred overseas. I am writing all these details in order to make it quite clear that Gaby and Ronnie are provided for as well as is humanly possible, and that therefore neither Siegfried nor the Krebsens need fear that they will become a liability on them in any circumstances. This might be of relevance in deciding what can be done for Vera.

If any documents are sent for Vera, it might be best to send them to Mr J.G. Flint, 7 Kingswood Avenue, Bromley, Kent, who is making arrangements on Vera's behalf.

Our children are leaving this week. They are very thrilled. We have just brought them home in order to get all their things in order.
With kindest regards to you all
Yours sincerely,

[Rudi]

[312] Genia Peierls to Hans and Rose Bethe

Birmingham 16.7.[19]40
(carbon copy)

My dear Bethes,
I am not quite sure which you will get first, this letter or the one which Gaby will write to you from Toronto. After about a fortnight of red tape and bother I am now writing to you in a whirl wind of socks, jerseys, suits and dresses, coats and shoes. Everything is very small, everything is marked, some bags are "wanted", others "not wanted", but I think

[1127] See letter [310].

one really wants everything for a journey on the high seas starting with a cool English July and finishing on a Canadian August.

We are going round the children looking for some more suitable spots on which they can be labeled. They are wearing identity disks, University badges, school blazers, and I think I shall just stick some labels over them and varnish them over to make them waterproof.

In the meantime we are stuffing Gaby with wisdom and knowledge. They are just home from school, and the time is very short, and Gaby gets slightly confused. For instance after I warned her against the dangers of constipation and ordered her to go to one of the mothers travelling with the party and ask for a remedy, she was also warned about the dangers of falling into the sea and the necessity of listening to what the sailors told you. Next morning she told me that: "If I or Ronnie have not been to the lavatory I must go and tell a sailor." Such a lot of pitfalls occur in every problem.

Rudi is writing you all the business side, I will tell you about the children.

Gaby is very jolly, very quick and very unsentimental. I think she has what is called a masculine mind. She is absolutely logical, very clear and unemotional in all her thinking. She loves arithmetic and one can explain to her everything. She will understand a very abstract problem, and she thinks a lot in this direction. For instance she told Rudi already a long time ago that every book must have an even number of pages because every leaf has two sides. Nobody has asked her about it. She is already, and will be even more so, a book-worm. She reads everything and is quite capable of understanding easy geography and history books. She is mentally much older than her age. She actually enjoys learning and thinking and solving problems. As other people play tennis, she plays arithmetic, for the sake of mental gymnastics. But I do not think she will be able to "invent" anything. She has not much imagination and is terrible terre-à-terre. She is never afraid or abashed in the mental or sentimental sphere. She will go and talk to everybody and to any amount of people; she is very self-confident, (even cheeky) and free. But she is a coward in a lot of physical things, she does not believe in herself in that respect at all. She cannot ride a bicycle or climb, but loves water and probably will soon learn to swim. She is bad at sewing, knitting and all handwork, and her handwriting is positively inherited! She is very definitely my daughter in very many respects (bicycle) but much more a woman than I ever was. She loves frocks and ribbons, can spend hours in front of a mirror, and adores babies and small children. She is actually very good with them and very reliable.

She tremendously enjoys life and is very easily contented. She is a chatterbox, but has a lot of tact and savoir-vivre.

Now about Ronnie or, as he is called at school, Podgkin (he is still very fat). He is very much Hans's spiritual son. First of all, food is very important for him, and he can always eat as long as there is something to eat. But when it is good, he enjoys it tremendously. Ronnie is a real gourmand.

Then he has the same sort of reckless and senseless bravery as you have. He will climb the most impossible tree, he will try and get on a bicycle until he is dark and green and ultra-violet. He screams in between, but wipes his tears and goes on. He was swimming in Yorkshire last year in a real surf so that he was turning somersaults in the waves for minutes on end, but he never minded it. But if he is afraid of something (dogs!) nothing can be done.

And then he needs a lot of petting, and he adores it. He is very affectionate and every woman loves him. But he must be dealt with very firmly and he must be encouraged to consider himself a grown-up.

He was extremely good when his tonsils were taken out. He stayed for three days in the nursing home and never even cried or rang the bell. Just because he wanted to show off. He can very easily be sorry for himself. He is very jolly and has a big sense of humour. He has a lot of imagination, but he is not as quick or as logical as Gaby.

If he will ever stay with you, your mother will adore him and spoil him. And he will love her and tell her that "she is the best woman in the world, that he loves her more than anybody else, and that he will marry her when he grows up." He says that to a lot of people. He can read and write a little, and can do a bit of adding, &c. But he does not like arithmetic. He is a little lazy and easygoing.

The two are very fond of each other. When I told Gaby that in Toronto they might be separated, she went pale (you need a lot to make her feel anything) and said "but he is still so young and I must look after him!" I hope they will put them together. They both are quite independent, can dress, wash and Gaby can make her bed and wash up a bit or dust a room. They will eat everything, sleep everywhere, and play and work in the middle of the most terrible din (Gaby particularly) they learned that at boarding school.

They had whooping cough and were inoculated against diphtheria. Ronnie had his tonsils out. He must be stopped when he eats too much; he does not stand too much of eggs and fat things.

Now you know about them nearly as much as I do.

Gaby is a very simple problem. She is clever enough to understand another point of view and if she will not be frightened by something or somebody, she will always be happy and adapt herself to every situation. But she must be kept busy and set hard problems to think about. She did very badly at school when she was put in too low a form and work was too easy for her.

Ronnie is more complicated. He must be loved and petted or he will be unhappy, but not too much or he will not grow up. He must be treated and talked to as a "big boy" and he will learn to play up and live up to it.

Thank you very much once more.

Yours sincerely,

[Genia]

[313] Rudi Peierls to Hans and Rose Bethe

Birmingham 17.7.[19]40
(carbon copy)

Dear Bethes,

Genia has promised that I would write about the "business side", but the joke is that the business side is practically non-existent.

I am sending with the children a letter addressed to their hostess asking her to let you know her address and to approach you in the event of any difficulty. I have made a will by which, in the event of Genia's or my death you would be appointed guardian of the children if they are then overseas. The executor under this will is Prof. Roy Pascal, of Birmingham University. He has absolute discretion in the way he uses the money on behalf of the children (it was very hard to persuade my solicitor to draw up a will giving so wide powers, that was quite against his feelings). He would obviously try and get permission to transfer the money (consisting only of my superannuation insurance) to America and if he succeeds, would delegate his powers of absolute discretion to you so that you would not be tied in any way by regulations of trusteeship, or by restrictions as to what share of the money to use for which child, &c. But I do not see at the moment that permission for a transfer would be given and thus the question will not arise.

Meanwhile I also enclose an official authority for you which may be handy if anything has to be done for the children where it would cause too much delay to get our signature for, or if there is any delay in getting the will approved, or you appointed as guardian. The document has no actual legal value, but it may help in the case of an application on behalf of the children or any other papers which would normally have to be signed by a parent or guardian.

You will understand that the intention of all this is not to give you more responsibility than you wish to take, but merely to avoid technical complications and bother.

I still hope that there will be a chance of showing you our gratitude.
Yours very sincerely,

[Rudi]

[314] Hans Bethe to Rudi and Genia Peierls

location unspecified, 15.8.1940

My dear Peierls',
Our telegram to you was so much the obvious thing to do that we hardly needed to think before we sent it.[1128] It might just as easily have been the other way round, and you would have done the same thing.

[1128] See letter [310].

So far, we have not yet heard from the children. We wrote a letter to Toronto University to tell them that we would take Gaby and Ronnie if there were any difficulty in finding a suitable family for them to live in Toronto. We have not yet had an answer from Toronto, but our letter was written only about 2 weeks ago, and it takes quite a long time from here to Toronto. But if there is no answer by the beginning of next week, and if we do not hear from Gaby in the meantime, we shall write again to Toronto.

Of course we shall go and see how the children are taken care of as soon as we are back in the East — i.e. probably early in October. I believe it may be best to convince ourselves personally whether all is well, and then decide whether it would be better to take them to Ithaca. For the purpose of bringing them over to America, it was certainly the best to have them go with the big transport of the other university children — everything goes so much more smoothly if there is an official organization behind it.

I remember enough of the children to know that they are about the nicest children one could acquire. All the real education, it is said, is done between 0 and 6 years; what follows is mostly the consistent development of the character built up in the first six years, and any outside influence contrary to the child's character provokes only antagonism and has an effect opposite to the intended one. Well, if this is true, nothing can spoil Gaby and Ronald — the first six years were too good. I have hardly ever seen children who were at the same time so natural and happy and so well-balanced and understanding as they were in 1938. It will not be a hard task for us — or for anybody else — to bring them up, and I know it will be a lot of fun.

Your description of them, Genia, was as good as if they had been with us for months. We are repeating all the time: "If I have not been to the lavatory I tell a sailor"; and it is if they were playing here in this room. I am already giving Gaby a problem in arithmetic and climbing up the big tree in our lawn with Ronnie.

Thanks very much for the business arrangements. I hope they will never become effective but if the situation arises the free hand arrangement will simplify everything tremendously.

We are practically living with you all this time. Whenever there are news in the radio we listen anxiously. There is seldom much reason for jubilation. Still, we are happy enough to know that so far the German planes have been repulsed every time with great losses, and that the R.A.F. obviously has learnt a great deal from the mistakes in the Belgian and French War. We are happy to know that British ships still are everywhere on the sea and that England's sea power has not been seriously challenged, that the French fleet did not fall in Hitler's hands and many other things. And we hope for the best.

Yes, this life was beautiful and full, and we are enjoying it at the moment in spite of everything. The last year alone was worth living for (this concerns my personal affairs only), and if there should be an end soon, we still have had our life. If England should be defeated, our turn will be next — everybody knows that here. And I am very much surprised at how indifferent I am to the possibility of death. I dislike the idea of dying mostly because there would then be one less person less to fight Hitler. If somebody had told me that 3 years ago, I should have considered him crazy.

Good luck — and I hope we can soon write to you something more positive about the children.

Yours,

Hans

[315] Rudolf Peierls to Else and Heinrich Peierls

Birmingham, 23.8.[19]40
(carbon copy)

Dear Parents,
We received you cable yesterday from which I saw that you had apparently been without news for some time, and that some letters of ours must have been delayed. I sent a cable in reply which, I hope, has reassured you that there is nothing to worry about.

It is very nice of you to suggest that Edgar or Hermann should go to see our children, but I would like to ask you not to send them. We have seen people who have taken refugee children here (including ourselves) and we know what nuisance it is to receive visits from relatives who come to "inspect" the children. We shall have frequent information about them from our friends who are with the same party and who have known the children for years. Edgar's style of life is so different from our own that the impression of how the children live would not convey very much information to us, and he might only give the children's hosts the wrong idea about our family. It is of course quite another matter of you should later happen to be in Toronto yourselves, or if the children should come to the States later, I am sure they will always be very happy to see you; but any distant relative whom they have never seen in their lives will neither cause nor derive much pleasure from such a visit.

Incidentally, we received a few days ago a parcel from New York sent by a shop there containing some coffee, tea and sugar, but it did not say on whose instructions it was sent. We suspect that this comes either from you or from Siegfrieds, and if it does, I would like to thank you (or whoever sent it) most sincerely for the kind thought. It is only fair to this country to say that, although we very much appreciate the chance of tasting American coffee and tea, the food situation here is very much better than your newspapers seem to think. There is plenty of coffee here (which is unrationed) at pre-war price; although tea is rationed, our ration amounts to more than our normal consumption; sugar is limited, but this does not amount to more than a mild nuisance. We are not starving just yet!

However, this does not mean we are ungrateful for the kindness of the sender whoever it was.

I am working fairly hard now, but I still spend some time on fire duty, although I have not so far seen a real fire. Nor have I, for that matter seen any signs of damage done by raids, although I had to travel a good deal.

This morning the papers published amended regulations for the release of internees and this has somehow increased our hopes of getting Alfred and Nina out.

With kindest regards to all of you
Yours sincerely,

[Rudi]

[316] Rudolf Peierls to Else and Heinrich Peierls

Birmingham, 16.10.[19]40
(carbon copy)

Dear Parents,
To-day I have good news: Nina has been released from internment. She is at present in Bromley, where I saw her on Sunday, but she will come to stay with us for the present as soon as she gets the police permission which, for technical reasons, will cause a few days' delay. She has been released because of her health, which has suffered from the discomfort and worries of internment, but we strongly hope that in more normal conditions she will soon recover again. In the circumstances, we have given up the idea of sending Vera to the school at Attingham, but Vera will go to the other branch of the same school which is just across the road from our house.

Also, the chances of Alfred's release in the near future have increased considerably, since a new committee has been formed which will especially deal with the cases of engineers, and this removes some of the difficulties that we always had with the applications for his release.

Genia has just recovered from a bad attack of tonsillitis. Fortunately the drugs which have been discovered in the last two years have made this a much more harmless thing than it used to be.

We hear regularly from the children and they are getting on alright.
I am extremely busy and have to travel a lot.
With kindest regards
Yours sincerely

[Rudi]

[317] Siegfried Peierls to Rudolf Peierls

New York, 6.11.1940

Dear Rudi,
We were glad to see from your last letter to your father[1129] that you and Xenia are well after the latter had recovered from a slight attack of tonsillitis and that Nina and presumedly also little Vera are with you. I hope that Nina's health will improve under good care.

We cannot understand why Vera has never written to your father nor to us. She used to be such a good correspondent.

We now hope that Alfred will also be with you in the very near future and that he will be able to take up his career.

I am glad to report that we are all well here and particularly pleased that your parents like the life in this country very much and feel entirely at home. They have a very nice little apartment and live near the Krebses that they can see each other very frequently and we visit them of course, as often as possible as they come very rarely to the City. Before the World's Fair closed, I took your father there to see the most interesting parts of it and I think he enjoyed his visit very much.

We think of you all frequently and will be glad to get your good news either through your parents or direct.
With love to you all from all of us, I am, very sincerely yours

Siegfried

[1129] Letter [316].

[318] Egon Orowan[1130] to Rudolf Peierls

Cambridge, 29.1.1941

Dear Peierls,
I apologize in being so late about sending you the note about the net; I had terrible troubles with polishing my little beads and therefore very little time for other work. Besides, I have married in the meantime[1131] and that takes up more time than you would think.

Enclosed you find the stuff prepared ready for the patent application but not yet applied for; therefore please, treat it as confidential (not towards Oliphant, of course).[1132]

Further enclosures refer to our talk on predictors and latent time; these are old patent specifications of mine from pre-Munich days; I had sent them to the War Office and they thanked me and said that they had already a predictor. Owing to their age, there are some primitivities in them; but I am convinced more than ever that they represent the sound principle and that all predictors of to-day are just pre-diluvial monsters. You will notice that the principle laid down in them can be applied for ultra-short-wave ranging.

The most interesting of the enclosures would have been No. 4; this, however, got stuck and had been authoritatively declared as confidential. So I am sorry that I cannot send it now.

My address is 1, Brookfield, Newnham Walk (it is the flat where Bhabha used to live; we have found an old hat of his in a drawer). In the meantime a Czech couple called Czerny lived there for 18 months,

---

[1130]Egon Orowan (1901–1989) studied physics, chemistry, mathematics, astronomy in Vienna and later electrical engineering in Berlin under Richard Becker and Imre Bródy in Budapest. In 1937 he accepted a post at Oliphant's institute in Birmingham before moving to Cambridge in 1939. After eleven year there he moved on to M.I.T. where he stayed until his death in 1989.

[1131]Orowan had married Jonal Schonfeld, a pianist from Hungary who had fled to Paris and later to England.

[1132]Egon Orowan had devised an aiming device for anti-aircraft-artillery, the provisional specifications of which are included as an enclosure to the letter. Furthermore, Orowan intended to apply for a patent in connection with improvements to the construction and use of torpedo nets.

without having cleaned up once, but not without smashing all furniture and everything.

Very soon I shall write to Genia thanking and replying to her two letters full of valuable and invaluable advices about marriage. Only one of these advices remains enigmatic to me: that in married life love should be dynamic and not static. Here I am not quite clear about the variables.

With kind regards,
Yours sincerely,

E. Orowan

[319] Egon Orowan to Rudolf Peierls

Cambridge, 6.2.1941

Dear Peierls,
I am sorry that you caught a cold in Cambridge, and that even in bed you were troubled by my letter. Thank you very much, on the other hand, for the careful attention you gave to the matter, in spite of your horizontal position.

Before going into the matter, I apologize that I was unable to stay at home longer than I did on the last day of your Cambridge tour; I had an appointment, arranged by Bragg, with a man whom I did not know and whom we are going to ask for things.

Our admiration for the tea set and the tray[1133] still goes on increasing towards the saturation value. Many thanks for the increment of the admiration that took place since your departure.

As to the problems concerning the 3 specifications: I shall send you a description of the matter with the protective net. As a matter of fact, I do not quite seriously want to spend one pound for a patent application. I am convinced that the stuff is the right one, but I am still more convinced that the people who in the last instance would judge

[1133]This was a wedding gift from the Peierls.

on its acceptance are not the right ones, and that would be probably a waste to spend good money on the matter. I think I can send a description for Oliphant on Monday (to you).

Now I shall review your remarks.

A) The net.

1. Whether the scheme is practicable is a financial, not a technical question.

2. I know no reliable data about the frictional drag is a plate moving parallel to its plane in water; in the 24th ed. of the Hütte[1134] it is emphasized that the displacement resistance cannot be separated from the frictional resistance on the ground of what is known with certainty. But I think more accurate calculations are, first, impossible, because the drag depends so much on the roughness of the surface which is a matter of design; and second, they are unnecessary for the present purpose, because we know that the depth of the net is of the order of half of the submerged depth of the ship (somewhat less), and that the depth of the ship is roughly 0.4 of its width. Thus the total surface of the net is roughly half of the frictional resistance of the ship which is probably a fraction of its total resistance.

3. As mentioned in the specification, the weight of the net can be compensated; its connection with the ship is a simple engineering problem; and I think that any flutter can be sufficiently prevented by the frame in which the net is fixed. I think that the strength of the structure which holds the net will be determined by the requirement that the net must withstand the impact of the waves in storm.

B) The aiming devices

1. Three-dimensional cams (which are, by the way, just one of the many ways which can be used) present no difficulty whatsoever

[1134] S. Hütte, glider designer.

in mass-manufacture; certainly not more than the manufacture of shoe-lasts. It can be done with any degree of precision by Kopierfräsmaschinen, Kopierschleifmaschinen, or special lathes with guided tool holders.

2. Certainly the point at which you must fire might be quite far from the position of the aircraft in the moment of firing; but the whole of the third specification is devoted to this ("Compensating device etc.") If you read that you will also see that a method is described there to take into account the second derivative of the position of the aircraft with respect to time. This cannot be done with the present predictor, and, on the other hand, it would be senseless to compensate for higher derivatives.
In fact, the two specifications refer to the same device; they are separated merely because you cannot claim more than one invention in one patent. If you read the second, you will see that the rest of your objections find their answer there. Those specifications are certainly more antiquated than the aeroplanes built when they were written but I am sure that their idea, that of a kinematic connection between observing device and the guns, with electric compensation for the speed and acceleration of the target, is that of the future, and that those impossible clumsy calculating predictors are just a kind of predictossaurus.

I wish you speedy recovery and hope that this wish will arrive after fulfillment. With many thanks,
and kindest regards,
Yours sincerely,

E. Orowan

Dear Genia,
I am still preparing a long letter as a reply to your really most valuable and admirable letters with the advices; I am still in a narrow and tight corner with my time. So pardon me, if I write a few words only in this letter. Many thanks for your visit, and for the beautiful set (cp. above).

Please keep razor and torch for the present; not worth while sending. Many apologies for the lapsus linguae; I am sure it is as tight as if the lapsus had not happened. Our chairs are in repair now; as soon as they are here, we ask Jacobsons. Do you remember my troubled state of mind in Tullivers? The method that occurred to me there has been the real solution. Pascoe is quite enthusiastic at work[1135] now that we have got that method. Married life is not as bad as I thought, mainly owing to some of Jolan's virtues. Main drawback is that landladies have husbands who clean their shoes, but my wife hasn't.
With kindest regards,
Yours sincerely,

E. Orowan

[320] Rudolf Peierls to Else and Heinrich Peierls

Birmingham, 7.2.[19]41
(carbon copy)

Dear Parents
We received both copies of your letter to Alfred and sent one on to him. We had taken it for granted that the arrangement that part of the money which Hermann had, was at our disposal would automatically end with your arrival, and Hermann would put the money at your disposal. It is a pity he did not, but now, as you will understand, it is too late and I have no choice but to call it in. I enclose a copy of my letter to Hermann on this subject. It sounds perhaps a little rude, but Hermann is fond of plain speaking.[1136]

[1135]K.J. Pascoe, Orowan's student at Cambridge.

[1136]In the 1930s, Heinrich Peierls had put equal amounts of money at his children's disposal for financial emergencies. The money had been deposited in the United States. There had already been a lengthy exchange of letters, when Rudolf Peierls' brother-in-law, Herman Krebs, had been reluctant to transfer Rudolf's share of the deposit to England at the time of his parents' emigration to the United Kingdom. (See *Peierls Papers* Supp. A 109–110.)

As you will have seen from my letters we had a disappointment that Vera was not released from hospital when we had hoped she would be; instead she had a relapse always just before coming out and there was later a complication with her tonsils which had to be taken out and which rather upset her. Now, however, she is truly better and we expect her home to-morrow. She actually paid us a short visit today, in the highest of spirits, and would have stayed except that her papers had not been ready yet at the hospital.

We hope to have Alfred here, too, for a few days when he gets leave; so far he has not had any leave yet, but we saw him briefly when he passed through Birmingham on business. There are now hopes that very shortly his position will be considerably improved, but there is nothing definite yet.

I am afraid your parcel has not arrived yet, but we have not given it lost yet. Meanwhile we are appreciating your good intentions.

Things are looking rather more cheerful here now and just today the fall of Benghazi comes as a pleasant surprise. Our optimism is increased by the long immunity from air raids. Nobody knows, of course, how long that may last, and we are ready for anything that may come, but while it lasts, it certainly is pleasant.

As you have probably read, we had quite arctic conditions here lately, and travelling was very difficult. I got rather behind with my work on that account, and then I caught the flu and was in bed for a few days so now I have got a busy time ahead.

With kindest regards from all of us.

[Rudi]

[321] Rudolf Peierls to Egon Orowan

Birmingham, 11.2.1941
(carbon copy)

Dear Orowan,
Thank you very much for your letter.[1137] To return the compliment, I might say that if you had read my last letter, you would have seen that I was quite aware that the inventions No.2&3, belonged together, and that I actually dealt with an objection to the last one. The point is that on page (2) of the third patent, you suggest that the rotation of auxiliary gear should be proportional to the speed of rotation of the telescope. This would only be correct if the equation of motion in polar co-ordinance were linear and that would only be approximately correct if the angle through which the line of sight moves during the time of flight of the shell, were small. You can easily see that this angle is in fact large, and that therefore the necessary rotation is not simply proportional to the angular velocity of the target: moreover, it is also not sufficiently accurate to assume the range to the proportional to the time of flight of the shell, hence your statement on page (3) that the [· · ·][1138] is accurate if the curvature of the path is small does not correspond to the facts. It is actually the object of the protector to work out the necessary correction and on this problem your invention does not touch. I think patent number 2 is of value for targets other than aircraft, and should be presented for such a purpose rather than for A.A. work.

As to the first, and I think most important problem, I had in mind just the kind of argument which you explained in your letter, and I thought you ought to have given indication that you have thought about these points in making the suggestion. I think one might even go a little further and discuss for example, the factor by which the brake resistance can be reduced, if one [· · ·][1139] by a smooth surface.

I any case, I hope when you send a revised manuscript or instruct me to hand the present one on to Oliphant, you will answer the question

[1137] Letter [319].
[1138] Missing in manuscript.
[1139] Missing in manuscript.

whether your request to treat the matter confidentially does not include Oliphant from passing it on to the proper authorities.
Yours sincerely,

[R.E. Peierls]

[322] Heinrich Peierls to Rudolf Peierls

Montclair, 14.2.1941

Dear Rudi.
Thanks a lot for your kind letter of Jan. 15th. and for your congratulations to my birthday, which we celebrated together with Annie (Hermann was out of town) Edgar and his nice wife[1140] and Lala, our guest, very nicely. Ilse and Henry came in the afternoon after school and we enjoyed very much having them here. Two days later we went to town to see Siegfrieds who gave us tickets for the Metropolitan Opera to see "Fideleo" conducted by Bruno Walther[1141] and Kirsten Flagstadt[1142] as Leonore, which is due in about one week.

What is your special job now? And what sort of trips do you make?

We got a very sad letter from Vera. We feel very sorry that you did not give her our invitation-letter, so that she saw that we wanted her here. We answered immediately and we do hope, she has left hospital by now. Perhaps you can give her that letter now, you can explain it to her, that you had reason not to show it to her when it arrived. We all send our best wishes and love the Krebs family included.
Fondly yours,

father

[1140]Edgar had remarried after splitting up with his first wife. See also letter [141].
[1141]Bruno Walther (1876–1962), Berlin-born musician, conductor and composer, had worked in Breslau, Riga, Berlin, Vienna, Munich, Leipzig and Paris before moving to the United States in 1939.
[1142]Kirsten Flagstadt (1895–1962), Norwegian-born soprano who sang in Scandinavia and Bayreuth before succeeding Frida Leider as the leading Wagnerian soprano at the New York Met.

Will you please give these pictures and enclosed letter to Vera and tell her that we sent a geographical Magazine to her, which we hope she will enjoy. From your children we have the nicest letters and news from Mrs. J. [1143]

[323] Egon Orowan to Rudolf Peierls

Cambridge, 22.2.1941

Dear Peierls,
I am sorry for being so late in replying to your letter.[1144] Exactly one week ago I sat here writing the enclosed letter to your wife, and hoping that within 24 hours I could write this and then post them together. In the meantime I tried several times to write, but in vain. The last time was very busy; there were menacing signs that suddenly I would have to abandon plasticity work and take up work for a Ministry which went through different Committees during the last two months and was started to be "very urgent". So I had to accelerate the lab work, and at the same time to think over the new work which involved a large high-tension and optical arrangement, apart from the mechanical part. Now this menace has passed; a letter was shown to me by Miss Smith in which a University board was sceptical about the qualification of Pascoe (the young student who is helping me) for the new work, and in the talk about the letter I learnt from Sir Lawrence that the new work should have been done by Pascoe, with my plans and my advisory co-operation of, say, half an hour a day. After a few illuminating remarks the matter was dropped, and now I can go on with plasticity, although probably very soon without Pascoe.

I am sorry to have misunderstood you on the predictor. I did not care much about the technical details of the gear between the acceleration recorder and speed recorder, and the gun, and assumed in the specification that the corrections may be linear. If they cannot, then another suitable kinematic transmission must be put between them, e.g. a cam. I think no serious difficulty can arise from this source. The

---

[1143] Mrs Jephcott, Canadian host of Ronnie Peierls.
[1144] Letter [321].

essence of the matter is that 1) a kinematical transmission can be used instead of a calculating machine between range-finder, speed-finder, and acceleration-finder on one side, and the adjusting gear of the gun on the other; 2) this works instantaneously; 3) it is much faster and more flexible in use against low and high flying targets; 4) it is probably very much lighter and cheaper than calculating predictors.

As regards plasticity, it seems that we have had to put all our nice theoretical considerations on gliding in cold storage. The bead-photography method has shown that it is not the slip between neighbouring planes, but the propagation of the slip to neighbouring planes which is essential. To be more explicit, I know now with a high degree of probability that

1. The jerky part of the extension runs along a fraction (e.g., a quarter) of the crystal and does not produce slip bands, or only on very fine and faint ones;

2. Slip bands are produced by the creep part of the extension.

Thus we can understand now a great deal of the hitherto mysterious things about glide. Slip bands are fine or absent at low temperatures, because there is no creep. The creep part is, with its huge temperature dependence, most probably an amorphous type of plasticity, due to atomic rearrangements at heavily disturbed parts of the lattice; the creep itself produces disturbances and thus it is a self-maintaining process, feeding on and producing disturbances. This explains why slip bands exist: creep is an instability, if the stress is high enough to produce new disturbances at a sufficiently high rate. The critical stress is at yield point: if there are initially no disturbances, the yield point can be considerably exceeded, as was systematically observed already in my old Zn-experiments. Now I am going to measure the speed of propagation of the jerky deformation along the crystal; more difficult but just as necessary is the measurement of the activation energy of the creep (the energy barrier on interchanges of place between atoms).

Another matter that keeps me busy is that of our platelet. We have not yet got our furniture back from repair; and when I asked for the other repairs fixed in our contract, Hockeys flatly refused, saying that the landlord has given permission for 10 pounds worth of repairs only, irrespectively of the repairs the performance of which he undertook in agreement which is kept there; he found that the schedule of repairs which was attached to the contract and to which reference was made there has been removed and could not be found. Hockeys are, on the other hand, sending people to us to see the flat for lease (I have rented it for four months, of course).

Thank you and your wife very much for sending me the letter of Gaby which I enjoyed so much. It is a superior piece of well-considered penmanship, plus charm. In addition I am glad that Anderson and the Arabic Nights have not been torpedoed.

I do not know whether you know the story of our stove. Achurch wrote me that the lorry people were there but refused to take goods to London and Cambridge. Then I rang up their agent here, and was told that, owing to bombing, there is no service now. I insisted that he has accepted the conveyance, and, and without obtaining any promise, I obtained the stove four days later. It was not aluminium coloured, but dirty white and grey with rust, and the vanish came off with a terrible smell at the first heating. But it works magnificently, Jolan is now quite another person, and we are blessing our names in the morning when we awake and in the evening when we lie down.

Sir Lawrence[1145] will go to Canada in the first days of March for 6 months. Fröhlich is here for the weekend. There is measles here; among others Kowarski's daughter has got i[t]. Schönberg's daughter Anne is 6 days old, she was 3600 gramms. Jolan has tea with Joan Thompson and Mrs. Dirac now. Taylor's nice mechanic, Thomson, broke his collar-bone, as he suddenly jumped off his cycle to take cover when a bomb fell near him in Cherry Hinton Road. It demolished two houses. I have not seen Jacobsons since.

With kindest regards.

Yours sincerely,

E. Orowan

[1145]William Lawrence Bragg (1890–1971) had just been knighted.

Addendum: The enclosed letter was written exactly 6 days, not a week ago.[1146]

[324] Rudolf Peierls to Egon Orowan

Birmingham, 2.3.1941
(carbon copy)

Dear Orowan,
Thank you very much for your letter with the exciting news about the slip process.[1147] It looks really as if by means of your new method one will at last learn what the process is.

As regards your inventions, your letter reminds me very much of the explanation of the telegraphy by means of a dog whose tail is pulled and of wireless telegraphy as being the same thing without the dog. Seriously, I have understood that the idea of patent No. 2 consists of a scheme whereby the image of the path of the shell is produced in a stereoscopic telescope (or its equivalent) and the job of aiming reduced to make a point of this coincide with the target. This, I think, is a very elegant idea for a slow-moving target, but for the reasons I gave, no good for a fast-moving one. Now, apparently you say that this is not essential, but what is essential is that the whole should be done by means of "kinematical transmission instead of a calculating machine". Next you will have to explain what exactly you mean by the term kinematical, since I should have considered any calculating machine the very type of a kinematical connexion. Your patent gives a particularly ingenious instance of a connexion which is simple as long as it is applicable, but if it is not I do not see your contribution to the problem. If I have misunderstood you, please give further details to help clear up the situation. [M]eanwhile I have not passed on the papers 2 and 3 to anybody since I felt that as long as I do not understand myself what they are about I cannot attempt to convince anybody else of their importance.

[1146] Orowan enclosed a letter to Genia.
[1147] Letter [323].

As regards No. 1, this is in Oliphant's hands unless he has already passed it on higher up. He was interested, but worried about the effect on the ship's resistance, in spite of your assurances. A very large part of the resistance is wave resistance, and I gather from somebody who seems to know the subject that a good shape for a ship is one in which the waves emerging from the front and the rear end of the ship are in opposite phase and combine to form a weak wave at large distances. These conditions would, of course, be upset in the net. At low speeds on the other hand, viscous drag is of greater importance. Would it not be worth while to discuss the matter with somebody who knows something about ships (e.g. G.I. Taylor)? In theory of course, if such an invention were passed on to the right authority, it ought to be their duty to consult the experts and get all these questions decided, but things don't work this way in practice, and one gets nowhere if one presents the idea and lets them take it or leave it. If you could produce quantitative figures on the probable losses &c it would very much improve your chances of it getting the proper consideration. Meanwhile I hope that Oliphant can be convinced of the importance of the matter on the available evidence.

I am glad you managed to get your stove eventually. Did Genia not tell you that the aluminium paint on it was rather inferior. But it is not a large surface to revarnish with a heat resistant paint, or to blacklead, or to look at in its present condition ("there is a war on") The idea that an ordinary solicitor can deal with Hockney & Son strikes me as exceedingly optimistic. Why not accept the £10 in cash and use it to refurnish the flat?

I just found out that Genia has written all this already, so there you have another example of the unanimity of our family.

Sincerely yours,

[R.E. Peierls]

[325] Rudolf Peierls to Else and Heinrich Peierls

Birmingham, 16.3.1941
(carbon copy)

Dear Parents,
Thank you very much for your letter of the 14th Febr[uary][1148] which arrived here a few days ago, more quickly than most air mail letters. We were very interested in the pictures of your flat and are very glad to know you in such an attractive home. The Geographical Magazine for Vera arrived about the same time.

Life is getting more cheerful here with spring starting, and longer days, and just now a splendid full moon which reconciles one with the black-out. I have taken more to travelling by car again to save time, and I had to be out late at night twice last week, but the moon made it a pleasure to drive at night, rather than a heavy strain.

The comparative peace which we enjoyed since Christmas has been broken by a few air raids, but there was no serious attack on our district, and even elsewhere, the raids do not seem to have been very heavy. But in any case, we are ready for them. To-night, it is our turn for fire-watching, i.e. to keep awake if there is a warning and to watch a number of houses in this road for incendiary bombs; and tomorrow night it is again my turn on the fire station.

Vera seems to be catching up very rapidly with what she missed at school because of her illness. Nina's health has improved very much. Alfred is still waiting for a decision, and the continued delays are naturally very disappointing for him.

I enclose a copy of a letter to Hermann, in reply to his cable. I am sorry to have to continue the correspondence with him in the present tone, but you will probably understand my position.[1149]
With kindest regards from all of us
Yours sincerely,

[Rudi]

[1148] Letter [322].
[1149] See letter [320], note 1136. Letter to Herman Krebs in *Peierls Papers*, Supp. A.112.

[326] Egon Orowan to Rudolf Peierls

Cambridge, 17.3.1941

Dear Peierls,
Many thanks for your letter of 10th. As regards your question about the difference of my present point of view from previous work, I suppose it refers to the result of last year's experiments rather than to the note to Nature[1150]which has no direct bearing on the questions I was concerned with during the last years.[1151] It (the letter to N[ature]) touches, on the other hand, on problems which kept me busy seven–eight years ago, i.e., on those different theories of real–, Locker,– secondary–, and other structures which were usually connected with experience by that mysterious 10^{-4}-s existing in spite of.

As regards the changes of my point of view due to the experiments of the Cambridge time, they are very substantial. When I began, I thought that the static interpretation of glide is wrong, and therefore the Becker one must be more or less correct. Now, as you know, things are more complicated than either of these theories had supposed and neither of them is even approximately correct. The only conception that survives is that of the dislocation. I think it is best to tell you how things appear now in dim and somewhat uncertain outlines.

When a crystal is loaded, glide starts somewhere and spreads probably over a considerable volume; this spreading is sometimes rather difficult with virgin crystals but easy with annealed ones. The great problem is how this happens. There are 3 or 4 possibilities, but I do not know which is the important one. This self-breeding of dislocations is the dominant thing in the first stage of the deformation; the older theories either said nothing about that (Becker[1152]) or led to the opposite conclusion that glide always produces harden-

[1150]E. Orowan and K.J. Pascoe, 'An X-ray criterion for distinguishing between lattice curvature and fragmentation', *Nature* **148**, 467–468 (1941).
[1151]E. Orowan, 'Problems of plastic gliding', *Proc. Phys. Soc.* **52**, 8–22 (1940).
[1152]Prof. Richard Becker (1887–1955), professor of theoretical physics at Berlin. Orowan had worked at his assistant in the late 1920s.

ing (Polanyi, Taylor).[1153] Now it appears that glide produces new dislocations in the immediate neighbourhood but hinders the movement of neighbouring dislocations, much in the sense of Taylor. The result is that glide zones arise which stop gliding in a certain neighbourhood and thus produce glide bands, this time in contrast to what Taylor assumed to happen. Now comes the thermal influence. It deblocs blocked dislocations and produces a creep which, of course, is mainly localized in the glide zones and thus gives rise to coarse glide packets which never occur without creep, i.e. at low temperatures. This is very much according to a short remark Andrade[1154] made in his Bristol paper.[1155] But this creep is very far from being a viscous one, because it shows a yield point, and thus the activation energy is strongly dependent on the stress, just as is the case with Becker's formula.

You see how things are now. Instead of a simple process, plastic glide certainly consists of two to three different processes, and these resemble different models used in the old theories. Each theory seems to be right in one point, and wrong in others. The main problems are now, how can dislocations arise at such low stresses, and how can they multiply. I am beginning to believe that the usual estimate of the theoretical shear strength is quite wrong for metals (I mean of a wrong order), because this is a case of extreme polarization where the stress is a function of shear displacement is everything but sine-shaped. Could we possibly

---

[1153] Michael Polanyi (M. Polanyi, 'Form of lattice distortion that may render a crystal plastic', *Z. Phys.* **89**, 660–64 (1934)) and G.I. Taylor (G.I. Taylor, 'Mechanism of plastic deformation of crystals, I. Theoretical', *Proc. Roy. Soc.* **A145**, 362–87 (1934)), roughly contemporarily with Orowan (E. Orowan, 'Plasticity of Crystals', *Z. Phys.* **89**, 605, 614, 634 (1934) and **97**, 573–95 (1935)) had realised in 1934 that plastic deformation of ductile materials could be explained in terms of Vito Volterra's theory of dislocation.

[1154] Edward Neville da Costa Andrade (1887–1971), had received his Ph.D. from Heidelberg before working with Rutherford at Manchester in 1914. Between 1928 and 1950 he was Quain Professor at University College, London, before becoming Fullerian Professor of Chemistry and Director of the Faraday Research Laboratory and Senior Research Fellow at Imperial College, London.

[1155] E.N. da Costa Andrade and Tsien, 'The Glide of Single Crystals of Sodium and Potassium', *Proc. Roy. Soc.* **163**, 1–15 (1937).

make an estimate by calculating an increased value of electron density wherever two atoms come near another during glide?

Many thanks for Ronnie's charming letter; we enjoyed it like many other people too.

I think I should not continue our discussion on the aiming device. It is difficult to discuss technical subjects if one is not clear about every detail one never knows where unforeseen snags arise and where reasonable objections can be avoided by a small con[s]tructive step.

Oliphant was here and I met him on Friday; he said he will see me on Saturday but probably had no time. We did not talk about the ship protector. I think the matter is just a comparatively simple constructive problem. If the Admiralty think they can manage U-boats with the means of the last war, let them go on. Probably they will finally resort to anti-torpedo protection instead of hoping in anti-Uboat measures. The question is only whether it is adopted a few hundred ship-sinkings earlier or later. Or are they deliberately refraining from the proper method to raise hopes in the Germans? I think this is perhaps the most probable explanation. But certainly Uboat war is the greatest danger because it does not menace the Germans who, therefore, will not invent anything against it which we could take over as is usually done.

Would you please give my kindest regards to your wife; I shall reply to her letters soon.

With kind regards,

Yours sincerely,

E. Orowan

[327] Rudolf Peierls to Raymond Priestley[1156]

Birmingham, 1.4.1941
(carbon copy)

Dear Vice-Chancellor,
You were good enough to offer to make enquiries whether my wife would be eligible to volunteer for the scheme mentioned in the recent letter of Dr. Snow.[1157] I am giving particulars of her qualifications below.

Mrs Eugenia Peierls was a Russian by birth and lived in Russia until 1931, when she became a German subject by marriage. She has lived in this country since 1933 and became a naturalized British subject in 1940. Her age is thirty-two.

She took a Russian University degree equivalent to a Bs.C. (Physics) and after that did a year's research in a Geophysical laboratory, but gave up this work in 1931, on getting married. Since then she has not done any scientific work, and her physics is obviously somewhat rusty, but she has been in touch with recent developments throughout and could probably get into the work fairly easily.

She joined the Civil Nursing Reserve on the outbreak of war and worked, first part-time and later full-time, in a hospital until 31 December 1940. Then, on the appeal for remunitions workers, she felt she could be more useful there and transferred to one of the Ministry of Labour training centres, where she is at present undergoing training for an engineering workshop.

She is quite satisfied with this work, but on hearing of Dr. Snow's appeal, we felt there might be work in which people with her qualifications were even more urgently required than in the Munitions industry.
Yours sincerely,

[Rudolf Peierls]

[1156] Raymond Priestley, distinguished Antarctic explorer and Vice-Chancellor of Birmingham University 1938–1952.

[1157] The University of Birmingham attempted to encourage people with a science background to get involved in science teaching in order to compensate for the loss of academics to war-related projects which prevented them from teaching.

[328] Rudolf Peierls to Klaus Fuchs

Birmingham, 10.5.[19]41

Dear Fuchs,
I have for some time been busy with research for a committee of the Ministry of Aircraft Production and I need somebody to help me with this work. I should be very glad if you could come to Birmingham to work with me on this and I have reason to believe that I could get your appointment for this purpose approved. I can only make a definite application to this effect when I have your readiness, in principle, to accept.

I cannot now disclose the nature of the purpose of the work, but it is theoretical work involving mathematical problems of considerable difficulty, and I have enjoyed doing it, quite apart from its extreme importance. There may be a little numerical work, but this is not the main thing and mainly it is the sort of work for which a good theoretical physicist is the best qualified, though it has no connection with quantum theory.

The job would be of a temporary nature, probably for six months at a time, to be extended as required (and I do not see an end to the work at present).

If you want to talk things over before deciding, could you come to Birmingham some time next week (expenses paid). In this case, please send me a wire suggesting a date.

Otherwise, please send me, as soon as possible, the following particulars:

A. A curriculum vitae, including the date of your arrival in this country, and the wording of your release from internment, also references.

B. Your aliens' registration number and number of identity card.

C. Salary required (I think £275 would seem a reasonable figure but I want your views).

How long notice would you require for coming up to Birmingham? There may be some delay with getting the approval of the Ministry for giving you information about the nature of the work, but it may be possible, meanwhile, to start here without being told what the whole thing is about. It is admittedly unsatisfactory to work on something of which you do not know the whole story, and I would oppose any such arrangement for any length of time, but it might do for a few weeks, in the interests of speed and in order to get you acquainted with some of the mathematical methods. Would you be ready to start temporarily on that basis?

I hope you will see your way to accept, and I am looking forward to the chance both of collaborating on this work and of discussing some real physics occasionally in our spare time.

Yours sincerely,

[R.E. Peierls]

[329] Rudolf Peierls to Francis Simon

Birmingham, 15.9.1941
(carbon copy)

Dear Simon,

I managed to find fairly complete information in the location of physicists by means of the lecture lists published in the "Physikalische Zeitschrift".[1158] I shall be reporting to you about our conclusions when Fuchs and I have had time to compare our notes. At the moment I am writing about two points: first, the lecture lists I mentioned do not give lectures on Physical Chemistry or Chemistry. Do you perhaps know whether those lecture lists are published regularly in any of the chemical journals? I am trying to arrange to have the physical ones copied

[1158] Peierls and Klaus Fuchs tried to deduce from publications of scientists and information contained in lecture lists, what kind of nuclear research was being done where in Germany. See letters (A. Brown.), vol. II, of the *Peierls Correspondence*, chapter 11.

or photostated, and would like them to be sent with the chemical ones, if available.

Then I also found in an article to Bodenstein's 70th birthday[1159] a remark that because of the war he was now again at the head of his former laboratory. This seems to show that his successor is either on war work or has been called up. Who is his successor?[1160]

Yours sincerely,

[R. Peierls]

[330] Rudolf Peierls to James Chadwick [1161]

Birmingham, 23.9.[19]41

Dear Chadwick,

I heard from Simon that you suggested looking at German periodicals, and Fuchs and I had just been in London for this purpose. I enclose a report on our conclusions which is in some respects incomplete. I hope Simon will be able to fill the gaps as regards Physical Chemistry and Chemistry,[1162] and we hope to look up a few people whose names occured to us only afterwards, when we are in London next.[1163] If you can suggest any further directions in which one should look we could do this when visiting the libraries next.

I am not having this report circulated to the members of the Committee,[1164] as I think its circulation should be restricted. I have shown

[1159] Max Bodenstein (1871–1942), had been the successor of Walther Nernst at the Department of Physical Chemistry at the University of Berlin. He retired in 1936 but returned to his department to teach in 1939.

[1160] Bodenstein's successor was Paul Günther (1862–1962), who moved to Karlsruhe after the war.

[1161] Enclosed with this letter is a report on German scientists

[1162] See also letters [329] and [331].

[1163] See correspondence between Peierls and Francis Simon in September 1941, NA/PRO: AB1/576.

[1164] This refers to the MAUD committee or its technical subcommittee, the government committee investigating the potential implications of the Frisch-Peierls memorandum for weapon production. See letter [304], note 1074.

it to Simon and Frisch, and perhaps you will arrange for it to be sent to anybody else who you think ought to see it.

Frisch tells me that the Bristol report on cross-sections has at last arrived.[1165] If, when you have studied it, you find it presents new factions, I shall be grateful to hear about it.
Yours sincerely

R. Peierls

[331] Francis Simon to Rudolf Peierls

Oxford, 24.9.1941

Dear Peierls,
I looked through the Zeitschrift f. Phys. Chemie, division (A) and (B), the only German periodical we have up to present in our library. The general impression is also that surprisingly many people are still working at their own places and on their own subject. I did not find anything of specific interest. Clusius is publishing on the separation of neon, and also on low temperature physics. It might interest you that C.F. von Weizsäcker has published a review of Kopfermann's book on Nuclear Moments, in January 1941.

I am going to-morrow to London to see Perrin, and on Friday I might drive over to Bristol to see how London is getting on.
Yours sincerely,

F. Simon

[1165] See extensive correspondence of Otto Frisch and Rudolf Peierls in PRO: AB1/574.

[332] Rudolf Peierls to James Chadwick[1166]

Birmingham, 14.10.[19]41

Dear Chadwick,
When I saw you last, we talked about the impossibility of using an electrolytic method on a large scale, and I stated then that I had worked out the size and power consumption of the apparatus required if its performance were similar to that of Brewer's apparatus. I have meanwhile discussed this question further, after noticing that from the data given about Brewer's apparatus and from simple theory, this apparatus is very far from equilibrium, and is, therefore, being used very inefficiently. One can, however, determine from Brewer's results the elementary constants governing the process with K, and from these work out the performance of a different type of apparatus used efficiently. The result of this calculation is naturally very favourable, and I enclose an outline of the arguments. You will see that if it be assumed that the elementary separation factor is the same for both elements, the power consumption is reasonably small but the equilibrium time very large. The latter can be reduced by increasing the field intensity, which may be technically possible if one correspondingly reduces the concentration. Unless the K ions are heavily hydrated, we would expect the separation factors to be in the ratio of the masses, that is, to differ by a factor of about 6. As the square of the separation factor enters both into the power consumption and into the equilibrium time this would make the results very much worse than I have stated.

Altogether it still seems to me extremely unlikely that this method will be suitable for large-scale work or will compete favourably with the others, but this does not follow with as much certainty as it seemed to do before. I just wanted to correct this in case I have given you the wrong impression.

[1166] Enclosed is a note on the Electrolytic Method.

My cold (which I certainly did not acquire in Liverpool) took me longer than I thought, and I shall only now be able to see London.[1167] I shall, of course, ask his opinion of this new aspect of the problem.
Yours sincerely,

R. Peierls

[333] Rudolf Peierls to James Chadwick

Birmingham, 17.10.[19]41

Dear Chadwick,
I had London here yesterday and talked things over with him. As regards the question of the possibility of large-scale separation by this method, he agrees in the main with my conclusions, but he is carrying similar arguments rather further by taking into account the question of the maximum permissible field strength which would not make the ions move too fast, so as to upset the transverse diffusion equilibrium. This limit involves the diameter of the capillary used or the size of the sand grains, and, in principle, one could increase the field intensity and thereby cut down the equilibrium times if one could go to fine capillaries or use fine sand. There is, therefore, no theoretical limit, excluding the use of the method on a large scale, but London thinks there may yet be insurmountable technical difficulties. He believes that in the sand method inhomogeneities of temperature may yet make the method unworkable, and in the capillary method one would want a tremendous number of capillaries, each containing its own ion boundary, and each having to be adjusted and emptied individually. It seems thus unlikely, but not impossible that this method may be made to work on a large scale, and this, of course, makes it important to carry out further research on the problem.

[1167] Heinz London (1907–1970), studied at Bonn, Berlin, Munich and finally Breslau, where he worked under Francis Simon on low temperature physics. In 1933 he moved, together with his colleagues N. Kurti and K.A.G. Mandelstam, to the Clarendon/Oxford, before taking up a post at Bristol University.

London would, of course, be only too glad to carry on at increased speed if sufficient means were put at his disposal, and as far as I can judge, that would mean research students and possibly a boy, as well as help from a glass blower and materials and instruments. At the moment, I understand he is carrying out all analyses himself, and is making himself the pure compounds he requires.

I do not know whether Simon has mentioned to you one possible complication that may arise regarding the financial arrangements. London has an invalid mother who is in a nursing home in Bristol, where, owing to the kindness of the matron, she is being looked after at very favourable terms. He would scarcely be able to leave her behind, because she has nobody else to talk to her, and both he and his wife spend an hour or so each day with her. To move her anywhere else would probably mean paying the normal charge in a nursing home, and that would of course increase significantly his expenses.

If it is planned to move London to a protected area, or to a factory which is classed as a protected establishment, and if his nationality should, therefore, be an obstacle, it might be useful to know that he has applied for naturalisation in spring, 1939, and that his application is an at advanced stage (he has already been interviewed, which is normally done a few months before the application is finally decided). I understand that, while applications for naturalisation are not normally proceeded with at present, they are considered in cases of people doing essential war work, and it might in his case be the shortest way out of the difficulties.

While he was here he saw Haworth,[1168] and they discussed the facilities available here, from the point of view of the possibility of London moving to Birmingham, but no definite conclusion was arrived at.

As regards the question of detecting the separation, London has not yet developed any method other that the photographic. Do you think it might be at all feasible to deposit electrolytically a controlled quantity

[1168]Sir Norman Haworth (1883–1950), graduated from Manchester University in 1906 and received his Ph.D. in chemistry from Göttingen in 1910. After teaching at St. Andrews and Durham, he moved to Birmingham in 1925 where he stayed for the remainder of his academic career.

of the ions on a plate and put the plate into a counter to measure the α activity? This would, of course, require that the deposit should form a uniform thin layer, which could probably be done. It would require that the quantity should be in direct relation to the current used, which is all right if no other positive ions are common to the process, and that the counting rate should be reasonable. The quantities available are of the order of a few milligrams, and I suppose it might not be hopeless to use them for counting.
Yours sincerely,

R. Peierls

[334] James Chadwick to Rudolf Peierls

Liverpool, 20.10.[19]41
(carbon copy)

Dear Peierls,
Many thanks for your letter about your talk with London.[1169] It seems that we should push on with the electrolytic method in order to find out whether it is practicable or not on a large scale. I gathered from Oliphant that Urey[1170] is not optimistic and I think that a decision about the facilities I asked for for this work will have to wait until we have discussed the matter with Urey.

I wrote to London some time ago asking about the amount of separated material which he obtains, for I think that the photographic method of measuring the separation is not good. The easiest and most reliable method for us is the counting of the α-particles from a known weight of material.

[1169] Letter [333].
[1170] Harold Clayton Urey (1893–1981) received his Ph.D. in chemistry from University of California in 1923, studied with Bohr, then taught at John Hopkins, Columbia from 1934 as professor. During the war he was director of war research, heading the atomic bomb project at Coloumbia University.

I did not know about London's domestic difficulties. These make it desirable for him to stay in Bristol, and I shall require about the help he could get there.

I have heard today that Pegram[1171] and Urey may arrive this week. I hope to see them before they start on a tour of inspection in order to discuss the general position and to focus our attention on the important points. I think it would be very desirable that you should join Simon if they wish to discuss with him the separation plant. I will write about this later.

I have also today received a list of points on which information is required in the States. The list has been prepared by Professors Pegram, Urey and Dunning. A copy has been sent to Haworth and Simon. The important enquiries concern the separation plant, and you will be closely interested in many of the points. Will you ask Haworth to show you the document and get into touch with Simon if you find it necessary?

Yours sincerely

[Chadwick]

[335] Rudolf Peierls to James Chadwick

Birmingham, 22.10.[19]41

Dear Chadwick,

Thank you for your letter.[1172] I was surprised to hear what you say about Urey, because I gathered from Oliphant that Urey was very optimistic indeed about he electrolytic method. However, as we can ask him directly soon, we shall soon know.

[1171] G.B. Pegram (1876–1958) graduated from Trinity College North Carolina and after a short high-school teaching spell became assistant teacher at Columbia University from where he received his Ph.D. in 1903 and where he remained for the rest of his working life, taking a particularly active role in administration. In the autumn of 1941, Urey and Pegram led a diplomatic mission to England to establish co-operation on the development of atomic weapons.

[1172] Letter [334].

I have obtained from Haworth a copy of the list you mention, and I am going to Oxford today to discuss with Simon any questions that arise from it. I shall keep myself available to join in the discussions with Simon when Urey and Pegram come to see him.
Yours sincerely,

R. Peierls

[336] James Chadwick to Rudolf Peierls

Liverpool, 28.10.1941
(carbon copy)

Dear Peierls,
Pryce's address is: H.M. Signal School Extension, Eastney Fort East, Portsmouth.[1173] He was intending to carry on some calculations, but when I saw him a month ago he had made very little progress. I am afraid he may now be too busy to do very much for us, and we may have to get someone else to do the work. In fact, I think the time has now come when we must consider our programme in detail and decide how many men we need.

I was glad to get your letter about London,[1174] for his domestic circumstances must be taken into account in any decision about the location of his work. I think he will probably have to leave Bristol, for it does not seem possible to provide him with the necessary facilities there. Any extra expenses will I think be met without difficulty.

I am expecting Pegram and Urey today.[1175] They will probably leave for Birmingham tomorrow to see Haworth, and then proceed to Oxford. As they will be discussing chemical problems with Haworth I think it would be best if you met them at Oxford if that is convenient to

[1173]Maurice Pryce was supposed to do some calculations in connection with isotope separations, and Peierls had asked Chadwick for a contact address. Letter Peierls to Chadwick, 27.10.1941, Churchill Archives Centre, Chadwick Papers, CHAD I 19/6.
[1174]Letter [333].
[1175]See letter [334].

you. I shall mention this point when discussing the arrangements with them.

Pegram and Urey will be accompanied by Perrin and Jackson,[1176] who are acting as secretaries under the new arrangement.[1177] I have been informed by Appleton[1178] that Perrin will take a full record of our discussion with Pegram and Urey for use both here and by the U.S. authorities. Appleton also adds: "At this stage of the work I think it will be sufficient to confine the discussion to the scientific possibilities and the results. The questions of larger issues have not yet been definitely settled and I feel it would be wiser not to discuss them at present." I take this to mean that we do not discuss the possibility of any development beyond the 20-stage model.

Yours sincerely,

[Chadwick]

[1176]J.F. Jackson, I.C.I., M.W. Perrin, I.C.I.

[1177]In September 1941, the Defence Services Panel of the Scientific Advisory Committee considered the MAUD Reports on the scientific feasibility of an atomic bomb. On the basis of the Panel's report, nuclear weapons research was reorganised as "Directorate Tube Alloys", lodged in the Department of Scientific and Industrial Research, headed by W.A. Akers, then Research Director at Imperial Chemical Industries (I.C.I.). A technical committee under chairmanship of Akers consisted of Chadwick, Simon, Halban, Peierls and Dr. Slade of I.C.I. See letter Chadwick to Mark Oliphant, 10.11.1941, Churchill Archives Centre, Chadwick Papers, CHAD I/19/3.

[1178]Edward Appleton (1892–1965), graduated from St. John's College, Cambridge in 1913/4 and continued research in Cambridge after the First World War. After further research in London, he took up the Chair of Natural Philosophy at Cambridge. He became a member of the Scientific Advisory Committee of the War Cabinet, and as head of the D.S.I.R. he assumed administrative control of British work on nuclear weapons.

[337] Rudolf Peierls to James Chadwick

Birmingham, 29.10.1941

Dear Chadwick,
I am writing to Pryce, but it was also my impression that we shall have to make arrangements without regard to him. In considering a programme of theoretical work, I think I shall need at least one full-time computer and possibly more. I propose to ask for one at this stage as probably, if necessary it would not take long to get another one. My impression is, however, that even with one computer to help us, Fuchs and I are going to spend most of our time for some time to come, on the separation problems, and we need somebody else, therefore, to carry on with Pryce's calculations on the expansion. I thought one might possibly get Dirac to come in and help with this, as it is a type of work he would be very good at, and I think that even tough it is work below his usual standard he would be quite willing to help. Before doing anything officially in this matter, I shall find out from Dirac how he is placed, and whether he would undertake a job of this kind.

I shall meet Pegram and Urey in Oxford tomorrow. I had a similar letter from Appleton concerning the restrictions of the subjects to be discussed.[1179] It is somewhat surprising, but I shall do my best to conform to the instructions.[1180]
Yours sincerely,

R. Peierls

[1179] See letter [336].

[1180] Anglo-American collaboration on nuclear matters was characterised by a change between co-operation and conflict at the highest political level, largely determined by whether or not unconditional, or even conditional, exchange of information was perceived to be in the national interest of either country. After a lukewarm reception of the results of British research and a distinct lack of sense of urgency in the US in 1940/1, by the autumn of 1941, particularly after a visit of Oliphant and Cockroft in August 1941, US interest in co-operation with British scientists grew. At this stage, however, British policy makers became more reluctant to share nuclear secrets unconditionally with the Americans. Appleton's letter was written in this political context.

[338] James Chadwick to Rudolf Peierls

Liverpool, 15.11.1941
(carbon copy)

Dear Peierls,
I have written to Rosenhead[1181] to ask him if he could help on the calculations concerning the turbine, and I have also spoken to him on the telephone. He has been authorised to help, and he is most anxious to do anything he can. I think the next step is for you to get into touch with him directly. His address is: Professor L. Rosenhead, P.D.E., Cilwendeg Park, Boncath, Pembs

The only information I have given him is that the calculations are connected with some form of turbine and that all we need is two or three hours work occasionally.
Yours sincerely,

[Chadwick]

[339] Rudolf Peierls to James Chadwick

Birmingham, 20.11.1941

Dear Chadwick,
I saw Akers yesterday and he mentioned that he has written to you about a lot of people into whose activities enquiries should be made. You remember that Fuchs and I made a search in the literature some weeks ago and I sent you a summary of the results.[1182] It might be an advantage if you could take those results into account in drawing up your list rather than sending in two independent reports. I enclose photostatic copies of lecture lists published in the Physikalische Zeitschrift

---

[1181]Louis Rosenhead (1906–1984), at the time Professor of Applied Mathematics at Liverpool University.
[1182]See letter [330]. Since the autumn of 1940, Fuchs and Peierls carried out regular surveys of German scientific literature to establish the nature of nuclear research in Germany. Copies in National Archives, PRO, AB3/111.

which may be of some help in getting names and localities. Unfortunately they do not cover chemistry at all, and physical chemistry only in a few places. Would you be good enough to return these lists when you have finished with them, as I have to send them back to M.A.P.[1183]

In one of the recent numbers of Naturwissenschaften, which contains a collection of papers from the various Kaiser Wilhelm Institutes, I found a very suggestive paper of Fleischmann,[1184] which deals with the use of a Clusius-Dickel tube as a means of separating chemically aggressive gases from a light buffer gas such as helium. He uses this to measure the vapour pressure of bromine, by having the bromine at the bottom of the Clusius-Dickel tube whereas the top containing helium is connected to a mercury gauge. He only quotes bromine as an example and it is easy to guess what other heavy aggressive gases he might have in mind. It is an accessory investigation which he might well make if he were interested in our problems, and which he might well publish if he were required to publish something that does not reveal his main line of work. His address is given as Heidelberg, as before.

R. Peierls

[340] Hans Halban to Rudolf Peierls

Cambridge, 20.12.1941

Dear Peierls,

Kowarski and I are both very thankful to you for your attitude and intervention in this awkward matter.[1185] Kowarski has just showed me

---

[1183] Michael Perrin.

[1184] Rudolf Fleischmann (1903–2002) was director of the Strasbourg Institute, created by the Germans after the occupation of France. It was patterned after the KWImF (Kaiser Wilhelm Institute for Medical Research) in Heidelberg. Fleischmann, formerly collaborator of Bothe's at Heidelberg, continued to work on isotope separation, and, among others, built a 3-storey high Glas tube separation device, based on designs for the Clusius-Dickel tube.

[1185] Hans Halban and Leo Kowarski has fled France after Germany's invasion, and they initially settled in Cambridge to continue their work on heavy water. See letter [309], note 1123. They later had major disagreements over the direction and organisation of their joint research. See letters [353] and [354].

the letter he received from you. He takes the whole thing still more seriously than I and he points out that Akers knowing me very well and him not at all, might — if taking the matter seriously which is not at all likely — deduce from Oliphant's remark that Kowarski is not a very loyal friend for me. Kowarski would very much like to talk the matter over with Oliphant who will be here over Christmas, but of course he needs your agreement for this. Personally I think that Chadwick is not very likely to take the matter too seriously. I think so especially since Frisch told me that when receiving Oliphant's letter he did not believe the story. Akers has had too many proofs of my attitude towards Kowarski to believe any real difficulties exist between the two of us. It would, however, seem to me that if Oliphant has also talked to Appleton, although very friendly towards me does not know me well and has seen Kowarski for only a few minutes. In this case I think one would have to try to make Oliphant "reshape" his statement, just as he did with a few other foolish statements he has made recently. Please write me your opinion as to this. Of course, I must respect in the whole matter your point of view.

As to the statement Kowarski is alleged to have made he said that it might be very possible that in a conversation with Oliphant or Bretscher he pointed out that he personally would prefer our work to go on with the Government and not with I.C.I. since the government work gives you an academic status whilst with I.C.I. you become more or less a paid servant. Even if he said this, this statement would not contain any more about the difficulties between Kowarski and myself.

Yours sincerely,

Hans Halban

[341] Rudolf Peierls to James Chadwick

Birmingham, 31.1.1942

Dear Chadwick,
In thinking over our conversation on Thursday, I feel that in discussing the question of efficiency with which the work is handled by Metro-Vickers,[1186] I may have given a somewhat wrong impression. I was not, of course, trying to convince you that the work is proceeding as quickly as it should. I feel fairly certain that the statements you have been given on the delays in the work of glands, and on the work not being in the hands of the most expert people, are at any rate very exaggerated, and I cannot imagine the glands being a serious difficulty (the leak and diffusion through the glands can be estimated theoretically to within, say, a factor 3 either way, and while experiments are desirable, to determine this exactly, they will not have a decisive effect on either the construction of the glands or the order of magnitude of the auxiliary pumps, etc.).

On the other hand, I have been seriously worried about the times given in the estimate at the last Committee Meeting,[1187] and also on the question of whether these times will be kept (they are estimates, as far as I understand, not part of the contract), but I do not know what can be done about it. It is probably true that M.V. are pushing this job through more quickly than any other job of a comparable size and difficulty they have ever handled, but this may not be quite enough. I do not think, though, that at this stage one could shorten the time by bringing in any other firm, who would first have to acquire all the experience already collected by Metro-Vickers; and I do not see by what

[1186]In May 1941, a contract for the design of a twenty-stage isotope separation plant was placed with Metropolitan Vickers, and the plant was estimated to be completed by the end of 1941. Already in the late spring and early summer of 1941, technical difficulties arose, initially with the production of membranes necessary for the separation. It soon became evident, that a large-scale industrial size separation as required for the production of a weapon would stretch Britain's resources beyond capacity. See Gowing, *Britain and Atomic Energy*, pp. 68–69, 217–30.
[1187]Scientific Advisory Committee to Tube Alloys. See letter [339].

means one can make them put even more pressure on the job than is being used at present.

In trying to convince you that the trouble was not where it had been reported to you, I hope I have not given you the impression that, in my opinion, everything was going as quickly as it should.
Yours sincerely,

R. Peierls

[342] Rudolf Peierls to Genia Peierls[1188]

New York, 15.3.[19]42

My dear grass widow and mother,
This is my third letter, but it looks as though it is going to overtake both the other by several lengths, so I have got to start at the beginning again, in other words tell you all about everything. I don't know when exactly my postman is going to leave so I better start with what you want to know most of all, my visit to Toronto.[1189]

You will have seen from my cable that I have already been there, but it was only a foretaste really, because I stayed only a day. It so happened that in travelling from Chicago to New York I could get in a day in Toronto without missing anything, and so of course I did. I had phoned Mrs Jephcott from Ottawa, and she was of course terribly thrilled she called Ronnie to the phone but he did not understand me as I expected. So I phoned the Sandersons from Chicago, and arranged to go to them first thing on arriving in Toronto. Gaby was in bed with a sniffle so it was best to see her in the morning and Ronnie in the afternoon. Gaby has grown a lot and is still quite pretty (that last picture did not do her justice) but has lost some of her childish charm. Her teeth are very big but do not protrude at all, so they do

[1188]Peierls was among a group of scientists who were sent to the US in the spring of 1942 to reinvigorate Anglo-American scientific, and in particular nuclear collaboration.
[1189]Gaby and Ronnie Peierls were still evacuated near Toronto.

not distort her mouth and are not visible when it is shut, (which still does not happen very often. She still chatters as always and giggles as much as she used to and her cold did nothing to dampen her spirits — you know Mrs Sanderson is terribly conscientious in keeping them in bed for the slightest cause. She does have an American accent but less so than I expected, and she has picked up a lot of slang words mostly because she is still so interested in language, she proudly showed off all her new words, and she still likes to find nice ways of expressing herself — she told me that Nancy[1190] was called pudding because she was so fat, and now even though she was slimming she was still called that "in memoriam" to the amused horror of Mrs S. The funniest thing is that Gaby has now acquired a passion for telling jokes. She really has read all the books we sent her, and many more from the school library, she has also read a lot of adventure stories and the type of stories you get in comics and in the many children's broadcasts. This I imagine is partly the influence of the wireless, and partly of her school friends. She was of course very excited to see me, and said I had changed a lot, which means really she had begun to forget me. I asked her did she want any special present if I came again, and she said she first wanted a bicycle. When I said I did not know whether one could still get one she said her next wish was a camera. Apparently Douglas[1191] has a box camera, and she wants one too. I shall get her one, although I do not think she will be able to make much of it yet. When I suggested a treat for my next visit, she said she would like to go up in a 'plane. I said again I did not know whether that would be possible, as the shortest trip is to Ottawa and that is a bit too far, there might be no sightseeing trips, so then she wanted to go to the movies and that at any rate can be done. Actually, Mrs S[anderson] is giving both girls bicycles for Nancy's birthday (Gaby's is an advance present for her birthday) and I shall make enquiries about the 'plane question.

Mr. S. was quite a surprise being so much younger than I imagined him to look from the pictures. Otherwise he is much the same as we imagined the cheerful and hearty business man, perhaps with less accent

[1190]Nancy Sanderson, daughter of the host family.
[1191]Douglas Sanderson, Nancy's older brother.

on prosperity than I imagined. Mrs. S. I liked a lot better than I expected, she is really very intelligent and sensible and un-fussy. Nancy is an extremely funny child, very nice, but I did not see much of her as she was at school, I only saw her for a few minutes before dinner. It was her bath night and she made elaborate arrangements for letting the bath fill, so obviously they do not each have a bath every day as I would have imagined, but I did not discover what the routine was. Douglas is really a dwarf, he is not much taller, and many pounds lighter than Nancy, and he is very shy. Gaby warned me if he did not want to see me at all, because he was so very shy with strangers, but actually we got on quite well. He showed me his model airplane which is quite an achievement. He knows a lot about planes and is very good with his fingers.

The Sandersons' house lies right in the open country, where Yonge Street has become a country road and there are just a few houses along the old street. In town, Yonge Street is rather a dreary affair not quite as bad as the Coventry Road, but almost as prosaic. The house is not quite as grande as we thought, and the garden appeared on the pictures enormously bigger (as gardens always do) than it is. It is really about the same as ours. The school which I saw in passing, is about $\frac{1}{4}$ of the house in size. Other familiar things that I saw with one glance were the neighbours' houses, (George Blackstock &c.) the "valley" the Jolly Miller. "Stephen our maid" turns out to be the male part of a Czech married couple, the other half being Elisabeth, the cook. Sheridan was in bed with a bad cold and did not come into action, nor did Mrs. Sheridan.

For lunch, Mrs S. and Mr. Jephcott took me to the "Carlton Club" which is an athletic club with swimming pool, badminton hall &c. and a good restaurant. Mr. J. was the biggest surprise of all. He looks very much older than I ever thought, and I am afraid if you thought him very handsome you would be disappointed. He has a very lined face, type somewhat between the family doctor and the don, rather serious but very kind and intelligent. Then, in the afternoon, Mr. J. and Mrs. S. took me down to the Jephcott house. Mr. S. went to the office for a short while and turned up later, otherwise he had taken the whole day off and J. had taken the afternoon off. For the whole afternoon while we were there, there were a continuous stream of laughter and excitement. Ronnie took me off to show me his room and all his things to the last detail but the girls were all the time after me, shouting "Ronnie's daddy" or just "daddy" for brevity and I had to remember all the old tricks of throwing Ronnie into the air (he is getting a shade heavy for that) and turning him upside down, &c. and performing all this on the three children in succession. Ronnie has grown much taller and more sensible and has a wider vocabulary, but otherwise he is just the same boy as when he left. He is on very friendly terms of equality with Ann and rather fatherly and protective towards Claire.[1192] He told me confidentially (and not at all in a spirit of complaint) that he had been spanked by Mrs. J. on three occasions, once when he nearly let Claire run across the road in front of a car, once when he made all three of them put out their heads into the rain with the effect of getting not only their heads, but all their clothing soaking wet, and I have forgotten the third occasion. He quite agreed that on all three occasions he had thoroughly deserved a spanking. Our fear of his becoming a bookworm seem quite unfounded. When he showed me his things he said he had far too many books, and said, "If you do send me things, don't send any more books, but send me games of something." His knife did arrive but it was taken away from him after he had once cut himself with it.

For tea there was ice cream as a special treat, and the fact that they had tea in the dining room with us was in itself a treat and Ronnie's face when he was consuming the ice cream was still the same as always, still the same gourmand.

Mrs. J. is just as charming as her letters. It was a surprise to me that her hair was not fair but quite white as is also Mrs. Sanderson's. In the general whirl of excitement I really had hardly a chance to talk to her properly. Claire is a darling. When I entered the house she greeted me with the question: "Why do you always make jokes?" and I had to do my best to live up to my reputation. Ann is very nice, too, and very happy, but I somehow feel she might later become a "difficult child". The Jephcott house is rather small, with a tiny garden, in a

[1192] Ann and Claire Jephcott, daughters of Ronnie's host family.

street rather like, say, the modern streets in Harborne, or like Barrow Road in Cambridge.

As I shall certainly be able to spend another two or three days in Toronto I did not visit any of the other children or mothers, but I did phone up Pauline,[1193] who did not sound too happy. I am planning to go and have quite a good long chat with her.

All the other news in brief: my trip was in fact quite uneventful, clothing, oxygen and food supply were quite adequate, the only drawback being that one was not allowed to sleep during the non-stop journey. On arrival I found that I had to call at Ottawa, so I flew there and the next day I flew from there to New York. The commercial airliners are very comfortable and, as they fly fairly low, give marvellous views. But the last lap to New York was very bumpy and made me decide that after all flying was not always pure pleasure.

Since then I have been in New York, Washington, Chicago, Princeton, and amongst other places I still have to go to California. On the way out there I shall arrange to stop at Toronto, and on the way back I shall spend a day in Ithaca. I have already spoken to Hans[1194] on the phone, and he was very thrilled to hear from me.

Father was again very ill and had an attack rather similar to the one he had in London, though perhaps not quite as severe, and things were so bad that when I phoned on my first evening in New York they told me to come out at once. I found him very weak but quite clear, and as quiet as anyone can be with Else running round in circles. I went out again last Sunday, and he was much better and obviously recovering, in fact the doctor, who had the first time taken a very serious view, was now quite confident. He is worrying all the time about Alfred and is quite upset that he cannot help him in any way.[1195]

I am working fairly hard, and all the personal contacts are extremely interesting and very much worth while. In fact, if I were to go back now, it would have been worth the trouble already, but I have not yet seen

[1193]Pauline Shapiro, family friend who had accompanied the children's exchange to Toronto.
[1194]Hans Bethe.
[1195]Alfred Peierls was still interned as enemy alien.

half the people I want to see, and there is still much more to talk about and to do. I expect I shall be ready to start back about the 9th April or a few days after, but the actual date of departure depends, of course, on transport conditions.

15.3.

I got that far by about 2 a.m. on Friday and my spelling is probably evidence of my state of mind. Meanwhile, the "postman" is leaving a little later, so I can carry on. It is very hard even to start describing my impressions of America, or of New York. I was completely overwhelmed at first, because I somehow always had thought only of the work here, and had not realized that, while I am not doing any definite sightseeing here, I could not help seeing a lot of the country. I never expected New York to be that beautiful. It is much less frightening than I thought. One gets used very quickly to the scale of the buildings downtown, and the effect of the Empire State Building with its 102 floors is merely to make a modest 25-storey building look like a cottage. But the enormous straight streets, the rocky ground which makes you come up against unexpected views on every step, the water all round with the large number of elegant bridges, the amount of light at night (impressive even if one does not come from a blacked-out place) makes the city so beautiful that it takes one's breath away.

Quite unexpectedly, Chicago is very attractive. It is given style by the enormous lake front, and by many parks which make its total area greater than London. Of Washington I have not seen much except for the Capitol which is conveniently built right opposite the station.

I shall go to Montclair again to-morrow, before leaving New York for the present. Apparently father is still very tired and Else begins to lose patience with the long period of convalescence, but objectively he seems to be getting on quite satisfactorily and the doctor is very pleased. I saw the Krebses for the first time I was out. Annie had just fallen downstairs or something or had hurt her leg, so she had to stay in bed and was dying from anger, that she could neither see father, nor meet me. So I went up to their house for a moment. Both children were up waiting for me. Ilse had just come back from a dance. She is quite good-looking but very badly-brought up, and I gather they are having great difficulties about her. I thought I could diagnose the trouble at once as one of the obvious textbook-cases, but I could not stay long enough to see if my diagnosis was right. The boy is quite straightforward and very pleasant.

I spent one evening with Siegfried, who has become a little deaf but otherwise seems ten years younger than father. They lamented very much the new taxes which make it rather difficult to run their 12-room apartment (they resent particularly a special tax on people employing four servants). Obviously they cannot move into a smaller apartment, much as they would like to, since there would not be room for their furniture, carpets, pictures, tapestries, and other things, (of which they have an amazing collection, arranged in very good taste) and one cannot sell anything nowadays as nobody wants to pay anything for it. So they feel like beggars (I am afraid I was not able to feel much pity for them.)

One evening I spent with the Pekeris'[1196] who asked a lot about you and the children, and seemed to know a lot about us from the Fan's[1197] who were their neighbours at M.I.T. They are very charming people. She has still trouble with her health and spends about half her time in hospitals.

I saw also Mancy's brother and got on with him very well, his wife is very nice, not quite young, but very sensible and a very steadying influence on him.[1198] Everybody says that since his second marriage he has become much less polite. His attitude to me, like everybody else's was quite what I wanted it to be.

Well, that is all the news I can remember. I shall send another letter about a fortnight after this, and then there will be no point in writing

---

[1196]Chaim Leib Pekeris (1908–1993), geophysicist and friend of Isidor Fankuchen's at M.I.T.

[1197]Isidor Fankuchen (1904–1964), crystallographer; he and his wife were good friends of Genia and Rudolf Peierls when they were in England.

[1198]Eugène Wigner had married Amelia Frank, a physics student, at the University of Wisconsin in 1936. She died of cancer in 1937, and in June 1941 Wigner remarried Mary Annette Wheeler, a physics teacher from Vasar College.

any more. I had your first cable, which greatly amused everybody at the office, and I have just heard another letter is on the way for me from Washington which, I believe, contains either a cable or a letter from you. I very much hope it is either one or the other — it seems to me ages since I left you and even since I had your cable.

I am frightfully envious at Simon having his wife with him here,[1199] although I know of course that this is not fair as they were separated for 18 months, but I just can't help imagining how much fun it would be to take you up to the top of the Empire State Building (where in fact I have not been myself) or just to walk along the Riverside Drive or in Times Square at night, and most of all to go to Toronto together! And on the way out west, when I shall have two nights and a day to collect my thoughts, I shall probably collect mainly thoughts on how much nicer it would be to travel in a dublet state and how much fun you are missing and how much more I would see through your eyes and through your intense enjoyment of things.

However, it is not exactly enjoyment that matters at the moment, with the news being what they are. We have just heard of the naval losses in the battle of Java, and everybody is rather gloomy.

It seems pretty clear I won't be home for Easter so you can have your holidays quite your own way. I did not quite know, this morning, whether to send you a short cable of congratulations for successfully completing 11 years,[1200] but decided there was not much worth saying over such a distance.

Love,

[Rudi]

[1199]Simon's wife was among the adults accompanying the children to Toronto. Therefore she could join her husband while he was travelling the US in the spring of 1942.

[1200]It was Rudolf and Genia Peierls' 11th wedding anniversary on 15th March 1942.

[343] Rudolf Peierls to Genia Peierls

location unspecified, US, 21.3.[19]42

My very dearest,
My second postman is about to leave so I better start writing what I hope will be my last letter from this side. I am at the moment on the way to Toronto for two days, Sunday and Monday, and then to the West.

Since I wrote you last I had a lot of interesting discussions in New York, spent an evening with father, who is still very tired, though very much better. (Else is very run down for working so hard looking after him, and one is not quite clear whether her sighs refer to the long time of reconvalescence ahead or the fact that he did recover). Then I caught the night train to Washington and met for breakfast there Simon who had come by another train, and then we were driven out to Virginia by one of the Washington people. Charlottesville, the seat of Virginia University, is a most charming place, many of the buildings very much in the early Georgian brick style, adapted to the Virginian climate (the University was built by Jefferson and so is his house nearly on the top of the highest hill which we also saw). The atmosphere changes very abruptly between Washington and Virginia, the division lines marked by the battlefields of the Civil War. It may have been imagination, but I felt that the type of negro one meets in Charlottesville is rather different from those in the North (it doesn't mean in racial type, but in manner, intelligence and attitude) (Many of the older ones are of the "faithful family servant" type.) The man we were travelling with confirmed my impression and said he thought the negroes in the south were much better because they had much more respect for the white man. That may be true but it is a kind of respect that makes me very uncomfortable.

Our friend insisted on staying the night there and driving us back in the morning along the "sky-line drive", i.e. a new road leading along the top of the a National Park on a 4000 ft. mountain ridge with marvellous views into the Virginian hills and the plain in the north.

Back in Washington I found your third cable which cheered me a lot. Were the happy returns meant literally or with reference to the

15th (coming rather as an afterthought)? Anyway, 11 years is not much of a milestone. What about a bet whether we shall make the 15 years or will you have lost patience?

One evening I went to see Rosa G. who seems to get on very well and has a very interesting job. She was very pleased to see me except that she resented my not being Sidney. Joe I could not see as he is moving and his telephone is disconnected and at the university there was not sign of him.

There was a lot of hard work in Washington, so when I returned to New York, I was rather worn out having had about seven hours sleep every night for the last week or so, and decided to go early to bed. Simon teased me because I had made that decision every evening and never carried it out and indeed when I got to New York and phoned the Tellers I found that Hans Bethe was staying with them for one night so I went out at once. It was great fun seeing him again, he has not changed at all, but we could not talk much, as there were many other people. Then I stayed in a hotel downtown as I had a meeting downtown early this morning and so I finally got to bed at 1 a.m. The trouble is I always want to do too many things and see too many people, as it always seems to be a pity to miss an opportunity. The hotel, which was a fairly respectable hotel in Pennsylvania Station proved a disappointment to my belief in American efficiency and service. I just loved the many ideas and gadgets, particularly the suggestion that on demand one could have a typewriter sent to one's room. I asked for one this afternoon as this seemed just the right thing to write a long letter to you and some official reports, but after waiting for about an hour they brought up a most decrepit looking monster which could not produce a single letter. Hence I wasted a lot of time and I am writing only now in the train.

At this point I fell asleep and I am now continuing from Toronto. I am actually lying in bed at the York Mills with Gaby in the room next-door.

This day was not as frantic as the first visit, and I begin to feel that by tomorrow night I really shall know something of the two families. Gaby's cold has, of course, disappeared (it did the day after I had been here) and it is now Claire's turn to stay in bed, but she is very sweet about it. Today, it turns out, is Mr. J[ephcott]'s birthday, and two days ago was Mrs. J[ephcott]'s. This, with Claire's in between, makes rather an accumulation. I sent Claire a very nice fluffy Dumbo from New York, which found approval.

I spent the morning here, had dinner with the Jephcott's (with Gaby) went in the afternoon to see Pauline, (my pen deserves its name) Oliver Hartree, John Polanyi and the Nimmo children. Dinner with the Jephcott's again here. Tomorrow Gaby and Ronnie up here in the morning, lunch with Mr. J. visit Ronnie's school and Dr. Bernhardt[1201] in the early afternoon, then call for Gaby and Nancy at their school. See Mrs. Gans there, then take the two girls and the young children out for tea. Dinner at the Sandersons, calling at J.'s after dinner and departure at 10.

To describe my impression of this second visit is much more difficult as it is mostly deepening impressions that had formed before. I did see Gaby and Ronnie together, and they were just as if they belonged together like children of the same family, but not more so than with the other children. There is apparently quite a family feud about education, or rather theories of education, Mrs. J. being in a minority of one in favour of the mother-child psychology. The joke is really on Mr. J. who as a bachelor used to visit the Sandersons when Douglas was a baby and used to say upon the slightest sound from him: "This child is not properly brought up. If I ever have children they will be seen and not heard!" Now he has in effect three children and while I don't know about their being seen, they can certainly be heard three blocks away and they are being encouraged by Mrs. J. (and in fact of course by him as well). (Ronnie let drop a remark that Uncle Clare was a woman-hater when he was young and I gather there may be a truth.) He evidently was not very young when he married and he was just swept off his feet by Mrs. J., which in fact is easily understandable. Ronnie had his report which was better than any he has ever had before, full of 's' (satisfactory) and quite complimentary remarks.

At this stage last night I must have gone off to sleep. I am now writing from the train after another full day in Toronto. The first time

[1201]Chairman of the Placement Committee.

I had just been rushing through but this time I got more settled and so it was much harder to leave, and I feel almost like crying. In fact, I am not sure that by the time I get back to my first-class bedroom I shall not have a good cry. I am quite afraid of going there again for fear that next time I have to leave I shall just not be able to.

The day went off more or less according to programme. Gaby missed school and I had a long chat with her in the morning. She is really very sensible and only kept asking why she could not come with me to California. She displayed no sign of wanting to come back to England with me, she probably realises that that cannot be done. In talking about school she said that she is getting rather fed up with the arithmetic lessons as they are doing the same thing over and over again. They are now starting fractions. Her taste in books, I am sorry to say, is not high-class, she loves the usual type of "Brownie's holidays" or words to that effect and said she hated "Helen's children", because there were so many pages just telling you what happened and not talking. She seems to enjoy dialogue, particularly pointful dialogue and her greatest joys are jokes. Then she enjoys puzzle books very much and chose one in exchange for Orovan's book.

Then Ronnie came up with Mrs, J. and we had quite a time roughing around. Ronnie looks much changed and grown-up in his braces and schoolboy-cap. His clothes don't seem to keep very good discipline, which his tie crooked and a gap showing between his blouse and his trousers. I felt he was strongly handicapped from both sides (Even Mrs. J. by the way is a kindred spirit in more senses than I expected: though dressed with the best of taste, her stockings showed a deplorable number of holes, some of them mended by no means all). Then we went down to the Forest Hill and Mr. J. showed me Clare's school which is absolutely the last word in nursery schools and we went to have dinner at the J.'s "Sister" felt very bad about having to stay in bed on such an exciting day, so part of the activities were transformed to her room. The old Mrs Curtis[1202] was keeping her company. Did I say she has just come out of hospital and seems a lot better. She looks the ideal grandmother and she is quite the type of old lady that would fall in love

[1202] Mrs Jephcott's mother.

with you at first sight, and the children are very fond of her. Anne is in some ways like you, in living with enormous intensity and never being neutral to anything or anybody either accepting or rejecting strongly. Her energy is such that her father suggested once buying her a treadmill and letting her run it for two hours every morning so she would be easier to live with. After lunch we took Ronnie to school and saw his school and his teacher. He had moved up to grade 3 only three weeks ago so his teacher did not know him very well yet, but she was very pleased with him and said he was managing the grade very well. His report apparently referred already to the work in his new grade. Seeing the big lads in his form, though, he looked a little tiny to be on a level with them. When his class had started we looked over a good bit of the school and saw the headmaster and I liked the school very much. Life is too short to put all my impressions of the school down, but I shall tell you all about it. Then with a stop for some toys on the way, we went down to Mr. Bernhardt's office. I met there his secretary, a very nice lady who told me all about the difficulties they had had — there were not many — and gave me some greetings to put to the Birmingham committee. Then there was Dr. Blatz who had just been in England and who is coming to Birmingham in the summer with a group of people to organize training for women to take charge of day nurseries. Dorothy Millescharp, Mrs. J.'s friend, is going with them. We were waiting for Dr. B. all the time, but he was kept somewhere and did not show up. This threw us somewhat out of gear, but Mrs. J. had time to rush me up to York Mills and I swapped drivers, Mrs S. took me down to Baron Renfew school where the children and Mrs. James were already waiting. Mrs James is very nice, not quite as girlish as her letters and very free and straightforward in manner. She took me to Gaby's schoolroom which is in the basement and is a recent innovation, so that Mrs James now only has two grades in the same room. With the extra space they have made the best of the existing possibilities and the room is quite attractive, decorated with children's drawings, cuttings, posters etc. By the way, I discovered the secret of "Social Studies" it is a new subject which represents history and geography in one.

Then we went for cake to Jacqueline and family and took them out for tea. Judy is as sweet as ever. Jacqueline is very grown-up and sensible and happy. The tea was a marvellous affair with hot chocolate and sardines on toast and ice cream and chocolate eclairs and whatever they fancied, they all seemed to know their limits, however, and I hope there won't be any tummy aches tomorrow.

24.3.

At this sleep interfered again and I had to delay the letter again until now. I am now sitting in the observation deck of the super streamliner "City of Los Angeles" and I shall ride in this marvellous affair for two nights and one day. To take up the interrupted account, there is not much left of the days in Toronto except meeting Mrs. S. who had done shopping while we were at tea delivering the young children home (Sheridan at the wheel – in spite of having quite the look of the old family chauffeur who in England would drive the family Rolls, he used his brakes almost as temperamentally as a New York cab driver.) Then dinner at Sandersons, good-bye to Gaby, who suddenly remembered a lot of questions she wanted to ask about the why and how of my work and my journey, then down for half an hour's chat to the Jephcott's house — without getting Ronnie out of bed to make my departure less traumatic. (Much better if he feels it is in order for me to show up one day and go away again without much ceremony.) Then down to the station with the two foster fathers, a few minutes with Bill Lambsden who had come to the station to see me off, and so to bed.

I did not cry myself to sleep after all.

30.3.

How annoying! I was sure my first postman had left long ago and you had got my letter, and now I suddenly hear that he was held up and for all I know he may still now be waiting. To extrapolate from this, heaven knows when Simon will be able to go and the letter I sent home will be stale like the rolls in the dining-cars here, before it ever gets to you. So I better write a few more lines to make it less disappointing.

Well, I am on the way back from the west and, short of seeing the children, this certainly was the climax of the whole trip as regards sights. As I wrote you I was travelling out by streamliner "City of Los Angeles" and spent a whole day riding through the barren hills of Wyoming which [?] on the very ferocity, the size of this country and the emptiness of the west. For relief a few snowy mountain ranges here and there, crowding together towards the Salt Lake which, however, we reached only after dark. We were due in L.A. at 8 a.m. but when I woke up that morning I found myself in the middle of the desert, with the train being four hours late owing to engine trouble. The descent is very beautiful with quite mad colours and old rock formations. Coming down, the water beginning gradually as one approaches Los Angeles and with it unlimited fertility, with palms and groves and all kinds of strange flowers. At the time the whole place is unbelievably green, more so than even England unless it is only the sun and the blue sky showing it up more. Everything is so new from the houses, which range from corrugated iron sheds through concrete or wooden bungalows to the most fantastic Spanish castles, to the people about labourers of Mexican, Indian, Filipino, so many other types that I wish I had five pairs of eyes.

On arriving in L.A. I found my plane to San Francisco gone and all others booked up. So cursing a little about the hours of the day and about being cheated out of an air trip, down the coast, I booked a berth on the evening train and spent the afternoon looking at the centre of Los Angeles, Hollywood, and then walking up to the top of the hills above Hollywood from which one has a marvellous view of the whole place. Los Angeles and Hollywood are pretty neat as I would have imagined them (but closer together) people very cheerful and well-dressed, marvellous shops (an enormous amount of jewelers working on the credit system!) Spanish influence noticeable mainly in the architecture and in the types one sees, not in the restaurants, among them there is a chain of places called "Pig'n Whistle", trying to be rather English but without much success.

Next day Berkeley, which is a dream. It is a suburb of Oakland, which is across the water from San Francisco and lies itself on a hillside along the bay. From the campus and from the houses of most of the scientists there you have the whole bay and the Golden Gate sight in front of you and particularly at night there is an inexhaustible sea of lights. The campus buildings are of insufficient quality but in this setting everything looks good.

It is a university of enormous size, and the place reminds me of Cambridge with its crowds of young healthy- and wealthy looking casually but well-dressed people. Except of course that the old part of your Bikini picture is missing.

As I walked through the hotel door there came through the other door a messenger with your cable. It could not have been better timed! The second night I did not remain at the hotel but stayed with a man I had not seen for 13 years and who has a marvellous house up the hill, built in the style of an old Spanish ranch house, evidently the equivalent of the Olde-World-Cottage-cum bathroom and refrigerator. On Sunday morning Bloch drove over from Stanford, to take me into San Francisco for a sea-food lunch at one of the fisherman-type of inns. There was not much time to see much of San Francisco, except, crossing the bay again by ferry, a picture of the general skyline, the bridges, prison island, and the island of the Exhibition.

Again with the streamliner (in which I am writing this now) and an evening amongst the wooded hills of Northern California, the train often running over the top with the villages low down in the valleys. Luckily there was nearly a full moon and I lay awake in my "roomette" for a long time while we climbed up through the snow-crowned woods, up to Reno (I nearly got out there, but then decided otherwise!) First thing in the morning was going along the dam through the salt lake, and then we joined the line by which I had come west and we are now again running through the barren country the only change being that since last week, some cattle have appeared as it apparently has turned warm enough now for them to be up on the hill pastures.

The view will now be the same until tonight. To-morrow Chicago, I just wonder whether there might be a letter from you there? There just might be.

Tons of love,

Rudi

2.[4].[1203] I am just back in N.Y. after a nice day in Ithaca, and heard of Alfred's cable. How thrilling! I hope this time it is real!

[344] Rudolf Peierls to Robert Oppenheimer

Montreal, 21.4.1942

Dear Oppenheimer,
I cannot resist the temptation to follow the example set by my boss[1204] in the way of bread and butter letters, though I hope I shall not emulate him as regards duration of stay. But conventions and precedents apart, I shall not forget the delightful day in your place and I hope Mrs Oppenheimer will not think too badly of me for intruding on a family weekend![1205]

I heard in Chicago that they have had correspondence with you about your collaboration, and I need hardly say that that makes me feel much happier about the prospects of the theoretical work and a satisfactory interpretation of the experimental work, although I do not know the precise form which this collaboration is to take.[1206]

During the rest of my stay I learnt that the approximation of constant γ used by Taylor is probably much worse than he — and I — had assumed and one of my next jobs will be to have this whole question reconsidered properly.
With kindest regards to all of you, including Peter.[1207]
Yours,

R.P.

[1203] This was misdated 2.3.
[1204] Mark Oliphant.
[1205] Peierls had met Oppenheimer in Zurich when working as Pauli's assistant. During Peierls's visit to the US they renewed their acquaintance and Peierls stayed with the Oppenheimers, when he was in San Francisco.
[1206] While travelling through the US, Peierls was summoned back to Chicago to discuss fast neutron work and nuclear weapons research. In these discussions, he and Arthur H. Compton, one of the senior Americans in charge of nuclear research planning, agreed about the desirability of putting Oppenheimer in charge of the work.
[1207] Peter Oppenheimer, Robert Oppenheimer's son, was born in 1941.

[345] Rudolf Peierls to James Chadwick

Birmingham, 22.6.1942

Dear Chadwick,
As Frisch has probably told you, Moon has returned to this country,[1208] and if you are still considering asking for his transfer, it might be best to do so fairly soon before he has settled down to a definite job here. I am not, of course, mentioning anything to him, unless you should want me to sound him out on the subject.

I called recently at the Central Register in connection with getting a computer, and also about some junior physicists for Simon, and as a result of my experience there it seems that if you still want a junior physicist of the type of Marshall,[1209] or with a full degree, it may still be possible to get one from the Central Register now, because they are getting lists of graduates from different Universities as the examinations finish and there is an interval before they get definitely allocated; but several Government departments are taking them on in [great] numbers, and once they are allocated it is very difficult to get them out again. They are not very efficient in correspondence, and I found the best plan is to write to Robinson or Hoff beforehand stating the requirements (they deal with different kinds of candidates), and then visit them to look at a selection they have prepared from their files. If you will be in London in the near future that would be the most efficient way; otherwise Jackson could go there, but he would not, of course, know precisely what you want.

Yours sincerely,

R. Peierls

[1208] Philip Moon (1907–1994), graduated from Cambridge in 1928 and continued research there under Oliphant. After another seven years of research at Imperial College, London, he followed Oliphant to Birmingham where, with the exception of the war years, he remained until his retirement in 1974.

[1209] Chadwick had corresponded with Peierls about hiring a number of people, among others a young graduate, Marshall. See letter Peierls to Chadwick, 14.5.1942, Churchill Archives Centre, CHAD/1/19/6.

[346] Rudolf Peierls to James Chadwick

Birmingham, 29.8.1942

Dear Chadwick,
Thank you very much for your letter; I am very glad you are getting on well. I shall then try to come and see you at Liverpool when it is convenient to you after your return. Amongst the points I was hoping to discuss with you was the position of the Cambridge part of your team and the question whether at present people in Cambridge, in particular Kinsey[1210] and the other full-time workers, had all the information they needed for efficiency. Meanwhile I hear from Akers about the further developments which rather complicate matters, but it is perhaps more desirable to talk these questions over.[1211]

There is one point arising from this: I feel even more strongly in favour of having some Physics panel, or perhaps more accurately "fast neutron panel" under your chairmanship, which would ensure adequate information for everybody, and which might sometimes also act as a safety valve.

Another point on which I would like to get your views is the position of yourself and the teams working under your direction both in Liverpool and Cambridge in the impending transfer. There seemed to be many arguments in favour of including fast neutron experimental teams in such a transfer, and I was glad to see from a remark you made, I believe to Perrin, that you share this view in principle. Any transfer will, of course, necessarily involve some waste of time, and also loss of contact when different teams are transferred at different stages, and in trying to think out a programme for my own work, which of course is largely linked to the work of the diffusion plant, I think it will be useful to get a preliminary idea of the times you had in mind, so as to arrange things with as little loss of contact as possible.

[1210]Bernhard Bruno Kinsey (1910–1995) graduated from Cambridge in 1932. After studying under Lawrence and Oppenheimer at Berkeley (1933–1936) he took up a lectureship at Liverpool while completing his Ph.D. at Cambridge. In 1939 he returned to Cambridge for war work.
[1211]See correspondence Peierls-Akers in PRO: AB3/111.

In connection with the transfer, you will have gathered from the minutes of the last meeting of the Technical Committee[1212] that there was some controversy, and this is another matter which I hope to discuss with you some time; I think you probably would approve of the line we took at that meeting, but I think it is desirable you should know all the details of the discussion. I feel fairly confident now that things will be all right in the respect.
With best wishes,
Yours sincerely,

R. Peierls

[347] Rudolf Peierls to James Chadwick

Birmingham, 12.9.1942

Dear Chadwick,
I enclose report CF 159 giving a summary of the present state of the American fast neutron work. The CF series (C for internal Chicago reports, F for fast neutrons) is very restricted in circulation and Perrin tells me we have been asked to limit circulation here as much as possible.[1213] At present all C reports are kept in Cambridge, so would you return it to Placzek when you have finished with it. If you want to keep a copy permanently it should go first to Perrin for reproduction.

It is a pity the notes are so scrappy that one cannot learn much from them. In particular, in the Washington report it is not clear whether one should reduce b[y] a factor 1.57 Heydenburg's value of the cross section obtained by accepting Amaldi, Tuve and Hafstad's result for the neutron intensity of the beam (this gave a cross section of about 2.3) or that obtained by accepting the A-T-H- calibration relative to a Ra-Be source, and taking Fermi's new value for the Ra-Be standard, which gives of course, about twice the cross section.

[1212]Technical Committee of the Scientific Advisory Board.
[1213]See correspondence Peierls-Perrin, PRO: AB3/111.

The "three-compartment" ion chamber is probably another form of Frisch's neutron spectrometer, which I mentioned to the Chicago group.[1214]

In connection with your report on the energy spectrum it occured to me that, if one assumes that the neutron is emitted from either of the fission fragments, so that its velocity is the superposition of the fragment velocity and its relative motion, it is possible to calculate the velocity distribution for the relative motion even from part of the neutron velocity distribution, and hence one can get an independent check on the absence of large numbers of low-energy neutrons which would complicate the interpretation. I hope I have this method ready quite soon.

Yours sincerely,

R. Peierls

[348] James Chadwick to Rudolf Peierls

Liverpool, 18.9.1942
(carbon copy)

Dear Peierls,

I must apologise for not acknowledging the receipt of the American report CF.159 before this. I had to go to Manchester just after receiving your letter and I handed it to Frisch.

The notes are certainly very scrappy and sometimes obscure. I have received today from Akers a later report which makes some points a little clearer. On the other hand, this later report is also obscure in places, and I think we shall have to ask the American workers to tell us just a little more so that we know better what they have done and get some indication of the reliability of their results.

[1214]Peierls had visited the Chicago group during his trip to the US in the spring of 1942.

In the last report Heydenburg[1215] of Washington gives a result which is rather disturbing if correct. He finds of 235 for neutrons of energy 2.6 – 4 MEV only $0.4 \times 10^{24}\text{cm}^2$.

Would you mind giving me your opinion of the note on inelastic scattering by Feather[1216] which I enclose. He wants to have it circulated not only here but in the States. I think there is something to be said for this, but I feel rather reluctant to circulate suggestions unless they are new and important. In my opinion circulation should be mainly confined to actual results.

As regards Feather's note, one could also raise objections (1) that there is nothing new in relating inelastic scattering to the presence of excited states, (2) that the correlation noted by Feather is not strong. I have tried to get Feather to modify the note, but with little success.

I would have no objection whatsoever to circulation here, but I think that circulation in the States might give a bad impression. Many of the workers in the States must be fully aware of the point of Feather's note and they might feel that we are adopting a somewhat schoolmasterly attitude towards them.

Yours sincerely,

[Chadwick]

[349] Rudolf Peierls to James Chadwick

Birmingham, 22.9.1942

Dear Chadwick,

Thank you very much for your letter[1217] and for letting me see the new version of Feather's report, which I herewith return. I feel that

---

[1215]Norman P. Heydenburg, at the time working at Washington, later professor of physics at Florida State University.

[1216]Norman Feather (1904–1978) graduated from Trinity College, Cambridge and became a fellow (1929–1933). After research at John Hopkins, he worked as a demonstrator and lecturer at Cambridge before taking up a professorship of Natural Philosophy at Edinburgh where he remained until his retirement in 1975.

[1217]Letter [348].

it would well be worth while to have a discussion of this problem sent over, because it is quite possible that no-one in America has thought of this approach. If they are aware of these arguments, no harm is done, provided, of course, the report is sent and worded in the right way. Even if the evidence is weak the discussion may lead to somebody else in America finding new evidence either in favour or against Feather's argument, or even re-analysing existing evidence so as to come to a definite conclusion.

I do not think we should assume that everything we are sending is necessarily a new contribution and implies that the Americans know nothing about it. They are now sending us all their reports which are written in the first instance for internal consumption, and it may be sent in the same sense. If you want to be quite sure of avoiding a misunderstanding on that score, could you not send it with a covering letter, to, say, Oppenheimer, so as to make the purpose of the paper quite clear. As to the actual wording, I agree that it might be more careful in places, and that the whole thing could be somewhat shorter. Would it not save time if you were to indicate to Feather explicitly the alterations in the wording you would suggest, as I think it extremely likely he would accept them immediately.

Yours sincerely,

R. Peierls

P.S. I understand from Placzek that at present Feather is not receiving any of the CF type reports, in particular the last one by Manley, because, following the wishes of the Americans these should be rather restricted, and the position is apparently that it would be for you to say if you considered that Feather should see any or all of those reports. I am mentioning this merely because there was apparently earlier a misunderstanding about a similar point concerning progress reports.

[350] James Chadwick to Rudolf Peierls

20.10.1942

Dear Peierls,
I should like to propose you for election to the Royal Society. If you have no objection to my doing this, will you provide me with a list of your published papers as soon as convenient.
Yours sincerely,

(Chadwick)

[351] Rudolf Peierls to James Chadwick

Birmingham, 21.10.1942

Dear Chadwick,
Thank you very much for your letter.[1218] I am very grateful to you for your offer to propose me for the R[oyal] S[ociety]. A few weeks ago Mott suggested he would propose me and I saw no reason not to accept this. He asked for a list of my papers and is, I believe, going ahead.

I have just had a cable message from Simon, who has been able to contact the American theoreticians at last. The message is very brief, but it appears that they are working on very similar lines to ours. A point of interest to you is that they are proceeding with experiments on the effect of fast neutrons in unseparated metal. Simon gives no details but he refers to point 10 of my August progress report, and I imagine the experiments are done from a similar point of view to the one expressed there.

From what Simon says, it appears that it would be desirable for me to go over on a short visit fairly soon. The American theoreticians are meeting early in November and then again in December. The December

[1218]Letter [350].

date may be too late for me, because then I hope the work in Valley[1219] will be under way and I should like to be here to help discuss the experiments. It seems probable, therefore, that I had better try to go in November.* Is your own programme more definite already? If somehow it could be arranged that we could be there together, or that our visits to some extent overlap, that would I think have advantages, particularly as the discussion on the theoretical aspects of the fast neutron work is very closely linked up with the experimental programme, and questions of programme are therefore bound to arise in the discussions.

I have written up the argument about nuclear distribution of fission neutrons and I enclose a copy. I am still a little worried about the result; it would, of course, make no difference to us if we knew the neutrons are not emitted isotropically, but it is a little surprising, and perhaps one could learn something by following up this point. Would you think, with this aim in view, any of the experiments suggested by Feather dealing with coincident measurements, e.g. coincidences between fission particles and neutrons, would be practicable?

Yours very sincerely,

R. Peierls

[Handwritten addition]

*I have just heard from Perrin that Simon is unlikely to be back before the second week in November. I may therefore not be able to go before late November.

[1219]Peierls and Francis Simon had continued to work on the theoretical aspects of the gaseous diffusion method of isotope separation. In early 1942 it was decided to use a former Ministry of Supply factory at Valley in North Wales for the purpose of carrying out the relevant experimental work. I.C.I. was to be responsible for the erection of the models and for running these; the research programme remained under the direction of Peierls and Simon.

[352] James Chadwick to Rudolf Peierls

Liverpool, 24.10.1942
(carbon copy)

Dear Peierls,
I am sorry I am too late with my suggestion about the R.S. for I have had it in mind for a long time, but I think it is perhaps more appropriate that you should be proposed by someone on the mathematical side, for it is the Mathematics Committee which will consider the matter.[1220] Would you let me have Mott's address, for I should like to sign the proposal form and I want to let him know that Fowler will support you in case Mott has not approached him.

I have heard from Simon through Perrin that some separated material is available — a relatively large amount, but the concentration is no better than before. I should like to examine a good sample before I go to the States, but I do not think I should wait very much longer. We have sent a cable about the sample which I hope will produce results.

Thank you for your note about the fission neutrons.[1221] I should be very surprised if our experiments are much in error, but I shall have to think over all the implications. We tried some time ago a fairly simple experiment which would have helped if it had been successful, but so far we have not been able to get the right conditions. Feather has promised to tackle the delayed neutron experiment seriously, but I have little hope of any result for some time. It is a difficult experiment.
Yours sincerely,

[Chadwick]

[1220] See letters [350–351].
[1221] Appendix to letter [351].

[353] Rudolf Peierls to James Chadwick

Birmingham, 7.12.1942

Dear Chadwick,
When I was in Liverpool we discussed the question of Massey and Hulme,[1222] but I am not quite sure whether we came to a definite conclusion as to what should be done about it. I believe you had written to Fowler about one of the points that came up in connection with Massey and were expecting to hear from him. I suppose unless this produced any new point of view we can assume that it will not be possible to get Massey, and perhaps the next step would be to approach Fowler to see whether he would take up the same attitude about Hulme. Do you see any objections to my going to see Fowler and discussing this question with him, or do you think it would be wiser if you wrote to him about it?

Simon and I have jointly sent a message to Halban[1223] via Perrin in which we strongly urge him to accept the suggestion that Kowarski should become a member of the Technical Committee and should bring his family.[1224] I do not think either of these claims can reasonably be refused. I gather from Perrin that you are of the same opinion and that

[1222]Henry R. Hulme, had published on nuclear physics, at the time he was working for the Admiralty; Harrie Stewart Wilson Massey (1908–1983), received his Ph.D. from Cambridge in 1932, became lecturer in Belfast in 1933, before taking up a professorship at University College London where he stayed for the remainder of his academic career.

[1223]As part of the restructuring of the Anglo-American nuclear research effort, Hans van Halban was hoping to convince the Americans to host his Cambridge team as an independent British unit at Chicago where all the American work had been concentrated under the leadership of Arthur Compton. Washington was opposed to such an independent unit and offered Halban a position in the Chicago group as responsible for the heavy water work. Halban refused and eventually his team was transferred to Canada as the nucleus of a much larger Anglo-Canadian outfit.

[1224]The British-Canadian co-operation was hampered by organisational and administrative problems, but also by clashes of nationality and personality. Among these was the rift between Hans Halban and Lew Kowarski. These culminated when Hans Halban only allocated a minor role to Kowarski in the team of British scientists working under him within the Canadian project. Kowarski's disappointment in this move was shared by many of his Cavendish colleagues who initially refused to go to

you would also like to see Kowarski made deputy leader of the team or something to that effect. We have not included this point in our own message as it is really for Auger as head of the team to arrange the work, and he will hardly be able to decide anything about this before the team has arrived and he can judge for himself how best to run the thing. This of course is purely our own personal opinion. I do hope, however, that Halban will agree to this and that a settlement will be reached which will allow people to think of more important problems.
Yours sincerely,

R. Peierls

[354] Hans Halban to Rudolf Peierls and Francis Simon

Montreal, 21.12.1941

Dear Simon and Peierls,
Please excuse my not answering your cable concerning Kowarski earlier.[1225] I certainly appreciated your taking this personal step to inform me of your opinion of the situation in Cambridge, but I always hoped that the arrival of Akers' letter as well as his telephone conversation with Perrin might clear the situation somewhat.

You will know by now that in Akers' cable of December 20th we have come to rather drastic decisions, and I would like to explain to you why we did, in spite of your recommendation.

We all (Placzek, Akers, Newell and I) fully agreed that we ought to make arrangements to avoid further discussions of this matter, to enable everybody to get to work. However, after all that happened in the past I was not of the opinion that this could be obtained by concessions. It is, of course, impossible in such a letter to rediscuss the whole deplorable case, but I would like to outline just one point.

Canada to join Halban's group. Akers, Chadwick, Peierls and Simon tried to move Halban into a more conciliatory position, and Appleton eventually persuaded most of the Cavendish team to accept their Canadian contracts. See letter [354]. Also Gowing, *Britain and Atomic Energy*, p. 191.

[1225] See letter [353].

I was in full sympathy when the people in Cambridge made it known that Kowarski ought to continue collaboration with us. As you know I was of this opinion for six months after Kowarski had informed me he wanted to discontinue collaboration, and I had only given up any hope of securing further collaboration when I saw that his reactions to every concession were further requests. I regretted very much that the people of the lab. did not approach me in the first instance, but this could be partly explained by my sudden departure. I regretted equally that they had the impression that I tried to postpone the final decision on the Kowarski affair until my departure, in order to avoid discussion. This was certainly not the case.

After the above you will understand that the request of the physicists of Cambridge was not too big a surprise to me, since I had myself only a short time before changed my attitude. I told Akers in our first conversation over here, of serious misgivings I had about the re-entrance of Kowarski into the team, but I told him that as a concession to him and to the Cambridge group, I was quite prepared to revise my attitude and give Kowarski a new chance.

However, I was at once of the opinion that we ought to avoid certain things. One was to give Kowarski the position of second in command under Auger, since at that time I had already formed the opinion that Laurence has, taking it on the whole, at least the qualification of Kowarski, and especially since we very much hope to obtain Pontecorvo, whose whole standing as a scientist is very much higher than Kowarski's. I also wanted to avoid Kowarski being a member of the Committee, since this would naturally raise the question of membership for Pontecorvo, Bauer and Gueron. Gueron was, for the moment prepared to renounce his membership, even if Kowarski became a member, but this would not change the fact that he was, in my eyes at least, as entitled to it as is Kowarski. I say this because membership of the committee ought to be given to those whose collaboration might be useful in the work of the Committee, and not as a recompense for merits in the past.

Another point seemed important to me: I think that members of the lab are certainly entitled to express their feelings as to the use Kowarski has for them as a teacher, and their feelings as friends when they think injustice has been done to him. I think, however, that we should get into the impossible situation by allowing all of them to decide the major policy of the Lab.

Unfortunately it has come to that, and the fact that some of them refused to embark before the Kowarski situation was cleared up goes certainly too far. I would not feel too well in handing over to Auger a physics team consisting of people who do not put their duty to hurry on with the work above everything else, and who do not trust that Akers and myself will not ask of Kowarski something completely unfair.

In writing this letter I have a feeling that I am not succeeding very well in telling you what I mean, and I hope you will read between the lines. Whilst this letter contains nothing which could not be shown to every member of the committee, and even to everybody concerned in Cambridge, it has not been worded with the care I would use in writing it for anybody but you or Perrin.

I telephoned the other day to Mrs Simon, in order to tell her that Newell had seen Simon in good health, and she told me that everybody was all right, and, exceptionally, nobody had even a flu.

You would both be very kind in encouraging my wife sometimes. She will need it very much because of the delay in her departure.
Yours,

(Signed) HANS HALBAN

[355] Max Born to Rudolf Peierls

Edinburgh, 1.1.1943

Dear Peierls,
Many thanks for your kind New Year's wishes which we cordially return. It was very kind of you to send me a copy of your memorandum about the teaching of Math-Phys at your university. I can subscribe it word by word. Concerning the title I have succeeded against Barkla's violent opposition to re-Christen my department Mathematical Physics (Applied Mathematics). It was not possible to suppress the traditionalist words. I should like to put forward similar suggestions here, but there are two obstacles.

1. I have not been asked to do so by my faculty;
2. I am tired of fighting against the discrimination of the Scottish university authorities with respect to modern science, and the resistance of Barkla against any change of routine.

There is further Whittaker's catholicism which leads him to an attitude rather inimical to experimental research; he wants my department representing science in the Art's faculty, or better "natural philosophy" with the weight on "philosophy" and very little empirical background.

But I shall grasp any opportunity to improve of the situation.

Should it not be possible to extend your model to a wider field, comprising all British universities? Could we not try to bring together the theoretical physicists of all places (Cambridge, Oxford, London, Manchester, Liverpool, Birmingham, Edinburgh a.o.) to develop a common plan consisting in a minimum request for staffing, research, scholarships, syllabus, examinations? Could we not ask Simon or Lindemann: pardon me, Cherwell, to take the lead? Or are you afraid that they are all too conservative? Your idea seems to be by improving things in Birmingham to compel the others by competition to follow your example. I am pretty sure the Scots won't follow.

I am looking forward to hearing from you. We are all well. My wife has recovered from her operation. I am trying to understand Dirac's paper's on the electromagnetic field[1226] but it is difficult. Schrödinger writes about new results about merging Einstein's gravitational theory with Infeld-B's electrodynamics, and very enthusiastic about it. I can hardly do any research work in term time, but try to read the literature in these holidays.

With kind regards to your wife from all of us.

Yours,

M. Born

[1226] P.A.M. Dirac, 'Quantum Electrodynamics', *Communications of the Dublin Institute of Advanced Studies*, **A1**, 1–36 (1943).

[356] Rudolf Peierls to James Chadwick

Birmingham, 16.2.1943

Dear Frisch,
I enclose now the results of the calculations on the density distribution near a point source. The curve which I sent you earlier is a special case for no absorption. From the remark in the last paragraph you will see that the curve has a vertical tangent for $r = 0$, and this has not been taken into account in drawing the earlier curve which we sent you. The error caused by this is, however, very small.

I believe your original idea was to apply this curve to the correction for an absorber of finite thickness. If we can be of any assistance in this, will you let us know?
Yours sincerely

R. Peierls

[357] James Chadwick to Rudolf Peierls

[Liverpool], 17.2.1943
(carbon copy)

Dear Peierls,
Thank you for your letter and for the calculation of a point source in an absorbing medium,[1227] and also for your earlier letter concerning the effect of a refractive index on the absorption in a sphere (which I believe I had not yet acknowledged). I am not at present interested on the correction for an absorber of finite thickness and I doubt, furthermore, whether any calculation of the correction would be very reliable, in view of the effects, such as persistence of velocity.

The model experiments have been stopped at present while we are waiting for the plastics, which ought to be here soon.

[1227] Letter [356].

I shall be in London this weekend, possibly including Monday, on the phone at Putney 4712 as usual.
Yours sincerely,

(Chadwick)

[358] Rudolf Peierls to James Chadwick

Birmingham, 23.2.1943

Dear Chadwick,
Thank you for your letter about the energy spectrum.[1228] This seems to determine the shape of the curve beyond any doubt, and the discrepancy with the elementary theory seems, if anything, worse than before. This conclusion depends, of course, mainly on three experimental points (the ones at .65, 1.1 and 1.3 mv), but they are so far away from would be required and so consistent with each other that unless there is an unsuspected systematic error it seems now absolutely out of the question to get agreement.

As regards A.C. electrolysis, I never quite understood on what principle this was supposed to work. Unless it is known that the effect depends only on the mass difference is it not very likely a priori that any effects usually observed would depend on other properties and therefore not give any effect on the case of isotopes.
Yours sincerely,

R. Peierls

[1228] Letter [357].

[359] James Chadwick to Rudolf Peierls

Liverpool, 27.2.1943
(carbon copy)

Dear Peierls,
The formula Pryce worked out for me for the reduction in neutron density, averaged of a circular disc of absorbing matter, caused by the presence of this matter is

$$d = \frac{4}{\pi}\frac{tr}{\lambda}\left(1 - \frac{3\pi r}{8.L}\right)$$

(λ = m.f.path, L = diff. length, r and m radius and mass of disc, $t = \sigma N m)/(\pi r^2 A)$ its thickness in terms of the l/e thickness).

On checking the experimental results, I found that the discrepancies I mentioned were not as big as I thought, and I eliminated some numerical errors thus further improving the agreement. The one experiment was with an indium disc $r = 0.7$ cm., $m = 134$ mg., $A = 115$, $\sigma = 167 \times 10^{24}$ cm^2, where Pryce's formula gives $d = 0.15$, while the experiment indicates $d = 0.14$ which is perfectly good enough. The other is a disc of 50 mg Dy_2O_3, A(Dy) = 166, $\sigma = 10^{-21}$, r = 0.6. Here the formula gives .22, the experiment .19. This is probably as good as one can expect. I must therefore apologise for libelling Pryce and the slow neutrons. If it is too difficult it might still be useful to extend the formula to cases where t is not $\ll 1$ and r not $\ll L$ and $\gg \lambda$.
Yours sincerely

[Chadwick]

[360] Rudolf Peierls to James Chadwick

Birmingham, 4.3.1943

Dear Chadwick,
The Billingham people[1229] have written a very short report on the present state of the technical problems associated with the separation plant. While I know that you generally have not the time to follow all the details of the developments on that side, I think you might be interested to see the report, which is comparatively readable and gives an outline of the most important points. I therefore enclose it in case you would like to see it, but would like to get the report back when you have finished with it.

I saw Oliphant for a short moment on Monday, and he still seemed very interested in the progress of the T.A. work, but I gathered that he is very anxious to start work on some problems more immediately in contact with the Services and with operations problems.[1230] I do not think, however, this means that he has made up his mind with any finality.
Yours sincerely,

R. Peierls

[1229]The I.C.I. plant at Billingham was the site of much of the technical work on slow neutron and the production of heavy water.
[1230]Oliphant was interested in the possibility of isotope separation by an electromagnetic method, a method that was being developed by Lawrence at Berkeley. However, he was also involved in the important war work on radar.

[361] James Chadwick to Rudolf Peierls

Liverpool, 5.3.1943
(carbon copy)

Dear Peierls,
Many thanks for sending me the report on the separation plant. While I can contribute very little to this side of the work, I am none the less extremely interested, and want to keep in touch as far as possible. I think, too, that we have something to give on the vacuum side, for we have more experience here on high vacuum work than most people in this country.

I wrote to Oliphant last week-end, but have not yet had a reply. My feeling is that if it were decided to go ahead on a good scale with the electromagnetic method I could persuade Oliphant to come in.
Yours sincerely,

[Chadwick]

[362] James Chadwick to Rudolf Peierls

Liverpool, 10.3.1943
(carbon copy)

Dear Peierls,
Thanks for your letter and the paper on hemisphere.[1231] This is indeed a comfortable way of becoming a co-author. We have not measured our models f[a]r enough to observe the crossing-over, but as far as our measurements go, they agree well with our curves.

At present we are studying the optical properties of two sample plastics. The one is coloured pink and has about twice the absorption coefficient in blue as in red; nevertheless, it gives a fairly straight logarithmic absorption plot even for white light, and may be very suitable

[1231] R. Peierls, 'Light absorption by two black hemispheres', DTA report, (1943), NA/PRO: AB4/915.

for your trick of varying the abs(orption) coeff(icient) by coloured filters. The other stuff is milky and looks very nice, but was found to give rather pronounced forward scattering, as shown in the enclosed curve.

I am rather doubtful whether this milky material is good enough and I should like to have your opinion on it. Or would it be possible for you to come over and discuss these similar questions with Marshall and myself. Any day next week (except Monday) would suit us, and I think it would be a great help if you could come.
Yours sincerely,

[Chadwick]

[363] Rudolf Peierls to Akers

Birmingham, 1.5.1943
(carbon copy)

Dear Akers,
I am writing to make some remarks about the general state of our project. Where these are critical, I hope you will no think they are directed against you. I know you are as impatient as everybody else about the delays. Quite possibly you are in agreement with all I say. Nevertheless I think it would be as well to state my opinions quite explicitly and frankly. I was hoping for an opportunity when I saw you in London but there was no time.

Our impatience may partly be due to the fact that this is not the first time we are waiting for a decision. Even without going back to the days of the MAUD committee and the time before it, there was the period of waiting for the decision of the Hankey committee in the summer of 1941,[1232] when everything was in a state of suspended animation for months, no move being made that involved any commitments, since

[1232] The MAUD Committee Report had been received by the Ministry of Aircraft Production on 29th July 1941, and was eventually passed on to Lord Hankey with a request for the Scientific Advisory Committee to consider the report and recommend each time the policy decision was due next week, so that it was just as well to wait for it. (I need hardly say that we have every reason to be satisfied with the nature of the decision when it did come).

From then onwards, the question of collaboration with America cropped up, and with it the uncertainty about the location of a plant if it was to built. When later the Valley models had got from the design stage into production, it would have been the obvious thing to go on designing a new model incorporating the experience gained in the meantime and suitable for higher pressure then becoming feasible. Instead, design work stopped since Elce argued that further work should be done in close contact with whoever was to build the final plant. His terms of reference did not include design of any further models and they were never extended.

At the same time, I believe, a chance of getting the collaboration going was missed through delay. If the policy committee had made up their mind in time to enable you to go to the States on your first visit with full authority it is likely that, with OSRD[1233] still in charge, a satisfactory scheme could have been found, and it would have had a chance of weathering the present storm. Ever since work on the engineering aspects has been held up because of the uncertainty whether, when, where and by whom a plant, if any, would be built. You know that Elce has expressed this view repeatedly and the same seems to be implied by J.D. Brown's letter to Simon of April 23rd.[1234]

I know, of course, that the present delay was caused by the unexpected and in the long run unpredictable difficulties that arose last autumn. By now, these have, however, been evident for about six months

---

further action. At the time a joint Anglo-American project was being considered, but UK scientists were neither informed about progress in the political decision making process nor about planning of nuclear research, which caused a serious delay in the progress of their research.

[1233]Office of Scientific Research and Development. Initially, the OSRD, under civilian leadership (at the time under the directorship of Vannevar Bush), oversaw nuclear research. In the autumn of 1942, as plans for a new laboratory which allowed the concentration of theoretical and experimental work on one site were agreed upon, the project was put under military supervision of General Leslie Groves.

[1234]Letter could not be located.

or more. As soon as there was a serious possibility that collaboration would fail, it was, in my view, the duty of the Policy Committee to consider this possibility and make up their mind what to do if it failed.

They could have decided that, without collaboration, there is no chance of completing the job in this country, and that we could only hope to get satisfactory collaboration going or drop the whole project. But as far as I know this decision has not yet been taken. If at all it should be taken consciously and not allowed to go by default.

As long as there is a possibility that one will want to go on with the project every week lost in preparing for it is a week's postponement on its completion. When I saw you a fortnight ago I had the impression that your preliminary work for intensifying the scale of T.A. research was now going on as quickly as physically possible, i.e. as quickly as it would if it were decided to go ahead. To judge from Brown's letter to which I referred, he does not seem to think that this is the case. The commitments required at this stage are not large measured by the scale of the project and are worth while even if a little later it is decided that we cannot afford to build a plant or if collaboration still becomes possible. An insurance premium is not usually considered wasted if the house does not catch fire!

In quoting past history I am not trying to criticize past mistakes. That is always easy, and it would be equally easy to point out, in retrospect, instances in which the scientists and engineers through errors of judgement, have caused delays. But there is a danger that this particular mistake will be repeated too often. There will still be several occasions when vital decisions on policy will have to be taken about our project. If each time there is going to be the same delay, the same waiting for definite information, the same method of taking the next step when the outcome of the preceding one is known, we may as well give up now, as the work will never get done in any reasonable time. The work is still on a comparatively small scale, but the many scientists, engineers, etc. whom it employs could be used well in other parts of the war effort; their work is wasted unless policy decisions are taken in time to get on as quickly as possible. This is one of the situations where quick action may be the only safe course, and where a great responsibility is incurred by doing nothing.

To put the same argument in a less serious form: Do you remember the standard Parliamentary reply to awkward questions during the pre-Munich period: "Her Majesty's Government cannot be expected to consider hypothetical situations"? I think the failure to face hypothetical situations, and to decide how to meet them before they arise, has already caused enormous delays in our project.

In writing this letter I can speak for myself only, although I imagine that Chadwick and Simon have rather similar views. I am sending them copies and no doubt if they wish to add any comments they will let you know.

Yours sincerely

R. Peierls

[364] Rudolf Peierls to Otto Frisch

Birmingham, 25.6.1943

Dear Frisch,

We have discussed with Wilson further the problem of the "safe factor" in ordinary metal. We have in mind measurements of the neutron flux F from a source of given intensity surrounded by a sphere of metal of varying radius. The flux is supposed to be measured outside by means of a fission chamber or some other indicator of similar response. The simplest case is then that in which F increases always as the radius of the sphere increases. This is the case to be expected if multiplication is possible by means of fast neutrons acting on 238 alone. In this case evidently a sufficiently large mass is unstable and any experimental indication of a rise therefore settles the issue.

We are trying to get the answer to the question of a lower limit to the critical size if no rise is found for a certain mass within a given accuracy. For some simple assumptions Wilson has found that F is proportional to $\frac{x}{sin x}$ where x is proportional to the radius i.e. $\frac{1}{8}$ the critical mass, F should have risen by 50 % , and hence a negative result to within 20% or 30% would certainly allow a safe increase in the mass by a factor 10 or so on this model. We have yet to convince ourselves that this result is not dependent on the detailed nature of the model and no reliance should therefore be placed on the actual value but it is certainly clear that the factor is considerable.

It is, however, very likely that fast neutrons by themselves are insufficient to give multiplication and then the curve will first drop and it may rise again at much larger size when the influence of slow neutrons (probably of the order of 100 k V) becomes felt. The definition of the safe factor for this case depends on the accuracy to which this curve can be followed to somewhere near its minimum, and on this point we would like some information from you as to the measurements you would regard as possible.

If it were possible to detect the rise of F over its minimum value by say 30% or 50% regardless of how small that minimum is then the safe factor is again considerabe although probably not as big as 10.

One particular model e.g. gives safe factors of the order of 4 in the mass. The minimum may, however, be as low as 2% of the unscreened source and I do not know how such small quantities could be measured.

I discussed this problem with Rotblat[1235] yesterday and he suggested that one could use a cyclotron as a source since in this case one would be dealing with considerable masses of metal and therefore may well afford to have a channel a few centimetres diameter to let in cyclotron target, and with high neutron intensities obtained in this way it is perhaps not hopeless to study the neighbourhood of the minimum.

If, on the other hand, it were only possible to detect the rise in the curve for F when F is again comparable with the value for no metal, one would have to expect extremely small safety factors and such experiments would indeed be very difficult to arrange.

In this case one might perhaps get a more sensitive test by using sources of slower neutrons but the results would not then be as easy to interpret as in the other cases.

[1235] Josef Rotblat, (1908–2005), obtained his doctorate from Warsaw University in 1937, a Ph.D. from Liverpool in 1950 and a D.Sc. from London in 1953. He joined the Manhattan Project, and became the only scientist to leave it on moral grounds when it became evident that Germany would not manage to build a nuclear weapon.

All this discussion implies, of course, that there will never be large masses of hydrogenous or other slowing-down material near. If this is not excluded the safety factors might come down a great deal.
Yours sincerely,

R. Peierls

P.S. I enclose for your information a draft report by Wilson giving some more details. We have not had it copied yet so will you please return it fairly soon with any comments you care to make. It is proposed to extend this to cover various other cases.

[365] Rudolf Peierls to Genia Peierls

Washington, 9.9.1943[1236]

My dear Genia.
This is the hottest place I have ever been in, but am now sitting almost in an ice box having a lovely room in the most modern hotel in the place. Each room here has its individual air-conditioning control, and it is quite ideal except that the control in Simon's room got out of order and he nearly froze to death the first night.

It begins to get rather monotonous to write all over again everything to do with the children, and yet this seems to be necessary since it interests you most and I have no idea about the order in which my letters will reach you.

Well, to start from the rear end: I went today to make all sorts of enquiries about the chance of bringing the children back with me, but so far I have drawn a complete blank. Even the chance of taking one in lieu of luggage on the clipper does not seem possible. There

[1236] In August 1943, the Quebec Agreement between the UK and the US had regulated the Anglo-American nuclear relationship for the duration of the war. In essence, Roosevelt had traded nuclear co-operation with the British in return for Churchill's assurance of a British cross-Channel invasion during 1944. As part of the renewed collaboration, Rudolf Peierls and some of his colleagues were visiting the US and Canada. He left mid-August for the US, and visited his children who were evacuated in Toronto while he was on the East Coast.

are quite definite rules against this and there seems no way of avoiding them. The story of "Margaret" is just a film story[1237] and in any case things are more difficult now. There is still one possible approach I have not tried so far. On the other hand, a small number of people are travelling back by ordinary North Atlantic Sea route, and although we may have to wait some time, I think they will eventually be able to come. I am trying to get their passports etc. in order now and to get them registered for a passage. All this of course, is assuming that there is no new development that would make us change our mind.

About the urgency of getting the children back, my present ideas are as follows: Gaby is perfectly happy here and particularly thrilled with Brankesome. She was not too happy at Moulton, partly because of the separation from Nancy, partly because she was hearing from Nancy all the time about all the amenities of Branksome that did not exist at Moulton, maybe also because she was prejudiced by Mrs. S.'s opinion of Moulton. Apparently there was quite a row with the head mistress when Mrs. S. removed Gaby, since this was considered an insult to the school, and the Committee would have liked to leave Gaby there. They think it is quite a good school. This may be right, but once Gaby felt about it as she did, I am glad the change was made. If you ask Gaby she says she would rather come home, but I do not think she feels strongly about it now. She may have done at Moulton. What I consider personally the most important argument in favour of bringing her back is that she is easily influenced; at the moment she is so entirely under Nancy's spell that she will usually turn to Nancy for anything involving a choice or a decision. Nancy is a nice enough girl, but decidedly low-brow, and I suspect she is the source of Gaby's taste in books. My feeling is, if Gaby is to be strongly under someone's influence, I should much rather like it to be ours. This, however, is evidently not an urgent matter.

As regards Ronnie, he is perfectly happy and there would be no need to accelerate his return. He says himself he would like to come home, but this is the result of some subtle propaganda by Mrs. J. who

---

[1237] 'Journey for Margaret', 1942 film about London war orphan who crosses the Atlantic.

was rather shocked by his refusal, last year, to consider coming back. He is certainly quite happy where he is, and he is such a personality that he will develop in just the same way whether he stays here or returns. I think, however, that when he hears that Gaby is going home this may catch his imagination and he may want to come home too. At the moment he realizes there is just no possibility of coming back. Therefore, if we can get Gaby back, I think he ought to come too, if at all possible.

11th September

Darling, since I started writing this letter I had your cable and am perfectly thrilled with the news.[1238] Now more than ever I am quite mad about being so far away from you and impatient to see you again. I got, together with your cable, your first airgraph which reached Montreal in about 10 days but took, of course, a little longer to reach me here. It was very timely to hear what the doctor said. By the way I was delighted to find every word in your airgraph perfectly legible, you must have been working very hard over it. Don't say I never praise your efforts!

Cables are so inadequate and I am just dying to know how you feel and whether your long hours of work are beginning to be a serious problem now! I see no prospect of my returning in under 3 or 4 weeks, but there is no point in speculating about it because by the time you get this letter I shall probably have given you a more precise date by cable.

To return to the question of bringing the children back I have changed my mind a little since I started writing and I think it may be a bit premature. I am continuing however, to get on with their passports and, and with exploring all possibilities of travelling back with them, but this latter seems very difficult indeed.

Now for a brief repetition of news about the children in case this letter gets ahead of the others. I saw Ronnie at Lorne Park on August

[1238] Genia sent a cable on 11 September 1942 which reads: "Lovely holiday expecting Jessica doctor very pleased health home soon everything ok." The cable indicated that Genia was pregnant, but she had a miscarriage not long after.

28–29, then went on to Wheatley, stayed there until September 1, travelled back to Toronto with Mrs. N. and the girls, spent two or three evenings at York Mills, went on a shopping (and photographing) expedition with Gaby on the 2nd, spent the 3rd afternoon and the 4th morning at Lorne Park and the 4th afternoon again at York Mills. I therefore really had plenty of opportunity to see both children alone and in the company of others although not together.

Gaby is very good-looking, (I shall try to enclose with this some small photos and the proofs of the big one) she looks on the whole rather like you or perhaps like your mother. Her teeth are very regular, they look very big, but I think that is natural for the second set of teeth at her age. They certainly do not protrude at all but if anything she has a very determined lower lip. She has, however, inherited my nose, the poor child. She still has her long dark eyelashes and that gives her a lot of charm. Her eyes can sparkle almost like yours, but she can also look bored, and I did not manage to get her best expression caught on any of the photographs. She does not always sit or stand very straight, but this is not as strong as I suspected, and she can move with a lot of grace in running about or on roller skates. She is better on skates than Nancy and very proud of it. She went swimming in Lake Erie, and while neither of the girls can swim properly yet they are both very confident in the water and will dive and play tricks with great delight. Gaby has not yet the perfectly unconscious social grace that she used to have but that is natural at her age. She still has her old tact and can talk freely with adults if she has anything to say, but more often she has not. She was rather thrilled with my visit, and I noticed that she was rather fidgety, always chewing her lips, playing with her belt, etc. but I learnt from Mrs Naylor that that was not usual but was just excitement of my being there. It was not easy to get her to talk about herself but I did finally get her to talk. When I asked her whether she would like to come back, she said yes she wanted to, so I explained the transport difficulties and promised we would let her come as soon as possible. I said also that if she felt in any way unhappy and was really keen to come at once we would try and get something exceptional done and get her back without waiting for her turn. She immediately said no, it was not like that at all, and she was quite happy to stay. She must, of course, be very careful, because she is so tactful and would not easily ask for something that would be difficult to get, but I believe that she was speaking quite genuinely in this case. She did say again could she not travel to Washington with me and all the other places, but did not insist. On another occasion I asked her whether she was able to talk to Mrs. S. about everything or whether there was anything she could not talk to her about but would talk over with you or me. She said no, there was nothing like that at all, but there were, of course, things she would talk about neither to Mrs. S. nor to us. When I asked what sort of things she said "You know, little things that Nancy and I think out. I talk about everything to Nancy, of course!" This answer I found very reassuring. On another occasion I asked what she usually thought about when she was not sleeping etc. After a lot of leading questions she finally said she often thought out adventures that might happen to her and in which she was the hero.

If my psychology is right, this means she is something of an introvert, and if we got her back one ought to do quite a bit to get her interested in more active things. She is, however, very observant and she gets on with people and with other children extremely well. I heard compliments from everybody about her, very much more so than ordinary politeness requires. Of the 800 inhabitants of Wheatley (who of course all know Gaby and knew at once that I was her father) think at least half must have spoken to me to say what a charming daughter I had.

The Naylors are very charming people. Mrs N. was a trained nurse, and a school teacher at some time and is a very efficient person. She runs her house without help and without fuss and somehow despises Mrs. S. for her helplessness. She told me the whole history of the Jephcott family which is very interesting.

This letter must, I suppose, end somewhere so I better turn to Ronnie. Ronnie has now all the social grace that Gaby had when she was small. He talks very intelligently and is still the same philosopher. When I asked him was it not an enormous piece of good luck that he got into the house of such charming people he agreed but added: "But I suppose if I had gone somewhere else far away and never known the Jephcotts, we might have thought, you know ..." The sentence did not end, but the idea was clear, and of course this does not mean that he is not extremely fond of the Jephcotts. At Lorne Park there is an enormous number of children, dogs, cats, etc. and they all drift about from one house to another, the Jephcott house being particularly popular, except when on occasion Mr. Jephcott gets fed up of trying to work in the din and throws them out. Ronnie was rather afraid he might not recognize me when I got out of the bus. To the great joy of the family the only people who got out of the bus were an old coloured laundress and I, so he had no trouble of picking me. He still has his old appetite and still is asleep every night when his head touches the pillow. He often wakes up early, and is then allowed to read in bed, and he is very fond of books. In fact, in talking to him about birthday presents, I could not get him to express any other wish than books. He is also extremely clever about putting things together and cardboard models that have to be assembled are his greatest joy, provided they need no great physical skill as his hands are still rather clumsy. He is extremely good at school, I gather, but rather slow, and that is why he does not show up well in exams. He knows his own mind and if Ann who is the bully of the family, will insist on doing something, he does not like, he will let her go off and stay by himself. By the way when talking to the Jephcotts I got the conversation on the Jewish question and they said that Ronnie knew very well that he was Jewish, as of course did Gaby. This rather surprised me because I had not realised that Ronnie was capable of understanding this when he came out.

By the way, I have the Moulton school magazine, and since I do not know if I can take or send it here is a poem by Gaby which was printed there:

Last night when I'd counted thirteen sheep
All of a sudden I went to sleep.
And do you know what I dreamt of all night?
Sheep, here and there, on left and right!
The bed had the head of a sheep soft and nice
The end had a tail that wound around twice
The blankets were made of nice sheep's fluff –
Fuzzy all over, but not at all rough.
I began to admire the tail and the head,
But the next thing I knew I fell out of bed.

I found myself lying right on the floor,
And the bed was not made of sheep anymore.
I gather this was edited slightly. It seems to be her first and last poem so far.

19.9.1943

My postman is leaving only now, so this letter has not gone off and is mostly obsolete. One thing that has come out very clearly these last few days is that the children ought to stay definitely where they are for the present. (not that it would be possible to bring them back very quickly if we wanted to!)[1239] We have been very busy lately and are very tired, but on the whole pleased with our work. I think I shall be through by about the 29th and shall then come home as soon as I can get a passage. This may, of course, take some time so do not worry if I do not turn up right away. It is very clear that, when I get back, there will be very many things to be done in a very short time and I will have to do a great deal of running about to do. I do not know how many weeks I shall have at home altogether. So will you please get plenty of sleep before I get back — I do not know how much you will get when I get there. There will be so many things we will want to talk about! Now that I can form a definite idea when I shall get back, it still seems such an enormous time and I do not know at all how I shall survive the next two or three weeks. I so much want to get home to you.

Else has got her visa at last to visit her parents. She will give up her apartment here and go up to New York next week so I shall meet her there again. In another week or so she will go for her six weeks' vacation out west.

I spent a few very pleasant hours with father and hope to see him again next week. He is very interested in everything and thinks very clearly but is very weak physically. I have not seen Annie or Hermann yet, except for a moment when they all collected at father's house to inspect me. I spent a very amusing evening in Washington with Rose

[1239]By the end of his visit to the US and Canada, it had become clear that Peierls would be continuing his war work on the other side of the Atlantic, which meant that a return of the children to the UK was no longer a desirable proposition.

Goldstein. I shall be back in New York tomorrow and shall stay there until I leave, except for the next weekend which I hope to keep free to go to Toronto. I have not yet got Ronnie's picture and I cannot come home without that, of course! Even Simon says he cannot go home if I have not got Ronnie's picture! Love to Vera and the twins. Do you know how I love you?

Rudi

[366] Rudolf Peierls to Klaus Fuchs

Montreal, 20.9.1943

Dear Fuchs,
You and Wilson must be getting pretty impatient at not getting any message about what we are doing here. You will have guessed that the reason why I have not so far given you any definitive information is not due to any lack of good intention on my part. Even now I cannot tell you very much because things are very largely in the melting-pot.

I can say, however, that very definite progress has been made which will lead to a satisfactory arrangement which will enable work to go on efficiently. In the plant problems my advice for the present is to continue everything that is in hand, I think everything we are doing is going to be useful at some stage, although some reorganization may become necessary. Simon and I have got full information on the proposed scheme here, but it is not certain as yet what changes in the scheme are practicable at this stage and therefore it is premature to map out any plan for this national work for this purpose. It is also not settled as yet what changes, if any, will be decided on in the programme of the British diffusion project.

On the other side, I have found, as expected, that the Oppenheimer group have made very considerable progress. They find that in the efficiency problem, radiation becomes of importance at a much earlier stage than was apparent from their previous data, and it has therefore more influence on determining efficiency than we believed. They have not done much on Miss Gray's problem.

Their method is to solve the equation for $\gamma = 1$ which is [?] and then to expand in a power series as regards $\gamma - 1$ This leads in first approximation to a manageable linear equation and they claim that this is of sufficient accuracy. I do not know enough details to form an opinion, but am somewhat doubtful. They have used for this purpose a smoothened equation of state, apparently they never received our report on the equation of state. I think Miss Gray's work, if satisfactory, should go on, if only as a check on their method.

I found that the spontaneous effect is somewhat larger than we believed, and hence questions of shape [ha]ve become more important again. They have nothing corresponding to the model [?] or the method of chords.

On the question of rate of multiplication with non-uniform density they have based most of their work on the differential equation of diffusion (infinite medium eq[uation]) and they claim to have established by using an iteration method in special cases that this is extremely good. Again I have no details, but feel lots of doubts.

I hear that a letter for me has arrived in Washington, which I imagined may be from you, I shall get this tomorrow on returning to New York. I shall spend most of the remaining time in New York. Simon and I have asked for the first available passage on or after 29th September, but to tell from this when we shall arrive needs evidently quite unjustified extrapolation.

All the information in this letter (with the exception of the last paragraph) scanty as it is, is for personal information of yourself and Wilson only, and in particular nobody outside the Birmingham groups in any circumstances.

With best wishes to all of you.

Yours sincerely,

R.P.

[367] Francis Simon to Rudolf Peierls

Oxford, 25.4.1944

Dear Peierls,
I believe you have been informed through our letters to Kurti about the work which has been done here and to-day Kurti himself has written you a very long report. However, you would probably like to know more about the general position and the ideas governing the work in England.

By necessity the situation could not have been very clear before we knew about the developments in the other fields, and particularly how the liaison over there was going to work. It was made still more vague by Oliphant's proposal, which he brought up during the few days he was back in this country, for the design of an improved electromagnetic plant. While it may well be that something on the lines which he expounded will be the final solution, it is just as probable, if not more so, that some snags may turn up and make the whole thing unworkable. If his suggestion is to play any part in our considerations, then a prototype of his unit should be built with very high priority. It does not seem impossible to build a solenoid in this country, but the development of the more important parts will have to be done over there, in addition to the work for the present plant, and I am not too hopeful that we could get any definite reply within say a year or so. Nevertheless a very optimistic attitude remained here after his visit: "Why worry about other methods of separation; the modified electromagnetic field one will do the trick!"

To give you my impression about the situation, perhaps it is simplest if I enclose a copy of a letter I wrote to Akers shortly after Urey's visit.

Not long afterwards, we had to take a decision on the policy of membrane development, and I wrote about this in my letter to Kurti of April 1st. The general idea is that very probably we cannot be of any short term assistance to the Americans (of course I exclude the work that you and your group and Kearton[1240] are doing over there). I believe,

[1240]C.F. Kearton, initially from I.C.I. at Billington had moved to America as resident technical liaison officer with Kellex in March 1944.

however, that we can very likely be helpful when the Americans have run into trouble after they have started up their plant. For instance, in the membrane field, I think we should pursue vigorously the loaded type of membrane, including also aluminium backings. If some unforeseen trouble arises with the present K.1 and we then are in a position to offer the Americans a fully developed membrane of this type, this might make all the difference. Intimately connected with all this is, naturally, the development of flat membrane assemblies. These two things would of course also be essential for a future British diffusion plant. If we also look into the question of gas bearing compressors, we would have all the essentials well developed if a British diffusion plant were after all to be built after the war.

A final decision about these points has however been postponed at the last moment, as — after some exchange of telegrams with Chadwick — Akers and Perrin favour the idea that the final solution for us would be the 94 or 233, and that should concentrate on these and stop all the work on the separation of 235, except to the extent to which it should continue for the Americans. They also believe that if it should turn out that one has to revise this decision, if for instance these two substances could not be used as a weapon after all, one could still switch back to the separation schemes without having lost too much time. It may well be possible that this is the right attitude, but my point is that we have not enough information here on this whole rather complex matter to be able to take a decision on such fundamental importance right now, and that the Technical Committee should meet in order to discuss the existing information. You will have heard that it was decided that Akers, Perrin and myself should come to Canada and meet you all there. Although it is now more than a week, we have not yet had a reply from over there, but I still hope that a visit will come off and that I will see you before this letter arrives. As, however, the reasonable things are not always the ones which take place, I would not bank on this, and am therefore writing to you.

A few other points:

a) Pickavance: Kurti has written about him already, and therefore I can be short. I visited him in Liverpool and he came to see us not long ago. I had a good impression of him and his collaborators. As he is new in this field, however, I do not think he is wrong when he says that he would not expect his mass spectrometer to reach the present standard of the performance of the Americans in less than a year or so, and even then we would still necessarily lag behind them. Whatever line we are going to follow, the mass spectrometer will always come in as an essential tool, which we must have at our disposal. You will remember that we had agreed at the last Technical Committee before our departure from the United States that Pickavance should come to the United States and learn and get familiar with Nier's technique, and I must say that I am rather depressed about the fact that there must obviously be some resistance on the American side to this visit. As his visit to Columbia would not interfere with the American work, I can only interpret their reluctance to agree to this visit as a very bad omen for future collaboration. [handwritten addition]]26.4. I just see in London that Pickavance is going out soon. I am very glad.

b) Tolansky: I assume that Urey has informed you of his visit to Tolansky. You will see from my letter, that I do not give any chance of success to a photochemical method of separation, but I understand that the knowledge of the nuclear spin of 235 will be of the interest to the people in Y. In order to get a really sound result from his experiments, it is necessary that he gets a small amount of enriched material, and Urey promised to take steps to procure this in the States. Perhaps you could enquire from Urey how far he has got in this matter. In the meantime Tolansky is preparing a method to determine the spectra of X without using more than 100 mgrs or so.

c) Thermal diffusion: Have you access to Cohen's manuscript on the experiments on the thermal diffusion in the liquid? If so, could you please let us know the essential data, as it may prove a great help to Arms, who is starting on such experiments.

d) Some progress has been made in the technique of oxidation as a method of stabilising nickel barriers, and it now looks very promising. Jones is just writing a report for which Emmett asked, and it will be sent on in a few days' time.

With best wishes to your family and to Kearton,
Yours sincerely,

F. Simon

[368] Rudolf Peierls to A.H. Wilson

location unspecified, 3.5.1944
(SECRET)

Dear Wilson,
I have just come back from another visit of four days to Y and am writing to give you an account of some of the things that I learned there and that have a bearing on problems you are working on. As before, I cannot write on certain interesting aspects of their work.

1. Critical Size
   Apart from the methods mentioned in my earlier letter, progress has mainly consisted in a perturbation theory developed by Serber. This is similar to Dirac's perturbation theory for the expanding boundary layer on which you will have seen my last letter to Dirac.[1241] The way in which Serber wants to apply this theory is to problems of odd shapes such as finite cylinders in which he believes it is a good approximation to replace the body by one in which the density is averaged over any spherical shell above its centre. He obtains, in this way, a problem of spherical symmetry but with non-uniform density and he replaces this first by a problem of a sphere of uniform density and then applies perturbation

[1241] Dirac, Fuchs, Preston and Peierls had collaborated on this problem. See their paper on the 'Approximate rate of neutron multiplication for a solid of arbitrary shape and uniform density: Part II application to the oblate spheroid, hemi-sphere, and oblate hemi-spheroid', NA/PRO: AB1/963.

theory. He tried to test this method for the case of a cylinder in which the length equals the diameter, for which an accurate solution is known from variation theory and he found that in this case the unperturbed case was already so close to the actual value (better than the numerical accuracy of either calculation) that he could not in fact test his perturbation theory. Personally I doubt very much whether his averaging over angles does not introduce as big an error as the assumption of uniformity and I advised him to test this point by trying the method for an extreme case such as a very long cylinder or a very flat slab.

2. Efficiency

Work on the small expansions is very similar to what Dirac has been doing except for the small differences in connection with perturbation theory. They had not at first noticed the point made by Dirac and confirmed in Davison's MS99 that the thickness of the shock wave in the container is negligible. They used instead an elaborate procedure integrating the hydrodynamic equations both for the container and inside and fitting them by a joint boundary condition. The result showed, as was to be expected, that one might as well use Dirac's form for the condition.

The numerical integration of the equation for the interior was done by using the equation in Fuchs' MS 63 which avoids the assumption of a rapid variation of the neutron density in the expansion wave which is employed in Dirac's form of the equation. In the numerical work a trick was used by means of which one obtains a more convenient expansion of the equation which holds over an appreciable part of the range and not merely in the immediate neighbourhood of the singular point. In fact, Fuchs has used another trick which involves getting a first integral for the equation, but this was not reported in MS 63. It would be of some interest to obtain Fuchs' notes on that which are contained in a typescript headed "Appendix to MS 63" and of which only a single copy exists amongst Fuchs' papers of which I believe he explained the general arrangement to you. If you would be good enough to extract this document and send it over to us, it would, I think, be of interest to the Americans.

3. Equation of State

I discussed with Bethe, the equation of state of the metal including the point stated in Davison's latest MS 108. I knew from the progress report that Davison was working on this, but the actual paper with the results has only come in after my return to New York. Bethe appears less afraid than we were of assuming constant γ for the following reason: At low temperature, the specific heat is small, but rises rapidly in the ionisation range eventually to fall off again and it is clear that if one plots the curve of E against T there must be a point at which the tangent goes through the origin. Near that point the assumption of E proportional to T is evidently a good approximation. They have done some numerical work on the equation of state and find that over a very large range, i.e. the range over which ionisation takes place gradually, the curve remains very close to the tangent through the origin. The slope of this tangent corresponds approximately equals 1.4. At still higher temperature, the specific heat would decrease so as to increase γ to 5/3, but there are two reasons why this is of no practical importance. The one is that the actual temperatures reached do not go far from the region where γ is 1.4. The other is that at these temperatures the heat capacity due to radiation begins to come in and again increases the specific heat and it more or less compensates the decrease.

As a result, Bethe believes that the assumption $\gamma = 1.4$ is an adequate approximation for the whole range.

4. Large Expansion

On the case of the later states in which the expansion wave was penetrated appreciably and in particular the stages after it has reached the centre, practically no work has been done other than the simple theory assuming uniform pressure inside of which I gave an account in my earlier letter. They are not at the moment planning to tackle this problem and the question was raised whether our group here might later, when immediate pressure of work on our present project subsided, make an attempt to discuss this problem. I do not yet know whether and when we may be in a position to undertake this.

5. Diffusion Approximation

   As I stated before, the simple theory for large expansion assumes diffusion theory for the multiplication in the expanded sphere. Serber is now proposing to make calculations using the integral equation in order to check the validity of the earlier work.

6. Shock Wave in Air

   I explained earlier the method Bethe had thought out for the case where γ is near 1 and the perturbation theory resulting from this idea. No work has yet been done on this beyond showing that in the case of the Neumann theory already the first approximation gives a very good result. It is now proposed to get Chandrasekhar who will join the team shortly to apply this method to the real case, based I imagine on our earlier work on the equation of state for air.

7. Partial Differential Equations

   I reported previously that they were struggling with a partial differential equation which was of the same general nature as the one tackled by Miss Gray, but in many ways easier. During my last visit, I suggested using our numerical method and this was done utilising a set of Hollerith machines (here called I.B.M. for International Business Machines) for such purpose. It turns out that they are extremely suitable for such work at least in the case arising at Y. With a problem containing about 20 space intervals, but only elementary boundary conditions and no discontinuity they were reaching a speed of half an hour to an hour for a complete cycle. An extension of their problem will introduce a discontinuity of the shock wave type, and they are now struggling with the problem of finding a suitable method for doing this. Their problem is easier than ours insofar as they can afford to take smaller time intervals, their total time range being rather short, but it is more difficult because the medium is in motion on both sides of the shock wave and they therefore lose the advantage of the density, pressure and velocity ahead of the shock front being independent of position.

   Von Neumann has made an interesting contribution by pointing out that if one ignores the discontinuity and takes the difference equation rigorously, one will obtain an instability which will make itself felt as an irregular motion behind the shock front. This simulates the increase in entropy following the shock and may in fact be interpreted as a molecular model of the heat motion behind the shock wave only with the Avogadro number of 20 or so instead of its real value. Neumann has tried this in a simple case in which the answer is known and found that the method works very well. I do not, however, believe it to be of practical value because by making such a simple model, one assumes in fact a wrong equation and one must include very complicated forms in the equation to put this right and also because the highly irregular motion forces one has to use rather smaller time intervals (by a factor of 2) than the stability of the equation would allow. The method has, however, the advantage that one can use the Hollerith machines right through without stopping to solve boundary conditions at a discontinuity.

This, I think covers all theoretical problems on which I have to report. The list is not complete (a) because I cannot report on some of the items and (b) because there was not enough time to discuss all of them.

I am not very well informed on the latest position concerning the experimental data, but I think it is fairly certain now that Bretscher's results on the cross sections are still the best available since some higher values found at Y have recently been reduced. The number of particles per fission tends to go up above the value originally quoted and may be as high as 2.5.

Measurements on the energy distribution of the emitted neutrons still do not agree with the relationship outlined in my MS 65. The deviations are in the same direction as the older Liverpool curves. There is no satisfactory explanation for this deviation. I have discussed with Weisskopf the possibility of the neutrons being emitted before the fragments separate which I think is not very likely. Another possible explanation is that the fragments are really rotating about an axis correlated with their direction of motion and that this causes an anisotropic distribution, but it would also be hard to understand the facts on this basis.

As a matter of general interest, you might like to know that the suggestion has now been made that I should go out to Y as soon as possible to take charge of the theoretical division dealing with one of the problems on which I am not able to report to you, but a decision on this depends on many factors which are not as yet certain, including the future of the work in New York. I hope that the impending visit of Akers and others will lead to a clarification of this point. In this connection it would be of great help to know your views, about whether or not you require Davison to remain with you. If he is to continue to work on chain reaction problems, which judging by his reports appears to be his main occupation, I believe his best place would be at Y provided he would be acceptable there, of which I am not yet sure. I appreciate, however, that you cannot make up your mind on this before the future of the plant work in England is settled and that is not the only reason why I am extremely impatient (as no doubt you are) that this question has not yet been decided. I would, however, be very grateful to have even an interim report from you on this question with your likely reaction in various possible cases.

This seems a good opportunity to give you a brief summary also of our work here but for convenience of circulation, I shall do this in a separate letter which I hope will go out simultaneously with this one or a day or two after it.

With best regards from all of us here and to the whole Birmingham team.

Yours sincerely,

Dr. R. Peierls

[369] Rudolf Peierls to Chadwick

location unspecified, 19.5.1944

Dear Chadwick,

We have drawn up a list of names. The following are extremely likely to be concerned with work in the direction mentioned — if any such work is going on.

1. W. Heisenberg. As you know, he is in charge of the Institute in Dahlem at which research work would most likely be located.

2. K. Wirtz — at the same laboratory. He has studied the theory of thermal diffusion, particularly in liquids. He has published papers on this, but not recently. There is evidence that he has been interested on all aspects of the T.A. problems.

3. Waldmann has published papers recently on separation methods and he has partly collaborated with Wirtz on theoretical problems.

4. It is of course likely that any major effort would have some connection with I.G. Farbenindustrie.

5. I need not say that knowledge of movement of large quantities of oxide would give useful information.
   The following names are less certain but might be connected with the work:

6. M. von Ardenne. His laboratory has recently developed high voltage work but was also evidently interested in T.A. problems in general.

7. Houtermans, who works with Ardenne, has published papers on separation as well as on nuclear problems, and he is quite possibly mixed up with such organisation.

8. Harteck is, or was, active in the T.A. field. He was concerned with polymer and may still work on that line but he must obviously have contact with any central organisation in the T.A. field and, if polymer has been abandoned, he might be concerned with other methods.

9. Jensen worked with Harteck and the same applied to him.

10. Walcher has written some time ago, a critical review of separation methods. He evidently knows a fair amount about separation and might have been brought in.

11. In connection with production and properties of hex, the Department of Inorganic Chemistry of the Technische Hochschule at Breslau is likely to be concerned as the substance has been made there first and substantial work done on it. ...[1242]

[370] Rudolf Peierls to Wallace A. Akers

location unspecified, 26.6.1944
(carbon copy)

Dear Akers,
I am writing to amplify the remarks made in a cable which I have asked Webster to forward to you today through his channels. I shall deal with the points in the same order as in the cable.[1243]

1. The general experience in work of this kind that will arise from the proposed diffusion project is that the person who has spent some years in thinking about the nature of such a plant as the relations of its various parts and the factors which are relevant or otherwise to the design is a very important asset that is hard to replace by other means. The amount of further fundamental theoretical work to be done on such a plant would no doubt be small and what there is could be done by fairly junior people with suitable guidance, but this in my opinion will not obviate the need for getting somebody like Fuchs in a position where he can exercise influence on the design work. This, I think, would [be] particularly important during the early stages when the fundamental features of the design are settled, and if the decision as to whether or not the project will go ahead is to be based on a detailed estimate, the period of preparing such an estimate is one during which his presence will be particularly important.

[1242] Second page of the letter missing.
[1243] Cable could not be located. At the time, discussions about the post-war future of British nuclear research included the possibility of a low-separation plant, and it was contemplated to ask for the release from the joint Anglo-American project of some key British personnel. See also letter [372].

I would like to make it quite clear that I imagine Fuchs would add to and not replace the contributions made by Bosanquet. I am sure that it would be of greatest advantage to have Bosanquet associated with any such project on a full-time basis, but I believe that with Fuchs and Bosanquet the chances of success would be very much higher than with only one of them.

It is true that if Y and this new project were to compete on an equal basis, Y should have priority because it is an immediate war project, but while the people out there were very keen to have Fuchs and while I personally would very much prefer him to join me out there, there is a theoretical team of considerable size working out there and his presence or absence there will be very much less decisive factor than on the LSD project.

I could not, of course, discuss the problem with Bethe but I did explain that Fuchs might be wanted at home and I don't believe that he will make a case so strong that it would be impossible to refuse.

All this, however, depends in time on the attitude of Kellex, since Chadwick is very insistent and I think rightly so, that one should avoid at all costs any impression that we are trying to withdraw from K-25 any assistance they may still consider desirable. I shall see Benedict on this matter to-day and if this leads to any new information I shall let you know by cable or letter. I feel certain from previous conversations with Benedict that he will agree that the usefulness of a theoretical team here is coming to an end. The difficulty is merely to make sure that this statement travels through the right channels from him to Keith, Felbeck and Groves, so that the latter will also be convinced that it is reasonable to discontinue our theoretical group here.

All this really is trying to answer your questions, namely, what would be the best way of ensuring success of the British LSD project if it were to proceed. On the other hand, it is also quite clear that if it were decided not to proceed with this scheme, then in my opinion, the best place for Fuchs would be at Y, although of course this depends on what kind of case Oppenheimer will make out to Groves on this matter. As long as the fate of the LSD project has not been decided, the question is obviously a difficult one and depends on one's estimate of what will happen. It is quite impossible to judge the prospects from here and in that case I can only pass the decision back to you.

One more thought that might help is that if the decision will only be taken on the basis of a month or two's concentrated design work which is starting now, one may consider calling on Fuchs for this where he will be very valuable, and in the event that the final decision will be against continuing the work or for continuing it at low pressure, one could again reopen the question of his transfer to Y. It is quite possible that that might make it too late — that in the meantime they will have taken on other people at Y and will not claim Fuchs in these circumstances — but this is a risk that is worth taking, provided the proposal of an LSD plant is really taken seriously.

2. Skyrme's case is much easier because we have not got a three-cornered problem. The difficulty remains that he should not be moved before the Americans are satisfied they have no further need for him in New York, and while the problem has not been formally raised it appears likely that nobody will raise any objection to his going to Y where I feel he would certainly be most useful.

3. You will have seen from my cable that the situation as regards Davison is somewhat complex since Placzek is in some doubt about the nature of the work his division will do in the future. The object of my question was therefore in the first instance to ascertain whether, from the point of view of health, Davison ought to go to Montreal or to Y. My impression is that Y would be ideal for him from this point of view, although I have heard that in some instances there are difficulties on returning to a low altitude. If, for example, it should turn out that it would not be wise to send him to Montreal, we need not worry further about that problem and it would then be a matter of Chadwick finding out in conversation with the Americans whether they would raise any objections on account of his past.

4. The question of Wilson's further programme is one on which, naturally, I could not have any definite views from here. It would be a pity to lose him from the project and my impressions of the remarks made at Washington were that he should transfer to nuclear physics in the event that our suggestion to strengthen the nuclear research in England is accepted, as they will then require some theoretical help. I am not clear whether this strengthening of the experimental research will in fact take place, but no doubt you will bear this factor in mind.

It is also fairly evident that until the question of Fuchs' return is definitely ruled out we should keep together the computers and junior staff now available, since they will be very valuable to Fuchs, if he returns. I may say that after our experience of trying to recruit and train computers here, the team at Birmingham has risen considerably in our estimate. It is, of course, quite feasible that Wilson would want to keep these people on in any case and, if so, the question of how to share them will have to be discussed between him and Fuchs when it arises.

Lastly, I would like to remark on the general nature of the planning and design work for the LSD scheme. The view taken at the Washington meeting that this is a fairly simple scheme was based mainly on specific ideas which I explained in a letter to you and which were derived in conversations between Simon, Kearton, Fuchs and myself. I believe that the latter three particularly have experience in the kind of flowsheet consideration that forms the basis of such a scheme and it may be inefficient to base even the estimate as to the feasibility of the scheme on work in which none of us have taken part. This forms part of the reason why I think it will be important to have Fuchs return as soon as he can be released from New York, but I also believe that Kearton ought to be brought into this unless there are strong reasons which make this impossible. This is, of course, a delicate matter since it is clearly impossible to withdraw Kearton against a strong objection on the part of the Americans, and I am not in a position to have any views as to the best way of ascertaining their reaction or what attitude to take, but clearly the first question to decide is whether in the event of the work on LSD going ahead you want Kearton back, and I think it very important that a decision on this point should be made as soon as possible, and that Chadwick should know the answer.

I have been very satisfied with the first few weeks spent out at Y. It is certainly an enormous pleasure after six months in New York to be at a place where no distinction is made between members of the American and the British organization and where work is guided by the necessity to get the best answer in the shortest possible time, rather than by questions of formal organization and prestige. I have learnt a great deal in those weeks and have in some measure been able to contribute to the work going on there. Naturally the place is not ideal — there are obstacles in the work as well as difficulties with regard to the living conditions — but both remain within reasonable limits and do not make it impossible to carry on useful work, chiefly because the technical people know what they want and carry enough weight to prevent the red tape growing beyond the limits that would stifle the work.

With best wishes to everyone at Old Queen Street.

Yours sincerely,

R. Peierls

[371] James Chadwick to Rudolf Peierls

Washington, 14.7.1944

Dear Peierls,

I have now had talks with both Kearton and Fuchs about the future of the New York section and in particular about their own positions. As a result, Kearton will approach Keith and Benedict with the object of getting a letter by one or both of them to Groves to say that the services of Fuchs and Skyrme are no longer required. It is possible that this matter was raised by Groves on a visit to New York earlier in the week, but I have had no news from him so far.

The position of Skyrme is quite clear. Bethe or Oppenheimer should write to Groves asking for his services in Y. Groves has provisionally agreed and there should be little delay over his transfer.

Fuchs' future is not so clear. I gave you the gist of a cable from Akers in my letter of July 11. I did not agree with the suggestion made in this cable that Fuchs was not required in England, but I wished to discuss the question with Kearton before I made up my mind. Kearton was very strongly of the opinion that Fuchs was quite necessary in England if work on any kind of diffusion plant is to continue ...

I have now had a talk with Fuchs himself. He feels that he has a special contribution to make in England, whereas in Y he would be one of a number and can make no really significant difference to the work.

I agree completely with these views of Kearton and Fuchs, and I feel sure you also agree at least in principle.

I come now to the point of this letter It would put me in a very awkward position if a request for Fuchs' services in Y were to be sent to Groves. If Groves were to agree I also should have to consent, for the consequences of refusing, on the grounds that he was needed in England for work which can have no significance for the war, might be quite serious. It would certainly cause great resentment in some quarters and our relations with the U.S. on this project would be impaired. I should attempt to justify his return as being useful for the New York project, for after his experience here he could interpret their requests and help to direct U.K. work into directions of immediate interest to them. This argument would of course not be valid if a low-separation diffusion plant were to be started in England.

I therefore do not want Bethe to ask for Fuchs. Further than that, I want Bethe to say that Fuchs would not be specially useful in Y, if Groves asks if they want him, as he may. This means some tactful work on your part and I hope you will be able to do what is necessary by suggestion rather than direct action.

I have prepared the ground here and I think the matter can be arranged. I have stated that Fuchs could be useful in Y but that his special qualifications are not on the nuclear side but on the diffusion plant.

Until I know something of what is happening in London I want to keep the New York position as fluid as possible.

Yours sincerely,

J. Chadwick

[372] Klaus Fuchs to Rudolf Peierls

New York, 17.7.1944

Dear Peierls,
I saw Chadwick on Friday and discussed with him the future of Skyrme and myself.

As regards Skyrme, Chadwick said that it would not be necessary for him to take the matter up personally with the General, since it had already seemed settled in principle — all that was required was a letter from Keith and Felbeck stating that Skyrme was no longer required in New York, also it would be necessary to make sure that the request from your place had gone through to the General's Office. It would probably take 2 or 3 weeks to get this settled.

My own position is rather more difficult but I think I had better leave it to Chadwick to tell you about our discussion.

Kearton is going to ask Keith and Felbeck to write the required letter about Skyrme and myself as soon as he manages to see them.

Our work here is practically finished. Skyrme has concluded the work on the effect of the pressure oscillations arising from the reciprocating pump, and finds that the effect is negligible. I finished all our outstanding problems and I am just writing up the last of the MSN-series!
Yours sincerely,

K. Fuchs

[373] Rudolf Peierls to Raymond Priestley

location unspecified, c/o W.L.Webster, 20.9.1944
(carbon copy)

Dear Vice-Chancellor,
It seems ages since I disappeared from Birmingham and now that Wilson's team has left, too, I feel that yet another link has been broken.

So I am writing this letter mainly in order to say that I am still alive and longing to come back.

The news is so good now and events move so fast that war in Europe may be over quite soon, and the time will soon come when the university will want to put into effect a part of its reconstruction (or as they say here, reconversion) programme, and will have to cope with the crowds of new students. I would give a lot to be sure of getting back in time for this. Apart from everything else, it is the best time to get new teaching programmes started and to get my department the way I want it. Unfortunately, however, I am now on a very short rope, and the date of my return does not depend on me. I cannot, of course, say much about my work here, except that it is war work and it is a question of policy whether for the war in the Pacific, this kind of work is to be given a high priority.

If I cannot be released immediately after the end of the war in Europe I shall try to get permission to return at least for a short visit, but even that will depend on the state of the work at the time.

In any case, I hope you will appreciate that I am not staying out here just for my own amusement, much as we enjoy the climate and the country of the American South West, to which I have moved from New York this summer. We find it very hard to sit in such a nice spot and to read about robot raids and other troubles at home. I still think I am incredibly lucky to have all my family together again.

Probably the return of quite a few members of the Edgbaston staff will raise similar questions to mine. One hope is that it may be much easier, since after the end of the war in Europe, the Pacific war may take much less time than we used to think, the progress made in the Pacific now is very spectacular, and the losses of planes and ships in the invasion seemed comparatively light, so there will be plenty of equipment available against Japan, as soon as its transport is organized.

With very best wishes,

Yours sincerely,

[Rudolf Peierls]

[374] Raymond Priestley to Rudolf Peierls

Birmingham, 25.10.1944

Dear Peierls,

We shall be very sorry if your return is delayed after the end of the German War but I personally shall not put up any barrage if you are wanted for the Pacific War. I know that you will let me know if you are of the opinion that the case for your retention is not really a good one. If it is we must put up with it and let you stay. I am sorry I did not see any of you when I was in the States this summer but I was only in Washington three days and I had not the faintest idea where you all were for secrecy has been well preserved.

We are already conducting negotiations about ways and means of getting staff back and I think procedure will be fairly simple as regards men and women in the Services or in civil occupations over here.

The University goes along quietly enough though there is more than enough work to do and everyone is getting a little jaded and some people a little irritable. That is inevitable, I fear, after a long strain like this has been. I agree with you that there is a good prospect of an early end to the Pacific War but we have just had a lesson here where the German War shows every sign of dragging on into the next summer though many of us were convinced that with ordinary fortune it might have been settled this November. We have been most unfortunate in this autumn's weather. I have vivid memories of October–November 1918 when we campaigned in France in weather that was perfect. I had many a gallop across the downs in the Cambrai area that would have been very hard going this autumn.

I am glad Mrs. Peierls and the children and yourself are all enjoying the country and the climate. It is delightful I know from experience and I hope it will not be too long before I pay another visit myself.

We are getting on with development plans and already have promises of help totalling about a quarter of a million pounds. I am not really much worried about the money aspect but I can see national building priorities holding things up until after I have had to retire if we are not fortunate.

All good wishes,

Yours sincerely,

Raymond Priestley

[375] Rudolf Peierls to James Chadwick

location unspecified, 28.3.1945

secret

Dear Chadwick,

Last night Fermi talked in the colloquium on the future of T.A. work and possible applications. He made the nice point that everybody who had been asked about non-military applications of fission had been firmly convinced that this was very important but that nobody had produced any clear idea about the specific applications. In this he included himself.

He mentioned some figures about power production that may interest you. He quoted the total amount of uranium in deposits of reasonable yield as 20,000 tons. If one takes the total electric power used in the United States as 5×10^7 KW, this would equal the energy obtained from burning all the 25 contained in the above amount over ten years. Since the conversion of the heat produced in piles to electricity will be inefficient the actual figure would be rather more like one year. If it was possible to burn the 28 also via 49 the figure would go up by a further 140, but it is clear that to replace existing power sources entirely by piles is out of the question.

This leaves the use of power in circumstances where the weight of fuel was a very important factor, and he quoted as a somewhat fantastic possibility the problem of sending a large rocket, say, to the moon; he did not quote somewhat more realistic cases such as power stations in very inaccessible places, say for an airfield near the north pole.

He then discussed briefly the possibility of a self-regenerating pile in which more 28 was turned into 49 than needed to replace the 49 burned up, and he seemed to regard this problem as soluble, though not yet solved. He dismissed the use of thorium, merely with the remark that too little was as yet known about the physical constants involved.

He mentioned briefly the applications of radio elements for medical and other purposes, and the use of new isotopes as tracers.

He lastly discussed possible contributions of piles to fundamental physics, but described only two experiments which he himself regarded as impossible, one being to observe the beta decay of the neutron by having an evacuated vessel inside a pile and searching it for hydrogen after a long exposure. The other concerned finding effects produced by the neutrinos emitted in the beta decay taking place in the pile.

At the co-ordinating council Weisskopf spoke on measurements at Trinity on which he has lately been acting as a consultant. He did not deal with the measurements of blast and other mechanical quantities but covered only the nuclear aspects. Of these the delayed gamma rays would seem a particularly attractive possibility, and as you know Moon is actively engaged in preparing such a measurement. The difficulties are that in the early stages the gadget material is still at too high density and its absorption not negligible whereas at later stages soil thrown up by the blast may interfere with the measurement. Delayed neutrons are another possibility, but there are two difficulties; one is that the prompt neutrons are slowed down on the H.E. and it will be some time before their intensity has dropped below the level of the delayed neutrons. The other point is that the scattering in air and reflection from ground require a lot of theory and make the observation somewhat indirect.

He also discussed the recovery of samples of soil containing some of the remaining 49 with fission products. The difficulty will be to make sure that the extent to which different elements are retained by the soil may be very different. However, if one looks for a number of different fission products giving weight particularly to those which contain no volatile element in their family tree, one might overcome this trouble. In the discussion the possibility was suggested of mixing with the gadget material a small amount of a stable isotope of a suitable fission product. The proportion of this found alongside 49 would then given an indication of how much of the radioactive isotope would have been retained.

Lastly, he discussed an experiment which looks of prohibitive difficulty but which Bob Wilson believes can be done. This is to measure the neutrons found during the first fifty generations (about 0.2.5 microseconds) while the nuclear reaction has not as yet affected the motion of the material. If it were possible to measure the multiplication rate i.e. the logarithmic derivative of the neutron number with respect to time, this would give a good opportunity to test in the event of poor efficiency whether this is due to a poor implosion to the initiator operating at the wrong instant, or to faults in the theory of efficiency.

A more extensive report on tests at Trinity will be given by Bainbridge next week, and I hope Moon will then be able to give a more adequate account — he is now on a visit of a few days at Trinity.

Very encouraging results have recently come from Froman's group. He has measured the velocity of a thin steel shell inside a two to one aluminium hemisphere imploded by an explosive system which is a scale model of the final design on one-third scale, except that it uses half a twelve-lens system in place of 32. The calculated initial velocity of the steel is 5,000 meters per second, whereas Froman finds 5,200.

Similar measurements with tube alloy gave 3900. The theoretical figure is 3900. This very satisfactory agreement tests really a number of assumptions that up to now did not seem quite certain and makes the chances for a Christy gadget with lenses look very good.

I reminded Bob Wilson again of his promised letter about the RF system and he promised to get it written very soon now.

In connection with the plans for Alberta Ramsey told Frisch that he had put Frisch's name down as one of the men he definitely wanted on the team. Frisch explained of course that there might be formal complications but Ramsey decided there was no harm in any case in asking for him. Incidentally Oppie asked me on two occasions if I had written to you on the question of British people going to Alberta, so evidently he is anxious to have the matter settled.

Yours sincerely,

R.E. Peierls

[376] Rudolf Peierls to James Chadwick

location unspecified, 5.4.1945
secret

Dear Sir James,
Your will doubtless like to hear some details of the arrangements being made for the Trinity Test; Bainbridge spoke about it at the colloquium on Tuesday, as will be described in the usual weekly letter to you. In the present letter I would like to tell you about the sides of the work with which I am particularly concerned. When Wilson's division was brought on to take charge of the measurements (nuclear and otherwise) on the test gadget it was arranged that I should act as consultant to Wilson but should also take charge of some specific sections of the work. These are:

1. Measurement of ionisation after the event for the protection of personnel, these measurements to be done with instruments that record at the base stations so that the ionisation level at the various points in the field is known before any men are sent into the area. The measurements are supplementary to those that will be made with ordinary ionisation meters by Hempelmann's people. We have made about 20 gadgets for the purpose, working on the principle that the ionisation current in a chamber of about a litre capacity charges up an argon discharge tube to its break down potential, when a short discharge passes until the extinction potential is reached, after which the process repeats. The number of pulses per minute is proportional to the ionisation intensity, and since the pulses though short are quite big they are ideal for sending as signals to the base station. We have had some of them working in the field and they seem to behave well; it will not take very long to complete the bulk construction and installation.

2. Measurements of the intensity of delayed gamma radiation at short distance (100/200 metres) and at short times (1 or 2 milliseconds) after the nuclear reaction. The reason for choosing short distances is that the correction for scattering and absorption by ground

> and air is small and therefore comparatively certain, while at the short times contemplated the blast wave will not have reached the ground and so there will be no question of any active material being mixed up with the ground or of any rock or sand having been thrown up into the air, thus causing uncertain amount of absorption. The ionisation will of course be extremely intense but it seems possible to handle it with a shallow chamber filled with argon. A rather elegant alternative way of measuring the ionisation is to make use of the heating effect due to these gamma rays; we had realised for some time that the air in the neighbourhood of the gadget would get hot instantaneously when the gamma pass through it, but it was left to Fermi to point out to us that solid bodies would similarly be heated; it then became obvious that resistance thermometers should be used to measure this heating effect without there being any need to worry about the conduction of heat from the air to the thermometer. At 100 metres the delayed gammas will heat the thermometer by several degrees during the first few milliseconds so there should be no difficulty as regards sensitivity. I am hoping to use both these methods.

The purpose of measuring these delayed gammas quantitatively is, of course, to deduce the number of fissions taking place in the gadget; it is therefore necessary to make laboratory measurements of the delayed gammas as a function of time. Frisch's dragon gave a good opportunity to make such measurements for 25; as the width of the pulse was several milliseconds it was not possible to measure as early as desired, but I was able to get a pretty good idea of the absolute intensity from about 10 milliseconds to 10 seconds after the fission pulse. The greatest uncertainty is due to the self-absorption of the core and tamper of the dragon; unfortunately the dragon was operating for so short a period that it was not even possible to make absorption experiments and determine the hardness of the gammas that I was measuring. The results are indicated on the accompanying graphs, and I think they are correct within a factor of 2, which is enough to go upon in designing the apparatus used at Trinity; since the Trinity gadget will work with 49 it is necessary to make similar but more accurate experiments on the early delayed gammas from that substance. This will be done with the co-operation of John Williams by irradiating a lump of 49 with 600 keV. neutrons produced by his von der Graaf machine. They will install an electrostatic beam-reflection system so that the neutrons will come in pulses of durations which can be made a small fraction of a millisecond, which will give ample time resolution for the purpose that I need. These experiments will not be done until after the "100 ton shot" that is to be made at Trinity early in May in order to test such instruments as blast gauges.

In addition to my measurements on short period delayed gammas, Segre is going to measure them at times ranging from .01 sec. to a few seconds after the event. He will necessarily have to make such measurements at a good many hundred metres from the gadget so that he gets his records before the blasts damages his apparatus; the corrections for air and ground scattering will be large and he proposes to use a mixture of theory and field experiments to determine them. While I think it is a good thing to measure at these longer times I am personally quite sure that the shorter times give a much cleaner experiment; I had to be quite stubborn about this, because when the measurements programme was re-organised it was at first proposed to make Segre responsible for all these measurements, and Segre decided (without consulting anybody who had thought about the problem in detail) that he was not going to measure earlier than .01 seconds; he refused to alter his plans and so I said I would do it myself. Oppenheimer and Weisskopf (who is looking after the theory of these measurements) agree that the earlier times are best and I am very glad to be doing that side of the work.

Yours,

R.E. Peierls

[377] Rudolf Peierls to Robert Oppenheimer[1244]

location unspecified, 2.5.1945
confidential

At the request of General Groves I enclose two copies each of our last three letters to Dr. Chadwick, one copy being for the transmission to the General.[1245]

These letters are written by Dr. Moon and myself for the primary purpose of collecting information that would help to keep Chadwick in contact with developments here. I would like to emphasise that for this reason the selection of material is entirely subjective and also that we are not trying to collect information other than that happens to come to our knowledge.

We are preparing a summary of the previous letters in the series and will send copies of these as soon as possible.

R.E. Peierls

[378] Rudolf Peierls to James Chadwick

location unspecified, 16.5.1945

Dear Chadwick,
We feel that on this occasion we ought to add an informal postscript to our weekly letter, not because the remarks we want to make should be particularly secret from anyone, but because they would be inappropriate in a letter that with our knowledge goes to Groves.

As you may have gathered from our description of Bainbridge's talk, the 100 ton shot did not go very well. As far as can be told, the real trouble was that not enough time was allowed for rehearsal and paradoxically the shortage of time was mainly caused by Bainbridge's insistence

[1244] Inter-office memorandum, Oppenheimer papers, LoC.
[1245] Explain Groves' increasing control of British communications and attempts to limit information being passed to member of the British mission.

on starting rehearsals when only a fraction of the apparatus had been installed. Since the area of the shot was closed for several hours at each rehearsal, the people whose apparatus was not yet ready were delayed by each rehearsal and the series barely converged by the actual time of the shot. Thus a considerable fraction of the apparatus and of the arming operations had not been properly rehearsed and it is not surprising that the original belief that most of the records would be good has proved unjustified. Bainbridge (as indicated in Moon's letter of last week) realises quite clearly that exhaustive rehearsals must be made prior to the gadget shot. It is obviously necessary that the cross-talk between neighbouring lines be investigated thoroughly and quantitatively a long while before the shot, so that troubles can not only be observed but can also be corrected where possible.

There were other organisational failures, the most striking of which was that the M.P.'s were not rehearsed in their duties and various errors in the instructions given them were therefore not discovered until some members of the staff were refused permission to go and turn on their apparatus. Since many of the guards had no radio, it was impossible to have the instructions corrected except in a few instances.

There were other difficulties such as a shortage of transport, but this particular difficulty is clearly realised and will probably be overcome.

The real troubles about the Trinity arrangements are (1) the refusal to give the job enough priority months ago so that it need not be done as a terrible rush at the last moment, and (2) Bainbridge's vague way of handling matters of organisation. It is probably too late to do anything about these, but if nothing is done one may find that a week or so before the scheduled date of the gadget test there is evidence that things are not in a good and reliable state, and then it will be necessary to make a very serious decision as to whether the shot should therefore be deferred or not.

Yours sincerely,

R.E. Peierls
P.B. Moon

[379] Rudolf Peierls to Robert Oppenheimer

location unspecified, 28.5.1945

We have now had copies made of the relevant parts of our letters to Dr. Chadwick. I enclose two copies, one of them for transmission to General Groves. This completes the series to date.

R. Peierls

Bibliography

A complete bibliography of Rudolf Peierls' works can be found in

R.H. Dalitz and Rudolf Peierls (eds.), *Selected Scientific Papers of Sir Rudolf Peierls, with Commentary*, Singapore: World Scientific, 1997, pp. 787–805

or in

R.H. Dalitz, "Complete bibliography for Prof. Sir Rudolf Peierls", *Nuclear Physics A*, vol. 604, pp. 7–23 (1996).

Archival Sources

The documents reproduced in this volume are taken from several archives in the UK and the US as well as from a collection of material held by the Peierls family.

The archival collections used are:

Papers and correspondence of Sir Rudolf Ernst Peierls, 1907–1995. Owning Repository: Department of Western Manuscripts, Bodleian Library, University of Oxford; cited as Peierls Papers.

Papers of P. A. M. Dirac, 1902–1984. Owning Repository: Florida State University, Paul A. M. Dirac Science Library, Tallahassee, FL 32306, USA; photocopies held at Churchill College Archives, Cambridge, England, CB3 0DS; cited as Dirac Papers.

Papers of Hans A. Bethe, 1906–2005. Owning Repository: Cornell University, Carl A. Kroch Library, Division of Rare and Manuscript Collections, University Archives, 2B Carl A Kroch Library, Ithaca, NY 14853; cited as Bethe Papers.

Papers of Sir James Chadwick, 1891–1974. Owning Repository: Churchill College, Archives Centre, Cambridge CB3 0DS, England, UK; cited as Chadwick papers.

Papers by Arnold Sommerfeld, 1868–1951. Owning Repository: Deutsches Museum, Museumsinsel 1, 80306 Munich, Germany; cited as Sommerfeld Papers.

Papers of J. Robert Oppenheimer, 1904–1967. Owning Repository: Manuscript Division, Library of Congress, Washington, D.C.; cited as Oppenheimer Papers.

Papers relating to Tube Alloys. Owning Repository: The National Archives: Public Record Office, Kew, Richmond, Surrey, TW9 4DU; cited as PRO.

The following list summarizes information about the provenance of the material reproduced in this volume. This includes information about when and where the letters were written, sender and addressee, whether the copy reproduced is the original or a carbon copy and which language the original was written in as well as where the reproduced copies are located. Where letters contain a mixture of languages, the predominant one is given first, the secondary language is given in square brackets thereafter.

Where no source is given, the letters are taken from the Peierls Papers in the Bodleian.

Key:

pc	—	postcard
g	—	German
e	—	English
r	—	Russian
RP	—	Rudolf Peierls
loc. unsp.	—	location unspecified
date unsp.	—	date unspecified

LIST OF CORRESPONDENCE

No.	date	from	to	place	original/ carbon copy	language	source
1	1.8.1922	Heinz Rudolph	RP	Essen	o	g	
2	8.11.1926	Else Peierls	RP	Berlin	o	g	
3	23.11.1926	Heinrich Peierls	RP	Berlin	o	g	
4	1.12.1926	Heinrich Peierls	RP	Berlin	o	g	
5	23.2.1927	Else Peierls	RP	Berlin	o	g	
6	16.7.1927	Heinrich Peierls	RP	Berlin	o	g	
7	3.8.1927	Heinrich Peierls	RP	Berlin	o	g	
8 (pc)	27.8.1927	Hans Bethe	RP	Bad Gastein	o	g	
9	19.9.1927	Hans Bethe	RP	Frankfurt/M	o	g	
10	1.10.1927	Hans Bethe	RP	Frankfurt/M	o	g	
11	15.12.1927	Heinrich Peierls	RP	Berlin	o	g	
12 (pc)	24,12,1826	Hans Bethe	RP	Baden-Baden	o	g	
13	25.1.1928	Heinrich Peierls	RP	Berlin	o	g	
14	28.1.1928	Hans Bethe	RP	Baden-Baden	o	g	
15	9.2.1928	Heinrich Peierls	RP	Berlin	o	g	
16	2.6.1928	Heinrich Peierls	RP	Glion	o	g	
17	5.6.1928	Hans Bethe	RP	train Baden–Frankfurt/M	o	g	
18	20.7.1928	Heinrich Peierls	RP	Berlin	o	g	
19	13.8.1928	Heinrich Peierls	RP	Berlin	o	g	
20	25.8.1928	Heinrich Peierls	RP	Berlin	o	g	
21	14.9.1928	RP	Paul Dirac	Hampton	o	e	Dirac Papers
22	27.9.1928	Heinrich Peierls	RP	Berlin	o	g	
23	15.10.1928	RP	Paul Dirac	Leipzig	o	e	Dirac Papers
24	2.12.1928	Hans Bethe	RP	Frankfurt/M	o	g	
25	17.2.1928	Hans Bethe	RP	Frankfurt/M	o	g	

Key: pc – postcard; g – German; e – English; r – Russian; RP – Rudolf Peierls; loc. unsp. – location unspecified; date unsp. – date unspecified

No.	date	from	to	place	original/ carbon copy	language	source
26	18.4.1929	Heinrich Peierls	RP	Rome	o	g	
27	14.5.1929	Heinrich Peierls	RP	Berlin	o	g	
28	13.6.1929	Heinrich Peierls	RP	Berlin	o	g	
29	23.7.1929	Else Peierls	RP	St. Moritz	o	g	
30	26.7.1929	Heinrich Peierls	RP	St. Moritz	o	g	
31	27.7.1929	Hans Bethe	RP	train Stuttgart–Heidelberg	o	g	
32	7.10.1929	Heinrich Peierls	RP	Berlin	o	g	
33	15.10.1929	Heinrich Peierls	RP	Berlin	o	g	
34	14.11.1929	Heinrich Peierls	RP	Berlin	o	g	
35	2.1.1930	Heinrich Peierls	RP	Berlin	o	g	
36	28.1.1930	Heinrich Peierls	RP	Berlin	o	g	
37	30.1.1930	Hans Bethe	RP	train	o	g	
				Baden–Munich	o	g	
38	6.2.1930	Lev Landau	RP	Leipzig	o	g	
39	10./11.2.1930	Heinrich Peierls	RP	Berlin	o	g	
40	6.3.1930	Lev Landau	RP	Berlin	o	g	
41	12.3.1930	Heinrich Peierls	RP	Berlin	o	g	
42	20.3.1930	Lev Landau	RP	Berlin	o	g	
43	8.5.1930	Heinrich Peierls	RP	Berlin	o	g	
44	17.5.1930	Lev Landau	RP	Haarlem	o	g	
45	28.5.1930	Heinrich Peierls	RP	Berlin	o	g	
46	3.6.1930	Werner Heisenberg	RP	Leipzig	o	g	
47	10.6.1930	Heinrich Peierls	RP	Berlin	o	g	
48	13.6.1930	Werner Heisenberg	RP	Heligoland	o	g	
49	4.7.1930	Heinrich Peierls	RP	Berlin	o	g	
50	11.7.1930	Hans Bethe	RP	Munich	o	g	

No.	date	from	to	place	original/ carbon copy	language	source
51	24.7.1930	Hans Bethe	RP	loc. unsp.	o	g	
52	25.7.1930	Heinrich Peierls	RP	Berlin	o	g	
53 (pc)	12.8.1930	Hans Bethe	RP	Zuoz	o	g	
54	26.8.1930			Grusjia	o	g	
55	28.8.1930			Grusjia	o	g	
56	13.9.1930	Genia Peierls	RP	Gagry	o	e	
57	24.9.1930	RP	Genia Peierls	Berlin	o	e	
58	26.9.1930	Genia Peierls	RP	Leningrad	o	e	
59	29.9.1930	RP	Genia Peierls	Berlin	o	e	
60	2.10.1930	RP	Genia Peierls	Zurich	o	e	
61	3.10.1930	Genia Peierls	RP	Leningrad	o	e	
62	8.10.1930	RP	Genia Peierls	Zurich	o	e [r]	
63	9.10.1930	RP	Genia Peierls	Zurich	o	e	
64	10.10.1930	Genia Peierls	RP	Leningrad	o	e [r]	
65	11.10.1930	Vladimir Fock	RP	Charkow	o	g	
66	14.10.1930	Genia Peierls	RP	Leningrad	o	e [r]	
67	14.10.1930	RP	Genia Peierls	Zurich	o	e [r]	
68	15.10.1930	Genia Peierls	RP	Leningrad	o	e [r]	
69	16.10.1930	RP	Genia Peierls	Zurich	o	e [r]	
70	22.10.1930	Genia Peierls	RP	Leningrad	o	e [r]	
71	24.10.1930	RP	Genia Peierls	Zurich	o	e [r]	
72	28.10.1930	Genia Peierls	RP	Leningrad	o	e [r]	
73	1.11.1930	Heinrich Peierls	RP	Berlin	o	g	
74	4.11.1930	RP	Genia Peierls	Zurich	o	e	
75	6.11.1930	Genia Peierls	RP	Leningrad	o	e [r]	
76	8.11.1930	Vladimir Fock	RP	Leningrad	o	g	

No.	date	from	to	place	original/ carbon copy	language	source
77	16.11.1930	RP	Genia Peierls	Zurich	o	e [r]	
78	17.11.1930	Genia Peierls	RP	Leningrad	o	e [r]	
79	30.11.1930	RP	Genia Peierls	Zurich	o	r [e]	
80	5.12.1930	Werner Heisenberg	RP	Leipzig	o	g	
81	6.12.1930	Genia Peierls	RP	Leningrad	o	e [r]	
82	6.12.1930	RP	Genia Peierls	train Zurich–Tübingen	o	r [e]	
83	11.12.1930	Genia Peierls	RP	Leningrad	o	r [e]	
84	12.12.1930	Werner Heisenberg	RP	Leipzig	o	g	
85	14.12.1930	RP	Genia Peierls	Zurich	o	r [e]	
86	20.12.1930	Vladimir Fock	RP	Leningrad	o	g	
87	20.12.1930	Heinrich Peierls	RP	Berlin	o	g	
88	21.12.1930	RP	Genia Peierls	Zurich	o	r [e]	
89	23.12.1930	Genia Peierls	RP	Leningrad	o	e [r]	
90	25.12.1930	RP	Genia Peierls	Arosa	o	r [e]	
91	1.1.1931	Genia Peierls	RP	Leningrad	o	r [e]	
92	6.1.1931	RP	Genia Peierls	Arosa	o	e	
93	10.1.1931	RP	Genia Peierls	Zurich	o	r [e]	
94	15.1.1931	Genia Peierls	RP	Leningrad	o	r [e]	
95	15.1.1931	Vladimir Fock	RP	Leningrad	o	g	
96	18.1.1931	RP	Genia Peierls	Zurich	o	r [e]	
97	22.1.1931	Genia Peierls	RP	Leningrad	o	r [e]	
98	22.1.1931	Heinrich Peierls	RP	Berlin	o	g	
99	26.1.1931	Werner Heisenberg	RP and Lev Landau	Leipzig	o	g	
100	27.1.1931	Genia Peierls	RP	Leningrad	o	r [e]	
101	28.1.1931	RP	Genia Peierls	Zurich	o	r [e]	
102	2.2.1931	Genia Peierls	RP	Leningrad	o	e [r]	
103	3.2.1931	RP	Genia Peierls	Zurich	o	r [e]	

No.	date	from	to	place	original/ carbon copy	language	source
104	4.2.1931	Heinrich Peierls	RP	Berlin	o	g	
105	7.2.1931	Genia Peierls	RP	Leningrad	o	r [e]	
106	9.2.1931	RP	Genia Peierls	Zurich	o	r [e]	
107	11.2.1931	Heinrich Peierls	RP	Berlin	o	g	
108	13.2.1931	Genia Peierls	RP	Leningrad	o	r [e]	
109	14.2.1931	RP	Genia Peierls	Zurich	o	r [e]	
110	18.2.1930	Genia Peierls	RP	Peterhof	o	r [e]	
111	19.3.1931	Heinrich Peierls	RP	Berlin	o	g	
112	31.3.1931	Heinrich Peierls	RP	Berlin	o	g	
113	23.4.1931	Hans Bethe	RP	Rome	o	g	
113	2.5.1931	Genia Peierls	RP	Detskoye	o	r [e]	
115	6.5.1931	Genia Peierls	RP	Leningrad	o	r [e]	
116	10.5.1931	Genia Peierls	RP	Lenongrad	o	r [e]	
117	10.5.1931	Genia Peierls	RP	Detskoye	o	r [e]	
118	21.5.1931	Genia Peierls	RP	Leningrad	o	r [e]	
119	24.5.1931	Genia Peierls	RP	Leningrad	o	r [e]	
120	28.5.1931	Genia Peierls	RP	Leningrad	o	r [e]	
121	2.6.1931	Genia Peierls	RP	Leningrad	o	e	
122	24.6.1931	RP	Genia Peierls	Zurich	o	g [e]	
123	25.6.1931	Genia Peierls	RP	Leningrad	o	r	
124	29.6.1931	RP	Genia Peierls	Zurich	o	e [r]	
125	1.7.1931	Wolfgang Pauli	RP	Ann Arbor	o	g	
126	1.7.1931	Wolfgang Pauli	RP	Ann Arbor	o	g	
127	3.7.1931	Wolfgang Pauli	RP	Ann Arbor	o	g	
128	4.7.1931	RP	Genia Peierls	Zurich	o	r [g]	
129	7.7.1931	RP	Genia Peierls	Zurich	o	r	
130	11.7.1931	RP	Genia Peierls	Zurich	o	r	
131	14.7.1931	Genia Peierls	RP	Pavlovsk	o	r	

No.	date	from	to	place	original/ carbon copy	language	source
132	16.7.1931	RP	Genia Peierls	on train Stuttgart	o	r	
133	19.7.1931	Genia Peierls	RP	Leningrad	o	r	
134	19.7.1931	RP	Genia Peierls	Leipzig	o	r	
135	21.7.1931	RP	Genia Peierls	Leipzig	o	r	
136	23.7.1931	Genia Peierls	RP	Leningrad	o	r	
137	26.7.1931	RP	Genia Peierls	Leipzig	o	r	
138	26.7.1931	Genia Peierls	RP	Leningrad	o	r	
139	30.7.1931	RP	Genia Peierls	Berlin	o	r [e]	
140	31.7.1931	Genia Peierls	RP	Leningrad	o	r	
141	3.8.1931	RP	Genia Peierls	Berlin	o	r	
142	6.8.1931	RP	Genia Peierls	Berlin	o	r [g]	
143	9.8.1931	RP	Genia Peierls	Berlin	o	r	
144	29.9.1931	Wolfgang Pauli	RP	New York	o	g	
145	12.11.1931	Hans Bethe	RP	Munich	o	g	
146	29.1.1932	RP	Arnold Sommerfeld	Zurich	o	g	Sommerfeld Nachlass
147	17.3.1932	Heinrich Peierls	RP	Berlin	o	g	
148	2.4.1932	RP	Genia Peierls	Moscow	o	r	
149	29.5.32	Genia Peierls	RP	Leningrad	o	r	
150	1.6.32	Genia Peierls	RP	Leningrad	o	r	
151	6.6.32	Genia Peierls	RP	Leningrad	o	r	
152	9.6.32	Genia Peierls	RP	Leningrad	o	r	
153	[date unsp.]	Genia Peierls	RP	Leningrad	o	r	
154	24.10.1932	Wolfgang Pauli	RP	Zurich	o	g	
155	26.10.1932	RP	Paul Ehrenfest	Rome	cc	g	

No.	date	from	to	place	original/ carbon copy	language	source
156	31.10.1932	Paul Ehrenfest	RP	loc. unsp.	o	g	
157	10.11.1932	RP	Paul Ehrenfest	Rome	cc	g	
158	21.4.1932	RP	Genia Peierls	Moscow	o	r	
159	22.5.1933	Wolfgang Pauli	RP	Zurich	o	g	
160	14.6.1933	RP	Hans Bethe	Cambridge	o	g	Bethe Papers
161	28.6.1933	RP	Hans Bethe	Cambridge	o	g	Bethe Papers
162	6.7.1933	RP	Hans Bethe	Cambridge	o	g	Bethe Papers
163	17.7.1933	RP	Wolfgang Pauli	Cambridge	cc	g	
164	18.7.1933	Heinrich Peierls	RP	Berlin	o	g	
165	24.7.1933	RP	Hans Bethe	Cambridge	o	g	Bethe Papers
166	10.8.1933	Else Peierls	RP	St. Moritz	o	g	
167	16.9.1933	Heinrich Peierls	RP	Berlin	o	g	
168	17.11.1933	Heinrich Peierls	RP	Berlin	o	g	
169	8.12.1933	RP	Paul Dirac	Manchester	o	g	Dirac Papers
170	16.1.1934	Heinrich Peierls	RP	Berlin	o	g	
171	22.1.1934	Heinrich Peierls	RP	Berlin	o	g	
172	27.1.1934	Niels Bohr	RP	Copenhagen	o	g	
173	28.1.1934	RP	Arnold Sommerfeld	Manchester	o	g	Sommerfeld Nachlass
174	2.2.1934	Arnold Sommerfeld	RP	Munich	o	g	
175	7.2.1934	Heinrich Peierls	RP	Berlin	o	g	
176	8.2.1934	RP	Paul Dirac	Manchester	o	e	Dirac Papers
177	14.2.1934	Heinrich Peierls	RP	Berlin	o	g	
178	17.2.1934	RP	Paul Dirac	Manchester	o	e	Dirac Papers
179	6.3.1934	Heinrich Peierls	RP	Naples	o	g	
180	20.3.1934	RP	Paul Dirac	Manchester	o	e	Dirac Papers
181	3.7.1934	Carl-Friedrich von Weizsäcker	RP	Leipzig	o	g	
182	19.7.1934	Heinrich Peierls	RP	Berlin	o	g	

No.	date	from	to	place	original/ carbon copy	language	source
183	21.9.1934	Heinrich Peierls	RP	Berlin	o	g	
184	25.9.1931	Heinrich Peierls	RP	Berlin	o	g	
185	5.10.1934	Heinrich Peierls	RP	Berlin	o	g	
186	10.3.1934	Heinz Rudolph	RP	Berlin	o	g	
187	10.6.1935	Wolfgang Pauli	RP	Zurich	o	g	
188	23.4.1936	Heinrich Peierls	RP	Meran	o	g	
189	2.7.1936	Else Peierls	RP	Berlin	o	g	
190	1.8.1936	Hans Bethe	RP	MS Queen Mary	o	g	
191	17.8.1936	RP	Niels Bohr	Cambridge	cc	g	
192	19.8.1936	Hans Bethe	RP	Seelisberg	o	g	
193	25.8.1936	RP	Hans Bethe	Cambridge	o	g	Bethe Papers
194	27.8.1936	Hans Bethe	RP	Seelisberg	o	g	
195	30.8.1936	RP	Hans Bethe	Cambridge	o	g	Bethe Papers
196	4.9.1936	Hans Bethe	RP	Seelisberg	o	g	
197	9.9.1936	Niels Bohr	RP	Copenhagen	o	g	
198	11.10.1936	RP	Niels Bohr	Cambridge	cc	g	
199	17.10.1936	Niels Bohr	RP	Copenhagen	o	g	
200	1.12.1936	Heinrich Peierls	RP	Berlin	o	g	
201	4.12.1936	Nevill Mott	RP	Bristol	o	e	
202	9.12.1936	RP	Nevill Mott	Cambridge	cc	e	
203	14.12.1936	Heinrich Peierls	RP	Berlin	o	g	
204	16.12.1936	Max Born	RP	Edinbrugh	o	g	
205	14.1.1937	Max Born	RP	Edinburgh	o	g	
206	13.2.1937	RP	Hans Bethe	Cambridge	o	g	Bethe Papers
207	2.7.1937	RP	Paul Dirac	Portscatho	o	e	Dirac Papers
208	12.8.1937	Heinrich Peierls	RP	Berlin	o	g	
209	12.10.1937	RP	Wolfgang Pauli	Birmingham	cc	g	

No.	date	from	to	place	original/ carbon copy	language	source
210	15.10.1937	Wolfgang Pauli	RP	Zurich	o	g	
211	28.10.1937	RP	Niels Bohr	Birmingham	cc	g	
212	6.11.1937	George Placzek	RP	Copenhagen	o	g	
213	25.11.1937	RP	Ralph Fowler	Birmingham	cc	e	
214	25.11.1937	Wolfgang Pauli	RP	Zurich	o	g	
215	1.12.1937	Ralph Fowler	RP	Cambridge	o	e	
216	7.12.1937	Walter Heitler	RP	Bristol	o	g	
217 (pc)	7.12.1937	Wolfgang Pauli	RP	Zurich	o	g	
218	13.12.1937	Walter Heitler	RP	Bristol	o	g	
219	27.1.1938	George Placzek	RP	Copenhagen	o	g	
220	1.2.1938	Niels Bohr	RP	Copenhagen	o	e	
221	8.2.1938	RP	Niels Bohr	Birmingham	cc	g	
222	3.3.1938	Heinrich Peierls	RP	Berlin	o	g	
223	19.3.1938	Wolfgang Pauli	RP	Cambridge	o	g	
224	24.3.1938	RP	Wolfgang Pauli	Birmingham	cc	g	
225	4.4.1938	George Placzek	RP	Copenhagen	o	g	
226	18.4.1938	Wolfgang Pauli	RP	Zurich	o	g	
227	22.4.1938	Nevill Mott	RP	Bristol	o	e	
228	22.4.1938	RP	George Placzek	Birmingham	cc	g	
229	23.4.1938	RP	Wolfgang Pauli	Birmingham	cc	g	
230	27.4.1938	George Placzek	RP	Copenhagen	o	g	
231	27.4.1938	RP	Nevill Mott	Birmingham	cc	e	
232	27.4.1938	Wolfgang Pauli	RP	Zurich	o	g	
233	7.5.1938	Niels Bohr	RP	Copenhagen	o	e	
234	11.5.1938	RP	George Placzek	Birmingham	cc	g	
235	14.5.1938	RP	Wolfgang Pauli	Birmingham	cc	g	
236	16.5.1938	George Placzek	RP	Copenhagen	o	g	
237 (pc)	21.5.1938	Wolfgang Pauli	RP	Zurich	o	g	

No.	date	from	to	place	original/ carbon copy	language	source
238	25.5.1938	Wolfgang Pauli	RP	Zurich	o	g	
239	3.6.1938	RP	Wolfgang Pauli	Birmingham	cc	g	
240	8.6.1938	RP	Niels Bohr	Birmingham	cc	g	
241	9.6.1938	Wolfgang Pauli	RP	Zurich	o	g	
242	27.6.1938	Niels Bohr	RP	Copenhagen	o	e	
243	28.6.1938	RP	Niels Bohr	Birmingham	cc	g	
244	14.7.1938	RP	Niels Bohr	Morwenstow	cc	g	
245	17.7.1938	George Placzek	RP	Peracaliente	o	g	
246	18.7.1938	Wolfgang Pauli	RP	Zurich	o	g	
247	19.7.1938	Niels Bohr	RP	Copenhagen	o	e	
248	21.7.1938	RP	Niels Bohr	Morwenstow	cc	g	
249	22.8.1938	Hans Bethe	RP	loc. unsp.	o	g	
250	24.8.1938	RP	Hans Bethe	Birmingham	cc	g	
251	25.8.1938	Else Peierls	RP	Zurich	cc	g	
252	26.8.1938	RP	Else Peierls and Heinrich Peierls	Birmingham	cc	g	
253	27.8.1938	RP	George Placzek	Birmingham	cc	g	
254	27.8.1938	Hans Bethe	RP	Seelisberg	o	g	
255	27.8.1938	Niels Bohr	RP	Copenhagen	o	e	
256	30.8.1938	RP	Siegfried Peierls	Birmingham	cc	e	
257	31.8.1938	RP	Wolfgang Pauli	Birmingham	cc	g	
258	2.9.1938	Annie Krebs	RP	Birmingham	cc	g	
259	4.9.1938	George Placzek	RP	Pasadena	o	g	
260	8.9.1938	Jenny Peierls	RP	Lake Placid	o	e	
261	24.9.1938	RP	Niels Bohr	Birmingham	cc	g	
262	24.9.1938	RP	Jenny Peierls	Birmingham	cc	e	
263	24.9.1938	RP	Else Peierls and Heinrich Peierls	Birmingham	cc	g	
264	17.10.1938	George Placzek	RP	Paris	o	g	

No.	date	from	to	place	original/ carbon copy	language	source
265	22.10.1938	RP	George Placzek	Birmingham	cc	g	
266	5.11.1938	RP	Nevill Mott	Birmingham	cc	e	
267	5.11.1938	RP	Hans Bethe	Birmingham	cc	g	
268	[date unsp.]	Nevill Mott	RP	Bristol	o	e	
269	[date unsp.]	Nevill Mott	RP	Bristol	o	e	
270	[date unsp.]	Nevill Mott	RP	Bristol	o	e	
271	12.11.1938	Nevill Mott	RP	Bristol	o	e	
272	10.12.1938	RP	Niels Bohr	Birmingham	cc	e	
273	10.12.1938	RP	Hans Bethe	Birmingham	cc	g	
274	12.12.1938	Hans Bethe	RP	Ithaca	o	g	
275	25.12.1938	RP	Hans Bethe	Birmingham	cc	g	
276	27.12.1938	RP	Heinrich Peierls	Birmingham	cc	g	
277	30.12.1938	Siegfried Peierls	RP and Alfred Peierls	New York	o	e	
278	1.1.1939	RP	Alfred Peierls	Birmingham	cc	g	
279	18.1.1939	David Shoenberg	RP	Cambridge	o	e	
280	23.1.1939	RP	Else Peierls and Heinrich Peierls	Birmingham	cc	g	
281	24.1.1939	RP	David Shoenberg	Birmingham	cc	e	
282	9.2.1939	Heinrich Peierls	RP	Berlin	o	g	
283	12.2.1939	David Shoenberg	RP	Cambridge	o	e	
284	19.2.1939	RP	Siegfried Peierls	Birmingham	cc	e	
285	15.4.1939	RP	Else Peierls	Oslo	cc	g	
286	6.6.1939	RP	Else Peierls and Heinrich Peierls	Birmingham	cc	g	
287	6.6.1939	Niels Bohr	RP	Copenhagen	cc	e	
288	10.6.1939	RP	Niels Bohr	Birmingham	cc	e	
289	16.6.1939	RP	Niels Bohr	Birmingham	cc	e	

No.	date	from	to	place	original/ carbon copy	language	source
290	7.7.1939	RP	P. L. Kapur	Birmingham	cc	e	
291	18.7.1939	RP	German Embassy	Birmingham	cc	g	
292	31.7.1939	German Embassy	RP	London	o	g	
293	22.8.1939	RP	Hans Bethe	Birmingham	cc	g	
294	31.8.1939	RP	Home Office	Birmingham	cc	e	
295	1.9.1939	Else Peierls	RP	Folkestone	o	g	
296	6.9.1939	Else Peierls	RP	Folkestone	o	e	
297	8.9.1939	Annie Krebs	RP	loc. unsp.	o	g	
298	13.9.1939	RP	Edward Appleton	Birmingham	cc	e	
299	13.9.1939	RP	Niels Bohr	Birmingham	cc	e	
300	15.9.1939	RP	Else Peierls	Birmingham	cc	e	
301	6.10.1939	RP	Else Peierls and Heinrich Peierls	Birmingham	cc	e	
302	6.12.1939	RP	W. L. Bragg	Birmingham	cc	e	
303	20.3.1940	RP	Else Peierls and Heinrich Peierls	Birmingham	cc	e	
304	22.4.1940	RP	Maud Committee	Birmingham	cc	e	PRO: AB4
305	5.6.1940	James Chadwick	RP	Liverpool	cc	e	Chadwick Papers
306	12.6.1940	RP	James Chadwick	Birmingham	o	e	Chadwick Papers
307	17.6.1940	Genia Peierls	Hans Bethe	Birmingham	cc	e	
308	11.7.1940	RP	Else Peierls and Heinrich Peierls	Birmingham	o	e	
309	12.7.1940	RP	James Chadwick	Birmingham	o	e	Chadwick Papers
310	12.7.1940	RP	Rose and Hans Bethe	Birmingham	cc	e	
311	14.7.1940	RP	Else Peierls and Heinrich Peierls	Birmingham	cc	e	

No.	date	from	to	place	original/ carbon copy	language	source
312	16.7.1940	Genia Peierls	Rose and Hans Bethe	Birmingham	cc	e	
313	17.7.1940	RP	Rose and Hans Bethe	Birmingham	cc	e	
314	15.8.1940	Hans Bethe	Genia Peierls and RP	loc. unsp.	o	e	
315	23.8.1940	RP	Else Peierls and Heinrich Peierls	Birmingham	cc	e	
316	16.10.1940	RP	Else Peierls and Heinrich Peierls	Birmingham	cc	e	
317	6.11.1940	Siegfried Peierls	RP	New York	o	e	
318	29.1.1941	Egon Orowan	RP	Cambridge	o	e	
319	6.2.1941	Egon Orowan	RP	Cambridge	o	e	
320	7.2.1941	RP	Else Peierls and Heinrich Peierls	Birmingham	cc	e	
321	11.2.1941	RP	Egon Orowan	Birmingham	cc	e	
322	14.2.1941	Heinrich Peierls	RP	Montclair	o	e	
323	22.2.1941	Egon Orowan	RP	Cambridge	o	e	
324	2.3.1941	RP	Egon Orowan	Birmingham	cc	e	
325	16.3.1941	RP	Else Peierls and Heinrich Peierls	Birmingham	cc	e	
326	17.3.1941	Egon Orowan	RP	Cambridge	o	e	
327	1.4.1941	RP	Raymond Priestley	Birmingham	cc	e	
328	10.5.1941	RP	Klaus Fuchs	Birmingham	cc	e	PRO: AB4
329	15.9.1941	RP	Francis Simon	Birmingham	cc	e	PRO: AB4
330	23.9.1941	RP	James Chadwick	Birmingham	o	e	PRO: AB4
331	24.9.1941	Francis Simon	RP	Oxford	cc	e	
332	14.10.1941	RP	James Chadwick	Birmingham	o	e	Chadwick Papers

No.	date	from	to	place	original/ carbon copy	language	source
333	17.10.1941	RP	James Chadwick	Birmingham	o	e	Chadwick Papers
334	20.10.1941	James Chadwick	RP	Liverpool	o	e	PRO: AB4
335	22.10.1941	RP	James Chadwick	Birmingham	o	e	Chadwick Papers
336	28.10.1941	James Chadwick	RP	Liverpool	cc	e	Chadwick Papers
337	29.10.1941	RP	James Chadwick	Birmingham	o	e	Chadwick Papers
338	15.11.1941	James Chadwick	RP	Liverpool	cc	e	Chadwick Papers
339	20.11.1941	RP	James Chadwick	Birmingham	o	e	Chadwick Papers
340	20.12.1941	Hans Halban	RP	Cambridge	cc	e	PRO: AB4
341	31.1.1942	RP	James Chadwick	Birmingham	o	E	Chadwick Papers
342	13.3.1942	RP	Genia Peierls	New York	o	e	
343	21.3.1942	RP	Genia Peierls	loc. unsp.	o	e	
344	21.4.1942	RP	Robert Oppenheimer	Montreal	o	e	Oppenheimer Papers
345	22.6.1942	RP	James Chadwick	Birmingham	o	e	Chadwick Papers
346	29.8.1942	RP	James Chadwick	Birmingham	o	e	Chadwick Papers
347	12.9.1942	RP	James Chadwick	Birmingham	o	e	Chadwick Papers
348	18.9.1942	James Chadwick	RP	Liverpool	cc	e	Chadwick Papers
349	22.9.1942	RP	James Chadwick	Birmingham	o	e	Chadwick Papers
350	20.10.1942	James Chadwick	RP	Liverpool	cc	e	Chadwick Papers
351	21.10.1942	RP	James Chadwick	Birmingham	o	e	Chadwick Papers
352	24.10.1942	James Chadwick	RP	Liverpool	cc	e	Chadwick Papers
353	7.12.1942	RP	James Chadwick	Birmingham	o	e	Chadwick Papers
354	21.12.1942	Hans Halban	RP and Francis Simon	Montreal	o	e	PRO: AB4
355	1.1.1943	Max Born	RP	Edinburgh			
356	16.2.1943	RP	James Chadwick	Edinburgh	o	e	PRO
357	17.2.1943	James Chadwick	RP	Liverpool	cc	e	Chadwick Papers
358	23.2.1943	RP	James Chadwick	Birmingham	o	e	Chadwick Papers
359	27.2.1943	James Chadwick	RP	Liverpool	cc	e	Chadwick Papers
360	4.3.1943	RP	James Chadwick	Birmingham	o	e	Chadwick Papers

No.	date	from	to	place	original/ carbon copy	language	source
361	5.3.1943	James Chadwick	RP	Liverpool	cc	e	Chadwick Papers
362	10.3.1943	James Chadwick	RP	Liverpool	cc	e	Chadwick Papers
363	1.5.1943	RP	Wallace Akers	Birmingham	cc	e	
364	25.6.1943	RP	Otto Frisch	Birmingham	cc	e	
365	9.9.1943	RP	Genia Peierls	Washington	o	e	
366	20.9.1943	RP	Klaus Fuchs	Montreal	cc	e	PRO AB4
367	25.4.1943	Francis Simon	RP	Oxford	o	e	PRO AB4
368	3.5.1944	RP	A. H. Wilson	loc, unsp.	cc	e	PRO AB1
369	19.5.1944	RP	James Chadwick	loc. unsp.	cc	e	PRO AB1
370	26.6.1944	RP	Wallace Akers	loc. unsp.	cc	e	PRO AB1
371	14.7.1944	James Chadwick	RP	Washington	cc	e	PRO AB1
372	17.7.1944	Klaus Fuchs	RP	New York	cc	e	PRO AB4
373	20.9.1944	RP	Raymond Priestley	loc. unsp.	cc	e	
374	25.10.1944	Raymond Priestley	RP	Birmingham	o	e	
375	28.3.1945	James Chadwick	RP	loc. unsp.	cc	e	PRO AB1
376	5.4.1945	RP	James Chadwick	loc. unsp.	cc	e	PRO AB1
377	2.5.1945	RP	Robert Oppenheimer	loc. unsp.	cc	e	
378	16.5.1945	RP	James Chadwick	loc. unsp.	o	e	Chadwick Papers
379	28.5.1945	RP	Robert Oppenheimer	loc. unsp.	cc	e	

Name Index